13th International Conference on Electrorheological Fluids and Magnetorheological Suspensions

(ERMR2012)

Journal of Physics: Conference Series Volume 412

Ankara, Turkey
2-6 July 2012

ISBN: 978-1-62276-990-2
ISSN: 1742-6588

Printed from e-media with permission by:

Curran Associates, Inc.
57 Morehouse Lane
Red Hook, NY 12571

Some format issues inherent in the e-media version may also appear in this print version.

Printed by Curran Associates, Inc. (2013)

For permission requests, please contact the Institute of Physics
at the address below.

Institute of Physics
Dirac House, Temple Back
Bristol BS1 6BE UK

Phone: 44 1 17 929 7481
Fax: 44 1 17 920 0979

techtracking@iop.org

Additional copies of this publication are available from:

Curran Associates, Inc.
57 Morehouse Lane
Red Hook, NY 12571 USA
Phone: 845-758-0400
Fax: 845-758-2634
Email: curran@proceedings.com
Web: www.proceedings.com

TABLE OF CONTENTS

ELECTRORHEOLOGICAL MATERIALS

MAGNETORHEOLOGICAL MATERIALS

Journal of Physics: Conference Series **412** (2013) 011001 doi:10.1088/1742-6596/412/1/011001

Preface

13th International Conference on Electrorheological Fluids and Magnetorheological Suspensions | Gazi University Ankara, Turkiye 2 - 6 July, 2012

The 13th International Conference on Electrorheological Fluids and Magnetorheological Suspensions (ERMR2012) was held in Ankara, Turkey at Gazi University in the Architect Kemaleddin historical hall on 2–6 July 2012. The first International Conference on Electrorheological Fluids and Magnetorheological Suspensions took place nearly 25 years ago and this conference continued the same tradition by providing an arena for researchers around the world to present their new research findings in these fields, and gave them the opportunity to learn about the latest research and technology and to renew their acquaintances. The meeting brought together scientists and engineers in multidisciplinary areas such as chemical engineering, mechanical engineering, materials science and engineering, physics, chemistry and polymer science and technology, to explore the state-of-art technology, identify key areas to be focused on and discuss their problems/issues. All oral presentations were held in a single session to enhance the interactions between the scientists and engineers. The ERMR2012 Conference included plenary lectures given by prominent leaders in their respective fields.

About 130 participants from more than 50 organizations attended the conference and 15 plenary speeches, 64 oral presentations and 57 poster presentations took place in the following areas: (i) synthesis, characterization and processing of novel ER/MR materials, (ii) dynamics, chain and structure formation of ER/MR materials, (iii) ER/MR elastomers, ferrogels and their characterizations, (iv) rheological techniques and measurements of ER/MR materials, (v) modeling and simulations of ER/MR materials, (vi) device development and control techniques and (vii) applications of ER/MR materials.

The ERMR2012 International Conference began with Turkish classical music performed by the musicians of the Ministry of Culture and Tourism. Rector Professor Dr R Ayhan welcomed the participants and the conference chair Professor H I Unal delivered an opening talk before the plenary lectures.

Magnetorheological (MR) materials and their applications featured prominently at the ERMR2012 Conference. C A Evrensel from the University of Nevada, USA, opened the first session with a plenary speech on 'Magnetorheological Fluids for Cancer Therapy'. In the field of magnetorheological elastomers (MRE), S Odenbach from TU Dresden, Germany highlighted new developments in data management which enable single particle analysis and statistical analysis of the structures in an MRE in his plenary lecture. Another plenary lecture in the MRE field was given by F Gordaninejad (University of Nevada, USA) on new developments for integrated systems with thick MREs. D Barber (LORD Corporation, USA) gave details of the latest developments of LORD® MR fluids at extremes in his plenary speech. R Hidalgo Alvarez from the University of Granada, Spain, focused on particle shape effects in MR fluids in his plenary talk. D J Klingenberg (University of Wisconsin, USA) illustrated how the increased extent of layer formation increases the yield stress by increasing the stresses transmitted by both the magnetizable and nonmagnetizable spheres in his plenary lecture. S Smoukov from Cambridge University, UK, demonstrated magnetically responsive bi-stable hierarchical materials with novel functionality, which can be effectively used to design sustainable and energy-responsive systems. J de Vicente from the University of Granada, Spain, highlighted true yield stress in magnetorheology with the aid of simulations and experimental techniques for a wide range of MR fluids, inverse ferrofluids and ionic liquids-based MR fluids in his plenary speech. H Böse (Fraunhofer Institute, Germany) gave his plenary speech on magnetorheological torque transmission devices with permanent magnets. W Li (University of Wollongong, Australia), M Nakano (Tohoku University, Japan) and S Xuan (University of Science and Technology of China) gave plenary talks on novel MR shear thickening fluids, MR rubber composites and MR plastomers, respectively.

H J Choi (INHA University, Korea), X Zhao (Northwestern Polytechnical University, China) and R Tao (Temple University, USA) delivered plenary lectures on issues relating to the area of electrorheological (ER) fluids. In their talks, Choi and Zhao covered novel ER materials with core-shell structured microspheres and micro/nano hierarchical structured titania particles, respectively, and Tao focused on the reduction of the viscosity of liquid suspensions for energy applications in transporting crude oil via pipelines.

Attendance of the presentations was exceptionally high. Poster presentations were divided into two sessions and held in the afternoons after the close of the sessions, which were filled with stimulating discussions. The award for the best student research in the area of electrorheological fluids, sponsored by the Winslow family, was given to Y D Liu of INHA University, Korea on research entitled 'Copolyaniline coated monodisperse polystyrene microparticles and their electrorheological response'. The award for the best student research on magnetorheological suspensions, sponsored by the LORD Corporation, was given to S Kaneko of Keio University, Japan on research entitled 'Effect of a magnetic field on sloshing pressure in a magnetic fluid'.

Besides the purely scientific program during the five days, some special events were also organized. A guided tour of the old city and some important landmarks of the capital city of Ankara were organized before the Gala Dinner. Special samples of Turkish classical music and folk dancing were performed by a group from the Republic of Turkey Ministry of Culture and Tourism to exhibit Turkish culture to the delegates at Gala Night, which was truly appreciated. Following the conclusion of this successful meeting, the next conference is expected to be organized at the University of Granada, Spain by Professor F Gonzalez-Caballero in 2014.

It is expected that during the next conference, the interaction between polymer chemists synthesizing the new ER/MR materials, experimentalists from the rheological side characterizing their rheological properties, theoretical physicists describing the electric field and magnetic field dependent phenomena in ER/MR fluids rheology, and mechanical engineers conducting vibration damping tests will give rise to a deeper understanding of the ER/MR phenomena and will result in new findings in this field.

The conference was sponsored by Gazi University, Anton-Paar GmbH, Kurimoto Ltd., Anamed Analitik Grup, TA Instruments, LORD Corporation, Turkish Powder Metallurgy Association (TTMD), LiKrom Ltd, Atomika Ltd., Turkish Patent Institute (TPI), Berkecan Ltd., Kurukahveci Mehmed Efendi Mahdumlari and Turkish Science-Research Foundation (TUBAV). The organizing committee truly appreciates the support from these organizations.

Special appreciation is also due to my students O Erol and H C Gullu. I would also like to thank members of the Local Organizing Committee and International Advisory Board.

Guest Editor

H Ibrahim Unal
Gazi University
Science Faculty
Chemistry Department
06500 Ankara/Turkey
E-mail: hiunal@gazi.edu.tr

Journal of Physics: Conference Series **412** (2013) 011002 doi:10.1088/1742-6596/412/1/011002

Conference Committees

Local Organizing Committee

H Ibrahim Unal Chair	Gazi University, Department of Chemistry
Ozlem Erol Conference Secretariat	Gazi University, Department of Chemistry
Duran Altiparmak	Gazi University, Department of Automotive Engineering
Yucel Ercan	TOBB University of Economics and Technology, Department of Mechanical Engineering
Hasim Yilmaz	Nevsehir University, Department of Chemistry
Mustafa Yavuz	Suleyman Demirel University, Department of Chemistry
Mehmet Cabuk	Suleyman Demirel University, Department of Chemistry
Turgay Ergin	Hacettepe University, Department of Mechanical Engineering
O Yunus Gumus	Nevsehir University, Department of Chemistry
Ismail Ozenturk	Gazi University, Department of Chemistry
Seyma Ozkan	Gazi University, Department of Chemistry
Seyda Cabuk	Gazi University, Department of Chemistry
Evrim Sever	Gazi University, Department of Chemistry
Halil C Gullu	Gazi University, Department of Mechanical Drawing and Construction

International Advisory Board

Andrejs Cebers	Institute of Physics, Latvia
Anezka Lengalova	Tomas Bata University, Czech Republic
Daniel Barber	Lord Corporation, USA
Daniel J Klingenberg	University of Wisconsin, Madison, USA
Evguenia Korobko	A V Lykov Heat and Mass Transfer Institute, Belarus
Faramarz Gordaninejad	University of Nevada, Reno, USA
Georges Bossis	University of Nice, France
Hyoung J Choi	Inha University, Korea
Holger Bose	Fraunhofer Institute, Wurzburg, Germany
James E Martin	Sandia National Laboratory, USA
Kunquan Lu	Chinese Academy of Science, China
Luwei Zhou	Fudan University, China
Masami Nakano	Tohoku University, Japan
Norman M Wereley	University of Maryland, USA
Rongjia Tao	Temple University, USA
Seung-Bok Choi	Inha University, Korea
Stefan Odenbach	TU Dresden, Germany
Weijia Wen	Hong Kong University of Science and Technology, China
Xinglong Gong	University of Science and Technology of China, China

Conference e-mail: ermr2012@gazi.edu.tr
Conference web page: www.ermr2012.gazi.edu.tr

Conference Sponsors

Gazi University, Anton-Paar GmbH, Kurimoto Ltd., Anamed Analitik Grup, TA Instruments, Lord Co., Turkish Powder Metallurgy Association (TTMD), LiKrom Ltd, Atomika Ltd., Turkish Patent Institute (TPI), Berkecan Ltd., Kurukahveci Mehmed Efendi Mahdumlari and Turkish Science-Research Foundation (TUBAV).

ERMR2012 Conference venue, Main Building of Gazi University, Ankara

ERMR2012 Conference photo

Conference Chair, Professor Dr Unal delivers welcome speech, discussions during the ERMR2012 Conference and a view from the poster area

Views from the city tour (Hamamonu old city, Anitkabir and Kocatepe Mosque)

Views from the conference Gala dinner

Conference chair, Dr Unal presents plaquettes to plenary lecturer Dr H.J.Choi and to sponsors, Turkish Patent Institute and Kurimoto Ltd

Best poster award winners on ER materials (left: Y D Liu, INHA University, Korea) and MR materials (right: S Kaneko, Keio University, Japan)

List of participants

Participant	Institution	Participant	Institution
Aadıl Elmoumnı	TA Instruments - Waters LLC, USA	Lingyu Sun	Beihang University, China
Ahmad Isnikurniawan	Keio University, Japan	Liu Degang	Academy of Armored Forces Engineering, China
Akio Inoue	Chuo University, Japan	Luwei Zhou	Fudan Unviersity, China
Alexsandra Karabko	Belarusian State University of Informatics and Radioelectronics, Belarus	M. Ashraf Imam	Naval Research Laboratory, USA
Ali Tadayoninejad	Shiraz University, Iran	Ma Ning	Dalian University of Technology, China
Ansgar Wiehe	Ostwestfalen-Lippe University, Germany	Marco Jackel	Fraunhofer Institute, Germany
Antonio José Faria Bombard	Universidade Federal De Itajubá, Brazil	Martin Obst	TU Dresden, Germany
Aslıhan Orum	Sabanci University, Turkey	Masaaki Motozawa	Tokyo University of Science, Japan
Aysegul Derme	Undersecretariat For Defence Industries, Turkey	Masami Nakano	Tohoku University
Barkan Kavlicoglu	Advanced Materials and Devices, Inc., USA	Mehmet Cabuk	Suleyman Demirel University, Turkey
Barnabas Horvath	University of Pannonia, Hungary	Miao Yu	Chongqing University, China
Basim Abu-Jdayil	UAE University, UAE	Miao Guo	Univeristy of Wollongong, Australia
Ben-Xiang Ju	Chongqing University, China	Michal Sedlacik	Tomas Bata University in Zlin, Czech Republic
Bogdan Sapinski	AGH University of Science and Technology, Poland	Mikalai Zhurauski	Heat and Mass Transfer Institute, Belarus
Cahit A. Evrensel	University of Nevada, USA	Min-Sang Seong	Inha University, Korea
Cécilia MAGNET	Université De Nice-Sophia Antipolis Parc Valrose, France	Miroslav Mrlik	Tomas Bata University in Zlin, Czech Republic
Chaoyang Guo	University of Science and Technology of China, China	Mohammad Mehdi Naserimojarad	Shiraz University, Iran
Chiranjit Sarkar	Indian Institute of Technology Delhi, India	Monica Moral Muñoz	University of Granada, Spain
Dan Baiasu	Atelierele CFR Grivita SA, Romania	Mustafa Yavuz	Suleyman Demirel University, Turkey
Daniel E. Barber	LORD Corporation, USA	Nafel Dogdu	Gazi University, Turkey
Daniel Klingenberg	University of Wisconsin, USA	Omer Y. Gumus	Nevsehir University, Turkey
Dingxin Leng	Beihang University, China	Ozlem Erol	Gazi University, Turkey
Dirk Güth	Ostwestfalen-Lippe University, Germany	Pavel Kuzhir	University of Nice, France

Dmitry Borin	TU Dresden, Germany	Petra Svrcinova	Institute of Hydrodynamics Academy of Sciences of The Czech Republic, Czech Republic
Elif Besel	Undersecretariat For Defence Industries, Turkey	Rongjia Tao	Temple University, USA
Evguenia Korobko	Lykov Heat and Mass Transfer Institute of NAS Belarus	Roozbeh Aayani	Shiraz University, Iran
Evrim Sever	Gazi University, Turkey	Roque Hidalgo-Alvarez	University of Granada, Spain
Faramarz Gordaninejad	University of Nevada, USA	Rui Li	Chongqing University of Posts and Telecommunications, China
Fernando Gonzalez-Caballero	University of Granada, Spain	S.H. Hashemabadi	Iran University of Science and Technology
Gang Sun	Chinese Academy of Sciences, China	Seok-Jae Ha	Inha University, Korea
Gennady Stepanov	State Scientific Research Institute of Chemistry and Technology of Organoelement Compounds, Russia	Seyda Cabuk	Gazi University, Turkey
Gergely Kronome	University of Pannonia, Hungary	Seyma Ozkan	Gazi University, Turkey
Gjergj Dodbiba	The University of Tokyo, Japan	Shawqi Almajdalawi	Tomas Bata University İn Zlin, Czech Republic
Guifré Julio	Université De Sherbrooke, Canada	Shinya Yamanaka	Muroran Institute of Technology, Japan
Guillermo Ramon Iglesias Salto	University of Granada, Spain	Shouhu Xuan	University of Science and Technology of China, China
Guo Jinmao	Academy of Armored Forces Engineering	Shunsuke Kaneko	Keio University, Japan
Guojiang Liao	University of Science and Technology of China, China	Stefan Odenbach	TU of Dresden, Germany
H. Ibrahim Unal	Gazi University, Turkey	Steffen Schneider	Bundeswehr Research Institute For Materials, Fuels and Lubricants, Germany
Hanna Lagger	Fraunhofer Institute For Mechanics of Materials IWM, Freiburg, Germany	Stephan Ulrich	Helmut-Schmidt-University, Germany
Hasim Yilmaz	Nevsehir University, Turkey	Stoyan Smoukov	Cambridge University, UK
Hideaki Masuda	Keio University, Japan	Sukran Katmer	Yalova Community College, Yalova University, Turkey
Hiroki Tomori	Chuo University, Japan	Sung Hoon Ha	Inha University, Korea
Hiroyuki Iwashina	Keio University, Japan	Takehito Kikuchi	Yamagata University, Japan
Hitoshi Nishida	Toyama National College of Technology, Japan	Tamer Turkucu	Gazi University, Turkey
Holger Böse	Fraunhofer ISC, Germany	Taro Nakamura	Chuo University, Japan

Hua Yan	Logistic Engineering University, China	Tatsuo Majima	Chuo University, Japan
Hui Juan Yao	Chongqing University, China	Tatsuo Sawada	Keio University, Japan
Hyoung Jin Choi	Inha University, Korea	Thomas Gundermann	TU Dresden, Germany
Ismail Ozenturk	Gazi University, Turkey	Tibor Medvegy	University of Pannonia, Hungary
István Szalai	University of Pannonia, Hungary	Tongfei Tian	University of Wollongong, Australia
Jakub Roupec	Brno University of Technology, Czech Republic	Turgay Ergin	Hacettepe University, Turkey
James E. Martin	Sandia National Lab., USA	Wei Li	Beihang University, China
Janusz Gołdasz	BWI Group, Technical Center Kraków, Poland	Weihua Li	University of Wollongong, Australia
Javier Alonso Lopez Medına	Universidad Del Valle, Colombia	Wen Ling Zhang	Inha University, Korea
Jean-Francois Desrosiers	Université De Sherbrooke, Canada	William Kordonski	QED Technologies International, USA
Jean-Philippe Lucking Bigué	Université De Sherbrooke, Canada	Wojciech Horak	AGH University of Science and Technology, Poland
Jessica Alves Marins	Universidade Federal De Itajubá, Brazil	Won-Ho LEE	Gangneung-Wonju National University, Korea
Jianbin Zeng	University of Technology, Sydney, Australia	Xiangpan Zheng	Fuzhou University, China
Jianbo Yin	Northwestern Polytechnical University, China	Xiaojie Wang	University of Nevada, USA
Jianwei Tu	Wuhan University of Technology, China	Xiaopeng Zhao	Northwestern Polytechnical University, China
Jie Fu	Chongqing University, China	Xinglong Gong	University of Science and Technology of China, China
Jile Jiang	Tsinghua University, China	Xuezheng Jiang	Nanjing University of Science and Technology, China
Jiong Wang	Nanjing University of Science and Technology, China	Xufeng Dong	Dalian University of Technology,China
Joanes Berasategui	Mondragon University, Spain	Yanceng Fan	University of Science and Technology of China
Jong-Chul Lee	Gangneung-Wonju National University, Korea	Yancheng Li	University of Technology Sydney, Australia
Jong-Seok Oh	Inha University, Korea	Yasuhiro Kakinuma	Keio University, Japan
Jose A. Ruiz-López	University of Granada, Spain	Yasushi Ido	Nagoya Institute of Technology, Japan
Juan Pablo Segovia Gutiérrez	University of Granada, Spain	Ying Dan Liu	Inha University, Korea
Juan De Vicente	University of Granada, Spain	Yoon-Hyun Kim	Changsung Corporation, Korea
Jung-Won Lee	Inha University, Korea	Youguang Guo	University of Technology Sydney, Australia
Ken Plunkett	North Carolina State University, USA	Yu Tian	Tsinghua University, China
Ken'ichi Koyanagi	Toyama Prefectural University, Japan	Yuhiro Iwamoto	Doshisha University, Japan

Kohki Tsuchiya	Tohoku University, Japan	Zbyněk Strecker	Brno University of Technology, Czech Republic
Konstantin Krivenkov	Helmut-Schmidt-University, Germany	Zekeriya Parlak	Sakarya University, Turkey
Kwang-Pyo Hong	Inha University, Korea	Zhang Jinqiu	Academy of Armored Forces Engineering, China
Laura Rodriguez-Arco	University of Granada, Spain	Zhaochun Li	Nanjing University of Science and Technology, China
Li Guoqiang	Academy of Armored Forces Engineering	Zhi-De Hu	Logistic Engineering University, China

Journal of Physics: Conference Series **412** (2013) 011004 doi:10.1088/1742-6596/412/1/011004

Peer review statement

All papers published in this volume of *Journal of Physics: Conference Series* have been peer reviewed through processes administered by the proceedings Editors. Reviews were conducted by expert referees to the professional and scientific standards expected of a proceedings journal published by IOP Publishing.

Preparation, microstructure and electrorheological activity of nanosized modified titanium dioxide and titaniferous fillers

E V Korobko[1,3], Z A Novikova[1], M A Zhurauski[1], A N Murashkevich[2] and O A Alisienok[2]

[1]A.V. Luikov Heat and Mass Transfer Institute of NAS Belarus, 220072, 15 P. Brovka str., Minsk, Belarus
[2]Belarusian State Technological University, 220050, 13a Sverdlova str., Minsk, Belarus

E-mail: evkorobko@gmail.com

Abstract. Nanodisperse powders of pure, superficially and structurally modified titanium dioxide as well as composition materials SiO_2–TiO_2 with different morphology are synthesized. Electrorheological properties of dispersions on their basis are investigated. It is established, that doping of TiO_2 by heterovalency replacement of the titanium with aluminum and chrome ions allows us to raise essentially electrorheological activity of dispersions in comparison with pure titanium dioxide.

1. Introduction

Micro- and nanosized poorly conducting, capable to be polarized in the electric field particles are used as a dispersed phase of electrorheological fluids (ERF). Among inorganic materials nanooxides of titanium, possessing high dielectric permeability, and of silicon, having low density, that provides sedimentation stability of ERF are a subject of active researches recently [1].

Investigating influence of microstructure (type of crystal structure, pore size, surface area) of pure TiO_2 on electrorheological properties of dispersions, Shang *et al.* [2] have arrived at a conclusion that TiO_2 particles of anatase structure show larger electrorheological (ER) activity in comparison with TiO_2 of rutile structure. However pure TiO_2 does not give a sufficient increase in viscosity in the electric field. Viscosity of the dispersion, containing 25 wt.% of filler, at the electric field intensity of 3.2 kV/mm and shear rate of 300 s^{-1} increases only by 7 times. It results from the fact that the fast ionic or atomic polarization inherent crystalline TiO_2 does not provide the optimum dielectric and polarizing properties necessary for revealing high ER-activity.

Hao *et al.* [3] consider that the greatest contribution to electrorheological effect (ERE) is brought by the interfacial polarization caused by presence of a surface charge. Therefore, structural and surface modification of nanodisperse inorganic oxides by using reproducible methods of synthesis is a perspective way to increase the quantity of charge carriers, to intensify interfacial polarization, and, consequently, to create effective ERF fillers with high ER-activity [4].

2. Experimental

Nanodisperse pure titanium dioxide, and composition materials SiO_2 – TiO_2 with various morphology:

[3]Corresponding author.

co-precipitated powders and the materials having morphology core SiO_2 – shell TiO_2, and also superficially and structurally modified titanium dioxide were synthesized and investigated. Co-precipitated materials SiO_2 – TiO_2 have been obtained by combined hydrolysis of silicon and titanium containing components. The materials with morphology core SiO_2 – shell TiO_2, have been obtained by formation of titanium dioxide shell on nanodisperse microspheres of silicon dioxide by the mechanism of electrostatic interaction. As initial silicon and titanium containing components titanium and silicon alkoxides: titanium tetraisopropoxide (TTIP), tetraethoxysilane (TEOS), liquid glass (l. g.), titanium tetra- and trichlorides were used. Surface modification of TiO_2 was carried out by introduction of urea or sodium dodecyl sulfate (SDS) during the process of sol-gel synthesis. Cetyltrimethylammonium bromide (CTAB) was used as a template. TTIP was a source of titanium dioxide. Samples were dried at temperature 120 °C for 5 h. Structural modification of TiO_2 was carried out by doping with cations of aluminum, chrome, lanthanum, yttrium. Synthesis was realized at the presence of templates – dodecylamine (DDA) and triethanolamine (TEA). Samples were dried at 120 °C for 5 h and calcined at 600 °C for 3 h. Doping components and templates were introduced both into initial solutions of titanium containing components, and in colloidal solutions of titanium dioxide, obtained by peptization of freshly precipitated titanium dioxide from solutions of titanium tetrachloride in isopropyl alcohol. The obtained materials were tested by means of methods of the physical and chemical analysis: adsorptive and thermogravimetry, chemical and X-ray phase analysis methods.

For determination of ER-activity of samples, dispersions on the basis of the obtained powders with weight concentration of 5 % in transformer oil are tested by means of the measuring cell consisting of two coaxial cylinders.

3. Results and discussion

Some conditions of synthesis and properties of obtained fillers are tabulated below. Pure titanium dioxide and nanocomposition materials SiO_2 – TiO_2 have the developed specific surface area. Dispersity of fillers, their phase structure and surface properties depend on preparation conditions and processing temperature. Samples after drying at 120°C represent amorphous materials characterized by appreciable hydrophilicity. Calcination of them at temperatures between 200–1000°C is accompanied by the processes of dehydration (weight losses at calcination in the range of temperatures 200–1000°C are 15 – 25% and they are caused by removal of physically adsorbed, coordinated bound water and surface hydroxyl groups), crystallization of titanium dioxide, considerable reduction of a specific surface area.

It is established that doping of titanium dioxide with compounds of Cr, Al, La, Y within 5–10 mol. %, and also introduction of templates slows down crystallization process of titanium dioxide at heat treatment and allows to receive nanocrystal product with a crystallite size d = 10–15 nm and a specific surface area S = 80–150 m^2/g. Titanium dioxide obtained in similar conditions and not containing doping additions, has d = 45–50 nm and S = 30–50 m^2/g.

According to the chemical nature and technology of production, the samples 1 – 4 of the table 1 concern to containing water materials, and samples 5 – 11 on a basis of nanodisperse doped titanium dioxide concern to waterless fillers.

The fillers, containing partially physically adsorbed and coordinated bound water (samples 1 – 4), have shown appreciable electrorheological activity (figure 1a). However, some of them (samples 1 and 4), as seen in figure 1b, have rather high conduction current densities 20–35 $\mu A/cm^2$. The sample of titanium dioxide modified by urea, obtained by template synthesis (figure 1a, curve 3), showing appreciable ER-activity, has a rather low values of density of conduction currents 0.01–2.5 $\mu A/cm^2$ (figure 1b, curve 3), in spite of the fact that heat treatment of this sample was carried out at 120°C. The ERF on the basis of the titanium dioxide sample modified with DDS (the sample 4 of the table 1), has shown ER-activity, comparable with the activity of the sample 3. However, heat treatment of water containing samples at temperature 200°C results in full loss of ER-activity because of the removal of physically adsorbed and, probably, coordinated bound water which is an activator of ERE in this case.

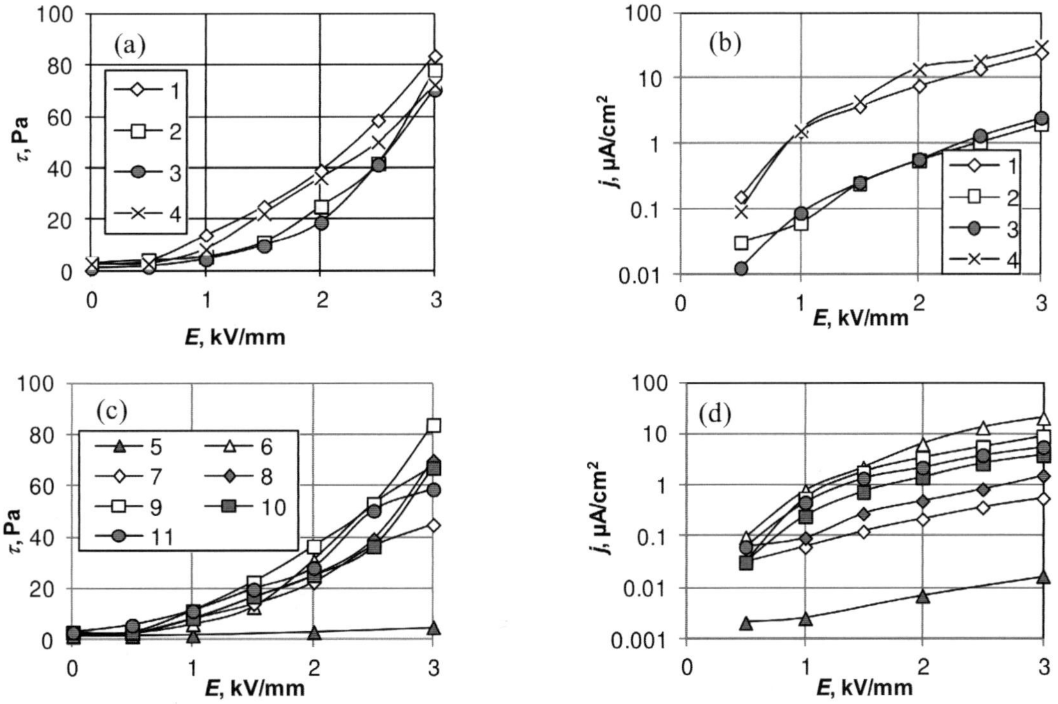

Figure 1. Dependence of shear stress (a, c) and conduction current density (b, d) of ER fluids on the electric field intensity (shear rate is 32 s^{-1} for samples 2, 5, 6; 17.2 s^{-1} for the rest of samples); filler concentration is 5 wt. %; numbers of samples correspond to numbers in the Table 1.

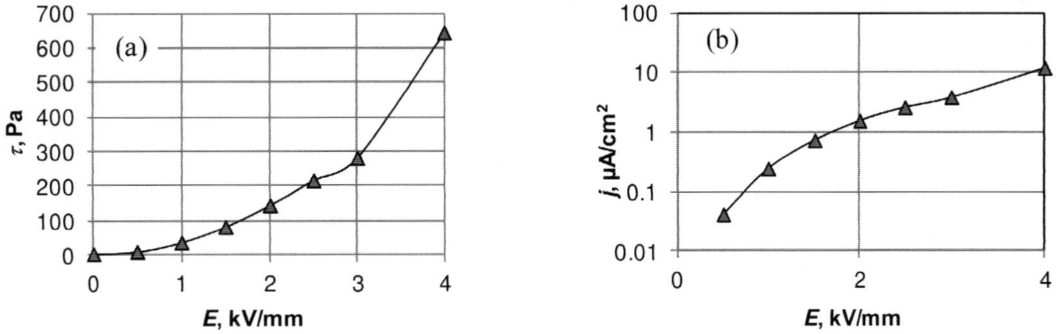

Figure 2. Dependence of shear stress (a) and conduction current density (b) of ERF with 30 wt. % (sample 8) on the electric field intensity (shear rate is 17.2 s^{-1}).

The samples doped with aluminum and chromium (6, 8, and 9), having the developed specific surface area 120–135 m^2/g, were the best among waterless fillers. Increase of shear stress τ of ERF on the basis of these fillers reach 30–40 times at E = 3 kV/mm at current density of 7–21 $\mu A/cm^2$ while pure mesoporous titanium dioxide (the sample 5) has shown increase only by 3.5 times.

Considering positive characteristics – high ER-activity and low values of current density of the sample TiO_2 doped by aluminum (the sample 8), the fluid containing 30 wt. % of filler in mineral oil is prepared on its basis. In figure 2 dependences of shear stress and current density on the electric field intensity are presented. In the electric field of intensity of 3 kV/mm shear stress increases more than by 2 orders. At that, values of current density do not exceed 12 $\mu A/cm^2$ (at the electric field intensity of 4 kV/mm). In the range of E up to 3 kV/mm, the dependence j (E) practically coincides with the dependence for the sample doped with lanthanum (10), which has a smaller dispersed phase concentration of 5 wt. % (figure 1d). The samples doped with chrome and yttrium (6 and 11) give even larger values *j* (at 5 wt. %). Therefore it is necessary to expect, that current density for these samples at a bigger concentration of a dispersed phase will be essentially higher and will not allow to reach high ER-activity. Thus, aluminum seems to be the most perspective doping element among the investigated metals.

Table 1. Samples of fillers and their properties.

No.	Chemical nature and morphology	Parent substances	Treatment conditions T (°C)	Treatment conditions Time (h)	Template	S (m^2/g)
1	Co-precipitated composite $SiO_2 - TiO_2$	L. g., $TiCl_3$	120	5	–	355
2	Core SiO_2 – shell TiO_2	L. g., $TiCl_4$	120	5	–	206
3	TiO_2 + urea + TEA	TTIP, urea, TEA, NaOH	120	5	CTAB	300
4	TiO_2 + DDS	TTIP, CH_3COOH	120	5	DDS	160
5	TiO_2	TTIP, TEA	120 450	5 3	CTAB	37
6	TiO_2 + Cr	TTIP, $CrCl_3$, TEA, NaOH	450	3.5	CTAB	135
7	TiO_2 + Al Ti/Al=90:10	sol TiO_2, $Al(NO_3)_3$	150 600	5 3	DDA	79
8	TiO_2 + Al Ti/Al=93:7	sol TiO_2, $Al(NO_3)_3$	150 600	5 3	DDA	120
9	TiO_2 + Al Ti/Al=94:6	sol TiO_2, $Al(NO_3)_3$	150 600	5 3	DDA	130
10	TiO_2 + La Ti/La=94:6	sol TiO_2, $La(NO_3)_3$	150 600	5 3	DDA	112
11	TiO_2 + Y Ti/Y=94:6	sol TiO_2, $Y(NO_3)_3$	150 600	5 3	DDA	142

Note: TiO_2 content (wt. %) in the samples: 1 – 32.0; 2 – 7.0; the rest samples – 82–90

4. Conclusion

Thus, it is established that structural modification of titanium dioxide (production of the mesostructured product in the presence of urea), and also doping TiO_2 by heterovalent replacement of the titanium with ions of some transition and rare-earth elements allows to raise essentially ER-activity of dispersions on its basis in comparison with pure TiO_2. For promotion of titanium dioxide as effective filler of industrial ERF it is necessary to continue investigations in wider interval of structures, temperatures, to optimize the production technology of such fillers by using the available initial reagents of the inorganic nature.

References

[1] Zhao X P and Yin J B 2006 *J. Ind. Eng. Chem* **12** (2) 184–98
[2] Shang Y, Ma S, Li J, Li M, Wang J and Zhang S 2006 *J. Materials Science & Technology* **22** (4) 572–6
[3] Hao T, Kawai A and Ikazaki F 1998 *Langmuir* **14** (5) 1256–62
[4] Wu Q, Zhao B Y, Fang C and Hu K A 2005 *Eur. Phys. J.* E **17** 63–7

Electrorheological behaviour under oscillatory shear of TiO_2 rod-like particles prepared via microwave-assisted molten-salt synthesis

M Sedlacik[1,2,4], M Mrlik[1,3], V Pavlinek[1,3], Z Kozakova[1,3] and P Saha[1,3]

[1] Centre of Polymer Systems, University Institute, Tomas Bata University in Zlin, Nad Ovcirnou 3685, 760 01 Zlin, Czech Republic
[2] Department of Production Engineering, Faculty of Technology, Tomas Bata University in Zlin, T. G. Masaryk Sq. 275, 762 72 Zlin, Czech Republic
[3] Polymer Centre, Faculty of Technology, Tomas Bata University in Zlin, namesti T. G. Masaryka 275, 762 72 Zlin, Czech Republic

E-mail: msedlacik@ft.utb.cz

Abstract. Titanium dioxide (TiO_2) rod-like particles were synthesized by a simple and rapid microwave-assisted molten-salt method. The X-ray diffraction analysis and electron microscopy provided information on particle composition and morphology, respectively. It was found that during the synthesis process the crystalline phase of TiO_2 transformed from anatase into rutile while the morphology changed from nanospheres into micrometer sized rod-like particles. The electrorheological (ER) properties were investigated via oscillatory shear tests. It was found that TiO_2 rod-like particles based silicone oil suspensions exhibited higher ER activity than those of original anatase TiO_2 nanoparticles probably due to side-by-side solid friction between particles as well as shorter time of their polarization. The changes in ER properties of rod-like particle based suspensions as a function of the applied electric field strength and particles weight fraction were also investigated.

1. Introduction

Electrorheological (ER) fluids are usually two-phase systems that undergo a tunable and reversible liquid to solid-like state transition upon external electric field application [1-3]. Typical ER fluids are composed of electrically polarizable particles dispersed in a non-conducting carrier liquid. In the absence of an electric field, ER fluids typically exhibit Newtonian behaviour. When the electric field is applied, the interfacial polarization of particles occurs resulting in the aggregation of particles into chain-like structures aligned in the direction of the field streamlines [4, 5]. Such a response usually takes place in a millisecond scale and is accompanied with the increase of rheological properties including viscosity, shear stress, and viscoelastic moduli. This phenomenon called ER effect is moreover completely reversible when the electric field is switched off. Hence, ER fluids could be used in various technical applications such as in automotive industry for clutches, brakes or damping systems [6].

[4] Corresponding author

Basically, the particle geometry can play a significant role in their ER response, since one-dimensional particles, whose major axis is aligned according to the field direction and have the same relative permittivity and volume as spherical particles, will form stronger internal structures [7,8]. Great influence of particles anisotropy was examined for example in ER fluids based on conducting polymers in which the ER performance was enhanced in comparison with fluids based on microspheres [9,10]. It is also obvious that anisotropic particles exhibit better sedimentation stability due to the hindered settling, easier redispersion as well as lower concentration necessary for the formation of chain-like structures spanning the operating gap [7]. However, the ER performance of anisotropic particle based ER fluids has been scarcely studied probably due to the lack of suitable methods for preparation of such particles.

In this study, with a view to prepare a novel anisotropic material with a good ER performance, the rod-like particles of titanium dioxide (TiO_2) were synthesized via a simple and rapid microwave-assisted molten-salt method. This technique is based on the use of low-melting salts to accelerate diffusion and thus formation of required structures with high crystallinity [11]. Moreover, a novel principle of microwave heating was used instead of classical heating in an oven that shortened the reaction time significantly. Furthermore, a comparison with the viscoelastic behaviour of suspensions of spherical TiO_2 particles was performed.

2. Experimental

2.1. Materials
Starting material for the synthesis titanium (IV) dioxide powder consisted of anatase crystalline phase (99.8% trace metal basis) was purchased from Sigma-Aldrich (USA). Salts used as molten environment were sodium chloride (NaCl, Penta, Czech Republic) and sodium phosphate dibasic dodecahydrate ($Na_2HPO_4.12H_2O$, Penta, Czech Republic) and they form eutectic mixture at 735°C. All chemicals were used as received.

2.2. Synthesis of TiO_2 rod-like particles
Titanium dioxide rod-like particles were prepared by a solvent-free and rapid microwave-assisted molten-salt method in a ceramic kiln with microwave-absorbing layer which enables fusing at high temperatures while the kiln is exposed to microwaves in a microwave oven [12]. Briefly, original TiO_2 of anatase crystalline phase was homogenized with eutectic mixture of NaCl / $Na_2HPO_4.12H_2O$ in a ratio of 4:5. The as prepared powder mixture was placed into corundum crucible, covered by a lid and placed into the ceramic kiln. The kiln was subsequently placed into microwave oven (Hyundai, Korea) and exposed to microwaves at standard frequency 2.45 GHz for 30 minutes at 750 W. After the reaction, the whole kiln was allowed to cool down to room temperature prior opening the corundum crucible. The obtained product was washed several times with distilled water and vacuum dried at 80°C for 12 hours.

2.3. Particle characterization
The crystallinity and phase composition of starting TiO_2 with anatase crystalline phase and prepared powder was carried out with X'Pert PRO (XRD, Philips, the Netherlands) X-ray diffractometer, fitted with copper target, Kα ray, scanning rate of 4° min^{-1} for the recording data in the range of $2\theta = 20°$ – 90° (360 kV, 20 mA). The morphology of starting TiO_2 and prepared powder was determined with TEM (Transmission Electron Microscopy JEOL 1200, JEOL Ltd., Japan) and SEM (Scanning Electron Microscope VEGA II LMU, Tescan, Czech Republic) operated at 10 kV, respectively.

2.4. Electrorheological measurements
The 5, 10, 15, and 20 wt.% ER fluids were prepared by dispersing the original anatase TiO_2 or prepared rod-like TiO_2 particles in a corresponding volume of silicone oil (Lukosiol M200, Chemical

Works Kolín, Czech Republic, viscosity $\eta_c = 200$ mPa·s). Before each measurement, the fluid was stirred first mechanically and then in an ultrasonic bath for 60 s.

Rheological measurements were performed using a rotational rheometer (Bohlin Gemini, Malvern Instruments, UK) modified for ER experiments, with a coaxial cylinder geometry. The fluids were placed into a Couette cell with a rotating inner cylinder of 14 mm diameter and stationary outer cylinder separated by 0.7 mm gap. DC high-voltage source TREK (TREK 668B, USA) was connected to the rheometer to generate the electric field strength 0–3 kV·mm^{-1}. Before each measurement at new electric field strength used, the formed internal structure within the suspension was destroyed by shearing of the sample at a shear rate 20 s^{-1} for 150 s. Further, dynamic viscoelastic tests were carried out by dynamic strain and frequency sweeps. The strain sweep was performed with applied strain in the range from 10^{-4} to 10^{-2} at the frequency of 6.28 rad·s^{-1} under an electric field to determine the linear viscoelastic region (LVR). The viscoelastic moduli were subsequently obtained from the frequency sweep tests (0.63 – 62.83 rad·s^{-1}) at a fixed strain amplitude in the LVR. The temperature in all experiments was kept at 25°C.

2.5. Dielectric properties

The frequency dependence of dielectric properties (relative permittivity, ε', and dielectric loss factor, ε'') were obtained with an impedance analyzer (Agilent 4524, Japan) in the frequency range 4×10^1 – 5×10^6 Hz. The dielectric spectra were characterized with a Cole–Cole model [13,14] (Equation 1).

$$\varepsilon^* = \varepsilon' - i\varepsilon'' = \varepsilon_\infty + \frac{\Delta\varepsilon'}{(1 + i\omega\lambda)^{1-\alpha}} \tag{1}$$

where polarizability, $\Delta\varepsilon'$, is the difference between static, ε'_0, and high frequency, ε'_∞, relative permittivities, ω, is angular frequency ($=2\pi f$), λ, is dielectric relaxation time ($=1/2\pi f_{max}$), in which f_{max}, is defined by a local maximum of the dielectric loss of ER fluid, and, α, is the scattering degree of relaxation times.

3. Results and discussion

3.1. Particles characterization

According to the XRD analysis (Figure 1), the microwave-assisted molten-salt synthesis transformed the original anatase crystalline phase of starting TiO_2 almost completely into the rutile crystalline phase in prepared particles, i.e. although there are presented some low diffraction peaks of anatase phase in the XRD pattern of prepared TiO_2 particles. The narrow and most intensive peaks show very good crystallinity and can be indexed to the tetragonal rutile phase.

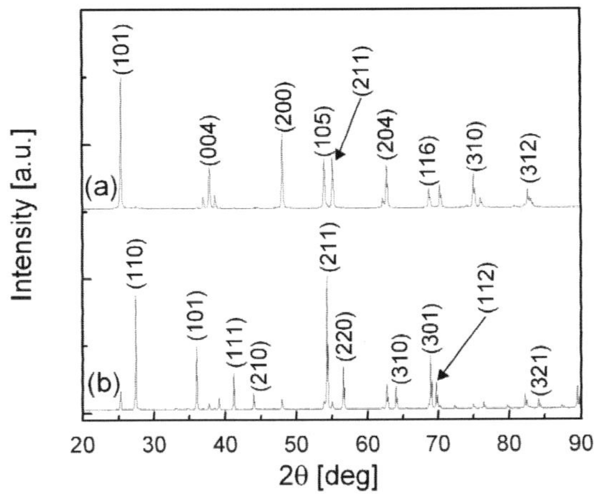

Figure 1. Powder XRD patterns of original anatase TiO_2 particles (a) and prepared TiO_2 rod-like particles via microwave-assisted molten-salt synthesis (b).

The changes in TiO_2 particle size and shape resulting from the use of microwave-assisted molten-salt synthesis were determined by TEM and SEM image analysis shown in Figure 2. Particles of original anatase TiO_2 are approximately spherical with mean particle size of about 200 nm. On the other hand, the successful transformation from spherical nanoparticles to rod-like microparticles is evident from Figure 2b. The length of rod-like particles is ranging from 5 to 10 μm while the diameter from 0.5 to 2 μm. The geometric aspect ratio, L/D, of prepared TiO_2 particles is correspondingly ranging from 5 to 10.

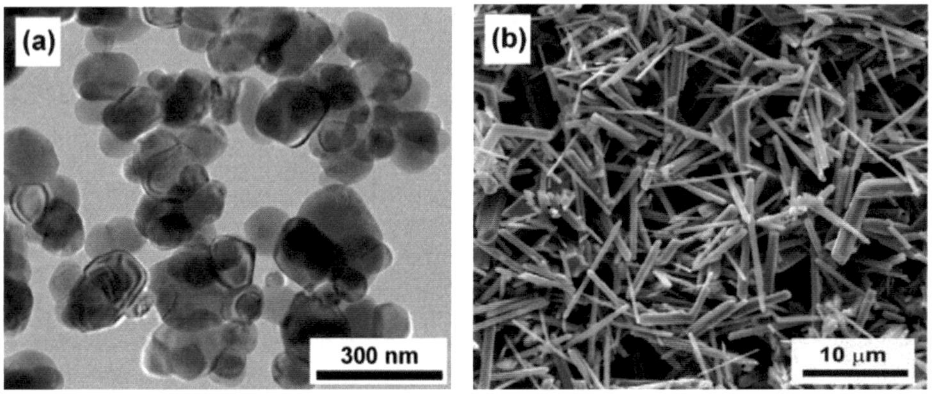

Figure 2. Electron micrographs of original anatase TiO_2 (a) and prepared rutile TiO_2 rod-like particles (b).

3.2. ER activity of prepared ER fluids

In order to check the structure formation within the ER fluid (Figure 3), dilute suspension of prepared rod-like TiO_2 particles was placed between microscope slides (the gap was fixed to 1 mm) connected to DC high-voltage source KEITHLEY (2400, USA). As can be seen in Figure 3a, in the absence of external electric field strength the particles are randomly dispersed. However, although the suspension was carefully mixed right before the observation, some particles are gathered together into aggregates. Such phenomenon could be caused by short-range attractive forces (Van der Waals or electrostatic) or mechanical cohesion between rod-like particle surfaces due to the solid friction between particles [13,14]. When the electric field is applied (Figure 3b), particles start to be polarized and the chain-like structure from connected particles is created between electrodes. Agglomerates formed in the absence of the field remain gathered resulting in incomplete alignment of individual particles with the field.

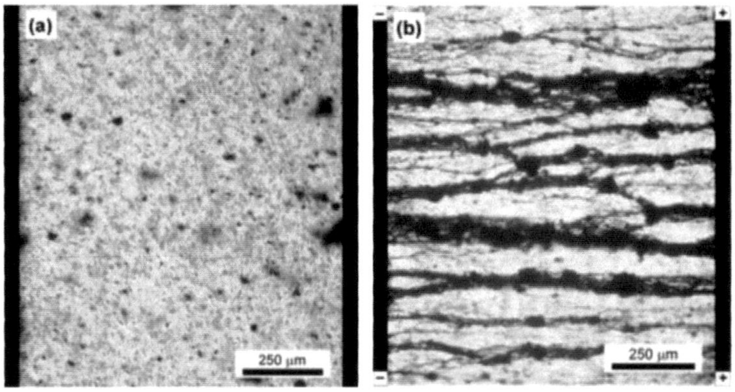

Figure 3. Optical micrographs of dilute ER fluid of TiO_2 rod-like particles in the absence (a) and in the presence of external electric field strength of 0.5 $kV \cdot mm^{-1}$ (b).

It is generally accepted that interfacial polarization of particles connected with the formation of chain-like structures in the presence of electric field is responsible for rheological changes of ER fluids (Figure 4). Hence, parameters obtained from equation 1 were used to help understand the ER activity of ER fluids based on TiO_2 spherical or rod-like particles (Table 1). Although, polarizability of particles, $\Delta\varepsilon'$, is relatively low, the time of their relaxation, λ, is much lower in the case of rod-like particles than those reported in ER fluids of various spherical particles [15,16], which indicates faster response to an electric field. It means that TiO_2 rod-like particles of rutile crystalline phase are polarized in shorter time, which may, consequently, contribute to their better ER activity in comparison with TiO_2 nanospheres of anatase crystalline phase.

Table 1. Dielectric parameters in equation 1 for original anatase TiO_2 particles and prepared rod-like TiO_2 particles based ER fluids (5wt.% concentration).

Parameter	Anatase TiO_2	TiO_2 rod-like
ε'_0 [–]	3.09	3.30
ε'_∞ [–]	2.93	3.07
$\Delta\varepsilon'$ [–]	0.16	0.23
λ [s]	4.45×10^{-5}	9.06×10^{-6}

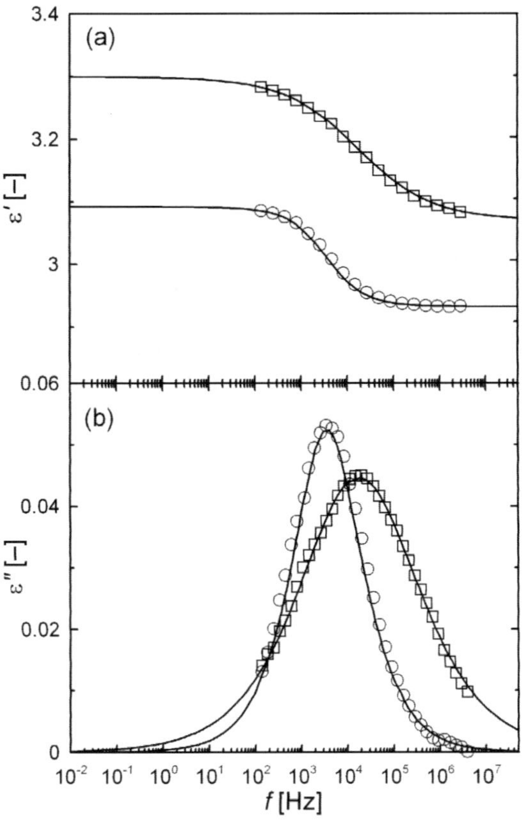

Figure 4. Relative permittivity, ε', (a) and dielectric loss factor, ε'', (b) as a function of frequency, f, for 5 wt.% suspension of original anatase TiO_2 particles (○) and prepared TiO_2 rod-like particles (□).

The stiffness of internal structures formed within the ER fluid in the presence of external electric field can be determined via the dependence of complex viscosity, η^*, on the angular frequency, ω. Figure 5 compares ER activity of TiO_2 rod-like particles with original TiO_2 nanoparticles at the same concentration. In the absence of electric field, rod-like particles based ER fluid exhibits almost Newtonian behavior while the suspension of original anatase nanoparticles is slightly pseudoplastic. On the other hand, strong electrostatic interactions in the presence of external electric field cause chain-like structure formation within the suspension which leads to a considerable enhancement of system stiffness. A further increase in pseudoplasticity occurs with increasing electric field strength applied. Moreover, the shear thinning behavior is also clearly observed with increasing frequency. Evidently, much stronger structures (higher viscosity) are formed in the rod-like particle based ER fluid at all electric field strengths applied due larger polarizability (Figure 4) and higher solid friction between particles [13]. The lower field-off viscosity and higher stiffness in the electric field consequently indicate higher ER efficiency of TiO_2 rod-like particle suspensions.

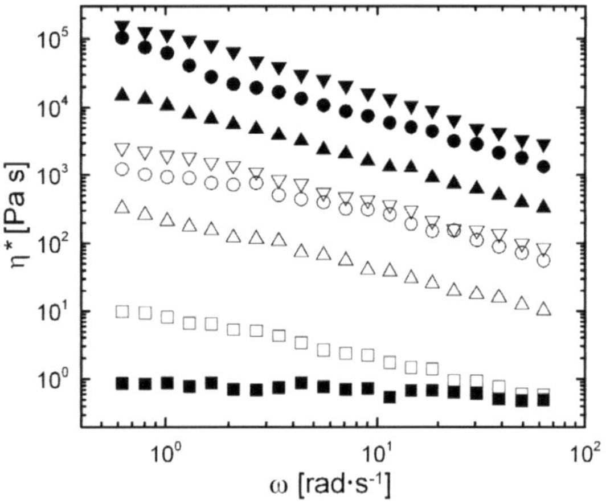

Figure 5. Dependence of complex viscosity, η^*, on angular frequency, ω, for 20wt.% suspensions of TiO_2 rod-like (solid symbols) and anatase TiO_2 (open symbols) particles in silicone oil at 25°C under various electric field strengths (kV mm^{-1}): 0 (■ □), 1 (▲ △), 2 (● ○), 3 (▼ ▽).

The formation of the internal chain-like structures of polarized particles in the presence of electric field is also accompanied by a change of viscoelastic moduli. Thus, Figure 6 shows the dependence of storage, G', and loss, G'', moduli on the strain amplitude, γ, in oscillatory flow for 20 wt.% ER fluid of TiO_2 rod-like particles. In the absence of electric field, the system exhibits liquid behavior when viscous portion is dominant over elastic one, i.e. G'' is larger than G'. When the electric field is applied G' becomes higher than G'' in the viscoelastic region and both moduli increase rapidly in several orders of magnitude from their field-off values. Especially in the case of electric field strength $E = 3$ kV·mm^{-1}, G' is higher than that at zero electric field by five orders of magnitude.

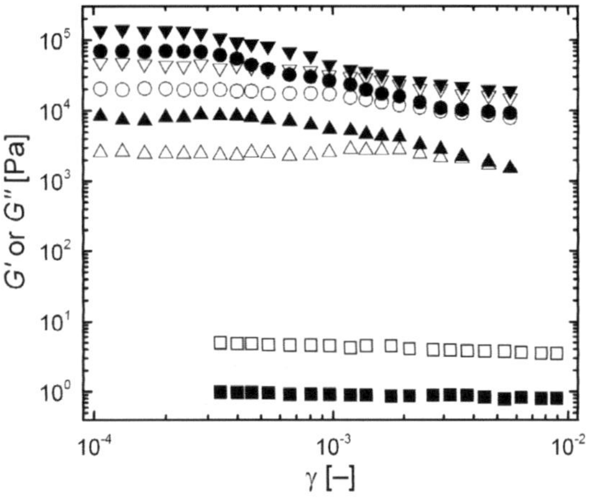

Figure 6. Storage, G', (solid) and loss, G'', (open) viscoelastic moduli as a function of strain amplitude, γ, for 20 wt.% TiO_2 rod-like particles suspension in silicone oil at 25 °C under various electric field strengths (kV·mm^{-1}): 0 (■ □), 1 (▲ △), 2 (● ○), 3 (▼ ▽).

For practical applications it is important to know the dependence of viscoelastic moduli on angular frequency, ω, in the linear viscoelastic region ($\gamma = 3 \times 10^{-4}$). As can be seen in Figure 7, the rod-like TiO_2 particles based ER fluid exhibits liquid-like behavior in the absence of external electric field since G'' dominates over G' in the whole frequency range. When the electric field is applied, polarized particles attract each other resulting in the increase of both moduli in several orders of magnitude. The G' starts to be higher than G'' indicating liquid to solid-like state transition and the stiffness of the system increases with increasing electric field strength. Moreover, G' becomes nearly independent of applied angular frequency which is characteristic for stiff three-dimensional network formed within the system which is sufficiently strong to transmit the elastic forces in such system [17].

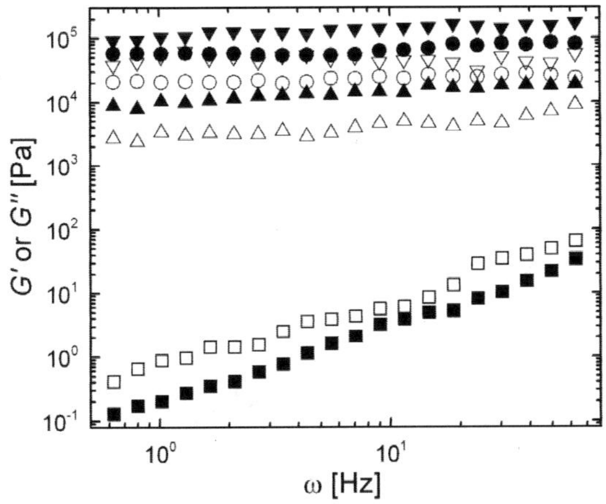

Figure 7. Storage, G', (solid) shear and loss, G'', (open) viscoelastic moduli as a function of angular frequency, ω, for 20 wt.% TiO_2 rod-like particles suspension in silicone oil at 25 °C under various electric field strengths (kV·mm^{-1}): 0 (■ □), 1 (▲ △), 2 (● ○), 3 (▼ ▽).

The ER activity of rod-like TiO_2 particles based fluids with different particle weight fraction is shown in Figure 8. Stiffness of the system significantly increases with the electric field strength applied. The ER effect rises also with particle weight fraction since the formed chain-like structures are more robust with stronger electrostatic interactions as well as with higher solid friction between particles.

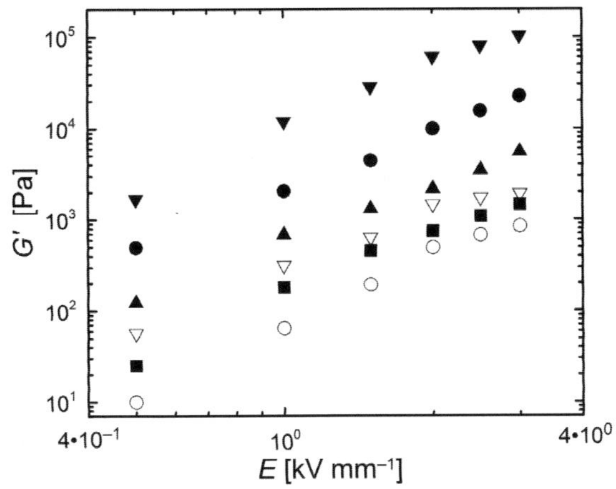

Figure 8. The dependence of storage, G', modulus on the electric field strength, E, at $\omega = 1$ rad·s^{-1} for 5 (■), 10 (▲), 15 (● ○), and 20 (▼ ▽) wt.% suspensions of rod-like (solid symbols) and spherical (open symbols) TiO_2 particles in silicone oil at 25°C.

4. Conclusions

Rod-like TiO_2 particles of rutile crystalline phase with high crystallinity were prepared via microwave-assisted molten-salt method as a novel one-dimensional dispersed phase of ER fluids. The oscillatory shear measurements revealed that silicone oil suspensions of prepared rod-like particles exhibit much higher ER activity than those based on TiO_2 nanoparticles of anatase crystalline phase due to side-by-side solid friction between particles. Moreover, the higher ER activity correlated well with dielectric spectroscopy measurement since the polarization of rod-like particles in the direction of the long axis was reflected in shorter relaxation times.

Acknowledgments
The authors wish to thank the internal grant of TBU in Zlin No. IGA/FT/2012/039 funded from the resources of specific university research for financial support. This article was written with support of Operational Program Research and Development for Innovations co-funded by the European Regional Development Fund (ERDF) and national budget of Czech Republic, within the framework of project Centre of Polymer Systems (reg. number: CZ.1.05/2.1.00/03.0111).

References

[1] Hao T 2002 *Adv. Colloid Interface Sci.* **97** 1
[2] Fang F F, Choi H J and Joo J 2008 *J. Nanosci. Nanotechnol.* **8** 1559
[3] Yin J B and Zhao X P 2011 *Nanoscale Res. Lett.* **6** 256
[4] Wen W J, Huang X X and Sheng P 2008 *Soft Matter* **4** 200
[5] Parthasarathy M and Klingenberg D J 1996 *Mater. Sci. Eng. R-Rep.* **17** 57
[6] Stanway R, Sproston J L and ElWahed A K 1996 *Smart Mater. Struct.* **5** 464
[7] de Vicente J, Segovia-Gutierrez J P, Andablo-Reyes E, Vereda F and Hidalgo-Alvarez R 2009 *J. Chem. Phys.* **131** 194902
[8] Mrlik M, Pavlinek V, Saha P and Quadrat O 2011 *Appl. Rheol.* **21** 52365
[9] Yin J B, Xia X, Xiang L Q, Qiao Y P and Zhao X P 2009 *Smart Mater. Struct.* **18** 095007
[10] Mrlik M, Pavlinek V, Cheng Q L and Saha P 2012 *Int. J. Mod. Phys. B* **26** 125007
[11] Li H L, Du Z N, Wang G L and Zhang Y C 2010 *Mater. Lett.* **64** 431
[12] Kozakova Z, Mrlik M, Sedlacik M, Pavlinek V and Kuritka I 2011 Preparation of TiO2 powder by microwave-assisted molten-salt synthesis *NanoCon: Proc. NanoCon 2011 (Brno, Czech Republic, 21–23 September 2011)* pp 71
[13] Lopez-Lopez MT, Kuzhir P and Bossis G 2009 *J. Rheol.* **53** 115
[14] Mason S G 1950 *Tappi J.* **33** 440
[15] Fang F F, Sung J H and Choi H J 2006 *J. Macromol. Sci. Part B-Phys.* **45** 923
[16] Fang F F, Kim J H, Choio H J and Seo Y 2007 *J. Appl. Polym. Sci.* **105** 1853
[17] Sedlacik M, Pavlinek V, Saha P, Svrcinova P and Filip P 2011 *AIP Conf. Proc.* **1375** 284

Synthesis and electrorheological effect of Cr doped TiO_2 nanorods with nanocavities in silicone oil suspensions

S Almajdalawi[1], V Pavlinek[1,2,3], M Mrlik[1,2], Q Cheng[1] and M Sedlacik[1,2]
[1]Polymer Centre, Faculty of Technology, Tomas Bata University in Zlin, Namesti T. G. Masaryka 275, 762 72, Zlin, Czech Republic
[2]Centre of Polymer Systems, University Institute, Tomas Bata University in Zlin, Nad Ovcirnou 3685, 760 01, Zlin, Czech Republic

E-mail: pavlinek@ft.utb.cz

Abstract. Titanium dioxide (TiO_2) nanorods with nanocavities doped with chromium (Cr) were synthesized by hydrothermal method. The morphology of prepared nanorods was determined by means of scanning electron microscopy (SEM), transmission electron microscopy (TEM), and energy dispersive X-ray spectroscopy (EDX). The electrorheological (ER) behaviour of suspensions based on TiO_2 was investigated under the steady shear, and the yield stress was chosen as a suitable tool for a comparison of ER performance. Some optimum in level of Cr doping of TiO_2 was found.

1. Introduction

In principle, the semi-conducting particles dispersed in non-conducting liquid can be reversibly organised into the chain-like or columnar structures after the application of an external electric field. This electrorheological (ER) phenomenon which sets in the order of milliseconds is reflected in a dramatic change of the rheological properties and thus attracts considerable attention in both academic and technical fields since the pioneer study was presented by Winslow in 1947 [1–4].

For preparation of ER fluids, various materials including organic, inorganic or composite were employed [5–14]. Among them, suspensions based on titanium oxide (TiO_2) seem to be very promising mainly because of a relatively high ER performance. TiO_2 nanotubes were synthesized by hydrothermal synthesis in alkali solution and their ER properties in silicone oil suspensions were measured [15]. The ER behaviour of sea-urchin-like hierarchical doped with chromium (Cr) was also investigated [16]. TiO_2 particles coated with polyaniline and polypyrrole were synthesized and effect of coating on ER response was studied [17, 18]. In another study [19], positive influence of rare earth doping of TiO_2 particles on ER performance was pointed out. ER effect of TiO_2 particles doped with different elements was compared. Suspensions of TiO_2 particles doped with aluminium showed a shear modulus (G_0, $G_0^2 = G'^2 + G''^2$) of hundreds of kilopascals at electric field strengths $E = 3.0$ kV mm^{-1} [20].

In the present study, TiO_2 nanorods doped with Cr were synthesized as a novel material for ER fluids. Cr was doped into the particle structure, nanocavities, and oxidized at the surface by the annealing process.

[3] Corresponding author.

2. Experimental

2.1. Chemicals
For the preparation of the ER particles, TiO_2, P25 powder, sodium hydroxide (NaOH), hydrochloric acid (HCl) and hydrated chromium chloride ($CrCl_6.6H_2O$) were employed. All chemicals were used as received without any further purification.

2.2. Synthesis of Cr doped TiO_2 nanorods with nanocavities
The synthesis was carried out according to literature [21] with some modification. TiO_2 (0.75 g) was dissolved in 10 M NaOH (70 mL) and placed into 100 mL Teflon autoclave. A given amount of $CrCl_6.6H_2O$ was added during the mixing (for 5 min). The mixture was heated for 48 h at 180°C. Product was washed with HCl and distilled water to pH $\approx$ 7. Finally, TiO_2 nanoparticles were dried in an oven and annealed for 1 h at 650°C.

2.3. Structure characterization
The morphology of synthesized nanoparticles was characterized by scanning electron microscopy (SEM) VEGA II LMU (Tescan, Czech Republic) and transition electron microscopy (TEM) JEOL 1200 (JEOL, Japan). Elemental analysis was done by energy dispersive X-ray spectroscopy (EDX) using SEM VEGA II LMU (Tescan, Czech Republic).

2.4. Preparation of ER fluids and rheological measurements
ER fluids (5 wt.%) were prepared by mixing of Cr doped TiO_2 nanoparticles with corresponding amount of silicone oil (Fluid 200, Dow Corning, UK; viscosity $\eta = 100$ mPa s, density $\rho = 0.965$ g cm^{-3}). Rheological parameters were measured by means of a rotational rheometer Bohlin Gemini (Malvern Instruments, UK) equipped with coaxial cylinders (14 mm and 15.4 mm in diameter) modified for ER measurements. The ER cell was connected to the DC high-voltage source TREK 668B (TREK, USA) providing the electric field strengths in the range of 0 – 3 kV mm^{-1}.

ER fluids were sonicated for 60 s before measurements. The built-up particulate structure (in the presence of electric field) was always destroyed by shearing at the shear rate 40 s^{-1} for 60 s in the absence of electric field. The temperature during all experiments was kept at 25 °C.

2.5. Dielectric measurements
Dielectric properties (the relative permittivity and the dielectric loss factor) of 5 wt. % suspensions were monitored using Impedance Material Analyzer Agilent 4294-A (Agilent, Japan) in frequency range of 50 Hz – 1 MHz.

3. Results and discussion
The morphologies of synthesized nanorods are presented in figure 1. SEM analysis shows rod-like structure and size of TiO_2 nanoparticles (figure 1a). The presence of small cavities in the surface of the nanorods was confirmed by TEM (figure 1b). According to literature [22], these cavities can be in the size of tens of nanometres (10 – 20 nm) in diameter with different structures – circular, hexagonal or rectangular.

The quantitative analysis of the TiO_2 nanorods performed by EDX confirmed the presence of Cr by two characteristic peaks at 5.40 and 5.95 keV (figure 2). These peaks result from energy emitted by the excited electrons during their return to the original level (stable state). The amount of energy depends on the difference between the energy levels inside the atoms.

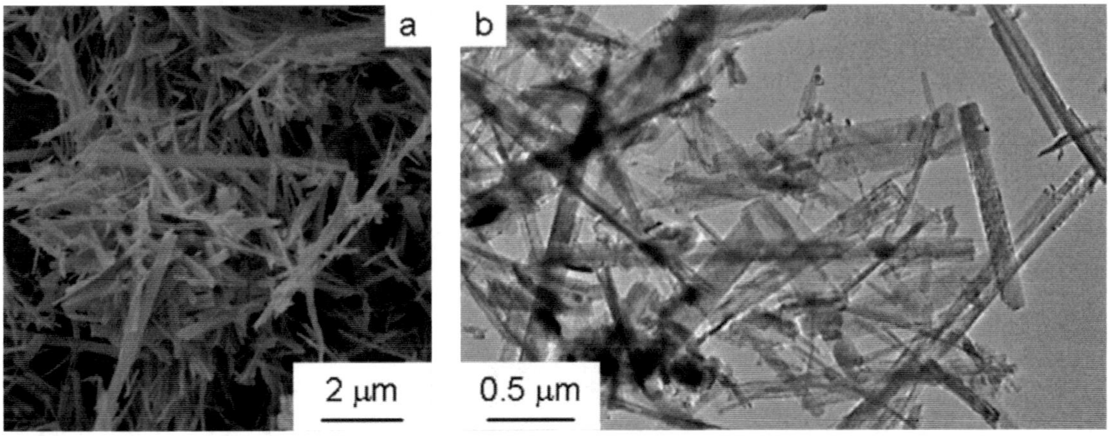

Figure 1. SEM image (a) of TiO_2 nanorods and TEM image (b) of 0.7 mol % Cr doped TiO_2 nanorods.

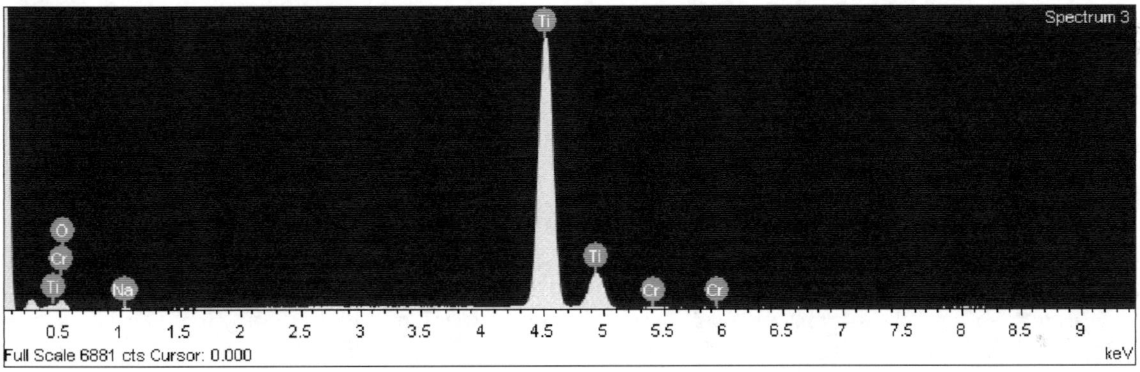

Figure 2. EDX analysis of 0.7 mol % Cr doped TiO_2 nanorods.

The stability of developed ER structures is moderated by reorganization and destruction of chain-like structures given by the competition between the electrostatic and the hydrodynamic interactions. The electrostatic interactions are responsible for the reorganization of polarized particles and hinder the flow, while the hydrodynamic interactions tend to destroy ER structures and promote flow.

Flow curves (the shear stress and the shear viscosity *vs.* the shear rate) of suspension (5 wt.%) prepared by dispersing TiO_2 nanorods doped with 0.7 mol% Cr in silicone oil are plotted in figure 3. In the absence of the electric field, suspension behaved almost as a Newtonian fluid. When the electric field was applied, the rheological behaviour dramatically changed as a result of particle chaining, especially at low shear rates. Higher intensity of the electric field was reflected in higher shear stress and higher viscosity corresponding to the stiffer ER structure. This suspension exhibited the strongest ER effect as will be discussed further.

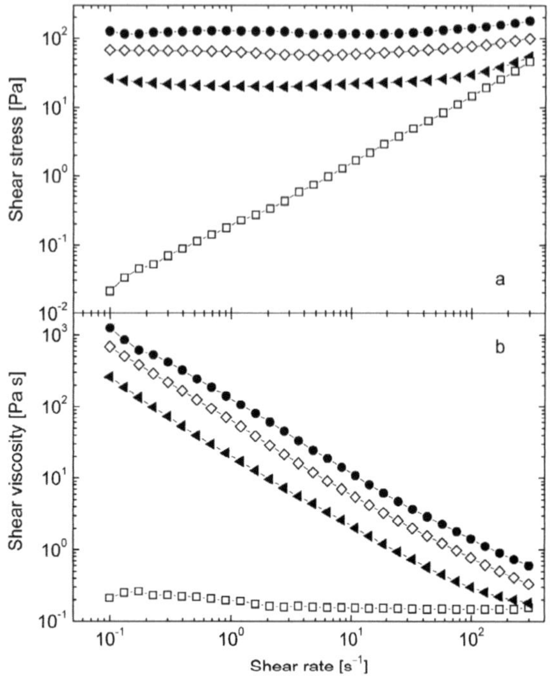

Figure 3. The shear stress (a) and the shear viscosity (b) *vs.* the shear rate dependence for 5 wt. % suspension of TiO_2 nanorods doped 0.7 mol % Cr in silicone at electric field strengths: (□) 0, (◀) 1, (◇) 2, and (●) 3 kV mm^{-1}.

Figure 4 shows the dependence of the yield stress of suspensions on Cr content (0 – 5 mol %) in doped TiO_2 nanorods at electric field strength 3 kV mm^{-1}. Evidently, very low content of Cr (0.7 mol %) in TiO_2 nanorods strongly supports ER effect and the maximal yield stress (130 Pa at 3 kV mm^{-1}) for 5 wt. % suspensions was observed. On the other hand, higher ratio Cr/TiO_2 results in a reduction of the yield stress because of increased conductivity of the TiO_2 nanorods doped to higher levels. As a consequence, the current density passing through ER suspensions increases and seems to be responsible for destroying of chain-like structure and thus significantly lower ER performance.

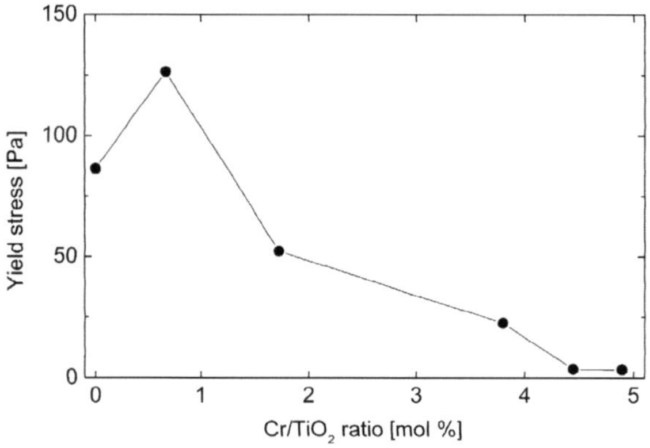

Figure 4. The yield stress of silicone oil suspensions (5 wt.%) *vs.* Cr content in doped TiO_2 nanorods at electric field strength 3 kV mm^{-1}.

The dielectric parameters of TiO_2 nanorods based suspensions were evaluated in the frequency range $10^2 – 10^6$ Hz. The relative permittivity and dielectric loss factor of the suspension with the highest ER effect are shown as an example in figure 5.

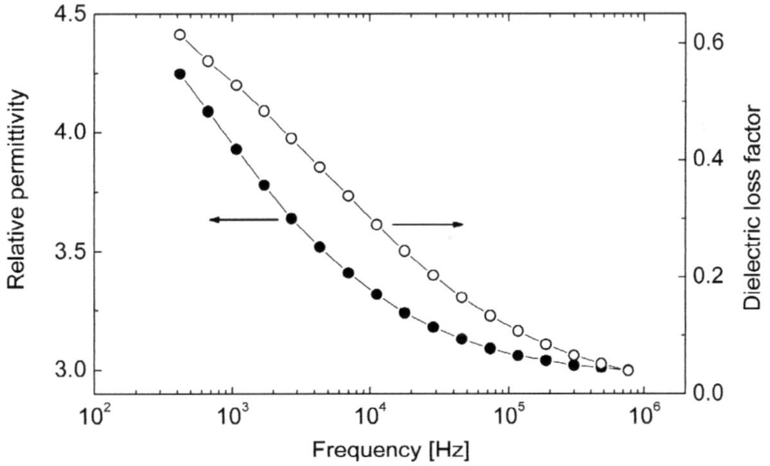

Figure 5. Dependence of the relative permittivity (●) and the dielectric loss factor (○) on the frequency for suspension of TiO_2 nanoparticle doped with 0.7 mol% Cr in silicone oil.

To obtain the ER effect, high polarizability (proposed as maximal difference in the relative permittivity approaching to zero and infinite frequency) as well as proper relaxation time (a reciprocal value of maximal frequency of dielectric loss factor) are necessary. Interfacial polarization showing relaxation maxima in the frequency range $10^2 – 10^5$ Hz was accepted as responsible for suitable ER activity [23].

4. Conclusions

TiO_2 nanorods doped with Cr in nanocavities were synthesized by solvothermal method for preparation of novel ER fluids. The influence of Cr amount on ER performance was investigated and optimum doping of TiO_2 nanorods with 0.7 mol % Cr was found. Higher or lower doping level resulted in low ER performance. Rather dilute suspension (5 wt. %) of TiO_2 nanorods doped with 0.7 mol% Cr in silicone oil showed interestingly large ER effect providing yield stress over 120 Pa.

Acknowledgements

The authors acknowledge the financial support of the Grant Agency of the Czech Republic (Grant no. 202/09/1626).

This article was also written with support of Operational Program Research and Development for Innovations co-funded by the European Regional Development Fund (ERDF) and national budget of Czech Republic, within the framework of project centre of Polymer Systems (reg. number: CZ.1.05/2.1.00/03.011.

6. References

[1] Winslow W M 1947 *US Patent* 2,417,850
[2] Block H and Kelly J P 1988 *J. Phys. D-Appl. Phys.* **21** 1661
[3] Parthasarathy M and Klingenberg D J 1996 *Mater. Sci. Eng. R-Rep.* **17** 57
[4] Hao T 2002 *Adv. Colloid Interface Sci.* **97** 1
[5] Otsubo Y and Edamura K 1994 *J. Colloid Interface Sci.* **168** 230
[6] Choi U S and Ahn B G 2000 *Colloid Surf. A-Physicochem. Eng. Asp.* **168** 71
[7] Kim Y D and Park D H 2002 *Colloid Polym. Sci.* **280** 828
[8] Lengalova A, Pavlinek V, Saha P, Stejskal J and Quadrat O 2003 *J. Colloid Interface Sci.* **258** 174
[9] Quadrat O and Stejskal J 2006 *J. Ind. Eng. Chem.* **12** 352

[10] Zhao X P and Yin J B 2006 *J. Ind. Eng. Chem.* **12** 184
[11] Yoon D J and Kim Y D 2006 *J. Colloid Interface Sci.* **303** 573
[12] Ahmari H and Etemad S G 2009 *Rheol. Acta* **48** 217
[13] Choi H J and Jhon M S 2009 *Soft Matter* **5** 1562
[14] Ramos Tejada M M, Espin M J, Perea R and Delgado A V 2009 *J. Non-Newton. Fluid Mech.* **159** 34
[15] Yin J B and Zhao X P 2008 *Colloid Surf. A-Physicochem. Eng. Asp.* **329** 153
[16] Yin J B, Zhao X P, Xiang L Q, Xia X and Zhang Z S 2009 *Soft Matter* **5** 4687
[17] Wei C, Zhu Y H, Jin Y, Yang X L and Li C Z 2008 *Mater. Res. Bull.* **43** 3263
[18] Lee I S, Lee J Y, Sung J H and Choi H J 2005 *Synth. Met.* **152** 173
[19] Zhao X P and Yin J B 2002 *Chem. Mater.* **14** 2258
[20] Tang H Z, He J H and Persello J 2010 *Particuology* 8 442
[21] Zhu H, Tao J and Dong X 2010 *J. Phys. Chem. C* **114** 2873
[22] Jesurani S, Kanagesan S, Vermurugan R, Thiruphati C, Sivakumar M, Kalaivani T 2011 *Mater. Lett.* **65** 3305
[23] Ikazaki F, Kawai A, Uchida K, Kawakami T, Edamura K, Sakurai K, Anzai H and Asako Y 1998 *J. Phys. D-Appl. Phys.* **31** 336

Solvothermal synthesis of different TiO_2 morphology and their electrorheological characteristics

S Almajdalawi[1], V Pavlinek[1,2,4], M Mrlik[1,2], Q Cheng[1], P Piyamanocha[1,2], M Pastorek[2,3] and M Stenicka[2,3]

[1]Polymer Centre, Faculty of Technology, Tomas Bata University in Zlin, Namesti T. G. Masaryka 275, 762 72, Zlin, Czech Republic
[2]Centre of Polymer Systems, University Institute, Tomas Bata University in Zlin, Nad Ovcirnou 3685, 760 01, Zlin, Czech Republic
[3]Department of Polymer Engineering, Faculty of Technology, Tomas Bata University in Zlin, Namesti T. G. Masaryka 275, 762 72 Zlin, Czech Republic

E-mail: pavlinek@ft.utb.cz

Abstract. Titanium dioxide (TiO_2) with different morphologies was prepared using solvothermal synthesis. The morphology and crystalline structure were characterized by scanning electron microscopy (SEM) and X-ray diffraction (XRD). Electrorheological (ER) effect and dielectric properties of 5 wt. % suspensions were evaluated.

1. Introduction

Electrorheological (ER) fluids are smart materials showing a dramatic change in the rheological properties upon the application of an external electric field [1]. A typical ER fluid consists of polarisable particles dispersed in an insulating liquid medium. After the application of the electric field, the polarized particles form columns in a very short time and thus change the consistence of ER fluid from a liquid-like to a gel-like. Once, the electric field disappears, particles can return to the original disordered state and ER fluid behaves as a liquid again. This is an attractive topic from research point of view [2–5] with an interesting possible application for brakes, engine mounts or shock absorbers.

Various materials have been employed for preparation of ER fluids [6–9]. However, ER performance can be significantly affected by the morphology of dispersed particles, too [10]. The solvothermal synthesis seems to be an easy method how to prepare particles with controlled morphology by changing of reaction temperature and time, concentration and type of the solvent, pH value of the solution, *etc.* [11].

In this work, TiO_2 particles with different morphologies were synthesized by controlling concentration of the reactant and duration of synthesis. The influence of these variables on the morphology and subsequently on ER behaviour and dielectric properties of prepared TiO_2 based suspensions was studied.

[4] Corresponding author

2. Experimental

2.1. Chemicals

TiO_2 anatase powder, sodium hydroxide (NaOH) and absolute ethanol were used as received without any further purification.

2.2. Synthesis of TiO_2 particles with different morphology

TiO_2 anatase powder, 10M NaOH solution and absolute ethanol were mixed in predefined ratio. Subsequently, the solution was transferred to 100 mL Teflon autoclave and heated at 200°C for different times (Table 1). The obtained product was filtered, washed with distilled water until pH = 7. Finally, the synthesized TiO_2 particles were dried at 80°C for 24h.

Table 1. Experimental conditions of TiO_2 particles synthesis.

Sample code	Weight of TiO_2 [g]	Time of heating [h]	Temperature [°C]
S1	0.3	24	200
S2	1.2	24	200
S3	1.2	48	200

2.3. Structure characterization

Morphology of synthesized TiO_2 particles was characterized by scanning electron microscopy (SEM) VEGA II LMU (Tescan, Czech Republic). The XRD spectra were obtained via X-ray diffractometer X´Pert PRO (PANalytical, The Netherlands).

2.4. Preparation of ER fluid and rheological measurements

ER suspensions (5 wt.%) were prepared by mixing of TiO_2 particles with corresponding amount of silicone oil Lukosiol M200 (Chemical Works Kolin, Czech Republic; viscosity $\eta = 200$ mPa s, density $\rho = 0.965$ g cm^{-3}, conductivity $\sigma \approx 10^{-11}$ S cm^{-1}).

A rotational rheometer Bohlin Gemini (Malvern Instruments, UK) equipped with coaxial cylinders (inner and outer diameters 14 and 15.4 mm) was used for steady shear measurements. The cylinders were connected to the external DC source providing electric field strengths from 0 to 3 kV mm^{-1}. All measurements were carried out under constant temperature of 25°C in the shear rate range of 0.5 to 200 s^{-1}.

ER fluids were stirred first mechanically and then for 1 min in ultrasonic bath before each measurement. Structures developed in the presence of electric field were destroyed by shearing at 40 s^{-1} for 60 s in the absence of electric field.

2.5. Dielectric measurements

Dielectric properties for 5 wt.% suspensions were measured using Impedance Material Analyzer Agilent 4294-A (Agilent, Japan) in the frequency range 50 Hz – 1 MHz.

3. Results and discussion

Various morphologies of TiO_2 particles were obtained when treated under different conditions like time of heating or concentration of the starting material (Figure 1). At low initial concentration of TiO_2, the synthesized particles tend to form rod-like structures (S1, Figure 1a), whereas higher content of TiO_2 in mixture is reflected in more flat morphology of particles (S2, Figure 1b). This structure has higher diameter, but nearly the same length as S1. However, when time of heating increases, the TiO_2 particles (S3, figure 1c) become longer and thinner in contrast to S1 and S2, respectively.

Figure 2 shows the XRD patterns of TiO_2 particles. The crystalline structure was nearly the same for all prepared samples. After solvothermal treatment and washing with water, the TiO_2 precursor is transformed into a mixture of crystals; ($Na_2Ti_3O_7$) and ($Na_2Ti_9O_{19}$). However, the main part is $Na_2Ti_3O_7$ crystal according to the database. The anatase crystalline phase, having diffraction lines of planes at 2θ values of 25.1°, 29.7°, 43.5°, and 47.7°.

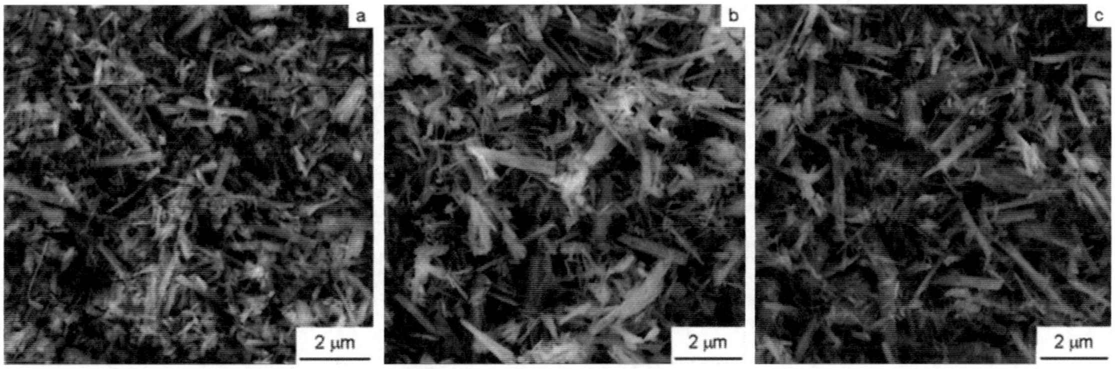

Figure 1. SEM images for S1 (a), S2 (b) and S3 (c).

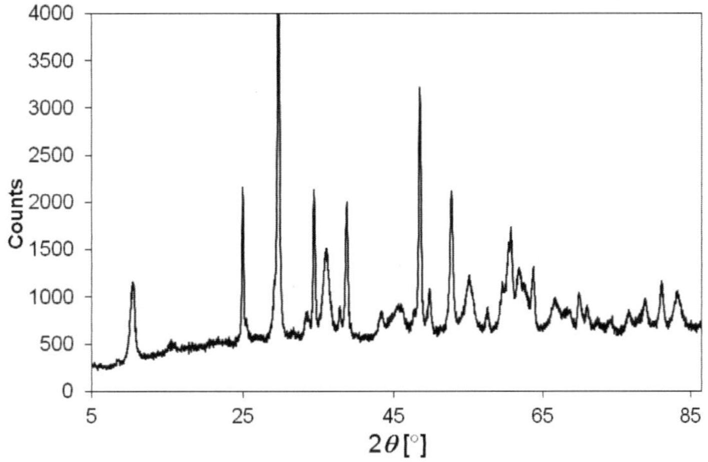

Figure 2. XRD-spectra of particles.

The rheological behaviour of 5 wt.% TiO_2 based suspensions under 0 and 3 kV mm^{-1} is presented in figure 3. In the absence of the electric field, suspensions behave nearly as Newtonian fluids – the shear stress linearly increases with shear rate. By application of the electric field, the shear stress dramatically increases and suspensions show a significant yield stress. The yield stress appears as a result of interaction between the particles induced by the external electric field (Figure 3a).

However, a slight decrease of the shear stress observed in the range of small shear rates (1 – 35 s^{-1}) can arise from a negative synergistic interaction between hydrodynamic and polarization forces. At high shear rates, where the hydrodynamic forces dominate, both the shear stress and the shear viscosity approach the values as in the absence of the electric field due to structural changes in the suspensions: the destruction rate of the structures become faster than the reformation rate.

Figure 4 shows the dependence of the yield stress on the external electric field strength. The yield stresses for suspensions of all the samples (S1–S3) show an increase with electric field strength. While S1 and S2 present comparable yield stresses, the yield stress of S3 is slightly lower.

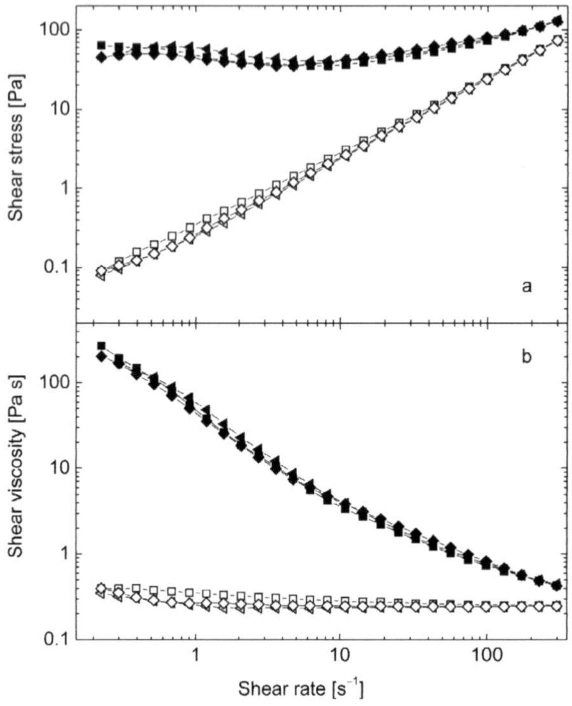

Figure 3. The shear stress (a) and the shear viscosity (b) *vs.* the shear rate dependence for 5 wt. % TiO_2 based suspensions in the absence of electric field strength (open symbols) and at E = 3 kV mm^{-1} (solid symbols). Sample codes: (□■) S1, (◁◀) S2, and (◇◆) S3.

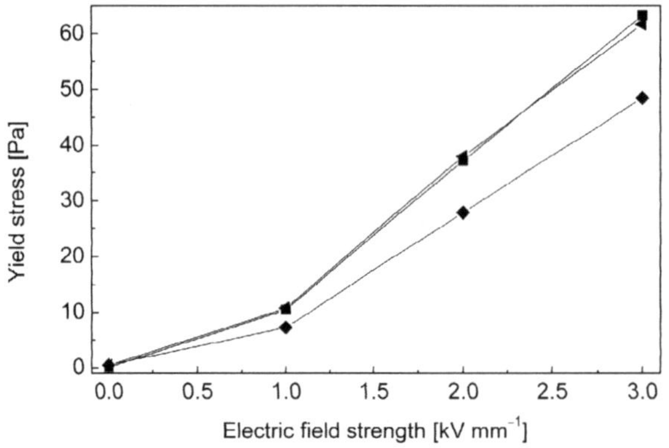

Figure 4. The yield stress *vs.* the electric field strength. Symbols as denoted in figure3.

As proposed in the literature [12], the ER response is significantly influenced by interfacial polarization. Therefore, the dielectric properties (relative permittivity and dielectric loss factor) of ER suspensions are very important for probing the mechanism that controls the electric properties of the materials [13]. Both parameters are plotted in figure 5 as a function of electric field frequency. The dielectric characteristics of suspensions were obtained from Havriliak-Negami equation [14]:

$$\varepsilon_s^*(f) = \varepsilon'_{s,\infty} + \frac{(\varepsilon'_{s,0} - \varepsilon'_{s,\infty})}{(1 + (i\,2\pi\,f\,t_{rel})^a)^b} \tag{1}$$

where $\varepsilon_s^*(f)$ is a complex suspension permittivity and $\varepsilon'_{s,0}(f)$ and $\varepsilon'_{s,\infty}(f)$ are the limiting values of relative permittivity at below and above the relaxation frequency, f is frequency, t_{rel} is a relaxation time, a is the scattering degree of relaxation times, and b is related to the asymmetry of the relaxation time spectrum (table 2).

Table 2. Parameters of Havriliak-Negami equation for TiO_2 based suspensions.

Sample code	$\varepsilon'_{s,0}(f)$	$\varepsilon'_{s,\infty}(f)$	t_{rel} [s]	a	b
S1	3.85	2.91	4.74×10^{-4}	0.34	1.65
S2	3.36	2.91	1.21×10^{-4}	0.32	1.75
S3	3.45	2.93	3.79×10^{-5}	0.39	1.75

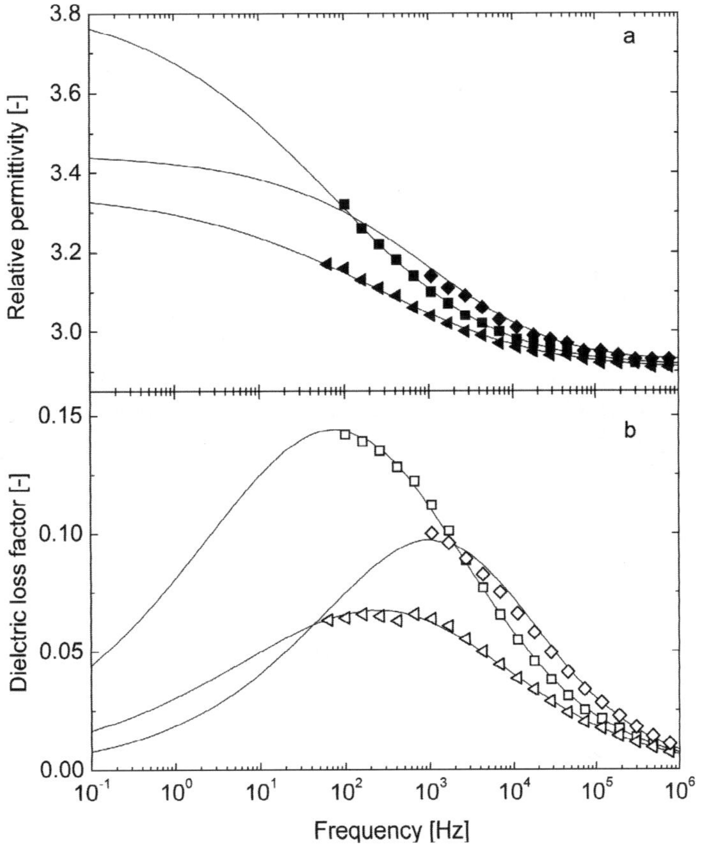

Figure 5. The relative permittivity (a) and the dielectric loss factor (b) *vs.* the frequency for 5 wt.% TiO_2 based suspensions. Symbols as denoted in figure3.

Suspension S1 shows the highest polarizability of the particles given by difference between $\varepsilon'_{s,0}(f)$ and $\varepsilon'_{s,\infty}(f)$ compared to other two suspensions. The difference in relative permittivity for S1, S2 and S3 was 0.94, 0.45 and 0.52, respectively as comes from table 2. In contrast to *e.g.* other ER systems [15], the polarizability of these ER fluids is rather low, but corresponds to the ER performance.

In addition to the polarizability of the particles, a proper relaxation time evaluated from the maximum of dielectric loss between $10^2 - 10^5$ Hz is necessary for a large ER effect [12].

Evidently, various morphology of prepared TiO_2 particles plays a role in particle organizations (Figure 5b and Table 2). ER fluid having higher relaxation time exhibited weaker ER effect because of slower chaining of particles.

4. Conclusions

TiO_2 particles with different morphology were synthesized using solvothermal synthesis method. These particles were mixed with silicone oil and ER performance was investigated. As pointed out, only 5 wt.% suspensions exhibited a moderate ER effect. In case of ER suspensions having similar polarizability, suspension with longer relaxation time exhibited weaker ER effect.

Acknowledgements

The authors wish to thank to the Grant Agency of the Czech Republic for the financial support of Grant no. 202/09/1626.

This article was written with support of Operational Program Research and Development for Innovations co-funded by the European Regional Development Fund (ERDF) and national budget of Czech Republic, within the framework of project centre of Polymer Systems (reg. number: CZ.1.05/2.1.00/03.011.

References

[1] Winslow W M 1947 *US Patent* 2,417,850
[2] Block H, Kelly J P, Qin A and Watson T 1990 *Langmuir* **6** 6
[3] Hao T 2001 *Adv. Mater.* **13** 1847
[4] Filisko F E 2002 *Electrorheological materials. Encyclopedia of Smart Materials. 1st ed. New York John Wiley and Sons* 1–2 376 ISBN 978-0-471-17780-7
[5] Quadrat O and Stejskal J 2006 *J. Ind. Eng. Chem.* **12** 352
[6] Choi H J, Park S J., Kim S T, Jhon M S 2005 *Diam. Relat. Mat.* **14** 766
[7] Cheng Q L, Pavlinek V, Lengalova A, Li, C Z, Belza, T and Saha P 2006 *Microporous Mesoporous Mat.* **94** 193
[8] Fang F F and Choi H J 2007 *J. Solid State Chem.* **180** 390
[9] Almajdalawi Sh, Pavlinek V, Cheng Q, Saha P, Mrlik M and Stenicka M 2011 *AIP-Conf. Proc.* **1375** 275
[10] Lengalova A, Pavlinek V, Saha P, Quadrat O and Stejskal J 2003 *Colloid Surf. A-Physicochem. Eng. Asp.* **227** 1
[11] Chang H Y, Lee H D, Wu M L and Lin L J 2010 *J. Lumines.* **130** 969
[12] Ikazaki F, Kawai A, Uchida K, Kawakami T, Edamura K, Sakurai K, Anzai H and Asako Y 1998 *J. Phys. D-Appl. Phys.* **31** 336
[13] Klingenberg D J, Dierking D and Zukoski C F 1991 *J. Chem. Soc., Faraday Trans.* **87** 425
[14] Havriliak S Jr and Havriliak S J 1997 *Dielectric and mechanical relaxation in materials. Analysis, interpretation, and application to polymers. 1st ed. Munich Hanser* 484 ISBN 1-56990-186-4
[15] Stenicka M, Pavlinek V, Saha P, Blinova N V, Stejskal J and Quadrat O 2009 *Colloid Polym. Sci.* **287** 403

Electrorheological response of polyindene/colemanite conducting composite

H I Unal[1], B Cetin and O Erol
Gazi University, Science Faculty, Chemistry Department, 06500 Teknikokullar, Ankara, Turkey

E-mail: hiunal@gazi.edu.tr

Abstract. In this study, electrorheological (ER) responses of colemanite and polyindene (PIn)/colemanite conducting composite dispersions in silicone oil (SO) were investigated. Antisedimentation ratios of the dispersions were determined. Some parameters which affect the ER properties of the dispersions such as electric field strength (E), shear rate, frequency and temperature were investigated. The rather unusual behavior known as "*negative ER effect*" was observed for colemanite/SO above E = 1.5 kV/mm and for PIn/colemanite/SO under all the values of electric field strength at volume fraction of 25%. This negative ER response was converted to positive with the addition of non-ionic surfactant.

1. Introduction

Electrorheological (ER) fluids are colloidal dispersions that commonly compose of polarizable powders dispersed in insulating liquids such as silicone oil (SO). ER fluids exhibit reversible changes (liquid-like to solid like) in their rheological properties as a function of the electric field strength (E). To enhance the ER effect and the stability of the suspension, sometimes, ER fluids require small amounts of additives such as polar compounds or surfactants. If the electric field induced viscous properties increase with the application of electric field, this phenomenon is termed positive ER effect. On the other hand, if the electric field induced viscosity of the ER fluid decreases as the applied electric field increases, this refers to negative ER effect [1].

Colemanite ($2CaO.3B_2O_3.5H_2O$) is the most important calcium and boron containing commercial borate mineral. The basic structure of colemanite contains endless chains of interlocking $BO_2(OH)$ triangles and $BO_3(OH)$ tetrahedrons with the calcium, crystalline water molecules and extra hydroxides interspersed between the chains. The sheets are held together by weak hydrogen bonds involving both water molecules and hydroxyl groups [2].

Polyindene (PIn) is one of the conducting polymers and because of its cyclic structure, phenyl rings of PIn are arranged horizontally along the main chain; hence, the repeating unit of PIn forms a nearly planar structure. This unique structure renders PIn having a high glass-transition temperature and special optical properties [3]. In the literature, it was reported that the viscosities of PIn/$CaCO_3$ [4] and PIn/kaolinite [5] composites showed positive ER effect and the ER activities of these system rose with PIn content.

In this study, colemanite and PIn/colemanite composites were used as ER active materials. Antisedimentation stabilities of these systems in non-aqueous medium (SO) were determined. The

[1] Corresponding author.

parameters which affect the ER properties of colemanite/SO and PIn/colemanite/SO systems were investigated such as electrical field strength, shear rate, frequency, temperature and surfactant. Full details of the study is given in our previous publication [6].

2. Experimental

2.1. Materials
Colemanite ($2CaO.3B_2O_3.5H_2O$) was kindly supplied by ETI Mining Co. (Turkey) and all the other chemicals were Aldrich products and used as received. Indene was used after distillation.

2.2. Synthesis and characterization of PIn/colemanite composite
PIn was in-situ polymerized in the presence of colemanite using $FeCl_3$ as oxidizing agent taking the monomer to initiator ratio as 2:1 in $CHCl_3$ and the obtained PIn/colemanite composite was subjected to various characterization techniques. The full details of the synthesis and characterization of PIn and PIn/colemanite composite are given in our previous publication [7].

2.3. Preparation of dispersions
Dispersions of colemanite and PIn/colemanite composite were prepared at a volume fractions of 25% by dispersing definite amount of solid particles in SO (ρ = 0.965 gcm^{-3}, η = 1.0 Pa s, ε = 2.61, Aldrich, Germany), which were both vacuum dried before mixing in an oven for overnight to remove any moisture present. The ionic-surfactant examined was Triton X-100 (*t*-octylphenoxypolyethoxyethanol) and the concentration of Triton-X in the dispersions was 5 wt.%. These surfactant containing dispersions were coded as T-colemanite and T-PIn/colemanite. All the prepared dispersions were allowed to equilibrate for overnight before ER measurements.

2.4. Antisedimentation ratio measurements
Antisedimentation stabilities against gravitational forces of the colemanite/SO and PIn/colemanite/SO dispersions were determined at T = 25±0.1°C. Glass tubes containing the above dispersions were immersed into a constant temperature water bath and formation of the first precipitates was recorded to be the indication of colloidal instability. During the neat eye observations, the height of phase separation between the particle-rich phase and the relatively clear oil-rich phase was recorded as a function of time by using a digital composing stick. The antisedimentation ratio was defined as the height of the particle-rich phase divided to the total height of dispersion.

2.5. Electrorheogical measurements
ER properties of the dispersions were determined with a Termo-Haake RS600 parallel plate torque electrorheometer (Karlsruhe, Germany). The gap between the parallel plates was 1.0 mm and the diameters of the upper and lower plates were 35 mm. The potential used in these experiments was supplied by a 0–12.5 kV (with 0.5 kV increments) dc electric field generator (Fug Electronics, HCL 14, Germany), which enabled resistivity to be created during the experiments.

3. Results and discussion

3.1. Antisedimentation stabilities of the dispersions
The antisedimentation stability of ER materials is one of the main criteria because the ER fluids are supposed to be homogeneously dispersed before applying external electric field strength.

The effects of dispersed particles on the antisedimentation ratios of the materials used in this study were examined. For the particle volume fraction of 25%, antisedimentation ratios at the end of 25 days were determined as 72% for colemanite/SO and 92% for PIn/colemanite/SO systems. It was observed that PIn/colemanite/SO dispersion system was significantly stable against gravitational sedimentation than colemanite/SO. Furthermore, increasing antisedimentation ratios were recorded with increasing average hydrodynamic particle sizes ($d_{0.5}$) of samples [$d_{0.5}$(PIn/colemanite) (2.2 μm)>$d_{0.5}$ (colemanite) (1.2 μm)] and decreasing apparent densities of the samples [$\rho_{colemanite}(1.69) > \rho_{PIn/colemanite}(1.02)$]. It was observed that the PIn/colemanite/SO dispersion system had higher antisedimentation ratio than colemanite/SO, although particle size of PIn/colemanite was bigger than colemanite. On the other hand, the density of PIn/colemanite was lower than colemanite and closer to the density of SO. Furthermore, in composite structure PIn chains form steric hindrance that holds the particles in suspended positions against agglomeration.

3.2. Electrorheological studies

3.2.1 .Viscosimetry results. The colemanite particles dispersed in SO were distributed randomly without applying E (Figure 1). When an electric field was applied, the particles became polarized and tended to attract each other, and consequently the fibril-like structures aligned to the direction of electric field were formed, which provide additional resistance against flow. For colemanite/SO system, this occurs under low electric field values and the formation of fibril-like structures could be weak (Figure 2). After E = 1.5 kV/mm, weak fibrillar structures between the electrodes broke-down because of the enhancement in electrophoretic mobility of particles with increasing applied electric field, and the colemanite particles moved towards the positive electrode, resulting in phase separation and decrease in viscosity (Figure 1). Similar migration of particles to one of the electrodes was reported for Teflon/SO system, which was attributed to the interactions between the oil and the Teflon particles [8].

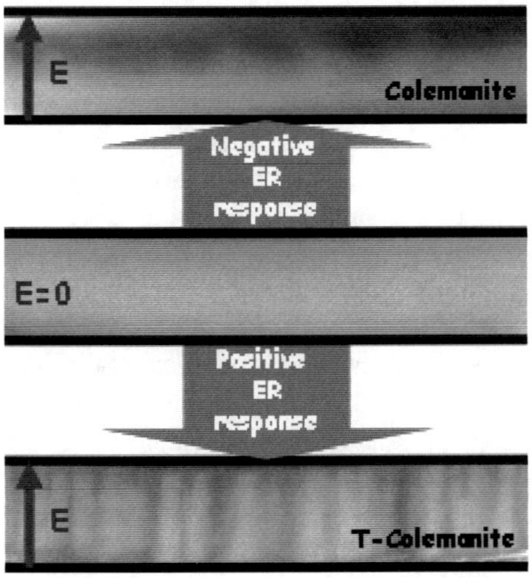

Figure 1. The ER response of colemanite/SO and T-colemanite/SO (φ = 25%).

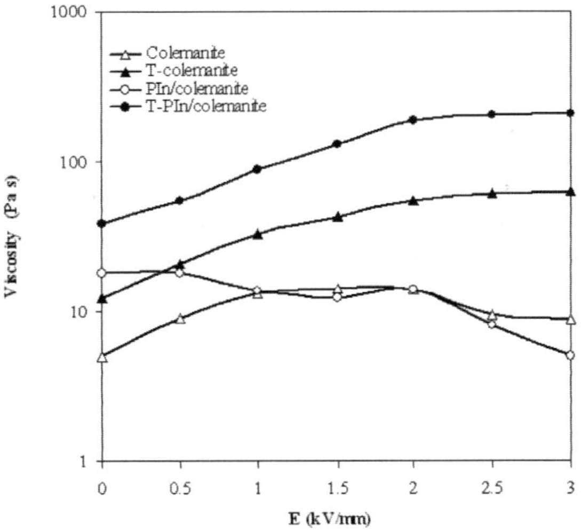

Figure 2. Change in viscosity of the materials with E (φ = 25%, T = 25°C) [6].

Sigma bonds between the hexagonal rings in colemanite structure facilitate the rotation of the rings, and this rotation prevents the formation of stronger fibrillar structures under electric field and also polyanion structure of colemanite contribute to the dispersed particles to migrate to one of the electrodes.

The PIn/colemanite/SO system showed slight changes in viscosity with increasing electric field. But, the mean apparent viscosity showed to be lower than that of the viscosity at E = 0 kV/mm condition which is an indication of negative ER behavior. In the presence of an electric field, structure of dispersion was destroyed by electrostatic forces acting on polarized particles, reorganizing them into fibrillar chains parallel to the electric field, whose resistance against the flow was lower than that of the structure of dispersion with the absence of electric field. Consequently, the negative ER effect was occurred.

After observing negative ER effect for the materials studied, especially at high electric field strengths, the effects of non-ionic surfactant (Triton-X) on ER behaviors of the materials were investigated. Significant increases in the electric field viscosities were observed with the addition of Triton-X. The presence of Triton-X surfactant observed to cause dramatic changes on the ER responses and the viscosities of T-colemanite/SO and T-PIn/colemanite/SO systems increased with increasing electric field, in contrast to pure colemanite/SO and pure PIn/colemanite/SO systems. It is believed that Triton-X interacted with colemanite and PIn/colemanite with the secondary forces, and prevented the particle migration to one of the electrodes. Triton-X surfactant provoked the modification of the interfacial properties of the dispersed particles and formed a kind of bridge between the adjacent particle surfaces. Thus, the formation of stronger fiber structure was provided and also fibrillar structures were observed easily by neat eyes for T-colemanite/SO dispersion system (Figure 1). Furthermore, the electric field viscosity enhancement of PIn/colemanite/SO system was higher than colemanite/SO system with the addition of Triton-X into these systems. This is due to the interaction between hydrophobic chains of Triton-X and polyindene chains in PIn/colemanite composite structure. Virtually, these interactions reversed the observed negative ER behavior of the materials to positive, as targeted.

3.2.2. Effect of shear rate on shear stress and viscosity. The effects of shear rate ($\dot{\gamma}$) on shear stress and viscosity for colemanite/SO and T-colemanite/SO systems were depicted in Figure 3(a, b), and for PIn/colemanite/SO and T-PIn/colemanite/SO systems were shown in Figure 4(a, b) for $E = 0$ kV/mm and $E = 1.5$ kV/mm conditions. The viscosities of the systems were observed to decrease exponentially at lower shear rate regions with typical curves of shear thinning non-Newtonian viscoelastic behaviors both under $E = 0$ kV/mm and $E = 1.5$ kV/mm conditions.

Non-Newtonian shear stress increments were observed with increasing shear rate for all the materials examined. But it was clearly seen that PIn/colemanite/SO system showed weaker structure with the presence of electric field than the absence against to the shearing forces as a results of negative ER effect (Figure 4a). On the other hand, with the application of electric field, Triton-X provoked the formation of chain-like structures in the dispersions which resulted in a resistance to flow with an abrupt enhancement of electric field induced viscosity at low shear rate regions. It can be concluded that Triton-X enhanced the formation of electric field induced fibrillar structures in dispersions as targeted.

Yield stress (τ_y) values of all the materials examined were calculated from the shear stress-shear rate curves obtained at control rate (CR) mode of the electrorheometer by extrapolating the flow curve to $\dot{\gamma} = 0$ s^{-1}. Yield stresses of the samples at $E = 3.5$ kV/mm are calculated to be as following: $\tau_{y(T\text{-}PIn/colemanite)}$ 408 Pa> $\tau_{y(T\text{-}colemanite)}$ 91 Pa> $\tau_{y(colemanite)}$ 6 Pa> $\tau_{y(PIn/colemanite)}$ 4 Pa. A tremendous yield stress increase was obtained for T-PIn/colemanite/SO system as targeted. It was concluded that, the interactions between the surfactant molecules and the dispersed particles caused the reduction of surface tension of the dispersion system; and also the surfactant molecules promoted the colloidal stability of the dispersed particles. This situation is in accordance with "water/surfactant bridge" model [9]. Besides, the surfactant molecules may overcome the negative ER effect via stabilizing the particles and preventing electromigration of the particles due to the charge injection at the electrodes [10] and preventing a possible phase separation between the upper and lower plates of the electrorheometer.

3.2.3. Effect of frequency on storage modulus and loss modulus. It is well known that the relationship between storage modulus (G') and loss modulus (G") indicates whether the material is solid-like (G'>G"), gel-like (G'≈G") or liquid-like (G">G') in the whole frequency range measured [11].

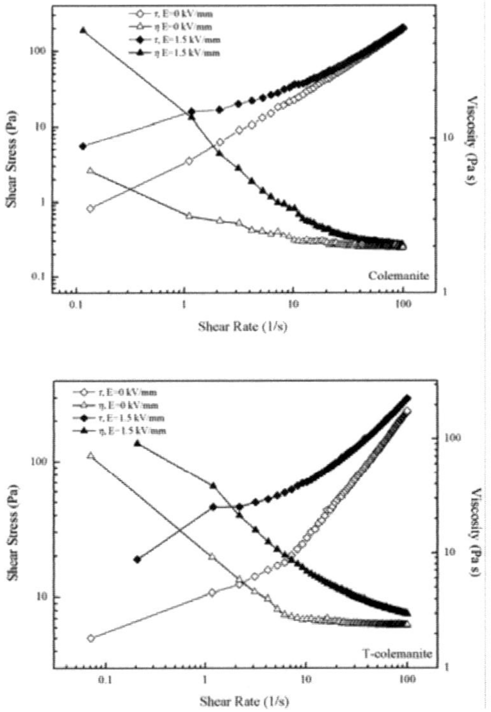

Figure 3. The effect of shear rate on shear stress and viscosity of (a) colemanite/SO and (b) T-colemanite/SO (T=25°C, φ = 25%) [6].

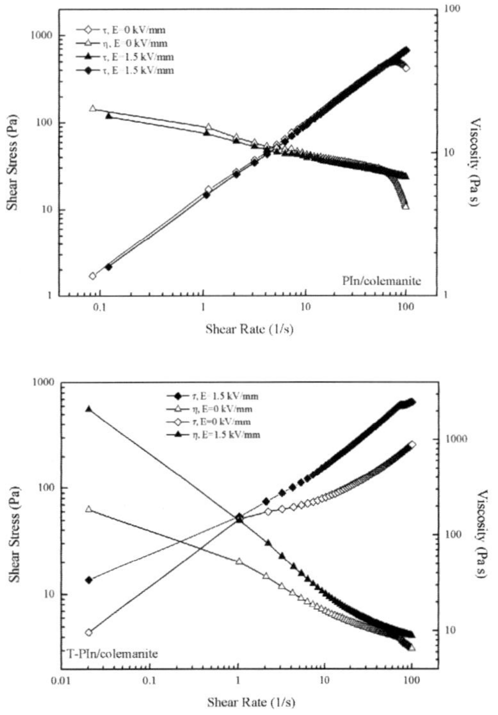

Figure 4. The effect of shear rate on shear stress and viscosity of (a) PIn/colemanite/SO and (b) T-PIn/colemanite/SO (T=25°C, φ= 25%) [6].

Figure 5 shows G' and G" as a function of frequency for colemanite/SO, T-colemanite/SO, PIn/colemanite/SO and T-PIn/colemanite/SO dispersion systems at E = 1.5 kV/mm. Stress sweep was first carried out to determine the proper stress values to ensure collecting data in the linear viscoelastic regime for the samples. It was observed that G' values of all the dispersions except PIn/colemanite/SO were higher than G" values and showed slight increments in the whole frequency sweep range and owing to not providing fully elastic solid-like structure in the dispersions. However, G' values of PIn/colemanite/SO were closer to G" values at moderate frequency values and also increased notably with increasing frequency. At lower frequencies the viscous behavior reflected by G" was dominant whereas, at the higher frequencies the elastic behavior reflected by G' was outweighing. This suggests that fibrillar structure between the electrodes could not occur at lower frequencies. Similar increment in G' were reported for goethite/SO suspension system at both lower volume fractions and electric field strengths [12].

3.2.4. The effect of temperature on shear stress. It is known that the temperature affects on various physical properties of the particles and dispersant liquid, such as the dielectric constant, conductivity and current density. Moreover, thermal motion and colloidal stability are influenced by change of temperature. The ultimate ER behavior depends on the balance attained among these factors [13].

To investigate the behavior of materials at elevated temperatures, temperature sweeps were carried out between 15°C and 75°C for all the materials studied (Figure 6). The shear stresses of all the samples, except colemanite/SO system, were recorded to decrease with increasing temperature. The overall shear stress losses ($\Delta\tau = \tau_{15°C} - \tau_{75°C}$) were in the order of: $\Delta\tau_{T\text{-PIn/colemanite}}$ (160 Pa)>$\Delta\tau_{\text{PIn/colemanite}}$ (15 Pa)>$\Delta\tau_{T\text{-colemanite}}$ (14 Pa) and shear stress increment for colemanite was calculated to be 3 Pa. With increasing temperature the thermal motions of the dispersed particles were enhanced; as a result the mobility of the particles raised and formation of stripes along the direction of the electric field got relatively difficult. Colemanite/SO system showed slight increment with raising temperature which was believed to cause by its crystal water.

On the other hand, the temperature effect on ER behavior became more complicated with the addition of Triton-X surfactant. As the temperature varies, beside change in the electrical properties of the dispersion, the adsorption and desorption behaviors of the surfactant molecules contribute to the ER behavior. From the results, it can be clearly seen that surfactant molecules caused to decrease of the ER performance of T-colemanite and T-PIn/colemanite dispersions significantly. A possible explanation to this behavior may be the formation of weaker chains as a result of desorption of the surfactant molecules between the adjacent particles that act as a bridge or aggregation of the surfactant molecules between the electrodes via phase separation.

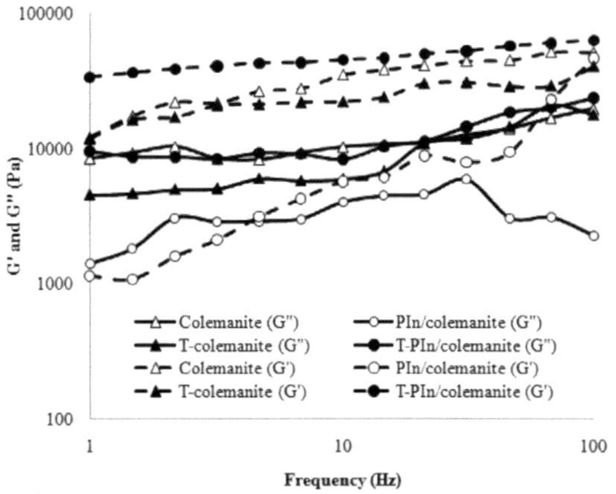

Figure 5. The effect of the frequency on the storage modulus and loss modulus of the samples ($\varphi = 25\%$, T = 25°C) [6].

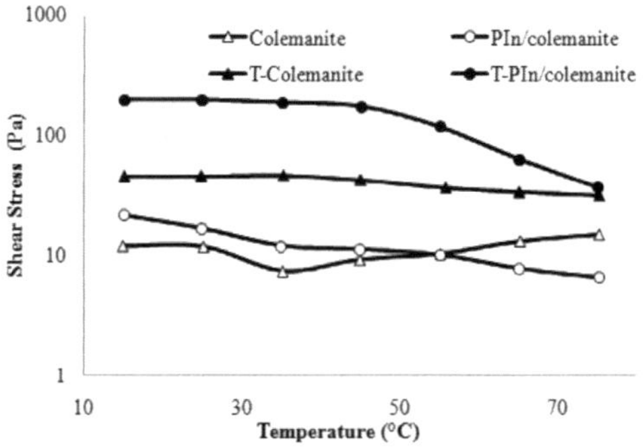

Figure 6. The effect of the temperature on the shear stress of the samples ($\varphi = 25\%$, T = 25°C) [6].

4. Conclusions

Colemanite/SO dispersion observed to show positive ER effect until E = 1.5 kV/mm and became negative ER active material when E>1.5 kV/mm was applied. On the other hand, PIn/colemanite/SO system showed negative ER activity under all the electric field conditions. Both behaviors were attributed to the migration of the dispersed particles to one of the electrodes. The use of Triton-X non-ionic surfactant in the dispersions observed to convert the negative ER effect to positive one by also enhancing the ER activity, which was explained by the water/surfactant-bridge formation model.

Acknowledgement

The authors thank to the Gazi University Research Fund (Grant No: 05/2011-05, 05/2012-34) and Turkish Scientific and Technological Research Council (Grant No: 111T637) and COST Action CM1101 for the financial support to this work.

References

[1] Hao T 2001 *Advanced Materials* **13** 1847
[2] Sari A and Tuzen M 2009 *Desalination* **249** 260
[3] Kanaoka S, Ikeda N, Tanaka A, Yamaoka H and Higashimura T 2002 *J. Polym. Sci., Part A: Polym. Chem.* **40** 2449
[4] Sarikaya S, Yavuz M, Yilmaz H, Unal H I and Sari B 2009 *Polym. Compos.* **30** 583

[5] Eristi C, Yavuz M, Yilmaz H, Sari B and Unal H I 2007 *J. Macromol. Sci. Part A: Pure Appl. Chem.* **44** 759
[6] Cetin B, Unal H I and Erol O 2012 *Smart Mater. Struct.* **21** 125011
[7] Cetin B, Unal H I and Erol O 2012 *Clays Clay Miner.* **60** 300
[8] Wu C W and Conrad H 1997 *J. Rheol.* **41** 267
[9] Hao T 2005 *Studies in Interface Science* **22** 475
[10] Ramos-Tejada M M, Arroyo F J and Delgado A V 2010 *Langmuir* **26** 16833
[11] Zou J, Su L, You F, Chen G and Guo S 2011 *J. Appl. Polym. Sci.* **121** 1725
[12] Ramos-Tejada M M, Espin M J, Perea R and Delgado A.V. 2009 *J. Non-Newtonian Fluid Mech.* **159** 34
[13] Yin J, Xia X, Xiang L and Zhao X 2001 *Smart Mater. Struct.* **20** 015002

Viscoelastic properties of electrorheological suspensions of core-shell (carbon/polyaniline) particles in silicone oil

M Sedlacik[1, 2, 5], S Almajdalawi[3], M Mrlik[1, 3], V Pavlinek[1, 3], P Saha[1, 3] and J Stejskal[4]

[1] Centre of Polymer Systems, University Institute, Tomas Bata University in Zlin, Nad Ovcirnou 3685, 760 01 Zlin, Czech Republic
[2] Department of Production Engineering, Faculty of Technology, Tomas Bata University in Zlin, T. G. Masaryk Sq. 275, 762 72 Zlin, Czech Republic
[3] Polymer Centre, Faculty of Technology, Tomas Bata University in Zlin, namesti T. G. Masaryka 275, 762 72 Zlin, Czech Republic
[4] Institute of Macromolecular Chemistry, Academy of Sciences of the Czech Republic, Heyrovsky Sq. 2, 162 06 Prague 6, Czech Republic

E-mail: msedlacik@ft.utb.cz

Abstract. Carbon/polyaniline particles with core-shell structure were synthesized as a novel dispersed phase for electrorheological (ER) suspensions in this study. Core of these composite particles was obtained by carbonization of polyaniline base in an inert atmosphere of nitrogen at 650°C and then coated with polyaniline shell. The morphology and composition of prepared particles were examined with scanning electron microscopy and Fourier transform infrared spectroscopy, respectively. The analysis revealed the conversion of polyaniline to carbon via ring-opening happened during the carbonization process and successful coating of carbonized particles with shell layer. The products retained the original granular structure after carbonization as well as after the coatings. The dielectric spectra analysis suggests high particle polarizability of carbonized material. Thus, the measurements performed under oscillatory shear flow showed a remarkably high ER intensity at relatively low electric field strengths. Coating of carbonized particles by polyaniline base changes compatibility of particle surface with silicone oil medium and, consequently, flow properties of suspensions in the absence of electric field, but does not influence the shear rate dependence of the complex viscosity in the electric field.

1. Introduction

Investigation of new materials with high efficiency still ranks among topical subjects in electrorheology (ER). Since Winslow's discovery [1] more than 50 years ago a number of studies have been performed. It appeared that not only high permittivity and low conductivity of materials provide high particles polarizability necessary for high ER performance. The results suggest that also particle structure and morphology may be very important. Various inorganic or organic powder materials have been used as a solid phase in ER suspensions. Among them, semi-conducting polymers such as polyaniline [2], polypyrrole [3], or paraphenylendiamine [4] proved to be very suitable.

[5] Corresponding author

Their chemical structure with conjugated system of π-bonds enables charge transfer and thus good particle polarization. Polyaniline (PANI) as the most interesting polymer in this group showed suitable thermal stability and possibility to control the conductivity via deprotonation.

This study deals with ER viscoelastic behaviour of silicone oil suspensions of core-shell particles in oscillatory shear. The particle core of carbonized polyaniline coated with polyaniline base shell provided material having various compatibility with oil medium and thus different field–off flow properties.

2. Experimental

2.1. PANI base preparation
The PANI base was prepared according to the report [5]. In this way 0.2 M aniline hydrochloride (Fluka, Switzerland) was oxidized with 0.25 M ammonium peroxydisulfate (Lach-Ner, Czech Republic) in water for 24 h at room temperature. The obtained particles were collected on a filter and washed with 0.2 M hydrochloric acid (Sigma, USA) and acetone (Sigma, USA) several times. Synthesized particles of polyaniline hydrochlorid (emeraldine) were thereafter deprotonated with excess of 1 M aqueous ammonia (Sigma, USA) to obtain PANI base. After washing with acetone the product was dried at 60°C in vacuum.

2.2. Carbonization of PANI base
PANI base particles were heated under a nitrogen atmosphere in an electric oven from the laboratory temperature at the rate of 15°C/min to the temperature 650°C which was maintained for 1 h. Then, the carbonization process was switched off and the sample was left to cool to laboratory temperature. Then, the prepared carbonized particles were sieved to obtain particle sizes lower than 45 μm (*Sample A*). According to the previous work [6] most of structural changes have already occurred at 650 °C. Thus, this temperature was used in the present study.

2.3. Coating of carbonized product with PANI layer
Sample B and *Sample C* of composite particles from the *Sample A* as a core coated with a PANI base layer of different thickness were prepared. In case of the *Sample B* the PANI content in the particle shall was the same as in the particle core before carbonization. The content of PANI base in the shell of the *Sample C* was half.

2.4. Scanning electron microscopy and FTIR spectroscopy
A scanning electron microscopy (SEM, VEGA II LMU, Tescan Ltd., Czech Republic), with an operating voltage of 30 kV, was adopted to observe the morphologies of prepared samples. On the other hand, the chemical composition of particles was examined with Fourier transform infrared spectroscopy (FTIR, Thermo Scientific, USA). The FTIR spectra were recorded using attenuated total reflectance (ATR) technique with Germanium crystal in the range 500 – 4000 cm^{-1} at 64 scans per spectrum at 2 cm^{-1} resolution.

2.5. Electrorheological measurements
Suspensions containing 10 wt.% of the *Sample A*, *Sample B*, or *Sample C* in silicone oil (Lukosiol M200, Chemical Works Kolín, Czech Republic; $\eta \approx 200$ mPa·s) were prepared. Measurements of their electrorheological properties were carried out on a rheometer Bohlin Gemini (Malvern Instruments, UK) with parallel plates 40 mm in diameter and 0.5 mm gap. The plates were connected to a DC high-voltage source TREK (TREK 668B, USA) with the applied voltage (0.125–0.5 kV) corresponded to the electric field strength 0.25 – 1 kV mm^{-1}. Dynamic viscoelastic tests were performed by dynamic strain and frequency sweeps. The rheological parameters were obtained from frequency sweep tests (0.1–10 Hz) at fixed strain amplitude in the viscoelastic region ($\gamma_A = 4 \times 10^{-4}$). The particle aggregated structures in the system were destroyed by shearing of the suspension at a shear rate 20 s^{-1} for 150 s

before each measurement at new electric field strength applied. The measurements were carried out at 25, 45, 65, and 85°C.

2.6. Dielectric properties

The frequency dependences of complex permittivity, ε^*, involving the relative permittivity, ε', and dielectric loss factor, ε'', of studied ER suspensions were measured with an impedance analyzer (Agilent 4524, Japan) in the range $f = 5\times10^3 - 5\times10^6$ Hz.

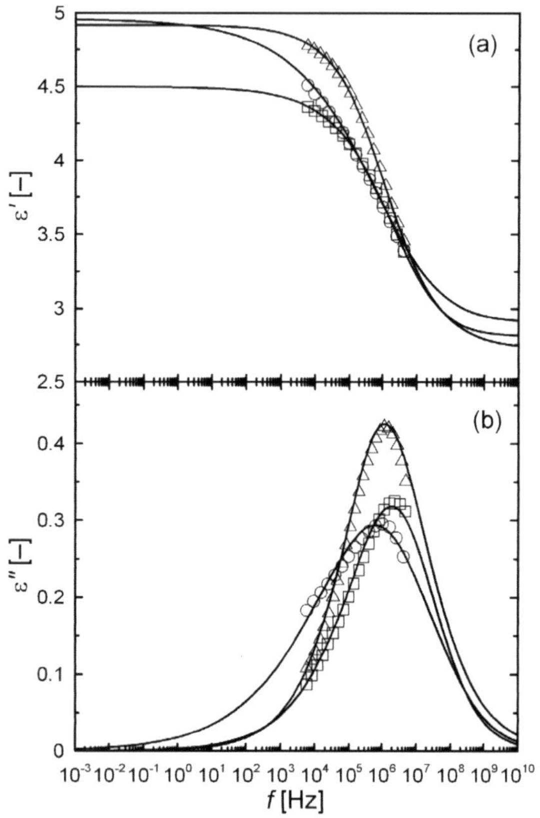

Figure 1. Relative permittivity, ε', (a) and dielectric loss factor, ε'', (b) as a function of the frequency, f, for 10 wt.% ER suspensions of *Sample A* (△), *Sample B* (□) and *Sample C* (○) in silicone oil at 25°C.

Table 1. Dielectric parameters in Eq. 1 for 10 wt.% ER suspensions of *Sample A*, *Sample B*, and *Sample C*.

Parameter	*Sample A*	*Sample B*	*Sample C*
ε'_0 [–]	4.92	4.50	4.96
ε'_∞ [–]	2.72	2.81	2.91
$\Delta\varepsilon'$ [–]	2.2	1.69	2.05
t_{rel} [s]	1.4×10^{-7}	2.1×10^{-8}	1.8×10^{-8}
a [–]	0.47	0.40	0.29
b [–]	0.98	1.75	2.24

The dielectric characteristics of particle suspensions (Table 1) were obtained from the Havriliak–Negami empirical model [7] fitting by least square method:

$$\varepsilon^* = \varepsilon'_\infty + \frac{(\varepsilon'_0 - \varepsilon'_\infty)}{\left(1 + (i\omega \cdot t_{rel})^a\right)^b} \tag{1}$$

Here, ε^*, is a complex fluid permittivity, the particle polarizability, $\Delta\varepsilon'$, is defined as the difference between, ε'_0, and, ε'_∞, which are the limit values of the relative permittivity at the frequencies below and above the relaxation frequencies, ω, is angular frequency ($= 2\pi f$), t_{rel}, is the relaxation time, a, is the scattering degree of relaxation times, and, b, is related to the asymmetry of the relaxation time spectrum. The relaxation frequency, at which the dielectric loss factor, ε'', has a maximum, is proportional to the rate of polarization of suspension particles.

3. Results and discussion

3.1. Characterization of prepared particles

The size and morphology of PANI base particles and *Sample A*, *Sample B*, and *Sample C* are shown in Figure 2. As can be seen in Figure 2b, after carbonization of PANI base particles the granular morphology was retained and only some shrinkage may be observed, which is in good accordance with literature [8]. After coating of carbonized particles the size of *Sample B* and *Sample C* slightly increased while their morphology became preserved.

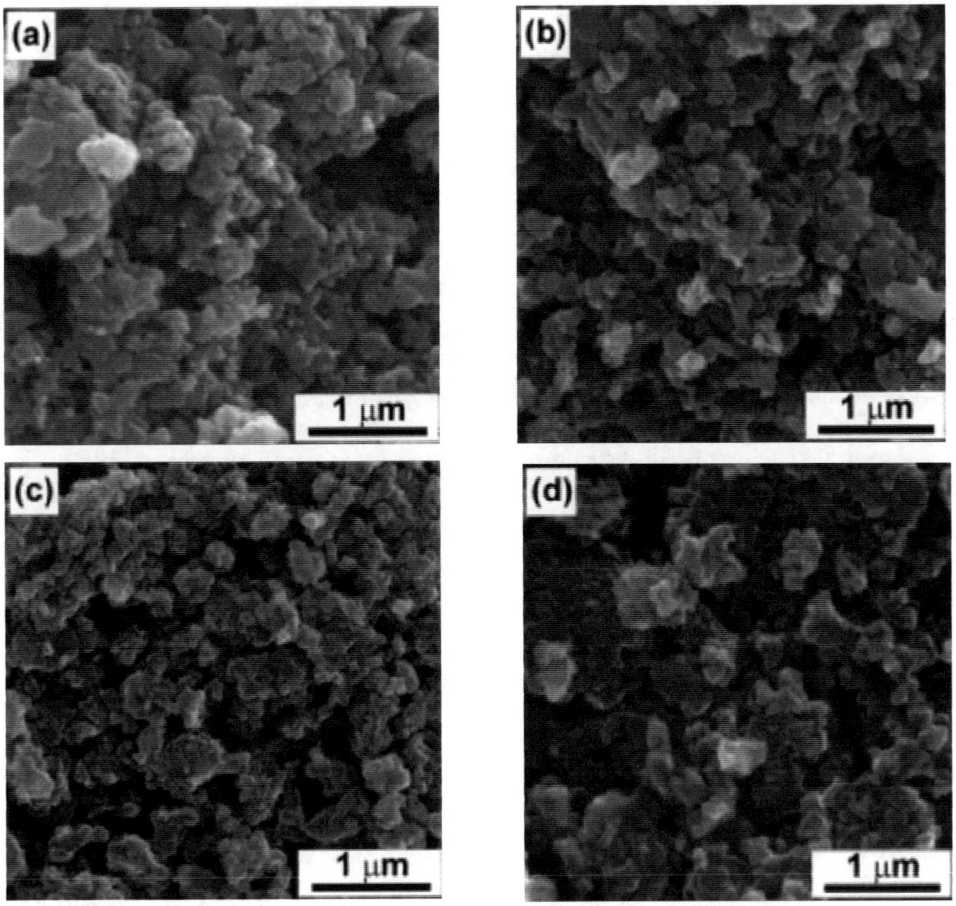

Figure 2. SEM images of PANI base (a), carbonized particles of *Sample A* (b), and coated particles of *Sample B* (c) and *Sample C* (d).

The infrared spectra showed the absorption around 1563 cm^{-1} in case of PANI base (Figure 3a), which is characteristic for the vibration of quinonoid rings, while the peak at 1485 cm^{-1} is attributed to the vibrations of C–C aromatic ring stretching [9]. The absorptions of these peaks were suppressed after carbonization (Figure 3b) indicating the ring-opening and possible crosslinking. The process of carbonization also reduced the absorption peak at 1298 cm^{-1} which is attributed to the C–N stretch of a secondary aromatic amine. On the other hand, the quinonoid peak was moved to higher wavenumbers and C–C aromatic ring stretching to lower wavenumbers after coating of carbonized particles (Figure 2c, d). The characteristic absorption band in the range of 1010 – 1170 cm^{-1} was assigned to the aromatic C–H in-plane bending modes. The relatively intensive peak at 1128 cm^{-1} indicates a certain amount of protonated polymer in PANI base probably due to the presence of highly organized chains which are resistant towards the deprotonation [10]. Deformations of out-of-plane C–H bonds, i.e. in 1,4-disubstituted rings, are located in range of 800–880 cm^{-1} with the maximum at 798 cm^{-1}. The absorption of this band was lowered after carbonization due to the ring-opening and moved to higher wavenumbers after a coating with PANI layer.

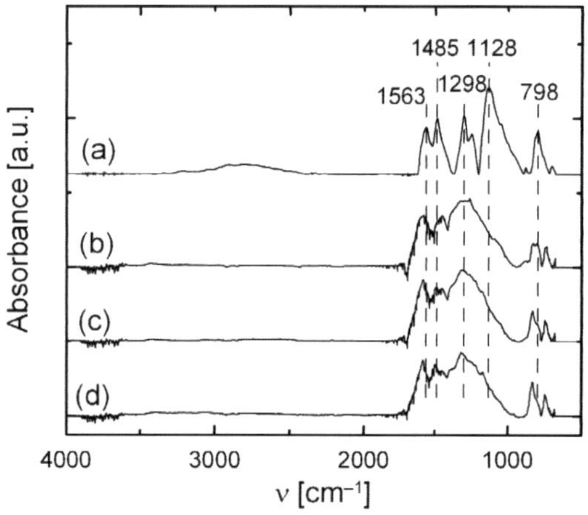

Figure 3. FTIR spectra of PANI base (a), carbonized particles of *Sample A* (b), and coated particles of *Sample B* (c) and *Sample C* (d).

3.2. ER effect and dielectric properties

Figure 4 illustrates the ER properties of the prepared suspensions under oscillatory shear. Only the suspension of the *Sample B*, i.e. carbonized particles coated with higher PANI base amount shows almost Newtonian behaviour in the absence of electric field. On the other hand, suspension of the *Sample C* coated with thinner PANI layer or suspension of uncoated *Sample A* exhibit slightly pseudoplastic behaviour probably due to particle aggregation caused by bad compatibility with silicone oil. The aggregates of the *Sample C* are less stable than that of the *Sample A* probably because of incomplete particle surface coating. Consequently the aggregates disintegrate with increasing frequency more easily.

When the external electric field was applied (Figure 4b) the internal chain-like structure of particles raised, the complex viscosity increased at relatively low electric field strength 0.5 kV mm^{-1} in several orders of magnitude and all suspensions under study reached very similar values regardless of electric field–off behaviour.

Figure 4c demonstrates that the highest efficiency as a relative increase in electroviscosity, $e = (\eta^*_E - \eta_0)/\eta^*_0$ [11] corresponds to the composite particles of *Sample B* due to lowest complex viscosity in the absence of electric field which suggests that the coating process of carbonized particles with PANI may significantly influence this quantity.

It is generally accepted that dielectric properties of ER suspension play important roles in the interfacial polarization of dispersed particles and formation of chain-like internal structures within the system. The dielectric spectra (Figure 1) and their characteristics in the Table 1 indicate that dielectric properties of all three studied suspension are very similar. This fact correlates well with ER behaviour

(Figure 4) in the presence of electric field and suggests that not polarization of only PANI base coating but dielectric properties of carbonized core control the ER behavior.

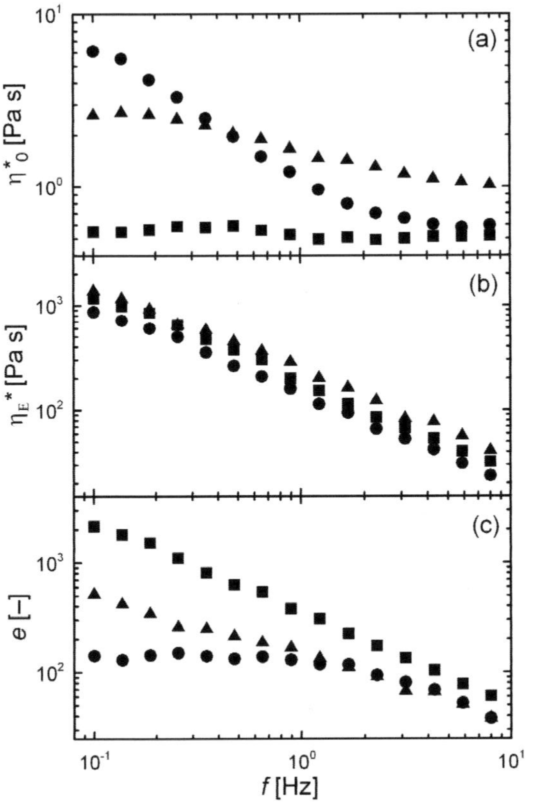

Figure 4. Complex viscosity in the absence of electric field, η^*_0, (a) in the presence of electric field strength of 0.5 kV mm^{-1}, η_E^*, (b) and electrorheological efficiency, *e*, (c) vs. frequency, *f*, for 10 wt.% ER suspensions of *Sample A* (▲), *Sample B* (■) and *Sample C* (●) in silicone oil at 25°C.

3.3. Viscoelastic characteristics

To illustrate the behaviour of viscoelastic moduli of ER materials in the electric field suspension of the *Sample B* has been used. In the absence of electric field the system exhibits liquid character and the loss modulus dominates (Figure 5). When the electric field was applied the elastic modulus became remarkably higher due to an increase in the system rigidity caused by the formation of internal chain-like structures of polarized particles.

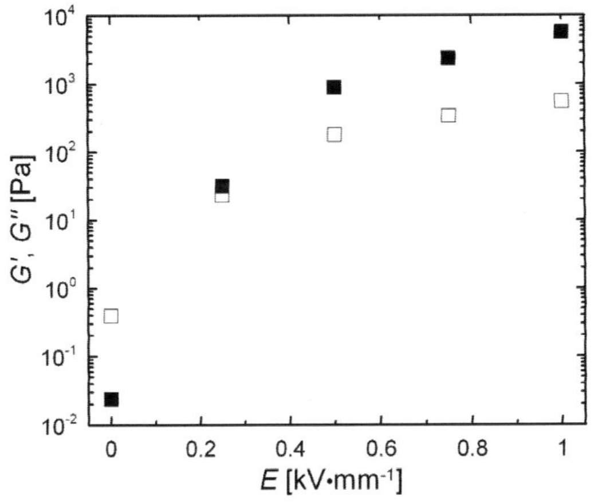

Figure 5. Storage, *G'*, (■) and loss, *G''*, (□) viscoelastic moduli vs. electric field strength, *E*, at *f* = 0.14 Hz for 10 wt.% ER suspension of *Sample B* particles in silicone oil at 25°C.

3.4. The temperature dependence

The wide temperature range for application is another important factor for the utilization of ER suspension in real device. Thus, PANI having the temperature stability approximately till 200°C [12] is a suitable candidate from the group of conducting polymers for the use as a dispersed phase in ER suspensions. The effect of temperature on the ER performance illustrates Figure 6. It is obvious that the elevated temperature increases the Brownian motion within the suspension reflected in a decreased viscoelastic moduli in the absence of electric field. In the presence of external electric field, the ER performance is lower at elevated temperatures since the internal chain-like structures of polarized particles are destroyed easily due to more intensive Brownian motion as well as lower silicone oil viscosity. However, the difference between viscoelastic moduli in the absence and in the presence of electric field is still higher than three orders of magnitude reflecting quite stiff internal structures formation even at high temperatures.

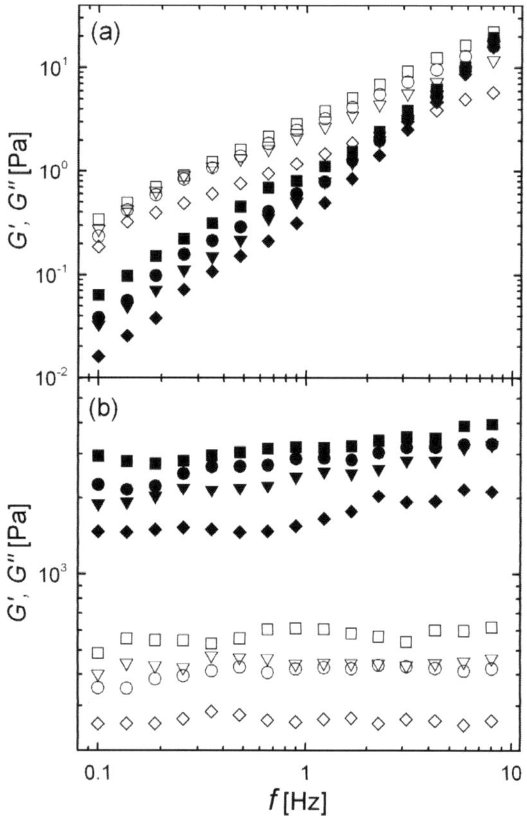

Figure 6. Storage, *G'*, (solid symbols) and loss, *G''*, (open symbols) viscoelastic moduli vs. frequency, *f*, in the absence of electric field (a) and in the presence of electric field strength of 0.75 kV mm^{-1} (b) for 10 wt.% ER suspension of *Sample B* in silicone oil at 25°C (■, □), 45°C (●, ○), 65°C (▼, ▽), and 85°C (◆,◇).

4. Conclusions

The results of the study suggest that the particles based on carbonized PANI can be interesting material with high polarizability suitable for suspensions providing high ER performance even at low electric field strengths. Coating of particles with PANI base layer changes compatibility of carbon surface with silicone oil medium. This affects electric field–off mutual particle interactions and changes flow behaviour in the absence of electric field. It appeared, however, that it did not influence markedly the suspension properties when the electric field was applied and thus very similar high values of shear dependence of complex viscosity of the chain structure of polarized particles of all studied suspensions were obtained.

Acknowledgements

The authors wish to thank to the Grant Agency of the Czech Republic for the financial support of Grant No. 202/09/1626. This article was written with support of Operational Program Research and Development for Innovations co-funded by the European Regional Development Fund (ERDF) and national budget of Czech Republic, within the framework of project Centre of Polymer Systems (reg. number: CZ.1.05/2.1.00/03.0111).

References

[1] Winslow W M 1947 US patent 2 417 850
[2] Stenicka M, Pavlinek V, Saha P, Blinova N V, Stejskal J and Quadrat O 2009 *Colloid Polym. Sci.* **287** 403
[3] Sedlacik M, Mrlik M, Pavlinek V, Saha P and Quadrat O 2012 *Colloid Polym. Sci.* **290** 41
[4] Plocharski J, Drabik H, Wycislik H and Ciach T 1997 *Synth. Met.* **88** 139
[5] Stejskal J and Gilbert R G 2002 *Pure Appl. Chem.* **74** 857
[6] Rozlivkova Z, Trchova M, Exnerova M and Stejskal 2011 *Synth. Met.* **161** 1122
[7] Havriliak S and Negami S 1966 *J. Polym. Sci.* C **16** 99
[8] Yin J B, Xia X A, Xiang L Q and Zhao X P 2010 *Carbon* **48** 2958
[9] Sedenkova I, Prokes J, Trchova M and Stejskal 2008 *J. Polym. Degrad. Stabil.* **93** 428
[10] Mu S L 2010 *Synth. Met.* **160** 1931
[11] Lengalova A, Pavlinek V, Saha P, Quadrat O, Kitano T and Stejskal J 2003 *Europ. Polym. J.* **39** 641
[12] Kulkarni V G, Campbell L D and Mathew W R 1989 *Synth. Met.* **30** 321

Electrorheology of aniline-oligomer suspensions under oscillatory shear

M Mrlik[1, 2, 4], **V Pavlinek**[1, 2], **S Almajdalawi**[2], **P Saha**[1, 2], **P Bober**[3] **and J Stejskal**[3]

[1] Centre of Polymer Systems, University Institute, Tomas Bata University in Zlin, Nad Ovcirnou 3685, 760 01 Zlin, Czech Republic
[2] Polymer Centre, Faculty of Technology, Tomas Bata University in Zlin, namesti T. G. Masaryka 275, 762 72 Zlin, Czech Republic
[3] Institute of Macromolecular Chemistry, Academy of Sciences of the Czech Republic, Heyrovsky Sq. 2, 162 06 Prague 6, Czech Republic

E-mail: mrlik@ft.utb.cz

Abstract. Preparation of the aniline oligomers by the oxidation of aniline with *p*-benzoquinone in the solutions of methanesulfonic acid (MSA) and the rheology of their suspensions in silicone oil are presented in this study. This synthesis provides particles of flake-like morphology and various conductivities depending on the molar concentration of MSA. Further, the electrorheological (ER) performance of the particles suspended in the silicone oil was measured as well as dielectric properties of suspensions. Finally, the effect of the temperature on the ER activity was investigated.

1. Introduction

Electrorheological (ER) suspensions are basically two-phase liquid systems usually consisting of solid particles (dispersed phase) and liquid medium (continuous phase) [1–3]. Solid particles have to be polarizable under application of the electric field while the liquid medium should be in this case an electric insulator (usually silicone oil). Suspensions having these parameters provide a unique property to change the state from liquid-like to solid-like after application of the external electric field [4–6]. This transition is connected to the development of the internal structures within the application of electric field and also leads to rapid increase in the basic rheological properties, such as viscosity and viscoelastic moduli [7–10].

Solid particles are mostly represented by conducting polymers such as polyaniline [11–13], polypyrrole [14–16], polyphenylenediamine [17], and also polythiophene [18]. The advantages of conducting polymers over other ER materials (based on inorganic particles mostly titanates and titanium dioxides) represent fast preparation, suitable density (enhanced sedimentation stability), no abrasivity and, finally, tunable conductivity, which enables to control the ER performance.

However, aniline oligomers which are represented by several aniline constitutional units linked together seem to be promising material for ER suspensions due to their suitable conductivity [19]. Faster preparation in comparison to polymers and also tunable conductivity predict promising applicability of this material in ER suspensions.

[4] Corresponding author.

In this study aniline oligomers were prepared by oxidation of aniline with an organic oxidant, *p*-benzoquinone. The synthesis was carried out in the presence of different molar concentration of methanesulfonic acid, thus providing samples of different conductivity. Dielectric properties of the sample silicone oil suspensions were measured and results were fitted with Havriliak-Negami model in order to obtain information about the polarizability of samples as well as their relaxation time. Finally, the effect of temperature on viscoelastic properties of suspensions was investigated.

2. Experimental

2.1 Synthesis of aniline oligomers

Aniline (0.2 M; Fluka, Switzerland) was oxidized with *p*-benzoquinone (0.5 M; Sigma-Aldrich, USA) in aqueous solutions of 0.1–0.5 M methanesulfonic acid (MSA). After mixing the solutions of aniline and oxidant, the mixture was left at room temperature for a few hours. The solids were isolated by filtration, rinsed with acid solution, then with acetone, dried in air, and then over silica gel. A part of each sample was converted to a base form by immersion in a large excess of 1 M ammonium hydroxide, and dried as above.

2.2 UV–visible spectra

Polyaniline base displays the typical absorption spectrum of emeraldine base with absorption maxima at 330 and 620 nm [19]. In the case of aniline oxidation by *p*-benzoquinone, the latter absorption maximum is practically missing. This is due to the presence of dominating aniline oligomer fraction, which absorbs strongly below 400 nm and has only a long absorption tail extending to longer wavelengths (Figure 1).

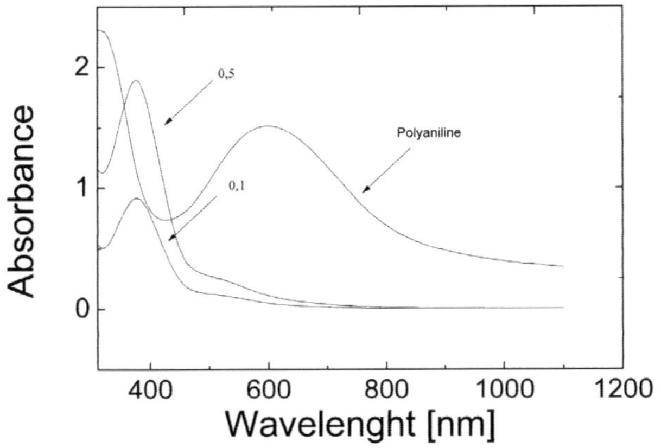

Figure 1. The UV–visible spectra of the oxidation products dissolved in *N*-methylpyrrolidone prepared by the oxidation of 0.2 M aniline with 0.5 M *p*-benzoquinone in the aqueous solutions containing various molar concentrations of MSA and standard polyaniline base.

2.3 Conductivity measurements

Powders of prepared aniline oligomers were pressed into the pellets of 13 mm diameter and 1 mm thickness at 7 MPa. The conductivity was measured via van der Pauw method at room temperature (Table 1).

Table 1. Conductivity, σ, and density, ρ, of aniline oligomers prepared in the solutions of 0.1 and 0.5 M methanesulfonic acid.

Medium	σ [S cm^{-1}]	ρ [g cm^{-3}]
0.1 M MSA	2.3×10^{-10}	1.36
0.5 M MSA	1.9×10^{-6}	1.35

2.4 Suspension preparation
Suspensions (10 vol.%) were prepared by mixing of aniline oligomers with corresponding amount of silicone oil (Fluid 200, Dow Corning, UK; viscosity η_c = 108 mPa·s, density $\rho_c = 0.965$ g·cm^{-3}). Before each measurement, the suspensions were stirred at first mechanically and then sonicated for 30 s.

2.5 Rheological and dielectric characterization
Rheological measurements were performed under controlled-shear-rate mode using rotational viscometer (Bohlin Gemini, Malvern Instruments, UK). The suspensions were placed into the parallel plate geometry with 40 mm in diameter and 0.5 mm gap. The instrument modified for ER experiments was connected to DC high-voltage source TREK (TREK 668B, USA) to generate electric field strengths 0–3.0 kV mm^{-1}. Before each measurement at new electric field strength, the built-up particulate structures were destroyed by shearing the sample suspension at the shear rate 50 s^{-1} for 80 s.

Dielectric properties of prepared suspensions were measured with Impedance Analyzer 4294A (Agilent Technologies, Japan) connected to 16482A cell for liquid materials. To investigate appropriate dielectric properties such as relative permittivity, ε', and dielectric loss factor, ε'', frequency range from 40 Hz to 5 MHz was selected. Measured dielectric spectra were fit with Havriliak–Negami (H–N) model (Eq. 1) [20].

$$\varepsilon^*_{HN}(\omega) = \varepsilon'_\infty + \frac{\Delta\varepsilon'}{\left(1 + (i\omega \cdot t_{rel})^a\right)^b} \tag{1}$$

Here, $\Delta\varepsilon' = \varepsilon'_s - \varepsilon'_\infty$ is particle polarizability, where, ε'_s, and, ε'_∞, are relative permittivities at zero and infinite frequency, f, respectively, ω, is angular frequency (=2 π f), t_{rel}, is the relaxation time, a,and, b, are shape-dependent parameters which describe the asymmetry of the dielectric function.

3. Results and discussion

3.1 Morphology of particles
The oxidation of the aniline with *p*-benzoquinone in the aqueous solutions of MSA provided samples with particle size around two micrometers (Figure 2). Morphology of the samples depends on the molar concentration of MSA. In 0.1 M MSA, the product of the synthesis was nearly globular particles. On the other hand, when the 0.5 M MSA was used, particles with mostly flake-like morphology were prepared.

Figure 2. SEM images of synthesizes particles (right) in 0.1 M MSA and (left) in 0.5 M MSA.

3.2 Dielectric characterization

In order to obtain information about polarizability (ability to react on the application of the electric field) and relaxation time (how fast particles respond to the electric field), data from the dielectric measurements were fit with H–N model (Table 2, Figure 3). The polarizability is higher and the relaxation time is lower for suspension of the sample prepared in 0.5 M MSA in comparison to suspension of sample prepared in 0.1 M MSA having lower conductivity. Thus, the enhanced ER activity of the suspension containing particles obtained from 0.5 M MSA due to higher conductivity connected to improved charge motion should influence viscoelastic properties under applied electric field.

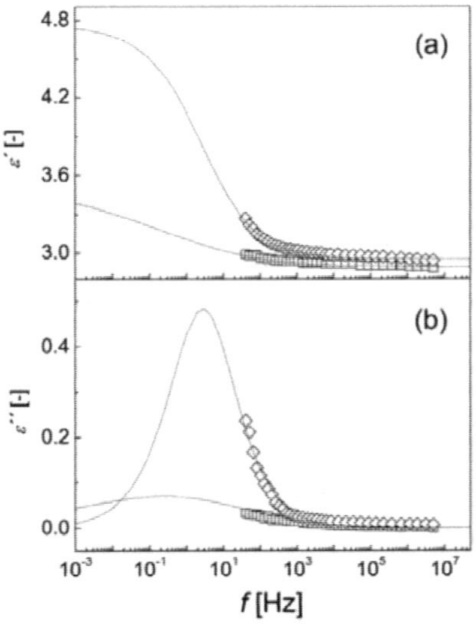

Figure 3. Frequency dependence of relative permittivity, ε', (a) and dielectric loss factor, ε'', (b) for 10 vol.% suspensions of aniline oligomers in silicone oil. Samples were prepared in 0.1 M MSA (□) or 0.5 M MSA (◇). Solid lines represent the H–N model fit.

Table 2. Parameters of the Havriliak–Negami model.

Medium	ε'_s	ε''_∞	$\Delta\varepsilon'$	t_{rel} [s]	a	b
0.1 M MSA	3.54	2.89	0.65	0.150	0.22	1.67
0.5 M MSA	4.74	2.95	1.79	0.048	0.62	1.12

3.3 Viscoelastic properties

Viscoelastic measurements provide information about dynamic characteristics of the prepared suspensions. Firstly, the linear viscoelastic regions from the dependence of the both viscoelastic moduli on the strain for the suspensions at various electric field strengths have to be established (Figure 4, left). Values of the strain will be further used for the frequency dependence of the moduli which is discussed below in the text.

In the absence of external field the viscous modulus dominates over elastic one confirming the liquid–like state of the suspensions (Figure 4, right). On the other hand, in the presence of the electric field the elastic modulus overcome viscous one for both cases, confirming creation of the internal structures. Toughness of these chain–like columns increases with increasing electric field strength. Further, larger ER effect was observed for the suspension that contains particles prepared in 0.5 M MSA, due to suitable conductivity and also dielectric properties of their suspension.

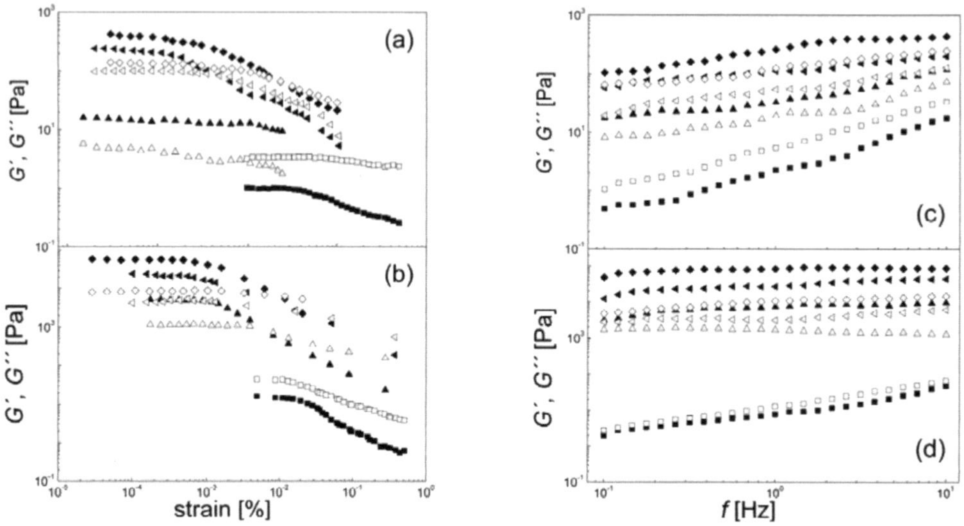

Figure 4. Strain dependence (left) and frequency dependence (right) of elastic modulus, G' **(solid symbols)** and viscous modulus, G'' **(open symbols)** for 10 vol.% suspensions of samples prepared in 0.1 M MSA (a, c) and 0.5 M MSA (b, d) under various electric field strengths, E (kV·mm^{-1}): (□, ■) 0, (△, ▲) 1, (◁, ◀) 2, (◇, ◆) 3, at 25°C.

3.4 Effect of the temperature on ER activity

In order to elucidate the influence of temperature on ER performance, the frequency dependence of the tan δ was plotted. **When tanδ (tanδ = G''/G') decreases, elastic behaviour of the suspension increases. In this case, tanδ** decreases with the increasing temperature (Figure 5). This trend is more pronounced at higher electric field strengths. Brownian motion stimulated at elevated temperatures inhibits the creation of the chain-like structures and decrease the ER activity.

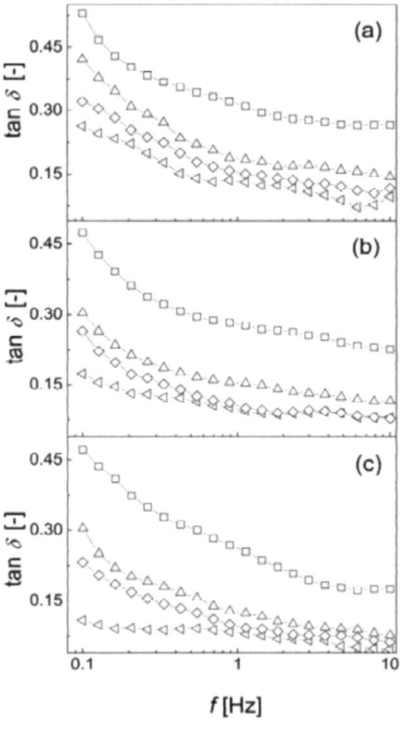

Figure 5. Frequency dependence of tanδ for 10 vol.% suspension of aniline oligomers prepared in 0.5 M MSA at (a) 1 kV·mm^{-1}, (b) 2 kV·mm^{-1} and (c) 3 kV·mm^{-1}, at various temperatures (°C): (□) 85, (△) 65, (◇) 45, (◁) 25.

4. Conclusion

Aniline oligomers were prepared by the oxidation of aniline with *p*-benzoquinone in the solutions of methanesulfonic acid (MSA). Various molar concentrations of MSA provided samples with different conductivities. Using the concentration of 0.5 M MSA enables to synthesize rather flake-like than globular morphology. The suspension of sample particles prepared in 0.5 M MSA provided the highest ER efficiency due to suitable conductivity. Dielectric properties such as polarizability of the particles and relaxation time obtained from the fitting of the dielectric spectra with H–N model also supported the enhanced ER effect of oligomers prepared in 0.5 M MSA. Finally, rheological investigation under various temperatures shows decreased ER activity of the suspensions at elevated temperatures.

Acknowledgement

The authors wish to thank the Czech Grant Agency for the financial support (202/09/1626). This article was written with support of Operational Program Research and Development for Innovations co-funded by the European Regional Development Fund (ERDF) and National Budget of Czech Republic, within the framework of project Centre of Polymer Systems (CZ.1.05/2.1.00/03.0111).

References

[1] Pavlinek V, Quadrat O, Saha P, Benes M J and Trlica J 1998 *Colloid Polym. Sci.* **276** 690
[2] Hao T 2001 *Adv. Mater.* **13** 1847
[3] Parthasarathy M and Klingenberg D J 1996 *Mater. Sci. Eng. R-Rep.* **17** 57
[4] Choi H J and Jhon M S 2009 *Soft Matter* **5** 1562
[5] Pavlinek V, Saha P, Perez-Gonzalez J, de Vargas L, Stejskal J and Quadrat O 2006 *Appl. Rheol.* **16** 14
[6] Lozano K, Hernandez C, Petty T W, Sigman M B and Korgel B 2000 *J. Colloid Interface Sci.* **297** 618
[7] Mrlik M, Pavlinek V, Cheng Q L and Saha P 2012 *Int. J. Mod. Phys. B* **26** 125007.
[8] Sedlacik M, Mrlik M, Pavlinek V, Saha P and Quadrat O 2012 *Colloid Polym. Sci.* **290** 41
[9] Mrlik M, Pavlinek V, Saha P and Quadrat O 2011 *Appl. Rheol.* **21** 52365
[10] Stenicka M, Pavlinek V, Saha P, Blinova N V, Stejskal J and Quadrat O 2010 *Appl. Rheol.* **20** 1
[11] Stenicka M, Pavlinek V, Saha P, Blinova N V, Stejskal J and Quadrat O 2009 *Colloid Polym. Sci.* **287** 403
[12] Quadrat O and Stejskal J 2006 *J. Ind. Eng. Chem.* **12** 352
[13] Yin J B, Zhao X P, Xia X, Xiang L Q and Qiao Y P 2008 *Polymer* **49** 4413
[14] Cheng Q L, He Y, Pavlinek V, Lengalova A, Li CZ and Saha P 2006 *J. Mater. Sci.* **41** 5047
[15] Zhang X T, Zhang J, Song W H and Liu Z F 2006 *J. Phys. Chem. B* **110** 1158
[16] Kim Y D and Song I C 2002 *J. Mater. Sci.* **37** 5051
[17] Wei C, Zhu Y H, Jin Y, Yang X L and Li CZ 2008 *Mater. Res. Bull.* **43** 3263
[18] Eyrik M, Yavuz M, Unal H I and Sari B 2009 *Polym. Compos.* **30** 1345
[19] Stejskal J and Trchova M 2012 *Polym. Int.* **61** 240
[20] Havriliak S and Negami S 1967 *Polymer* **8** 161

Microconvection heat transfer in electrorheological fluids in rotating electric field

A A Mokeev, S A Gubarev, E V Korobko[1] and N A Bedik

A.V. Luikov Heat and Mass Transfer Institute of NAS Belarus, 220072,
15 P. Brovka str., Minsk, Belarus

E-mail: evkorobko@gmail.com

Abstract. This work is dedicated to the study of heat transfer characteristics of electrorheological fluids in rotating electric fields. The problem of heat transfer in the electrosensitive heat carrier considering the microconvection transport mechanism is formulated and numerically solved. It is shown on the example of a flat layer, that internal rotations of elementary particles intensify heat transfer in ERF.

1. Introduction

It is known that in a static electric field, a spontaneous rotation of insulators placed in a low conductive medium [1–5] occurs. When electrorheological fluids (ERF) are used as a medium the rotation of the insulator is enhanced by 1 – 2 orders of magnitude compared to the pure fluid [6]. This effect lies in the basis of dielectric motors [7, 8]. Recently papers studying the dynamics of electrofluids in a rotating electric field [9] have appeared. In [10, 11] a principle of a centrifugal pump on that basis is proposed. In [12], based on differences in electrical properties of cell membranes and their size a method of their separation by applying a rotating electric field is suggested. The peculiarities of heat transfer in ERF when a rotating electric field is applied is not fully studied, in contrast to extensive studies in the field of microconvection heat and mass transfer in magnetic fluids under the influence of time-dependent magnetic fields [13–17]. This issue is particularly important in the aspect of controlled electric field devices for process intensification of heat exchange. This research investigates the features of the physical mechanism of microconvection heat transfer in electrorheological fluids and determines the influence of external alternating electric fields on heat transfer in them.

2. Physical and mathematical modeling of heat transfer in ERF in a rotating electric field

Let us consider the mechanism of heat transfer in ERF flowing in a flat slot channel under the influence of a rotating electric field. By analogy with the phenomenological approach proposed in [1], heat transfer, taking into account microconvection will be described using the effective thermal conductivity tensor, taking into consideration the energy transfer by two mechanisms – microconvection and conduction.

Electrorheological suspension is a quasi-homogeneous liquid dispersion in which the filler particles are evenly distributed throughout the volume. The electric field structures are formed, directed along the electric field vector.

[1]Corresponding author.

Macroscopic properties of ERF are expressed by values of physical quantities averaged over the local equilibrium subsystems <*F*>, located in a physically infinitesimal volume $\mathrm{d}V(\vec{r},t)$, and containing a sufficiently large number of particles $\mathrm{d}N(\vec{r},t)$. The values <*F*> differ little from the average of the neighboring elementary volumes $\mathrm{d}V(\vec{r}+\mathrm{d}\vec{r},t+\mathrm{d}t)$ and are continuous functions of the coordinates of the centers of mass of subsystems. According to the mechanics of heterogeneous media [18], averaging over the local equilibrium systems is carried out separately for the liquid carrier and solid dispersed phase, and the results are summed up. The process of heat conduction in the absence of macroscopic convection of ERF according to the first law of thermodynamics:

$$\rho_i \frac{\mathrm{d}u_i}{\mathrm{d}t} = \rho_i A_i + \rho_i Q_i, \qquad (2.1)$$

where ρ_i – density of *i*-th phase; A_i – the work of internal forces; u_i – mass density of the internal energy of the *i*-th phase; $\rho_i Q_i = -\mathrm{div}(\vec{j}_{qi}) + \sum_1^m Q_{ij}$ – the difference between incoming and outgoing heat fluxes due to heat $\vec{j}_{qi}$ and contact heat fluxes between the phases Q_{ij}.

In the absence of phase transitions of the contact heat flow across the surface with a temperature difference ΔT_{ij}:

$$Q_{ij} = C_V \mathrm{d}m \beta_{ij} \Delta T_{ij}, \qquad (2.2)$$

where C_V – heat capacity of perceiving the mass d*m*; β_{ij} – the coefficient of contact heat transfer.

Solid particles in a stationary ERF under the influence of a constant external electric field strength $\vec{E}$ are in the form of chains connecting the channel wall in the direction of the vector $\vec{E}$. In particular, if the potential difference is applied to the walls of the flat channel, the chain will be built across the channel. This structure is constantly disturbed by Brownian motion and reduced again due to the shear flow due to the strength of the electric interaction.

Let us consider the case when in a rotating electric field with intensity of $\vec{E}(\Omega t)$ an elementary volume of fluid in steady motion rotates with the rotation frequency Ω in the plane of rotation of vector $\vec{E}$ maintaining chain structures. The interaction between the particles varies with the rate of relaxation. Mathematical modeling of heat and mass transfer in such a system based on the solution of the equations of hydrodynamics and heat for individual particles and the vortex flow of the dispersion medium between them is very complex. To resolve this problem in application to magnetic fluids a phenomenological approach is proposed in [18], based on the use of the effective thermal conductivity tensor, taking into account the convective and conductive heat transfer mechanisms taking into consideration the anisotropy caused by the internal rotation of the particles.

We obtain similar expressions for the case of electrorheological fluids. For heat transfer in a stationary ERF it is possible to apply microscopic approach to the averaging physical quantities of the phase and volume [18]. In ERF a heterogeneous environment when using averaging microconvection flows of mass and heat is produced in cells that are released around each particle in the form of an almost spherical volume of radius ($b + a$) (figure 1). These cells in the case of ERF with the particles form a chain, from wall to wall, and the averaging is performed on a pair of adjacent cells and carrier phase fluid surrounding the entire chain of cells.

By turning the electric field vector at an angle $\Theta = \Omega \Delta t$ with the angular velocity Ω chains tend to be reconstructed along a new direction, but in the stationary liquid particles remain in the chains of the former direction. Acting on the mass of liquid electrical forces only occurs in an inhomogeneous dielectric, i.e. on the surface of discontinuity of the dielectric constant and conductivity – on surfaces of the particles. Changing the direction of the electric field vector does not affect the mass transfer of the carrier liquid. Its movement arises only because of the rotating surface of the particle entrainment.

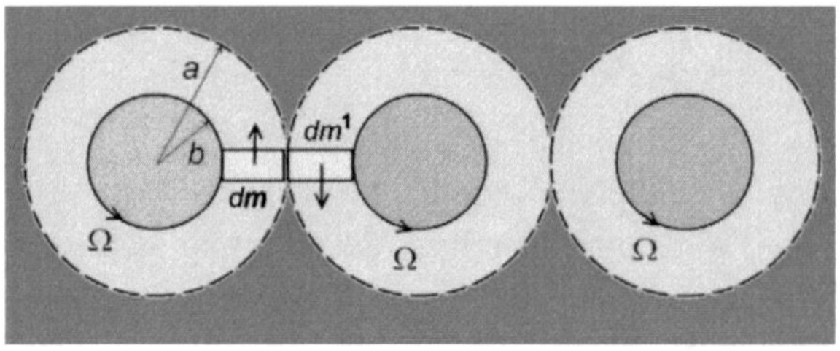

Figure 1. The model of heat transfer between two rotating neighboring particles.

The moving surface of the particle carries away the surface layer of the liquid in rotation with the same frequency. Neighboring particles according to the dipole interaction draws the liquid in the opposite direction. In the separating layer of liquid (figure 2) almost planar shear flow appears in which the speed of the middle layer is equal to zero. The flow between the particles is approximately taken with a rectilinear Couette velocity profile. The physical mechanism of heat transfer is as follows.

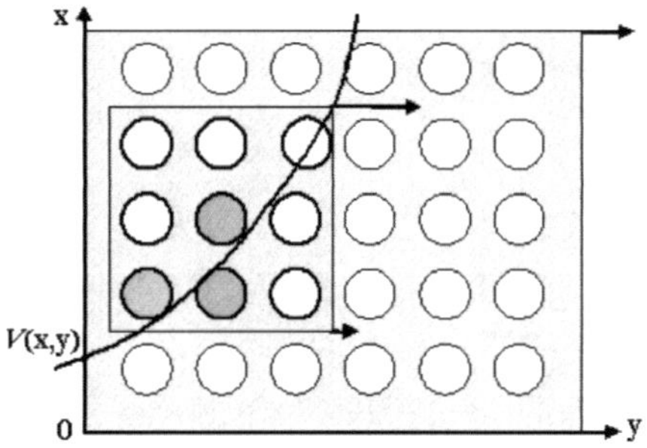

Figure 2. Separation of an infinitely small volume in a flowing ERF.

Element of the fluid mass dm, followed by rotation of the moving particle with a linear velocity $\vec{V} = \frac{1}{2}\vec{\Omega}\times\vec{b}$ in the time $\Delta t = \frac{T}{2} = \frac{\pi}{\Omega}$ receives from the heat:

$$\delta Q_i = k\,\mathrm{d}S\,\Delta T_1 = C_v\,\mathrm{d}m\,\beta_i\,\Delta T_1, \tag{2.3}$$

where ΔT_1 – temperature drop between the particle and fluid, °C; β_i – a dimensionless coefficient of heat transfer.

The following half-period element of mass dm is in contact with the corresponding element dm^1, moving along the surface of neighboring particles in the opposite direction and, in the case of steady-state heat transfer, it sends the same amount of heat. More precisely, different elements in series come in contact, but at the end of the half-period the element at the exit of the gap between the particles receives the amount of heat such as what it would have received under the continuous contact in the middle of the gap. Then the heat is transferred to the next link in the chain of particles, etc. to the opposite wall of the channel. On average, there is a closed stream.

In the steady state heat transfer process due to the symmetry of the system along the walls of the channel the amounts of heat transferred across the chains are mutually balanced. Heat is transferred only perpendicular to the walls. This average over the unit cell (particle and surrounding fluid) and the dispersion of phase (carrier fluid), the convective heat flux from the heated to the temperature T_1 to a wall cooled to a temperature T_2, the other wall, equal to half the width of the gap between the particles, depth of 2 $(a + b)$, covers the longitudinal chain of particles N_c, and 2 $(a + b)$ N_c. In the first approximation, the velocity profile in the flow line $V(y) = ky$, so that

$$Q_V = \frac{\mathrm{d}}{\mathrm{d}t}\int_0^{a+b}(a+b)V(y)\mathrm{d}t\mathrm{d}y = a\int_a^{b+a} aV(y)\mathrm{d}y = a(a+b)^2\Omega . \qquad (2.4)$$

Given the differences between the actual shape of the particles from the cylindrical coefficient 2 in the last factor is omitted. Mass flow rate Q_m and the mass flux density equal to:

$$Q_m = \frac{\mathrm{d}m}{\mathrm{d}t} = \rho\frac{\mathrm{d}V}{\mathrm{d}t} = \rho a(a+b)^2\Omega , \qquad (2.5)$$

$$j_m = \frac{Q_m}{S} = \frac{\rho\Omega a(a+b)^2}{a(a+b)} = \rho\Omega(a+b) \qquad (2.6)$$

Each element of mass dm transfers heat quantity from the wall:

$$\delta Q = j_{\mathrm{qk}}\,\mathrm{d}S\,\mathrm{d}t = C_V j_{\mathrm{m}}\mathrm{d}S\,\mathrm{d}t\,\beta\,\Delta T_1 , \qquad (2.7)$$

and the density of the convective heat flux j_{qk} is:

$$j_{\mathrm{qk}} = C_V j_{\mathrm{m}}\beta\Delta T_1 = C_{\mathrm{V}}\rho\Omega a(a+b)^2\beta\Delta T_1 . \qquad (2.8)$$

We assume that the number of links in the chain is the same in the steady heat transfer process, so the total difference in temperature between the walls is the sum of temperature differences at each of the N_c links in the chain of particles and the temperature gradient between the walls is equal to the gradient on the link:

$$\frac{\Delta T}{h} = \frac{\Delta T_1 N_{\mathrm{c}}}{(a+b)N_{\mathrm{c}}} = \nabla T , \qquad \Delta T_1 = (a+b)\nabla T . \qquad (2.9)$$

Heat flux density along the chain is equal to the flux density through the link:

$$j_{\mathrm{qk}} = C_{\mathrm{V}}\rho\Omega(a+b)\beta\Delta T_1 = C_{\mathrm{V}}\rho\Omega(a+b)^2\beta\nabla T = \lambda_{\mathrm{k}}\nabla T . \qquad (2.10)$$

The total density of heat flow is the sum of convective and conductive components:

$$j_{\mathrm{q}} = j_{\mathrm{qk}} + j_{\mathrm{qm}} . \qquad (2.11)$$

The molecular heat flux passes through a chain of particles with a density j_{qp} intervals and in parallel along the liquid layer between the neighboring chains with a density j_{qf}:

$$j_{\mathrm{qm}} = j_{\mathrm{qp}} + j_{\mathrm{qf}} . \qquad (2.12)$$

The flow along the chain in the same steady course through the particle and the layer of liquid between the particles: $j_{\mathrm{qp}} = j_{\mathrm{qf}} = \lambda_{\mathrm{p}}\nabla T_{\mathrm{p}} = \lambda_{\mathrm{f}}\nabla T_{\mathrm{f}}$, so

$$\nabla T_{\mathrm{f}} = \frac{\lambda_{\mathrm{p}}}{\lambda_{\mathrm{f}}}\nabla T_{\mathrm{p}} . \qquad (2.13)$$

On the other hand: $\nabla T_{\mathrm{p}} = \frac{\Delta T_{\mathrm{p}}}{2a}$, $\nabla T_{\mathrm{f}} = \frac{\Delta T_{\mathrm{f}}}{2b}$, so

$$\nabla T_{\mathrm{f}} = \frac{\lambda_{\mathrm{p}}}{\lambda_{\mathrm{f}}}\frac{b}{a}\Delta T_{\mathrm{p}} . \qquad (2.14)$$

Full temperature difference along the chain of particles N_c is equal to

$$\Delta T = (\Delta T_{\text{p}} + \Delta T_{\text{f}})N_{\text{p}}, \qquad N_{\text{p}} = \frac{h}{2(a+b)}, \tag{2.15}$$

so $$\Delta T_{\text{p}} + \Delta T_{\text{f}} = \frac{\Delta T \cdot 2(a+b)}{h} = 2(a+b)\nabla T, \ \Delta T_{\text{p}} + \frac{\lambda_{\text{p}}}{\lambda_{\text{f}}}\frac{b}{a}\Delta T_{\text{p}} = 2(a+b)\nabla T,$$

$$\left(1 + \frac{\lambda_{\text{p}}}{\lambda_{\text{f}}}\frac{b}{a}\right)\Delta T_{\text{p}} = 2(a+b)\nabla T, \ \Delta T_{\text{p}} = \frac{2(a+b)\lambda_{\text{f}}a}{\lambda_{\text{f}}a + \lambda_{\text{p}}b}\nabla T, \ j_{\text{qc}} = \lambda_{\text{p}}\frac{2(a+b)\lambda_{\text{f}}a}{\lambda_{\text{f}}a + \lambda_{\text{p}}b}\nabla T$$

and the average coefficient of thermal conductivity of the molecular chains of particles is

$$\lambda_{\text{c}} = \lambda_{\text{p}}\frac{2(a+b)\lambda_{\text{f}}a}{\lambda_{\text{f}}a + \lambda_{\text{p}}b}. \tag{2.16}$$

The molecular heat flux along the layer of fluid between adjacent chains of particles continuously from one side to the other channel is determined by the usual law of heat conduction $j_{\text{qf}} = \lambda_{\text{f}}\nabla T$, and added to the flow along the chains, so that the total molecular heat flux density is

$$j_{\text{qm}} = j_{\text{qp}} + j_{\text{qf}} = \left[\lambda_{\text{f}} + \lambda_{\text{p}}\frac{2(a+b)\lambda_{\text{f}}a}{\lambda_{\text{f}}a + \lambda_{\text{p}}b}\right]\nabla T \tag{2.17}$$

and the total coefficient of molecular thermal conductivity of ERF is

$$\lambda_{\text{m}} = \lambda_{\text{f}} + \lambda_{\text{p}}\frac{2(a+b)\lambda_{\text{f}}a}{\lambda_{\text{f}}a + \lambda_{\text{p}}b}. \tag{2.18}$$

Full thermal conductivity across the layer of ERF is equal to the sum of convective and molecular components

$$\lambda_{\lrcorner} = \lambda_{\text{k}} + \lambda_{\text{m}} = C_{\text{V}}\rho\frac{\Omega}{2}(b+2a)^2\beta_i + \lambda_{\text{f}} + \lambda_{\text{p}}\frac{2(a+b)\lambda_{\text{f}}a}{\lambda_{\text{f}}a + \lambda_{\text{p}}b}. \tag{2.19}$$

Coefficient of thermal conductivity along the layer of ERF parallel to the walls of the channel due to the symmetry of stationary ERF has the same meaning. In the direction perpendicular to the plane of rotation of the electric field, the thermal conductivity is equal to the molecular

$$\lambda_{\Uparrow} = \lambda_{\text{m}} = \lambda_{\text{f}} + \lambda_{\text{p}}\frac{2(a+b)\lambda_{\text{f}}a}{\lambda_{\text{f}}a + \lambda_{\text{p}}b}. \tag{2.20}$$

Stationary ERF in a rotating homogeneous electric field is thermally homogeneous and isotropic, i. e., its coefficient of thermal conductivity in each element's volume $\mathrm{d}V(\vec{r})$ is the same in all directions perpendicular to the axis of rotation of $\vec{E}$.

Total heat flux

$$\begin{aligned}\vec{j}_q &= \vec{j}_{qx} + \vec{j}_{qy} + \vec{j}_{qz} = \vec{e}_x(\lambda_{xx}\nabla_x T + \lambda_{xy}\nabla_y T + \lambda_{xz}\nabla_z T)\\ &+ \vec{e}_y(\lambda_{yx}\nabla_x T + \lambda_{yy}\nabla_y T + \lambda_{yz}\nabla_z T) + \vec{e}_z(\lambda_{zx}\nabla_x T + \lambda_{zy}\nabla_y T + \lambda_{zz}\nabla_z T),\end{aligned} \tag{2.21}$$

where $\lambda_{\text{xx}} = \lambda_{\lrcorner}$, $\lambda_{\text{xy}} = \lambda_{\text{xz}} = \lambda_{\text{yx}} = \lambda_{\text{zx}} = \lambda_{\text{zy}} = 0$, $\lambda_{\text{yy}} = \lambda_{\lrcorner}$, $\lambda_{\text{zz}} = \lambda_{\Uparrow}$.

Thus, the thermal conductivity tensor has the following form

$$\lambda_{\alpha\beta} = \begin{pmatrix} \lambda_{\lrcorner} & 0 & 0 \\ 0 & \lambda_{\lrcorner} & 0 \\ 0 & 0 & \lambda_{\Uparrow} \end{pmatrix}. \quad (2.22)$$

Substituting it into the equation of energy conservation

$$\rho C_V \frac{\partial T}{\partial t} - \operatorname{div}(\vec{j}_q) = f, \; j_q = \sum_{\alpha=1}^{3} (\vec{e}_\alpha \lambda_{\alpha\beta} \nabla_\beta T) \quad (2.23)$$

gives the equation of heat conduction

$$\rho C_V \frac{\partial T}{\partial t} - \lambda_{\lrcorner} \left(\frac{\partial^2 T}{\partial x^2} + \frac{\partial^2 T}{\partial y^2} \right) - \lambda_{\Uparrow} \frac{\partial^2 T}{\partial z^2} = f, \quad (2.24)$$

where f – the density of internal heat sources.

3. Convection heat transfer in ERF in a rotating electric field

With the shift of one of the electrodes in a layer of ERF Couette flow is created. It is divided into the physical motion of the infinitesimal volume elements dV, within each of which the velocity profile $V(x)$ can be considered linear in first approximation. ERF in each such microvolume flows as in a flat slot channel with the width h with a velocity profile $V(x) = k\,x$. In the flowing ERF chains of particles are virtual, i.e. they are broken by a continuous unidirectional shift and re-formed from other elements downstream in the direction of rotation of the electric field vector. With a steady flow at a constant rate [19] particles are fully entrained by a flow of carrying liquid.

With a steady flow – continuous unidirectional shift at a rate of the channel walls V the structural links of the particles in suspension are broken, the particles or small aggregates are almost independent.

If the fixed wall of the channel is heated to a temperature T_1, and moving to $T_2 < T_1$, then the channel creates a temperature field with a gradient $\nabla_x T$. Under the action of rotating with angular velocity in the plane with flow velocity Ω of the electric field with intensity

$$\vec{E}(\Omega\cdot t) = \vec{i}E_x + \vec{j}E_y \quad E_x = E_0 \cdot \cos(\Omega\cdot t) \quad E_y = E_0 \cdot \sin(\Omega\cdot t),$$

a steady motion of a particle in orbit with a period $T_\Omega = \frac{2\pi}{\Omega}$, with a frequency Ω with a linear velocity of the surface points $V_\Omega = \Omega\cdot b$ in the plane of rotation of intensity.

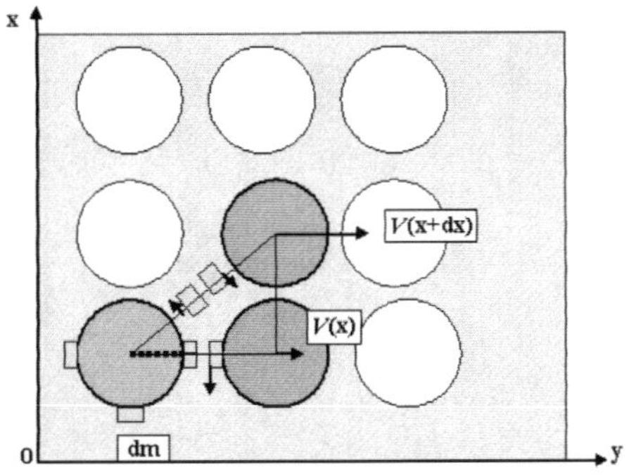

Figure 3. The location of the particles, rotating counterclockwise. Dotted line – the particle radius a, the rectangle – a half of the gap b.

Rotating particle carries the liquid carrier into the rotation at a speed $V_\Omega = \Omega\cdot b$, and the next – at the same speed in the opposite direction. The velocity of the middle layer of liquid between them is zero. Element of the fluid mass dm, adjacent to the particle surface, is entrained by it with this high speed. In the separating layer of liquid almost a planar shear flow appears in which the speed of the middle layer is equal to zero. The flow between the particles is approximately taken with a rectilinear Couette velocity profile. Elements of the fluid mass dm with density ρ with the specific heat C_V, entraining by the rotating surface of the particles are moving at an average linear velocity $\vec{V}_0 = \frac{1}{2}\vec{\Omega}\times\vec{b}$ and form microvortices, a compound of which creates a chain of micropumps, pumping the fluid, and with it the heat in the directions perpendicular and longitudinal to the flow.

Element of the fluid mass dm, isolated in the middle layer of fluid adjacent to the particle surface and carried along by it to meet the similar elements dm^1 of a neighboring particle, is in contact with them consistently. For a time equal to a half of the period of rotation of a particle $\Delta t = \frac{T}{2} = \frac{\pi}{\Omega}$ at the temperature difference of the wall and the element ΔT the amount of heat received from the wall of element of ERF of the mass dm = ρdV = $\rho\Delta Sa$, is determined by the density of heat flow j_q through the contact area $\Delta S = [2(a + b)]^2$. Over time the amount of heat transferred to the contact:

$$\delta Q = \langle j_q \rangle \Delta S \Delta t = \beta \cdot C_V \cdot \rho \cdot \mathrm{d}V \cdot \Delta T \qquad (3.1)$$

where β – dimensionless coefficient of heat transfer.

It is the product of the average density of heat flow through the average element of mass dm in the interval between the particles and the contact area of ΔS for the contact time Δt. The average heat flux into the element dm [5] is greater, the greater the heat element of mass C_V:

Heat flux density is equal to its amount in the volume of a single cylinder $\Delta S V_T \Delta t$ with a cross-sectional area ΔS, perpendicular to the velocity of the carrier energy V_T, with the length of the side of $V_T\Delta t$, transmitted through ΔS at a time Δt, per unit area and unit of time:

$$j_q = \frac{\delta Q}{\mathrm{d}S\mathrm{d}t} = \beta \cdot C_v \cdot \rho \cdot V_T \cdot \Delta T . \qquad (3.2)$$

Heat transfer coefficient β determines the velocity of propagation of thermal energy to a distance equal to the thickness of the layer of liquid between the particles.

The resulting amount of heat energy obtained by mass dm

$$\delta Q = j_q \cdot \Delta S \cdot \Delta t = \beta \cdot C_v \cdot \rho \cdot \Delta V \cdot \Delta T . \qquad (3.3)$$

is transferred by mass dm during time Δt along the axis OX with an average linear velocity $V_0 = \Omega(a + b)$ at a distance $\Delta x = V_0\Delta t = \Omega(a + b)\Delta t$. After a time Δt an element dm is in contact with others, consistently successive elements of the mass dm^1, each of which enjoys a neighboring chain in the structure of the particle temperature in the lower ΔT. During the contacts on the area of $\Delta S = 4(a + b)^2$, with a steady process of transfer the element dm transfers to the elements dm^1 the same amount of heat, as in (3.1).

At the same time this amount of heat is transferred along the axis OY to the next particle along the chain at a distance

$$\Delta Y_V = \left(V(X) + \frac{\mathrm{d}V}{\mathrm{d}X} \cdot \Delta X\right) \cdot \Delta t = \left[V(X) + \frac{\mathrm{d}V}{\mathrm{d}X} \cdot (a+b)\right] \cdot \Delta t$$

The total length of heat transfer is defined as:

$$\Delta L = \sqrt{\Delta X^2 + \Delta Y^2} = (a+b) \cdot \Omega \cdot \Delta t \cdot \sqrt{1 + \left[1 + \frac{V(X)}{(a+b)\cdot\Omega} + \frac{1}{\Omega}\frac{dV}{dX}\right]^2} . \qquad (3.4)$$

and the rate of transfer:

$$V_T = 2\cdot(a+b)\cdot\Omega\cdot\sqrt{1+\left[1+\frac{V(X)}{(a+b)\Omega}+\frac{1}{\Omega}\frac{dV}{dX}\right]^2}\,. \tag{3.5}$$

Heat flux density at the temperature difference $\Delta T = (a+b)\cdot\nabla_X T$ is

$$j_\mathrm{q} = \frac{\delta Q}{\mathrm{d}V}\cdot V_T = \beta\cdot C_V\cdot\rho\cdot(a+b)^2\cdot\Omega\cdot\sqrt{1+\left[1+\frac{V(X)}{(a+b)\Omega}+\frac{1}{\Omega}\frac{\mathrm{d}V}{\mathrm{d}X}\right]^2}\cdot\nabla_X T\,. \tag{3.6}$$

On the other hand

$$j_\mathrm{q} = \sqrt{(j_\mathrm{q})_X^2+(j_\mathrm{q})_Y^2}\,, \tag{3.7}$$

and components of the heat flux density along the coordinate axes are equal in magnitude:

$$(j_\mathrm{q})_x = \lambda_{xx}\cdot\nabla_x T\,, \tag{3.8}$$

$$(j_\mathrm{q})_y = \lambda_{xy}\cdot\nabla_x T\,, \tag{3.9}$$

Where λ_{XX} and λ_{XY} – the thermal conductivity tensor components, determined by heat transfer during shear deformation of the structure:

$$\lambda_{XX} = \beta C_V\rho(a+b)^2\Omega\,, \tag{3.10}$$

$$\lambda_{XY} = \lambda_{XX}\left[1+\frac{V(X)}{(a+b)\Omega}+\frac{1}{\Omega}\frac{\mathrm{d}V}{\mathrm{d}X}\right], \tag{3.11}$$

$$\lambda_{XZ} = 0\,. \tag{3.12}$$

If the electric field vector rotates in a plane perpendicular to the flow, the amount of heat δQ transported is the same as at the rotation in the plane of flow and transport distance is equal to

$$\Delta L = \sqrt{\Delta X^2+\Delta Y^2} = (a+b)\Omega\Delta t\left[1+\frac{V(X)}{(a+b)\Omega}+\frac{1}{\Omega}\frac{\mathrm{d}V}{\mathrm{d}X}\right]^{0.5}, \tag{3.13}$$

then the components of the conductivity tensor in this case are:

$$\lambda_{XX} = \beta C_v\rho(a+b)^2\Omega\,, \tag{3.14}$$

$$\lambda_{XY} = \lambda_{XX}\left[1+\frac{V(X)}{(a+b)\Omega}+\frac{1}{\Omega}\frac{\mathrm{d}V}{\mathrm{d}X}\right], \tag{3.15}$$

$$\lambda_{XZ} = 0\,. \tag{3.16}$$

The difference between the thermal conductivity λ_{XY} from (3.11) reflects the absence of heat transfer in the direction of rotation of the flow velocity.

When temperature drops $\nabla_Y T$ along the velocity vector expressions are obtained for the components of the heat flux density along the coordinate axes:

$$(j_\mathrm{q})_X = \lambda_{YX}\cdot\nabla_X T,\quad \lambda_{YX} = \beta\cdot C_V\cdot\rho\cdot(a+b)^2\cdot\Omega\,, \tag{3.17}$$

$$(j_q)_Y = \lambda_{YY} \cdot \nabla_X T, \quad \lambda_{YY} = \beta \cdot C_V \cdot \rho \cdot (a+b)^2 \cdot \Omega \cdot \left[1 + \frac{V(X)}{(a+b) \cdot \Omega} + \frac{1}{\Omega} \cdot \frac{dV}{dX} \right], \tag{3.18}$$

For an arbitrary direction of the temperature gradient in the plane xoy:

$$\nabla T = \vec{i} \nabla_X T + \vec{j} \nabla_Y T \tag{3.19}$$

heat flux density is:

$$\vec{j}_q = -\sum_{\alpha\beta} (\vec{e}_\alpha \cdot \lambda_{\alpha\beta} \cdot \nabla_\beta T) \tag{3.20}$$

Effective thermal conductivity tensor of a plane flow of ERF is asymmetric and is the sum of symmetric and asymmetric tensors

$$\lambda_{\alpha\beta} = \begin{pmatrix} \lambda_{xx} & \lambda_{xy} & 0 \\ \lambda_{yx} & \lambda_{yy} & 0 \\ 0 & 0 & \lambda_{zz} \end{pmatrix} \tag{3.21}$$

Substituting in the equation of energy conservation, that is, the equation of local first law of thermodynamics, the expression for the heat flux density, f – the density of heat sources

$$\rho C_V \frac{\partial T}{\partial t} + \operatorname{div}(\vec{j}_q) = f ,$$

gives the equation of heat conduction

$$\rho C_V \frac{\partial T}{\partial t} - \sum_{\alpha\beta} \left[\frac{\partial}{\partial x_\alpha} \left(\lambda_{\alpha\beta} \frac{\partial T}{\partial x_\beta} \right) \right] = f , \tag{3.22}$$

Given the expressions of the tensor components of the effective thermal conductivity:

$$\rho C_V \frac{\partial T}{\partial t} = \lambda_{xx} \frac{\partial^2 T}{\partial x^2} + \frac{\partial}{\partial x} \left(\lambda_{xy} \frac{\partial T}{\partial y} \right) + \lambda_{xx} \frac{\partial^2 T}{\partial x \partial y} + \lambda_{yy} \frac{\partial^2 T}{\partial y^2} + f , \tag{3.23}$$

If the temperature gradient is across the direction of flow, the equation (3.23) simplifies to:

$$\rho C_V \cdot \frac{\partial T}{\partial t} - \beta \cdot C_V \cdot \rho \cdot (a+b)^2 \cdot \Omega \cdot \frac{\partial^2 T}{\partial x^2} = f . \tag{3.24}$$

In the end channel at its edges $\frac{\partial T}{\partial y} \neq 0$, and the problem requires the solution of the previous equation with appropriate boundary conditions together with the equation of motion if the velocity profile $V(x)$ is not specified.

4. The numerical solution

To assess the influence of the parameters of the rotating electric field on heat transfer in ERF let us solve the equation (3.24). Let us consider the one-dimensional case in the absence of internal heat sources, i. e. at $f = 0$. At one boundary the condition is given of the first kind (the known and constant temperature), the second – the condition of thermal equilibrium. At the initial time the temperature of ERF is assumed known and constant throughout the channel. Thus, we will solve the problem:

$$\frac{\partial T}{\partial t} - \beta(a+b)^2 \Omega \frac{\partial^2 T}{\partial x^2} = 0, \tag{4.1}$$

$$T\Big|_{x=0} = T_1; \ \frac{\partial T}{\partial x}\Big|_{x=h} = 0; \ T(t=0) = T_0. \tag{4.2}$$

The problem (4.1) – (4.2) was solved numerically using the finite difference method. In the finite-difference form:

$$\frac{1}{\Delta t}(T_i^k - T_i^{k-1}) = \beta(a+b)^2 \Omega \frac{1}{\Delta x^2}(T_{i+1}^k - 2T_i^k + T_{i-1}^k), \tag{4.3}$$

$$T_0^k = T_1; \ T_N^k - T_{N-1}^k = 0; \ T_i^0 = T_0. \tag{4.4}$$

The problem (4.3) – (4.4) was solved by the sweep method [20]. The coefficients of the sweep:

$$A_i = \beta(a+b)^2 \Omega \Delta t, \ B_i = -\Delta x^2 - 2\beta(a+b)^2 \Omega \Delta t, \ C_i = \beta(a+b)^2 \Omega \Delta t, \ D_i = T_i^{k-1} \Delta x^2, \tag{4.5}$$

$$A_0 = 0, B_0 = 1, C_0 = 0, D_0 = -T_1; \ A_N = 0, B_N = -1, C_N = 1, D_N = 0. \tag{4.6}$$

Figure 4 shows the results of model calculations (4.1) – (4.2) at $T_0 = 10$ °C and $T_1 = 50$ °C after 1, 10, 30 and 60 s when applying the electric field. The thickness of the flat channel was assumed to be 1 cm. As electrosensitive carrier ERF prepared in the laboratory of rheophysics and macrokinetics of A.V. Luikov Heat and Mass Transfer Institute of National Academy of Sciences of Belarus was used, with the studied rheological, thermal and physical properties. The heating rate of a remote wall was considered at a given constant temperature of the wall heating.

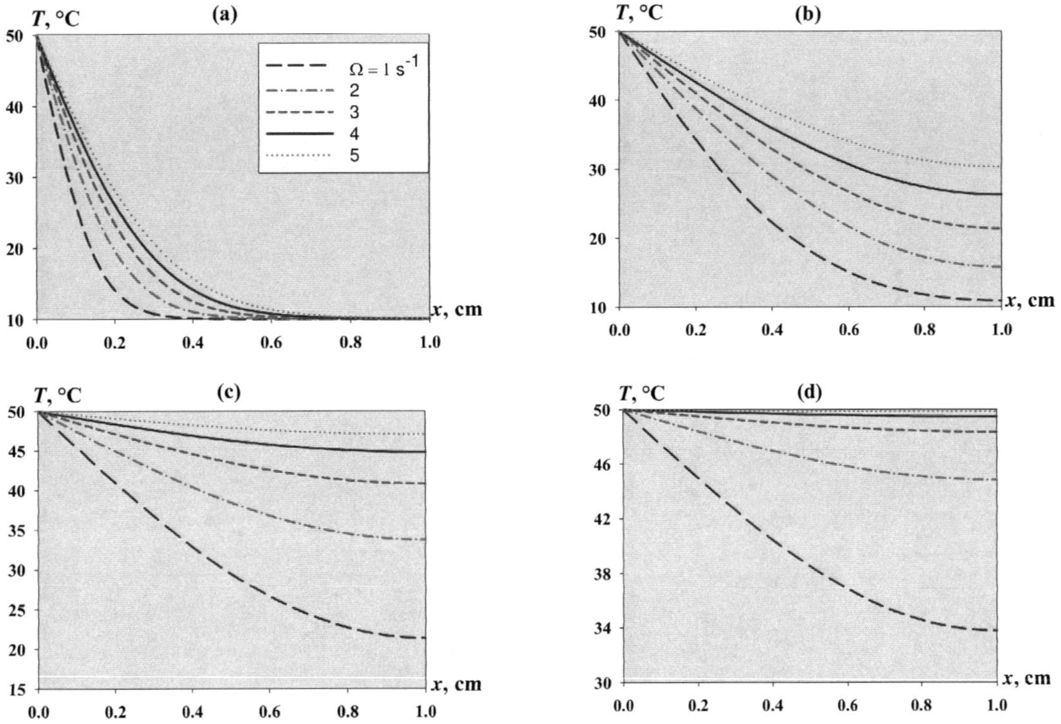

Figure 4. The kinetics of heating ERF in a flat channel at different intensities of the internal rotation through: (a) 1 s, (b) 10 s, (c) 30 s and (d) 60 s.

As seen from figure 4 the presence of the internal rotation leads to a significant acceleration of heat transfer from one wall of a flat channel to another: after 1 min after the electric field at $\Omega = 3$, 4 and 5 s^{-1}, channel wall temperatures were almost identical (up to 2 °C).

Let us consider the case of cooling. Let the temperature of ERF flat channel walls to be known at the initial moment of time. We will study the process of cooling the system, setting the distal boundary conditions of heat transfer by natural convection and radiation. In this case we will vary the intensity of the internal rotations. The results of these calculations are shown in figure 4. The initial temperature was 50 °C, the ambient temperature 10 °C.

The coefficient of heat exchange with the environment $\alpha_{env.}$ in the absence of external convection is

$$\alpha_{env.} = \alpha_{nat.conv.} + \alpha_{rad.}, \tag{4.7}$$

where $\alpha_{nat.conv.}$ – coefficient of heat transfer by natural convection of air from the outer surface of the absorber, W / (m^{2}K); $\alpha_{rad.}$ – coefficient of heat transfer by radiation from the outer surface of the absorber, W / (m^{2}K).

The value of $\alpha_{nat.conv.}$ was evaluated according to [21] for the horizontal plane: $\alpha_{nat.conv.}$ = 5 W / (m^{2}K). The coefficient $\alpha_{rad.}$ was calculated by the Stefan–Boltzmann law: $\alpha_{rad.}$ = 5.7 W / (m^{2}K).

The results of model calculations (2.25) – (2.26) for the described conditions are shown in figure 5.

The time during which a plane boundary of the channel at x = 0 was cooled to within 1 °C to ambient temperature was as follows: for $\Omega = 1$ s^{-1} – the simulation time exceeded 1000 s, 2 s^{-1} – 807 s, 3 s^{-1} – 538 s^{-1}, 4 – 404 s, 5 s^{-1} – 323 s.

Thus, the rotation of the particles under the influence of an alternating electric field has a significant influence on the rate of heat transfer in ERF. This effect can be used to intensify the process of heat transfer in channels of heat exchangers of different configurations.

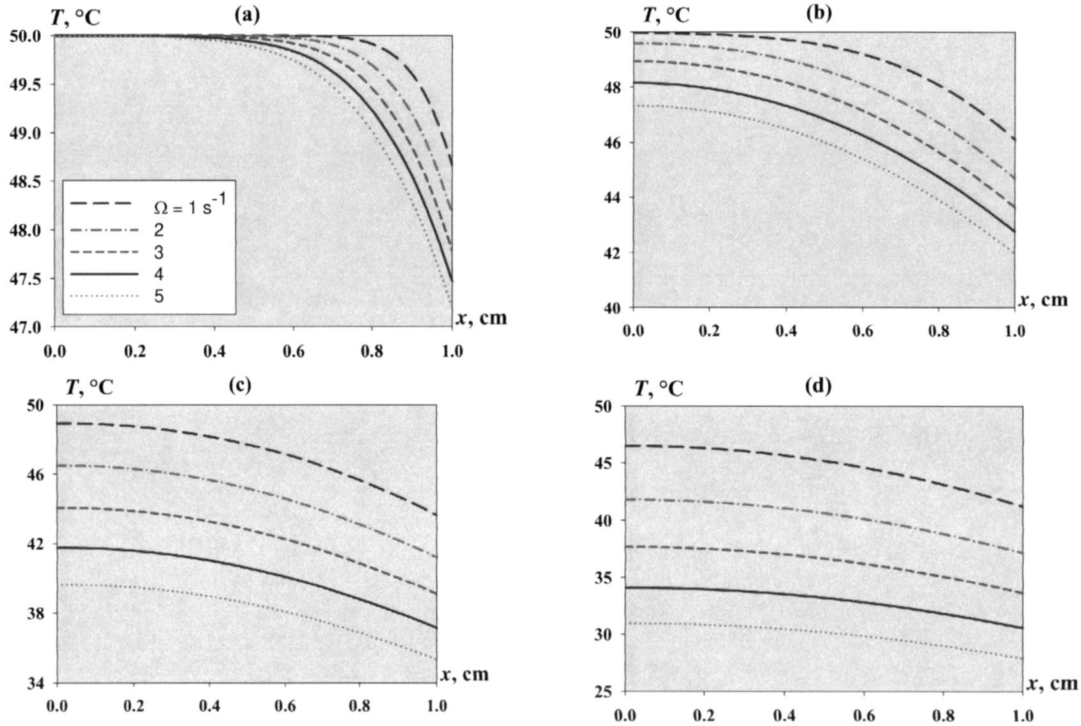

Figure 5. The kinetics of cooling ERF in a plane channel at different intensities of the internal rotation through: (a) 1 s, (b) 10 s, (c) 30 s and (d) 60 s.

5. Conclusions

A study has been performed on physical and mathematical modeling of microconvection heat transfer in electrorheological fluid in an external alternating electric field varying with a given angular frequency, which takes into account the occurrence of microvortices carrying the energy and the mass of the dispersion medium as a result of particles' rotation of the dispersed medium.

The problem of heat transfer in a flat layer of electrically sensitive fluid in an alternating electric field with the microconvection is formulated and numerically solved. Numerical simulations showed that the presence of internal rotation significantly intensifies the heat exchange process in electrorheological fluids, which can be used to create heat exchangers, controlled by an external electric field.

References

[1] Turcu I 1987 *J. Phys.* A **20** (11) 3301–8
[2] Rowe A D, Leake M C, Morgan H and Berry R M 2003 *J. Modern Optics* **50** (10) 1539–54
[3] Dolinsky Y and Elperin T 2009 *Phys. Rev.* E **80** (6) 066607
[4] Holzapfel C, Vienken J and Zimmermann U 1982 *J. Membr. Biol* **67** (1) 13–26
[5] Dolinsky Y and Elperin T 2009 *Phys Rev* E **79** (2) 026611
[6] Shulman Z P and Nosov V M 1985 *Rotation of Insulating Bodies in Electrorheological Suspensions* ed O G Martynenko (Minsk: Nauka i Tekhnika)
[7] Benecke W, Wagner B, Fuhr G, Hagedorn R, Glaser R and Gimsa J 1998 Patent US 5726509
[8] Otsubo Y and Edamura K 1997 *Appl. Phys. Lett* **71** 318–21
[9] Dolinsky Y and Elperin T 2007 *Physical Review* E **75** 066611
[10] Fuchimoto R and Tsukiji T 2009 *Proc. 10th Int. Conf. on Fluid Control, Measurements, and Visualization (Moscow, Russia, 17–21 August 2009)* 630–35
[11] Cheng L, Tsukiji T, Hayakawa K and Ruan X 2008 *Proc. 7th JFPS Int. Symp. on Fluid Power, (Toyama, Japan, 15–18 September 2008*) 833–6
[12] Fugr G, Hagedorn R and Göring H 1985 *Plant and Cell Physiology* **26** (8) 1527–31
[13] Bashtovoi V G, Vislovich A N and Kashevskii B E 1978 *Prikl. Mat. Tenh. Fiz.* (3) 88–93
[14] Kashevskii B E and Ivanova N I 1985 *Magnetn. Gidrodin.* (3) 48–52
[15] Shatalov A F, Popov A A and Kozhevnikov V M 1998 *Vestn. Sev-Kas GTU* **3** 68–70
[16] Ivanova N I 1986 *Development of the Methods of Investigation of Heat and Pulse Transfer Processes in Magnetic Fluids with Internal Rotations* Abstract of Candidate's Thesis (in Physics and Mathematics), A V Luikov Heat and Mass Transfer Institute, National Academy of Sciences of Belarus, Minsk
[17] Tsebers A O 1978 *Prikl. Mat. Mekh.* **42** 673–8
[18] Nigmatullin R I 1978 *Principles of the Mechanichs of Heterogeneous Media* (Moscow: Nauka)
[19] Mokejev A A, Korobko E V and Vedernikova L G 1992 *J. Non-Newtonian Fluid Mechanics* **42** 213–30
[20] Dul'nev G N, Parfenov V G and Sigalov A V 1990 *Application of PC to Solve the Problems of Heat Transfer* (Moscow: Vyssh. Shkola)
[21] Martynenko O G and Sokovishin Yu A 1982 *Free-convective Heat Transfer: Handbook* (Minsk: Nauka i Technika)

Response time of an ER-fluid under shear and flow modes

B Abu-Jdayil
Chemical & Petroleum Eng. Dept., UAE University, Al-Ain, UAE

E-mail: babujdayil@uaeu.ac.ae

Abstract. The response times of an ER-fluid in shear and flow modes under the effect of AC- and DC-fields were investigated experimentally. In shear mode, the response time was obtained under the effect of different shear rates and electrical field strengths by comparing the shear stress measured statically with that measured dynamically. The response times in shear mode decreased with increasing the field strength and decreasing the shear rate. In flow mode, the response times of the ER-fluid under various electric field strengths and flow rates were determined from the pressure measurements in channel flow. The response times in flow modewere exponentially decreasedwith the increase of the volumetric flow rate and the electric field strength. In both modes, increasing the frequency of the AC-field led to increase of the response time of the ER-fluid.

1. Introduction

The electro–rheological (ER) phenomenon is related to changes in rheological behavior of certain suspensions or solutions upon the application of an electric field [1]. The possibilities for employing the ER-effect in practical applications have long been recognized, e.g. in clutches [2], shock absorbers, hydraulic valves [3], vibration dampers, motor bearings [4], etc. The principle attraction of using ER-fluids in these devices is their fast response time. This feature makes possible feedback control systems for robotics and automotive applications, which are greatly improved over existing models. Therefore, an important parameter to characterize an ER-fluid is the response time, which has different definitions in the literature [5]. Most of the researchers have used the total time needed by the fluid to reach the saturated structure after the application of the electric field. The response time of structure formation in ER-fluids is thought to be controlled by a competition between the field-induced particle interactions and the viscous drag forces.

Wen *et al.* [6] have studied the response time of microspheres suspended in electrorheological fluids. They found that both the initial response time depends on the conductivity of the microspheres. Lee and Choi [7] identified that the response time of the ER-fluid to the step electric field is almost the same between the shear and flow modes. But a lot of parameters that can affect the response time of ER-fluids have been not studied. In shear mode, the response time was found to relate to shear rate and the applied electric field strength [8]. It was found that the rising time decreased with the increase of the shear rate. On the other hand, for constant shear rate, the applied field strength only slightly affected the time constant. The constant increased a little when the applied field strength was very strong [8]. Rejon *et al.* [5] found that the response time of different ER-fluids in DC field was proportional to $1/E^a$, where E is the applied field and a is an adjustable paratmeter. In flow mode, Nam *et al.* [9] found that the response times are exponentially decreased with the increase of the flow velocity and the decrease of the electric field strength. Ulrich *et al.* [10] approximated the effect of the electric field on the response time in a valve arrangement by a mathematical model. Recently, the study of Perera *et al.* [11] revealed that the response time of an ER-fluid developed for a medical device and measured by the dielectric analyserwas temperature dependent. However, most of the studies on the response time of ER-fluids did not consider the effects of electric field type and the frequency of the AC-field.

In this study, the response times of an ER-fluid under the effect of AC and DC electric fields in shear and flow modes were investigated. The effects of different operating parameters such as the volumetric flow rate, shear rate, electric field strength, and field frequency on the time response were examined.

2. Experimental

2.1. ER-fluid

The ER-fluid used in this study was supplied by Bayer Company (Leverkusen, Germany). It is an ER-fluid based on soft, crosslinked polyurethane particles in silicone oil (BAYSILONE OIL M – Polydimethylsiloxan).The ER-fluid is water-free and contains additional additives and emulsifier. The polyurethane particles contain dissolved metal ions which are necessary for fast polarizationand for the ER effect.

2.2. Rotational viscometer

The rheological data of the ER-fluid were measured with a Haake M5/SV1 conventional rotational viscometer (Searle type) with a fixed outer cylinder and a rotating measuring bob. The radius of the rotating cylinder is 10.5 mm, length of the cylinder 60 mm, and the gap width 1 mm. The software SROT of Haake Co. was used to record the rheological data.

2.3. Channel flow

A PVC rectangular channel (flat channel) was used to investigate the time response of the ER-fluid under the flow mode. To prevent sedimentation of the particles in the ER-fluid the channel is mounted vertically. The rectangular channel had a length of 280 mm with cross-sectional dimensions of 2 mm height and 20 mm in width. In this study, two stainless-steel smooth plates with 40 mm long were used as electrodes. These electrodes are flush with the channel walls and extending over the entire width. The volumetric flow rate of the ER-fluid was adjusted by a gear pump and bypass valve. With this arrangement it was possible to vary the volumetric flow rate (Q) from 0.316 up to 5.50 cm^3/s. In addition, the volumetric flow rate was measured with a mass flow meter (Type Mass 2100, Dan Foss, Denmark) at the outlet of the channel. The pressure drop (Δp) was determined in the midplane of one of the 74 mm long insets over a length of 60 mm, using two pressure transducers (Rosemount, type G1151) with Δp ranges from 0 to50 mbar and from 0 to 1000 mbar. The temperature of the ER-fluid was measured at the inlet of the channeland controlled by a heat exchanger at 25 °C. Further details on the channel flow set up can be found in ref. 12.

2.4. High voltage devices

The high voltage power supply of alternating current (AC) used in this work was a device developed according to the standards of the Society of German Electrician (VDE). Its working range is: Voltage, U: 0–6 kV (effective); Current, I: 0–10 mA (effective); Frequency, f: 50–500 Hz.

For direct current (DC) measurements a high voltage power supply unit from the company HEIZINGER (typeHNC 10000-1) was used. Its working range is: Voltage, U: 0–10 kV; Current, I: 0–1 mA.

3. Results and discussion

3.1. Rotational viscometer results

To investigate the response time of the ER-fluid under the shear mode, the shear stress of the ER-fluid was measured in the rotational viscometer at constant shear rate using the static and dynamic methods. In static measurements, the electric field was applied on the system before starting the shearing process, while in dynamic measurements the electric field applied after starting the shearing process. The response time was determined as the total time needed by the fluid to reach the saturated structure (i.e. the static shear stress) after the application of the electric field. Figure 1 shows the typical effect of the strength of DC-field at high shear rate. By considering the steady-state conditions, it is clear that the response time of the ER-fluid decreased with the field strength.

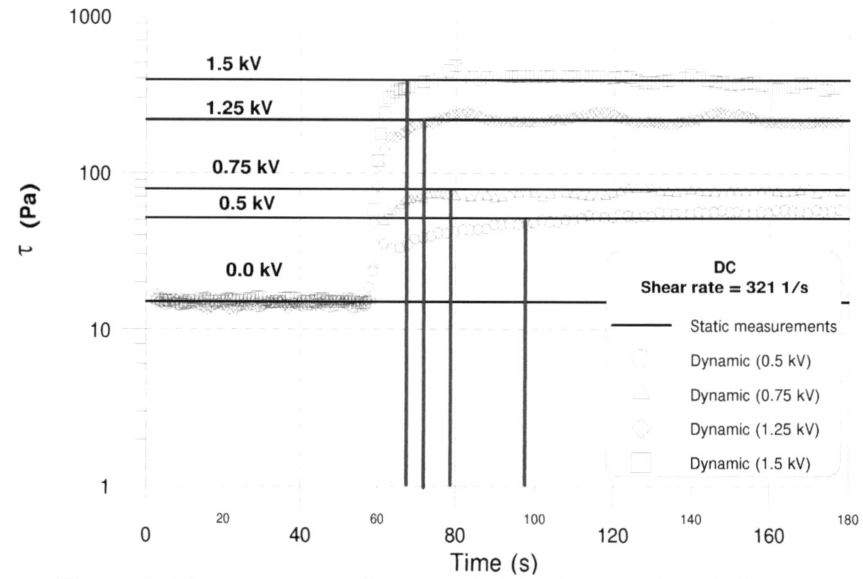

Figure 1. Rising response of the ER fluid in shear mode (DC-field).

On the other hand, it was observed that the response time of the ER-fluid increased by increasing the shear rate, see figure 2. This behavior can be explained by recalling the determination method of the response time. In static measurements the electric field is induced before shearing, which means that the cluster formation process is already completed prior to the cluster deformation due to the hydrodynamic force [13]. While in the dynamic measurements, where the electric field is induced after shearing, the cluster begins to form the quasi-stable structure as the electric field is induced on the ER fluid already subjected to the hydrodynamic force. In other words, the cluster formation and deformation are occurred in the dynamic measurements simultaneously. Therefore, the calculated response time here, which is the difference between the rising time of static stress and the rising time of the dynamic stress, can be considered as the cluster formation time [13]. The obtained relationship between the response time (cluster formation time) and the shear rate is in consistence with the results obtained by Choi *et al.* [13]. The increases of the cluster formation time is increased with the increase of shear rate because the collision chance among the aggregated ER particles and the deformation rate of the cluster structure are increased with the increase of hydrodynamic force.

As can been seen in figure 2, the response time of the ER-fluid used varied between 2.94 s at 1.5 kV/mm and shear rate of 10.19 s^{-1} and 34.9 s at 0.5 kV/mm and shear rate of 321 s^{-1}. The results are in accordance with those obtained by Rejon *et al.* [5] for different ER-fluids. The dependence of the response time (t) of the field strength (E) was modeled by the following equation:

$$t \propto \frac{1}{E^a} \quad (1)$$

where a is an adjustable parameter. This behavior was found by Klingenberg *et al.* [14] where a was approximated to 2. Rejon *et al.* [5] found an a parameter varying between 0.41 and 1.05. In this study, the parameter a in DC field varied with the shear rate between 0.85 and 1.40, see Table 1.

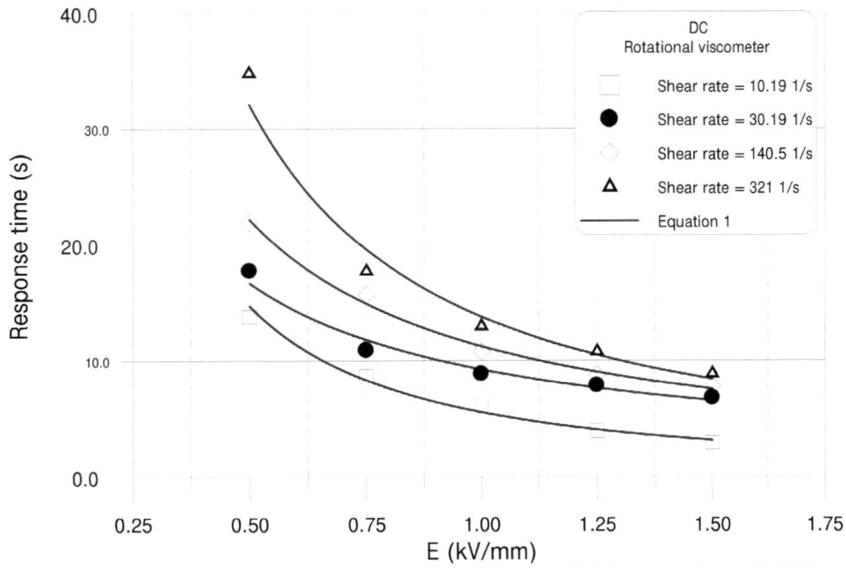

Figure 2. Response time for ER-fluid for different field strengths and different shear rates (DC-field).

Table 1. Parameter a of the relationship between response time and electric field (Eq 1) for DC- and AC-fields.

Shear rate	DC-field	AC-field			
(s^{-1})		f = 50 Hz	f = 100 Hz	f = 200 Hz	f = 300 Hz
10.19	1.40	0.56	0.31	0.31	0.33
30.19	0.85	0.37	0.26	0.28	0.31
140.50	0.98	0.38	0.28	-	-
321.00	1.22	0.40	0.27	0.30	0.26

The same dependence of the response time on field strength and shear rate has been also observed in the AC-field. On the other hand, a dependence of the response time on the field frequency has been detected. The effect of AC-field frequency (f) on the response time of the ER-fluid is illustrated in Figure 3. With increasing frequency (at constant field strength and shear rate) the time response of the ER-fluid increased. The response time increased with increasing field frequency because the time available for charge transport during polarization (in one directionper half cycle) decreases with increasing frequency, which has delayed the structure formation [15].

Figure 4 shows the effects of the field strength and on the response time fitted to equation 1. The regressed values of a parameter are reported in Table 1. It is clear that none of the values is close to 2.0, where they varied between 0.26 at high shear rate and field frequency and 0.56 at low shear rate and field frequency. On the other hand, it is noticeable in Table 1, that the a values of DC-field are greater than that of AC-field at the same shear rate. In other words, the response time of ER-fluids in DC-field is shorter than that in the AC-field for the same field strength and shear rate.

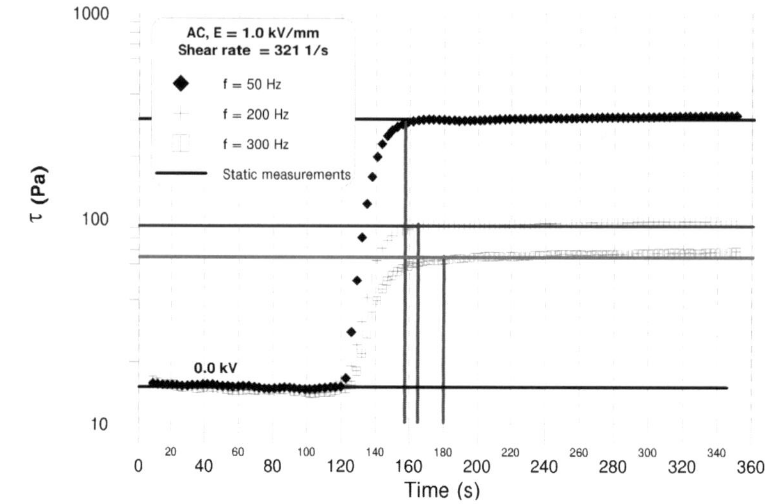

Figure 3. Rising response of the ER fluid in shear mode (AC-field).

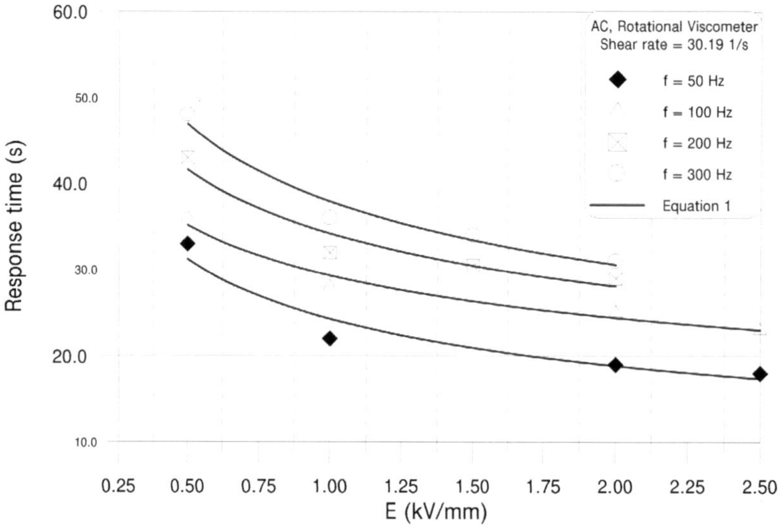

Figure 4. Response time for ER-fluid for different field strengths and different frequencies (AC-field).

3.2. Channel flow results

Most of the previous studies have adopted the shear mode experiments and the DC field in simulation to investigate the response time of ER-fluids. But the flow mode is no less important than the shear mode due to its applicability in different engineering devices such as valves, clamps and dampers. The study on the pressure response with time for ER-fluids is important for the effective design and control of such devices [9].

For flow mode measurements, the pressure drop $\Delta p = f(Q)$ was measured under steady-state conditions; firstly, a constant volumetric flow rate of the ER-fluid was fixed. Then after 60 s an electric field was applied and the pressure drop as a function of time was measured. The time response of the ER-fluid was determined as the time needed to reach a steady state value of Δp.

As can be seen in figure 5, the response time of the ER-fluid decreased slightly with the strength of the DC-field. This behaviour was also function of the volumetric flow rate, where the dependence of the response time on the field strength was very clear at low volumetric flow rate, see figure 6. The volumetric flow rate and the electric field strength were changed in the ranges of 0.316-5.50 mL/s and

0.5 – 2.0 kV/mm, respectively. Figure 6 shows also that the response times and their decreasing rates were decreased with the increase in the volumetric flow rate. At high field strength, and volumetric flow rate more than about 2.0 mL/s, the rise response time tends to be nearly constant. Compared with the previous studies on the dynamic response of the ER-fluid in flow mode [9], the similar inverse relationship between the pressure response time and volumetric flow rate was obtained. In addition, Nam *et al.* [9] have also observed the decrease of the pressure rising time with the field strength. The pressure rising times in DC-field were proportional to the volumetric flow rate ($t \propto Q^{-n}$) with the exponent of about -0.256 and -0.593 at low and high electrical field strength, respectively. An exponent of -0.412 has been found by Nam *et al.* [9] for pressure response time of an ER-fluid. This behaviour was explained by Nam *et al.* [9], that when the flow volumetric flow rate is increased at the same electric field, the densification process is promoted and the average size of the clusters in steady state is decreased. Therefore, the dynamic response time of the ER fluid in the pressure flow is exponentially decreased due to the densification period shortenedwith the increase of the volumetric flow rate.

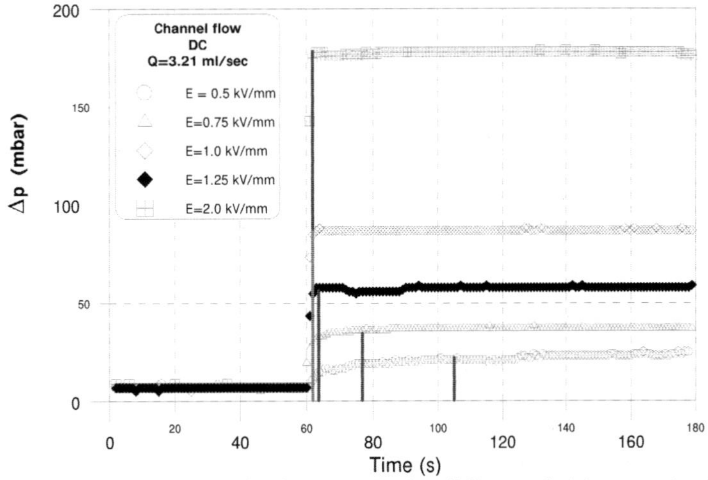

Figure 5. Rising response in flow mode for different field strengths (DC-field).

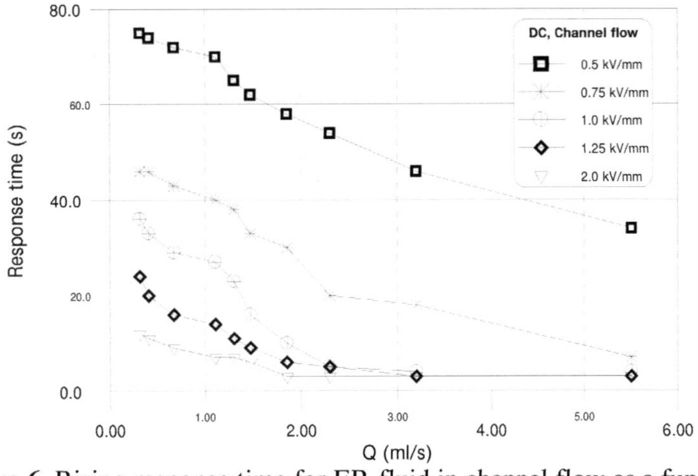

Figure 6. Rising response time for ER-fluid in channel flow as a function of volumetric flow rate and field strength (DC-field).

Figure 7 shows the typical pressure response of the ER-fluid under the effect of the AC-field. Similar trends have also been observed here as in the case of DC-field, with some dependence of the response time on the field frequency.In addition to its effect on the response time, the volumetric flow rate has affected the range of the pressure values at steady state conditions. As can be seen in figure 7, increasing of the volumetric flow rate has increased the fluctuations in the measured pressure drop. The degree of fluctuations also increased with increasing the field frequency at constant volumetric flow rate and constant field strength. When the volumetric flow rate is increased at constant field strength, the hydrodynamic force will increase in competition with the field-induced particle interaction forces, which leads to this fluctuated behavior.

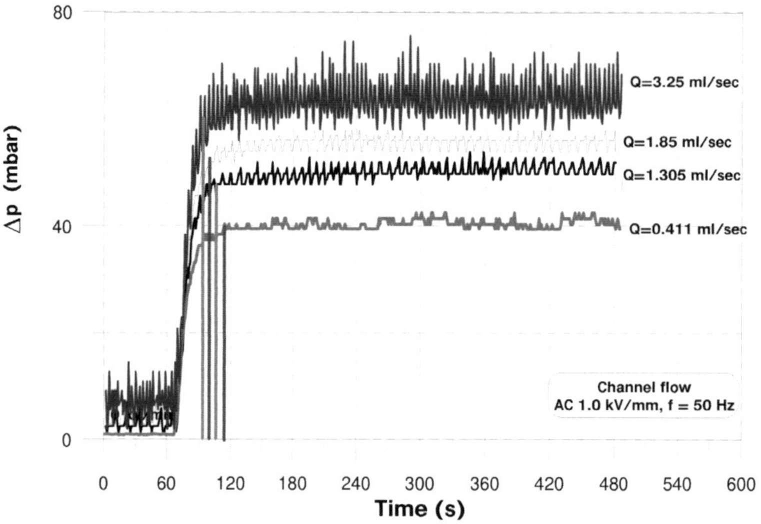

Figure 7. Rising response in flow mode for different volumetric flow rates (AC-field).

The effects of volumetric flow rate and field strength on the pressure response time at f = 50 Hz are illustrated in figure 8. For all field strengths, and volumetric flow rate more than about 2.0 mL/s, the rise response time tends to be nearly constant. The pressure rising times in AC-field were proportional to the volumetric flow rate ($t \propto Q^{-n}$) with the exponents of about -0.469 and -0.410 at low and high electrical field strength, respectively, which are in accordance with the previous findings of Nam *et al.* [9]. It was also observed that the response times under DC-field conditions were shorter than that under AC-conditions with 50 Hz for the same volumetric flow rate and field strength. An exception was the field strength of 0.5 kV/mm.

The main important difference between the shear and flow mode is that the response time has an inverse relationship with the volumetric flow rate in flow mode, while in the shear mode the response time is proportional to the shear rate. It was indicated in previous studies that there is a difference in the mechanism of the structure formation in both modes [9]. It is expected that the pressure response of the ER-fluid in channel flow is mainly determined by the densification process in the competition of the field-induced particle interaction forces and the hydrodynamic forces, unlike those in steady shear dominated by aggregation process.

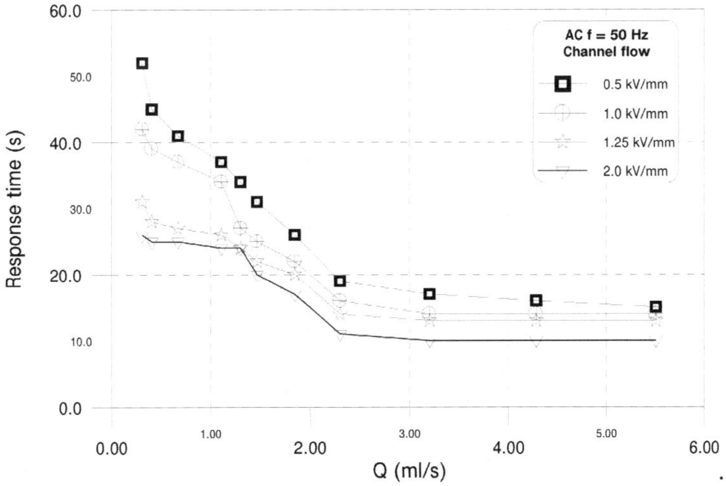

Figure 8. Rising response time for ER-fluid in channel flow as a function of volumetric flow rate and field strength (AC-field).

4. Conclusions

Response times of an ER-fluid in shear and flow modes were investigated experimentally. The effects of electrical field strength, field type, shear rate and volumetric flow rate on the response time of ER-fluid were determined. In shear mode, where the measurements were carried out in a rotational viscometer, high field strength and low shear rate correspond to quicker response. While in flow mode, where the measurements were carried out in a channel flow, high field strength and high volumetric flow rate correspond to quicker response. In both modes, it was observed that at the same field strength and flow conditions, the response times under DC-field are shorter than those under AC-field (50 Hz). In addition, increasing the field frequency of AC-field results in longer response time in shear and flow modes. For practical applications, where short response time is required in different ER devices, using of DC field or AC field with low frequency is recommended.

References

[1] Brunn P O, Abu-Jdayil B 2004 *Rheologica Acta* **43** 62
[2] Hartsock D L, Novak R F and Chaundy G J 1991 *J. Rheol.* **35** 1305.
[3] Strandrud H T 1966 *Hydraulic Peneum* Sept 139
[4] Duclos T G 1988 *Automotive Eng.* **96**, 45
[5] Rejon L, Ramirez A, Paz F, Goycoolea F M and Valdez M A 2002 *Carbohydrate Polymers* **48** 413
[6] Wen W, Zheng D W and Tu K N 1998 *Physical Review E* **57** 4516
[7] Lee H G and Choi S B 2002 *Materials and Design* **23** 69
[8] Tian Y, Zhang, Meng Y and Wen S 2005 *J. Colloid Inter. Sci.* **290** 289
[9] Nam Y J, Park M K and Yamane R 2008 *Experiments in Fluids* **44** 1196
[10] Ulrich S, Böhme G, Bruns R 2009 *J. of Physics: Conference Series* **149** 012031
[11] Perera N I, Maheswaram M P, Mantheni D, Perera H D, Matthews M E, Sam-Yellowe T and Riga A 2011 *American Journal of Analytical Chemistry* **2** 85
[12] Abu-Jdayil B, Asouid H and Brunn P O 2007*Materials and Design* **28** 928
[13] Choi B H, Nam Y J, Yamane R and Park M K 2009 *J. of Physics: Conference Series* **149** 012004
[14] Klingeberg D J, Zukoski C F and Hill J C 1993 *J. of Applied Physics* **73** 4644
[15] Abu-Jdayil B and Brunn P O 1997 *Chemical Engineering and Processing: Process Intensification* **36** 281

On the comparison of electrorheological measurements with different generation of an electric field

P Peer[1,4], M Stenicka[2], P Filip[1] and V Pavlinek[2,3]

[1]Institute of Hydrodynamics, Academy of Sciences of the Czech Republic, Pod Patankou 5, 166 12 Prague 6, Czech Republic
[2]Centre of Polymer Systems, University Institute, Tomas Bata University in Zlin, Nad Ovcirnou 3685, 760 01 Zlin, Czech Republic
[3]Polymer Centre, Faculty of Technology, Tomas Bata University in Zlin, namesti T. G. Masaryka 275, 762 72 Zlin, Czech Republic

E-mail: peer@ih.cas.cz

Abstract. The rheological measurements of PANI powders mixed (10 wt.%) in silicone oil were compared using two electrorheological (ER) apparatuses different in generation of an electric field: (i) a rotational rheometer Physica MCR 501 (Anton Paar Co.) equipped with the electrorheological cell, (ii) a rotational rheometer Bohlin Gemini CVOR 150 (Malvern Instruments) modified for ER experiments. It was shown that the data obtained for a plate-plate arrangement using the Physica MCR 501 provide the highest values exceeding both the plate-plate data for the Gemini 150 and the, mutually comparable, concentric cylinders data for both rheometers.

1. Introduction

Most of commercial available rheometers can be equipped with electrorheological (ER) devices fully using the functionality of the basic instruments. However, generation of an electric field in the individual devices substantially differs. The aim of this contribution is to compare the experimental data obtained from the following rheometers: a rotational rheometer Physica MCR 501 (Anton Paar Co.) and a rotational rheometer Bohlin Gemini CVOR 150 (Malvern Instruments). For both instruments two geometrical arrangements were used: parallel plates and concentric cylinders. Each system is covered with a hood ensuring safety regulations and temperature stability controlled by Peltier elements (Physica) or fluid circulator (Bohlin). The bottom plate for the parallel plate device and the cup in the concentric cylinder setup are grounded. Opening the hood automatically switches the power supply off. The rheometers themselves are fully isolated from an electric current. The substantial difference between the two electrorheologically adapted rheometers consists in completing an electric circuit through an upper shaft.

In the case of the MCR 501 a spring wire adhering to an upper shaft by an application of a spring completes a circuit loop, see figure 1. Intensity of an electric field is varied through an external DC high voltage power supply unit HCP 14-12500 (F.U.G. Elektronik GmbH, Rosenheim, Germany). It provides an electric field strength up to12.5 kV/mm at the condition that electric current does not exceed 1 mA (otherwise a voltage is proportionally reduced). Dimensions of the measuring systems

[4] Corresponding author.

used are as follows: P-PTD200/E parallel plates of a diameter 50 mm, C-PTD200/E a bob and cup arrangement with the inner and outer diameter 16.6 mm and 18 mm, respectively.

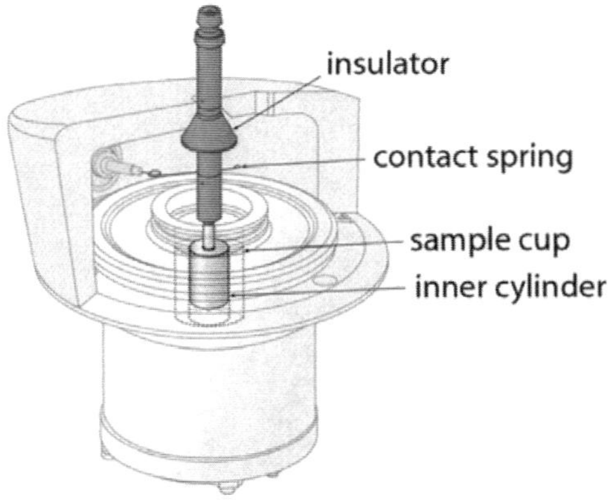

Figure 1. Geometrical position of a spring wire (contact spring) in the Physica MCR 501 equipped with the C-PTD200/E device (a bob and cup arrangement).

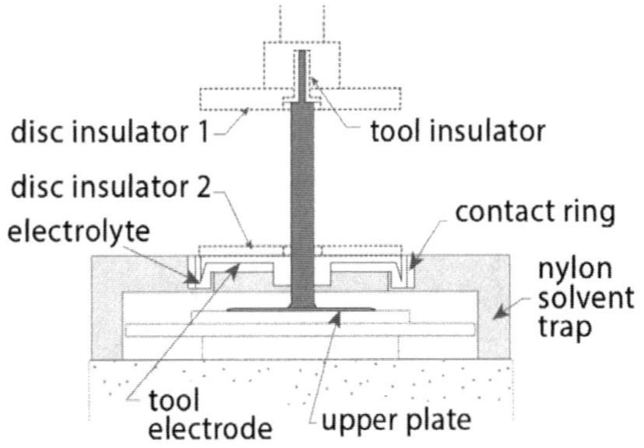

Figure 2. Geometrical location of electrolyte in the Gemini 150 device.

On the other hand the circuit loop in the Gemini 150 device is completed through an electrolyte housed in the circumferential gutter, see figure 2. Intensity of an electric field, influenced also by a density of electrolyte, is regulated by a DC high-voltage source TREK (TREK 668B, USA) providing the electric field strength up to 3 kV/mm with a maximum electric current 5mA. Dimensions of the measuring systems used are as follows: the diameter of parallel plates attains 40 mm and in the concentric arrangement the inner and outer diameter is 25 mm and 27.5 mm, respectively.
Prior to the ER experiments using either rheometer there is a necessity to carry out the measurements if mere air is applied. This enables to measure a friction between a spring wire and an upper shaft, or between a contact ring and electrolyte. Then these magnitudes have to be subtracted from the data obtained for ER materials.

2. Experimental

2.1. Material
Electrorheological suspension was prepared by mixing of PANI powder (Sigma Aldrich, USA, base, 50,000 g/mol) with silicone oil (Lukosiol M200, Chemical Works Kolin, Czech Republic) in 10 wt.% concentration. PANI powder was ground, sieved to obtain particle sizes smaller than 45 μm, and dried at 80°C in vacuum oven to a constant weight.
Electrorheological properties (shear viscosity, storage and loss moduli) of polyaniline particles suspended in silicone oil were measured using both devices and both geometrical arrangements. Each point in steady shear flow measurements as well as in small-strain oscillatory tests (dynamic amplitude sweep and frequency sweep) was measured at least three times.

2.2. Geometrical arrangements
The following gaps were used for the individual ER devices:
Physica MCR 501: PP 50 - gap 0.3 mm, CC 17 - gap 0.7 mm;
Gemini 150: PP 40 - gap 0.3 mm, CC 25 - gap 1.25 mm.

2.3. Temperature
The temperature was kept at 20°C during experiments.

2.4. Conductivity measurement
The DC conductivity of PANI base pressed into pellets (13 mm in diameter, 1 mm in thickness) was determined by a four-point method direct conductivity measurement in a Van der Pauw setup. The conductivity of PANI samples attained 7.72 x 10^{-9} S/cm.

2.5. Scanning electron microscope
Figure 3 illustrates morphology of PANI particles using a scanning electron microscope (SEM, Tescan, Czech Republic) with the accelerating voltage of 30 kV.

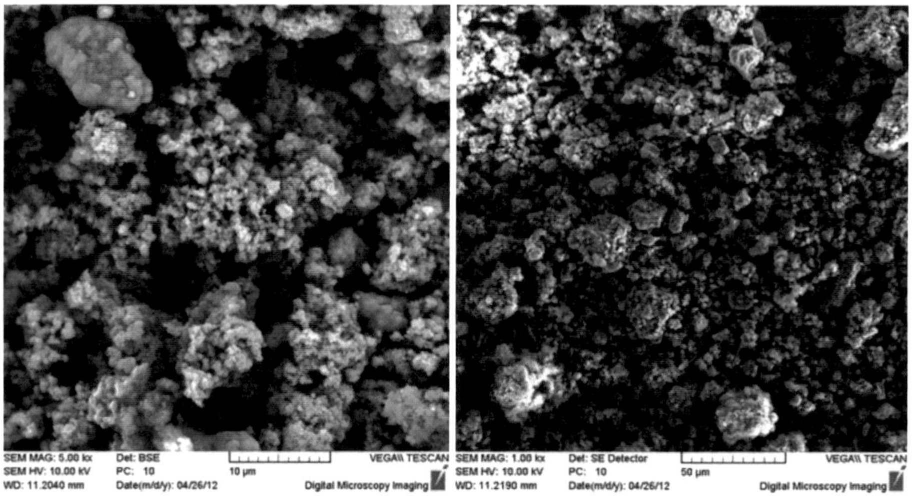

Figure 3. SEM pictures of PANI particles.

3. Experimental intercomparison of both generations of an electric field
First, flow curves of a carrier fluid (silicone oil) were obtained followed by ER measurements of the prepared ER suspension. The comparisons were always made for all four ER devices (two in either rheometer).

3.1. Measurements of silicone oil without electric field
Basic introductory comparison was carried out with a carrier silicone oil. This measurement is documented in figure 4. As can be seen, the flow curves for shear rate in the range 1-100 s^{-1} are identical.

3.2. Measurements of ER suspension
The samples (10 wt.% concentration of PANI powder in silicone oil) were stirred mechanically and than placed in an ultrasonic bath for 30 s before each measurement. Consequently, the intercomparisons were made for three different modes. The courses of measurements (time intervals, electric field strength (xxx V stands for the individual values of voltage as indicated in the figures), range of measured points) are summarised in table 1.

For each mode, figures 5-10 compare behaviour of flow curves consequently for the same geometry (plate-plate (PP) or concentric cylinders (CC)) of both rheometers, and both geometries of individual rheometers. The experimental data in these figures are always reduced by the values obtained for the corresponding measurements with mere air.

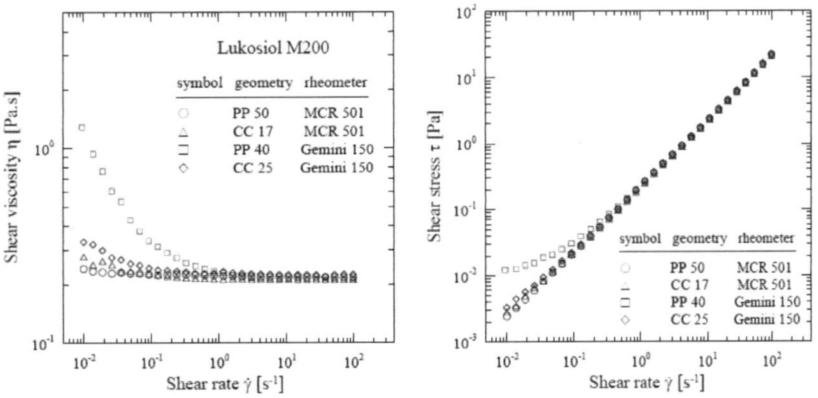

Figure 4. Flow curves of silicone oil (E = 0 kV/mm).

Table 1. Regimes of measurements for the individual modes.

T=20°C	Mixing	Pause	Pause	Measurement
Steady shear	10 s^{-1}	0 s^{-1}	0 s^{-1}	0.01 - 100 s^{-1}
	30 s	30 s	60 s	
	0 V	0 V	xxx V	xxx V
Amplitude sweep	10 s^{-1}	0 s^{-1}	0 s^{-1}	0.0001 – 1 / 0.01[-]
	30 s	30 s	60 s	5 Hz / 3 Hz
	0 V	0 V	xxx V	xxx V
Frequency sweep	10 s^{-1}	0 s^{-1}	0 s^{-1}	0.001 / 0.0006 [-]
	30 s	30 s	60 s	0.1 - 10 Hz
	0 V	0 V	xxx V	xxx V

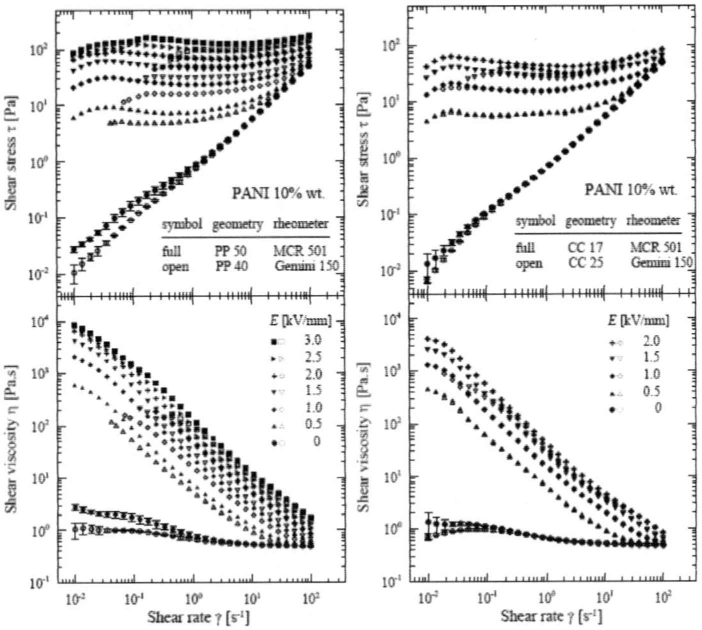

Figure 5. Shear viscosity and shear stress in dependence on shear rate, same geometrical arrangement (PP or CC), different rheometers.

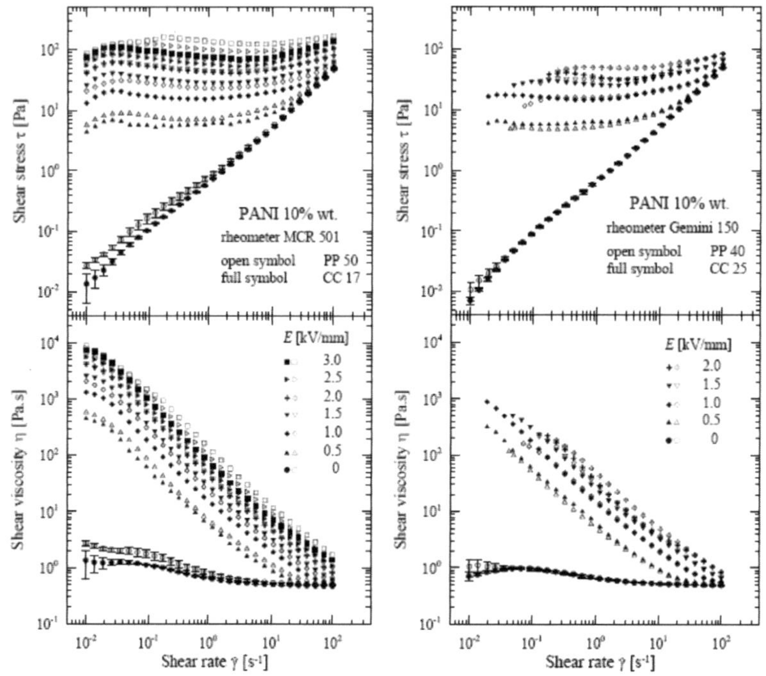

Figure 6. Shear viscosity and shear stress in dependence on shear rate, different geometrical arrangements, same rheometer.

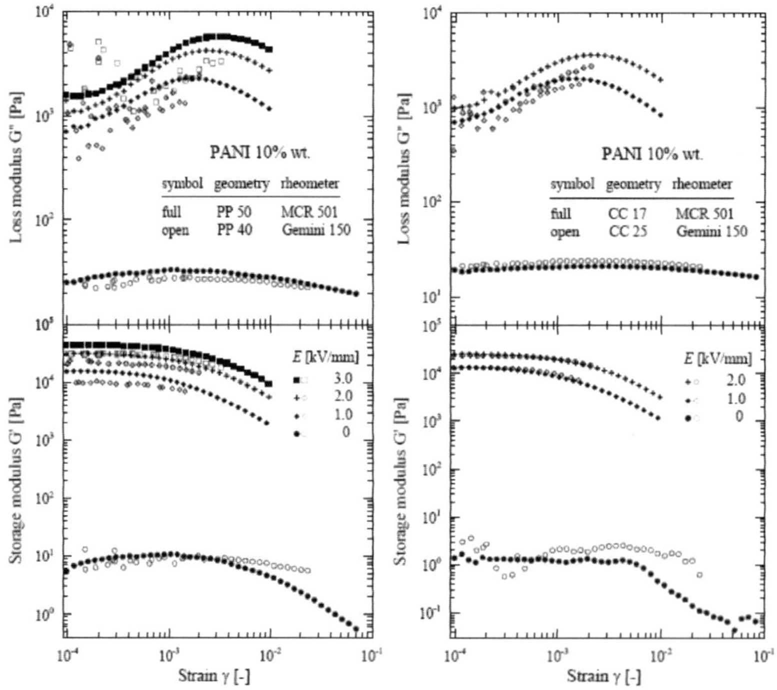

Figure 7. Storage and loss moduli in dependence on strain, same geometrical arrangement (PP or CC), different rheometers.

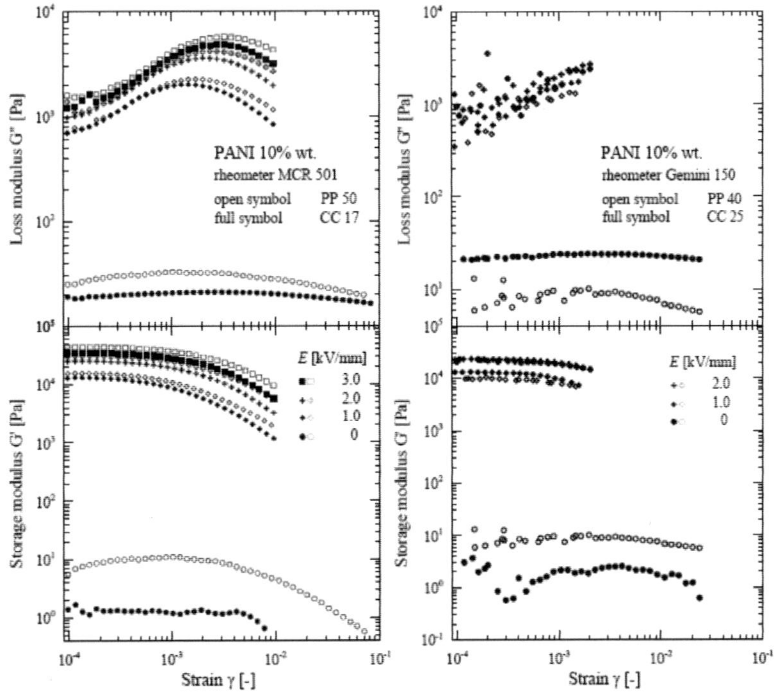

Figure 8. Storage and loss moduli in dependence on strain, different geometrical arrangements, same rheometer.

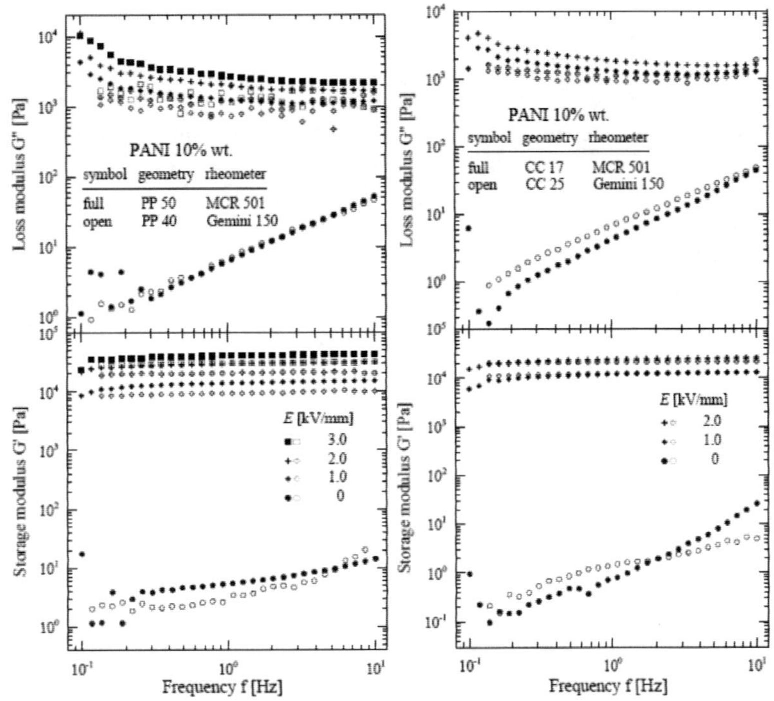

Figure 9. Storage and loss moduli in dependence on frequency, same geometrical arrangement (PP or CC), different rheometers.

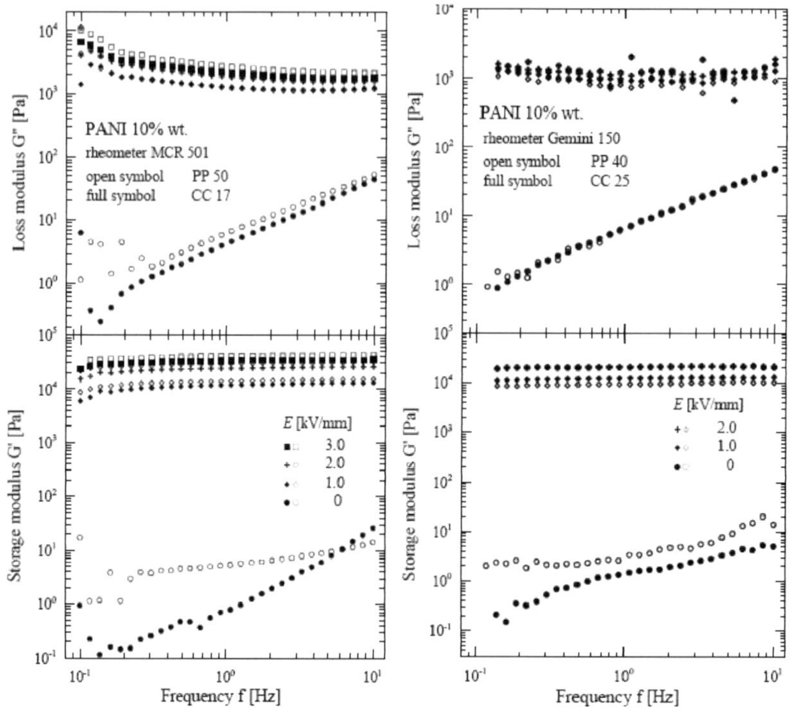

Figure 10. Storage and loss moduli in dependence on frequency, different geometrical arrangements, same rheometer.

4. Discussion and conclusions

The experimental data obtained for different geometrical arrangements and different rheometers exhibit non-negligible discrepancies. This fact is not surprising, cf. e.g. Rides *et al.* [1] who compared shear viscosity results obtained at high shear rates using extrusion rheometers (the same manufacturer) including an instrumented injection moulding machine.

There is a series of reasons causing this ambiguity: differences in construction of both rheometers (nevertheless in both cases all the parts in contact with the materials are made of stainless steel (1.4571 -MCR 501, 316 - CVOR 150)), anti-shock balancing in the laboratories (significant for lower values of independent variables), generation of an electric field, diameters of corresponding geometrical arrangements (e.g. 40 vs. 50 mm).

Based on figures 5-10 it is possible to come to the conclusions that

- the effect of completion of an electric circuit is different, but loosely speaking for mere air in both cases comparable e.g. with viscosity of water, however, when ER fluids are applied this value is not significant;
- the data from the Physica MCR 501 are smoother and behaviour of the corresponding curves is more continuous (at this point an age difference of the individual devices is also projected);in the case of the Gemini 150 the experimental data are closer for both geometries used;
- in the Physica MCR 501 usage of parallel plate geometry provides - on average - higher values in comparison with concentric cylinder geometry;
- when using parallel plate geometry, the data obtained by the Physica MCR 501 provide higher values (shear viscosity, storage and loss moduli vs. strain or frequency) than those obtained by the Gemini 150;
- when using concentric cylinder geometry, the data obtained by the Physica MCR 501 and the Gemini 150 are comparable in spite of the difference in corresponding gaps.

Acknowledgements
This work has been supported by the Grant Agency of the Czech Republic, Grant No. P105/11/2342 and Operational Program Research and Development for Innovations co-funded by the European Regional Development Fund (ERDF) and national budget of the Czech Republic, within the framework of Centre of Polymer Systems project (reg. No. CZ.1.05/2.1.00/03.0111).

References

[1] Rides M, Kelly A L and Allen C R G 2011 *Polym. Testing* **30** 916

Efforts to establish standardized methods to characterize electrorheological suspensions – An interim report focusing on flow mode characterization

S Schneider[1,4], K Holzmann[2] and S Ulrich[3]

[1] Bundeswehr Research Institute for Materials, Fuels and Lubricants, Institutsweg 1, 85435 Erding, Germany
[2] FLUDICON GmbH, Landwehrstrasse 55, 64293 Darmstadt, Germany
[3] Helmut-Schmidt-Universität Hamburg - University of the German Armed Forces, Holstenhofweg 85, 22043 Hamburg, Germany

E-mail: steffenschneider@bwb.org

Abstract. Intensive research, project development, and success in industrial applications based on electrorheological (ER) technology lead to considerations to develop standardized methods to characterize ER suspensions. Some independent work was completed in earlier projects and in 2006 a working group was founded in Germany consisting of seven different partners. In April 2011, the first part of DIN 51480 was issued addressing "Electrorheological suspensions – Requirements, testing and application – Part 1: Basics and field strength independent properties". In the future, two additional parts are planned to be released. Part 2 will describe methods to determine field strength dependent properties under laboratory conditions. Part 3 will describe methods to determine field strength dependent properties under application-like conditions. The majority of the work has already been completed for the development of Parts 2 and 3. However, some efforts still need to be completed before these two parts can be released.

1. Introduction

Intensive research, project development, and success in industrial applications based on electrorheological (ER) technology lead to considerations to develop standardized methods to characterize ER suspensions.

Some independent work was completed in earlier projects and in 2006 a working group was founded in Germany consisting of seven different partners:

- FLUDICON GmbH, Darmstadt, Germany
- Institute for Fluid Power Drives and Controls (IFAS), RWTH Aachen University, Germany
- Fraunhofer-Institut für Silicatforschung, Würzburg, Germany
- NMW Neue Materialien Würzburg GmbH, Würzburg, Germany
- Automation and Control Institute (ACIN), Vienna University of Technology, Austria
- Bundeswehr Research Institute for Materials, Fuels and Lubricants, Erding, Germany
- Helmut-Schmidt-Universität - University of the German Armed Forces, Hamburg, Germany.

[4] Corresponding author.

In April 2011 the first part of DIN 51480 was issued addressing “Electrorheological suspensions – Requirements, testing and application – Part 1: Basics and field strength independent properties” [1].

In the future two, additional parts are planned to be released. Part 2 will describe methods to determine field strength dependent properties under laboratory conditions. Part 3 will describe methods to determine field strength dependent properties under application-like conditions.

2. The new DIN 51480 specification

2.1. Part 1 of DIN 51480: Basics and field strength independent properties

One purpose of Part 1 is to give an introduction into ER technology. Furthermore, it gives advice on how to determine field strength independent properties using standardized and non-standardized test procedures.

The outline of Part 1 is as follows:

- Fundamentals, terms and definitions of typical properties of ER suspensions
- Determination of the dynamic base viscosity depending upon the shear rate
- Determination of the dynamic base viscosity depending upon the temperature
- Determination of the mean particle size and the particle size distribution according to DIN EN 725-5
- Determination of the density of an ER suspension according to DIN 51757 and DIN EN ISO 3838
- Determination of the relative solid content and the density of the solid particles following DIN EN ISO 3838
- Determination of the sedimentation
- Determination of the re-dispersing behaviour

2.2. Part 2 of DIN 51480: Field strength dependent properties

The purpose of Part 2 is to characterize ER suspensions with regard to their field dependent behaviour using laboratory equipment and standardized methods if available.

The outline of Part 2 is as follows:

- Determination of the shear stress at constant shear rate depending upon the electric field strength in rotational motion
- Determination of the yield stress depending upon the electric field strength in oscillatory motion
- Determination of the electric breakdown voltage following DIN EN 60156 (Figure 1)
- Determination of the relative permittivity ε_r, dielectric dissipation factor $\tan\delta$ and the d.c. resistivity ρ following DIN EN 60247

In the current state, an attempt will be made to determine the yield stress using already approved, standardized methods. Therefore, we will try to adopt one of the following specifications:

- DIN 51810-2: Testing of lubricants – Testing rheological properties of lubricating greases – Part 2: Determination of flow point using an oscillatory rheometer with a parallel-plate measuring system.
- E DIN 54458 (Draft): Structural adhesives – Determination of the flowability and application behaviour of viscoelastic adhesives using the oscillation rheometry.

If both are not applicable than it becomes necessary to develop a new test procedure capable of determining the yield stress using an oscillatory rheometer with a parallel-plate measuring system.

The electric strength of an insulator is the maximum electric field strength which the material can withstand without experiencing an electric breakdown (spark). The electric strength E of an insulating material is expressed in the same unit as field strength (specified in kV/mm or V/mm). Therefore, electric strength is also referred to as breakdown field strength and is calculated by using equation 1.

$$E = \frac{U}{l} \tag{1}$$

where U is the breakdown voltage in V or kV and l is the insulator thickness in mm.

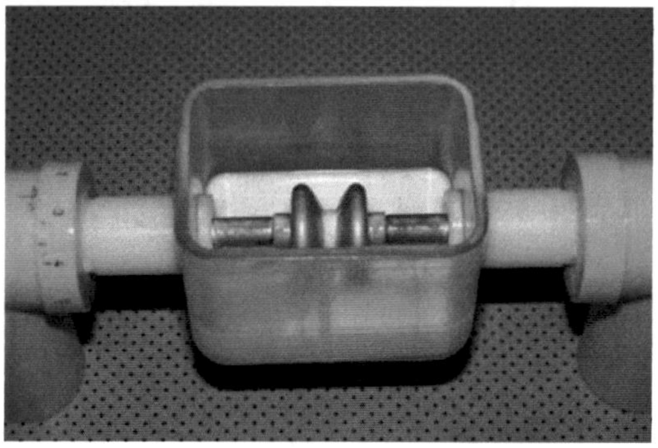

Figure 1. Determination of the electric breakdown voltage.

Figure 1 shows the test setup according to DIN EN 60156. The DIN specification requires a series of ten tests without interruption. In deviation from the specification, the ten measurements need to be carried out as single tests. Due to chain formation in each single test the material between the gap needs to be homogenized after each single test.

2.3. Part 3 of DIN 51480: Determination of field strength dependent, application-like properties

ER suspensions experience higher shear rates than obtainable with rheometer setups in dampers and other industrial applications. Depending upon the rheometer geometries for bob and rotor, shear rates are typically possible up to 2,000 s^{-1}. At higher shear rates turbulent flow, like Taylor vortices, may occur. In dampers ER suspensions often experience shear rates higher than 10,000 s^{-1}, a fact which led to considerations to design a hydraulic test bench. With this equipment, the field strength dependent behaviour of ER suspensions under application-like conditions could be tested.

The outline of Part 3 is as follows:

- Requirements to the hydraulic test bench (Figure 2, Figure 3).
- Measurement of field strength dependent stationary pressure differences, calculation of the dynamic base viscosity, and calculation of the field strength dependent yield stress [4].
- Determination of the field strength dependent step response behaviour [5, 6].

In principle, a distinction is made between the two design types of electrorheological valves, the annular gap design and the rectangular gap design.

Both types will be considered in part 3 of DIN 51480. However, since all measurements described in this paper are carried out with an annular gap valve, this type is subject of further descriptions here.

A design concept for the development of cylinder drives based on electrorheological fluids was introduced by WOLFF-JESSE, FEES, and ZAUN [2, 3]. The design of an annular gap was used in the cylinder drives itself as well as in electrorheological valves to evaluate the rheological behaviour under laboratory conditions. Such an annular gap is shown in Figure 2 and Figure 3. It is composed of an outer electrode (cylinder of radius Ra) connected to earth and an inner electrode (cylinder of radius Ri) connected to the voltage U, forming an annular gap. Since the height H = Ra − Ri of the gap is small compared to the mean radius Rm = (Ra + Ri) /2, the ER valve can be approximated by an equivalent flat channel with the length L and the width W = 2Rmπ. The ER valve can be considered as a hydraulic throttle with a pressure difference Δp calculated from the pressures p1 and p2 before and behind the gap (equation 2).

$$\Delta p = |p_1 - p_2| \tag{2}$$

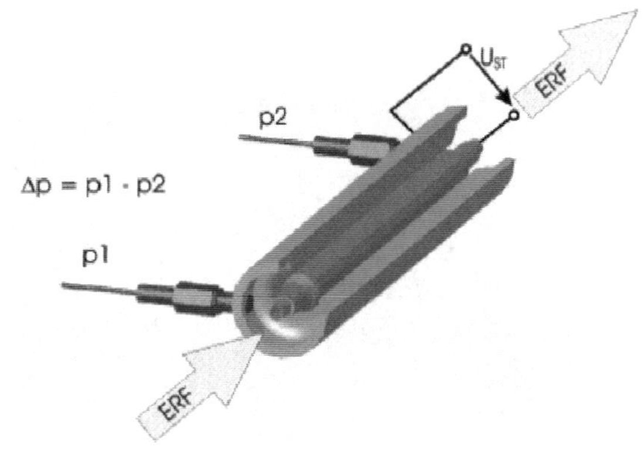

Figure 2. Operating mode of an ER valve [3].

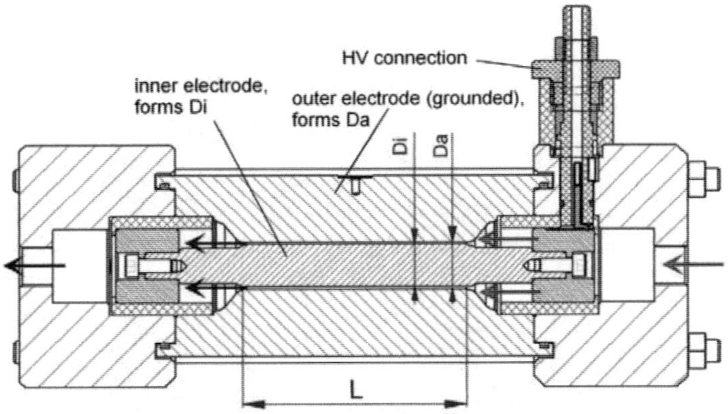

Figure 3. Sectional view of an annular ER valve.

3. Determination of the field strength dependent yield stress

3.1. Assumption

The yield stress τ_0(E, T) is the lowest shear stress beyond which plastic material shows the rheological behaviour of a liquid (see DIN 1342-1). For ER suspensions, this electrorheological yield stress τ_0(E, T) depends, among others, on the electric field strength E and the temperature T.

Under the assumptions that (i) the flow in the gap remains laminar, (ii) the fluid in the gap is incompressible and (iii) the temperature is constant, the constitutive equations of many ER fluids can be simplified to a "Bingham-like" material model (3)

$$\sigma_{12} = \tau_0(E)\operatorname{sign}(\dot{\gamma}) + \eta\dot{\gamma} \quad \text{for} \quad \dot{\gamma} \neq 0 \tag{3}$$

with the shear stress σ_{12}, the dynamic viscosity η, the electric field strength E, the field dependent yield stress τ_0 (E) and the shear rate $\dot{\gamma}$.

3.2. Test procedure

The stationary pressure difference Δp will be measured at constant volume flow q under variation of the applied electric voltage U.

With the pressure difference Δp_l from measurements without applied electrical field (U = 0 V) the dynamic base viscosity will be derived according to equation 4.

$$\eta = \frac{WH^3}{12L}\frac{\Delta p_1}{q} \tag{4}$$

From the voltages U_i the corresponding field strength E_i will be calculated according to equation 1:

$$E_i = \frac{U_i}{H} \tag{5}$$

In the following steps the pressure gradient P_i is calculated according to equation 6. Furthermore, an assistant variable Φi is introduced according to equation 7. In a final step the field dependent yield stress $\tau_{0,i}$ is calculated according to equation 8.

$$P_i = \frac{\Delta p_i}{L} \tag{6}$$

$$\Phi_i = \arccos\left(\frac{12\eta q}{WP_i H^3} - 1\right) \tag{7}$$

$$\tau_{0,i} = P_i H \cos\left(\frac{\Phi_i}{3} + \frac{4\pi}{3}\right) \tag{8}$$

As a result the measurements were visualized in a diagram showing the dependence between electrical field strength and yield stress. The entire algorithm is shown in Figure 4. The complete derivation is described in [4].

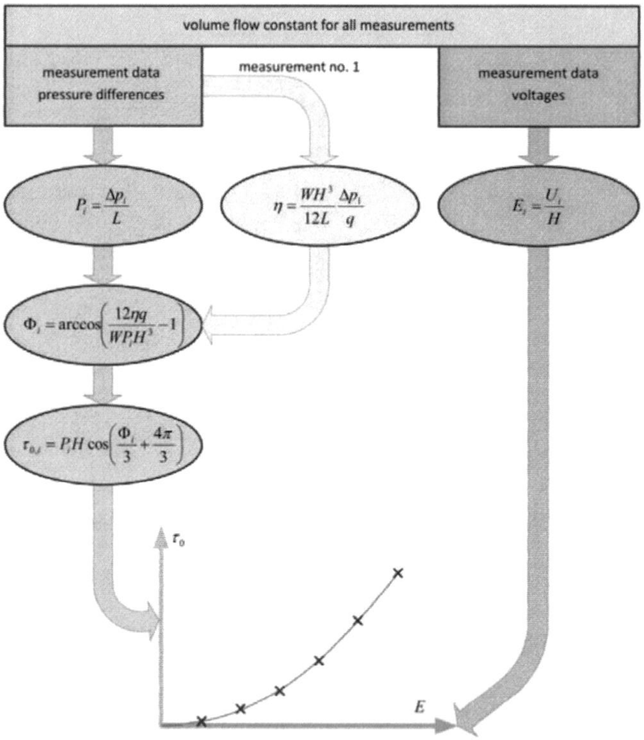

Figure 4. Algorithm to identify the field strength dependent yield stress.

3.3. Reproducibility

The acceptance of a test procedure depends on repeatability and reproducibility. The reproducibility is the difference between two single and independent results obtained by different operators working in different laboratories on identical test method. The measurements shown in Figure 5 and Figure 6 were carried out by three different operators working in the same laboratory with the same apparatus under constant operating conditions on the same ER fluid. The tests were carried out at constant volume flows q of 5 and 10 litres per minute. The geometry used was version V105 (Table 1).

The repeatability is the difference between successive results obtained by the same operator with the same apparatus under constant operating conditions on identical test material. The repeatability of test carried out by operator 2 is shown in Figure 7 and for test carried out by operator 4 in Figure 8.

These measurements are evidence that this test procedure shows an acceptable reproducibility when carried out in the same laboratory. Comparisons between other laboratories are planned but not available yet.

3.4. Influence of the volume flow

The test was carried double each at two different volume flows q, 5 and 10 litres per minute, and geometry version V105. Calculating the shear rate out of the volume flow and the geometry data according to Table 1 shear rates are 13,893 and 27,785 s^{-1}, respectively. The measurements were carried out by operator 2.

As one can see in Figure 7, there is an influence of the volume flow to the yield stress which can not be eliminated by the test procedure.

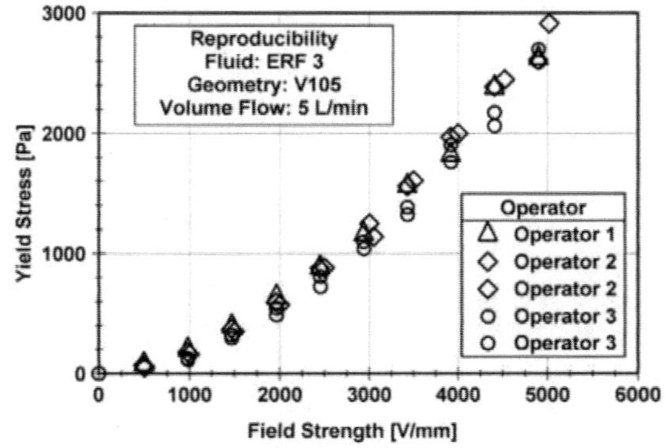

Figure 5. Reproducibility between three operators at 5 L/min.

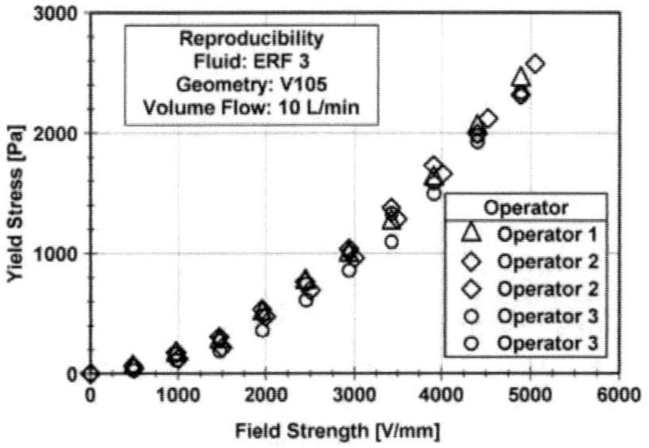

Figure 6. Reproducibility between three operators at 10 L/min.

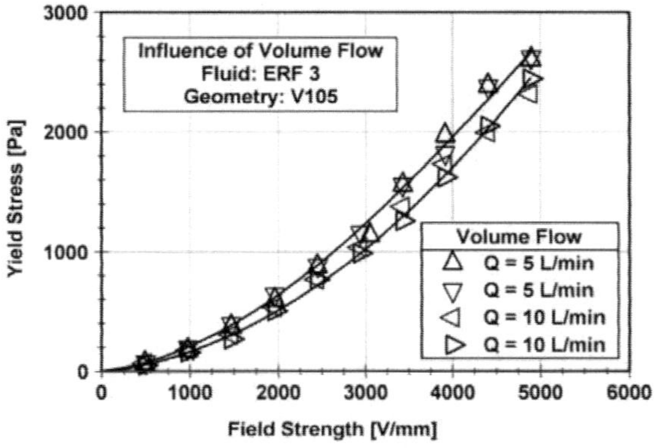

Figure 7. Influence of the volume flow.

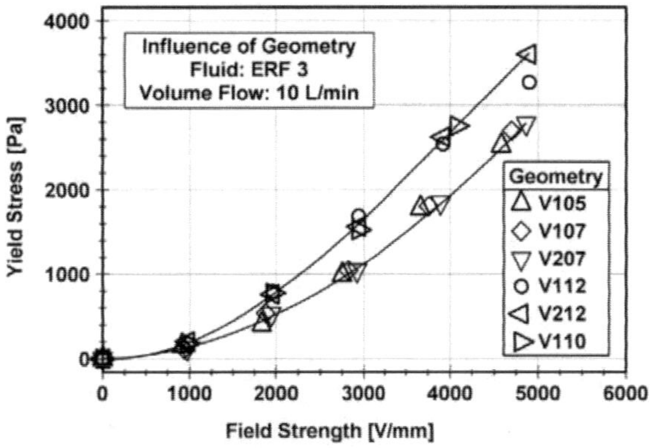

Figure 8. Influence of the gap geometry.

3.5. Influence of the gap geometry

As there are different flow channel geometries available, it was of interest if there is also an influence by the geometries. The measures of the geometries used are listed in Table 1. The measurements were carried out by operator 4 at a constant volume flow q of 10 litres per minute.

Table 1. Annular gap geometries data.

Geometry Version	**V105**	**V107**	**V110**	**V112**	**V207**	**V212**
Mean Diameter Dm [mm] / B [mm]	40.025	30.05	40.05	30.02	30.025	30.02
Gap Height H [mm]	0.535	0.73	1.01	1.20	0.705	1.20
Gap Length L [mm]	100	100	100	100	200	200
Gap Cross Section Area A [mm^2]	67.3	68.9	127.1	113.2	66.5	113.2
Inner Electrode						
Diameter Di [mm]	39.49	29.32	39.04	28.82	29.32	28.82
Outer Electrode						
Diameter Da [mm]	40.56	30.78	41.06	31.22	30.73	31.22
Shear Rate						
@ volume flow 10 litres per minute	4629	4521	2451	2752	4685	2752

It can be seen that the yield stress, whose calculation is described above, is also dependent on the geometry especially from the height of the channel. The simple Bingham model where the yield stress is only dependent on the electric field strength seems not to be sufficient to describe the flow behaviour of the ER suspension accurately. Other influences, like wall slip or non-Newtonian flow behaviour, have to be taken into account, which are known for these complex suspensions [7].

But a standardisation method has to be applicable for fluid investigators as well as for application engineers. Therefore, it has to be as simple as possible but reproducible and clear. For this reason it seems necessary to define the flow channel geometry and the conditions to gain the electric dependent rheological parameters of ER suspensions. The chosen geometry and conditions should be similar to typical application of ER systems.

4. Summary

Much effort was put in to develop and release Part 1 of the specification. The majority of the work has already been completed for the development of Parts 2 and 3. However, some efforts still need to be completed before these two parts can be released. Especially a standard flow channel to measure the electrical dependent rheological parameters has to be defined in Part 3.

References

[1] DIN 51480 2011 Electrorheological suspensions – Requirements, testing and application – Part 1: Basics and field strength independent properties

[2] Wolff-Jesse C and Fees G 1998 Examination of flow behaviour of electrorheological fluids in the flow mode *Proc. Instn. Mech. Engrs.* vol 212 part I

[3] Zaun M 2006 Design concept for the development of electrorheological cylinder drives *J. Intell. Mater. Syst. Struct* **17** 303-7

[4] Holzmann K 2011 Modellierung und Regelung von elektrorheologischen Dämpfern, Technical University Vienna, PhD thesis

[5] Ulrich S, Böhme G and Bruns R 2009 Measuring the response time and static rheological properties of electrorheological fluids with regard to the design of valves and their controllers *Journal of Physics: Conference Series* **149** 012031

[6] Gurka M, Petricevic R, Schneider S, and Ulrich S 2011 Characterization of step response time and bandwidth of electrorheological fluids *J. Intell. Mater. Syst. Struct* **22** 1745-48

[7] Schneider S 2012 Wall slip effects measuring the rheological behaviour of electrorheological (ER) suspensions *Int. J. Mod. Phys. B* **26** 1250006.

Development of a manipulator with an opposed-placement-type ER clutch contributing to collision force reduction

A Inoue, N Kanno, M Yoshikawa and T Nakamura[1]

Chuo University, Biomechatronics Laboratory, 1-13-27 Kasuga, Bunkyo-ku, Tokyo, 112-8551, Japan

E-mail: nakamura@mech.chuo-u.ac.jp

Abstract. In recent years, human and robots have begun to work together in the fields of medical treatment and welfare. In such environments, robots must be designed to operate safely; thus, reducing collision force when collision is occurred with their human colleagues is essential. In this study, a manipulator with opposed-placement-type electrorheological (ER) clutches is designed to reduce the collision force. The main feature of the manipulator is a structure that sandwiches the link between two ER clutches and equips the link with a pneumatic cushion. The input shaft of each clutch is rotated in the opposite direction via a bevel gear and a spur gear from an electric motor. Because the link can be reversed by switching the clutch, the manipulator can operate in high-speed reverse motion. Position control, reverse motion control and collision experiments were performed to test the effectiveness of the manipulator. The reverse motion control experiment revealed that the manipulator can reverse more rapidly than a conventional manipulator, which is only driven by a motor. From the collision control experiment, it was found that the manipulator can reduce the collision force more effectively than the conventional manipulator.

1. Introduction

Recently, humans and robots have begun to work together in the fields of medical treatment and welfare [1-3]. In such environments, robots must operate safely so as to avoid collisions or inadvertent contacts with their human colleagues. A manipulator with an electrorheological (ER) clutch and a pneumatic cushion sensor has been developed as a robot arm that satisfies this requirement [4, 5]. Because of its flexible joint, this manipulator can reduce collision and contact forces, but has proven difficult to control. To address this problem and further reduce the collision force, high-speed reverse motion control has been proposed [6].

In this study, we developed a manipulator that can be reversed at high speed. Figure 1 illustrates the manner in which the manipulator achieves the intended high-speed reversal motion. In stage 1, a collision occurs. A pneumatic cushion sensor detects the collision and pushes the colliding object. In stage 2, a reversal torque is instantaneously generated so that the impact force is reduced by an amount dependent on the reduction in pushing distance and contact time. In stage 3, the situation following operation of the reversal motion control is illustrated.

In this paper, we demonstrate the action of the developed manipulator with an opposed-placement-type ER clutch in terms of impact force reduction. The ability of the manipulator to reduce collision forces is tested in a series of collision experiments.

---

[1] Corresponding author.

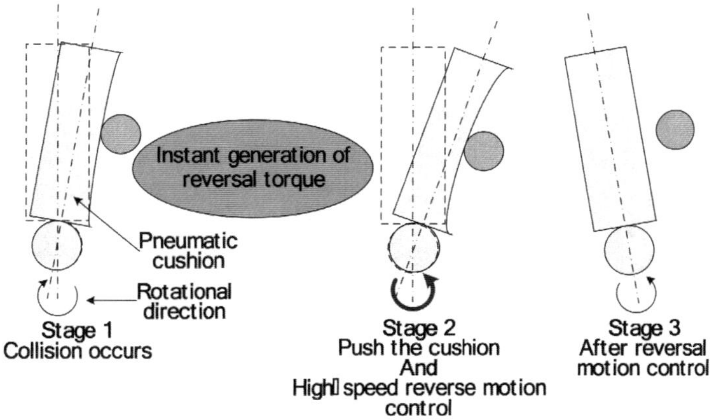

Figure 1. Schematic of idealized high-speed reverse motion control.

2. Manipulator with the opposed-placement-type ER clutch

The developed manipulator and its specifications are presented in Figure 2 and Table 1, respectively. This manipulator comprises two ER clutches, an electric motor and a pneumatic cushion.

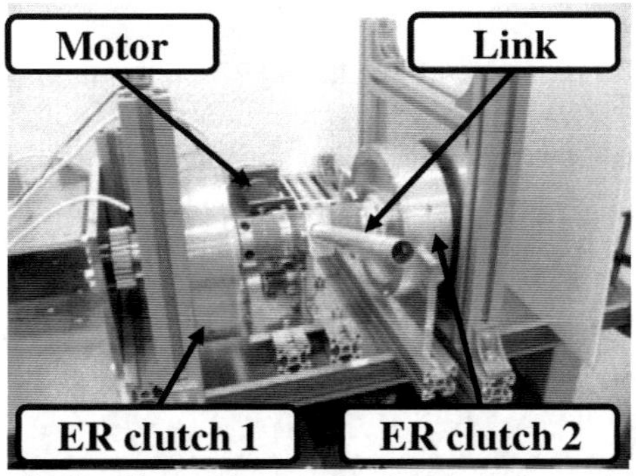

Figure 2. Configuration of the developed manipulator.

Table 1. Specifications of the manipulator.

Parameter	Value
Length of the link	0.3 [m]
Mass	0.2 [kg]
Rotation angle	215 [deg]
Center of gravity position	0.12 [m]
Inertia (Link + output shafts of ER clutches)	0.026 [kgm^2]
Human pain tolerance	13 [N]
Contact position of the base of the link	0.2 [m]
Initial pressure of the pneumatic cushion	5 [kPa]

2.1. Mechanism of the opposed-placement-type ER clutch

Figure 3 shows the mechanical design and the driving mechanism of the arm. The main feature of the manipulator is the structure that sandwiches the link between two clutches. With this structure in place, the input shaft of each clutch can rotate in the opposite direction via a bevel gear, a spur gear and a motor. The arm is attached to the output shafts of the ER clutches. High-speed reversal of the arm is achieved by switching to another input shaft that drives another ER clutch, not by reversing the motor. As a result of reversing the link by switching the clutch and ignoring the inertia of the motor and gear, the manipulator can reverse at high speed.

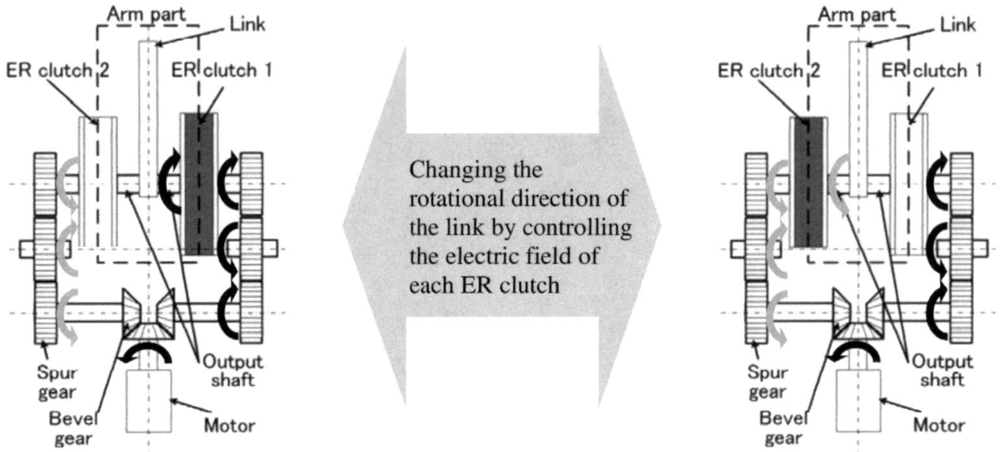

Figure 3. Mechanism of the developed manipulator.

2.2. ER clutch

Figure 4 and Table 2 display the configuration and specifications of the ER clutches, respectively. Figures 5 and 6 show the dependence of the torque transmitted to the ER clutches on the applied electric field and the angular rotation speed of the input shaft. In these figures, the applied electric field is increased from 0 to 2.0 kV/mm in 0.5 kV/mm increments, while the rotation speed is increased from 5 to 20 r/min in 5r/min increments. It is seen that the transmitted torque increases by applied electric field. In addition, we note that torque does not depend on the rotation speed.

The clutch is made by ER Tech. Multiple disks are mounted on the input and output shafts. The ER fluid enclosed between these disks transmits the torque from the input to the output shaft. The transmitted torque can be increased by applying an electric field independent of temperature. In addition, the torque rapidly responds to the applied electric field. Therefore, this high-speed reverse motion shows promising future applicability. ER clutches 1 and 2 generate a maximum transmitted torque of about 12 Nm and 7 Nm, respectively, under an electric field of 2.0 kV/mm. The maximum transmitted torque of the ER clutches is decided by calculating the torque required to reverse the arm in advance. In this study, the maximum transmitted torque performances of the two ER clutches are designed to be different to enable a differential investigation of impact force reduction and reverse speed.

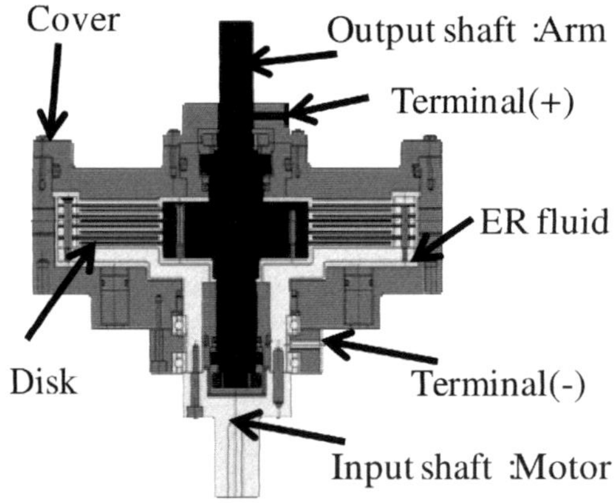

Figure 4. Configuration of the ER clutch.

Table 2. Specifications of the ER clutch.

Parameter	Value (ER clutch1)	Value (ER clutch 2)
Maximum torque [Nm]	12	7
Output shaft [mm]	ϕ 20	ϕ 15
Input shaft [mm]	ϕ 25	ϕ 20
Diameter [mm]	215	185
Weight [kg]	8.2	6.1
Thickness [mm]	56	56
Outer disk diameter [mm]	170	140
Inner disk diameter [mm]	90	70
Number of sheets of the disk	4	4
Disk thickness [mm]	1	1

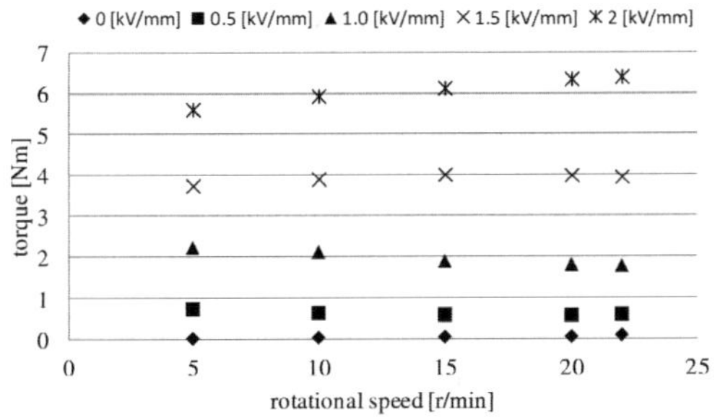

Figure 5. Relationship between the rotational speed and torque of the ER clutch (7 Nm) as a function of the electric field.

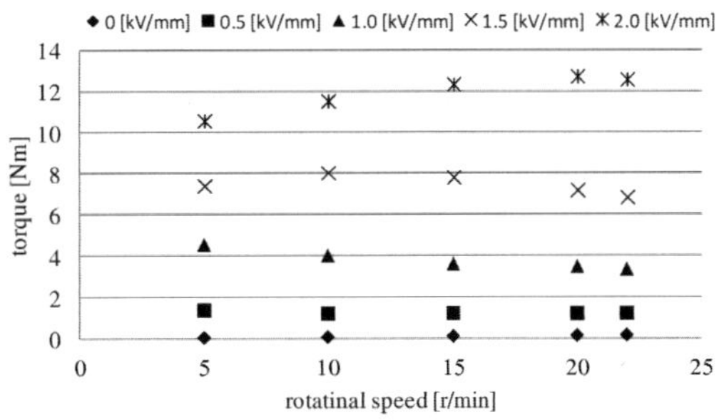

Figure 6. Relationship between the rotational speed and torque of the ER clutch (12 Nm) as a function of the electric field.

2.3. Pneumatic cushion

Figure 7 illustrates the pneumatic cushion attached to the link of the developed manipulator. The cushion specifications are listed in Table 3. The pneumatic cushion reduces the collision force that is measured by internal pressure changes, which are detected by a pressure sensor.

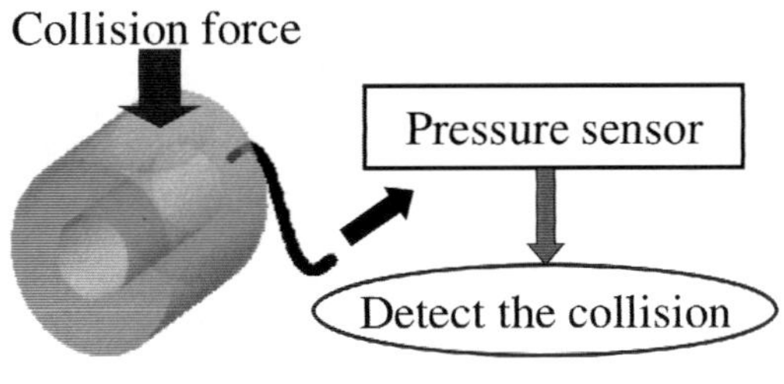

Figure 7. Schematic of the pneumatic cushion

Table 3. Specifications of the pneumatic cushion.

Parameter	Value
Outer diameter	100 [mm]
Inner diameter	20 [mm]
Length	140 [mm]
Thickness	0.05[mm]
Material	Polyurethane

2.4 Experimental system

Figure 8 illustrates the control system of the manipulator. The manipulator is driven by an electric motor. The torque generated by the motor is transmitted by the ER clutches, which in turn powers the link. The control system consists of a MATLAB and Simulink interface with dSPACE. The angle of the input shaft is measured by counting the output pulses by the encoder interfaced to a PC, while that of the output shaft is measured by a potentiometer.

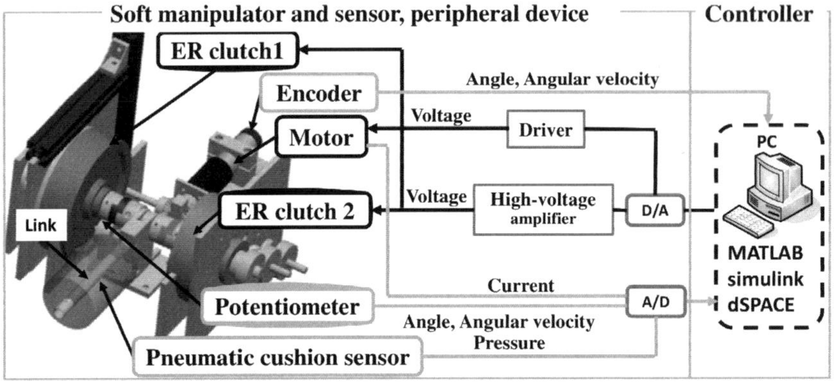

Figure 8. Configuration of an experimental system.

3. Experiment

3.1. Position control experiment

The aim of the position control experiment is to test the positioning accuracy when a changing electric field is applied to the ER clutch.

The target angle of the arm is set to 180°, and the electric field applied to ER clutch1 is increased from 0.5 to 2.0 kV/mm in 0.5 kV/mm increments. The target angle command is executed in the first second following the start of the experiment.

The results of the experiment are plotted in Figure 9. This figure shows that precise angle control is possible under electric fields exceeding 0.5 kV/mm. Low electric fields are insufficient to overcome the friction and base viscosity of the clutches; hence, the arm is not moved at applied electric fields below 0.5 kV/mm . The curve is static for the first few seconds following the increase in the target angle because of the delay in transmitting the torque from the motor to the output shaft of the ER clutch.

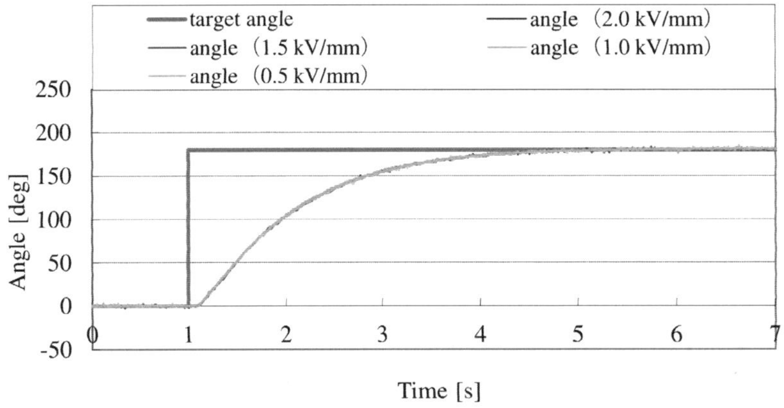

Figure 9. Results of the position control experiment.

3.2. Reverse control experiment

To determine whether the developed manipulator can achieve the desired level of high-speed reverse motion control, a reverse motion experiment was conducted under three separate regimes.

Regime 1: The arm is reversed by switching the ER clutch.

Regime 2: The arm is reversed by controlling the motor, but with the manipulator attached to the ER clutch.

Regime 3: The arm is reversed by controlling the motor.

Figure 10 shows the experimental setup. First, the input shaft of each ER clutch is rotated by the motor, and the arm is rotated such that the electric field is only applied to ER clutch 1. Once the angle of the arm has reached 180°, reverse motion control is implemented. In the first regime, the applied electric field is switched to the other clutch when the arm angle reaches 180°. At this time, the rotational direction of the motor is not changed. In the second regime, the electric field is always applied to ER clutch 1, but the rotational direction of the motor is changed, and the arm is reversed once its angle reaches 180°. In the third case, as the arm angle reaches 180°, the rotational direction of the motor itself alters. In this regime, the manipulator without the ER clutch is used.

The results of the reverse motion control experiment are plotted in Figure 11. The black solid line, black dashed line and grey line show the responses under regime 1, 2 and 3, respectively. The vertical dashed line in the figure indicates the time at which the arm reaches 180°. Switching of clutches (regime 1) is seen to be the fastest means of achieving angular reversal (cf. regimes 2 and 3). From the figure, the overshoot of regimes 1, 2 and 3 is 1.3°, 5.7° and 6.8°, respectively. Under regime 2, the angle reaches a maximum of 185.7° and does not immediately decline thereafter (note the horizontal section of the curve). This phenomenon occurs because of the slip between the input and output shafts of ER clutch 1 as the motor is reversed. Consequently, the response of the system under regime 2 following reversal is slower than that under regime 3.

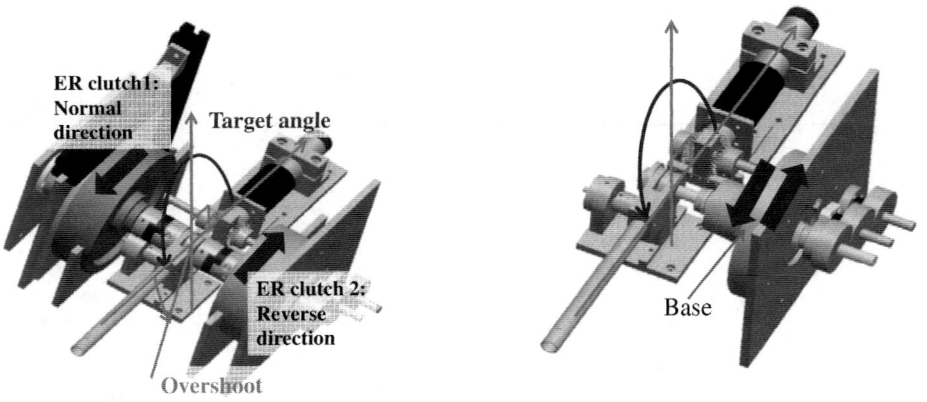

Figure 10. Left section shows the manipulator with the ER clutch used in experimental tests 1 and 2. However, under regime 2, ER clutch 2 is not used to reverse the arm. The right section shows the manipulator without the ER clutch tested in regime 3.

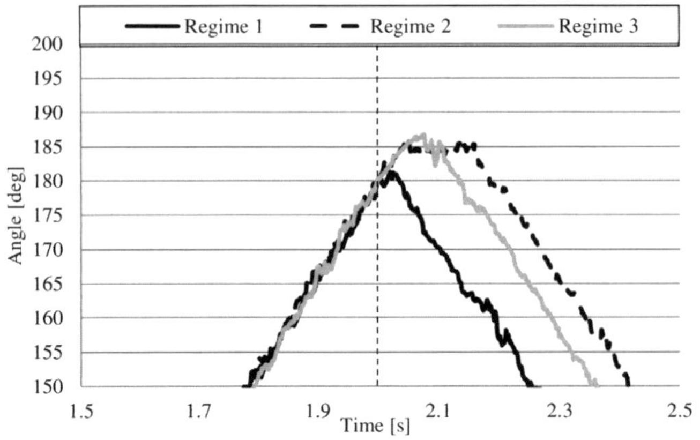

Figure 11. Results of the reverse motion control experiment.

3.3. Collision experiment

To confirm whether the developed manipulator can effectively reduce the collision force, a collision experiment is conducted, the setup of which is shown in Figure 12.

First, the arm is rotated, causing the link to collide with the force sensor. Next, when the collision force surpasses the threshold (the human pain tolerance [7]), this link is reversed by each of the three methods described in 3.2. Table 1 summarizes the experimental conditions. The following experimental designs were implemented:

Regime 1: Motion is reversed by changing the ER clutch. The motor is rotated in a constant direction after collision.

Regime 2: Motion is reversed by controlling the motor. The manipulator used this test is fixed to an ER clutch.

Regime 3: Motion is reversed by controlling the motor. The manipulator used in this test is not attached to the ER clutch.

Regime 4: This test is non-reverse motion control. Only the force sensor is pushed. The manipulator is not attached to the ER clutch under this regime. The arm is directly driven by the motor.

The results of this experiment are plotted in Figure 13. From the figure, it is seen that reversing the motion by switching the clutch is the most effective at reducing the collision force (the contact time is reduced in regime 1 relative to the other tests). The reason of the wavy response (regime 4) results from jumping phenomenon of the bevel gear.

From the figure, the maximum impact forces under all regimes are seen to exceed the threshold (the human pain tolerance). This is because that we set the threshold to be equal to that of human pain tolerance without taking into account the dynamic characteristics of the ER clutches and the pneumatic cushion sensor. In future, the threshold will be decided on these dynamic characteristics.

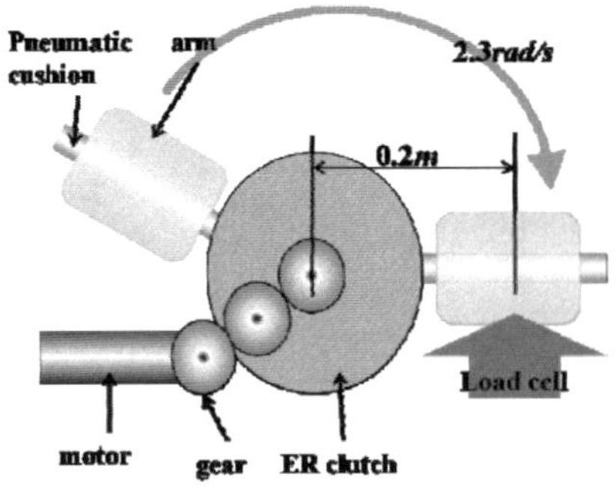

Figure 12. Schematic of the setup of the collision experiment.

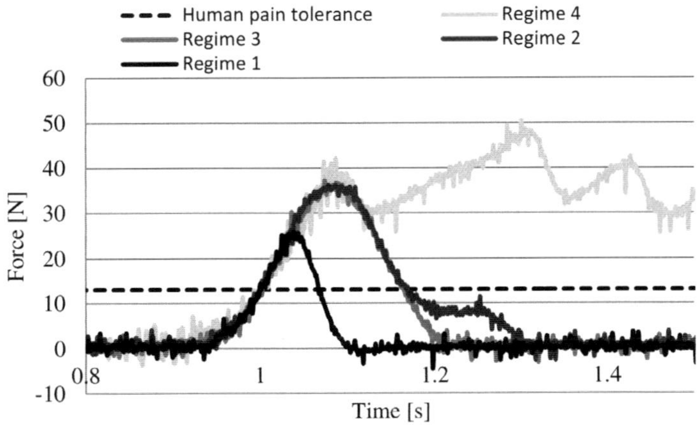

Figure 13. Measured impact forces for each buffer method.

8. Conclusion

In this study, we developed a manipulator with an opposed-placement-type ER clutch and tested its utility by position control, reverse motion control and collision experiments. From the position control experiment, the manipulator was found to exert precise positional control when an electric field exceeding 0.5 kV/mm is applied to the ER clutch. From the reverse motion control experiment, the developed manipulator was shown to reverse more rapidly than the conventional manipulator. The collision experiment revealed that the developed manipulator reduces collision forces more effectively than the conventional manipulator. However, the maximum collision force of the developed manipulator exceeds the threshold (the human pain tolerance) because of a time delay in the ER clutch and the pneumatic cushion sensor. In future studies, by accounting for the dynamic characteristics of the ER clutch and the pneumatic cushion sensor, we aim to reduce the maximum impact force to more desirable levels.

References

[1] Kawamura K and Iskarous M 1994 Trends in service robots for the disabled and the elderly *Proc. IEEE/RSJ Int. Conf. Intelligent Robots and Systems, 12–16 Sep 1994, Munich, Germany,* pp. 529–51

[2] Woosub L, Junho C and Sungchul K 2009 Spring-clutch: A safe torque limiter based on a spring and CAM mechanism with the ability to reinitialize its position 2009 *Proc. IEEE/RSJ Int. Conf. on Intelligent Robots and Systems*, 10–15 Oct. 2009

[3] Wand R-J and Han-Pang H 2010 An active-passive variable stiffness elastic actuator for safety robot systems *Proc. IEEE/RSJ Int. Conf. Intelligent Robots and Systems*

[4] Boku K, Kusaka Y, Akamatsu Y and Nakamura T 2008 Development and control of three degrees of freedom soft manipulators which considered a collision with the environment that was able to include a human being *J. Robot. Mechatron.* **20** 634–40 [Remark 9]

[5] Nakamura T, Akamatsu Y and Kusaka Y 2008 Development of soft manipulator with variable rheological joints and pneumatic sensor for collision with environment *J. Robot. Mechatron.* **20** 634–40

[6] Yoshikawa M, Boku K and Nakamura T 2010 High speed switching control of 1 DOF manipulator using ER clutch *Proc. 12th Int. Conf. Electrorheological Fluids and Magnetorheological Suspensions* (ERMR2010)

[7] Yamada Y, Suita K, Ikeda H, Sugimoto N, Miura H and Nakamura H 1997 Evaluation of pain tolerance based on a biomechanical method for human-robot coexistence *JSME* **63** 238–43

Journal of Physics: Conference Series **412** (2013) 012013 doi:10.1088/1742-6596/412/1/012013

Temperature and shear rate characteristics of electrorheological gel applied to a clutch

K Koyanagi[1,4], Y Takata[1], Y Kakinuma[2], H Anzai[3], K Sakurai[3], T Motoyoshi[1] and T Oshima[1]

[1]Toyama Prefectural University, 5180 Kurokawa, Imizu, Toyama, Japan
[2]Keio University, 3-14-1 Hiyoshi, Kohoku-ku, Yokohama, Kanagawa, Japan
[3]Fujikura Kasei Co., Ltd., 5-13-1 Sakurada, Kuki, Saitama, Japan

E-mail: koyanagi@pu-toyama.ac.jp

Abstract. This investigation reports the physical characteristics of electrorheological (ER) gels, which are a type of functional material having controlled surface friction. We previously developed slip clutches using ER gels sandwiched between electrodes, and verified their responses and controllability. We newly report the temperature and shear rate characteristics of ER gel in this study because the input and output electrodes of the clutch continuously slip past each other. While the temperature of ER gels increased when energized, the shear stress hardly changed. Instead, wearing and adaptation to the electrode affect the property. The shear rate hardly affected the shear stress in the high-shear-rate region. Conversely, the shear stress depended on the shear rate in the lower region.

1. Introduction

Electrorheological (ER) gels are a type of functional material with a surface that has controlled shear friction characteristics. We have developed slip clutches using ER gels sandwiched between electrodes, and linear actuators applying the clutch [1, 2]. The clutches can transmit the power of a motor with control provided by the magnitude of an electric field. The input and output electrodes of the clutch continuously slip past each other. Therefore, the shear rate of ER gels depends on the load.

ER fluid, another functional material, behaves like a Bingham fluid when energized. It is known that the shear stress of ER fluids depends little on the shear rate. Therefore, clutches using ER fluids are not required to control the transmitting torque in spite of variance in the shear rate.

In contrast, it is not well known how the shear stress of ER gels is affected by variation in the shear rate. When applications of ER gels require velocity or position control, the accuracy of the torque control is important for the actuator or the clutch using the gel. Therefore, the torque needs to be compensated if the gel depends on the shear rate. This paper reports on the relationship between the rotating speed and the torque of an ER gel clutch obtained experimentally.

[1] To whom correspondence should be addressed: Ken'ichi Koyanagi; Kurokawa 5180, Imizu, Toyama 939-0398, Japan; Tel: +81-766-56-7500; Fax: +81-766-56-8030; E-mail: koyanagi@pu-toyama.ac.jp
[4] Corresponding author.

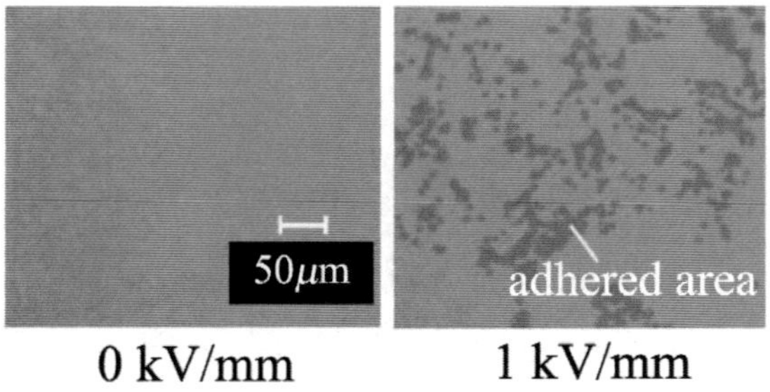

Figure 1. Adhesive effect of ER gel.

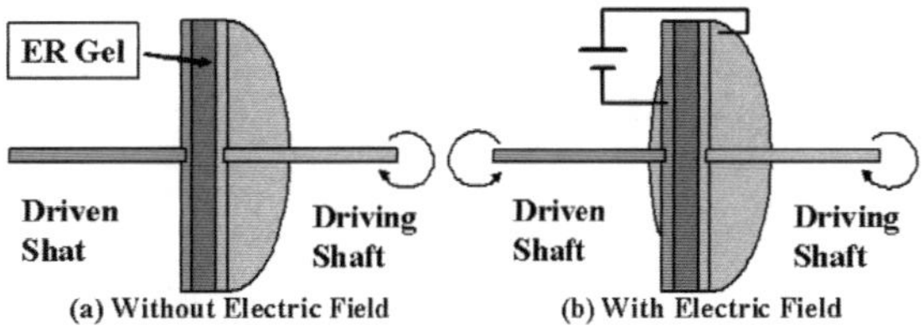

Figure 2. Image of an ER gel clutch.

Another factor critical to accurate and stable control is temperature. Because the ER gel slips continuously in the clutch, it generates heat over a long period of use. It has been reported that the shear stress of an ER gel under high temperature decreases [3]. However, the characteristics of devices such as an ER clutch remain unclear. This work therefore also investigates the temperature and wearing characteristics of an ER gel clutch that is energized for a long period.

2. ER gel and ER gel clutch

2.1. Outline of an ER gel clutch

ER gels are functional materials. Dielectric polymer particles about 20 μm in diameter are dispersed in an insulating oil, and a gelling agent is then added [4]. When an electric field is applied to electrodes sandwiching the ER gel, dielectric polarized particles and the gel component increase the shear stress of the surface of the ER gel several tenfold as shown in figure 1.

An ER gel can be applied to a working surface of a slip clutch as shown in figure 2. While the ER gel is de-energized, the clutch input force is not transferred because of minimal shear stress. When the ER gel is energized by an electric field, the increased shear stress results in a force transfer; that is, the clutch is engaged. The magnitude of the transferred force can be controlled by the magnitude of the electric field. Devices with ER gels do not require sealing because the gels are not liquid like ER fluids. The sealing is a mechanical design issue for devices that use functional fluids or powder.

2.2. ER gel drum

We developed a disk-type slip clutch using ER gel as the working surface as shown in figure 3; we refer to the system as an ER gel drum (ERGD). The inner structure of the ERGD is shown in figure 4.

The inner structure of the ERGD consists of the output disk engaging with the outside of the drum, the input disk connecting to the input shaft, and the ER gel. A sheet of ER gel is pressed on the output disk and sandwiched between the output and input disks, which stress the gel. The input shaft is rotated by the motor at constant speed. Because the shear stress of the ER gel changes according to the magnitude of the applied electric field, an arbitrary torque can be transmitted to the output of the ERGD.

The advantage of this structure is that it is easy to increase the output torque of the ERGD by changing the number of ER gels in the drum. To equip a drum with many ER gel sheets, thin input and output disks are installed; these thicknesses are 0.3 mm despite an ER gel sheet having a thickness of 0.5 mm [5]. Figure 4 is an image of eight sheets of ER gel inserted in parallel.

Figure 3. An ER gel drum.

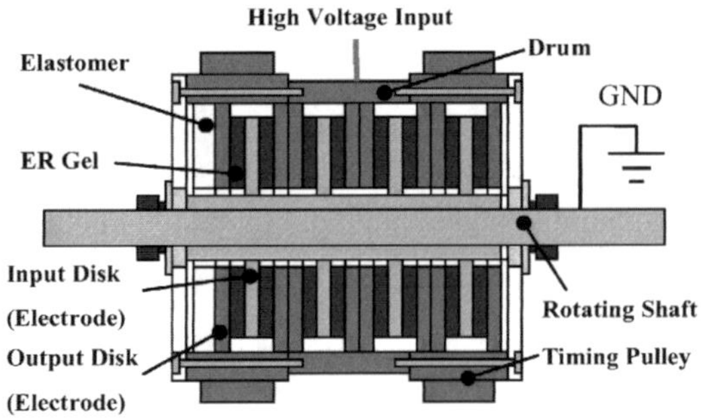

Figure 4. Structure of an ERGD.

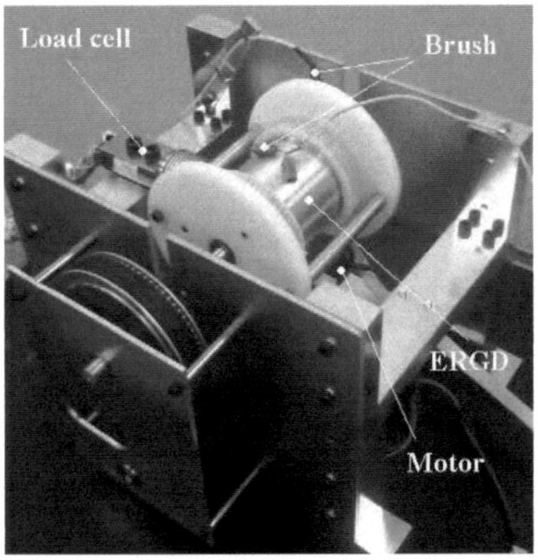

Figure 5. The ER gel drum test bed.

Figure 6. ERGD to measure temperature.

2.3. Experimental equipments

A test bed of the ERGD shown in figure 5 was used to obtain the basic characteristics of the ERGD.

The output torque of the ERGD was gauged with a load cell while the rotational velocity of the motor was controlled. The controller was constructed on a personal computer with ART-Linux as the OS. The sampling time was 1.0 ms for real-time sensing and motor control. The ERGD was supplied a high voltage up to 2.0 kV/mm by an amplifier through brushes. A 60-Hz notch filter to remove noise from the power source and a 10-kHz low-pass electrical filter were introduced before data analysis.

3. Temperature and Wearing Characteristics of ER gel

3.1. Objective and Method

It has been reported that the electro-adhesive effect of ER gel under a static condition is affected by temperature [6]. However, no experimental result has been reported for a shear condition in which the temperature tends to increase easily. In contrast, it is experientially known that the shear stress of the ER gel gradually stabilizes when the gel is used for a certain time under the shear condition. Thus, the variance of the output torque and heat generation needs to be verified for the ER gel clutch in use for a long period.

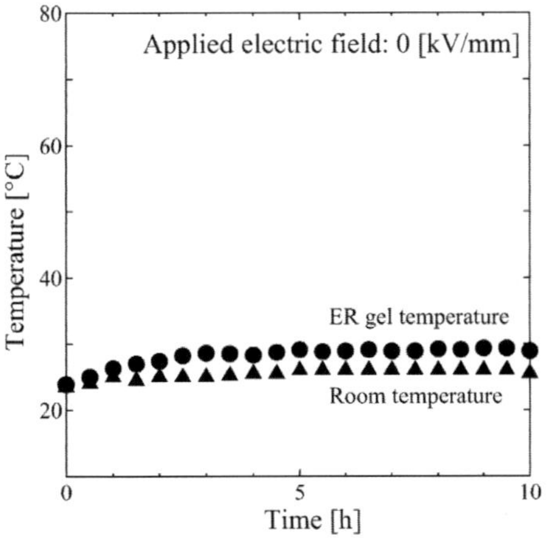

Figure 7. Temperature variation of the ERGD during 10 hours continuous testing (0 kV/mm).

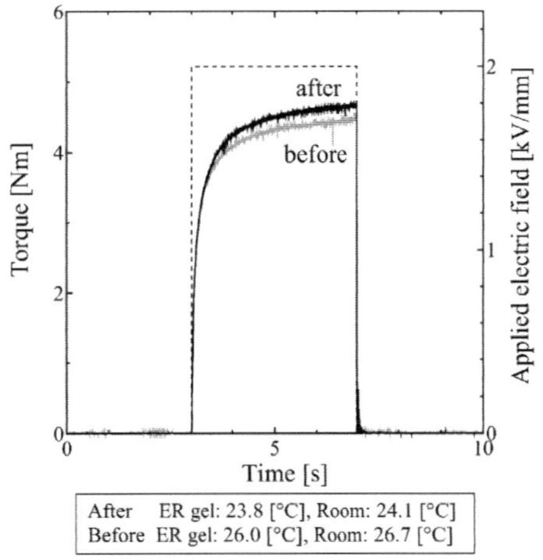

Figure 8. Step responses of torque before and after continuous testing (0 kV/mm).

Experiments to investigate the wearing and temperature characteristics of the ER gel were carried out using the test bed mentioned in 2.3. The ERGD was activated continuously for 10 hours and the output torque and temperature were measured. Two sheets of ER gel were installed in the ERGD. The magnitude of the applied electric field was 0 or 2.0 kV/mm. The rotational speed of the input of the ERGD was controlled to $\pi/2$ rad/s (15 rpm). The output torque was recorded every 200 ms.

The temperature of the ERGD was measured every 30 minutes by an infrared thermometer (FLUKE 561). Since a hole was cut in the side cover of the ERGD as shown in figure 6, the temperature of an output disk within the ERGD can be measured directly. Emissivity was supposed to be 1 in this experiment because frosted black tape was applied to the output disk.

3.2. Without an electric field

The transition of the temperature of the ER gel with no electric field applied over 10 hours is shown in figure 7. Torques of the ERGD before and after the experiment are shown in figure 8.

Figure 7 shows that the temperature only increased by several degrees during 10 hours shearing with no electric field. No wearing is seen after the experiment. The output torque increased by several percent after cooling to room temperature as shown in figure 8. This increase is due to the ER gel adapting to the input disk to some degree during the long shearing period.

3.3. With an electric field

The temperature of the ERGD with a 2.0 kV/mm electric field applied is shown in figure 9. Figure 10 shows the output torque of the ERGD during the test. Figure 11 shows the surface of an ER gel after the experiment. Torques of the ERGD before and after the experiment are shown in figure 12. Torques were measured after cooling the ER gel to room temperature in both cases.

Figure 9 shows that the temperature of the ER gel increased to around 66 °C about 2.5 hours after the start of the experiment, and then remained almost constant. The output torque of the ERGD also increased as shown in figure 10 according to the increase in temperature. Figure 12 shows that the torque remained high even after the ER gel cooled. As shown in figure 11, the

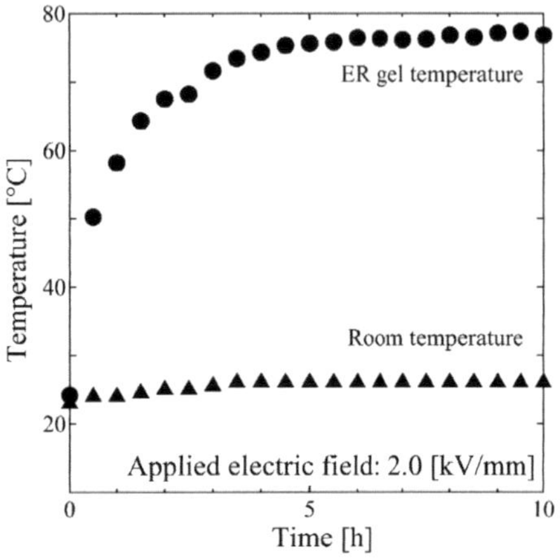

Figure 9. Temperature variation of the ERGD during 10 hours continuous testing (2.0 kV/mm).

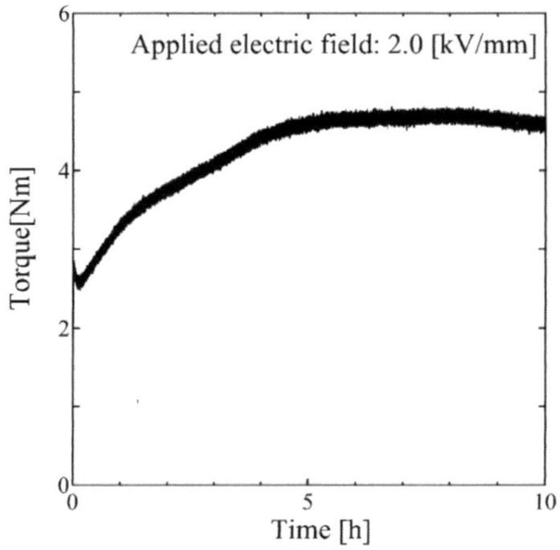

Figure 10. Torque during 10 hours continuous testing (2.0 kV/mm).

Figure 11. Surface of an ER gel after 10 hours continuous testing.

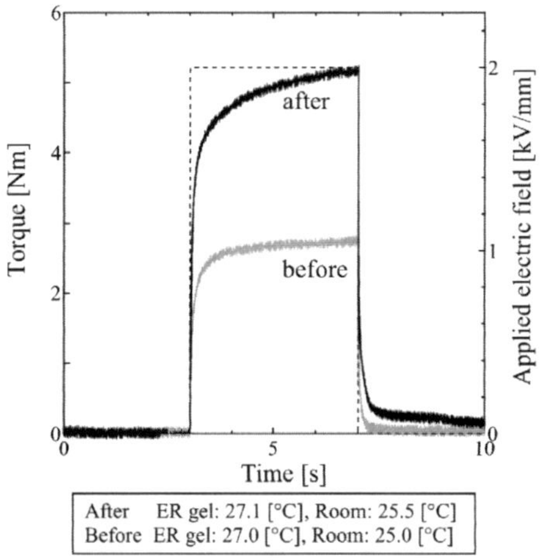

Figure 12. Step responses of torque before and after continuous testing (2.0 kV/mm).

surface of the sheared area of the ER gel seems a little abraded compared with that of the outer area of the gel, which was not in contact with the counter electrode.

3.4. Discussions on the effects of the temperature and the wearing

The reason why the output torque increased with time is considered as follows.

(i) The surface of the ER gel gradually adapted to the surface of the input disk while energized, and the contact condition of the surfaces thus improved.

(ii) The ER gel wore thin.

(iii) The shear stress of the ER gel increased according to its temperature (a variance of a physical property).

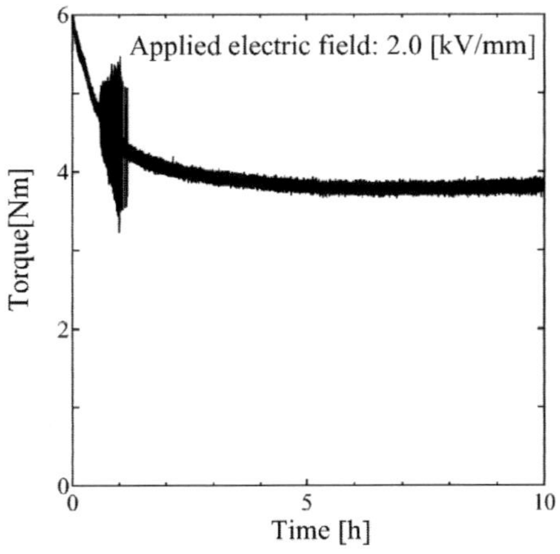

Figure 13. Torque during 10 hours continuous testing (2.0 kV/mm).

Case (i) We repeatedly confirmed that ER gels used to some degree have a high and stable shear stress characteristic. It is considered that this characteristic is due to a little wearing of the ER gel improving the geometric properties, especially the parallelism, between surfaces of the ER gel and the counter electrode, thereby improving their contact condition.

Case (ii) Large shear stress acts on the surface of the ER gel while the gel is energized. This will wear the ER gel during a long period of energizing, and the gap between electrodes will then decrease because the gel becomes thinner. The shear stress of the ER gel increases linearly as the gap reduces.
Figure 11 confirms the wearing of the surface of the ER gel over a long time period. However, a decrease in the thickness of the ER gel was not confirmed when the thickness was measured with a pair of vernier calipers (having a minimum measurement of 0.05 mm). Therefore, case (ii) cannot be considered as the main issue.

Case (iii) The construction of ER gels is such that many particles are mixed between silicon-bridging sites in the gel part, and the variances in their chemical and physical properties are not well known at high temperature. It is commonly stated that the surface friction resistance of viscoelastic materials decreases when the temperature of the materials increases above the glass transition temperature [7]. In addition, the property will revert when the temperature recovers. The shear stress of ER gels has been reported to decrease as the temperature increases under static friction conditions [3, 6]. Therefore, case (iii) also cannot be considered the main issue.

As deduced above, the main explanation for the experimental results presented in this chapter is case (i) that the improvement of the characteristics of the ER gel is due to the wearing of the gel, and not a change in gel characteristics with temperature.

The results differ from the results reported in [3], which were obtained from experiments in which the temperature was increased by heat inflow and the contact conditions between the ER gel and the electrode had yet not improved. However, our results are for the situation that the contact conditions have improved during long-duration activation; thereby, the conditions for the ER gels differ from those in [3]. This would account for the differences in experimental results.

In fact, when the same experiment with the same ER gels was conducted, the torque did not increase even as the temperature increased as shown in figure 13. While the torque in the earlier

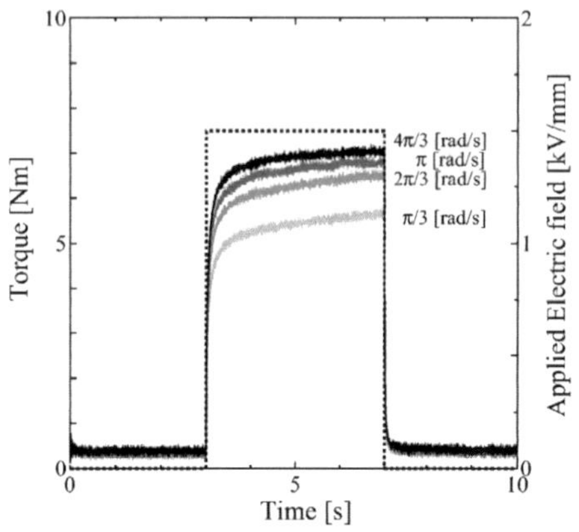

Figure 14. Step responses of torque for constant rotational speeds.

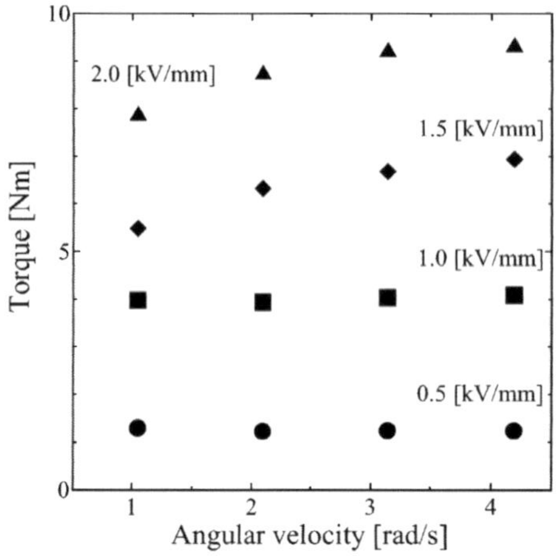

Figure 15. Torque vs. rotational speed of the ERGD.

phase of the experiment oscillated and was greater than that in figure 12, it is a future topic of investigation.

4. Shear rate characteristics of ER gel

4.1. Objective and Method

This chapter presents relations between the shear stress and shear rate of an ER gel. The number of ER gel sheets was fixed at 10. The magnitude of the applied electric field was 0.5, 1.0, 1.5 or 2.0 [kV/mm]. Each result was the average of five experimental data for the same condition.

4.2. At a constant shear rate

Torques of the ERGD were measured at constant rotation speeds of the clutch input of $\pi/3$, $2\pi/3$, π and $4\pi/3$ rad/s.

Figure 14 shows the step responses of the ERGD with a 1.5-kV/mm electric field applied. The torque increased approximately from 5 to 7 Nm. Figure 15 shows relations between the torque, rotation speed and magnitude of the electric field. A stronger electric field and higher shear rate result in larger torque. The effect, however, seems small at rotation speeds exceeding 2 rad/s.

4.3. At a variable shear rate

In the case that the rotational speed of the clutch input, or the shear rate, varied linearly, the results are shown in figure 16. Figure 17 shows the results of experiments in which the angular velocity of the input of the ERGD was varied stepwise.

Variance of the rotational speed seems to little affect the torque under 1.5 kV/mm. By contrast, high rotational speed increases the torque for an applied electric field exceeding 2.0 kV/mm.

4.4. Discussions on the effects of the shear rate

The variance in the output torque characteristics of an ER gel clutch depending on the rotational speed of the clutch input was investigated. When a weak electric field was applied, the torque was little affected by the variation in rotational speed if the speed exceeded 2 rad/s. In contrast, the torque depended on the input speed when a strong electric field was applied. Nevertheless,

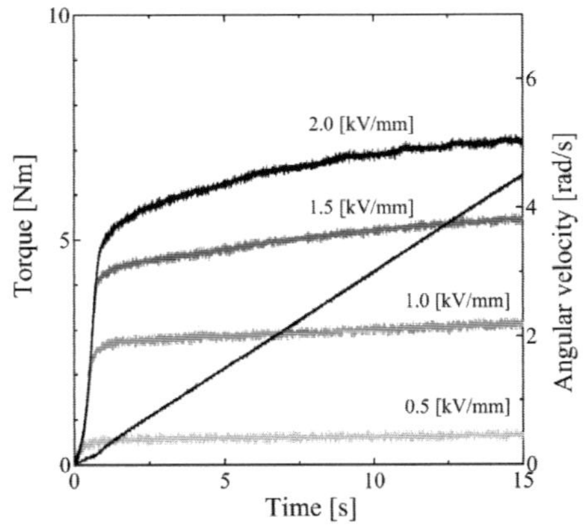

Figure 16. Step responses of torque for linearly increasing rotational speed.

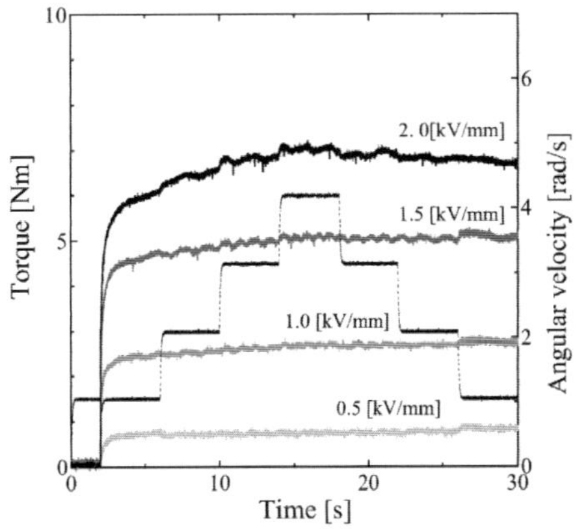

Figure 17. Step responses of torque for stepwise increasing rotational speed.

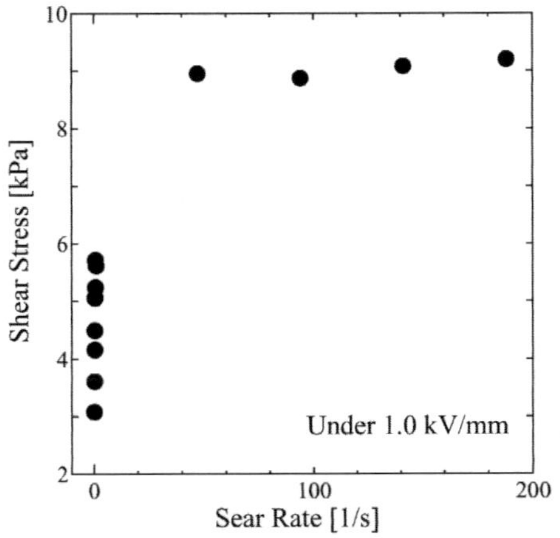

Figure 18. Relation of the shear stress and shear rate of the ER gel.

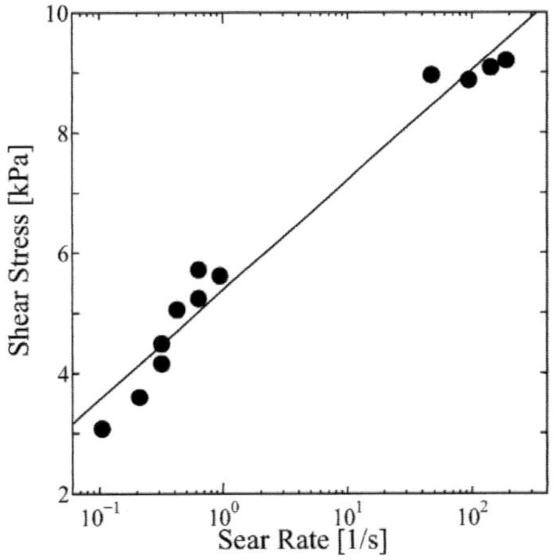

Figure 19. Relation of the shear stress and shear rate of the ER gel (semi-logarithmic).

there would be a wearing effect as mentioned in the previous chapter. Therefore, shaking down under a strong electric field for some hours might dissolve the dependence.

If results from other research on ER gels [8] are included, the characteristics of the shear rate and shear stress under a 1-kV/mm electric field are as shown in figure 18. Figure 19 shows the results plotted on a semi-logarithmic graph. The shear stress is hardly affected by the shear rate in the high-shear-rate region despite the shear stress being sensitive to variation in the shear rate in the low-shear-rate region. It would be important in the design phase to keep the shear rate high if ER gels are sheared constantly, like in this study.

5. Conclusions

This paper reported experimental research on the temperature characteristics of ER gel and their relation to wearing, and the shear rate characteristics of ER gel.

The temperature of ER gels increased while energized because they operated in a constant sheared situation. The shear stress hardly changed with increasing temperature but improved owing to wearing and adaptation. Thereby, several hours shaking down will improve the performance of ER gels.

The shear rate hardly affected the shear stress of ER gels in the high-shear-rate region exceeding 40 s^{-1}. Conversely, the shear stress depended on the shear rate in the lower region, with an increase in the shear rate increasing the shear stress. However, a high shear rate possibly affects the shear stress if a strong electric field is applied to the ER gel, which could be the focus of further investigation.

ER gels will perform well in mechatronic devices when they are used in such a shear rate region that the effects of variation in the shear rate are small after satisfactory shaking down.

Acknowledgement
This work was partially supported by MEXT/JSPS KAKENHI Grant Number 24650057.

References

[1] Koyanagi K, Kakinuma Y, Anzai H and Yamaguchi T 2010 *Advanced Robotics* **24** 1963–1981
[2] Koyanagi K, Kakinuma Y, Anzai H, Sakurai K, Yamaguchi T and Oshima T 2011 *Proceedings of the 2010 International Conference on Electro-Rheological Fluids and Magneto-Rheological Suspensions* pp 95–101
[3] Kakinuma Y, Aoyama T, Anzai H, Sakurai H, Isobe K and Tanaka K 2006 *Precision Engineering* **30** 280–287
[4] Kakinuma Y, Aoyama T, Anzai H, Sakurai H, Isobe K and Tanaka K 2005 *International Journal of Modern Physics B* **19** 1339–1345
[5] Takata Y, Koyanagi K, Kakinuma Y, Anzai H, Sakurai K, Motoyoshi T and Oshima T 2012 *Proceedings of the 2012 IEEE International Conference on Mechatronics and Automation* pp – (in press)
[6] Shida Y, Kakinuma Y, Aoyama T, Minami H and Anzai H 2011 *Proceedings of the 2010 International Conference on Electro-Rheological Fluids and Magneto-Rheological Suspensions* pp 81–87
[7] Vorvolakos K and Chaudhury M K 2003 *Langmuir* **19** 6778–6787
[8] Dohi M, Koyanagi K, Kakinuma Y, Anzai H, Sakurai K and Oshima T 2011 *Proceedings of SICE Annual Conference 2011* pp 2988–2992

Journal of Physics: Conference Series **412** (2013) 012014 doi:10.1088/1742-6596/412/1/012014

Dynamic model of the electrorheological fluid based on measurement results

K Krivenkov[1], S Ulrich and R Bruns

Institute of Machine Elements and Technical Logistics, Helmut Schmidt University, Hamburg, Germany

E-mail: konstantin.krivenkov@hsu-hh.de

Abstract. To develop modern applications for vibration decoupling based on electrorheological fluids with suitable control strategies, an appropriate mathematical model of the ERF is necessary. The devices mostly used have annular-shape electrorheological valves. This requires the use of flow channels to measure the static and dynamic properties of the electrorheological fluids in similar flow conditions. Particularly for the identification of the dynamic behavior of the fluids, the influences of the non-electrorheological properties on the overall system must be taken into account. In this contribution three types of parameters with several nonlinear dependencies for the mapping of the static and dynamic properties of the ERF are considered: electro-rheological, hydraulic and electrical. The mathematical model introduced can precisely demonstrate the static and dynamic behavior of the electrorheological fluid and can be used for the future design of real systems for vibration decoupling or other systems with high dynamic requirements.

1. Introduction

To map the dynamic properties of the ERF in a simulation program, many approaches are in existence [1-7]. Here mainly phenomenological models are used, which describe in general the behavior of the overall electrorheological system quite well. But for designing highly dynamic systems these models cannot be used. On the one hand, unrealistic fluids and operating conditions are used. On the other hand, these models do not cover all outcomes which are necessary for the design and development of highly dynamic systems [8-10].

In the present work, a mathematical model to predict the static and dynamic behavior of the commercially available electrorheological fluid is presented [11]. The breakdown of the whole system into electrorheological, electrical and hydraulic areas allows the modeling of all effects which are relevant to develop systems for vibration decoupling or other systems with high requirements to the dynamics.

The full simulation model consists of the hydraulic and electric equations with field- and pressure-dependent parameters and can predict the following behaviors of the real system: the superharmonic oscillations of pressure, the amplitude- and phase-frequency characteristics, the shape of the pressure signals with all important nonlinear phenomena and the drift of the mean pressure in the overall

[1] Author to whom any correspondence should be addressed

system. Moreover, the representation of the dynamic properties of the electrorheological fluid itself without any significant influences is possible.

2. Measurement components and dynamic parameters

To measure the static and dynamic properties of the electrorheological fluids in conditions similar to the real applications, flow channels are required [12, 13]. The static (stationary) parameters of the electrorheological fluid, like pressure drop dependent on field strength and shear rate, can be measured with only one flow channel.

However, by measuring the dynamic properties of the electrorheological fluids, the influence of the hydraulic parameters as hydraulic capacity C_H and inductance L_H of the overall system need to be considered. These parameters are dependent, for the most part, on the design of the hydraulic components of the system. To reduce this influence, two reciprocally controlled flow channels are required (figure 1) [9]. In spite of this arrangement of the active components, it cannot be assumed that the true dynamic behavior of the ERF has been measured.

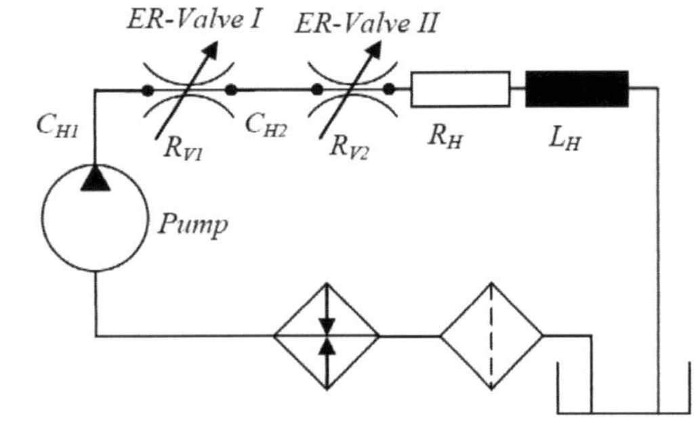

Figure 1. Schematic representation of hydraulic system for dynamic measurements and its dynamic parameters.

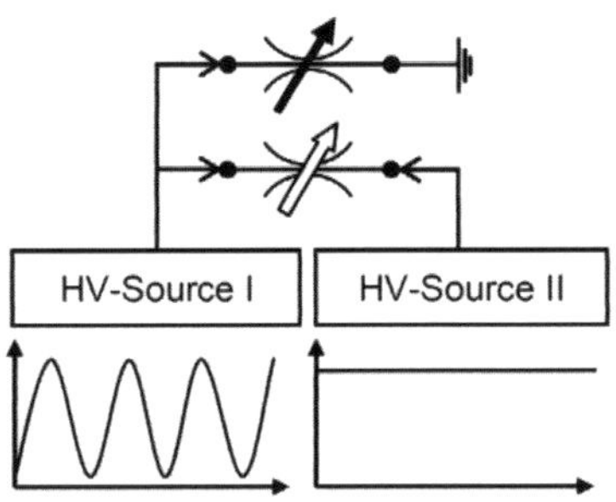

Figure 2. Connection of the high-voltage sources to the flow channels and required control signals.

For such kind of control of the electrorheological channels, two high-voltage sources are necessary. At least one HV-source must cover the dynamic of the measured system with its own dynamic characteristics. The second needs only the required electrical power which is necessary for the operation of the flow channel. The reciprocal dynamic control of the two valves is realized through the connection which couples the two channels electrically with each other. The connections and the required control signals are shown in figure 2. The required phase shift between the respective channels is realized by the phase shift in building up electric fields. Therefore, it makes use of the fact that for an electric field potential differences between the electrodes are required. Besides, the level of the voltage U_{stat} satisfies the following relationship, wherein U_0 is the activation voltage of the electrorheological fluid:

$$U_{stat} = 2U_{dyn.mean} - U_0 \tag{1}$$

As well the HV-sources as the wires, the connectors and the flow channel itself are electrically less than ideal and have a transfer function, internal resistances and capacities. It follows that, to identify the dynamic behavior of electrorheological fluids, the dynamic behavior of the hydraulic and electrical parts of the overall system must be identified first.

3. Modelling of the hydraulic system

The differential equations which describe the dynamic behavior of the hydraulic system with two reciprocally controlled flow channels form the basis of the model. The pressure $p_{1,2}$ upstream of each flow channel can be calculated from the different pressures as a result of the electrorheological effect

$p_{ER1,2} = k_{e1,2} \cdot E_{1,2}$, flow resistance $p_{R1,2} = R_{V1,2} \cdot Q_{1,2}$ and inertia of the electrorheological fluid $p_{L1,2} = L_{H1,2} \cdot \mathrm{d}Q_{1,2}/\mathrm{d}t$.

$$p_1 = k_{e1}E_1 + R_{H1}Q_1 + L_{H1}\frac{dQ_1}{dt} + p_2 \tag{2}$$

$$p_2 = k_{e2}E_2 + R_{H2}Q_2 + L_{H2}\frac{dQ_2}{dt} \tag{3}$$

The volume flow rates in each valve are different and are dependent on the compressibility of the fluid and the pressure lines. They can be determined from the continuity equation as follows:

$$Q_{1,2} = Q_{Pump,1} - C_{H1,2}\frac{dp_{1,2}}{dt} \tag{4}$$

To fill these equations electrorheological and hydraulic parameters are needed. The static electrorheological properties of the ERF, hydraulic resistance $R_{H1,2} = \mathrm{d}p/\mathrm{d}Q_{1,2}$ and linearization factor of the pressure drop $k_{e1,2} = \mathrm{d}p/\mathrm{d}E_{1,2}$, can be implemented in the model as nonlinear functions of the field strength and the volume flow rate (figure 3 (a), (b)). For this purpose, a modified Carreau-Yasuda equation can be used [14]:

$$\frac{\eta_s}{\eta_B} = 1 + \frac{(k_c \chi^{m_c} - \eta_B)}{\eta_B}\left[1 + (\lambda_c \chi)^{\beta}\right]^{\frac{n_c - 1}{\beta}} \tag{5}$$

The hydraulic parameter of the system is the compression modulus K_{ERF} of the electrorheological fluid. In comparison to conventional hydraulic fluids the electrorheological fluid has substantially lower compression modulus. Especially in the region of low pressures, where the flow channels are usually operated, it should be represented as exactly as possible (figure 3 (c)). The measurement of the compression modulus and its implementation in the model allows a much better calculation of the pressure and the compression flow rates in the overall system.

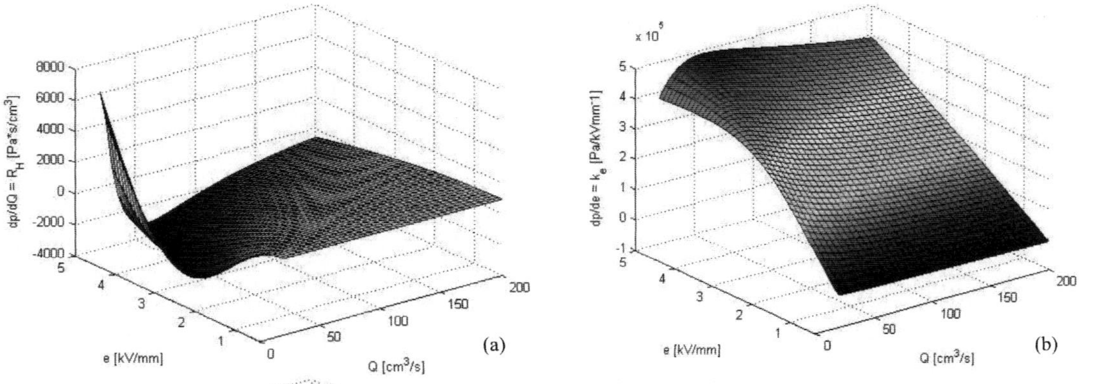

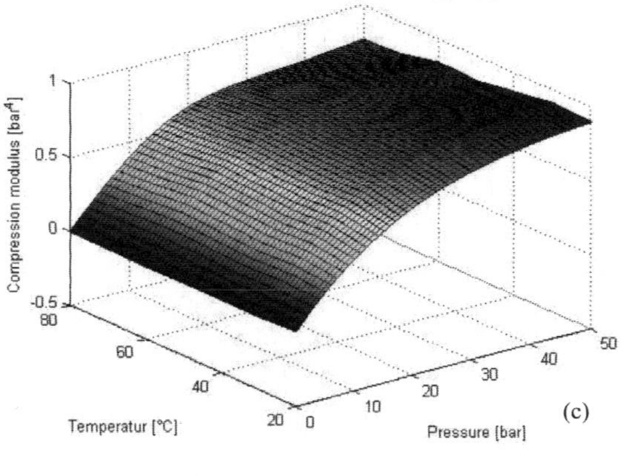

Table 1. Numerical values for equation (5)

$\eta_B\ (Pa\,s)$	0.027
$k_c\ (Pa\,s^{m_c+1} kV^{\alpha m_c} mm^{-\alpha m_c})$	314.1
$m_c\ (-)$	-0.893
$\lambda_c\ (s\,kV^{\alpha} mm^{-\alpha})$	1.745e-4
$\beta\ (-)$	0.677
$n_c\ (-)$	3.831e-8
$\alpha\ (-)$	1.5

Figure 3. Static parameters of a hydraulic system. Hydraulic resistance R_H (a), linearization factor k_e (b) at 60°C and compression modulus K_{ERF} (c).

The details of the geometric parameters of the flow channels, operating temperature, density and the basic viscosity of the fluid complete the hydraulic model, thus allowing calculation of the pressure drop in the channels as a function of the applied field strength, shear rate, frequency and operating temperature.

4. Modelling of the electrical system

Because of the strong coupling of the hydraulic and electrical parameters of the system, the derived hydraulic model with implemented nonlinear hydraulic and electrorheological parameters must be extended with equations to map the electrical properties of the overall system.

To describe the electrical properties, the system can be further subdivided into static and dynamic. The static properties describe the well-known, almost quadratic, dependence of the electrical current on the applied electric field strength [15]. This nonlinear static behavior of current-voltage characteristics is independent from the shear rate in the flow channel. Hence, the only differential electrical resistance $R_{el1,2} = dU_{1,2}/dI_{1,2}$, which is dependent only on the applied electrical field, can be derived. In this case Ohm's law has no validity.

To describe the dynamics of the electrical system the well-known phenomenological model of a dielectric under the influence of electric fields is used (figure 4). This model allows the calculation of important electric parameters like the capacity or loss factor of the system and can be mathematically defined as follows:

$$Z_{1,2} = \frac{U_{1,2}}{I_{g1,2}} = \frac{R_{el1,2}}{j\omega C_{el1,2} R_{el1,2} + 1} \tag{6}$$

Moreover, the frequency response of the introduced static electrical parameters at the operating frequency can be derived. For this purpose the impedance, which is dependent on the working frequency of the system, is determined from the measured current-voltage relationship (figure 5). However, this measurement requires a separate control of the flow channels.

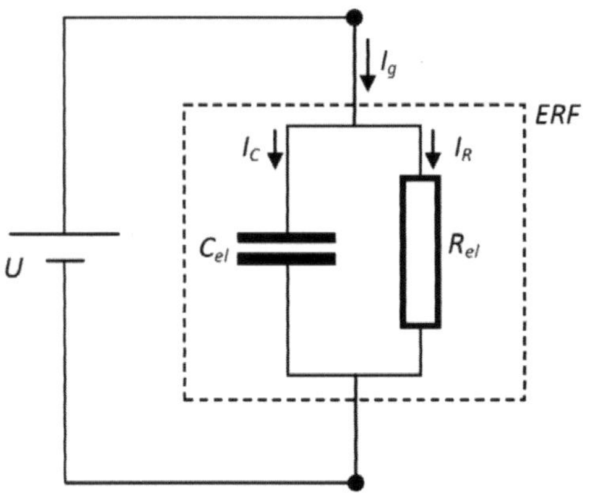

Figure 4. Phenomenological model of a dielectric with parallel connection of the electrical capacitance C_{el} and resistance R_{el}.

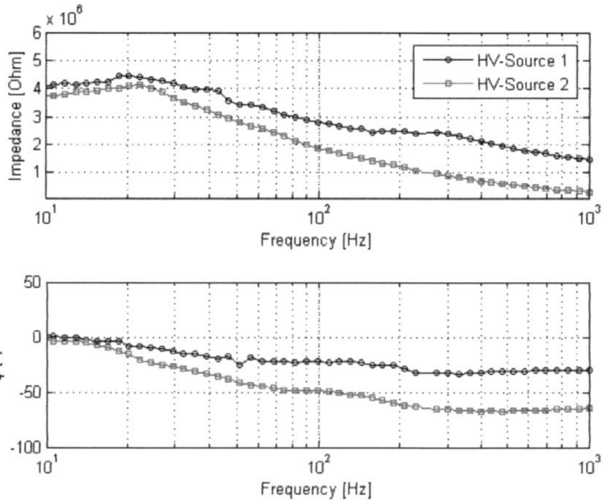

Figure 5. Measured frequency response of the impedance of the overall electrical system (HV-source – connections – flow channel - ERF).

The amplitude of the impedance has a nonlinear dependence on the strength of the applied electric field. It corresponds at very low operating frequencies to the values of the static electrical resistance. This allows the use of the Hammerstein model for the exact mapping of the measured results. The non-linear dynamic behavior can be described as a structure which consists of a series connection of a static non-linearity with the linear time-invariant transfer function. This transfer function corresponds to the first order system with lumped parameters. Therefore, the dynamic behavior of the electrical system with electrorheological fluid can be defined with a nonlinear Debye-like model.

5. Coupling of hydraulic and electrical model

In the presented models the mapping of the nonlinear static and the dynamic behavior of each system is performed by the addition of linear differential equations with required nonlinear parameters. The subsystems contain the electrorheological, hydraulic and electrical parameters with corresponding dependencies on the strength of the applied electric field, the volume flow rate (shear rate), the operating frequency and temperature. To obtain an overall dynamic model, the individual models will be coupled to each other.

Theoretical and experimental studies have shown that the overall system can be represented as a series connection of the electrical and hydraulic subsystems. It was found that the negative effects at higher frequencies are caused by non-measurable electrical interactions within the electrorheological fluid. For this reason a fictive voltage is introduced, which is responsible for the effective viscosity change in the fluid.

This voltage is also a connecting link between the two introduced subsystems. The electrical system is excited with a voltage from the high voltage source. In the existing electrical model the electric currents flowing through the electrorheological fluid can be determined. These currents are used for the excitation of the fictive auxiliary system to calculate a fictively effective voltage. The fictive auxiliary system consists of a parallel arrangement of a capacitor and electrical resistance. The values of these parameters are identical to the calculated quantities of the electrical subsystem. The fictively effective voltage is used for the excitation of the hydraulic subsystem. The resulting overall dynamic model is shown in figure 6.

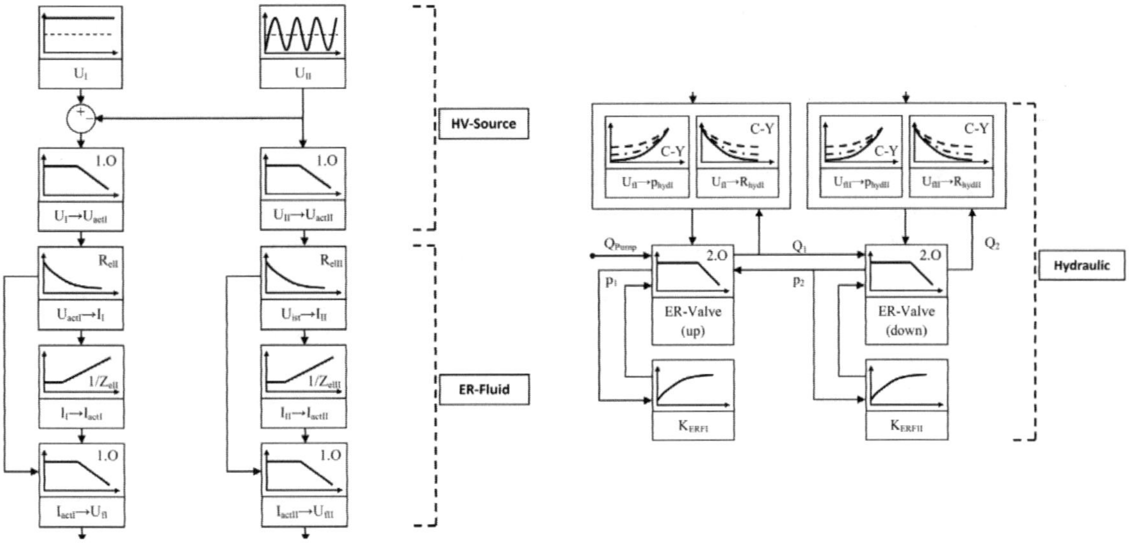

Figure 6. Schematic representation of the derived overall dynamic model.

The dynamic model of the high voltage source describes the behavior of the input voltage at the electrodes. In using high-voltage sources, the dynamics of the voltage build-up in the electrically active gap of the flow channels is load dependent. The higher the amplitude of high voltage, the slower is its rise and fall times. Consequently, this decreases the frequency bandwidth wherein the full amplitude of the voltage can be generated.

In addition, the entire system was provided with the characteristics of the pump and the associated control algorithm for keeping the flow rate constant at each operating point of the flow channels.

6. Results and discussion

The experimental data must now be compared with the simulation results of the new dynamic model. Firstly, for the design of highly dynamic systems the amplitude and phase response of the system is important. Figure 7 shows the measured and simulated dynamic behavior of electrorheological fluid in a frequency bandwidth of 10...1000 Hz. The excitation of the system is performed by a dynamic voltage of U_{II} = 2.5±2.3 kV/mm and static voltage of U_I = 4.8 kV/mm. The two flow channels are geometrically identical and are 100 mm long and 30 mm wide. The height of the electrically active gap is 1 mm. The flow rate is 5 L/min and kept constant by an integrated control algorithm. This corresponds to a shear rate of 16.7×10+3 s^{-1}.

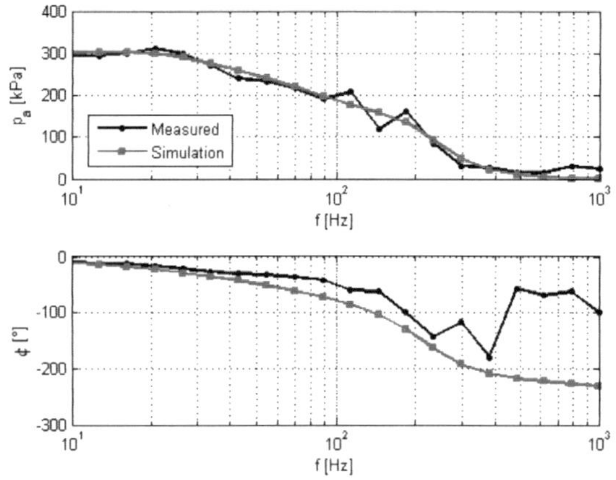

Figure 7. Measured and simulated frequency response of the electrorheological fluid.

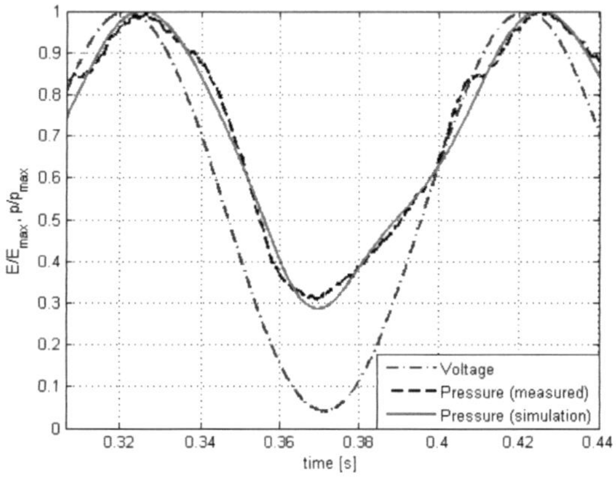

Figure 8. Time plot of the measured and simulated pressure.

The relatively large deviation between the measured and simulated frequency response occurs only at very high frequencies. In bandwidth, which can be used effectively, the mapping quality is about 90%. These deviations are due to the strong pulsations of the pump, which are relatively difficult to reproduce in the model.

Secondly, the time response of the pressure increase needs to be considered. As shown in figure 8, the pressure in the system increases about three milliseconds earlier than the voltage, which should theoretically be responsible for the increase in pressure. This effect is caused by the characteristic phase shift between the electrical voltage and current. It follows that not only the voltage, but also the electric current, which is passed through the fluid, is responsible for the increase in pressure. The simulation can reproduce with high accuracy the effect due to the presented auxiliary model and fictively effective voltage.

Furthermore, the model can represent with high accuracy the super harmonic vibrations of pressure in the flow channels. The problem of mapping the descending mean pressure at constant mean values of the voltage at higher frequencies is achieved with the new model.

7. Conclusions

The presented full-simulation model consists of hydraulic and electric equations with the most important process variables-dependent parameters and can predict the static and dynamic behavior of the electrorheological system. The effects, which are important for the design of highly dynamic systems and corresponding control algorithms, can be accurately predicted with the introduced model. These include the superharmonic oscillations of the pressure, the amplitude- and phase-frequency characteristics, the negative phase shift between pressure and voltage and the drift of the mean pressure in the system. Moreover, the representation of the dynamic properties of the electrorheological fluid itself is possible without almost any influences.

Acknowledgments
This research is supported by the Bundeswehr Technical Centre for Ships and Naval Weapons in Eckernförde, Germany.

References
[1] Bonnecaze R T and Brady J F 1992 Dynamic simulation of an electrorheological fluid *J. Chem. Phys.* **96** 2183-2202
[2] Kamath G M and Wereley N M 1997 A nonlinear viscoelastic-plastic model for electrorheological fluids *J. Intel. Mat. Syst. Str.* **6** 351-359
[3] Koyanagi K and Terada T 2010 Time response model of er fluids for precision control of motors *J. Intel. Mat. Syst. Str.* **21** 1517-1522
[4] Peel D J, Stanway R and Bullough W A 1996 Dynamic modelling of an ER vibration damper for vehicle suspension applications *J. Intel. Mat. Syst. Str.* **5** 591-606
[5] See H 1999 Advances in modelling the mechanisms nad rheology of electrorheological fluids *Korea-Aust. Rheol. J.* **11** 169-195
[6] Whittle M, Firoozian R, Peel D J and Bullough W A 1994 Decomposition of the pressure response in an ER valve control system *J. Intel. Mat. Syst. Str.* **5** 105-111
[7] Yen W S and Achorn P J 1991 A study of the dynamic behavior of an electrorheological fluid *J. Rheol.* **35** 1375-1384
[8] Gamota D R and Filisko F E 1992 Linear/nonlinear mechanical properties of electrorheological materials *Int. J. Mod. Phys. B* **6** 2595-2607
[9] Gurka M, Petricevic R, Schneider S and Ulrich S 2011 Characterization of step response time and bandwidth of electrorheological fluids *J. Intel. Mat. Syst. Str.* **22** 1745–1748
[10] Klingenberg D J 1998 Particle polarization and nonlinear effects in electrorheological suspensions *MRS Bulletin* **23** 30-34
[11] Gurka M, Johnston L and Petricevic R 2010 New electrorheological fluids—characteristics and implementation in industrial and mobile applications *J. Intel. Mat. Syst. Str.* **21** 1531–1535
[12] Brooks D A 1992 Correlation of measurement techniques in electro-rheological fluids *Theoretical and Applied Rheology* **2** 763-765
[13] Wunderlich T and Brunn P O 2000 Pressure drop measurements inside a flat channel – with flush mountes and protruding electrodes of variable length – using an electrorheological fluid *Exp. Fluids* **28** 455-461
[14] Ulrich S, Böhme G and Bruns R 2009 Measuring the response time and static rheological properties of electrorheological fluids with regard to the design of valves and their controllers *J. Phys.: Conf. Ser.* **149** 012031
[15] Conrad H and Sprecher A F 1991 Charakteristics and mechanisms of electrorheological fluids *J. Stat. Phys.* **64** 1073-1091

Journal of Physics: Conference Series **412** (2013) 012015 doi:10.1088/1742-6596/412/1/012015

Elastoviscoplastic behavior model of electrorheological fluids in various deformation modes

E V Korobko[1], M A Zhurauski and A A Makhaniok

A.V. Luikov Heat and Mass Transfer Institute of NAS Belarus, 220072,
15 P. Brovka str., Minsk, Belarus

E-mail: evkorobko@gmail.com

Abstract. The results of investigations of rheological properties of electrorheological fluids in various deformation modes: oscillatory, continuous flow, constant applied loading – are presented in the article. A generalized rheological model on the basis of a consecutive combination of viscoplastic, elastic and two viscoelastic elements with differing relaxation times, taking into account time dependence of retarded elastic deformation, is suggested.

1. Introduction

Numerous rheological investigations in different deformation conditions of the broad spectrum of electrorheological fluids (ERF) compositions with the components having the various chemical nature and structural characteristics have revealed distinctive features of mechanical behaviour, characteristic for the materials possessing properties both of Newton fluids, and quasi-solid bodies. Rheological behaviour of ERF in stationary or quasi-stationary conditions under the influence of large viscoplastic strains is most studied. In connection with complexity of the structural phenomena the elementary linear and power phenomenological models are applied to their description [1 – 5]. They do not allow to show real relaxation processes at change of ERF properties under the influence of shear forces in time, that excludes possibility of their adequate application in non-stationary dynamic conditions. Now there is no the conventional theory of ERF deformation based on basic researches and considering rheological, physical and strength properties of its structural skeleton and layers of dispersion medium or systems as a whole on macro- and micro-levels.

Depending on type of the structure created by the field, its pre-history, loading and additional external conditions (temperature), elastic, viscous or plastic properties can prevail in ERF. Features of process of deformation will be defined by a complex of characteristics of structural elements of a material: their arrangement in the bulk, character of their molecular and polarization interaction, distribution of the internal stresses depending on presence of heterogeneities, etc. Character of deformation is defined also by time of action of force and the electric field in comparison with time relaxation of medium. Taking into account the analysis of relaxation phenomena for various elastoviscoplastic materials the modeling analysis which is based on comparison of deformation behavior of real bodies and idealized mechanical models is used. As a whole, ERF is among complex structural materials and therefore they can be presented in the form of multi-element rheological elastoviscoplastic models consisting of four and more elements.

[1]Corresponding author.

The choice of adequate model by means of mathematical modeling of the mode of deformation of medium is based on use of results of rheological tests of specific compositions of ERF which allow to reveal the main features of mechanical behavior of medium. The object of present work is construction of the generalized model of viscoelastic behaviour ERF at various deformations on the basis of measurement of viscoelastic characteristics and the time mechanical response in non-stationary shear conditions.

2. Materials and equipment

The dispersion of goethite (goethite is iron needle ore α-FeOOH; rhombic crystal system, rhombobipyramidal kind of a symmetry) with content of a solid component 58 wt.% with the addition of 0.6 wt.% of a surfactant (monooleate of glycerin) was used for investigation of rheological properties. Transformer oil served as dispersion medium.

Rotational rheometer "Physica MCR301" by Anton Paar with measuring cell of plate – plate type with diameter of plates 50 mm and gap between plates of 1 mm and cell of coaxial cylinders type (diameter of inner cylinder is 26.7 mm, of outer one is 28.9 mm, gap between cylinders is 1.13 mm) was used for rheological experiments.

3. Experimental technique

Tests of sample were carried out in following deformation modes:

1) harmonic oscillations at fixed angular frequency ω=6.283 s^{-1} (f=1 Hz) with a variation of amplitude of oscillations from 0.01 to 100 % that has allowed to define components of complex shear modulus (storage modulus G' and loss modulus G'') depending on deformation amplitude, as well as limits of linear viscoelasticity area;

2) harmonic oscillations in a range of angular frequencies ω from 0.314 to 314 s^{-1} (a range of frequencies f from 0.05 to 50 Hz) at small amplitudes (0.1 %) that corresponds to linear viscoelasticity area;

3) constant shear rate in a range from 0.00001 to 1000 s^{-1}, that allowed to get flow curves on the measured values of shear stress;

4) tests for creep and relaxation: constant shear stress τ was applied, it was maintained some time, then τ=0 was established, dependence of strain ε on time was fixed (before shear the electric field with the same intensity as at shear was applied to ERF during 5 min for completion of processes of internal structure realignment). Electric field intensity E varied in range up to 2 kV/mm.

4. Results and discussion

ERF on a goethite basis under investigation show elastic properties in the range of strains to 1 – 2 % without electric field, at larger strains a transition into viscoplastic condition occurs. In the electric field the elastic limit is shifted to area of 0.1 %, the range of resiliency decreases by order, but at that the elasticity modulus increases more than by 3 order (to 1.2 MPa) at applying of electric field of 2 kV/mm (figures 1, 2). The structure becomes more rigid, however level of strain for transition to irreversible flow decreases. The components of complex shear modulus practically do not depend on deformation frequency in the field in the range of frequencies 0.05 – 50 Hz. In the absence of electric field loss modulus grows with frequency (from 27 to 250 Pa at E = 2 kV/mm).

Examples of creep curves are given in figures 3, 4. Initially there is an instantaneous elastic deformation, further deformation increases to some value, after loading removal there is an elastic restoration on value of instant deformation and relaxation process begins. After some time interval deformation decreases to residual value.

Rheological behavior of ERF can be described by complex model (figure 5), containing two viscoelastic elements of Kelvin – Voigt, having essentially differing characteristic relaxation times. This generalized model is seems to the most preferable to the analysis though the issue about kind of rheological model cannot be considered as closed, and the made choice as strictly founded. Further we check its adequacy to experimental data.

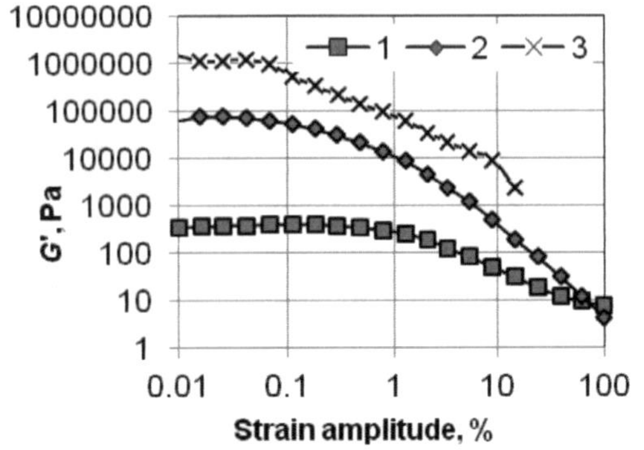

Figure 1. Storage modulus of ERF: 1 – E=0, 2 – E=0.5 kV/mm, 3 – E=2 kV/mm.

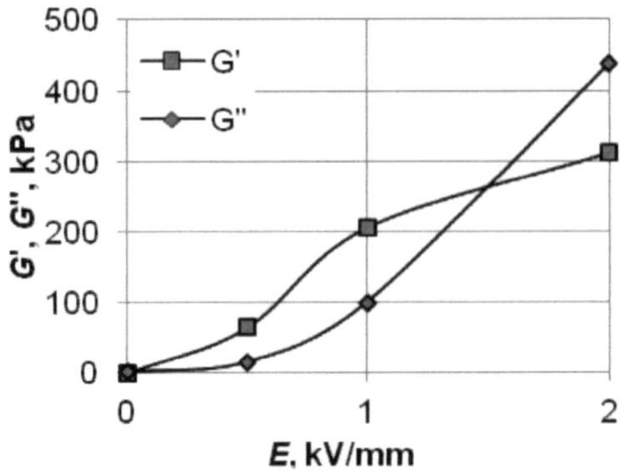

Figure 2. Dependence of storage modulus and loss modulus of ERF on electric field intensity.

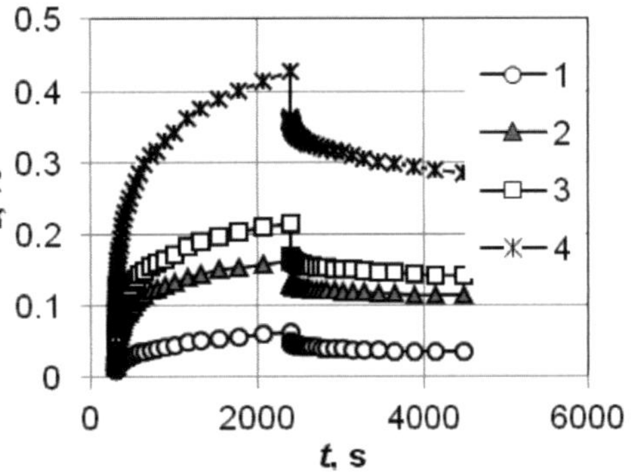

Figure 3. Dependence of ERF deformations on time at constant shear stresses (E=1 kV/mm): 1 – τ=40 Pa, 2 – 100, 3 – 150, 4 – 200.

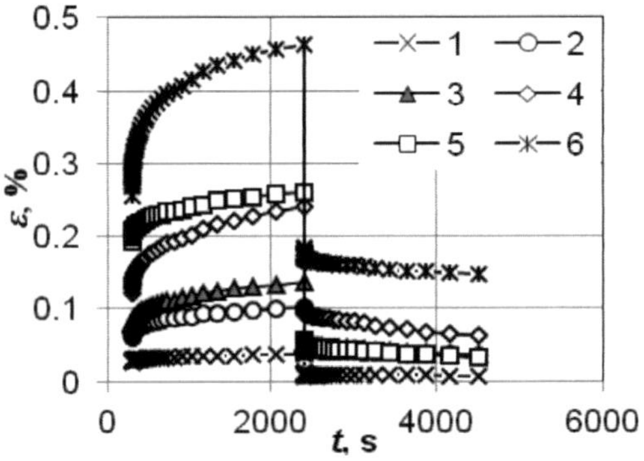

Figure 4. Dependence of ERF deformations on time at constant shear stresses (E=2 kV/mm): 1 – τ=100 Pa, 2 – 200, 3 – 300, 4 – 500, 5 – 750, 6 – 1000.

In the context of present work we confine oneself to research of only viscoelastic properties of ERF at various, but constant in time shear stresses and electric field intensities. The choice of values of shear stress was made so that they did not exceed yield stress for given level E. For determination of value $\tau_0(E)$ the experimental technique of linear growth of shear stress was used.

The received dependence $\tau_0(E)$ is shown in figure 6 and well approximated by exponential dependence (a continuous line)

$$\tau_0 = \tau_{00} \exp(-E / E_0), \tag{1}$$

at values of τ_{00} = 8.93 Pa, E_0 = 0.35 kV/mm.

Appreciable more precise definition of $\tau_0(E)$ at E >1.5 kV/mm is provided with small adjustment of dependence (1) in the form of

$$\tau_0 = \tau_{00} \exp\left(- \frac{E / E_0}{1 + E / E_1} \right) \tag{2}$$

at values of τ_{00} = 8.93 Pa, E_0 = 0.35 kV/mm, E_1 = 70 kV/mm (hatch line in figure 6).

Approximately and indirectly about structure of rheological model we can judge on correlation between values of relative deformation at loading stage $\varepsilon^{I}(t)$ and unloading stage $\varepsilon^{II}(\Delta t)$, where

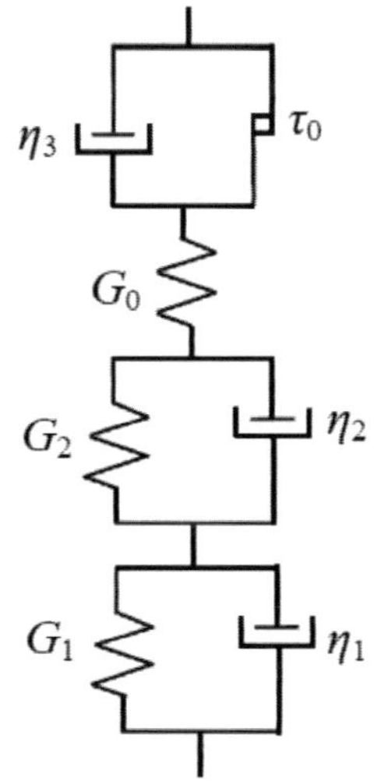

Figure 5. Rheological model of elastoviscoplastic behaviour of ERF.

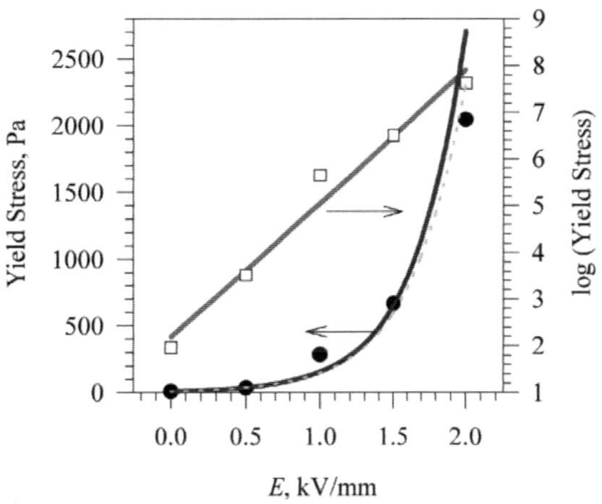

Figure 6. Dependence of yield stress on the electric field intensity.

$\Delta t = t - t_u$, t_u is absolute value of the moment of time at which loading was removed (unloading and retardation of system have begun). In figures 7(a)-(d), corresponding to various levels of loading, the correlation between $\varepsilon^{II}(\Delta t)$ and $\varepsilon^{I}(t)$ at $\Delta t \equiv t$ is illustrated by insets in each figure. With increase in value of constant loading relation between $\varepsilon^{II}(\Delta t)$ and $\varepsilon^{I}(t)$ at $\Delta t \equiv t$ comes nearer to linear as though at the larger τ action only of one element of Kelvin – Voigt starts to prevail. Continuous lines in figure 7 show result of approximation of experimental data with generalized (two-tier) model of Kelvin – Voigt at τ=const, i. e.

$$\varepsilon^{I}(t) = \varepsilon_0 + \varepsilon_1(1 - e^{-t/\Theta_1}) + \varepsilon_2(1 - e^{-t/\Theta_2}), \tag{3}$$

$$\varepsilon^{II}(\Delta t) = \varepsilon_{u1} e^{-\Delta t/\Theta_1} + \varepsilon_{u2} e^{-\Delta t/\Theta_2}, \tag{4}$$

$$\varepsilon_k = \frac{\tau}{G_k},\ k=0,1,2;\quad \Theta_k = \frac{\eta_k}{G_k};\quad \varepsilon_{uk} = (1 - e^{-t_u/\Theta_k}),\ k=1,2.$$

Adjustment of calculated values of relative deformation to the measured values was carried out by minimization of function

$$F(\varepsilon_0, \varepsilon_1, \varepsilon_2, \Theta_1, \Theta_2) = \sum_n [\varepsilon_{\exp}(t_n) - \varepsilon_{calc}(t_n)]^2 = \min.$$

Here indexes *exp* and *calc* designate accordingly measured and calculated values of deformations. Adjustment to experimental data was carried out on all curve of deformation. By separate consideration of curves $\varepsilon^{II}(\Delta t)$ and $\varepsilon^{I}(t)$ received parameters of model (3) – (4) would be different for each stage.

Results of data treatment without field are presented in figures 8 and 9. We will note approximate proportionality between ε_k and τ, argumentative of weak dependence of mechanical properties of ERF on value of shear stress at small deformations in the absence of electric field.

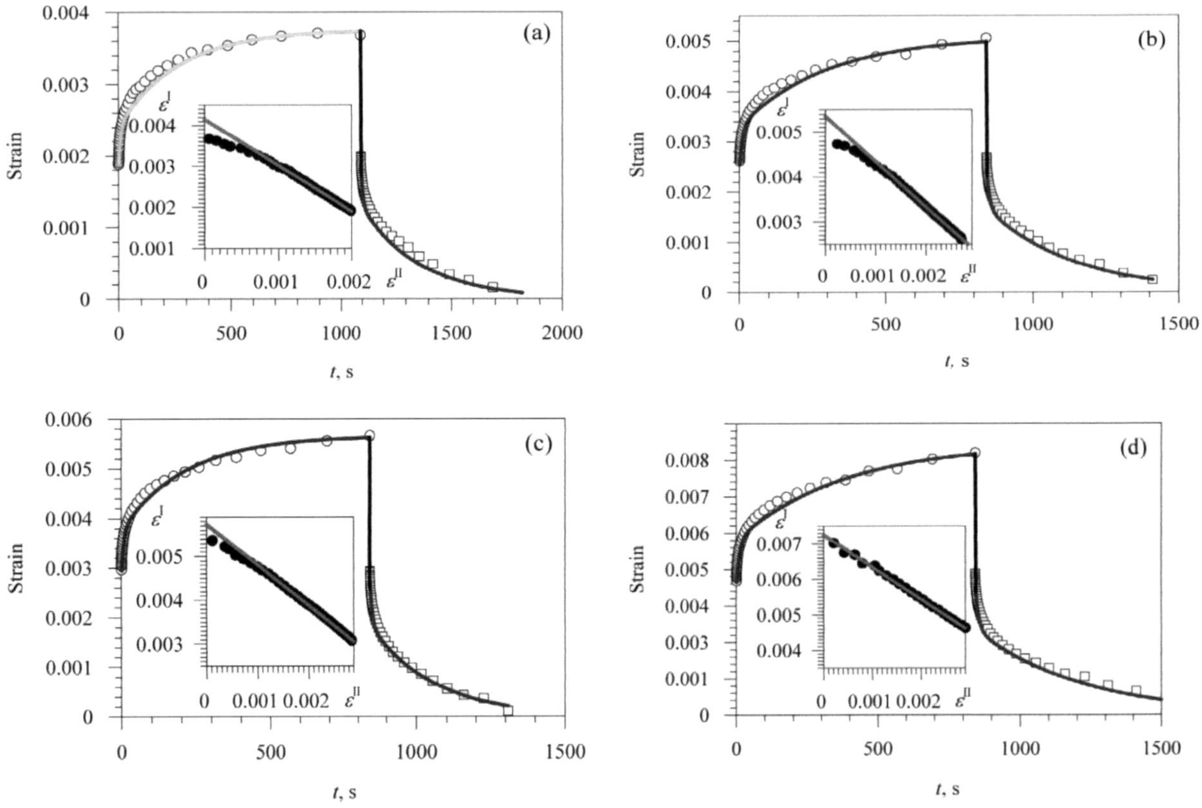

Figure 7. Deformation curves of ERF at different values of shear stress in the absence of electric field: (a) $\tau = 1$ Pa, (b) 1.5, (c) 2 and (d) 2.5.

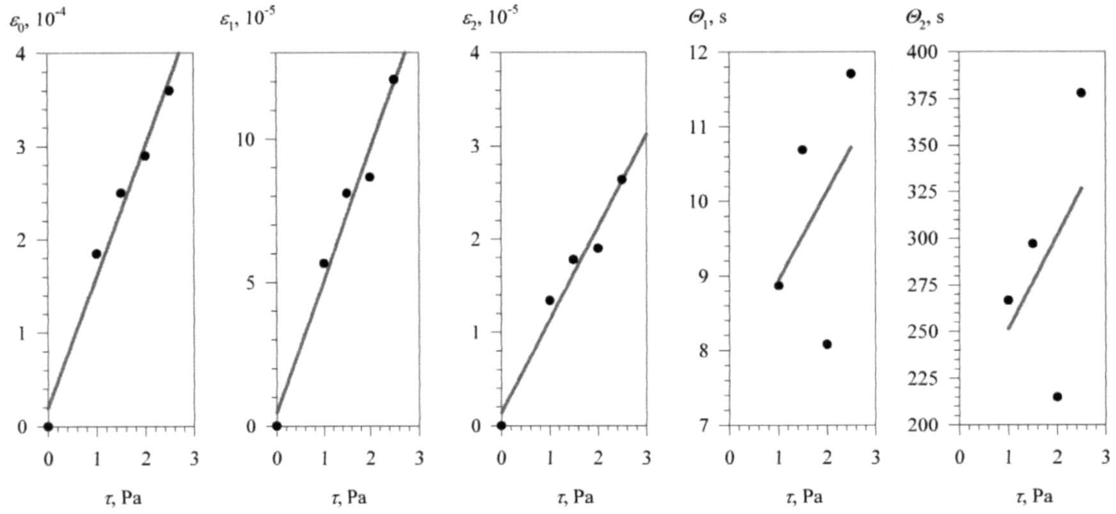

Figure 8. Dependence of parameters of model (3) – (4) on shear stress at $E = 0$ kV/mm.

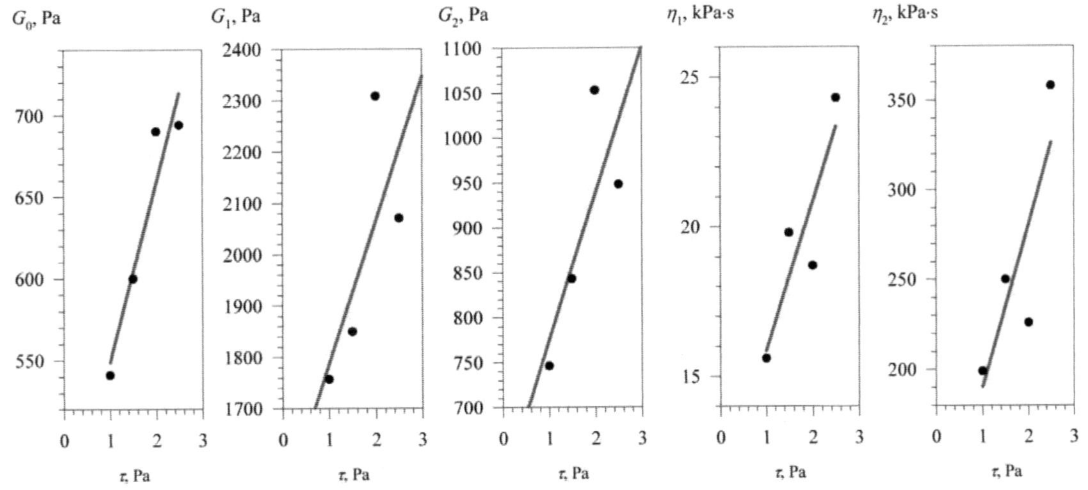

Figure 9. Viscoelastic characteristics of ERF at E=0 according to model (3) – (4).

Average values of mechanical constants are equal 631, 1997, and 898 Pa for G_0, G_1 и G_2, and also 19.6 and 258 kPa·s for η_1 and η_2 with the maximum relative deviations from average value 17, 14, 20, 26, and 30%.

In the presence of electric field it is possible to observe both quantitative and qualitative changes in behavior of deformation curves. In case of independence of mechanical properties on τ all curves $s(E,\tau,t)=\varepsilon/\tau$ at the same level of intensity of field E and different τ must coincide (as it follows from (3)–(4)). However it does not occur. It is possible to note only essential rapprochement between curves $s(E,\tau,t)$ at increase in intensity of electric field to 2 kV/mm. Among the considered levels of intensity of electric field only the data at E=2 kV/mm are close to a condition of a constancy of mechanical properties of ERF in time. Expressions (3) and (4) can be right only for this field intensity. We give the further analysis of curves of deformation for this reason only for case E=2 kV/mm.

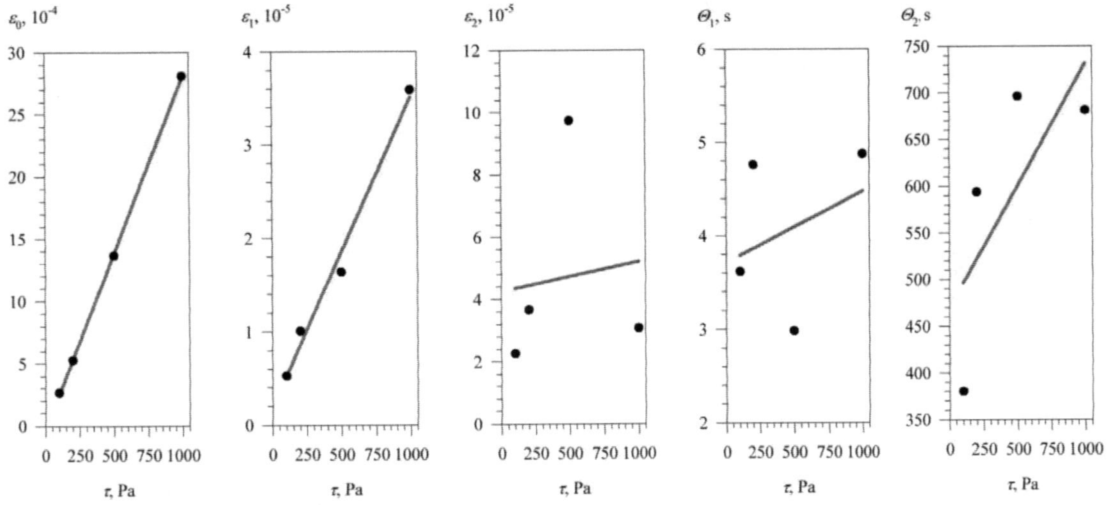

Figure 10. Dependence of parameters of model (3) – (4) on shear stress at E = 2 kV/mm.

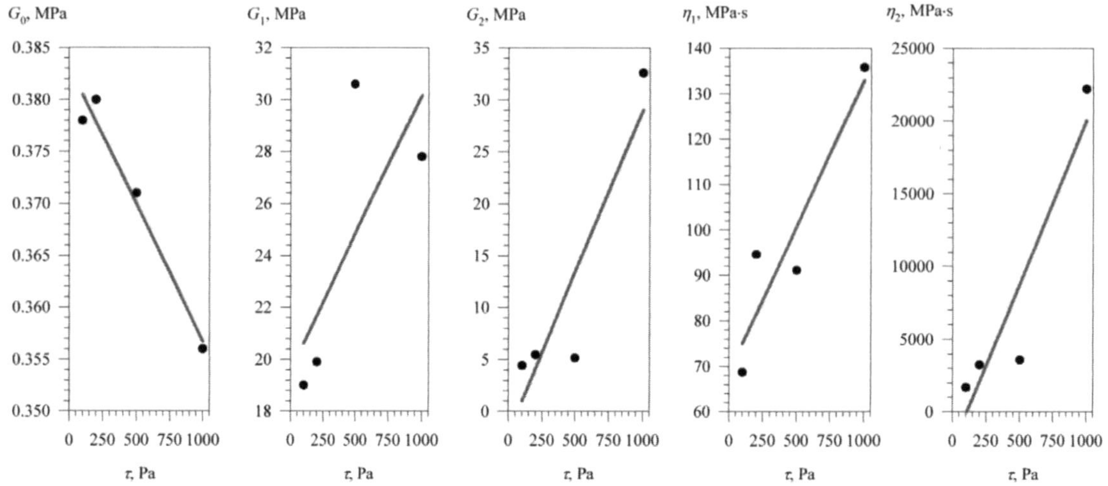

Figure 11. Viscoelastic characteristics of ERF at E=2 kV/mm according to model (3) – (4).

Final results of treatment of experimental data are illustrated in figures 10 and 11. They show that with increase of loading at a set value of electric field intensity rigidity of Hook element increases, and rigidity of units of model Kelvin – Voigt decreases. Internal viscosity of these units with growth of τ also increases.

5. Conclusion

Thus, on the basis of the model analysis it is suggested multi-element composite rheological model for calculation of mechanical behaviour of electrorheological fluids, considering a complex of structural properties and loading conditions. Its further development will allow to establish correlation between structural-mechanical properties (elasticity, viscosity, plasticity) and composition, physical and chemical structure of electrorheological fluids in the external electric field, will make possible calculation of flow of electrically controlled media in technical devices (the adaptive shock-absorbers, fixing devices).

Acknowledgment

The authors acknowledge Belarusian Fund for Fundamental Research for financial support of investigations (Project No. T10R-045).

References

[1] Klingenberg D J and Zukoski C F 1990 *Langmuir* **6** (1) 15–24
[2] Wang X and Gordaninejad F 1999 *J. Intel. Mater. Syst. and Struct.* **10** (8) 601–8
[3] Brunn P O and Abu-Jdayil B 2004 *Rheol. Acta* **43** 62–7
[4] Zhang Y L, Yu J Y, Li H L, Gan Y, Kong F S and Zhang J Q 2006 *Smart Materials and Structures* **15** 1719–24
[5] Wereley N 2008 *J. Intel. Mater. Syst. and Struct.* **19** (3) 257–68

Magnetorheology of carbonyl iron particles coated with polypyrrole ribbons: The steady shear study

M Mrlik[1, 2, 5], M Sedlacik[1, 3], V Pavlinek[1, 2], P Peer[4], P Filip[4] and P Saha[1, 2]

[1] Centre of Polymer Systems, University Institute, Tomas Bata University in Zlin, Nad Ovcirnou 3685, 760 01 Zlin, Czech Republic
[2] Polymer Centre, Faculty of Technology, Tomas Bata University in Zlin, namesti T. G. Masaryka 275, 762 72 Zlin, Czech Republic
[3] Department of Production Engineering, Faculty of Technology, Tomas Bata University in Zlin, namesti T. G. Masaryka 275, 762 72 Zlin, Czech Republic
[4] Institute of Hydrodynamics, Academy of Sciences of the Czech Republic, Pod Patankou 5, 166 12 Prague 6, Czech Republic

E-mail: mrlik@ft.utb.cz

Abstract. The aim of this study is a preparation and application of polymer coating on the surface of carbonyl iron particles. Oxidative polymerization of the pyrrole provides ribbon-like morphology. Compact coating of particles has a slightly negative impact on their magnetic properties (measured for magnetic field strength in the range from 0 to 300 mT); however, there is a significant increase in sedimentation stability. The effect of the particle concentration and temperature on the toughness of the internal structures was also investigated.

1. Introduction

Magnetorheological (MR) suspensions belong to the class of smart materials whose characteristic properties are changed upon application of the external stimuli [1-3]. In the case of MR suspensions this stimulus is represented by a magnetic field [4-6].

Generally, MR suspensions are two-phase system consisting of ferromagnetic particles with low value of coercivity dispersed in medium with negligible magnetic properties [7, 8]. This composition enables to create internal structures after application of the magnetic field. This phenomenon is fully reversible and thus after the demagnetization, internal chain-like structures are destroyed and suspension reverts to its original liquid state. Such behaviour has great potential and is used in real applications as e.g. in active damping systems and torque transducers [5].

The sedimentation is a very important factor which influences the applicability of these materials in industry. Magnetic particles are mainly based on the metal (iron, zinc, cobalt) having relatively high density, thus very fast sedimentation appears when they are dispersed in the liquid medium. There are several approaches to improve the sedimentation stability such as adding various additives (surfactants, thixotropic agents, nanofillers), or usage of bidispersed or bimorphic MR fluids [7, 9, 10]. Promising approach consists of preparation of the core-shell based magnetic particles, where a magnetic core is coated with a suitable layer decreasing the overall density of the core-shell particles and enhancing interactions between particles and liquid medium [11,12].

[5] Corresponding author

In this study, the carbonyl iron (CI) particles were coated with polypyrrole (PPy) ribbons synthesized via oxidative polymerization. The magnetic properties of the prepared particles were examined. Steady shear measurements in the absence and presence of magnetic field were performed and from the obtained flow curves a yield stress was evaluated for various strengths of magnetic field. The effect of the particle fraction and temperature was also analysed. Finally, the positive influence of the particle coating on the sedimentation stability was proved.

2. Experimental

2.1 Materials
Carbonyl iron (HS grade) consisting > 97% of iron particles, was produced by BASF (Germany), Pyrrole (Py, 98%, Aldrich Chemicals, USA) was distilled twice under reduced pressure before use. Both, oxidizing agent ammonium persulfate (APS, $(NH_4)_2S_2O_8$, 98%, Aldrich Chemicals, USA), and surfactant cetyltrimethyl ammonium bromide (CTAB, LACH-NER, Czech Republic), were used as received.

2.2 Synthesis of core-shell particles
First, CI particles were treated with 0.5M HCl following a procedure according to reference [13] and hydroxyl groups were created on the surface of the particles. Later, the particles modified in this way were sonicated with the surfactant CTAB water solution for 1 hour. Then the dispersion of CI particles was moved to the three-neck flask cooled down to 0-5°C and monomer Py was added under vigorous stirring. After 20 minutes, the oxidizing agent APS was added drop-wise. The reaction was carried out for additional 18 hours [14].

2.3 Characterization of the prepared particles
Morphology of the bare CI microparticles and the ones coated with PPy ribbon-like particles was studied using a scanning electron microscopy (SEM, VEGA II LMU, Tescan Ltd., Czech Republic), with an operating voltage of 5 kV. FT-IR spectra of the prepared samples were obtained using a Nicolet magna-550 spectrometer, USA, in the range of 4000-700 cm^{-1}. Further, magnetic properties of particles were examined using a vibration sample magnetometer (VSM, EG&G PARC 704, Lake Shore, USA) at room temperature.

2.4 Suspension preparation
Bare CI and CI/PPy ribbon-like particles were suspended in silicone oil (Dow Corning, Fluid 200, USA, η ~ 100 mPa s) with 20, 30 and 40 wt.% particle concentrations. Suspensions were stirred mechanically and than placed in an ultrasonic bath for 30 s before each measurement.

2.5 Rheological characterization
The rheological properties under external magnetic field in the range 0-300 mT were investigated using a rotational rheometer Physica MCR501 (Anton Paar GmbH, Austria) with the Physica MRD 180/1T magneto-cell. The true magnetic flux density was measured using a Hall probe and the temperature was checked with the help of an inserted thermocouple [15]. Rheological measurements at temperatures 25, 45 and 65°C were performed using a Viscotherm VT2 circulator with temperature stability ±0.02°C.

3. Results and discussion

3.1 Morphology of the particles and structure characterization
Morphology of the prepared particles is depicted in Figure 1. The bare CI particles (Figure 1a) exhibit nearly spherical shape with a diameter around 1 μm. The coating of CI particles with PPy ribbons (Figure 1b) changes their shape and size negligibly.

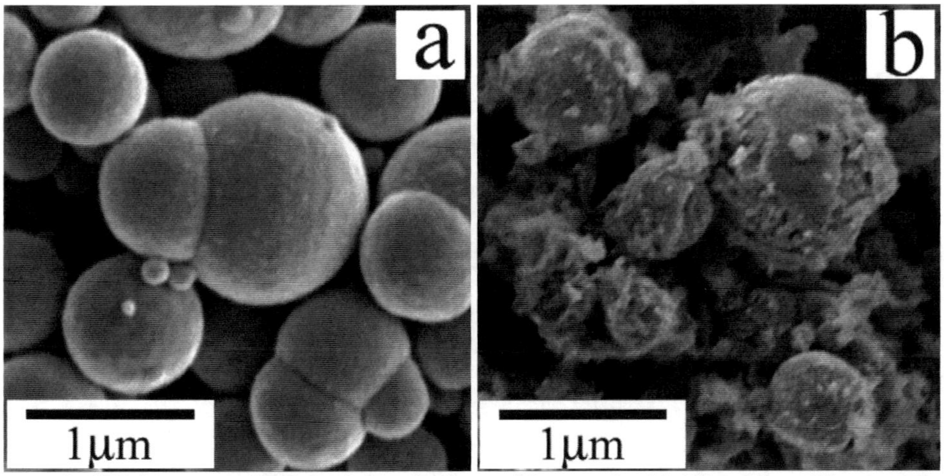

Figure 1. SEM images of CI particles: (a) bare, (b) coated with PPy ribbon-like particles.

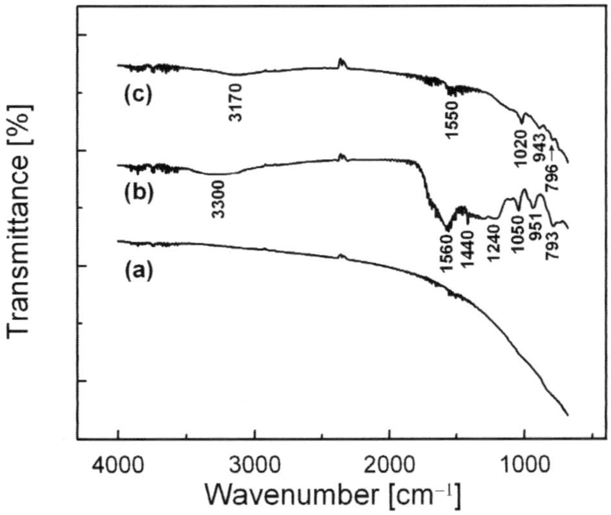

Figure 2. FT-IR spectra of (a) bare CI particles, (b) bare PPy ribbon-like particles, (c) coated CI with PPy ribbon-like particles.

Figure 2 shows FT-IR spectra of the bare CI particles, bare PPy ribbon-like particles and coated CI with PPy ribbon-like particles. As expected there are no visible characteristic bands for the bare CI particles due to their composition >97% of iron. Further, typical absorption of PPy is illustrated in Figure 2b where the characteristic bands are: for N-H at 3300 cm^{-1}, for C-N at 1560 and 1440 cm^{-1}, and the vibrations related to C-H at 1240 and 1050 cm^{-1}. Strong bands near 943 cm^{-1} indicate the doping state of PPy [16]. Finally, Figure 2c documents successful coating of the bare CI particles with PPy ribbons. The broad band at 3170 cm^{-1} is related to the N-H stretching mode from PPy. However, the bands become weaker due to presence of bare CI in the CI/PPy ribbon-like system. Analogously to the reference [17], probably due to polymerization of PPy in the presence of the CI particles some of the vibrations are shifted to lower wavenumbers.

3.2 Magnetic properties

Magnetic properties of bare CI particles and CI coated with PPy ribbon-like particles are shown in Figure 3. Both types of particles exhibit minimal remanence. Coating of CI particles with a polymer layer implies lower magnetization saturation in comparison with the bare CI particles. However, a reduction of magnetic properties is more than balanced by their improved sedimentation stability.

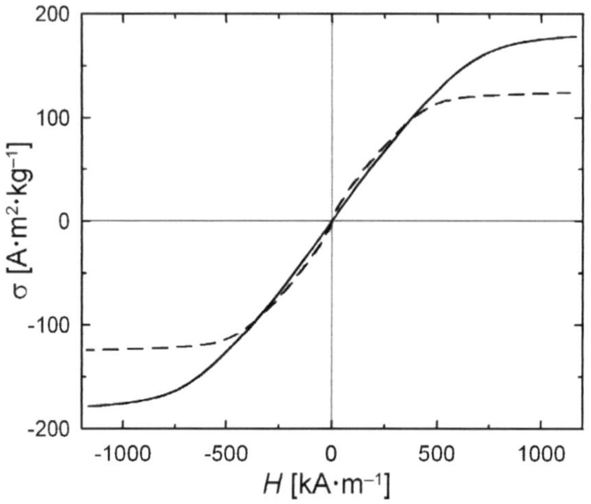

Figure 3. Magnetization curve of bare CI (solid line) and CI/PPy ribbon-like particles (dashed line).

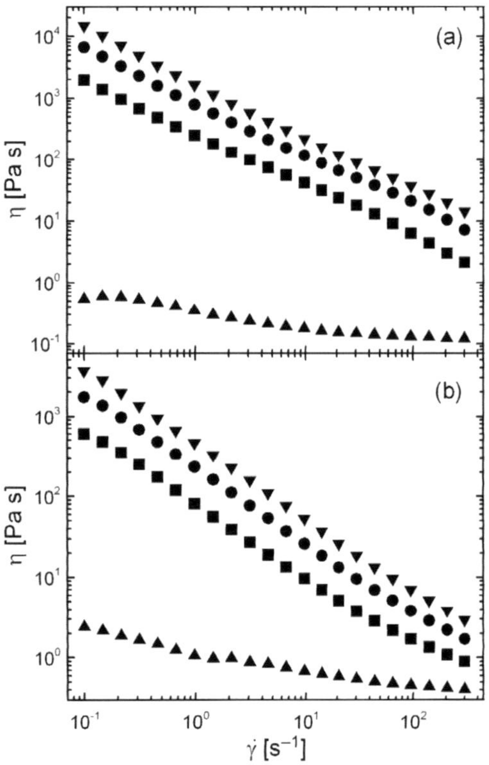

Figure 4. Dependence of shear viscosity, η, on shear rate, $\dot{\gamma}$, for 40 wt.% suspensions of (a) bare CI and (b) coated CI with PPy ribbon-like particles at various magnetic field strengths B (mT): (▲) 0, (■) 84, (●) 174, (▼) 263.

3.3 Steady shear behaviour

Figure 4 illustrates steady shear behaviour of the prepared suspensions. In the absence of magnetic field, both MR suspensions exhibit nearly Newtonian behaviour. This situation is dramatically changed when a magnetic field is applied. Its presence evokes chain-like structures in both suspensions resulting in an increase of shear viscosity by several orders of magnitude. Furthermore, the toughness of the internal structures increases with increasing magnetic field strength. As expected, PPy ribbon-like coating reflects in lower values of viscosity due to the suppressed magnetic properties in comparison to bare CI particles.

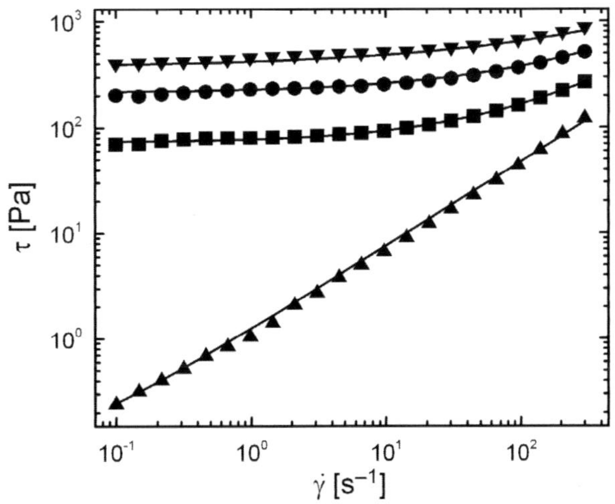

Figure 5. Dependence of shear stress, τ, on the shear rate, $\dot{\gamma}$, for 40 wt. % of coated CI with PPy ribbon-like particles in silicone oil at various magnetic field strengths, B (mT): (▲) 0, (■) 84, (●) 174, (▼) 263, where the solid lines represent the Herschel-Bulkley model.

In order to obtain an information about the internal structures, the dependence of shear stress, τ, on the shear rate, $\dot{\gamma}$, was evaluated using the Herschel-Bulkley model (as in [18])

$$\tau = \tau_y + k \cdot \dot{\gamma}^n \quad (1)$$

where, τ_y, is the yield stress, k, is the consistency parameter and ,n, is the flow behaviour index. The parameters in the Herschel-Bulkley model corresponding to the individual curves in Figure 5 are summarised in Table 1.

Table 1. Parameters of the Herschel-Bulkley model.

B [mT]	τ_y [Pa]	k [Pa·s^n]	n [-]
0	0.06	1.19	0.8
84	73.3	4.48	0.7
174	215.4	12.96	0.6
263	379.3	39.54	0.4

2.4 Effect of the particle concentration and temperature on the internal structure toughness

The yield stress obtained by fitting of the flow curves using the Herschel-Bulkley model was chosen as a measure of toughness of internal structures created under the application of magnetic field. As can be seen in Figure 6 the particle concentration represents a crucial parameter influencing the structure toughness. Toughness of the internal structures increases with increasing particle concentration. The increase from 20 to 40wt.% enhances a yield stress nearly by one order of magnitude.

The effect of temperature on toughness of the structures is depicted in Figure 7. Toughness of the structures increases with increasing temperature applied to the suspension under presence of magnetic field. Application of higher temperature results in an increase of the yield stress. Similar results were already reported by Kim *et al.* [19].

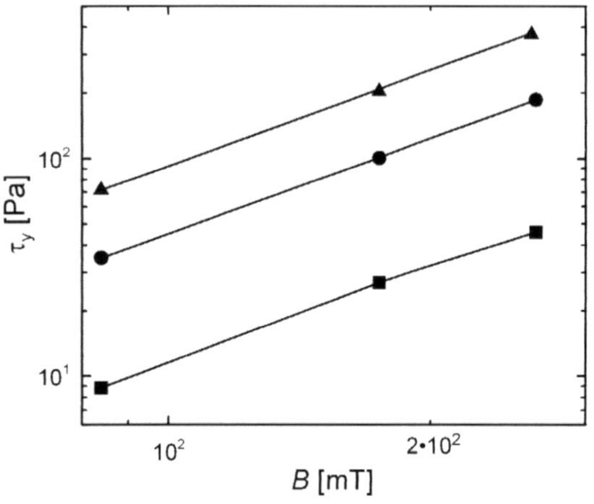

Figure 6. Dependence of the yield stress, τ_y, on the magnetic field strength, B, for suspensions in silicone oil of various concentrations (wt. %): (■) 20, (●) 30, (▲) 40 at temperature 25°C.

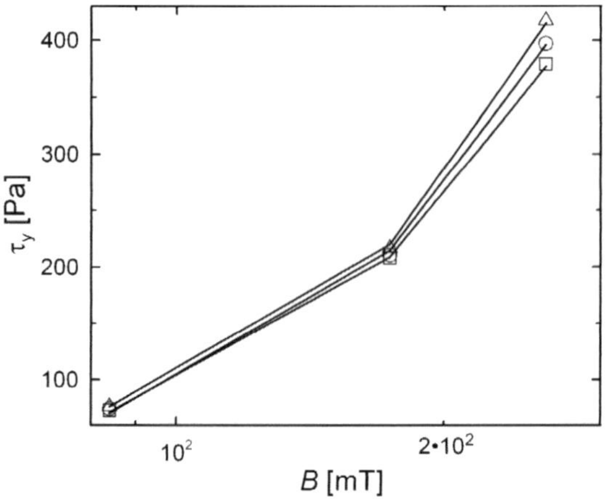

Figure 7. Dependence of the yield stress on the magnetic field strength for 40 wt.% particle suspensions in silicone oil at various temperatures (°C): (□) 25, (○) 45, (△) 65.

2.5 Sedimentation stability

Sedimentation strongly influences the use of MR suspensions in industry. The sedimentation test measured by a naked-eye method reveals that stability of bare CI particle suspension is poor (Figure 8). On the other hand, when the PPy ribbon-like structures were synthesized on the surface of CI particles, the sedimentation stability increased and the particles exhibited stable behaviour during 30 hours of test duration. Polymer coating on the CI particles with PPy ribbons partially decreases the particle density and probably increases the particle-silicone oil interactions [20] that contribute to enhancement of the sedimentation stability.

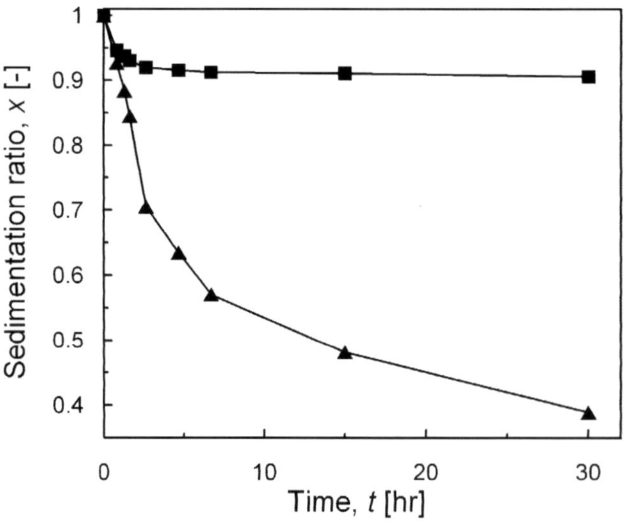

Figure 8. Sedimentation ratio of 40 wt. % suspensions in silicone oil for bare CI (▲) and CI/PPy ribbon-like particles (■) at temperature 25°C.

4. Conclusion

In the present study, the easy preparation of the CI/PPy ribbon-like particles was introduced. The coating of CI particles was realized via in situ oxidative polymerization of PPy. Magnetic properties of prepared particles are lower in comparison with the bare ones due to coating of CI particles with a polymer layer. When the magnetic field strength of 263 mT is applied, the magnitude of viscosity of the suspension of CI/PPy ribbon-like particles increases by three orders in comparison when no magnetic field is applied. Further, both particle concentration and temperature positively influence the toughness of the suspension structure (an increase in the yield stress). Finally, the sedimentation stability of the suspension consisting of coated CI with PPy ribbon-like particles was significantly improved, due to reduced particles density and enhanced particle-silicone oil interactions.

Acknowledgement

The authors wish to thank the Grant Agency of the Czech Republic for the financial support of Grant No. 202/09/1626.

This contribution was written with support of Operational Program Research and Development for Innovations co-funded by the European Regional Development Fund (ERDF) and national budget of Czech Republic, within the framework of project Centre of Polymer Systems (reg. number: CZ.1.05/2.1.00/03.0111).

References

[1] de Vicente J, Klingenberg D J and Hidalgo-Alvarez R 2011 *Soft Matter* **7** 3701
[2] Bica I 2006 *J. Ind. Eng. Chem.* **12** 501
[3] Bossis G, Lacis S, Meunier A and Volkova O 2002 *J. Magn. Magn. Mater.* **252** 224
[4] Park B J, Fang, F F and Choi H J 2010 *Soft Matter* **6** 5246
[5] Wang D H and Liao W H 2011 *Smart Mater. Struct.* **20** 023001
[6] Sedlacik M, Pavlinek V, Lehocky M, Mracek A, Grulich O, Svrcinova P, Filip P and Vesel A 2011 *Colloid Surf. A-Physicochem. Eng. Asp.* **387** 99
[7] Lopez-Lopez M T, de Vicente J, Gonzalez-Caballero F and Duran J D G 2005 *Colloid Surf. A-Physicochem. Eng. Asp.* **264** 75
[8] Sedlacik M, Pavlinek V, Saha P, Svrcinova P, Filip P and Stejskal J 2010 *Smart Mater. Struct.* **19** 115008
[9] Fang F F, Choi H J and Jhon M S 2009 *Colloid Surf. A-Physicochem. Eng. Asp.* **351** 46
[10] Wereley N M, Chaudhuri A, Yoo J H, John S, Kotha S, Suggs A, Radhakrishnan R, Loven B J and Sudarshan T S 2006 *J. Intell. Mater. Syst. Struct.* **17** 393
[11] Fang F F, Choi H J and Seo Y 2010 *ACS Appl. Mater. Interfaces* **2** 54
[12] Fang F F, Choi H J and Choi W S 2010 *Colloid Polym. Sci.* **288** 359
[13] Belyavskii S G, Mingalyov P G, Giulieri F, Combarrieau R and Lisichkin G V 2006 *Protect. Met.* **42** 244
[14] Zhang X T, Zhang J, Song W H and Liu Z F 2006 *J. Phys. Chem. B* **110** 1158
[15] Laun H M and Gabriel C 2007 *Rheol. Acta* **46** 665
[16] Cheng Q L, He Y, Pavlinek V, Li C Z and Saha P 2008 *Synth. Met.* **158** 953
[17] Somani P R, Marimuthu R, Mulik U P, Sainkar S R and Amalnerkar D P 1999 *Synth. Met.* **106** 45
[18] Wang X J and Gordaninejad F 2007 *J. Appl. Mech.-Trans. ASME* **74** 13
[19] Kim S G, Lim J Y, Sung J H, Choi J H, Seo Y 2007 *Polymer* **48** 6622
[20] Tsuda K., Takeda Y, Ogura H and Otsubo Y 2007 *Colloid Surf. A-Physicochem. Eng. Asp.* **299** 262

Temperature induced effects on the durability of MR fluids

A Wiehe[1], **C Kieburg**[2] **and J Maas**[1,3]
[1]Ostwestfalen-Lippe University of Applied Sciences, 32657 Lemgo, Germany, www.motion-ctrl.de,
[2]BASF SE, Carl-Bosch-Strasse, 67056 Ludwigshafen/Rhein, Germany

E-mail: juergen.maas@hs-owl.de

Abstract. Although commercial MR fluids exist for quite some time now and the feasibility as well as the advantages of the MR technology have been demonstrated for several applications by a variety of MR actuator prototypes, a sustainable market break-through of brake and clutch applications utilizing the shear mode is still missing. Essential impediments are the marginal knowledge about the durability of the MR technology. To overcome this situation, a long-term measurement system was developed for the durability analysis of MR fluid formulations within a technical relevant scale with respect to the volume of MR fluid and the transmitted torque. The focus of the presented series of measurements is given to the analysis of temperature induced effects on the durability. In this context four different failure indicators can be distinguished, namely an apparent negative viscosity, deviations in torque data obtained from different measurements as well as a pressure increase and a drop in the on-state torque. The measurement data of the present durability experiments indicate a significant dependency of the attainable energy intake density on the temperature. The aim of such durability tests is to establish a reliable data base for the industry to estimate the life-time of MR devices.

1. Introduction

The measurement of the durability is complex and often takes long periods of time. This situation becomes even more difficult, as the wear behavior of MR fluids has been rarely investigated so far, especially for the shear mode utilized in clutches and brakes. Durability tests have been performed with an MR clutch utilized as fan drive for trucks, [1]. Subsequently, the MR fluid was extracted from the MR fan drive after the durability test and analyzed in detail afterwards. The organics analysis [2] showed a degradation of the poly-alpha-olefin based fluid. Iron oxidation was also detected [3], which most likely caused the decrease of the MR effect during the durability test. The degradation of the base oil, in combination with an increasing pressure and a continuously decreased on-state torque, and in-use-thickening (IUT) [4] are important effects that were observed in some MR fluid formulation tests [5]. Lifetime dissipated energies (LDE) [4] of up to 5 MJ/mL were also achieved with optimized high shear stress MR fluid formulations for clutch and brake applications [6] using a rheometer-sized test cell.

To overcome this situation, a long-term measurement system was developed, which is suitable to analyze the durability of different MR fluid formulations within a technical relevant scale with respect to the volume of MR fluid and the transmitted torque. This system was initially presented in [7] and therefore only a brief description is given in section 2. Previous measurements with the commercial

[3] To whom any correspondence should be addressed.

MR fluid Basonetic® 5030 from BASF SE showed basic effects on the durability of MR fluids: A variation of the load cycle profile from constant to a square wave shaped power dissipation yielded no effects on the durability. This indicates that the life time dissipated energy (LDE) determined by the proposed durability tests can be extrapolated to real life MR brakes. Further measurements indicate a significant effect of the dissipated power per volume on the LDE, which decreases in the case of an increasing dissipated power resulting from increasing rotational speed and constant braking torque.

In the present contribution, the focus of a new series of measurements is given to the analysis of temperature induced effects on the durability. Life cycle test were performed by varying temperatures in the range from 20°C to 80°C, each with duration of several weeks. The potential thermal expansion of the MR fluid has to be taken into account when changing the temperature in a higher range. The analysis of the thermal expansion of the MR fluid is described in section 3 and the accordingly modified measurement process is presented in section 4. The experimental results of the new series of experiments are discussed in section 5.

2. Experimental

2.1. Long-term measurement system

The utilized torque test bench is equipped with a specially developed MR actuator for continuous loads (continuous load actuator, CLA) [7]. For this purpose, a cylindrical, horizontally oriented double shear gap (drum type) design was chosen to avoid the mass force induced phenomena of centrifugation and sedimentation (Figure 1). Homogenous shear conditions throughout the shear gap are an additional advantage of this design, which improves the transferability of the results. An easy and reproducible extraction of the MR fluid after the experiment is considered, which ensures a quick disassembling of the CLA and an easy cleaning of the shear gaps.

The lifetime of MR devices is affected by both, the durability of the MR fluid and in particular by the wear resistance of the utilized seals. Increased leakage due to worn seals almost inevitably causes a breakdown of MR devices. Thus, the choice of the seal type was one of the main aspects in the design phase of the CLA. Different kinds of seals were evaluated in extensive durability experiments described in [7]. The finally chosen combination of a spring steel seal and a RSS combines a good sealing capacity with comparably low friction and cost.

The influencing quantities temperature, rotational speed, and excitation current of the CLA are controlled in a closed-loop. The measured quantity is the torque generated by the CLA. Additionally, the pressure in the area of the outer seal can be assessed. A half section schematic of the CLA and the system structure of the long-term measurement system are shown in figure 1.

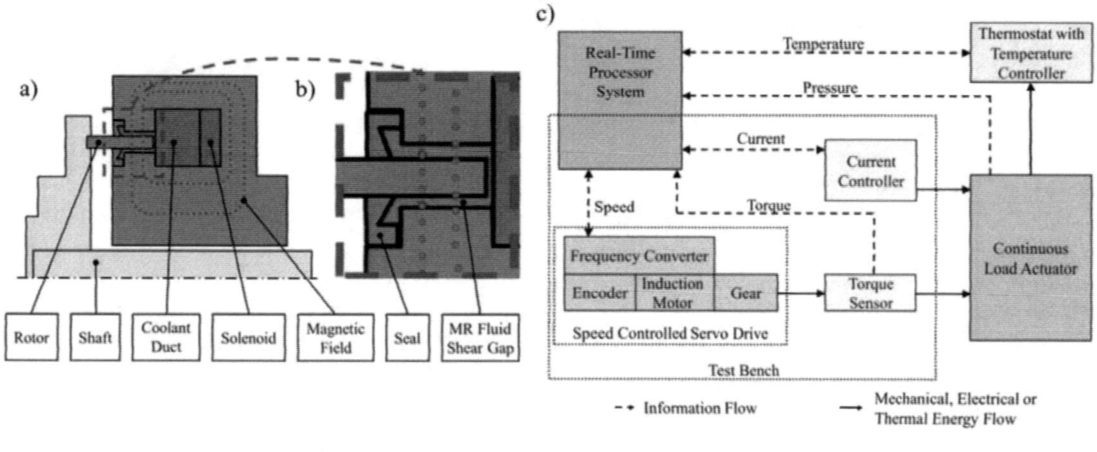

Figure 1. (a) Half section schematic of the Continuous Load Actuator (CLA), (b) detail view of the shear gap, and (c) system structure of the long-term measurement system [7].

2.2. Thermal expansion of MR fluids

In contrast to the previous experiments published in [7], the present series deals with the effect of the temperature on the LDE. Therefore greater changes in the MR fluid's temperature occur especially during the preparation phase of the durability experiment. The effect of the MR fluid's thermal expansion was analyzed in a preliminary study. In general, the density of the MR fluid is reduced by a temperature rise due to the thermal expansion effect. A free volume of MR fluid, which is not limited by external boundaries, would grow. The cubic expansion coefficient β is the relation between the volume expansion and the change in the temperature, e.g. [8]. The change in volume ΔV due to the change in temperature ΔT is calculated usually assuming a constant cubic expansion coefficient β, which is independent from the temperature. In the case of chemical mixtures, the expansion coefficient is not independent from the volume, which leads to the following expression:

$$\beta = \frac{1}{V}\left(\frac{\partial V}{\partial T}\right)_{P=const.} \xrightarrow{\beta \neq f(T)} \Delta T = T_1 - T_0 = \int_{V_0}^{V_1 = V_0 + \Delta V} \frac{1}{V \cdot \beta(V)} dV \tag{1}$$

For suspensions like MR fluids, the volumes of the single phases can be treated separately. Considering the values given in Table 1, it becomes obvious that the expansion β_{Fe} of the iron particles can be neglected compared with the carrier fluid β_{cf}. For small changes in temperature a linear approximation can be found, which leads to an expression easy to handle. In the relevant range of temperature the approximation error is small in relation to the assumptions already made.

$$\Delta T \approx \frac{1}{\beta_{cf}} \int_{V_0}^{V_0 + \Delta V} \frac{1}{V} dV \longrightarrow \Delta V = V_0 \left[\exp\left(\beta_{cf} \cdot \Delta T\right) - 1 \right] \approx V_0 \cdot \beta_{cf} \cdot \Delta T \tag{2}$$

Since the shear gaps of the CLA have to be treated as a confined volume the expansion of the MR fluid volume is impossible. An increase in the temperature leads to an increase in the pressure of the MR fluid instead. This pressure increase is related to the compressibility of the MR fluid. Based on the definition of the compressibility factor (e.g. [8]) a simple expression for the pressure change Δp can be found in a comparable way as used for equation 2. The compressibility of the iron κ_{Fe} can be neglected, in this case too.

$$\kappa = -\frac{1}{V}\left(\frac{\partial V}{\partial p}\right)_{T=const.} \longrightarrow \Delta p \approx \frac{\Delta V}{\kappa_{cf} \cdot V_0} \tag{3}$$

The expression for the temperature induced pressure increase can be obtained from equations 2 and 3 substituting the change in volume ΔV.

$$\Delta p \approx \frac{\beta_{cf}}{\kappa_{cf}} \Delta T \tag{4}$$

Using the values given in Table 1 and assuming a temperature increase of 80 K leads to a change in volume of approximately 9 % based on equation 2 and a theoretical pressure increase of 900 bar (13e3 psi) using equation 4. This theoretical pressure value cannot be observed in reality due to the seals' elasticity and leakage occurring inevitably. But it shows that the thermal expansion of an MR fluid cannot be neglected when changing the temperature in such a range. A measurement is shown in figure 2 indicating the pressure increase due to the thermal expansion of the MR fluid.

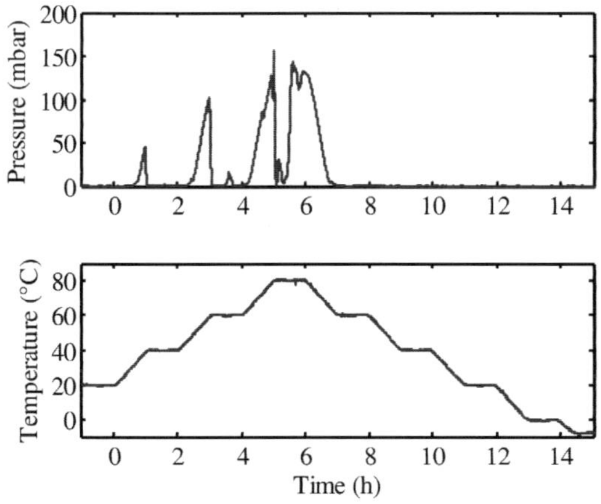

Figure 2. Measurement data indicating the pressure increase due to thermal expansion. The measured pressure is significantly smaller than the theoretical estimated value by equation 4. This deviation occurs due to the elasticity of the seals and leakage. The pressure drops occurring in the phases of a constant temperature are caused by leakage, too. Both effects are contrary with the assumption of a confined volume initially stated for equation 4.

Table 1. Parameters utilized for the calculation of the thermal expansion.

Parameter	Symbol	Unit	**Iron (Fe)**	**Cyclohexane ($C_{10}H_{12}$)**
thermal expansion	β	$°C^{-1}$	35.4e-6	1.15e-3
compressibility	κ	Pa^{-1}	16e-12	1.13e-9
typical for	-	-	particles	carrier fluid
assumed volume fraction	V	%	50	50

2.3. Modified measurement process for the examination of the temperature effect

The aim of the utilized testing process is the durability analysis of MR fluids by ensuring reproducible conditions, while the whole characteristic is observed in relation to the influencing quantities. The measurement process described in [7] was modified to regard the thermal expansion of the MR fluid when investigating the temperature effect on the durability.

At the preparation and the start of a durability experiment disturbing influences resulting from the assembly and filling process, from the bedding-in of the seals, and from the thermal expansion should be minimized. A flow chart of the modified measurement process is shown in figure 3. The parameters used for the present series of experiments are stated in table 2.

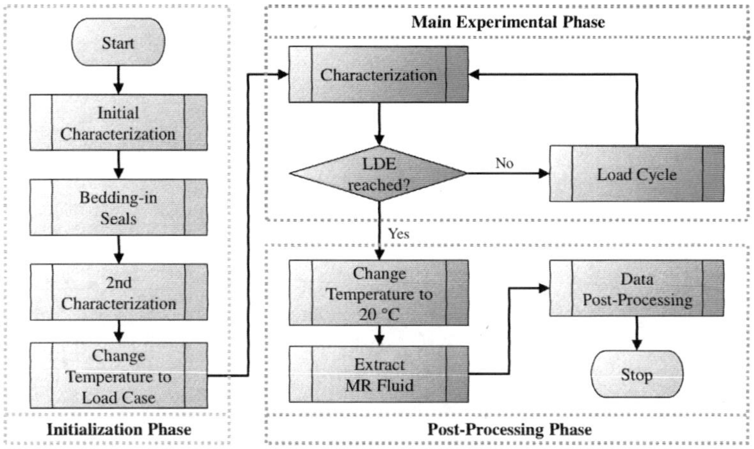

Figure 3. Modified measurement process.

The modified measurement process can be divided into three phases. The aim of the initial phase is to guarantee consistent conditions for each experiment. At the beginning, an initial characterization is performed by measuring the torque in relation to the current (On-State) and the rotational speed (Off-State). All characterizations are performed following the same principle: At first the off-state torque behavior is measured by changing the rotational speed and maintaining a current of 0 A and second the on-state torque behavior is measured by changing the current and maintaining a constant rotational speed. All characterizations during the experiments are performed with the same set of parameters except for the temperature, which is 20°C during the initialization phase. In the main phase, the characterizations are performed at the temperature defined by the load case.

Table 2. Parameters utilized for the examination of the temperature effect on the durability of MR fluids.

Sub-procedure	Parameter	Value	Unit	Comment
Characterization - Off-State	Current I	0	A	
	Speed n	10 : 20 : 70	rpm	Min. : Inc. : Max.
	Temperature ϑ	20 / Load Cycle	°C	Initialization Phase / Main Phase
Characterization - On-State	Current I	0 : 0.25 : 2	A	Min. : Inc. : Max.
	Speed n	5	rpm	
	Temperature ϑ	20 / Load Cycle	°C	Initialization Phase / Main Phase
Bedding-In Seals	Current I	0	A	
	Speed n	40	rpm	
	Temperature ϑ	20	°C	
Change Temperature to Load Case	Current I	0	A	
	Speed n	5	rpm	
	Initial Temperature ϑ	20	°C	
	Final Temperature ϑ	Load Cycle	°C	
	Change Rate	10	°C / h	
Load Cycle	Current I	0.7	A	
	Speed n	15	rpm	
	Temperature ϑ	20 / 40 / 60 / 80	°C	present series of experiments
	Load Profile	Constant		
	max. Duration	16	h	

The data of the unstressed MR fluid obtained during the first characterization can be utilized for the estimation of variations, which can result from the assembly and filling process. The bedding-in of the seal contact surfaces is performed afterwards. The parameters were chosen to enable a quick bedding-in and to minimize the load of the MR fluid. The seals do not provide the full sealing capacity before the sealing surfaces are bedded-in. Because of the higher viscosity of the carrier liquid the bedding-in is performed at room temperature to minimize leakage. After the seals' bedding-in, a second characterization is performed. Afterwards, the temperature is changed according to the load case maintaining a defined change rate to prevent high temperature gradients and resulting thermal stresses within the CLA. During this phase a volume compensation of the shear gaps is established to prevent a pressure increase and leakage by thermal expansion (section 3). The change to the temperature defined by the load case typically takes several hours. To prevent the sedimentation of the particles the rotor of the CLA is moved at a low rational speed. When the desired temperature is achieved the volume compensation is disabled so that the shear gaps are nearly a confined volume and the pressure increase due to the chemical degradation of the MR fluid becomes measureable.

In the first step of the main experimental phase a further characterization of the torque behavior in the on- and off-state is performed. A load cycle is started afterwards causing defined wear of the MR fluid. During this load cycle, a pre-defined workload profile is repeated (Table 2) and the time behavior of the torque and the influencing quantities is recorded until a desired energy intake or period of time is achieved. An additional characterization of the on- and off-state behavior is performed after every load cycle. This alternating procedure is repeated until the overall energy intake (LDE) is reached, indicated by an improper change of the MR fluid characteristics. A final performance characterization is carried out in the end. In the post-processing phase the temperature of the CLA is changed to 20°C maintaining a defined change rate prior to the extraction of the MR fluid. For the calculation of the energy intake density w during the data post-processing phase the time integral of the power dissipation per volume is calculated, [7]:

$$w(t) = \int_0^t \overline{\tau}(t) \cdot \overline{\dot{\gamma}}(t) \cdot dt = \frac{2 \cdot \pi}{60 \cdot V_{MRF}} \int_0^t \left(T_m - T_p\right) \cdot n \cdot dt \tag{5}$$

The measured torque T_m must be corrected by the parasitic torques T_p, which mainly result from the seals and can be estimated based on the characterization data. Further quantities are the volume of the shear gaps V_{MRF} and the rotational speed n.

3. Results and discussion

The objective of the presented experiments was the examination of the temperature effect on the durability of MR Fluids. The measurement process (figure 3, table 2) was used for all experiments presented in this section.

3.1. Results of a single experiment

At first, the measurement data of a single, exemplary experiment are described in detail. Already in the initialization phase, measurement results are obtained. The torque behavior over time shown in figure 4(a) represents the bedding-in process of the seals. The measured torque decreases while the surfaces of the seal point contact are adapting on each other. Afterwards the torque remains constant at a low level. The seal's bedding-in has a potential effect on the torque behavior of the CLA. So the on- and off-state torque was characterized before and after the bedding-in process showing the effect of the seals' bedding-in. The torque behavior remains stable apart from the seals' friction, which is reduced by the bedding-in. After bedding-in, the torque behavior in the off-state shows a clear viscous component superimposed by coulomb friction. The torque behavior regarding to the current shows a quite linear relation. The MR effect, which can be defined as the relation of the maximum to the minimum torque of an MR actuator, is remarkably enhanced by the low off-state friction of the utilized RSS-Nilos seal system [7].

The main phase is the essential part of the durability experiment, which is following the initialization phase. The torque data obtained during the load cycles and the on-state characterizations (by variation of the current) is compared in figure 5. The decrease in torque after reaching a certain energy intake density is caused by leakage of the MR fluid (particles and carrier liquid). This drop in the on-state torque is a distinct failure indicator preventing further proper operation of an MR device and hence of reaching the LDE. Before the drop in the on-state torque begins, a slight difference between the torque obtained during the load cycle and the characterizations can be observed. This indicating effect is caused by the MR fluid partly solidifying in the shear gap over time during the load cycle and the subsequent homogenization by mixing during the characterizations. "Solidifying" in this context is used for certain fluid volumes that suffer a loss of base fluid, that show strongly increased viscosity. Such an ambiguous behavior is undesirable for MR actuators and occurs only after a certain energy intake. Therefore, the deviation in the torque data can be used as an early indicator of reaching the LDE of the MR fluid.

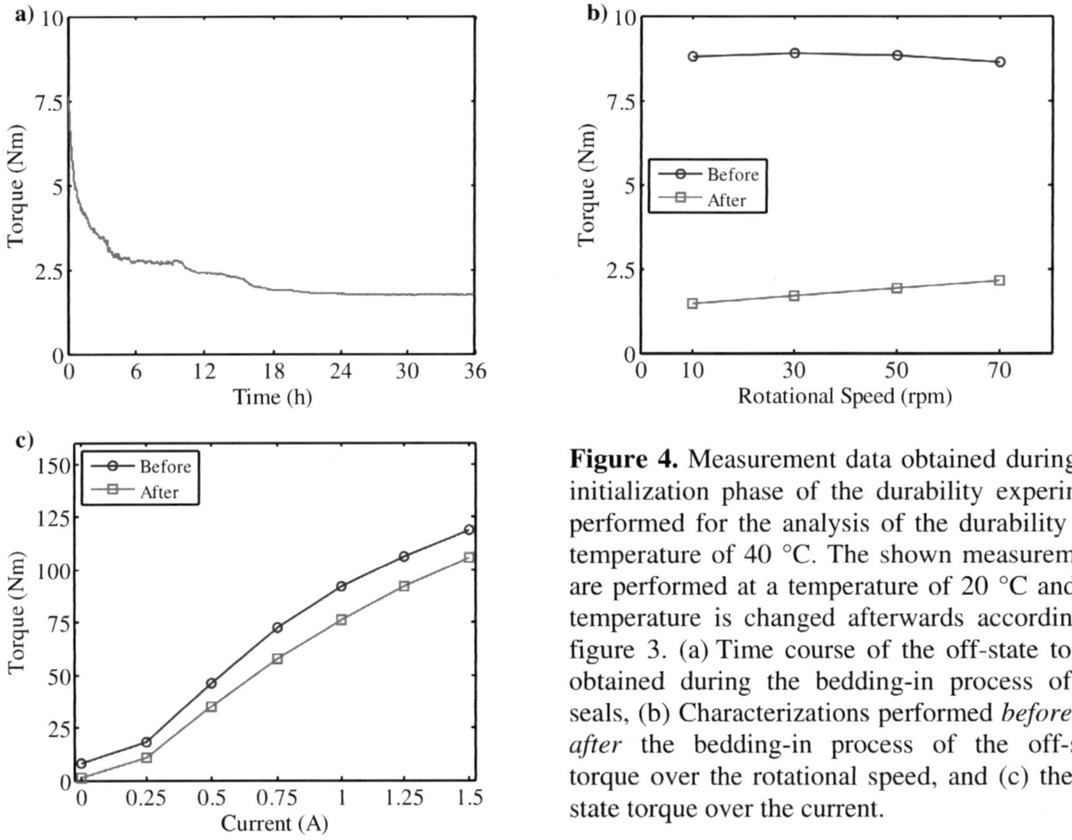

Figure 4. Measurement data obtained during the initialization phase of the durability experiment performed for the analysis of the durability at a temperature of 40 °C. The shown measurements are performed at a temperature of 20 °C and the temperature is changed afterwards according to figure 3. (a) Time course of the off-state torque obtained during the bedding-in process of the seals, (b) Characterizations performed *before* and *after* the bedding-in process of the off-state torque over the rotational speed, and (c) the on-state torque over the current.

The formation of solidifying structures of the MR fluid leads to a further effect. The torque curves of the off-state characterizations obtained at different rotational speeds are crossing. This suggests that an apparent negative viscosity – assuming a Newtonian flow behavior of the MR fluid in off-state mode – occurs. In fact, however, the solidified structure of the MR fluid was destroyed between the first measurement at the lower velocity and the second at the higher velocity by simple remixing. The apparent negative viscosity is a further early indicator.

An abrupt rise in the off-state torque occurs at a higher energy intake density, which is caused by the pressure increase also shown in figure 5. The pressure increase indicates the chemical degradation of the carrier fluid. Gases resulting from the degradation reaction can be the reason for the pressure increase. The abrupt change in pressure at approximately 3 MJ/mL is resulting from MR particles blocking the channel of the pressure sensor firstly and then suddenly breaking. Probably the pressure increase has begun earlier at approximately 2 MJ/mL. This increase in pressure leads to a larger contact pressing force of the seals and thus to a higher friction. From that point on typically a leakage of the MR fluid (particles and carrier fluid) can be observed. The leakage of the MR fluid is most often indicated by the drop in the on-state torque as stated before. Improving the seal capacity to minimize the leakage due to the pressure increase caused by a chemical degradation would result in a high seal friction, which would deteriorate the torque quality of an MR actuator in a not acceptable range.

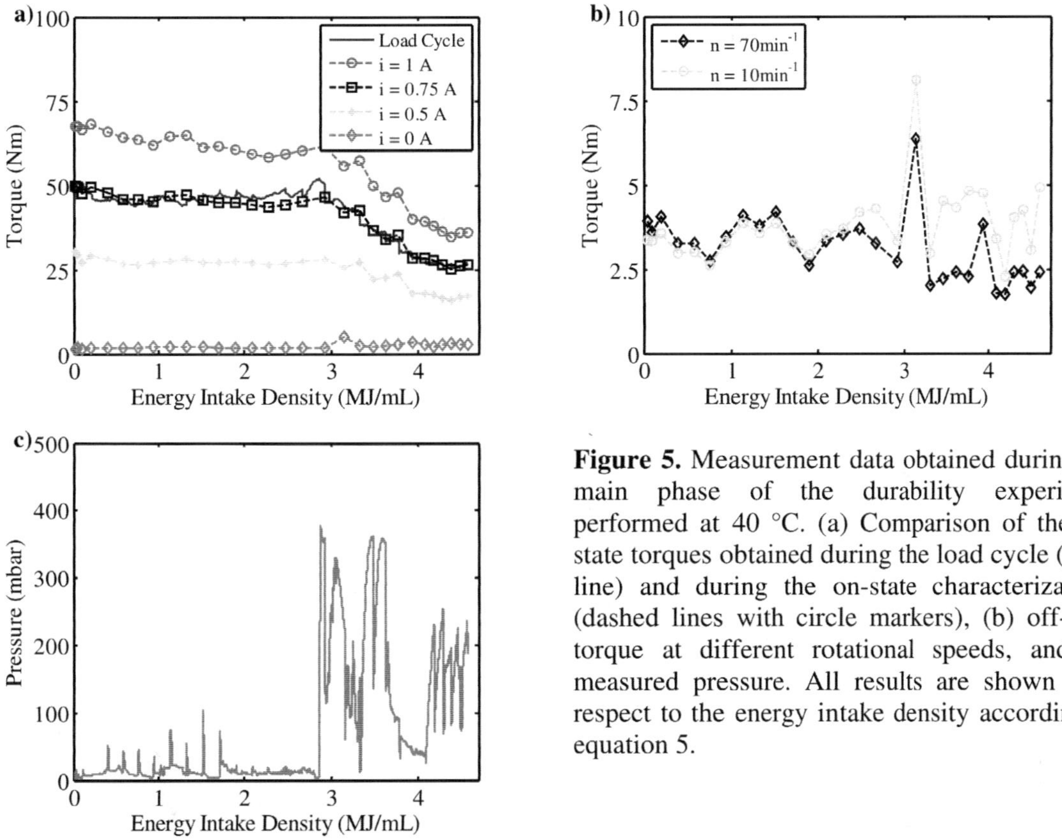

Figure 5. Measurement data obtained during the main phase of the durability experiment performed at 40 °C. (a) Comparison of the on-state torques obtained during the load cycle (solid line) and during the on-state characterizations (dashed lines with circle markers), (b) off-state torque at different rotational speeds, and (c) measured pressure. All results are shown with respect to the energy intake density according to equation 5.

In conclusion the following indicators of the degradation of the MF fluid can be stated:

- apparent negative viscosity (if our specific method of assessment is used)
- deviation in torque data obtained from different measurements
- pressure increase caused by a chemical degradation of the carrier fluid
- drop in the on-state torque

3.2. Temperature effect on the lifetime dissipated energy

Experiments with different temperatures are performed to investigate the effect on the durability of the MR fluid. The data of the bedding-in procedure as well as of the first main phase characterization are depicted in figure 6. The comparison of the different experiments does not result in significant differences. Varying the rotational speed the off-state torque shows only the typical temperature dependency of the viscosity. The data depicted in figure 6 show that there were equal conditions at the start of the main phase for all conducted experiments.

The data obtained during the main phases of the several experiments are presented in figure 7. In all the experiments a similar behavior of the MR fluid is observed, but the failure indicators appear at different energy intake densities regarding to the temperature. The mentioned effects on the durability are indicated by the measured torque T_m and pressure p obtained during the load cycles, as well as the MR torque T_{mr} and the parasitic torque T_p, which are calculated based on the characterization data.

The attainable LDE is decreasing in the case of an increasing temperature from 20°C to 60°C. Thus, an increased degradation of the MR fluid is caused at a higher temperature. The increase in the

LDE when changing the temperature from 60°C to 80°C is remarkable, which could be reproduced for several times. A fluctuation of the measurement thus appears implausible. The reason has to be clarified within further studies in detail. Reasons could be different degradation reactions at this temperatures as well as the temperature behavior of the used surfactants.

The characterizations were performed at a representative speed n according to table 2. Every durability experiment presented in this section took several weeks. The torques was calculated as follows [7]:

$$T_p = T_m(i)\,\big|_{i=0} \tag{6}$$

$$T_{mr}(|B(i)|) = T_m(i) - T_p \tag{7}$$

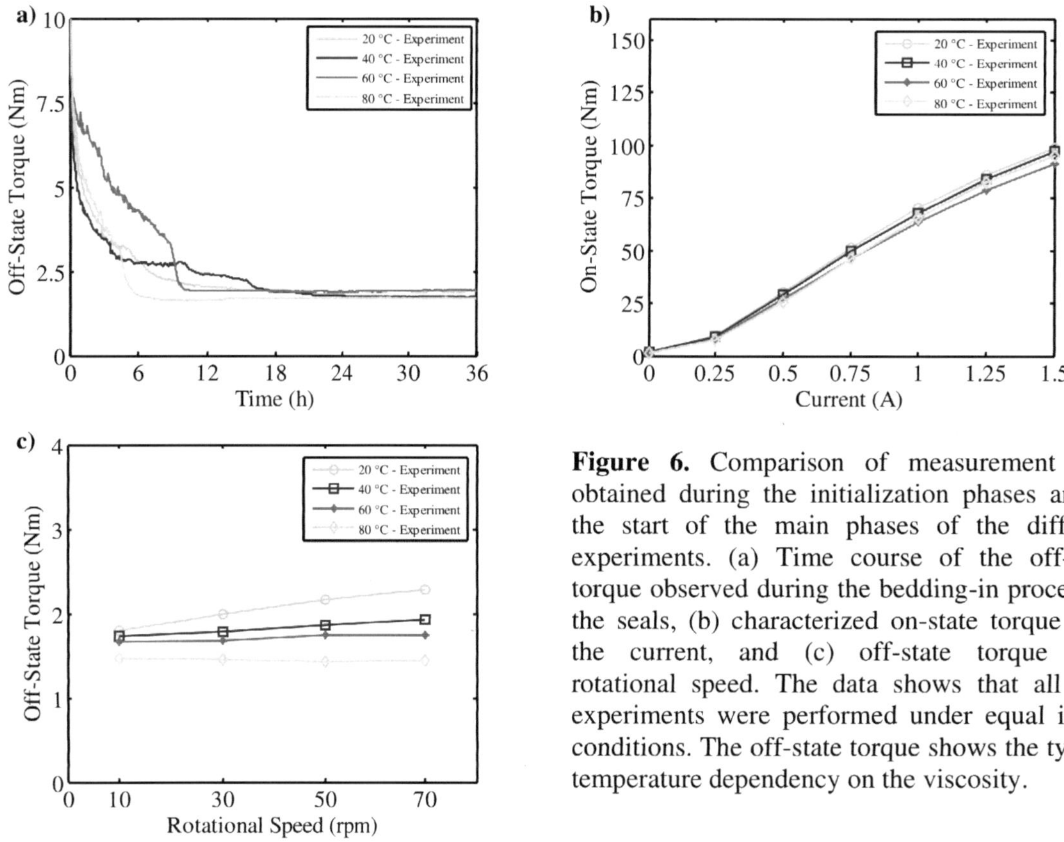

Figure 6. Comparison of measurement data obtained during the initialization phases and at the start of the main phases of the different experiments. (a) Time course of the off-state torque observed during the bedding-in process of the seals, (b) characterized on-state torque over the current, and (c) off-state torque over rotational speed. The data shows that all four experiments were performed under equal initial conditions. The off-state torque shows the typical temperature dependency on the viscosity.

In figure 8 the energy intake densities are compared at which the failure indicators (see subsection 5.1) occur. The dependency of the attainable energy intake density on the temperature becomes obvious with this depiction, too. The early indicators *apparent negative viscosity* and *deviation in torque data* occur portending the solidification of the MR fluid before a significant drop in the on-state torque states the final failure. Besides this, a pressure increase results due to the chemical degradation of the MR fluid.

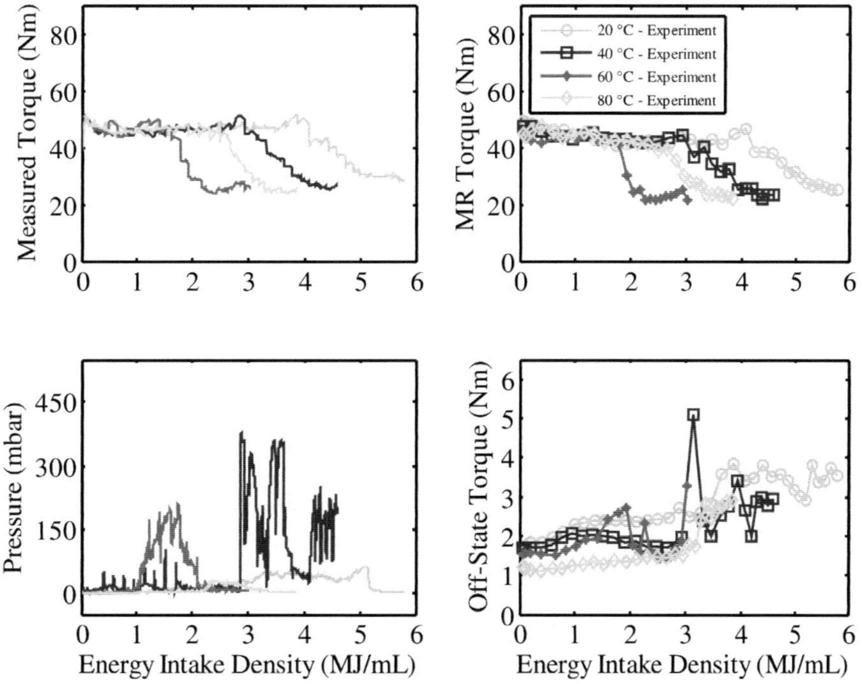

Figure 7. Effect of temperature on the durability of the MR fluid. The measured torque and the pressure were obtained during the load cycles. The MR torque at a current of 0.75 A is calculated by equation 6 and the off-state torque by equation 7. All results are shown with respect to the energy intake density according to equation 5.

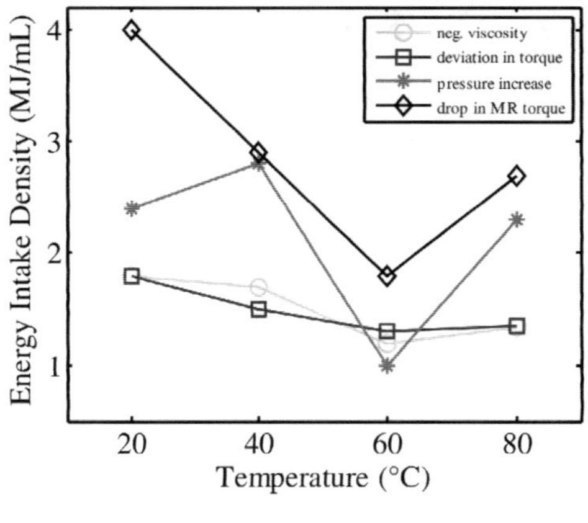

Figure 8. Temperature effect on the failure indicators discussed in subsection 5.1. Beside the pressure build-up inside the shear gaps, the apparent negative viscosity observed during the off-state characterization as well as the deviation in the torque data obtained during the load cycles and the on-state characterizations are appropriate early indicators of a failure. The failure is indicated by a significant drop in the MR torque finally. The deviation of the pressure increase of the 40°C – experiment is caused by MR particles blocking the channel of the pressure sensor.

4. Conclusions

The aim of this contribution was the investigation of the temperature effect on the durability of MR fluids to enable engineers to calculate the lifetime of MR devices from the scratch. An essential aspect for temperature durability experiments is the consideration of the thermal expansion of the MR fluid. The existing measurement process was modified accordingly. It can be divided into three phases: Initialization, main and post-processing phase.

When investigating the durability, four indicators of an occurring failure can be distinguished, namely an apparent negative viscosity, the deviation in torque data from different measurements, a pressure increase within the confined shear gaps and a drop in the on-state torque caused by leakage of the MR fluid. The deviation in torque data and the apparent negative viscosity are caused by a solidification of the MR fluid during the load cycle. Both can be used as early indicators of the degradation of the MR fluid as well as the pressure increase.

Durability experiments were performed by varying temperatures in the range from 20°C to 80°C, each with a duration of several weeks. The measurement data of the durability experiments indicate a significant dependency of the attainable energy intake density on the temperature. Lifetime dissipated energy decreases with increasing operating temperature. Thus, an increased degradation of the MR fluid is caused at a higher temperature. The increase in the attainable energy intake density when changing the temperature from 60°C to 80°C is remarkable. Its cause has to be clarified within further studies in detail.

References

[1] Smith A L, Ulicny, J C and Kennedy L C 2007 Magnetorheological fluid fan drive for trucks *J. Intelligent Material Systems and Structures* **18** 1131–36

[2] Ulicny J C, Hayden C A, Hanley P M and Eckel D F 2007 Magnetorheological fluid durability test – organics analysis *J. Materials Science and Engineering A* **464** 269–73

[3] Ulicny J C, Balogh M P, Potter M and Waldo R A 2007 Magnetorheological fluid durability test – iron analysis," *J. Materials Science and Engineering A* **443** 16–24

[4] Carlson J D 2002 What makes a good MR fluid? *J. Intelligent Material Systems and Structures* **13** 431–35.

[5] Kieburg C, Oetter G, Laun M, Gabriel C and Steinwender H 2008 MR all-wheel-drive prototype car driving tests and durability requirements for the MR fluids used *11th Int. Conf. on Electrorheological Fluids and Magnetorheological Suspensions – ERMR 2008* Dresden Germany 25–29 August

[6] Gabriel C, Oetter G, Kieburg C and Laun M 2010 Durability testing on magnetorheological fluids for clutch and brake applications *Conf. Proc. Electrorheological (ER) Fluids and Magnetorheological (MR) Suspensions – ERMR10 (ed R Tao)* Philadelphia USA 16–20 August 605–11 (World Scientific, Singapore)

[7] Wiehe A and Maas J 2012 Large-scale test bench for the durability analysis of magnetorheological fluids *J. Intelligent Material Systems and Structures* published online 17 July 2012 DOI: 10.1177/1045389X12453960

[8] Lide D R 2005 *CRC Handbook of Chemistry and Physics* 86th ed. (CRC Press)

Journal of Physics: Conference Series **412** (2013) 012018 doi:10.1088/1742-6596/412/1/012018

Effect of an applied magnetic field on sloshing pressure in a magnetic fluid

S Kaneko[1], **T Ishiyama**[2] **and T Sawada**[1,3]

[1]Department of Mechanical Engineering, Keio University, 3-14-1 Hiyoshi, Kohoku-ku, Yokohama 223-8522, Japan

[2]Department of Astronautical Engineering, University of South California, 854B Downey Way, RRB 225, Los Angeles, CA 90089-1192, USA.

E-mail: sawada@mech.keio.ac.jp

Abstract. Sloshing of a magnetic fluid can be controlled through altering the sloshing natural frequency by applying magnetic fields. The magnetic field is applied by two permanent magnets. The intensity of the applied magnetic field is changed by varying the position of the magnets and the magnet type. The dynamic pressure responses on the inner wall surface were used to gain an understanding of the sloshing phenomenon in this investigation. Several frequency response spectra are obtained in each experimental condition. These results indicated that the free surface disturbance caused by the sloshing was suppressed by the applied magnetic field.

1. Introduction

The sloshing phenomenon is the oscillation of the free surface of a liquid in a container because of external excitation [1]. Sloshing generates impact pressure and changes the centre of gravity of the container. When the container is excited to a frequency that is near the sloshing natural frequencies, resonance is reached and the free surface is disturbed.

A magnetic fluid is a colloidal liquid in which many ferromagnetic nanoparticles are suspended. It is a functional fluid that responds to an applied magnetic field [2]. The apparent gravity acting on the magnetic fluid varies due to the magnetic body force produced by the magnetic field. Thus it is possible to change the sloshing natural frequencies via the external magnetic field. Therefore, it is possible that sloshing can be controlled through the sloshing natural frequency by applying a magnetic field.

Sloshing of a magnetic fluid has been studied by only a few groups. Sudo et al. [3] investigated the vertical sloshing of magnetic fluid droplets using both rectangular and cylindrical containers. Sawada et al. [4] investigated lateral sloshing of a magnetic fluid in a rectangular container. Also, lateral sloshing of a magnetic fluid in a vertically-applied non-uniform magnetic field was studied by Sawada et al. [5], and the experimental results were compared with those obtained from nonlinear theory. There was an experimental investigation of pressure distribution related to water sloshing by Akyildiz and Unal [6].

In this study, two permanent magnets and an oscillating rectangular container filled with a magnetic fluid subjected to a horizontal magnetic field were used to investigate the influence of an applied magnetic field on sloshing. Small disc-type pressure transducers were placed on the

[3] Corresponding author.

inner wall surface of the container. This experimental approach was based on the measurement of the pressure at several points using pressure transducers, and the real-time computation of a discrete Fourier transform. The dynamic pressure response was thus determined, and the influence of the magnetic field on the sloshing was investigated.

2. Linear theory

A linear theoretical model is used for sloshing in a rectangular container. A two-dimensional fluid domain is considered, which is bound at the bottom and at both sides by the walls of a rectangular container and by the fluid free surface at the top. The container oscillates laterally with a single-frequency excitation $a \cos \omega t$, where a and ω are the amplitude and frequency of the excitation, respectively. The sloshing analytical model is illustrated in figure 1. Here L is the width of the tank, h is the height of the fluid and η is the displacement of the free surface.

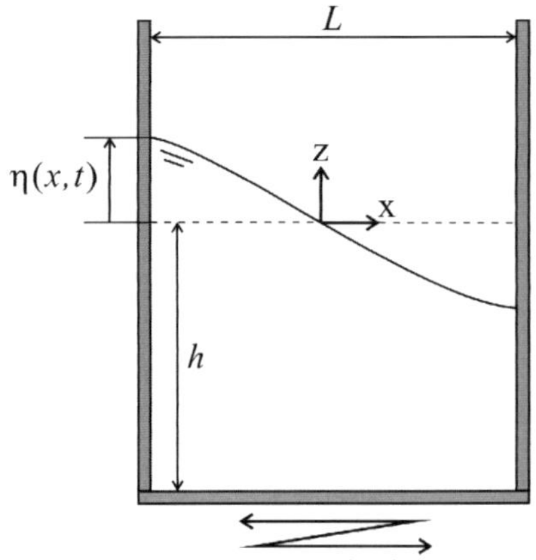

Figure 1. Sloshing analytical model.

The fluid is assumed to be incompressible and inviscid, and the flow is assumed to be irrotational. The mass conservation for an incompressible fluid is written as

$$\nabla \cdot \boldsymbol{V} = 0 \tag{1}$$

where $\boldsymbol{V}$ is the velocity vector. Equation (1) is referred as the continuity equation. Under the assumption of an irrotational flow, the following velocity potential $\Phi\,(x, z, t)$ exists:

$$\boldsymbol{V} = -\nabla\Phi \tag{2}$$

Substituting equation (2) into equation (1) leads to the Laplace equation

$$\nabla^2\Phi = 0 \tag{3}$$

Equation (3) is the governing differential equation for the flow, which is satisfied in the whole domain. To obtain a solution for the sloshing problem, the Laplace equation (3) has to be integrated into the fluid domain.

In order to integrate the governing equation, four boundary conditions have to be considered. The first is the wall boundary condition:

$$\left.\frac{\partial\Phi}{\partial x}\right|_{x=-\frac{L}{2}} = \left.\frac{\partial\Phi}{\partial x}\right|_{x=\frac{L}{2}} = 0 \tag{4}$$

The second is the bottom boundary condition:

$$\left.\frac{\partial \Phi}{\partial z}\right|_{z=-h} = 0 \tag{5}$$

The third is the kinematic free surface boundary condition. After linearization, the boundary condition assuming a small surface wave can be written as

$$\boldsymbol{V} \cdot \boldsymbol{n} = \frac{\partial \eta}{\partial t} \tag{6}$$

where $\boldsymbol{n}$ is the unit vector normal to the surface. Finally, the fourth is the dynamic free surface boundary condition. An unsteady irrotational Bernoulli equation is considered to be:

$$-\frac{\partial \Phi}{\partial t} + \frac{1}{2}\left(u^2 + w^2\right) + gz + \frac{p}{\rho} = C\left(t\right) \tag{7}$$

where $C\left(t\right)$ is a time-dependent integration constant referred to as Bernoulli constant, ρ is the density and g is gravitational acceleration. It is necessary to consider the inertia force due to external excitation. Pressure p can be treated as constant because p is equal to the atmospheric pressure at the free surface. Furthermore, the x and z components of velocity u and w are assumed to be minute. Then, the dynamic free surface boundary condition can be written as

$$\frac{\partial \Phi}{\partial t} + gz - ax\omega^2 \cos \omega t = 0 \tag{8}$$

After solving the Laplace equation (3) using separation of variables, the following dispersion relation is obtained.

$${\omega_1}^2 = gk \tanh kh \tag{9}$$

where k is the wave number associated with the sloshing mode. ω_1 is the first resonant angular frequency for the sloshing mode k. The experimental results are described using dimensionless excitation oscillation frequency ($f^* = f/f_1$), where f_1 is the first resonant frequency obtained from dispersion relation.

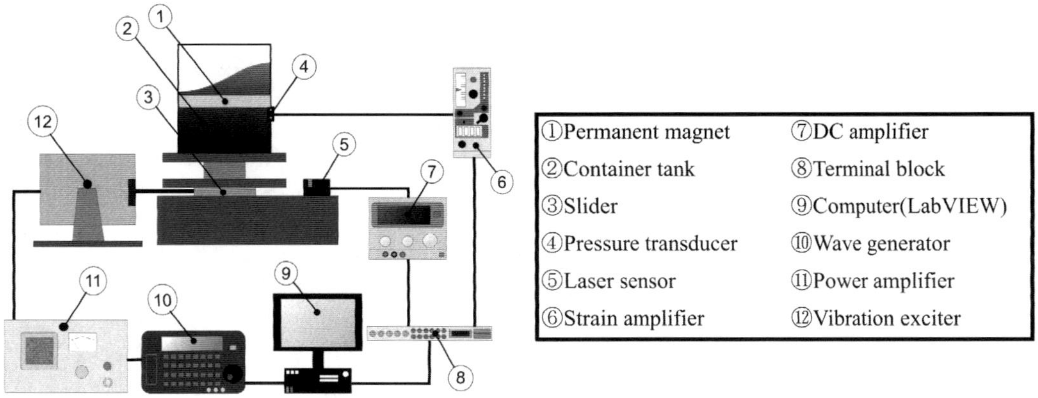

Figure 2. Experiment apparatus.

Table 1. Characteristics of the magnetic fluid.

Density at 25 °C	1.16×10^3 kg/m^3
Viscosity at 27 °C	2.0 mPa · s
Saturation magnetization at 25 °C	16.3 mT

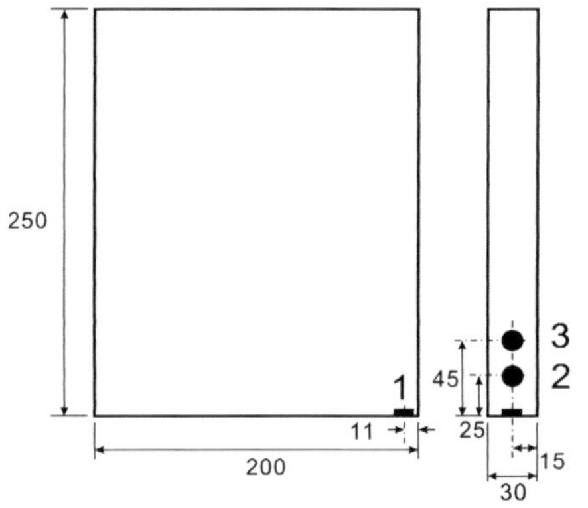

Figure 3. Positions of three pressure transducers.

3. Experiment

3.1. Experimental set-up

Figure 2 shows a schematic diagram of the experimental apparatus. The inner dimensions of the rectangular container are 200 mm × 30 mm × 250 mm, and it is made of a transparent acrylic resin with a thickness of 4 mm. The container is filled with a water-based magnetic fluid. The characteristics of the magnetic fluid used for the experiments are shown in table 1. A slider mechanism under the rectangular container is connected to a vibration exciter via two joints and a stainless cylindrical bar. The vibration exciter is controlled by measurement control software (*LabVIEW*) on a computer. When the vibration exciter generates a vibration, the container oscillates and the free surface of the magnetic fluid is disturbed. There are three small disc-type pressure transducers on the inner wall surfaces of the container, as shown in figure 3. Sensor 1 is placed on the bottom plate and sensor 2 and 3 are placed on the side wall. Pressure signals measured by the transducers are sent to the computer, and a fast Fourier transform (FFT) is used to obtain the amplitude of the pressure signal.

The magnetic fluid depth h was 100 and 60 mm for all the experiments and the excitation frequency f ranged from 0.5 to 3.5 Hz in increments of 0.01 Hz. The amplitude of the oscillation a was 0.5, 1.0, and 1.5 mm for all experiments. The magnetic field was applied by two rectangular parallelepiped permanent magnets of the same type, composed of neodymium or ferrite. These magnets were maintained inside a box made of transparent acrylic resin with a thickness of 3 mm. Figure 4 shows the arrangement of the two permanent magnets. We used couples of neodymium and ferrite magnets. There were two spatial parameters for changing the applied magnetic field intensity. y_m represents the distance between the inner wall surface of the container and the surface of the permanent magnet, and was varied by inserting 5 mm thick acrylic boards in

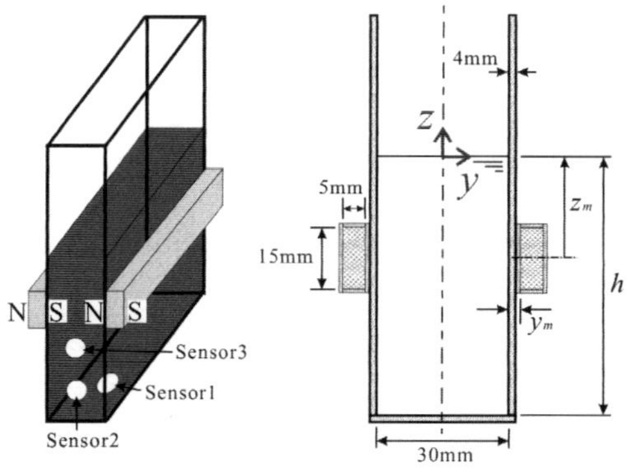

Figure 4. Application of the two permanent magnets.

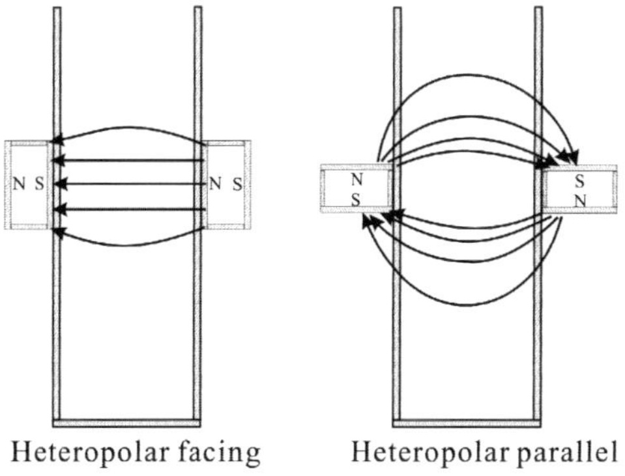

Figure 5. Arrangement of the two magnets.

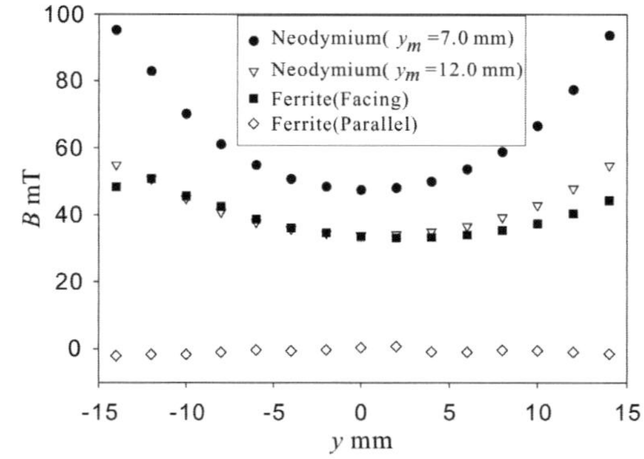

Figure 6. Distribution of the magnetic field on the centre line ($z = 0$).

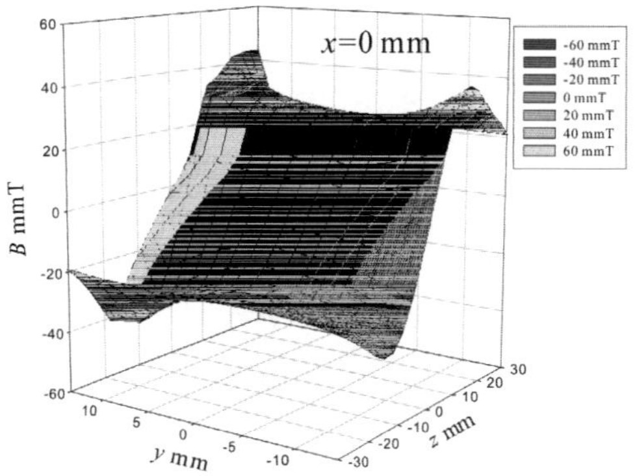

Figure 7. Distribution of the magnetic field (Ferrite, Heteropolar parallel).

between the wall and the magnet. z_m represents the displacement from the free surface of the magnetic fluid at rest to the centre of the magnet. When the permanent magnet is placed below the free surface, the sign of z_m is negative. In the experiments, the excitation frequency, magnetic fluid depth, excitation amplitude, magnet type, parameters y_m and z_m were varied.

3.2. Distribution of the magnetic field

Four types of magnetic fields generated by the neodymium and ferrite magnets were used in the experiments. The neodymium magnets were in a heteropolar facing arrangement with different y_m values of 7.0 mm. The ferrite magnets were in a heteropolar facing or heteropolar parallel arrangement, and both had y_m values of 7.0 mm and 12.0 mm. The arrangements of the two magnets are shown in figure 5. The distribution of the magnetic field intensity between two permanent magnets along their centre line for the four types of magnetic fields is illustrated in figure 6. Here y is the displacement from the centre point of the two magnets, and B is the magnetic flux density. The magnetic field intensity increases near each permanent magnet, and decreases with increasing distance from the permanent magnet. The weakest point of B appears at the centre point between the two magnets. The magnetic flux density is almost zero in the magnetic field generated by two ferrite magnets placed in a heteropolar parallel arrangement.

Figure 7 shows the 3-D distribution of magnetic field for ferrite magnets in a heteropolar parallel arrangement. The distribution of the magnetic field in the z direction is asymmetrical.

4. Results and discussion

The frequency responses of the dimensionless pressure amplitude on the inner wall surface of the container are shown in figure 8. Here the following dimensionless pressure is introduced.

$$P^* = \frac{P}{\rho g L} \tag{10}$$

P is the pressure amplitude. P is taken from the root means square (RMS) amplitude of the signal wave form. The FFT harmonic is considered to be the same as the excitation vibration. The results without the magnetic field are represented by a solid line in figure 8. The relationships between positions of the permanent magnets z_m and either the resonance frequency f_r or dimensionless pressure amplitude at resonance $P^*{}_r$ were obtained from the frequency response. Figure 9 shows the resonant frequency shift as a function of z_m for h = 100 mm and a = 1.0 mm at Sensor 1. Figure 10 shows the dimensionless pressure amplitude

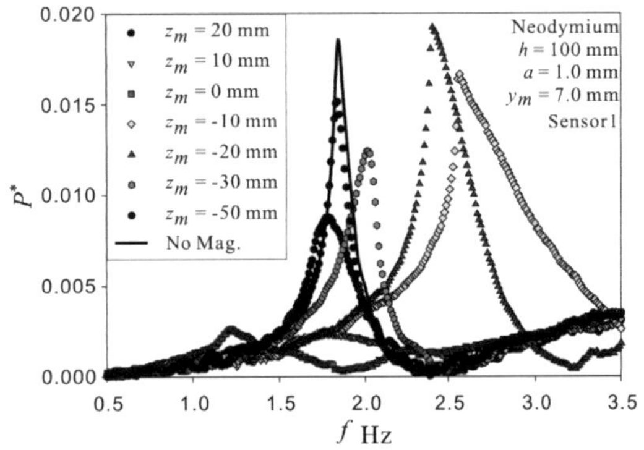

Figure 8. Frequency responses of the pressure amplitude at Sensor 1 (Neodymium).

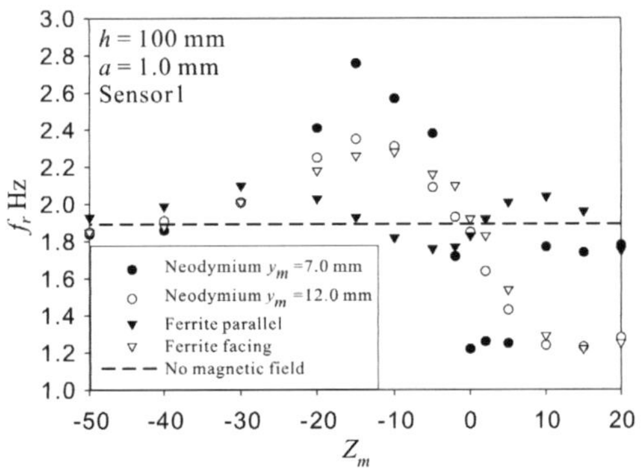

Figure 9. Resonant frequency shift with z_m at Sensor 1.

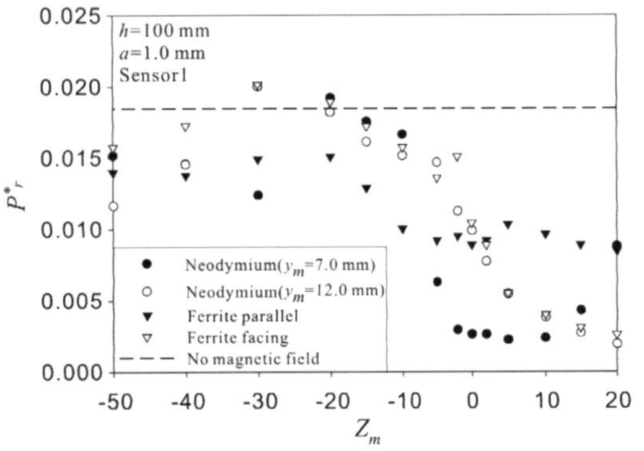

Figure 10. Dimensionless pressure amplitude at resonance with z_m at Sensor 1.

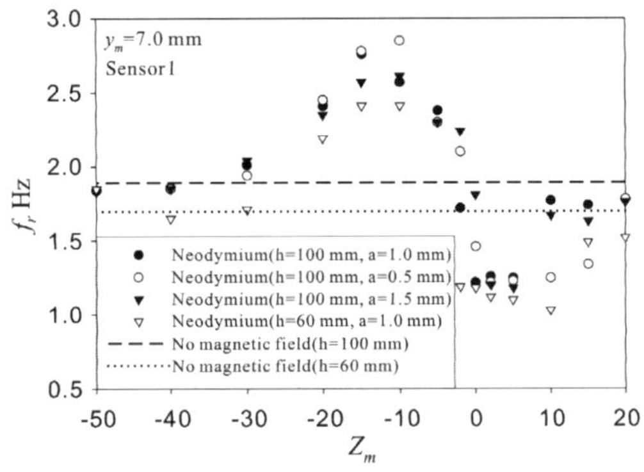

Figure 11. Resonant frequency shift with z_m at Sensor 1

at resonance for $h = 100$ mm and $a = 1.0$ mm at Sensor 1, and figure 11 shows the resonant frequency shift as a function of z_m for $y_m = 7.0$ mm at Sensor 1. The broken line in these figures indicates the resonant frequency without a magnetic field. In the previous study, the difference in f_r and $P^*{}_r$ between $y_m = 7$ mm and 12 mm under a magnetic field generated by neodymium magnets was investigated [7]. In this study we focus on the difference in magnet type, excitation amplitude, and the depth of the magnetic fluid.

When the magnets are above the free surface, the resonant frequency is lower than that without a magnetic field for each condition, as shown in figures 8 and 9. On the other hand, the resonant frequency becomes greater than that without a magnetic field when the magnets are placed below the free surface of the magnetic fluid. These different results are caused by a change in the apparent gravitational force. However, the resonant frequency shift when the ferrite magnets are in a parallel arrangement shows an opposite tendency to the other results because of the distribution of the magnetic field. This was caused by difference in the distribution of magnetic flux between the two magnets and difference in direction of the magnetic flux above and below the magnets. Further detailed analysis of the distribution of the magnetic field is necessary in order to explain this behaviour.

The dimensionless pressure amplitude at resonance $P^*{}_r$ shows lower values than those without a magnetic field, as shown in figure 10. This is because the movement of a magnetic fluid is restricted by a magnetic field. Therefore, it can be said that sloshing in a magnetic fluid can be controlled by a magnetic field from the point of view of the dynamic pressure variation.

The influence of the excitation amplitude a and the magnetic fluid depth h on the resonant frequency are shown in figure 11. There are no large differences caused by excitation amplitude. There are different values of f_r near $z_m = 0$ mm; however, these are caused by instability of feedback control of excitation amplitude. On the other hand, f_r is changed with the magnetic fluid depth. When the fluid depth is large, the resonant frequency is also large with or without magnetic fluids. Therefore, the magnetic fluid depth has no relevance to controlling sloshing by a magnetic field.

5. Conclusion

The dynamic pressure change of a magnetic fluid sloshing in a rectangular container under lateral excitation was investigated experimentally. The resonant frequency was changed by varying the positions of the magnets, the type of magnetic field and the arrangement of magnets. Notably, the range of the resonant frequency and the dimensionless pressure amplitude at resonance was changed significantly by varying z_m. Based on these results, it can be concluded that a magnetic field applied near the free surface of a magnetic fluid can be used to control the sloshing phenomenon of the fluid. It is important to note that sloshing does not depend on the excitation amplitude or on the magnetic fluid depth.

Acknowledgements

We would like to thank Mr. Jean Klingler of Souriau Japan K. K. for his experimental help. This work was partially supported by a Grant-in-Aid for Scientific Research (C) from the Japan Society for the Promotion of Science.

References

[1] Ibrahim R A 2005 *Liquid sloshing dynamics* (Cambridge: Cambridge University Press) chapter 1
[2] Fertman V E 1990 *Magnetic fluids guidebook* (Washington: Hemisphere Publishing) chapter 1
[3] Sudo S, Hashimoto H, Iked A and Katagiri K 1987 *J. Mag. Mag. Mater.* **65** 219-222
[4] Sawada T, Kikura H, Shibata S and Tanahashi T 1993 *J. Magn. Magn. Mater.* **122** 424-427
[5] Sawada T, Kikura H and Tanahashi T 1999 *Exp. Fluids* **26** 215-221
[6] Akyildiz H and Unal E 2005 *Ocean Eng.* **32** 1503-16
[7] Kaneko S, Klingler J, Ishiyama T and Sawada T 2012 *J. Jpn. Soc. Appl. Electromag. Mech.* **20** 416-421

Investigation of cluster growth in MR fluids using ultrasonic wave propagation

A Isnikurniawan[1], J Kuroiwa[2], S Abell[1] and T Sawada[1,3]

[1]Department of Mechanical Engineering, Keio University, 3-14-1 Hiyoshi, Kohoku-ku, Yokohama 223-8522, Japan
[2]Panasonic Corporation, Circuit Components Business Unit Industrial Devices Company, Toyo-oka City, Hyogo 668-0298, Japan

E-mail: sawada@mech.keio.ac.jp

Abstract. This research reports the experimental results of a study on cluster growth in Magneto-rheological (MR) fluid. The experiments were conducted by applying different magnetic field (M/F) sweep rates to MR fluid. A magnetic field was swept from 0 to 400 mT at different times. Since the MR fluid is opaque, an ultrasonic measurement method was applied. Ultrasonic propagation velocity (V) in MR fluid is dependent on magnetic field and is strongly related to cluster formation in this fluid. Therefore, ultrasonic propagation velocity was measured and the change of ultrasonic propagation velocity ($\Delta V/V_0$) was calculated. Based on the experimental results, cluster growth in MR fluids can be analyzed experimentally. Effect of volume fraction in MR fluid on cluster growth is presented. Furthermore, the relation between the change of ultrasonic propagation velocity and M/F sweep rate is discussed.

1. Introduction

Magnetorheological (MR) fluids are a type of magnetic functional fluid that changes from a free-flowing fluid to a semi-solid form on the application of a magnetic field. These fluids are formed from micron-sized magnetic particles and a carrier liquid. It is well-known that the magnetic particles in MR fluids will coagulate and form clustering structures parallel to magnetic force lines [1]. These clustering structures restrict the motion of the fluid, thereby increasing the viscous characteristics of the suspension [2]. The most important advantage of these fluids over conventional mechanical interfaces is their ability to achieve a wide range of viscosity (several orders of magnitude) in a fraction of millisecond [3].

The growth of clusters in a ferrofluid has been investigated by Józefczak et al. [4]. They reported that the ultrasonic propagation velocity changes as a function of the magnetic field sweep rate and the temperature of the ferrofluid. Cluster growth in MR fluids is rare. However, understanding the growth of clusters in MR fluids is important for developing applications of these fluids. Therefore, it is important to investigate cluster growth in MR fluids.

Several experimental methods have been used to investigate the inner structure of MR and magnetic fluids. López-López and Vicente [5] proposed a new method by measuring the change in the electromotive force induced in a sensing coil which was placed around the settling suspension. Their method can be used to analyze the effect of the magnetite volume fraction on the stability

[3] Corresponding author.

of the MR Fluid. Jolly et al. [6] conducted magneto-induction experiments to measure the microstructure response of quiescent MR fluid to an applied magnetic field. They measured the time scales associated with microstructure development by recording the evolution of fluid polarization in response to a step change in applied magnetic field. Sawada et al. [7] have applied another measurement method. They applied Rayleigh light scattering technique to visualize the inner structure of a diluted magnetic fluid with and without magnetic field.

In this study, experiments were performed by applying different magnetic field (M/F) sweep rates to MR fluids. Because MR fluids are opaque, an ultrasonic measurement method was utilized. The ultrasonic propagation velocities in MR fluids [8] were changed under the influence of an external magnetic field. Therefore, the ultrasonic propagation velocity was measured, and the change in the ultrasonic propagation velocity was calculated for various M/F sweep rates. Based on the experimental results, an analysis of cluster growth in the MR fluids is presented. In addition, the relationship between the change in the ultrasonic propagation velocity and the M/F sweep rate is discussed.

2. Experimental

A block diagram of the experimental apparatus is illustrated in figure 1. The MR fluid was placed in a rectangular container made from acrylic plates. Two ceramic oscillators were attached to the container. The ultrasonic frequency of these oscillators was 2 MHz. One oscillator acting as a transmitter was connected to the pulse generator, while the other, which acted as a receiver, was connected to the oscilloscope. A pulse generator was used to send an ultrasonic wave to the oscillator, while the digital oscilloscope was used to monitor the ultrasonic propagation between these oscillators. The oscilloscope was also used to measure the ultrasonic propagation properties such as the ultrasonic propagation velocity.

An electromagnet was used to generate the magnetic field in the test cell. The magnetic field was applied in a direction parallel to the ultrasonic propagation direction, as shown in figure 1. The magnetic flux density of the electromagnet was varied from 0 to 400 mT for different lengths of time, and the density of the magnetic flux was adjusted by changing the current in the power supply. A function generator was applied to adjust the power supply in order to produce

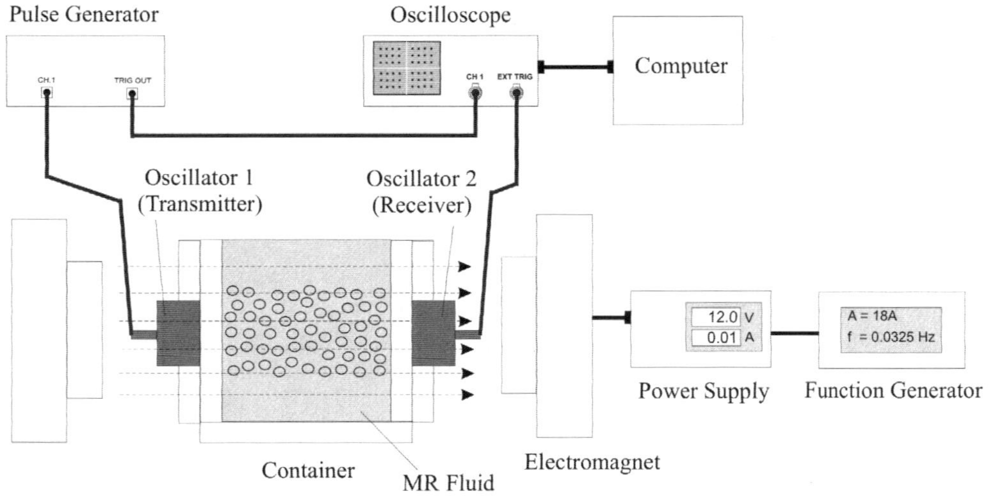

Figure 1. Block diagram of experimental apparatus.

Table 1. Properties of MR fluids at 40 °C.

Property	Value	
Serial name	MRF122-EG	MRF132-DG
Particle material	Iron	Iron
Mean particle size	3 - 10 μm	3 - 10 μm
Volume fraction	0.22	0.32
Carrier liquid	Hydrocarbon oil	Hydrocarbon oil
Viscosity	42 ± 20 mPa · s	92 ± 15 mPa · s
Density	2.28 - 2.48 × 10^3 kg/m^3	2.98 - 3.18 × 10^3 kg/m^3

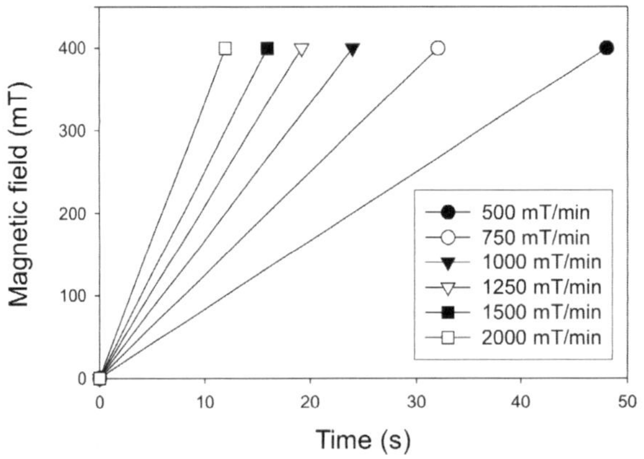

Figure 2. Magnetic flux density (mT) vs. time (s) for various M/F sweep rate.

different M/F sweep rates.

In this experiment, two types of MR fluids, MRF122-EG and MRF132-DG from Lord Corporation, were used. The main difference between these MR fluids is the volume fraction of magnetic particles, which is 22 % for MRF122-EG and 32 % for MRF132-DG. The detailed properties of these fluids are given in Table 1.

For all the experiments, the temperature of the test cell was maintained at 25 °C, and M/F sweep rates of 500, 750, 1000, and 1250 mT/min were applied. For example, at an M/F sweep rate of 500 mT/min, 400 mT of magnetic flux density was swept in 48 sec. Figure 2 presents more detailed information on how the magnetic flux density was swept.

The ultrasonic propagation velocity was measured, and the change in the ultrasonic propagation velocity ($\Delta V/V_0$) was calculated using the following equation:

$$\frac{\Delta V}{V_0} = \frac{V - V_0}{V_0} \tag{1}$$

where V and V_0 are the ultrasonic propagation velocity with and without the applied magnetic field, respectively.

3. Result and discussion

Figure 3 shows the values of $\Delta V/V_0$ for (a) MRF122-EG and (b) MRF132-DG at various M/F sweep rates. At different M/F sweep rates, the value of $\Delta V/V_0$ increases with an increase in the magnetic field. It is well known that this change is due to the formation of clustering structures; when the magnetic field increases, more clusters are formed [1].

Figures 3 (a) and (b) show that the influence of the volume fractions of the MR fluid on $\Delta V/V_0$ is significant. At a lower volume fraction (MRF122-EG), the value of $\Delta V/V_0$ is lower than that for the higher volume fraction, and the cluster size is proportional to the volume fraction. As a result, the value of $\Delta V/V_0$ increases when the volume fraction of the MR fluid increases. In other words, the size of the cluster increases for higher volume fractions. This result suggests that the volume fraction and M/F sweep rate affect the formation of the clustering structures in MR fluids.

Figure 4 shows the maximum values of $\Delta V/V_0$ for the two types of MR fluids at various M/F sweep rates. From the figure, it can be seen that the maximum value of $\Delta V/V_0$ decreases at higher applied M/F sweep rates. It appears that the magnetic particles in the MR fluids do not have sufficient time to form clusters of greater size at high M/F sweep rates. Therefore, the

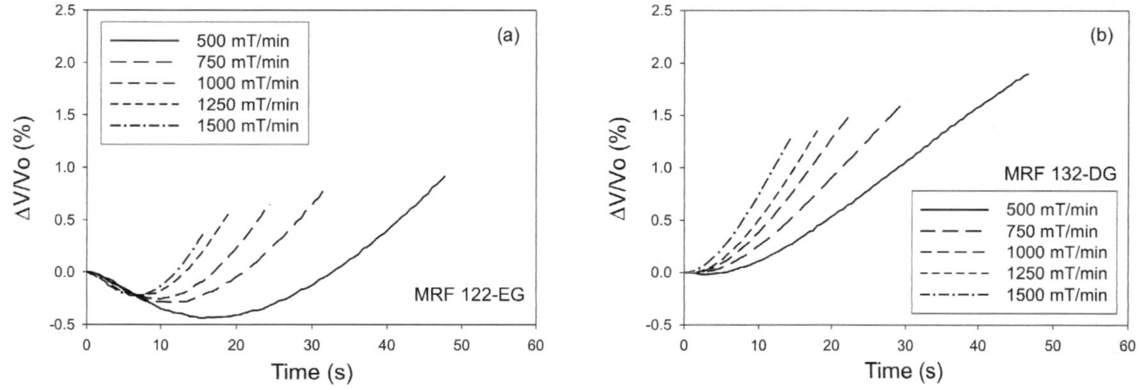

Figure 3. Change of ultrasonic propagation velocity in (a) MRF122-EG and (b) MRF132-DG vs. time for various M/F sweep rates.

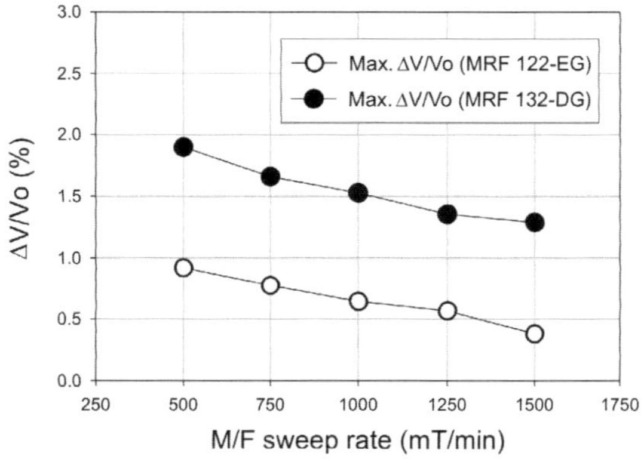

Figure 4. Maximum value of change in ultrasonic propagation velocity vs. M/F sweep rates.

maximum size of the cluster decreases at higher M/F sweep rates. This result indicates that the M/F sweep rate affects the maximum size of the clusters in MR fluids.

4. Concluding remarks

Experiments were conducted to investigate cluster growth in MR fluid. Two different volume fractions of MR fluids were subjected to magnetic fields using different sweep rates. The size of the clusters increased for higher volume fractions. The maximum value of $\Delta V/V_0$ decreased at higher applied M/F sweep rates. The results indicate that the volume fraction and M/F sweep rate affect the formation of the clustering structures in MR fluids.

Acknowledgments

A. Isnikurniawan gratefully acknowledges partial financial support from the ASEAN University Network-Southeast Asia Engineering Education Development Network Project of the Japan International Cooperation Agency and KLL PhD. Program Research Grant of Keio University. This research has been partially supported by a Grant-in-Aid for Scientific Research (C) from the Japan Society for the Promotion of Science.

References

[1] Bramantya M A, Motozawa M and Sawada T 2010 *J. Phys.: Condens. Matter.* **22** 204117 (5pp)
[2] Jolly M R, Bender J W and Carlson J D 1998 *Proc. SPIE* **3327** 262-274
[3] Bossis G, Volkova O, Lacis S, and Meunier A 2002 *Lect. Notes Phys.* **594** 202-230
[4] Józefczak A, Skumiel A and Łabowski M 2003 *J. Magn. Magn. Mater.* **258-259** 474-476
[5] López-López M T and de Vicente J 2005 *J. Mater. Res.* **20** 874-881
[6] Jolly M R, Bender J W and Mathers R T 1999 *Int. J. Mod. Phys. B* **13** 2036-2043
[7] Sawada T, Kikura H, Yamanaka G, Matsuzaki M, Aritomi M and Nakatani I 1999 *Proc. SPIE* **3783** 389-396
[8] Motozawa M, Sawada T and Matsumoto Y 2007 *Int. J. Mod. Phys. B* **21** 4914-4921

13th Int. Conf. on Electrorheological Fluids and Magnetorheological Suspensions (ERMR2012) IOP Publishing
Journal of Physics: Conference Series **412** (2013) 012020 doi:10.1088/1742-6596/412/1/012020

MRF in a plate-plate magnetorheometer: Numerical insight into the particle-wall interface

H G Lagger[1], C Bierwisch and M Moseler

MikroTribologie Centrum μTC,
Fraunhofer Institute for Mechanics of Materials IWM,
Woehlerstrasse 11, 79108 Freiburg, Germany

E-mail: hanna.lagger@iwm.fraunhofer.de

Abstract. Particle-based simulations are a suitable tool to gain insight into the microstructural behavior of a magnetorheological fluid (MRF). For the application of MRF in clutches, the amount of torque transmission is a matter of particular interest. Concerning the contact between iron particles and clutch walls, several questions arise: Is a higher wall roughness beneficial for torque transmission? What is the influence of wall magnetism on torque transmission? What are the mechanisms on particle scale that lead to increased torque transmission? Inspired by a previous experimental study, we performed three-dimensional simulations based on the Discrete-Element-Method (DEM) with different wall roughnesses and different magnetic conditions to investigate the mechanisms of shear stress transmission at particle level. The simulations show that a higher wall roughness leads to higher torque transmission only in the case of non-magnetic walls. For ferromagnetic walls, no influence of wall roughness on torque transmission is observed. This is in qualitative agreement with the experimental results.

1. Introduction

Magnetorheological fluids consist of micron-sized iron particles suspended in a carrier oil. Upon application of an external magnetic field, chains of iron particles are formed along the magnetic field lines. Due to these particle chains, the MRF changes from a liquid-like to a near solid-like state. This adaptive behaviour makes MRF interesting for several industrial applications, such as dampers, clutches or actuators.

For the application in clutches, it is important to have a high amount of torque transmission. In this study, particle-based simulations were performed to gain insight into the mechanisms inside the clutch. With an understanding of the mechanisms on particle level, useful hints for the design of magnetorheological clutches with regard to high torque transmission can be derived.

Experimental studies by Laun *et al.* [1] on a plate-plate magnetorheometer (see Figure 1 A) show a significant effect of plate material (non-magnetic vs. ferromagnetic) and plate roughness on the measured shear stress and, thus, on the torque transmission. Different plate roughnesses were obtained by grooved plates with a defined groove geometry on the one hand (see Figure 1 B) and flat plates without grooves on the other hand. In the case of non-magnetic plates, an increase

[1] To whom any correspondence should be addressed.

of plate roughness results in a considerable increase of shear stress. For ferromagnetic plates, no influence of roughness on shear stress could be found.

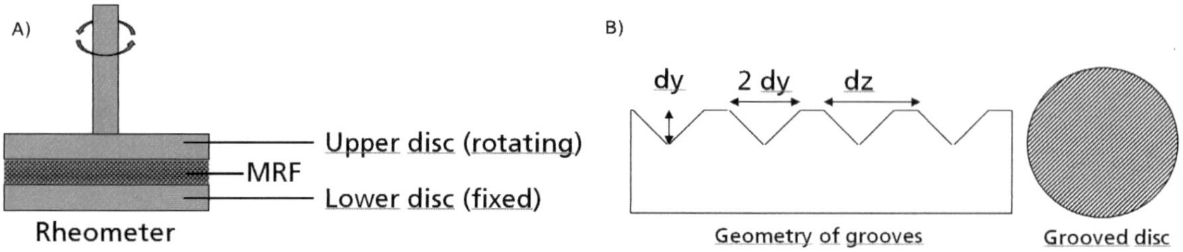

Figure 1. Schematic drawing of a magnetorheometer (A) and groove geometry (B) as used by Laun *et al.* [1].

These effects were explained by the differences of the flux density distribution in the MRF. In the case of non-magnetic plates, the flux density reaches a maximum inside the grooves. The iron particles always move to areas of higher magnetisation. Thus, the iron particles adhere magnetically to the grooves and act as an anchor for the particle chains. This leads to an increase in torque transmission with increasing groove size. In the case of ferromagnetic plates, the flux density inside the grooves is minimal. Therefore, no iron particles will be found inside the grooves in this case.

In this work, the role of varied disc roughness for magnetic and non-magnetic material is investigated numerically. The particle-based approach allows detailed insight into the dynamics of the iron particles. Anchor effects, if existing, can be observed directly.

2. The Discrete-Element-Model

The numerical description of the MRF is based on the Discrete-Element-Method, originally proposed by Cundall and Strack [2]. In this method, the forces on each particle are computed in each time step. Afterwards, Newton's equation of motion is integrated explicitly. Thereby, the positions of all particles are obtained for the following time step.

The forces acting on the particles are magnetic interactions, the hydrodynamic force of the fluid on the particles and a repulsive force, which describes collisions between particles and between particles and walls.

2.1. Magnetic interaction

The MRF particles are modelled as magnetic dipoles. The force between two dipoles i and j depends on the respective magnetic moments $\mathbf{m}_i$ and $\mathbf{m}_j$. The magnetic moments are calculated by the "mutual dipole model" as described in Keaveny *et al.* [3], i. e., for the calculation of the local magnetic field at a particle's position not only the external magnetic field is taken into account, but also the contribution of every other magnetic particle. The magnetisation $\mathbf{M}$ of a particle, which is defined as the magnetic moment per particle volume, is described by the following magnetisation curve [4]

$$M = |\mathbf{M}| = M_S \tanh\left(\frac{\chi}{M_S} H_{in}\right) \tag{1}$$

where χ is the dimensionless magnetic susceptibility of the material, M_S is the saturation magnetisation (in A/m) of the particles and H_{in} is the absolute value of the magnetic field (in A/m) inside the particle. For our simulations, we chose $\chi = 5.25$ and $\mu_0 M_S = 2.1\ T$. $\mu_0 = 4\pi \cdot 10^{-7}\ \frac{Tm}{A}$ is the permeability of free space.

The force on dipole i caused by dipole j is given by

$$\mathbf{F}_{ij}^{mag} = \frac{3\mu_0}{4\pi}\left[\frac{(\mathbf{m}_i\cdot\mathbf{m}_j)\,\mathbf{r}_{ij} + (\mathbf{m}_j\cdot\mathbf{r}_{ij})\,\mathbf{m}_i + (\mathbf{m}_i\cdot\mathbf{r}_{ij})\,\mathbf{m}_j}{r_{ij}^5} - 5\frac{(\mathbf{m}_i\cdot\mathbf{r}_{ij})\,(\mathbf{m}_j\cdot\mathbf{r}_{ij})\,\mathbf{r}_{ij}}{r_{ij}^7}\right]. \quad (2)$$

Here, $\mathbf{r}_{ij} = \mathbf{r}_i - \mathbf{r}_j$ is the vector connecting the centers of particle i and j and $r_{ij} = |\mathbf{r}_{ij}|$ is its norm.

In the case of inhomogeneous external magnetic fields, there are additional forces on the particles due to gradients in the external magnetic field. The force on particle i due to flux density $\mathbf{B_i}$ at the position of particle i is given by $\mathbf{F}_i^{mag} = \boldsymbol{\nabla}(\mathbf{m}_i\cdot\mathbf{B}_i)$, where $\boldsymbol{\nabla}$ is the spatial gradient. Thus, the total magnetic force on particle i reads

$$\mathbf{F}_i^{mag} = \sum_{\substack{j \\ j\neq i}} \mathbf{F}_{ij}^{mag} + \boldsymbol{\nabla}\left(\mathbf{m}_i\cdot\mathbf{B}_i^{ext}\right), \quad (3)$$

where $\mathbf{B}_i^{ext} = \mu_0\mathbf{H}_i^{ext}$ and $\mathbf{H}_i^{ext}$ is the externally applied magnetic field at the position of particle i. The required inhomogeneous magnetic fields were computed with the Finite-Element-Method and were imported into the DEM simulations.

2.2. Hydrodynamic interaction

The hydrodynamic interactions are modelled as a Stokes drag force of the carrier fluid on the magnetic particles,

$$\mathbf{F}_i^{Stokes} = 6\pi\eta R_i\left(\mathbf{v}_{fluid} - \mathbf{v}_i\right), \quad (4)$$

where $\mathbf{v}_i$ is the velocity of particle i, R_i is the particle radius, $\mathbf{v}_{fluid}$ is the velocity of the carrier oil at the particle's position and η its viscosity. The simulations were performed with a shear rate of 1000/s. The oil viscosity was chosen as $\eta = 5$ mPa s. The fluid velocity in the case of flat walls can be described as a linear Couette flow. For the simulations with rough walls, the required transient velocity fields were computed with the Finite-Element-Method and were imported into the DEM simulations.

2.3. Numerical rheometer

The three-dimensional model of a plate-plate magnetorheometer consists of two parallel walls confining the MRF in vertical direction. In the two horizontal directions, periodic boundary conditions are applied. The MRF particles are modelled as magnetic spheres. Each wall is modelled as an ensemble of overlapping, non-magnetic spheres (see Fig. 2).

All spheres have a diameter of 5.0 μm. The simulations were performed with 33 % volume fraction of particles in the suspension. To investigate the influence of wall roughness and magnetic flux density distribution, simulations with different wall profiles and with varying flux density profiles were performed.

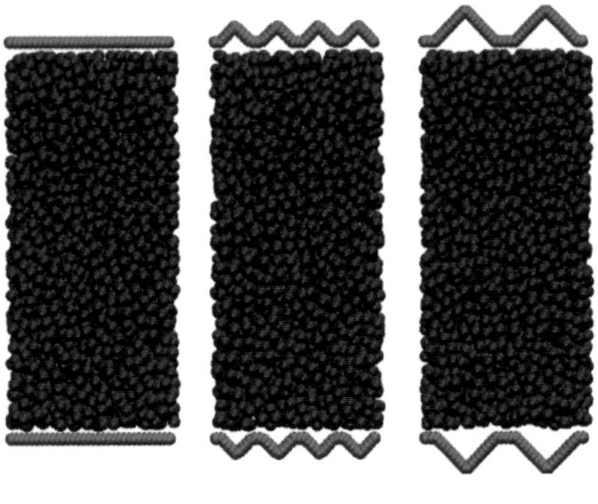

Figure 2. Numerical rheometer with flat walls (left), with small (middle) and large grooves (right). The MRF particles are modelled as spheres, the walls as ensembles of overlapping spheres. In horizontal directions, periodic boundary conditions are applied.

Variation of wall profiles:

- Walls without grooves (flat walls).
- Walls with small grooves. Wall profile as in Figure 1 B, with $dy = 6\ \mu m$ and $dz = 16\ \mu m$.
- Walls with large grooves. Wall profile as in Figure 1 B, with $dy = 12\ \mu m$ and $dz = 32\ \mu m$.

Variation of flux density profiles:

- Homogeneous flux density
- For grooved walls:
  - Flux density *minima* in the grooves (as in the case of ferromagnetic discs)
  - Flux density *maxima* in the grooves (as in the case of non-magnetic discs)
- For flat walls: Alternating flux density minima and maxima along the wall.

3. Results

3.1. Anchor effect

Visualisations of the simulations confirm that the particles move to the locations of highest flux density, as predicted by theory. The particle-based nature of the simulations allows to determine whether or not an anchor effect is present in the shear cell. We define that *no* anchor effect is present, if slip does occur directly at the interface between the walls of the shear cell and the nearby layer of MRF particles. On the other hand, we define that an anchor effect is present, if the region where slip occurs is at least one particle diameter away from the wall, not including the grooves.

In Figure 3, snapshots from simulations with homogeneous flux density are shown. The particles are coloured according to the initial horizontal position (Figure 3 A). If an anchor effect is present, the particle layers below the upper wall and above the lower wall should have the same colour as the adjacent wall, since the position relative to the wall did not change. In the case of homogeneous magnetic field, no anchor effect can be observed considering the differing colours of the wall and adjacent particles.

Snapshots from simulations with inhomogeneous flux density distribution are shown in Figure 4. In the case of flux density maxima in the grooves, MRF particles are pinned to the wall inside the grooves and act as an anchor to the other particles. In the case of flux density minima in the grooves, the particles are anchored at the flux density maxima outside the grooves. Thus, there is no slip at the particle-wall interface in both cases. Additionally, we simulated a case where the walls are flat, but where an inhomogeneous flux density profile is present nonetheless, with alternating minima and maxima along the wall. Also in that case,

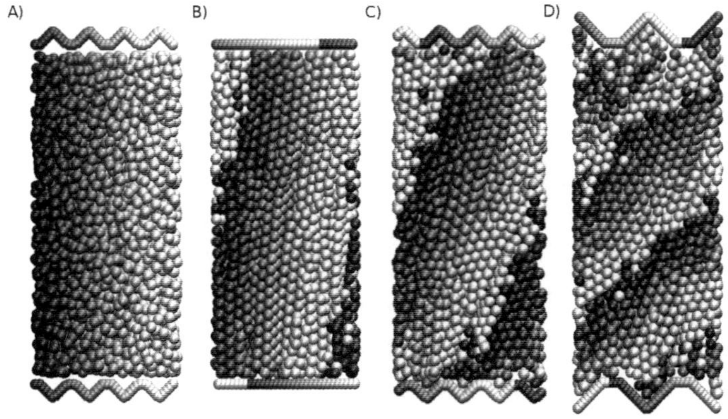

Figure 3. Snapshots from simulations with homogeneous flux density. A) Walls with small grooves, initial configuration. B) Flat walls, snapshot after 1.5 ms. C) Walls with small grooves, snapshot after 1.0 ms. D) Walls with large grooves, snapshot after 1.5 ms. Colouring according to initial horizontal position.

an anchor effect is observed, indicating that the presence of grooves is not necessary for the manifestation of an anchor effect.

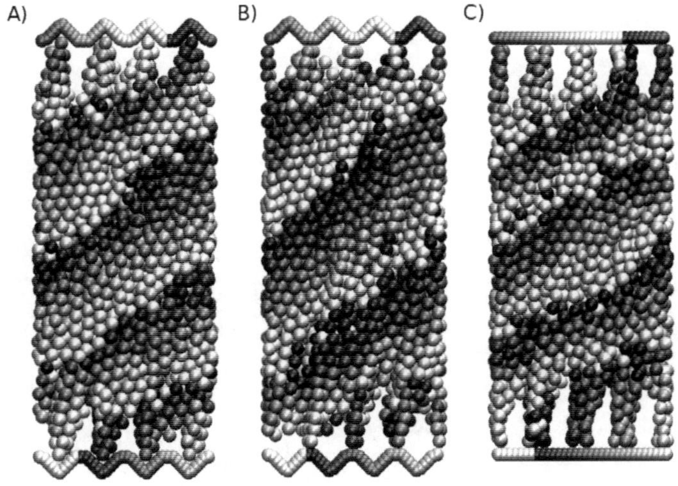

Figure 4. Snapshots from simulations with inhomogeneous flux density. A) Walls with small grooves; flux density maxima inside the grooves; anchor effect inside the grooves. B) Walls with small grooves; flux density minima inside the grooves; anchor effect outside the grooves. C) Flat walls; alternating flux density maxima and minima along the wall; anchor effect at regions of flux density maxima. Snapshots were taken after 1.5 ms simulation time. Colouring according to initial horizontal position.

The results are summarised in Table 1.

Table 1. The presence of an anchor effect in the simulations. Yes: Anchor effect was observed in the simulation. No: Anchor effect was not observed.

	Homogeneous flux density	Inhomogeneous flux density	
		Minima in the grooves	Maxima in the grooves
No grooves	No	Yes	
Small grooves	No	Yes	Yes
Large grooves	No	Yes	Yes

3.2. Shear stress
The calculation of shear stress in the simulation is based on the Kirkwood stress tensor [5,6]. The results are shown in Table 2.

Table 2. Shear stress in the simulation in kPa. Values are averaged from two independent runs.

	Homogeneous flux density	Inhomogeneous flux density Minima in the grooves	Inhomogeneous flux density Maxima in the grooves
No grooves	2.04	3.80	
Small grooves	3.97	4.01	4.09
Large grooves	4.13	3.94	4.82
Relative stress change from small to large grooves	4.0 %	-1.7 %	17.8 %

The shear stress is lowest in the case of no flux density gradient and no grooves. The change from small to large grooves only results in a significant shear stress increase in the case of flux density maxima in the grooves.

4. Discussion
4.1. Comparison with experimental results
The anchor effect was proposed by Laun *et al.* [1] to account for differences in the measured shear stress in the MRF shear cell.

In the simulations, the anchor effect can be observed both for flux density minima and for flux density maxima inside the grooves. Even in the case of flat walls an anchor effect can be produced by a suitable magnetic flux density profile. This demonstrates that the attractive magnetic force of the wall is sufficient to pin particles to the wall, irrespective of the wall profile. In the case of homogeneous flux density, no anchor effect can be observed.

In the case of flux density maxima inside the grooves, large grooves show an increase of the shear stress by 17.8 % compared to small grooves. This effect is not observable for a flux density profile with minima inside the grooves. These findings qualitatively are in agreement with the experimental findings of Laun *et al.* [1]. They observed that in the case of non-magnetic walls, corresponding to flux density maxima inside the grooves, a higher wall roughness leads to higher shear stress transmission. In the case of ferromagnetic walls where flux density is minimal inside the grooves, no such increase of shear stress was measured. This was explained by the presence of anchor effects in the former and by the absence of anchor effects in the latter case.

From the simulations, no correlation between anchor effect and shear stress can be deduced, since an anchor effect is present in all cases of inhomogeneous flux density distribution. Furthermore, in the simulations with homogeneous flux density, there is a significant increase of shear stress from flat walls to walls with small grooves. In both cases however, no anchor effect occurs. Switching from small grooves and homogeneous flux density to small grooves with flux density minima inside the grooves, no significant difference in shear stress is measured, although anchor effects occur in one case and not in the other. Thus, according to the present results, the anchor effect - as it was defined in this paper - seems to be neither sufficient nor necessary for an increase of shear stress in the simulations.

It remains to say that Laun *et al.* might have had a different definition of anchor effect in mind, e.g. only considering anchoring inside the grooves. In that case the explanations of the effects by Laun *et al.* are not contradictory to the simulation results presented here.

4.2. Predictive power of the model

The presented model reproduces correctly several aspects of magnetic particle dynamics: movement of the particles in direction of the flux density gradient, chain formation along the magnetic field lines, and shearing and breaking of the chains due to wall and fluid movement can be observed in the simulations. The scope of validity of the model can be limited due to the following aspects:

- Finite-Size-Effects: Due to the high computational effort, the simulations were performed with smaller systems than the experiments in Laun *et al.* [1]
- The fluid phase was treated as a background velocity field. There is force coupling from the fluid on the particles, but not vice-versa.
- The magnetic particles are modelled as magnetic dipoles. To treat the near field interaction more accurately, also higher multipoles could be taken into account (see Ref. [3]).

It remains to be investigated if the results presented here can be generalised to bigger systems and to what extend the mentioned approximations influence the simulation results.

5. Conclusions

In this work, we presented a model to simulate magnetorheological fluids at the particle-level based on the Discrete-Element-Method. The principal characteristics of magnetic particle dynamics can be observed in the simulations.

The shear stress in the simulations was analysed and discussed with respect to the presence of anchor effects. In the simulations with inhomogeneous magnetic fields, the MRF particles are pinned to the walls at the locations of maximum flux density. This so-called anchor effect can be observed in all cases: When the flux density maximum is located inside the grooves, but also when it is located between the grooves, and when there are no grooves at all. The magnetic field gradient alone is sufficient to keep particles pinned at the wall, independent of the geometrical details. In the simulations with homogeneous magnetic field, no anchor effect was found.

The analysis of shear stress shows, that for flux density maxima inside the grooves, an increase of wall roughness results in an increase of shear stress. Similar results were found in experiments with non-magnetic wall material, which corresponds to flux density maxima inside the grooves. For flux density minima inside the grooves, no significant difference in shear stress could be observed in experiments or simulations with varied wall roughness. Based on the observations of this study, these findings cannot be explained by the presence or absence of anchor effects, since anchor effects and high shear stresses were not found to be systematically related.

6. Acknowledgements

Part of this work was financially supported by the Federal Ministery of Education and Research, Germany.

7. References

[1] Laun H M, Gabriel C and Kieburg C 2011 *Rheol. Acta* **50** 141–57
[2] Cundall P A and Strack O D L 1979 *Geotechnique* **29** 47–65
[3] Keaveny E E and Maxey M R 2008 *J. Comput. Phys.* **227** 9554–71
[4] Jiles D C and Atherton 1986 *J. Magn. Magn. Mater.* **61** 48–60
[5] Larson R G 1999 *The Structure and Rheology of Complex Fluids* (New York: Oxford University Press) 360–87
[6] Bardet J P and Vardoulakis I 2001 *Int. J. Solids Struct.* **38** 353–67

Adhesion peculiarities at the contact of MRF with solid state in gradient magnetic field

V A Kuzmin[1], E V Korobko[1,4], V P Roizman[2], Z A Novikova[1], A P Dostanko[3] and A O Karabko[3]

[1]A.V. Luikov Heat and Mass Transfer Institute of NAS Belarus, 220072, 15 P. Brovka str., Minsk, Belarus
[2]Chmelnitski State University, 29016, Institutskaya str. 11, Khmelnytskyi, Ukraine
[3]Belarusian State University of Informatics and Radioelectronics, 220072, 6 P. Brovka str., Minsk, Belarus

E-mail: evkorobko@gmail.com

Abstract. The results of studying surface properties of MRF, obtained by experimental investigation of the influence of the magnetic field on adhesion indices and spreading characteristics of the finite volume (a drop) of the suspension are reported.

1. Introduction

Formation of the structure of magnetic fluids and magnetorheological suspensions from the magnetization of the elementary particles and its effect on flow patterns, despite the continuous study [1–7], is still not well understood. This is especially true of finite volumes of fluids during the spreading with the open outer boundary in the form of a layer on the surface of objects, in particular in terms of gradient magnetic fields. However, this problem is relevant when creating the technology of formation films with oriented structure. This work is undertaken to address this gap. The influence of external fields on the relaxation processes and surface free energy is investigated in relation to the more conductive metallic liquids. In fact, a number of papers are dedicated to the problem of determining the influence of the magnetic field on the spreading of metals on the surface. In [6, 7] a theoretical and experimental analysis of spreading mercury droplets on the surface of polycrystalline zinc is carried out. Several stages corresponding to different regimes of spreading that are determined in the current were discovered. The influence of various internal (surface tension, viscosity, diffusion, dissolution, chemical reaction) and external (gravitational and electric fields) factors on the metallic liquid spreading on a solid surface has been determined. In [6] it was theoretically demonstrated the possibility of the magnetic field to influence the kinetics of spreading of the metal drop. Assuming that the influence of the magnetic field can be carried out through the ponderomotive force, and taking as the driving forces of the spreading a surface tension force, a viscous friction and Lorentz force, based on the solution of the Navier-Stokes equation a continuity equation was obtained for the dependence of the spreading on the intensity of the magnetic field. Its solution gave a result that can be interpreted as an equivalent increase in the viscosity of the fluid.

There are also studies on the influence of external fields on the surface phenomena of structuring fluids – electrorheological fluids (ERF), that differ by a powerful force interaction of polarized

[4]Corresponding author.

particles of the filler [8, 9], as well as magnetic (colloidal) liquids [10, 11], that do not exhibit such significant rheological changes in magnetic fields.

Studies of these problems in relation to magnetorheological fluids are scarce and fragmented, both theoretically due to lack of closure relations to determine the force interactions of structuring particles and the surface free energy and experimentally, since most studies have been based on a detailed study of the magnetic parameters of the medium, due to peculiarities of their structure and surface phenomena, the change of which is affected by a non-stationary or non-uniform magnetic field.

It is known that in a homogeneous bulk phase there are no forces in the absence of external fields when describing it macroscopically. Points of the bulk phase are characterized only by the densities of the components present in it, the densities of the free and total energy and entropy. By microscopic description of the bulk phase the local forces (including through magnetic interactions) and molecular fields should be considered. By averaging over the volume, which includes a sufficient number of molecules and ions, they also vanish in contrast to the transition zone between the two phases in which the strength of interaction are damped further deep into each of the adjacent phases. These forces will be called surface forces. Accordingly, the density of the components that make up at least one of the phases, the densities of the free and total energy and entropy vary in a direction across the interfacial zone.

The present work is dedicated to the experimental determination of the characteristics of the magnetic field effect on the force components of the motion equation (Navier-Stokes equations), which solution together with the continuity equation will further allow to evaluate the characteristics of MRF spreading (time spreading, instantaneous values of the radius of spreading). The objectives of this paper are to establish patterns of the magnetic field influence on adhesive performance and characteristics of the spreading of the finite volume (a drop) of MRF suspension.

2. Experimental

As an investigated material a magnetorheological fluid (MRF) based on chromium dioxide powder and MOBIL oil (MRF-X) is taken where the volume content of dispersed phase ranges from 5 to 35 vol. %.

First of all, the experiments were conducted to study the magnetic properties of magnetorheological suspensions. Magnetization curves were obtained by means of measurement with two Hall sensors using the device shown in figure 1. Measurements of the magnetization is reduced to the measurement of the induction field in the sample cell 7 with a sample of MRF and in the air gap of an electromagnet consisting of a magnetic conductor 4 and coil 5, fed by the source 6. Hall sensor 9 in the sample registers the magnitude of the magnetic induction B, and the sensor 8 in the air gap – the value of $\mu_0 H$. For a sufficiently long sample (ratio of sample length to width is greater than 10) the distortion field in the sample is small.

Then the expression could be written $B = \mu_0(J+H)$ and magnetization of the sample in the field:

$$J = \frac{B - \mu_0 H}{\mu_0}. \tag{1}$$

Supplying the inputs 2, 3 of differential amplifier 10 by signals from two sensors we can obtain a signal on the exit 1, proportional to the magnetization of the sample J, and a signal from the sensor 8 to characterize the intensity of the magnetic field H.

When determining the magnetization it is necessary to bear in mind that a sensor 9 occupies in the suspension the layer of a thickness h=0.7 mm, dividing the cylindrical sample of MRF with a radius of R=3 mm in two parts. This layer is not filled by the suspension, that is equivalent to the appearance in it the magnetization $-J$. It results in that the induction of the magnetic field in the center of the layer where a sensor 9 is situated is lower than the induction in the volume of MRF by the value of

$$\Delta B = \frac{\mu_0 J}{2} \int_{-h/2}^{h/2} \frac{R^2 dh}{\left(R^2 + h^2\right)^{3/2}} = \frac{\mu_0 Jh}{\sqrt{h^2 + 4R^2}} .$$

Respectively the equation (1) gives the value of the magnetization, lowered by the value $\Delta J = \frac{Jh}{\sqrt{h^2 + 4R^2}}$. This correction is necessary to be considered when determining the real value of the magnetization. Measurement errors for the magnetic values were within 2 %.

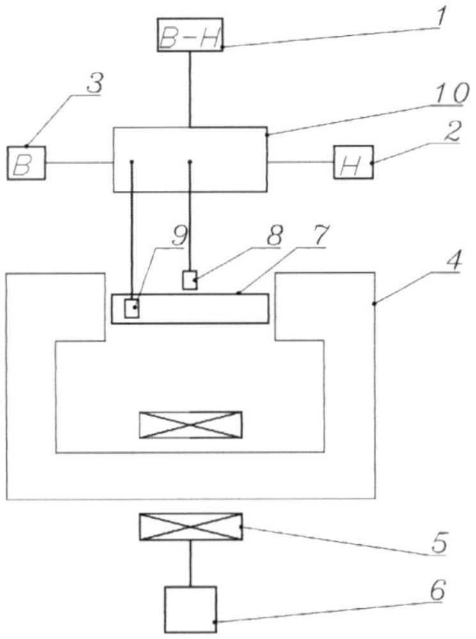

Figure 1. Scheme of the device to measure magnetic properties of MRF: 1 – output of differential amplifier, 2 – input from the sensor 8, 3 – input from the sensor 9, 4 – magnetic conductor, 5 – electromagnet coil, 6 – power source of the electromagnet, 7 – cuvette with a MRF sample, 8, 9 – Hall sensors, 10 – differential amplifier.

It is found that magnetic permeability is increased with the increase of dispersed phase concentration and is decreased with the increase of the intensity of the magnetic field which is due to the saturation in strong magnetic fields. Obtained dependence $\mu(H)$ for MRF-X with maximal concentration of the filler is present in figure 2. Chromium dioxide CrO_2 is a hard-magnetic material and has a lower magnetic susceptibility and intensity of saturation than that of the widely used powder of carbonyl iron.

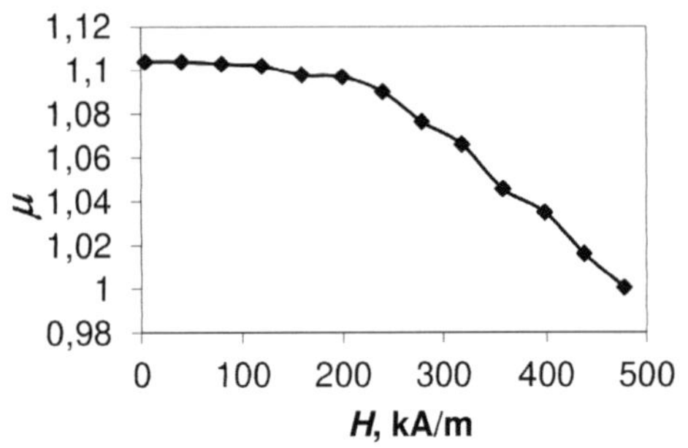

Figure 2. Dependence of magnetic permeability of MRF-X on the intensity of magnetic field.

Peculiarity of MRF as magnetic material is the mobility of the carriers of magnetic moment – the particles of the dispersed phase. Therefore suspensions are magnetized substantially because of the restructuring. Formation in the magnetic field of the structures, directed along field lines increase the susceptibility of MRF in this direction.

Investigations have shown that magnetization curves on the initial section are close to linear ones up to intensities of the field 30 kA/m, i. e. MRF in this area has a constant magnetic susceptibility.

More over the initial susceptibility increases practically linearly with the increase of ferromagnetic particle concentration. In the magnetic field of about 300 mT there is a saturation of MRF no matter what concentration of the dispersed phase is. It is found, that MRF susceptibility is rather sensitive to the transformation of the structure in the process of the flow. In fact, for resting MRF the value of the initial susceptibility does not depend on the field, in the shear flow it drops with the increase of the shear rate and the decrease of the intensity of the field. These changes of susceptibility are due to the changes of the size of aggregates and their orientation along the field direction, and also due to the relaxation time of the equilibrium structure and are determined by the ratio of magnetic and hydrodynamic forces.

The rheological characteristics were obtained by means of a rheometer Physica MCR 301 by Anton Paar using a measuring cell consisting of two coaxial cylinders (the diameter of the internal cylinder is 26.7 mm, of the external one is 28.9 mm, the gap between cylinders is 1.13 mm). Measurement errors for the rheological values were within 0.5 %. The rheological characteristics are shown in figure 3.

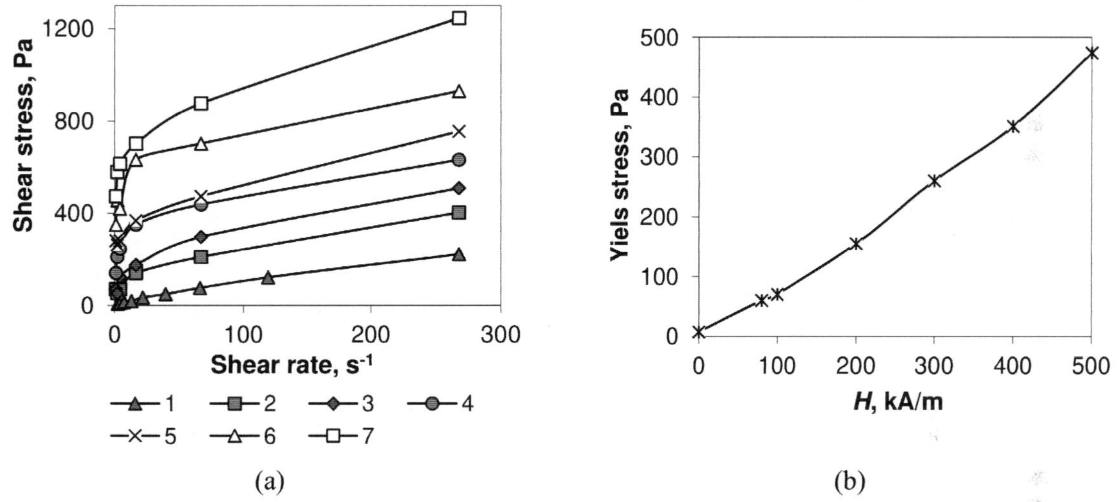

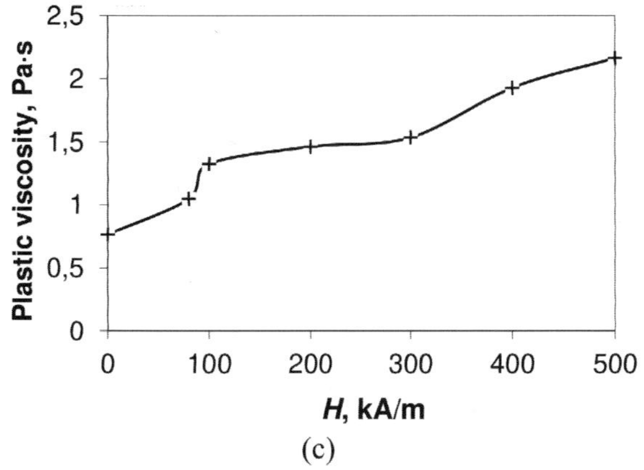

Figure 3. Rheological characteristics of MRF-X: (a) flow curves at the intensity of the magnetic field: 1 – 0, 2 – 80, 3 – 100, 4 – 200, 5 – 300, 6 – 400, 7 – 500 kA/m; (b) the dependence of the yield stress on the intensity of the magnetic field; (c) the dependence of the plastic viscosity on the intensity of the magnetic field.

3. Results and discussion

With the aim of determining formation conditions of a continuous compact MRF layer the influence of the inhomogeneous magnetic field on the shape of the surface, spreading characteristics and surface tension on its interface with the air is experimentally estimated on the example of a finite volume of MRF (a drop). Laboratory equipment (figure 4) has provided the measurement conditions for the wetting angle, radius and height of the MRF drop by means of video fixation of the spreading process on the base that is the inductor of the magnetic field. Also the shape and the surface of the drops were studied by means of Axio Lab A1 microscope by Carl Zeiss. In the experiments the composition of MRF, the size and mass of MRF drop, the induction of the external magnetic field, the bending angle of the base are varied. In figure 5 we show the dependence of the induction of the magnetic field on the distance *z* to the surface of the base, that was changed depending on the number of nonmagnetic additional layings (glass) with the aim of regulating the value of the intensity of the magnetic field on the surface, on which the drop fell (in its different points). The investigated suspension was put into a batcher with a diameter 1.5 or 2.5 mm. A drop of the suspension fell on the surface of the base from the given height (3.2, 11 or 20.3 cm). The height of the drop downfall did not influence substantially its parameters. At a bigger diameter of the batcher (2.5 mm) the diameter of the drop was bigger by 10 – 15% than that at a smaller diameter of the batcher (1.5 mm), for which the mass of the drop was fixed and was 28 g. It is necessary to note that the influence of the magnetic field on the process of drop formation at the end of the batcher has been noticed. In the magnetic field we needed more time for a drop creation and its downfall.

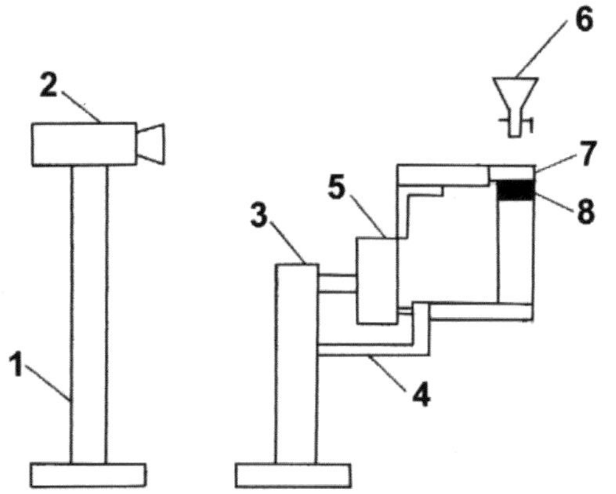

Figure 4. Equipment scheme for determining spreading characteristics of MRF: 1 – holder, 2 – camera, 3 – mount, 4, 5 – rotation device, 6 – batcher, 7 – base, 8 – magnet

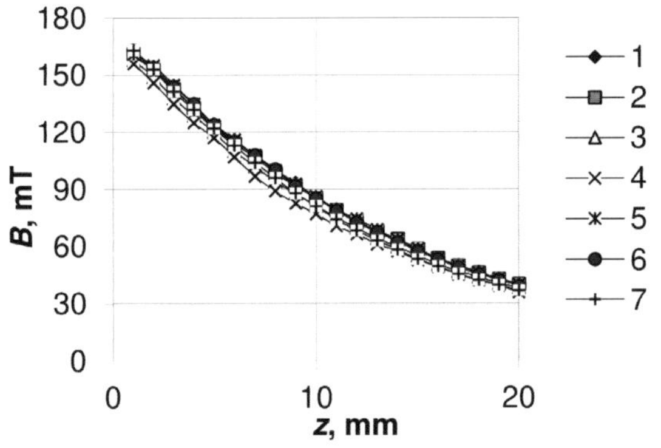

Figure 5. Areal distribution of the magnetic field on the base (*x* and *y* are rectangular coordinates in the plane of the base): 1 – x=0, y=0, 2 – x=2.5 mm, y=0, 3 – x=5 mm, y=0, 4, 5 – x=7.5 mm, y=0, 6 – x=0, y=2.5 mm, 7 – x=0, y=5 mm, 8 – x=0, y=7.5 mm.

Investigations of the influence of the magnetic field on the spreading and wetting of MRF were performed for two cases. In the first case a drop fell on the surface of the base in the absence of the field. Then the induction of the magnetic field has been raised up to 160 mT. The shape of the drop has been registered in two projections (side and frontal) by means of a digital camera. Pictures of the drops on the surface of the base in the absence of the magnetic field are presented in figure 6. The diameter and the height of the drop, its wetting angle was determined visually using the pictures. Observational errors were 0.2 mm in the diameter and the height, 0.5° in the wetting angle.

When the drop falls on the base in the absence of the field it has a classical shape of the drop on a lyophilic surface. By increasing the field up to 60 mT the shape of the drop is not changed substantially, and we can detect only small spreading (the increase of the diameter by 2 – 3% and the decrease of the height down to 25%).

If a drop falls on the base under the magnetic field with the intensity of 160 mT, then it had a much smaller diameter than in the absence of the field (in 1.5 times) and twice as bigger height. These observations can confirm the influence of the magnetic field on the process of drop spreading in terms of stopping this spreading.

An interesting peculiarity is the presence of peculiar fractures on the surface of the drop closer to its peak. It is explained by the distribution of the intensity of the magnetic field in its volume and on the surface in accordance to the force lines of the magnet-base.

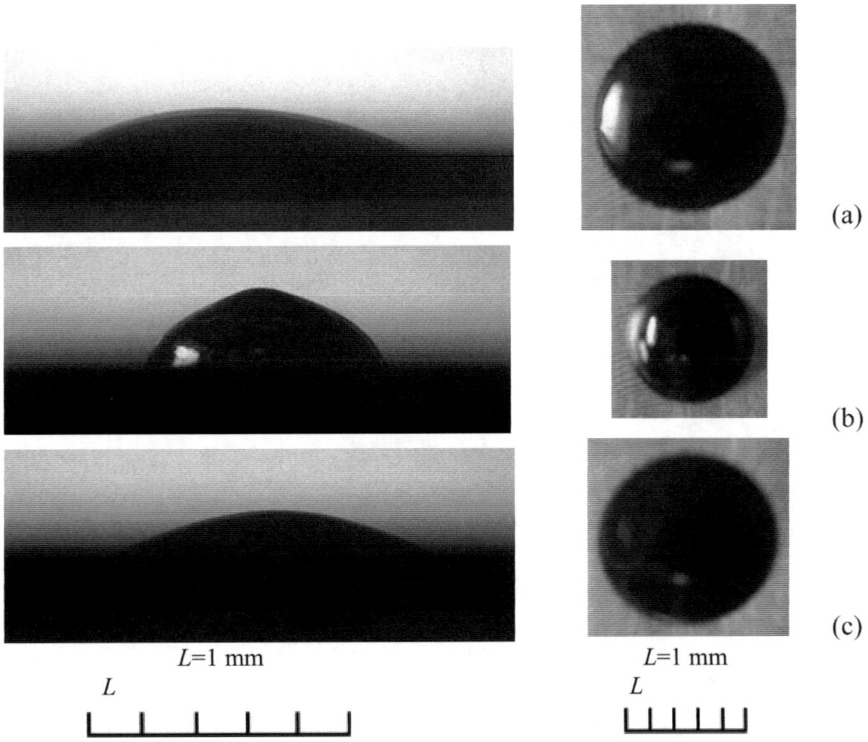

Figure 6. MRF drop on the surface of the base: (a) in the absence of the magnetic field, (b) under the influence of the magnetic field of 160 mT and (c) after switching off the magnetic field; side view (left), top view (right).

When decreasing the induction of the magnetic field from maximal value down to zero, the drop gets the same classical shape, as when the drop is registered in the absence of the field with a small difference in the size (the diameter of the drop without the field is 8.4 mm, the height 1 mm; when switching off the field the diameter of the drop is 7.5 mm, the height 1.1 mm). The dependence of the diameter and the height of the drop on the induction of the magnetic field is shown in figure 7.

The wetting angle of MRF θ grows with the increase of the intensity of the field (figure 8). On horizontal base the wetting angle linearly increases with the increase of the concentration of the dispersed phase. The bending of the base by 45° leads to some deformation of the drop by the gravitation force and therefore to the change of the frontal wetting angle to the side of its increase. The dependence of θ on the concentration of dispersed phase resumes from being linear.

It was found that the magnetic field leads to the decrease of the spreading rate and the setting time for reaching the stationary parameters of the drop (figure 9) of the investigated compositions. The main spreading process has been finished in all cases in time frame of 10 s. A further dwell up to 1 minute did not result in any substantial change of the spreading radius and height. However, by longer dwelling times, especially for a high concentration suspension we have observed a huge change of the shape of the whole surface of the drop – the formation of the bump in the center and the change of the wetting angle at the interface of three phases.

After analyzing experimental results, we have studied the dependence of the radius of spreading on time until the equilibrium values are reached $R = R_0 + kt^n$, where for each MRF composition the coefficients R_0, k, n depend on the intensity of the magnetic field.

It is known that in the surface layer fluid molecules have an additional potential energy in comparison with the molecules in the fluid volume. A surface layer on the whole has additional energy that is a part of the inner energy of the fluid. Coefficient of surface tension σ represents another additional free energy which a surface unit of the surface layer possesses. σ also has the meaning of the surface tension force, that is applied to a unit of length of the contour.

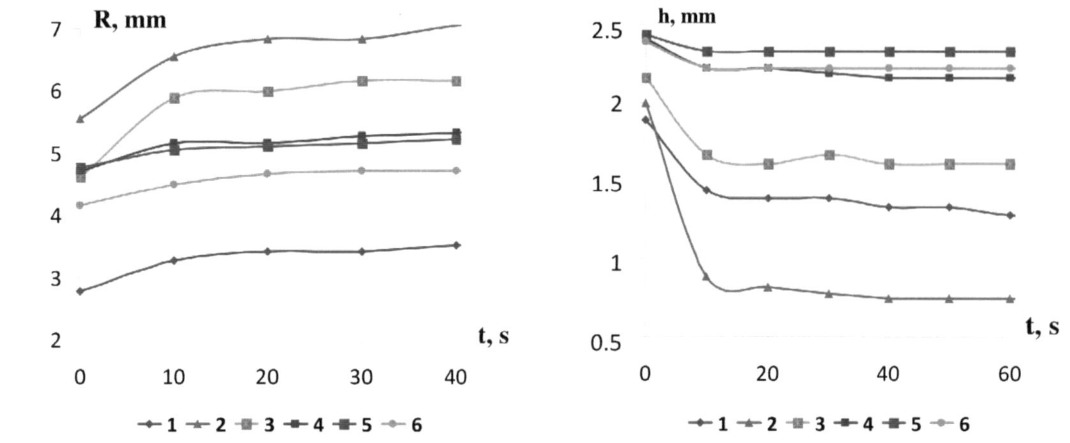

(a) (b)

Figure 7. The dependence of the diameter (a) and the height (b) of MRF drop on the time of spreading in the magnetic field: (a) 1 – 160 kA/m, 2 – 0, 3 – 20, 4 – 40, 5 – 80, 6 – 120 and (b) 1 – 20 kA/m, 2 – 0, 3 – 40, 4 – 80, 5 – 120, 6 – 160.

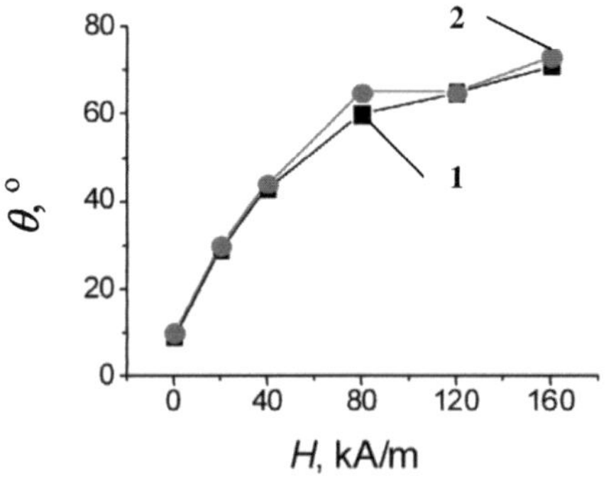

Figure 8. The dependence of the wetting angle of MRF on the induction of the magnetic field: 1 – θ for the diameter, 2 – θ for the height.

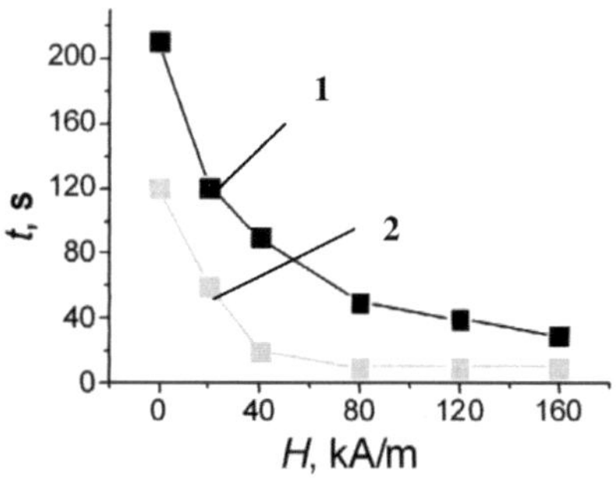

Figure 9. Dependence of setting time for reaching the stationary parameters of the drop when spreading: 1 – setting time for the diameter, 2 – setting time for the height.

The surface tension was determined by means of tearing a thin plate from the surface of MRF at the condition of full wetting of its butt end in the magnetic field. In this case when tearing it the film of the fluid is raised in which the gravitation force is equal to the applied force. The film of the fluid is broken and the plate starts to come off. The critical value of the force is fixed, and the surface tension is determined, that is connected to its value by means of calibration coefficient. It was stated that in maximal magnetic field the coefficient of the surface tension $\sigma = 0.0544$ N/m^2, that is only by 0.0011 N/m^2 more than without the influence of the field.

The interaction of the inner structure of MRF with the base (adhesion characteristics) was also evaluated by the measurement of the normal tearing forces of a glass part from the base with MRF under the magnetic field. A scheme of the base and the corresponding distribution of the magnetic field are shown in figure 10(a), and the dependence of the specific tearing force on the intensity of the field in figure 10(b).

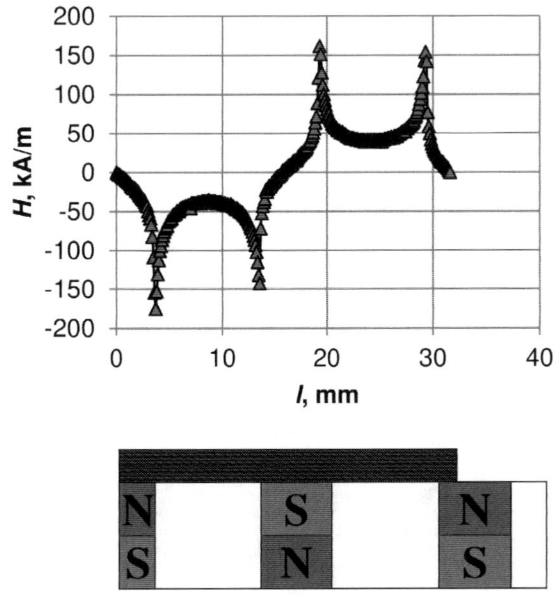

(a)

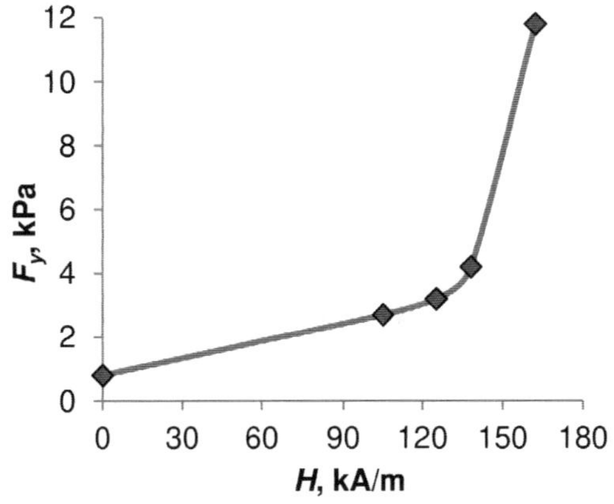

(b)

Figure 10. (a) Corresponding distribution of the magnetic field; (b) dependence of the specific tearing force on the intensity of the magnetic field.

4. Conclusions

On the basis of the obtained results it can be concluded that when spreading MRF on the solid surface in the magnetic field, in addition to common considered forces, Lorentz force also appears because of the gradient magnetic field is impacting the system, which leads to the apparent increase of viscosity and the force.

References

[1] Shukin E D, Summ B D, Gorunov Yu V, Perzov N V and Perzov A V 1963 *Colloid Journal* **25** 253

[2] Gorunov Yu V 1962 *Chemistry Successes* **23** 1064–69

[3] Elutin V P, Kosticov V I and Maurach M A 1965 *Surface Phenomena in Fusions and in Forming Solid Phases From Them* (Nalchik: Kab-Balk. Press)

[4] Sorokin Yu V, Hlinov V V and Esin O A 1968 *Surface Phenomena in Fusions* (Kiev: Navukova dumka)

[5] Bushovsky A I *Wetting and the Surface Properties of the Fusions and Solid State* 1972 (Kiev: Navukova dumka)

[6] Bekulov M T, Zadumkin S N and Hatazukov A S 1975 *On the Study of Surface Phenomena in Metallic Fusions* (Ordzhonikidze)

[7] Chernov V V 2006 *Spreading and Wetting of Conducting Fluid Phases of the Surface of Solid State in the Magnetic Fields* Thesis (01.04.14), Nalchik

[8] Korobko E V, Gorodkin R G, Goncharova N A and Baikov V I 2003 *Int. J. Applied Mechanics and Engineering* **8** 67–72

[9] Korobko E V, Gorodkin R G, Goncharova N A and Baikov V I 2003 *Injenerno-Fizichesky Zhurnal* **76**(4) 36–9

[10] Bashtovoi V G, Kuzhir P, Reks A A, Bossis G and Volkova O 2002 *Int. J. Modern Physics* B 2002 **16** (17–18) 2590–6

[11] Lavrova O, Matthies G, Polevikov V and Tobiskan L 2004 Numerical modeling of the equilibrium shapes of a ferrofluid drop in an external magnetic field *Proc. in Applied Mathematics and Mechanics (PAMM)* **4** (1) 704–5

Preliminary investigation of magneto-rheological fluid durability in continuous slippage clutch

J-F Desrosiers, J-P Lucking Bigué, M Denninger, G Julió, J-S Plante[1] and F Charron
Université de Sherbrooke, Sherbrooke, QC, J1K 2R1, Canada

E-mail: jean-sebastien.plante@usherbrooke.ca

Abstract. Magneto-rheological (MR) devices, such as dampers, engine mounts and clutches, are now appearing in the ground vehicle industry. Although important work has been directed to assess the longevity of MR dampers, few studies have targeted the aspect of magneto-rheological fluid (MRF) durability when used in continuous shear, such as in MR clutches. The objective of this research is to identify the degradation phenomena associated with MRF used in continuous shear and to understand the main causes and effects, in order to propose design improvements to enhance clutch durability. Experiments are conducted on two test benches in order to reproduce MRF aging in a controlled environment and to evaluate the proposed solutions. The effect of the operating conditions (shear rate, shear stress and temperature) on the long term degradation of the torque-current relationship is evaluated. Two degradation phenomena are identified: base oil expansion and particle oxidation. The dominant failure mode of the tested clutches is a MRF leakage resulting from the base oil expansion which occurs between 1.5 MJ/mL and 9 MJ/mL of dissipated energy depending of the operating conditions. Two solutions are proposed to extend clutch durability: 1- MRF circulation and 2- compliant elements.

1. Introduction

Magneto-rheological (MR) clutches offer highly controllable forces and rapid dynamics compared to conventional actuation approaches [1], as well as a high reliability potential due to their simplicity and absence of metal-to-metal contact. MR clutches can be categorized according to the relative slip between their input and output. Thus, when used as torque limiters, as per [2], no slippage occurs in the clutch unless the maximum transmissible torque is reached. MR clutches can also operate in continuous slippage [3, 4], where the input shaft rotates at a constant speed and the transmissible torque is constantly adjusted by the clutching action to control the output position or speed (see Figure 1) of the system.

This later clutch category has interesting advantages in terms of actuator bandwidth, since the input inertia is separated from the output [5]. However, these MR clutches must dissipate large amounts of energy through a small quantity (~10 – 100 mL) of MR fluid (MRF) that is subject to continuous shear stress. The durability of the clutches is therefore limited by the MRF operating life.

---

[1] Corresponding author.

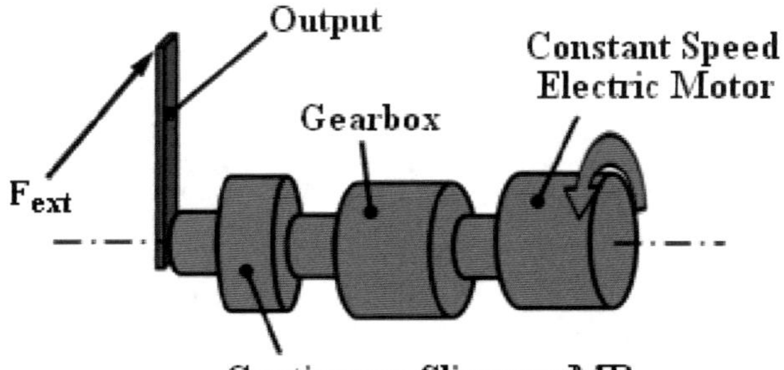

Figure 1. Example of a continuous slippage MR clutch.

Even though a great number of MR fluid devices have been proposed such as dampers [6], clutches [3, 7, 8] and brakes [9], MR-fluid durability has not been thoroughly addressed in literature. MR fluid durability becomes of prime concern as MR devices make their way to the market and more specifically, in economic sectors where reliability is of outermost importance such as aerospace. In damper applications, the maximum operating life of the MRF has been estimated to be 10 MJ/mL, where an increase in off-state viscosity of the MRF causes device failure [10]. For clutching applications, commercial MRF's have been shown to undergo an increase in oil volume around 6 MJ/mL [11] that leads to a significant torque decrease of the device. However, the occurrence of the expansion phenomenon, its exact nature, and relation with MR clutch design is not clear in the literature. A study makes no mention of this phenomenon up to 70 MJ/mL [12] and that device failure is associated to iron particle oxidation, slowly reducing torque output. Up to this day, a general comprehension of MRF degradation in clutch applications has not been given and thus, clear design guidelines for durability still do not exist.

The objective of this investigation is to lay the basis of a holistic description of MR-fluid degradation in clutch applications. In this work, two test benches are specifically developed to understand the degradation phenomena that characterize the aging of MRF submitted to different operating conditions of shear stress, shear rate and temperature. An accelerated aging test bench is developed to obtain large scale degradation signature of a MR clutch containing MRF that is aged up to 100 MJ/mL. Chemical and morphological analysis of the used MRF samples, in combination with pressure measurements performed with a full-scale clutch, reveal two MRF degradation phenomena occurring at specific energy levels: base oil expansion and particle oxidation. For all tested operating conditions, the base oil expansion is recorded between 1.5 MJ/mL and 9 MJ/mL of dissipated energy. This MRF degradation phenomenon is identified as the dominant failure mode of the tested MR clutches as it causes MRF leakage which decreases torque output. A fluid circulation mechanism is shown to extend clutch durability with regards to the dominant failure mode by allowing a larger amount of MRF to participate in the energy transfer process. A mitigation measure is also proposed to extend MRF life beyond the first failure mode, thus extending the transmissible energy per volume of commercially available MRFs.

2. Methodology

2.1. Experimental setup

An accelerated aging test bench, shown in Figure 2 (a), is designed at small scale to minimize the total testing time and rapidly obtain the degradation signature of MRFs. A full scale device, shown in Figure 2 (b), is used to identify the dominant failure mode observed in the degradation signatures and to validate proposed solutions to optimize durability.

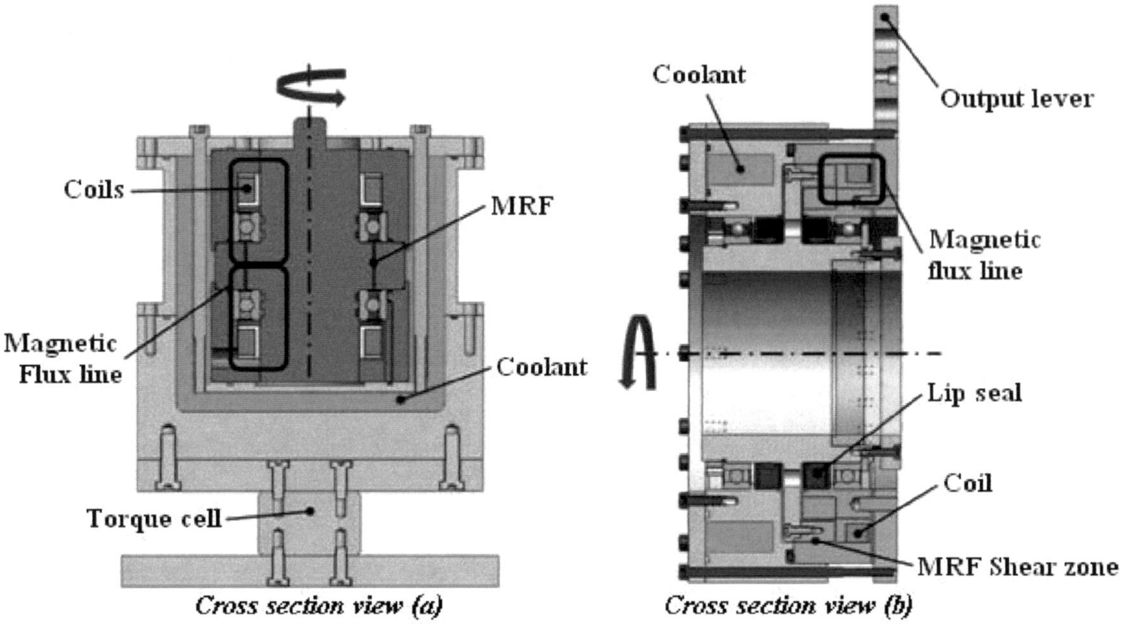

Figure 2. Test benches designed for (a) accelerated aging and for (b) full scale validation.

The design parameters of each test bench are summarized in Table 1. For the accelerated aging test bench, specific attention is given to the uniformity of the operating parameters. A drum type clutch is chosen to provide a uniform shear rate (<3500 s^{-1}), while double coils are used to provide a uniform magnetic flux density (<1.2 T). Bearing seals are used to contain the MRF in the shear gap while keeping the volume of MRF not submitted to magnetic fields (dead volume) as small as possible, here < 5%.

The full scale test bench uses radial lip seals, placed away from the shear gaps, in order to replicate the typical design of a MR clutch. This results in an additional volume of MRF which does not dissipate energy. Thus, the dead volume of the full scale clutch represents 90% of its 30 mL overall volume.

Table 1. Main characteristics of the test benches.

Characteristic	Test bench	
	Accelerated aging	Full scale
Clutch type	Drum	Drum
MRF volume	1.5 mL	30 mL
Dead volume	<5%	>90%
Uniform magnetic field	1.2 T max	1.2 T max
Shear rate	0-3500 s^{-1}	0-3000 s^{-1}
External cooling	1000 Watts	1000 Watts
Max aging rate	350 W/mL	350 W/mL
Sealing technology	Sealed bearing (6908 DDU)	Lip seal (CRx80x100x10 CRW1 V)
Torque sensor	Yes	Yes
Temperature sensor	Yes	Yes
Pressure sensor	No	Yes

The accelerated aging test bench is driven by a 2 hp motor, with variable output speed between 0 and 1500 RPM. The speed of the electric motor is controlled while the torque output of the clutch is measured with a torque sensor (Transducer Techniques model TRT-200). The test bench is equipped with a temperature sensor (Omega thermocouple No. TJ36-CASS-116G-6-SB-SMPW-M) while a Copper Core radiator (HX-360HL) is used to control the temperature of the clutch. The measurement and control are done with a National Instruments USB card 6211.

The full scale test bench is also driven by a 2 hp motor with variable output speed between 0 and 250 RPM. The torque output of the clutch is measured with a load cell (Tedea Huntleigh 1042) fixed on the clutch output lever. The test bench is equipped with two temperature sensors (Omega thermocouple No. TJ36-CASS-116G-6-SB-SMPW-M). The first one is located at the shear interface and the other one measures the MRF temperature in the dead zone. Two pressure sensors (SSI Technologies Inc. No. P51-50-G-B-I36-4.5v-R) are located on each side of the shear interface. A Copper Core radiator (HX-360HL) coupled with three fans is used to regulate the temperature of the clutch. The full scale test bench is equipped with a National Instruments Compact RIO-9022 embedded control and acquisition system for control and data logging.

2.2. Test sequence

Unless otherwise noted, each test described in this research is submitted to a two parts sequence (see Figure 3), and is performed with BASF Basonetic 5030 MRF. In the first one hour portion of the test sequence, a steady torque reference is set while the current command is adjusted automatically by a closed-loop controller, operating at the rate of 100 Hz. As this steady-state portion of the test sequence is used to gather a general degradation signature for the MRF, the torque output, acquisitioned at the rate of 100 Hz, is averaged and recorded at a rate of 20 Hz. After each hour, the steady-state test is stopped and the open-loop portion of the test sequence is performed. In this portion, deterministic current steps of specific amplitudes are imposed and the torque response is recorded at the acquisition rate of 1000 Hz. This test portion is used to qualitatively evaluate the current-torque relationship of the MRF aging from the off-state (no current) to the on-state (maximum current).

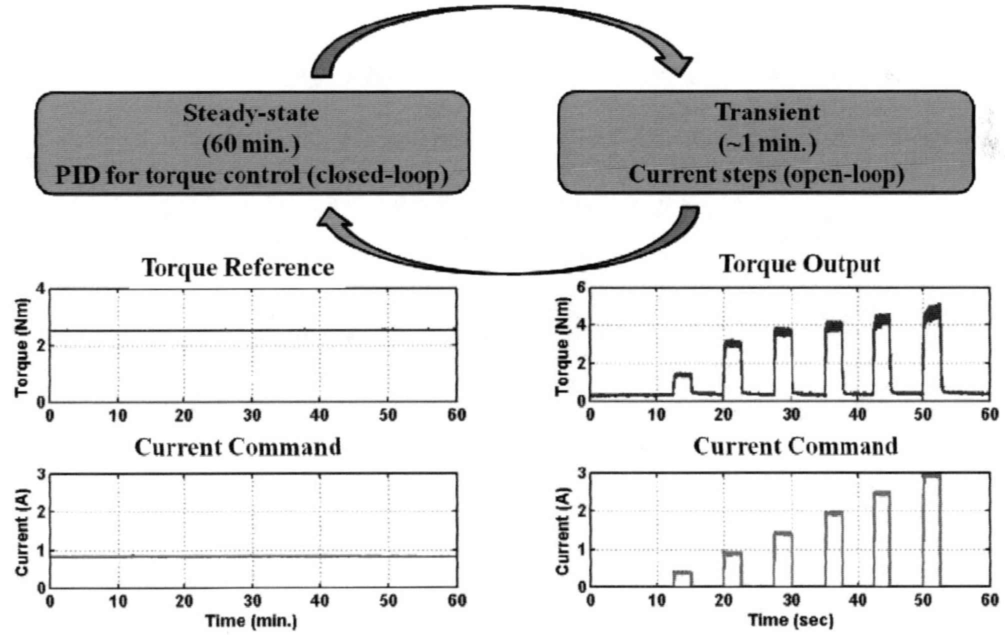

Figure 3. Test sequence used to age and to characterize the MRF (a) steady-state closed-loop control and (b) transient open-loop current steps.

2.3. MRF chemical and morphological analysis

Several analyses are performed with new and aged MRF in order to correlate the MRF degradation signature and behavioral changes with acute physiological variations in the MRF composition. Thermo gravimetric evaporation of volatile constituents is performed by temperature ramping up to 500 °C under a non-reactive environment (argon gas) to precisely determine the quantity of oil present in each MRF sample submitted to accelerate aging. Scanning electron microscopy (SEM) is used to observe the dry constituents (e.g.: iron) present in the MRF. Combined with X-ray fluorescence and X-ray diffraction, the SEM images are also used to evaluate the presence of oxygen (oxide) in the MRF samples. Finally, granulometry measurements are performed to quantify the particle size distribution.

3. Large scale degradation of the torque-current relationship

3.1. Typical degradation signature

Several tests are performed under the typical operating conditions, listed in Table 2 to identify the degradation signature of a MR clutch containing BASF Basonetic 5030 MRF.

Table 2. Typical operating parameters of a MR clutch.

Parameter	Value
Shear stress	30 - 60 kPa
Shear rate	1000 - 2500 s^{-1}
Temperature	40 – 80 °C
MRF	BASF 5030
Gap	0.5 mm

A typical closed-loop torque-current relationship measured from the steady-state test described in section 2.1 is shown in Figure 4, for operating conditions of 40 kPa, 2350 s^{-1} and 60°C. For each open-loop current step, the resulting mean torque is normalized by its initial value (unused MRF torque output at given current) and presented in Figure 5. The calculations are done over a one second period, starting when the torque reaches its stabilized value.

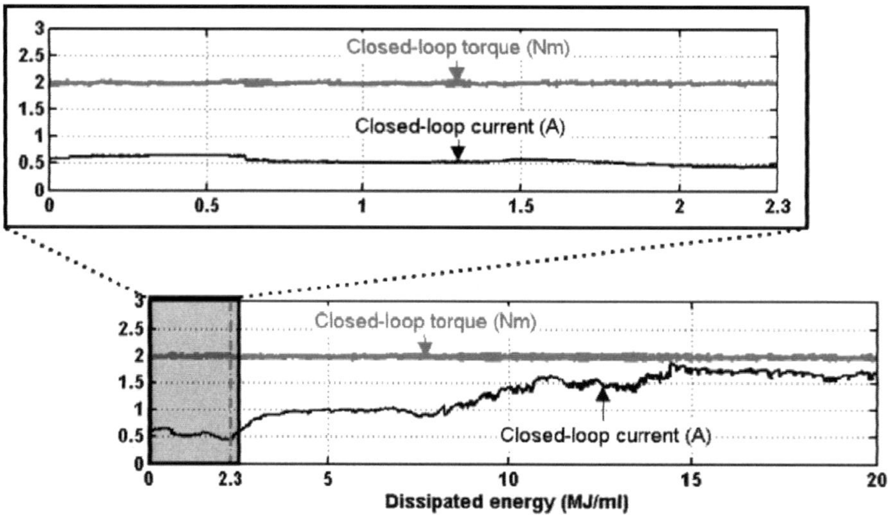

Figure 4. Closed-loop degradation signature of Basonetic 5030 in the accelerated aging test bench at 40 kPa, 2350 s^{-1} and 60°C.

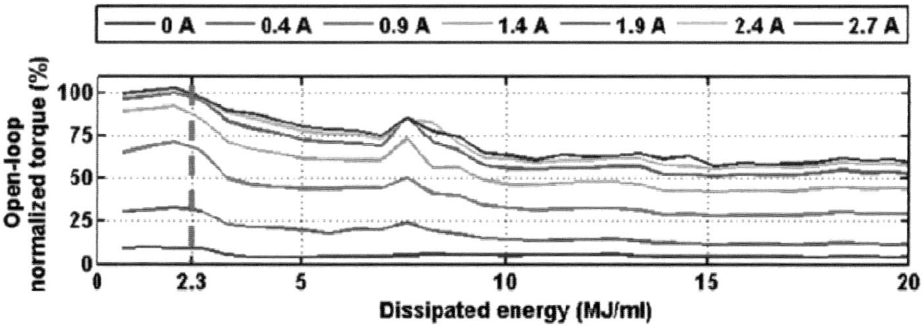

Figure 5. Typical degradation signal of BASF Basonetic 5030 submitted to the operating conditions of 40 kPa, 2350 s^{-1} and 60 deg C. Torque reference is set at 2 Nm.

Altogether, this particular test last up to 20 MJ/mL and two distinct parts are identified. Before 2.3 MJ/mL, both the closed-loop and open-loop current-torque relationship are found fairly steady (see Figure 4 andFigure 5). For this test, at 2.3 MJ/mL the first sign of degradation is noted through a sharp increase (0.5A to 1A) in closed-loop current, which is correlated with a ~18% decrease of the maximum open-loop torque.

3.2. Effect of operating parameters and MRF composition

The accelerated aging bench test is used to determine the occurrence of the first degradation sign as a function of the clutch operating conditions (shear stress, shear rate and temperature). As seen in Figure 6, maximum measured life dissipated energy is observed when the shear stress, shear rate, and temperature are set at minimum values (40 kPa, 1000 s^{-1}, 40ºC). The life dissipated energy is always found to decrease when aging tests are performed with one parameter increased at a time. Although these results do not show the influence of combined operating parameter modification, the results qualitatively demonstrate the relative influence of each parameter. Temperature is found to be the most influential parameter of MRF's life, while shear rate is the least influence. Note that these results are intended to represent the general tendency of an operating parameter and thus, must be used with care due to the limited number of operating points investigated.

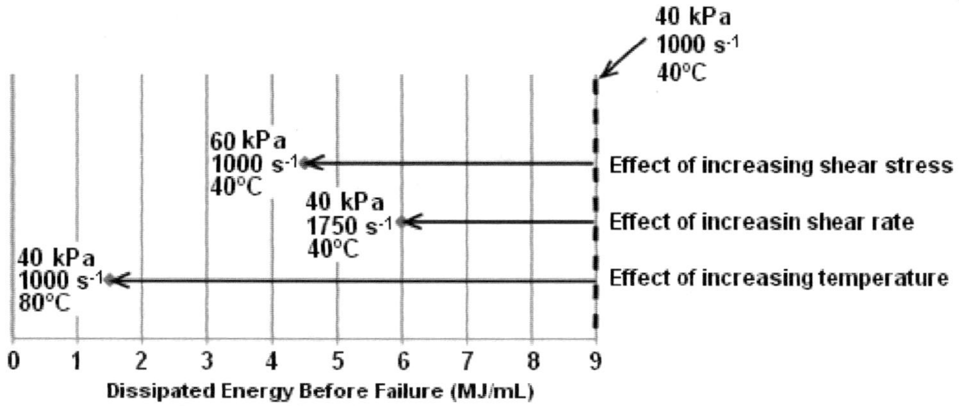

Figure 6. Effect of operating parameters on base oil expansion.

Aging tests are also performed on Lord MRF-140CG. As seen in Figure 7, the degradation signature of this MRF is found similar to that obtained with the BASF 5030 and for a similar level of dissipated energy (4.5 MJ/mL).

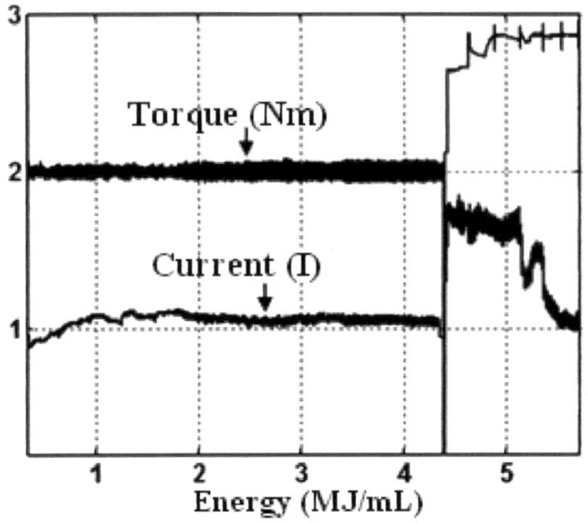

Figure 7. Degradation signature of Lord MRF 140CG.

Altogether, a first sign of degradation is observed between 1.5 MJ/mL and 9 MJ/mL for all test conditions and for both MRF fluid formulation (BASF 5030 and Lord MRF 140CG) tested with the accelerated aging bench test.

4. Detailed study of MRF degradation phenomena and resulting failure modes

4.1. Base oil expansion

The first sign of MRF degradation is characterized by a sharp increase in closed-loop current and a sharp decrease in open-loop torque between 1.5 MJ/mL and 9 MJ/mL. To confirm that this behavior is caused by base oil expansion as reported by [11], a test is performed on the full-scale test bench, which is equipped with pressure sensor. In this test, the current in the coil is set at a constant 1.5 A, and the operating conditions are set at typical values (25 kPa, 2700 s^{-1}, 70°C). As shown in Figure 8, the pressure inside the MR clutch as well as the open-loop torque output are both found stable up to 1.4 MJ/mL. After this energy input, a rapid increase in pressure (between t = 6 min and t = 20 min) is recorded, suggesting that the oil undergoes a transformation. However, since the torque output remains fairly steady (95% of its initial value) during the pressure rise, it is believed that the overall performance is not directly affected by this degradation phenomenon.

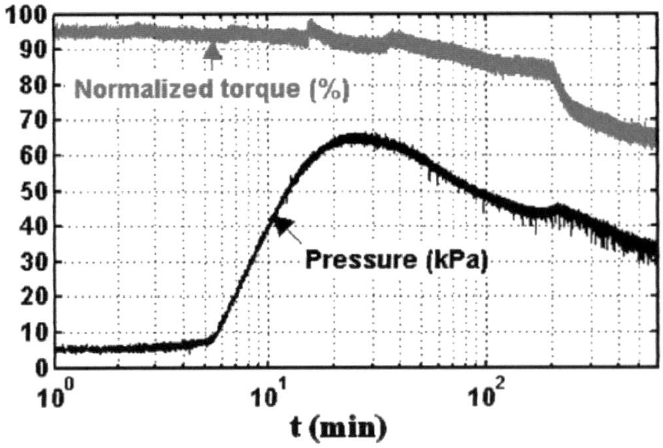

Figure 8. Effect of a sudden expansion of base oil on torque output and pressure measurement.

As the pressure approaches the maximum rated pressure of the seals (70 kPa), the torque output of the clutch starts to slowly decrease, and is accompanied by a decreasing pressure. The simultaneous decrease in pressure and torque is attributed to seal failure, followed by particle leakage. Loss of particle caused by oil expansion is therefore identified as the main clutch failure mode, which causes the observed degradation in the torque-current relationship expressed in Figure 4 and FigureFigure 5.

As shown in Figure 9, thermogravimetric analyses of the remaining MRF showed a lower volume percentage of base oil than in original MRF, suggesting that seal failure also caused severe base oil leakage. This could, in part, explain the erratic torque-current, measured in the accelerated aging tests.

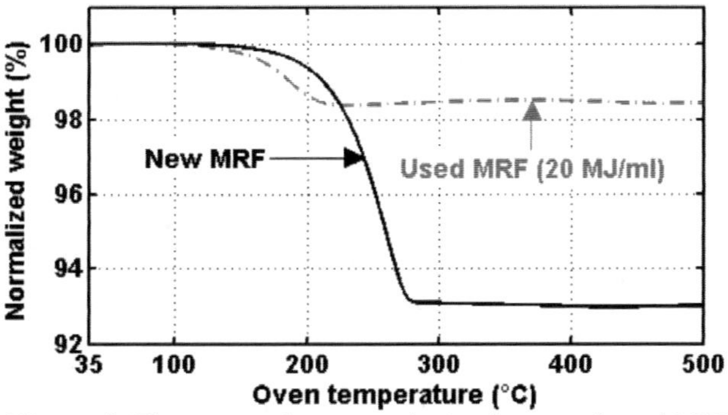

Figure 9. Thermogravimetry analysis of new and aged MRF.

4.2. Iron particle oxidation

Samples from the accelerated aging tests are submitted to x-ray diffraction (XRD) tests and scanning electron microscopy (SEM). As shown in Figure 10, MRF samples aged up to 10MJ/mL did not show any presence of oxide. However, at 100 MJ/mL, distinctive peaks corresponding to magnetite crystalline structure are observed on the XRD measurements.

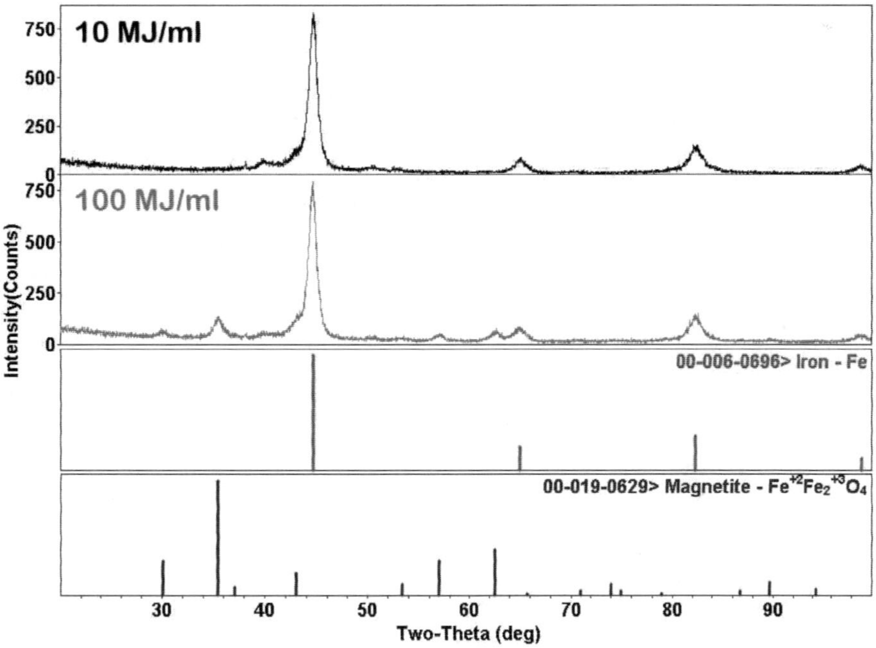

Figure 10. X-ray diffraction showing peaks corresponding to magnetite in MRF 10 MJ/mL and 100 MJ/mL aged sample.

The SEM images of new and aged MRF presented in Figure 11, show morphological changes of the particles. Compared to new MRF (see Figure 11(a)), Figure 11(b) and (c), show the evidence of oxide formation around the micron size particles when MRF is aged at 100 MJ/mL. Similar results were presented by Smith *et al.* (2007), who showed that an oxide layer surrounding the iron particles aged at a dissipated energy level of 70 MJ/mL, is responsible for a decrease in maximum output torque. In the current study however, the same conclusion cannot be drawn since oil expansion caused particle leakage before significant amounts of oxide can be identified in the MRF.

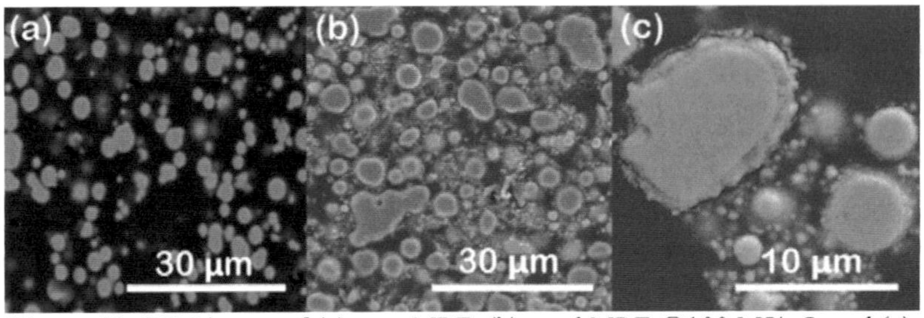

Figure 11. SEM image of (a) new MRF, (b) aged MRF @100 MJ/mL and (c) aged MRF @ 100 MJ/mL.

In Figure 11 (b) and (c), smaller particles can also be seen, mixed with the micron sized particles. A granulometry test, presented in Figure 12, shows that the used MRF sample (100 MJ/mL) contains a much larger percentage of small particles (sub-micron) than new MRF sample. The presence of small oxide particles has been reported by [10], where the authors suggested that the brittle oxide shells are broken from micron sized particles, increasing off-state viscosity of MRF, eventually leading to MR damper failure. In this study, no significant increase in off-state torque is measured. However, note that oil and particle leakage could have mitigated the effect of an increase in off-state MRF viscosity or yield stress.

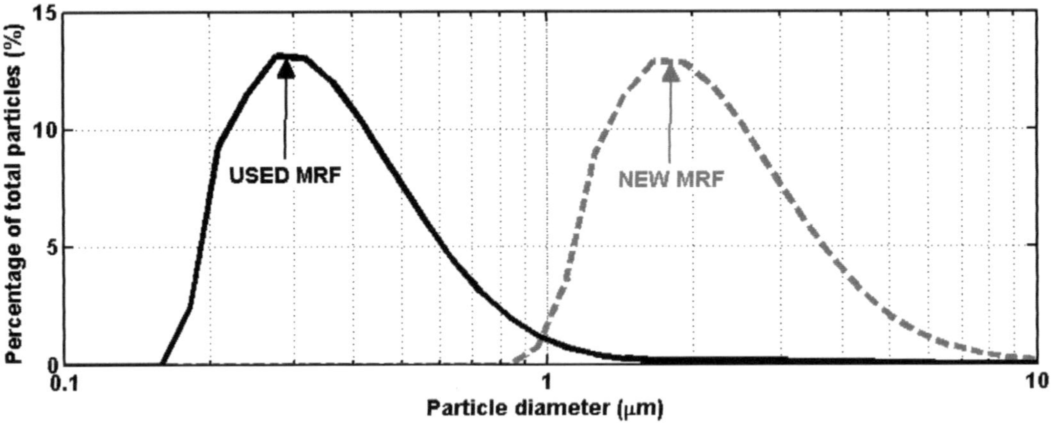

Figure 12. Granulometry of a new and aged (~100 MJ/mL) MRF samples.

5. Potential solutions to extend MRF life

For all tests performed in this study, particle and oil leakage, caused by oil expansion, is observed between 1.5 MJ/mL and 9 MJ/mL. Combined with the very small quantities of MRF required in typical MR clutches shear interfaces, these levels of dissipated energy limit the overall life of continuous slippage clutches to a few hours of operation. The following section proposes solutions to extend the life of MR clutches, by either increasing the total MR fluid present in such device or by mitigating the effect of base oil expansion on the failure mode.

5.1. Circulation of MRF

The first proposed solution is to add a circulation mechanism that pumps MRF from a reservoir into the shear zone of the MR clutch (Figure 13). This aims to increase the total amount of MRF in the clutch to extend clutch durability. To validate this hypothesis, two tests are performed with similar operating parameters (40 kPa, 2350 s^{-1} and 70°C). From the accelerated aging test bench, where the total amount of MRF is confined in the shear interface, the dissipated energy limit of the MRF corresponding to the specified operating conditions is measured at 1.5 MJ/mL. From the full-scale test bench, where only 10% of the MRF is present in the shear interface, a recirculation mechanism is used to circulate the MRF.

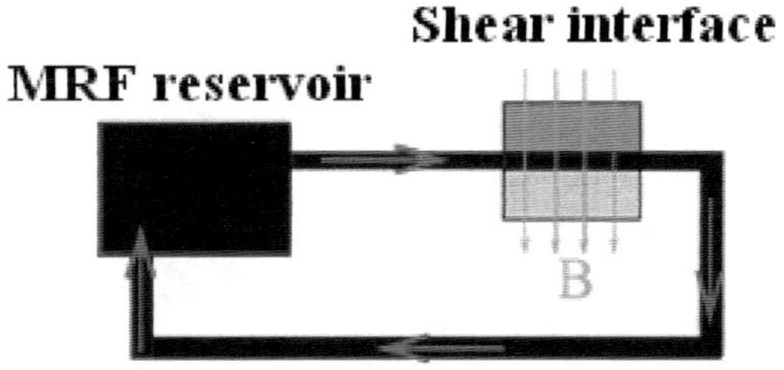

Figure 13. 30 mL of MRF circulate in a 4 mL shear zone of a MR clutch.

As seen from Figure 14, the total energy input to the full-scale test bench is measured to be 42 MJ, before a pressure increase is recorded and loss of torque is measured. This corresponds to a 1.4 MJ/mL of the total fluid present in the clutch, thus validating the feasibility of using recirculation to extend the life of a MR clutch. While this strategy delays the occurrence of base oil expansion, other phenomena could also lead to MRF leakage such as seal failure due to extended exposure to MRF abrasive particles.

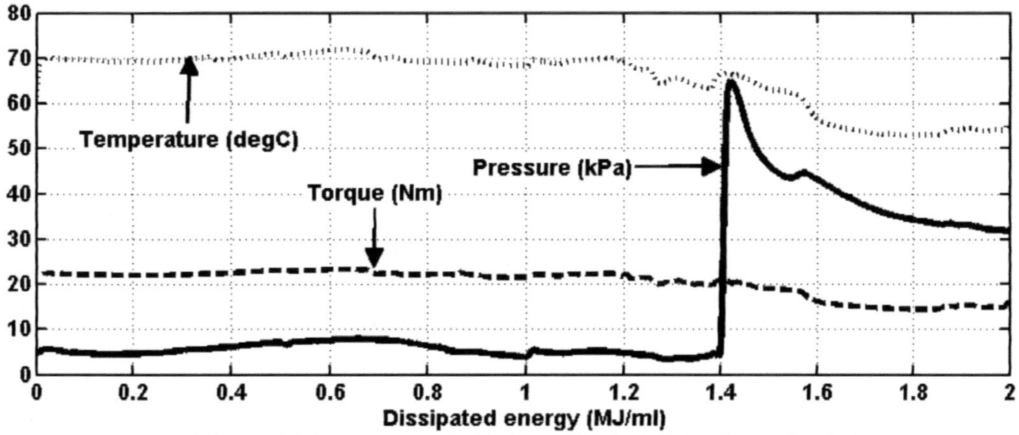

Figure 14. Energy per mL dissipated in the MRF using recirculation.

5.2. Compliant element

For applications where MR clutch costs and weight must be minimized, the addition of a MRF reservoir could not be a suitable option. To overcome this issue, compliant elements or voids could be used inside the MR clutch in order to mitigate the effects of base oil expansion, thus increasing the life of the MRF without the weight disadvantage.

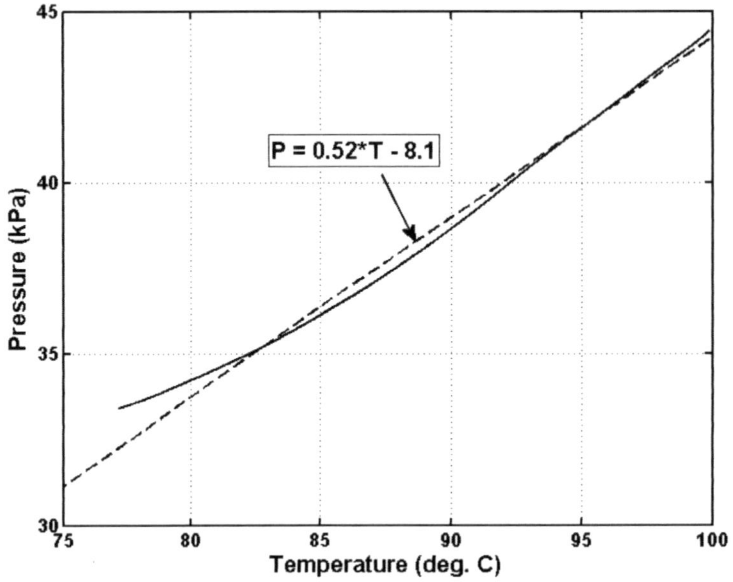

Figure 15. Relationship between the pressure and the temperature.

The percentage of volume expansion undergone by the MRF must be determined in order to design an appropriate compliant element. Hence, a method to approximate the expansion volume of MRF is presented by drawing an analogy with the coefficient of thermal expansion of the MRF. First, the full-scale prototype (containing no expansion chamber) is run at different temperatures and the corresponding pressures are recorded. As seen on Figure 15, the relationship between pressure and temperature is found to be linear in the range of interest (35 kPa to 45 kPa).

Knowing the coefficient of thermal expansion α of the MRF, here 0.00007 K^{-1} [13], it is possible to determine an equivalent bulk modulus (K) for the studied clutch with Equation 1, where V_0 is the initial MRF volume, ΔP is the recorded pressure variation and ΔV is the MRF volume variation.

$$K = -V_0 \frac{\Delta P}{\Delta V} \tag{1}$$

Assuming the equivalent bulk modulus value determined from equation 1 is valid up to the maximum recorded pressure (65 kPa), it is possible to correlate pressure variations to changes in MRF volume. Hence, the volume expansion caused by oil degradation, which corresponds to the pressure increase shown in Figure 8, is found to be 8%.

In order for the clutch to remain functional despite base oil expansion, this supplementary expansion volume must be taken into account in the design of expansion chambers, supplementary to the amount required for thermal expansion.

6. Conclusion

Results from this work lead to a quantitative description of MR fluid durability when used in shear clutch mode. The effects of operating conditions such as shear rate, shear stress and temperature on the large-scale degradation signature of the torque-current relationship were analyzed on two MR-clutch designs. Two MRF degradation phenomena are identified: base oil expansion and particle oxidation. Solutions to extend the life of MR clutches were also proposed: recirculation and expansion volume.

For all tests conditions, clutch failure is dominated by oil expansion and occurs at energy input levels comprised between 1.5 MJ/mL and 9 MJ/mL. Post-oil expansion behavior is characterized by a loss of transmitted torque for a given current due to MRF leakage and by a slightly erratic behavior in the torque vs current relationship. This is attributed to seal failure, caused by the oil expansion of ~10%. The exact nature of oil expansion is not known and remains to be investigated. When pushed beyond the oil expansion limit, oxidation appears at energy input levels of 10 MJ/mL or more. In this study, it is not possible to conclude on the effect of oxidation on clutch performance since the effects of oxidation cannot be isolated from the effects of oil expansion.

It was confirmed experimentally that adding a supplementary volume of MRF and circulating it through the shear interface allows to increase the maximal energy dissipated by the clutch in direct proportion with the supplementary volume. The introduction of an additional volume that can withstand both expansion due to temperature change and MRF aging could also be used to prevent excessive pressure on the seal. Experimental evidence of the impact of additional volume on durability remains to be provided.

Future work will include testing of clutches equipped with compliant elements in order to validate the extent to which the MRF life can be lengthened after the base oil expansion. Also, investigation of the base oil degradation causes and impacts on MR fluid properties will be conducted.

References

[1] Klingenberg D J 2001 Magnetorheology: Applications and challenges *AIChE Journal* **47** 246–9

[2] Saito T and Ikeda H 2007 Development of normally closed type of magnetorheological clutch and its application to safe torque control system of human-collaborative robot *Journal of Intelligent Material Systems and Structures* **18** 1181–5

[3] Hakogi H, Ohaba M, Kuramochi N and Yano H 2006 Torque control of a rehabilitation teaching robot using magneto-rheological fluid clutches *JSME International Journal, Series B: Fluids and Thermal Engineering* **48** 501–7

[4] Fauteux P, Lauria M, Heintz B and Michaud F 2010 Dual-differential rheological actuator for high-performance physical robotic interaction *IEEE Transactions on Robotics* **26** 607–18

[5] Shafer A S and Kermani M R 2009 On the feasibility and suitability of MR and ER based actuators in human friendly manipulators *Intelligent Robots and Systems, 2009. IROS 2009. IEEE/RSJ Int. Conf. on* pp 2904 –2909

[6] Sassi S, Cherif K, Mezghani L, Thomas M and Kotrane A 2005 An innovative magnetorheological damper for automotive suspension: From design to experimental characterization *Smart Materials and Structures* **14** 811

[7] Chen J Z and Liao W H 2010 Design, testing and control of a magnetorheological actuator for assistive knee braces *Smart Materials Structures* **19** 035029

[8] Kieburg C, Oetter G, Laun M, Gabriel C and Steinwender H 2008 MR All-wheel-drive prototype car driving tests and durability requirements for the MR fluids used *Proc. of the 11th Int. Conf. on Electrorheological Fluids and Magnetorheological Suspensions*

[9] Karakoc K, Park E J and Suleman A 2008 Design considerations for an automotive magnetorheological brake *Mechatronics* **18** 434–47

[10] Carlson J D 2002 What makes a good MR fluid? *Journal of Intelligent Material Systems and Structures* **13** 431

[11] Wiehe A, Güth D and Maas J 2011 Large scale test bench for the durability analysis of MR fluids *Proc. of the 12th Int. Conf. on Electrorheological Fluids and Magnetorheological Suspensions* (World Scientific Publishing) pp 620–6

[12] Smith A L, Ulicny J C and Kennedy L C 2007 Magnetorheological fluid fan drive for trucks *Journal of Intelligent Material Systems and Structures* **18** 1131–36

[13] Ay R, Golnaraghi M F and Khajepour A 2001 Investigation on a Semi-Active Hydro Mount Using MR Fluid *Defense Technical Information Center*

Resistance of a commercial magnetorheological fluid to penetration

K F Plunkett[1], MA Imam[2], B Rath[2] and H Conrad[1]
[1] Materials Science and Engineering Dept., North Carolina State University, Raleigh, NC 27695-7907, USA
[2] Materials Science and Component Technology Directorate, Naval Research Laboratory, Washington, D.C. 20375, USA

E-mail: kenfplunkett@gmail.com

Abstract. The stress, σ, required to penetrate a commercial magnetorheological (MR) fluid exposed to an applied magnetic field, B, in the range of 0 - 0.25 Tesla was determined for: (a) a constant, continuous field and (b) on-off field. With constant field application, it was found that σ vs. the penetration depth, x, of the cylindrical indenter into the MR fluid at a constant rate of 0.12-0.18 mm/s was parabolic in form, namely $\sigma = \sigma_o x^n$ where σ_0 = AH, A = 1.7 kPa/(kA/m) and n = 0.3-0.5. The magnitude of σ_o is approximately six times that of the shear yield stress, τ_y, of the MR fluid. A model is proposed to explain the resistance to penetration into the MR fluid with the constant field. The results obtained in the on-off field tests and the in-out tests are in keeping with the model.

1. Introduction

A new class of materials of interest is magnetorheological (MR) fluids for wide range of applications. These fluids consist of micron-sized ferromagnetic particles suspended in a dielectric liquid such as a hydrocarbon oil or silicone oil [1]. With application of a magnetic field the viscosity of the suspension changes in times of the order of milliseconds [2] from Newtonian behavior to Bingham behavior with a so-called yield stress. The shear yield stress, τ_y, is the stress required to rupture the chain-like arrangement of the particles aligned along the magnetic flux [3]. The shear yield stress of typical MR fluids ranges from 10 to 60 kPa for a magnetic field H_{MRF} = 100 kA/m [1].

Theoretical considerations by Ginder and Davis [4] give for the yield stress prior to magnetic saturation of the particles in an MR fluid:

$$\tau_y = K_1 H_{MRF}^{\alpha} \tag{1}$$

and upon saturation

$$\tau_y = K_2 M_s^{\alpha} \tag{2}$$

where K_1 and K_2 are functions of the volume fraction φ of the particles, H_{MRF} is the magnetic field within the suspension, α is a constant on the order of 1-2 and M_s is the magnetic saturation of the particles. Tang and Conrad [5] found reasonable agreement between experiment and theory for shear tests on a suspension of carbonyl iron in silicone oil.

Carlsen [1] proposed the following empirical expression for the yield stress of MR fluids

$$\tau_y = (2.72 \times 10^5 \varphi^{1.534} C)\tanh(6.33 \times 10^{-6} H_{MRF}) \quad (3)$$

where the constant $C \approx 1$ and H_{MRF} is given by

$$B = 1.91\varphi^{1.311}[1 - \exp(-10.97\mu_0 H_{MRF})] + \mu_0 H_{MRF} \quad (4)$$

B is the applied magnetic field in Tesla and μ_0 the vacuum permeability.

To the authors' knowledge, no prior study on the strength of MR fluids has been performed in which penetration techniques have been implemented in the presence of an applied field. The objective of the present investigation was to determine their behavior in a compression (squeeze) mode.

2. Experimental

2.1. MR Fluid

The MR fluid employed in the present tests was Lord MRF-132DG, which was provided gratis by the Lord Corporation, Cary, North Carolina. The specifications pertaining to this fluid can be found in the Lord Technical Data sheets, located on the company's website. These give that the fluid consists of ~0.30 volume fraction (φ) magnetic particles (presumably carbonyl iron) suspended in a hydrocarbon oil and has a yield stress in shear of 55 kPa at H_{MRF} = 150 kA/m. Typical magnetic properties of the fluid are given in Figure 1.

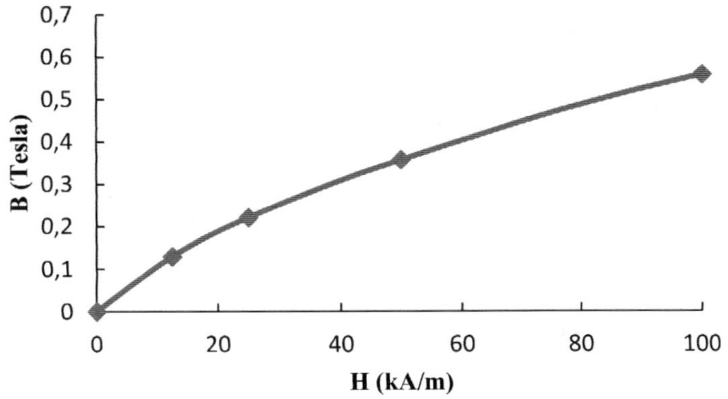

Figure 1. B vs. H_{MRF} given by Lord Corp. for their MRF-132DG.

2.2. Magnetic field apparatus

A schematic of the apparatus developed by us to apply the magnetic field during a compression test and the geometry and dimensions of the container and indenter are presented in Figure 2. The DC power supply consisted of two HP 6032A (60 V, 50 A) units provided by the U.S. Naval Research Laboratory and connected in series. The magnetic field in the air gap between the poles was measured with a F.W.

Bell IDR-329-T-UL: DC Kilogauss meter with an ultrathin transverse probe. A plot of the measured magnetic field in the airgap of 22 mm (outer diameter of the Al MR fluid container) between the curved pole pieces vs. the electric current, I, in the electromagnetic coil is presented in Figure 3. The flux is roughly proportional to the current up to 2 A, after which it increases at a decreasing rate.

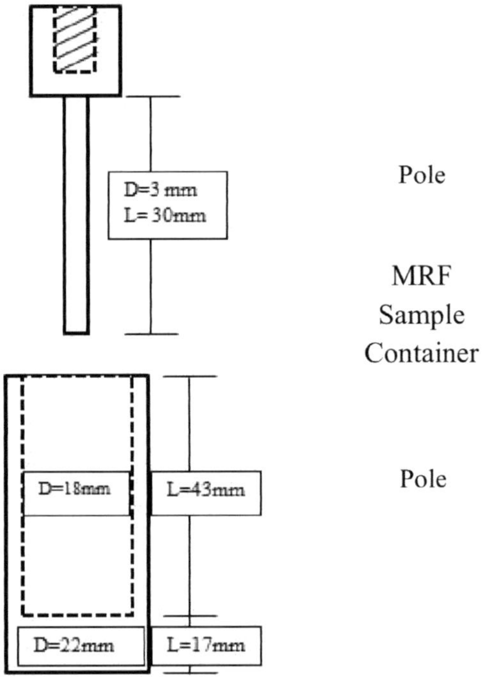

Figure 2. Schematic of the manner in which the magnetic field was applied to the MR fluid. Note the iron core poles were machined to fit around the Al MRF container. Also shown are the geometry and dimensions of the MR fluid container and the indenter. For testing, the indenter was located directly above the center of the sample container and was vertically lowered into the MRF.

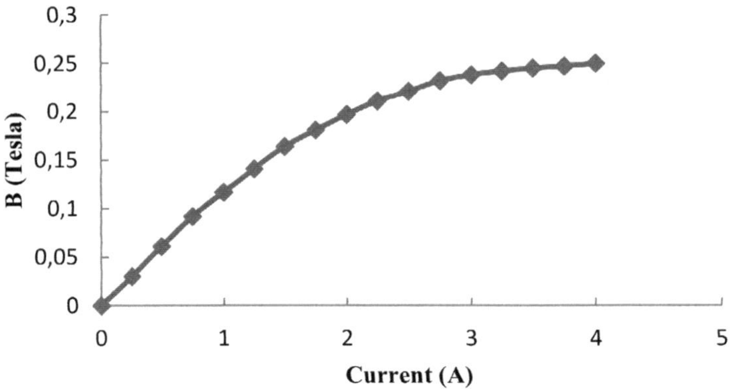

Figure 3. Measured magnetic flux in the 22 mm airgap (outer diameter of the Al MR fluid container) vs. the current in the electromagnetic coil.

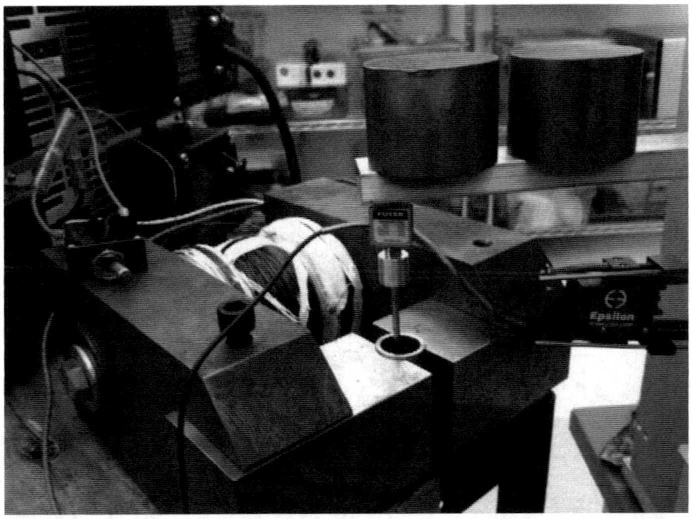

Figure 4. Photograph of the setup by which the compression tests were performed.

2.3. Compression (Penetration) tests

A photograph of the set-up by which the compression tests were performed is shown in Figure 4. The compressive load was applied at a fixed displacement rate of 0.12 to 0.15 mm/s by a commercial RSL Digital Displacement Loading Frame tension-compression mechanical test machine. Particular displacement rates are displayed on a per figure basis if applicable. The load was measured and recorded using a Futek LRM 200, 1 lb, JR S-Beam load cell.

Four types of compression (indentation) tests were performed:

- Tests with a fixed magnetic field throughout the entire test (Type A).
- On-off tests in which the field is alternately turned on and off during the test (Type B).
- In-out tests in which the indenter is applied with a fixed magnetic field, then retracted while the field is still maintained (Type C).
- Tests in which Type B was performed during a Type C test (Type D).

The Type A tests provide a measure of the resistance of the MR fluid to penetration; the Type B provides information on how quickly the resistance decreases following the removal of the field; the Type C provides information on the frictional or surface energy resistance imposed on the penetrator by contact of its cylindrical surface with the MR fluid, and Type D provides information on the effect of on-off cycles on the friction resistance. The resistance to penetration in each test is here reported as a nominal compression (or tensile) stress given by $\sigma = P/A$, where P is the load and A is the cross sectional area of the penetrating end of the indenter.

The following procedure was employed in performing a test with a constant field:

- The MR fluid was vigorously shaken and poured into the Al container.
- The indenter was centered and then lowered to within 0.5 to 1 mm above the surface of the MR fluid.
- The power supply was turned on to provide the magnetic field.
- The mechanical test machine was activated and the load vs. displacement was recorded.
- The test was terminated following a penetration of ~ 20 mm, and the penetrator withdrawn.

Similar procedures were employed in the on-off and in-out tests.

3. Results

3.1. Continuous magnetic field

Plots of the compressive stress vs. the displacement of the indenter as a function of the current passing through the electromagnetic coil are presented in Figure 5. The σ vs. displacement curves are parabolic in form. Incremental measurements using the gauss meter indicated a homogeneous field within the empty sample container. Field homogeneity could not be directly verified in situ due to the gauss meter's fragility, but it is inferred to be so. The aberrations in Figure 5's parabolic curves are the result of environmental vibrations being detected by the load cell. Converting the current to applied magnetic field, B, (Figure 3) the stress required for a displacement of 1, 5, 10, and 15 mm vs. B is given by Figure 6. It should be noted that for each displacement, x, the curvature of the plot is initially hyperbolic to B = 0.06 Tesla and then becomes essentially linear to B = 0.25 Tesla, the largest field attainable with our electromagnet setup. There is however some indication that for x = 20 mm, the stress becomes constant for B > 0.24 Tesla. Further work is required to determine whether this is in fact the case.

It should also be noted that during a compression test a brown translucent liquid, presumably the host hydrocarbon oil with no visible evidence of the magnetic particles, was squeezed up between the outer surface of the indenter and magnetized MR fluid. Moreover, the compressive stresses in the present tests are between 1 and 2 orders of magnitude larger than those normally obtained for the shear of MR fluids [1, 4, 5].

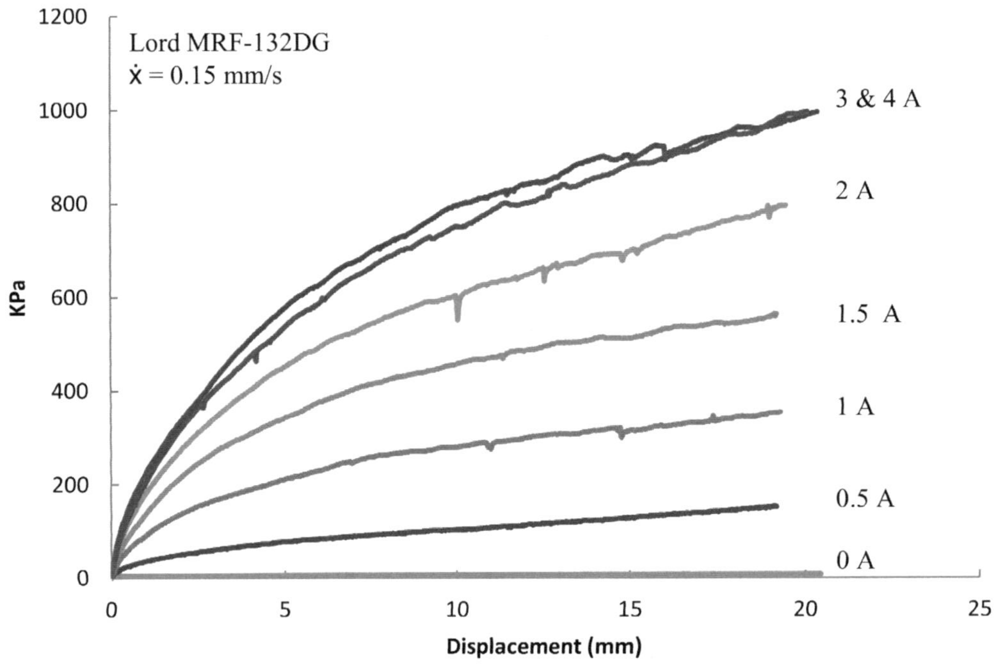

Figure 5. Compressive stress vs. displacement as a function of the current in the electromagnet in increments of 0.5 A. Stress curves for 3 and 4 A are provided to indicate the magnetic saturation of the electromagnet.

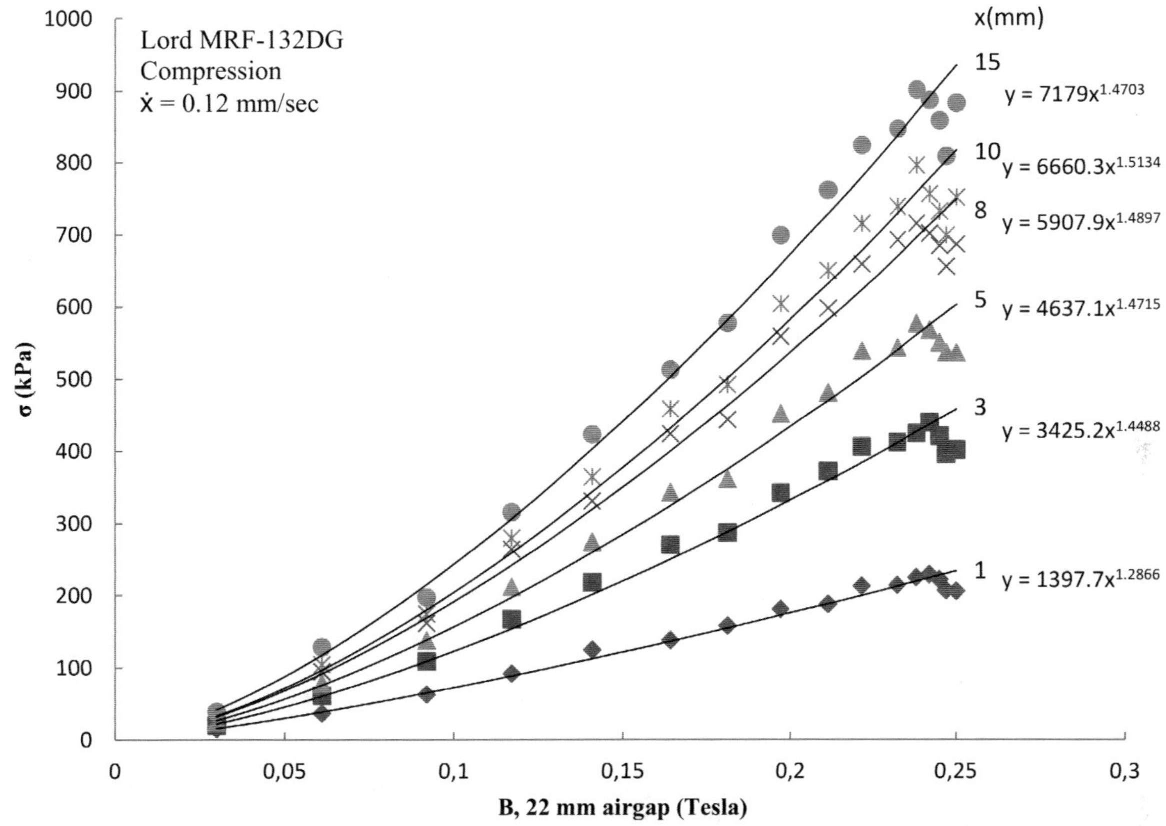

Figure 6. σ vs. B as a function of displacement, x; From the bottom, displacements are recorded at 1, 3, 5, 8, 10, and 15 mm. Equations for 1, 3, and 15 mm lines are $y = 1397.7x^{1.2866}$, $y = 3425.2x^{1.4488}$, and $y = 7179x^{1.4703}$ respectively.

3.2. On-off tests

The stress-displacement curves obtained in the tests in which the power supply to the electromagnet was periodically turned on and off are presented in Figure 7 for a current of 2 A. Analogous results were obtained for using a current of 1 A and 3 A. Included are the results for 2 cycles and 10 cycles. Also included in the figure is the σ vs. displacement curve obtained in the constant continuous field tests. To be noted in Figure 7 is: (a) the decrease in load when the field is turned off is abrupt and less than the response time of the recorder (< 0.25 seconds), (b) upon turning the field on, the initial increase in stress is abrupt up to a "yield stress" (σ_y) and then increased at a less rapid rate, (c) σ_y decreased the more frequent the number of on-off cycles, (d) the ratio σ_y/σ became smaller the greater the number of on-off cycles and the higher the applied magnetic field.

Journal of Physics: Conference Series **412** (2013) 012023 doi:10.1088/1742-6596/412/1/012023

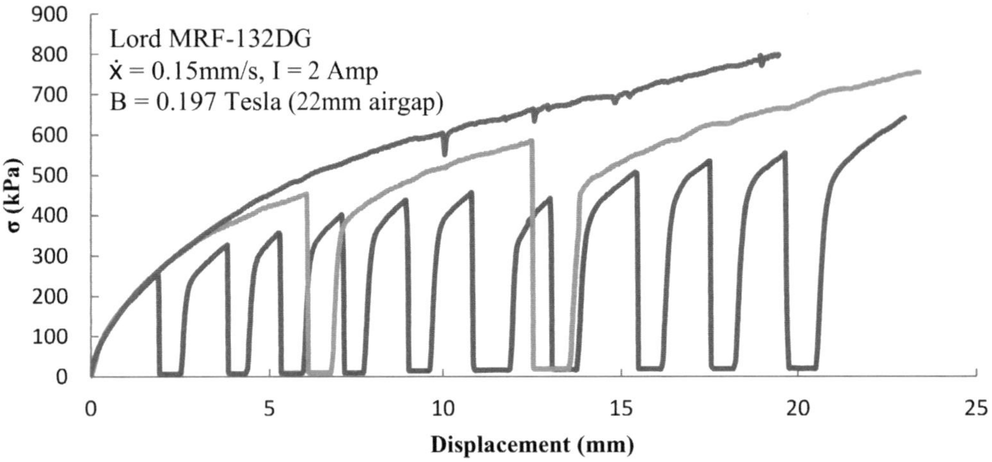

Figure 7. On-off field cycling tests (2 and 10 cycles) with electromagnet current, I = 2A, compared to the behavior for a continuous current of 2 A.

3.3. In-out tests

The behavior in the tests in which the indenter first penetrated the MR fluid with the field on, and then was retracted from the fluid, with the same field strength and displacement rate, is shown in Figure 8 for the field corresponding to 2A. To be noted is that immediately following the direction change of the indenter, the MR fluid exerts a resistive, pulling, force on the indenter, the magnitude of which is ~1/4 of the compressive force prior to retraction. This force is interpreted to result from shear friction on the indenter's walls. The negative pressure is similar when the field is turned on and off as well as when the field was held constant throughout the test. Also to be noted is that the tensile force decreases as the indenter is retracted from the MR fluid. Further, the decrease in tensile force which occurs when the field is turned off during the extraction becomes less with the extraction.

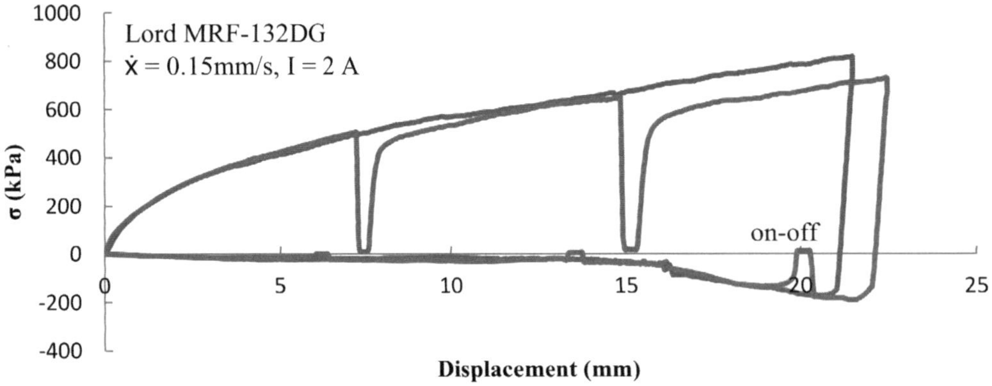

Figure 8. In-out tests with the on-off cycling test superimposed on a continuous test with an electromagnet current of I = 2 A.

4. Discussion

The parabolic nature of the σ vs. displacement curves in Figure 5represent equations of the form $\sigma = \sigma_0 x^n$, which equate to

$$log\, \sigma = log\, \sigma_0 + n \log x \tag{5}$$

where σ_0 is the compressive stress for x = 1 mm and n is the slope of the plot of log σ vs. log x. Log-log plots of the σ vs. displacement curves for several continuous electromagnetic currents are presented in Figure 9. The currents have here been converted to applied magnetic field, B, according to Figure 3. The results in Figure 9 are reasonably fit by straight lines with σ_0 increasing with magnetic field strength and n ≈ 0.3-0.5 increasing with field strength. A plot of σ_0 vs. H_{MRF} taken from Fig.3 is shown in Figure 10. There occurs a reasonable fit to a straight line with a slope of 12.3 (kPa)/(kA/m). This slope is ~ 6 times that given in the Lord Technical Data sheet for the shear stress, τ_y, in the same H_{MRF} range and is included in Figure 15. The linear fit for σ_0 vs H_{MRF} is in accord with theoretical considerations of τ_y [4,5], and the magnitude of σ_o is effectively six times larger.

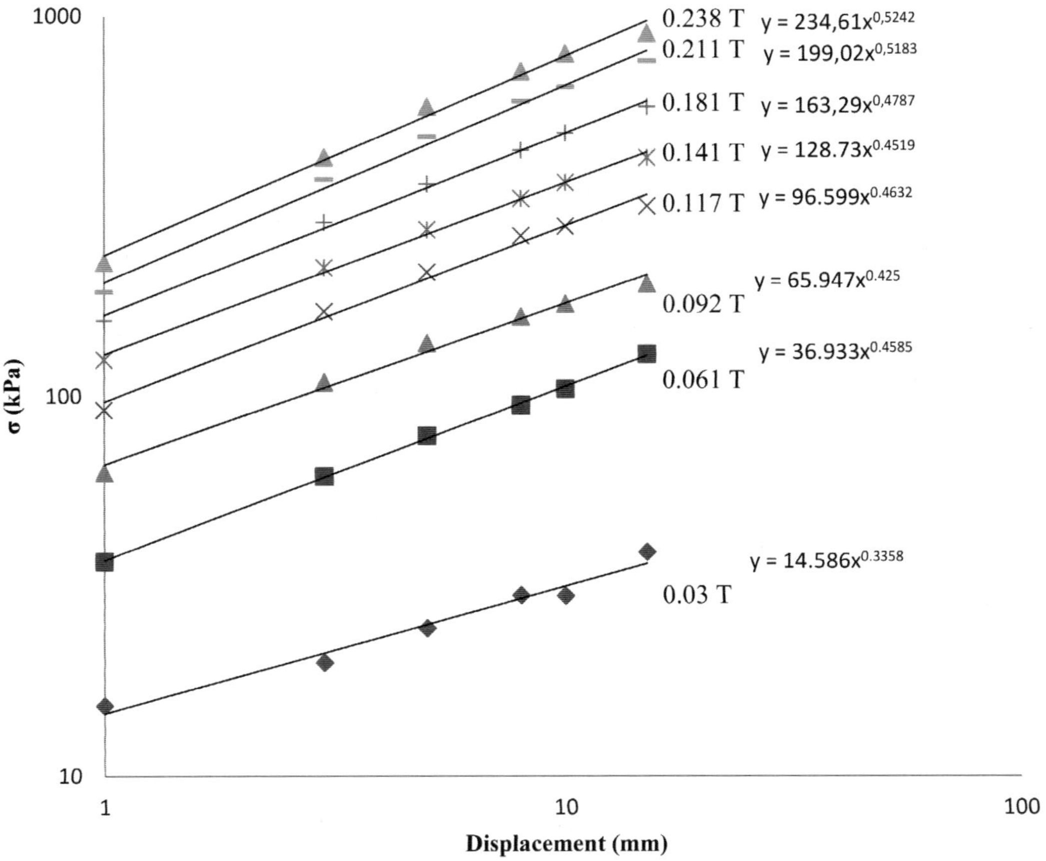

Figure 9. Log-log plots of the σ vs. displacement curves for a constant field as a function of the applied field, B, (Tesla); Field strengths for equation denoted lines are .03 T, .061 T, and .238 T, from the bottom respectively.

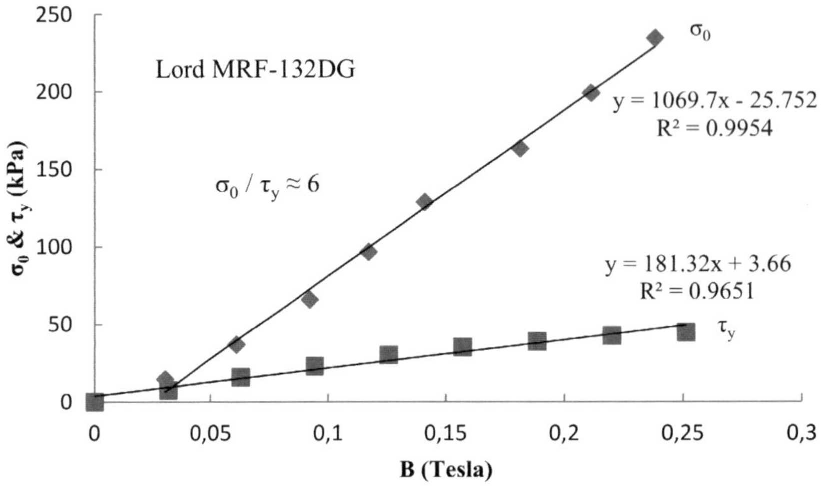

Figure 10. σ_o and τ_y vs. B (normalized to airgap).

Based on the observation that a translucent brown liquid with no evidence of particles (presumably the hydrocarbon oil host carrier fluid) appeared at the surface of the MR fluid during the compression, the authors propose the model shown in Fig.11 for the behavior in the present compression tests. In this model the host oil is squeezed out between the planar chain-like arrangement of the magnetic particles, thereby giving a slightly higher volume fraction of particles in the oil-depleted regions adjacent to and below the indenter. This in turn leads to an increase in stress required for further penetration of the indenter i.e. in the displacement, x. Due to the relative size difference between the indenter and sample container, in addition to the penetration depth, it is believed that interference from the walls or bottom of the container has a negligible effect on the experimental results.

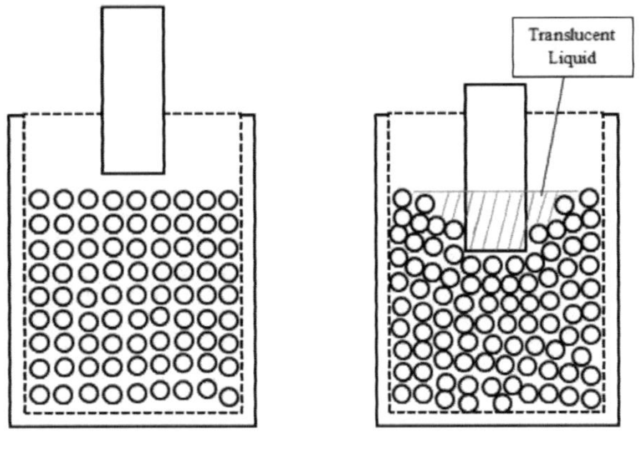

Figure 11. Model for the stress corresponding to the compression of the MRF. With penetration, the oil is squeezed out between the magnetic particles up along the side of the indenter, giving a higher volume fraction of particles below the indenter and in turn a higher stress with increasing penetration.

Accepting the above model, the decrease in stress which occurs when the field is turned on in the on-off tests can result from a disruption in the lower-energy, chain-like structure of the particles during the displacement when the field is off. The more frequently this on-off transition occurs, the greater is the disruption and the lower the resistance to further displacement.

According to our model, the resistive tensile pull which acts on the indenter immediately following the reversal of the indenter could result from the presence of elastic compression strain in the chains below the bottom of the indenter and from the interaction between particles at the side surfaces of the indenter.

Acknowledgements
This research was funded by the Naval Research Laboratory, Washington, D.C.

References

[1] Carlsen D 2005 MR Fluids and Devices in a Real World *Electrorheological Fluids and Magnetorheological Suspensions* K. Lu, R. Shen and J. Lu, eds., World Scientific, Singapore pp 531
[2] Goncalves F D and Carlson J D 2007 Investigating the time dependence of the MR effect *Int. J. Mod. Phys. B* **21** 4832
[3] Tang X, Chen Y and Conrad H 1996 Structure and interaction force in a model magnetorheological system J. *Intel. Mater. Syst. Struct.* **7** 517
[4] Ginder J M and Davis L C 1994 Shear stresses in magnetorheological fluids, Role of magnetic saturation *Appl. Phys. Lett.* **65** 3410
[5] Tang X and Conrad H 1996 Quasistatic measurements on a magnetorheological fluid *J. Rheol.* **40** 1167

The behavior of the MR fluid during durability test

J Roupec[1], I Mazůrek, Z Strecker and M Klapka

Institute of Machine and Industrial Design, Faculty of Mechanical Engineering, Brno University of Technology, Technická 2896/2, 616 69, Brno, Czech Republic

E-mail: roupec.j@fme.vutbr.cz

Abstract. The article describes results of durability test of a magnetorheological fluid (MRF), which was carried out in rheometer of own design. The rheometer design enables to measure the rheological properties of MR fluid and to expose it to a long-term loading simultaneously, without any manipulation of the measured sample. During the durability test a change of the two most important parameters of Bingham model describing the behavior of MR fluids can be followed – dynamic viscosity and yield stress. In this paper the yield stress and viscosity were evaluated depending on temperature in OFF-state. The results show a significant change of yield strength during durability test depending on temperature of loading. Dependence of yield stress on temperature was proved. The viscosity decreased by 36% from its initial value after the dissipation of 9÷20kJcm^{-3} from total 1.2 MJcm^{-3} and then has remained the same until the end of durability test. Viscosity was evaluated depending on temperature.

1. Introduction

Current magnetorheological (MR) fluids are suspensions formed by a carrying fluid and ferromagnetic particles, most commonly powdered iron. Upon the application of an external magnetic field the MR fluids can change their state from fluid to a semi-solid or plastic state, and back, in several milliseconds. These properties can be appropriately utilized for regulation of linear and rotary motion. The widest commercial application of these fluids is in mechatronic damping elements – MR dampers. MR dampers are frequently used in car suspension systems [1,2], suspension of driver seats in goods vehicles [3,4], vibration damping during seismic activity [5,6] or damping of cable bridge vibrations caused by wind and rain [7,8].

A flow curve of MR fluid exhibits Non-Newtonian behavior and is most commonly replaced by the Bingham model of plastic material, alternatively, for a more accurate solution, by so-called biviscous model (Figure 1). If the flow curve is described by the Bingham model, then the resulting shear stress in MR fluid can be described by the following equation:

$$\tau = \tau_y(H) + \eta \cdot \dot{\gamma} \qquad \tau \geq \tau_y, \tag{1}$$

where H is the magnetic field intensity, γ' is shear rate of the fluid, η is the viscosity in the off-state (i.e., viscosity at $H = 0$), τ is total shear stress and τ_y is yield stress. The total shear stress τ in Equation (1) is expressed as a sum of viscous part $\eta.\gamma'$ and yield stress τ_y.

[1] To whom any correspondence should be addressed.

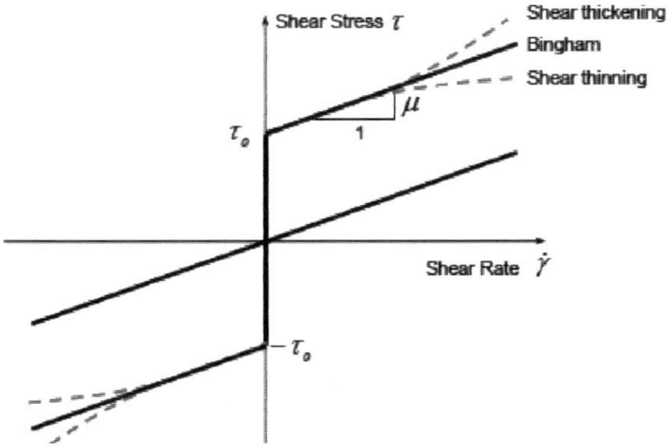

Figure 1. Rheological models of flow curve of MR fluid [9].

Although MR fluids have been used for several years, problems occurring during their long-term operation have not yet been satisfactorily resolved. In particular, the effect of temperature, shear stress, shear rate, and the number of cycles on the properties of MR fluids have not been addressed.

Already Carlson et al. [10,11] discovered that the damping force increases with the number of load cycles during long-term loading of magnetorheological damper. They called this phenomenon in-use-thickening (IUT). Since in their experiment the damping force was only sampled at constant speed of pulsator, it was not possible to reconstruct the flow curve from this single point. At that time it was only possible to determine the apparent viscosity. This resulted in a hypothesis that the IUT is demonstrated by viscosity increase. However, the apparent viscosity is not a material constant for Non-Newtonian fluids. It is dependent on shear rate or shear stress. It is therefore necessary to describe the behavior of non-Newtonian fluids by a flow curve in the desired range of shear rates. Data of apparent viscosity (published by Carlson) can serve only for rough comparison of MR fluid consistency and only under identical test conditions. The flow curve is mathematically described by two parameters (yield stress and viscosity) which are constant for given state of MR fluid in a wide interval of shear rates.

Carlson determined oxidation as a source of IUT and this was confirmed by Ulicny [12]. Ulicny performed a durability test, he described oxidation of Fe-particles and he also described the decrease of the power of the MR-clutch, in which the MR fluid was loaded. Figure 2a shows non-oxidized iron particles, Figure 2b shows oxidized particles after intensive loading of the MR fluid. A visible oxidized layer is both hard and brittle. Nanoparticles of oxides are separated from this oxidized layer due to particle interaction during the MR fluid flow. These nanoparticles increase fluid apparent viscosity in the off-state. Sunkara proceeded in his work and he described the dependence of decrease of the yield stress on oxidation [13].

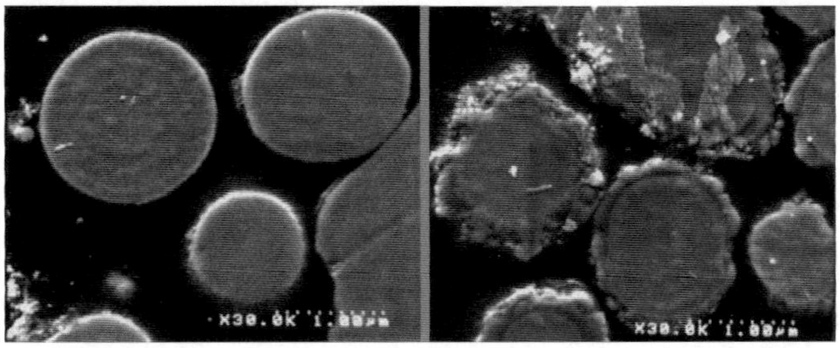

Figure 2. SEM pictures of scratch pattern (a) non-tested iron particles; (b) particles from 540h_MRF sample [12].

In Figure 3 “the viscosity increase” of MR fluid (according to Carlson) with the number of cycles is shown with the solid line. The dashed lines mark “viscosity” of new MR fluid with the addition of oxide nanoparticles. Although more efficient MR fluids have been developed, fluid durability as a function of temperature, shear stress and shear rate has not been published nor discussed yet. However, it is generally well-known that oxidation intensity grows with increasing temperature.

Likewise, it can be assumed that higher shear rate will result in a more noticeable collision of iron particles and thereby also in creation of oxide particles.

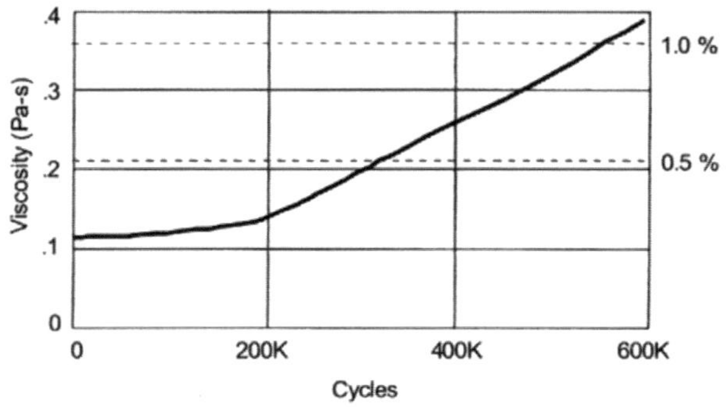

Figure 3. Off-state viscosity with added nanoparticles [10,11].

To date, nobody published complex description of MR fluid during long-term operation. The description of MR fluid behavior with the aid of an apparent viscosity is not appropriate. For complex description of MR fluid behavior, it is necessary to use the mathematical description of the whole flow curve in the wide range of shear rates. The best model for such behavior is probably a Bingham model.

2. Methods and experimental approach

For long-term tests, the special load unit was developed (Figure 4). This unit works like a slit-flow rheometer which allows loading the MR fluid and also measurement of flow curves without any manipulation with the sample.

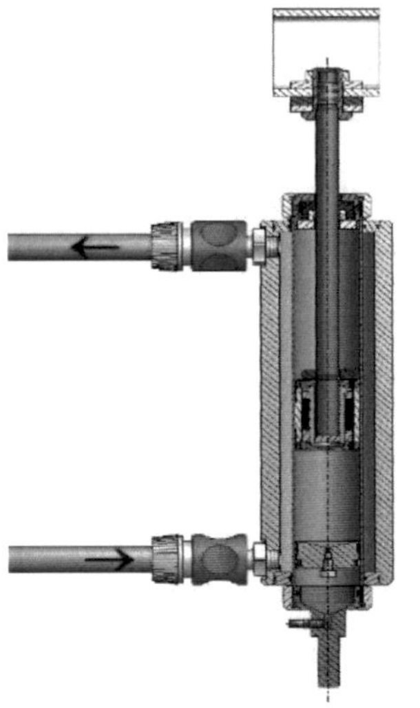

Figure 4. Cross section of slit-flow rheometer.

The real piston of the proposed rheometer has a specific slit with thickness h_1=0,75mm and width b=2πr=88mm divided into two identical sections of length l_1=9mm (Figure 5). Between them, the fluid passes through the slit with length l_2=19mm where is much weaker magnetic field with intensity H_2<<H_1. The geometric profile of this slit is also different (h_2>>h_1). The hydraulic resistance of this piston part is therefore very small and its influence was included in the correction function that is evaluated during the necessary calibration.

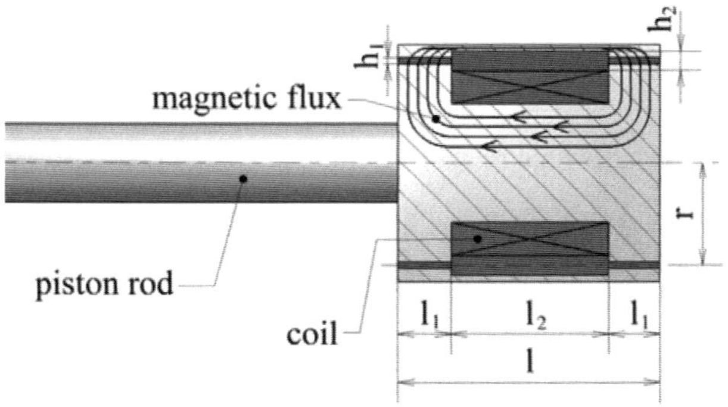

Figure 5. Simplified diagram of piston group.

Our load unit was installed to the mechanical pulsator with stroke 50mm (Figure 7). Tested MRF was 140CG by Lord Corporation. Volume of MRF was 100ml. The MR fluid was exposed to the long-term operation in ON-State (H_I=134 kA/m) with the maximum shear rate γ'=35 000 s^{-1}. Two durability tests with different temperatures (50 and 70°C) were conducted. The durability tests were conducted in accordance with the diagram in Figure 6. The tests were stopped after specific life dissipation energy LDE = 1.2 $MJcm^{-3}$.

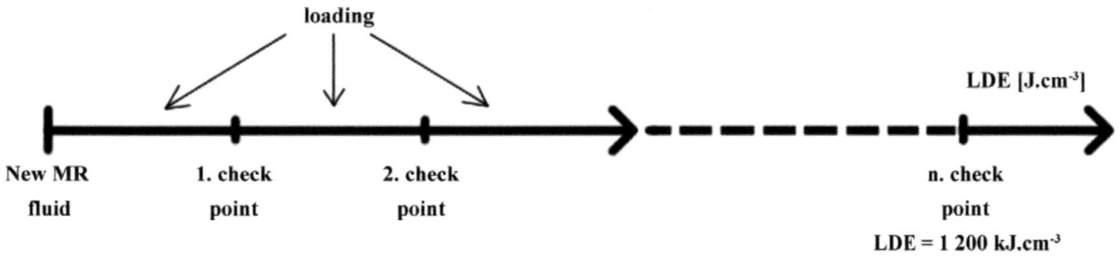

Figure 6. Scheme of durability test.

At each checkpoint of test, the flow curves in off-state and on-state within the temperature range 20 to 80°C were measured using our load and rheometric unit. Flow curves at control points were obtained using the start-up tests (Figure 7). From measured points of velocity and force there were selected only those, where the piston acceleration was close to zero (e.g. points F_I and v_I on Figure 7). This choice eliminated the influence of gas content elasticity of the load unit and inertia forces on the hysteretic nature of the point distribution. This also ensured a stable flow in the unit slit. F-v (Force-velocity) characteristic was obtained from these tests and was transformed into a flow curve using the appropriate mathematical relations. For presented results in OFF-state, an influence of apparent slip wall effect during the evaluation was neglected due to relatively low consistency of MRF.

The behavior change of MR fluid in off-state was evaluated from the Bingham's model parameters change (viscosity, dynamic yield stress) and the influence of temperature was observed too.

A repeatability of measurement could not be accurately determined due to unequal conditions during test. Especially temperature could not be kept on exactly the same value during temperature unstable start-up test for measuring flow curves. For this reason, the temperature dependence of the monitored parameters was measured in the wide interval 20-75°C. From the measured points of shear stress (at γ'= 10.000 s^{-1}) dependence on temperature and dynamic yield stress dependence on temperature (Figure 8), it is possible to assume that the repeatability is on a very good level.

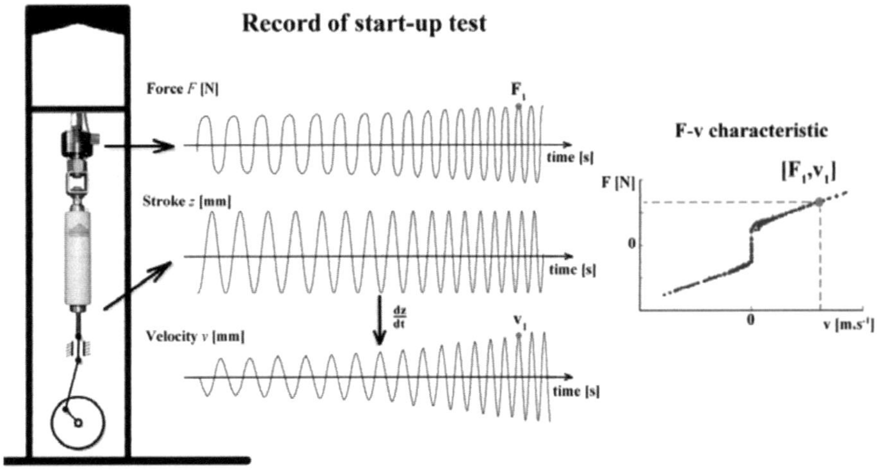

Figure 7. Schematics of the start-up test procedure and F-v characteristics derivation.

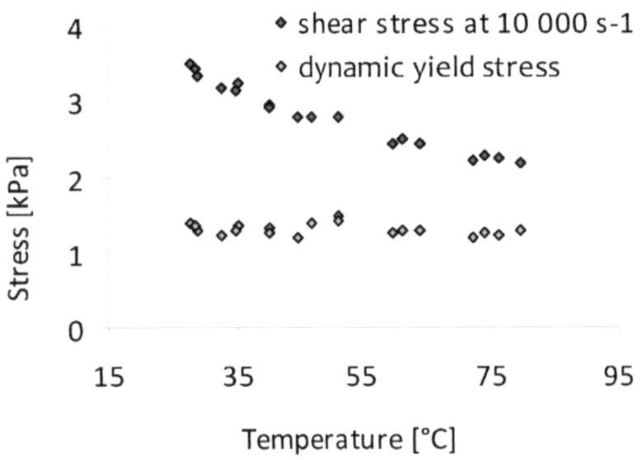

Figure 8. Presentation of repeatability of measurement on temperature (in) dependence of measured points at LDE=37 kJ cm^{-3}.

3. Results and discussion

All presented results are valid for the off-state. Figure 9 shows the change of consistence of the MR fluid before and after loading.

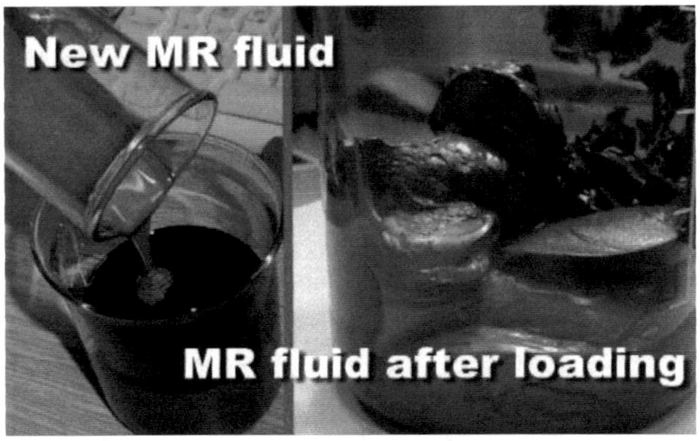

Figure 9. The change of consistence of the MR fluid before and after loading.

Figure 10 shows the significant viscosity decrease (up to 36%) after dissipation of 90 kJ cm^{-3}. After this initial decrease, the viscosity remained stable. The beginning of the second durability test was monitored more precisely and it was identified that the viscosity decrease came already when LDE = 9÷20 kJ cm^{-3}.

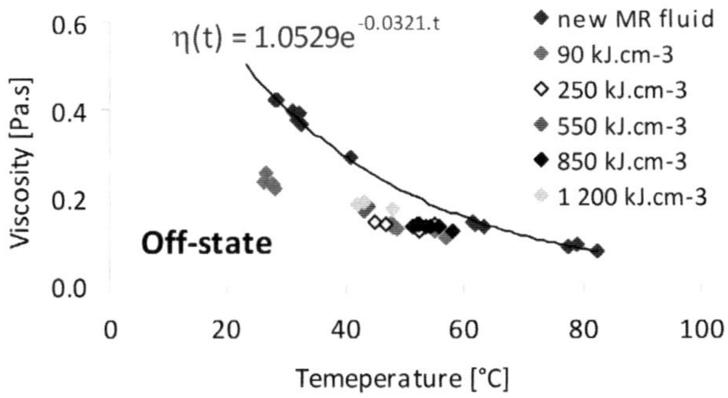

Figure 10. Decrease of viscosity during durability test at temperature 50°C.

The comparison of viscosity curve during loading with different temperatures (Figure 11) shows that the temperature did not influence the viscosity. Figure 10 shows all measured viscosity values during both tests except the beginning of the loading, where the viscosity decreased.

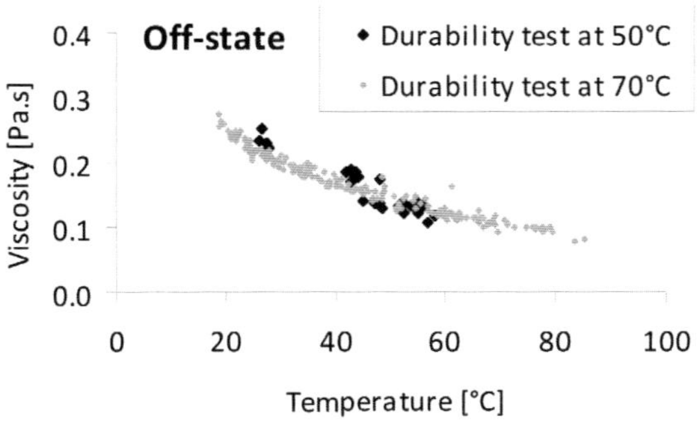

Figure 11. The comparison of viscosity curve during loading with different temperatures.

The second parameter of the Bingham model is a dynamic yield stress. Figure 12 shows the course of the dynamic yield stress curve during durability test at 50°C. There is a significant increase from 1 to 4kPa.

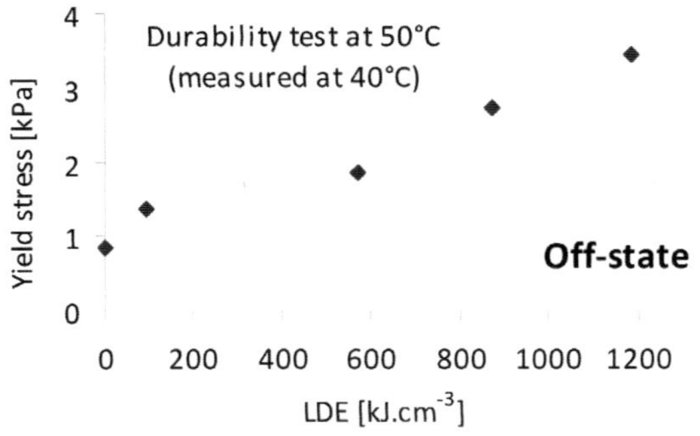

Figure 12. Increase of dynamic yield stress during durability test at temperature 50°C.

During durability test at 70°C the increase of dynamic yield stress was considerable lower from 1 to 2.5 kPa (Figure 13).

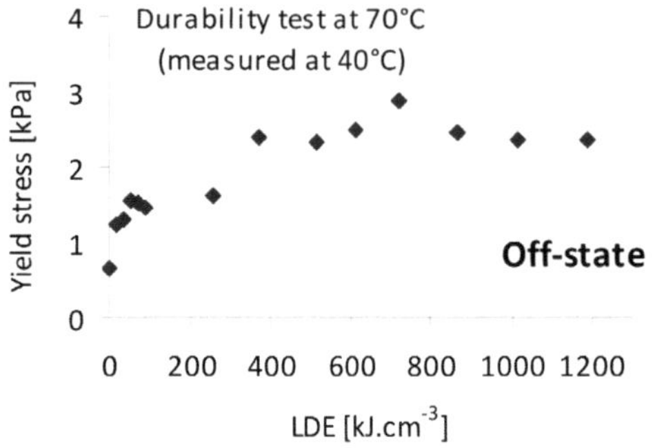

Figure 13. Increase of dynamic yield stress during durability test at temperature 70°C.

There was also observed a certain dynamic yield stress dependence on the temperature, which was considerably changing during durability test (Figure 14). In the beginning of the MR fluid test, the dynamic yield stress was independent on the temperature. With the increasing number of load cycles (and thus increase of LDE), the dependence on temperature started to be obvious and increasing in time.

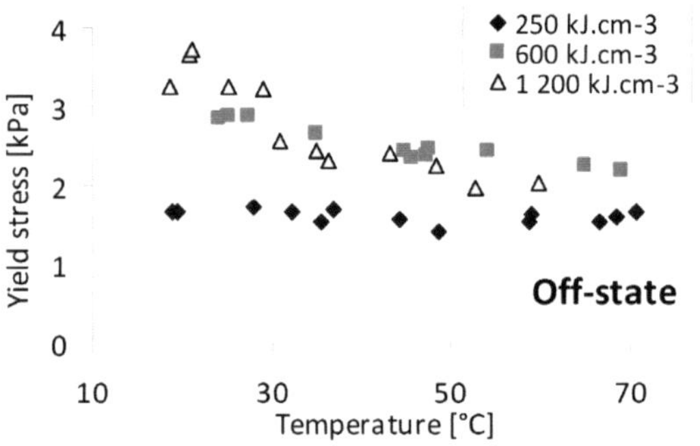

Figure 14. Increase of dynamic yield stress during durability test at temperature 70°C.

4. Conclusion

The significant MR fluid viscosity decrease during the beginning of the durability test was described. After that the viscosity was no more changing. Contrary the dynamic yield stress course had increasing character, which fully corresponds with the published knowledge that the consistence of the MR fluid is increasing because the oxidized layer is separated from the surface due to particles interaction. During the test with higher temperature, the increase of dynamic yield stress is less significant. This phenomenon is caused by lower load of the MR fluid, because the apparent viscosity is lower and thus interactions among particles are less significant.

Acknowledgments

This work was developed with the support of the research plan CZ.1.07/2.3.00/30.0039 and grant BD 13101006.

References

[1] Gehm R 2001 Delphi Improves Cadillac's Ride *Automotive Engineering International* **109** 32-33

[2] Wang D H, Bai X X, Liao W H 2010 An integrated relative displacement self-sensing magnetorheological damper: prototyping and testing *Smart Mater. Struct.* **19** 105008

[3] Choi S B, Nam M H, Lee B K 2002 Vibration control of MR seat damper for commercial vehicles *J.Intell. Mater. Syst. Struct.* **11** 936-44

[4] McManus S J, St Clair K A, Boileau P E, Boutin J 2002 Evaluation of vibration and shock attenuation performance of a suspension seat with a semi-active magnetorheological fluid *J. Sound Vib.* **253** 313-27
[5] Yang G, Spencer Jr. B F, Carlson J D, Sain M K 2002 Large-scale MR Fluid Dampers: Modeling and Dynamic Performance Considerations *Eng. Struct.* **24** 309-23
[6] Chen C, Ricles J M, Sause R, Christenson R 2010 Experimental evaluation of an adaptive inverse compensation technique for real-time simulation of a large-scale magnetorheological fluid damper *Smart Mater. Struct.* **19** 025017
[7] Ok S Y, Kim D S, Park K S, Koh H M 2006 Semi-active fuzzy control of cable-stayed bridges using magneto-rheological dampers *Eng. Struct.* **29** 776-88
[8] Kim I H, Jung H J, Koo J H 2010 Experimental evaluation of a self-powered smart damping system in reducing vibrations of a full-scale stay cable *Smart Mater.Struct.* **19** 115027
[9] URL: <http://www.lord.com> [cit. 2009-06-11]
[10] Carlson J D 2002 What makes a Good MR fluid? *J. Intell. Mater. Syst. Struct.***13** 431-35
[11] Carlson J D 2003 Critical factors for MR fluids in vehicle systems *Int. J. Veh. Des.* **33** 207-17
[12] Ulicny J C, Balogh M P, Potter N M, Waldo R A 2006 Magnetorheological fluid durability test–Iron analysis *Mater. Sci. Eng.* **443** 16-24
[13] Sunkara S R, Root T W, Ulicny J C and Klingenberg D J 2009 Iron oxidation and its impact on MR behavior *J. Phys.:Conf. Ser.* **149** 012081

Variation of forced convective heat transfer in rectangular duct flow of a magnetic fluid under magnetic field

M Motozawa[1, 3], **T Sekine**[1], **T Sawada**[2] **and Y Kawaguchi**[1]
[1]Tokyo University of Science, 2641 Yamazaki, Noda, Chiba, Japan
[2]Keio University, 3-14-1 Hiyoshi, Kohoku-ku, Yokohama, Kanagawa, Japan

E-mail: motozawa@rs.noda.tus.ac.jp

Abstract. Variation of forced convective heat transfer in a rectangular duct flow of a magnetic fluid under a magnetic field was investigated experimentally. Experiments were performed changing the magnetic field intensity, and this magnetic field could be varied from 0 mT to 600 mT. The Reynolds number based on the hydraulic diameter was set to 960, 1900 (laminar flow), and 2830 (turbulent flow). The results of the experiments show that in the case of laminar flow of the magnetic fluid, when a magnetic field is applied to a magnetic fluid flow, heat transfer locally increases in the region where the magnetic field exists. In contrast, in the turbulent flow of the magnetic fluid, heat transfer is not enhanced but reduced. In order to better understand this heat transfer phenomenon, we measured the velocity distribution of magnetic fluid flow by the Ultrasonic Velocity Profile (UVP) method. In the case of laminar flow, the result shows that the flow velocity at the center of the rectangular duct decreases and the velocity gradient in the near-wall region increases. Moreover, we calculated the flow resistance under a magnetic field by measurement of the pressure gradient, and the relationship between heat transfer and flow resistance was discussed.

1. Introduction

Magnetic fluid [1] is a stable colloidal solution which uniformly disperses ferromagnetic particles in solvents such as water or kerosene. These ferromagnetic particles are about 10 nm in size and coated with surfactant in order to avoid aggregation. Therefore, these particles can remain suspended because of Brownian motion. In addition, magnetic fluid has enough liquidity to be treated as a Newtonian fluid under no magnetic field. However, once a magnetic field is applied to magnetic fluid, several interesting flow phenomena, which are not seen in a Newtonian fluid, have been observed [2, 3]. Since magnetic fluid was developed about 50 years ago, many studies on the physical properties [4] and interesting phenomena of magnetic fluid [5] have been performed. In recent decades, magnetic fluid has been applied in mechanical engineering [6], medical engineering [7], space engineering, and so on.

In thermal engineering, magnetic fluid also has interesting thermophysical properties such as temperature-dependent magnetization. Therefore, the thermophysical properties of magnetic fluid have also attracted attention and have been applied to many technologies recently. One of the most successful applications of the thermophysical properties of magnetic fluid is the audio speaker [8]. In audio speakers, magnetic fluid is used to fill the space around the voice coil. Because the thermal conductivity of magnetic fluid is much greater than that of air, the fluid provides a lower heat resistance between the coil and the pole plate. As a result, the operating temperature of the voice coil

[3] Corresponding author.

can be reduced. On the other hand, magnetic fluid is also useful for heat transport applications. Iwamoto *et al.* [9] developed a self-driven heat transport device using temperature-sensitive magnetic fluid. As mentioned above, because magnetic fluid contains nano-order-size ferromagnetic particles in the carrier liquid, it can be considered a kind of nanofluid. Recently, there has been intensive research on the heat transfer of nanofluids and several studies on the heat transfer characteristics of nanofluid have been performed [10]. Therefore, the study on heat transfer of the magnetic fluid can be regarded as the combination of heat transfer of the nanofluid and active control of heat transfer by a magnetic field. Moreover, if heat transfer enhancement and reduction of flow resistance can be realized simultaneously, it is very useful for many engineering techniques. As a basic study of these backgrounds, we investigate the forced convective heat transfer characteristics in rectangular duct flow of a magnetic fluid.

Several studies on the heat transfer characteristics of magnetic fluid have been carried out attracting great engineering interest [11-13]. However, the heat transfer phenomena of magnetic fluid are very complicated and there is not enough knowledge about them. In our previous study [14], we performed an experimental investigation of forced convective heat transfer in the rectangular duct flow of a magnetic fluid in a laminar flow. The result showed the heat transfer was enhanced locally in the region in which the magnetic field existed. Generally, if heat transfer is enhanced, the flow resistance will increase in the same region. Therefore, it is necessary to investigate the relationship between heat transfer and flow resistance.

Moreover, in order to better understand the characteristics of the complicated physical phenomena of magnetic fluid, the internal velocity fields should provide useful information. However, because magnetic fluid is opaque, it is impossible to apply optical methods such as Laser Doppler Velocimetry (LDV) and Particle Image Velocimetry (PIV). To realize the measurements of the velocity profile of an opaque fluid, Takeda developed the Ultrasonic Velocity Profile (UVP) method more than 20 years ago [15]. The UVP method is a way of measuring a velocity profile along the ultrasonic beam from the ultrasonic Doppler shift frequency. After the development of UVP, Sawada *et al.* [16] and Kikura *et al.* [17] applied the UVP method to velocity profile measurement of magnetic fluid. These experimental studies have shown the efficiency of the UVP method for velocity profile measurement of magnetic fluid.

In this study, we performed an experimental investigation of variation of forced convective heat transfer in the rectangular duct flow of a magnetic fluid. The experiments were carried out in both a laminar flow and a turbulent flow. The magnetic field intensity could be varied from 0 mT to 600 mT. To discuss the heat transfer characteristics of magnetic fluid flow, we measured the velocity distribution of the rectangular duct flow of a magnetic fluid subject to magnetic fluid by UVP. In addition to this heat transfer experiment, the flow resistance of the magnetic fluid flow under a magnetic field was measured. On the basis of the results of these experiments, the effect of magnetic field on the forced convective heat transfer, and the relation between heat transfer and flow resistance in the rectangular duct flow of magnetic fluid was discussed.

2. Experiment

2.1. Experimental apparatus

Figure 1 shows a block diagram of the flow system of the experimental apparatus. The experiment was performed with a closed-circuit loop with the rectangular duct as a test section. The fluid was circulated by a pump and the flow rate could be adjusted by a bypass. An impeller type flow meter was installed downstream of the pump to measure the flow rate. Two thermocouples were set at the inlet and outlet of this duct. Therefore, we could measure the temperature of the flow liquid at inlet T_{in} and outlet T_{out}. A storage tank on the flow loop was equipped with a heater and a cooler to keep T_{in} at a constant level. An electromagnet was located at the center of the test section. This electromagnet could apply a uniform magnetic field to the magnetic fluid flow and the magnetic field could be varied from 0 mT to 600 mT.

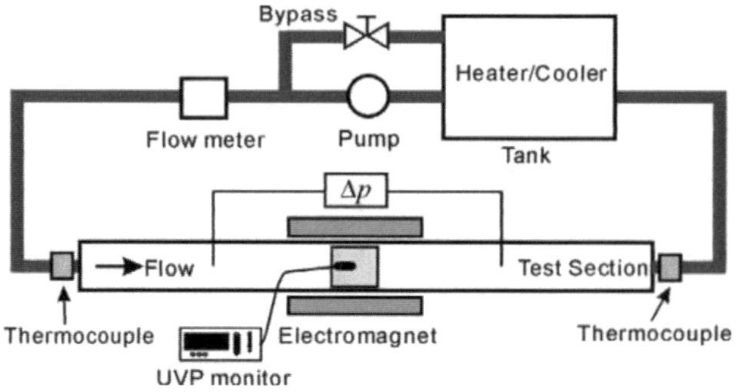

Figure 1. Flow system.

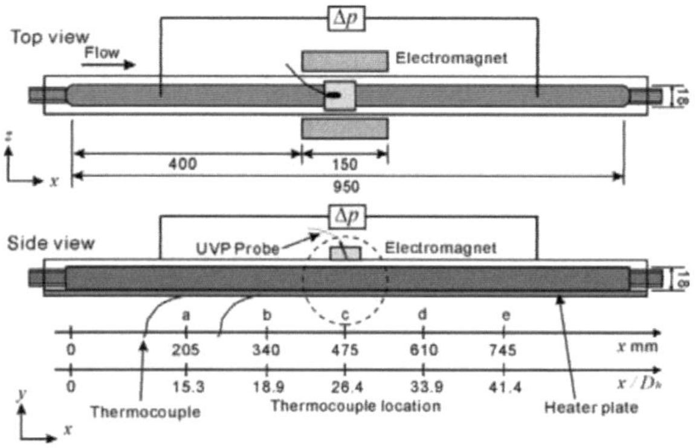

Figure 2. Detailed structure of the rectangular duct.

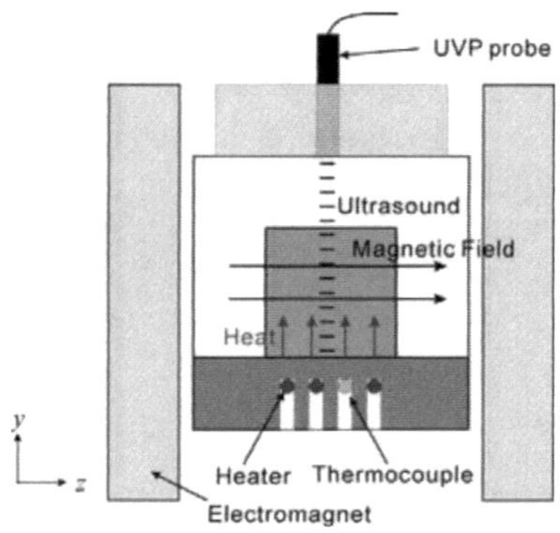

Figure 3. Cross-section of the duct.

The detailed structure of the rectangular duct and the cross-section of this duct at the center position are shown in Figs. 2 and 3, respectively. This rectangular duct was made of a transparent acrylic resin except for the heater plate, and was 18 mm × 18 mm in cross-section and 950 mm in length. The hydraulic diameter D_h of this duct is 18 mm. Two pressure taps were located on the top of the rectangular duct, 700 mm apart, and the static pressure gradient between the two pressure taps was measured. There was an electromagnet between these pressure taps. Therefore, we could measure the change in the flow resistance under the magnetic field. We defined the Cartesian coordinates as shown in Figs. 2 and 3 (i.e. x: Streamwise direction, y: Direction of normal to the heater plate, z: Spanwise direction). In this experiment, the Reynolds number based on the bulk mean velocity and hydraulic diameter was set to 960, 1900 and 2830. As a result of the measurement of the velocity distribution by the UVP method, we confirmed that the magnetic fluid flow was a laminar flow at Re = 960 and Re = 1900, but the magnetic fluid flow was a turbulent flow at Re = 2830.

The heater plate was attached to the bottom of the rectangular duct. This heater plate comprised the heater and a copper plate as shown in figure 3. The flow liquid in the rectangular duct could be heated by this heater plate with a uniform heat flux. The heat flux of this heater plate was 41 kW/m^2 in this experiment. We defined the entrance of the duct as $x = 0$. Five thermocouples were installed in the heater plate and located at x = (a) 205 (x/D_h = 15.3), (b) 340 (x/D_h = 18.9), (c) 475 (x/D_h = 26.4), (d) 610 (x/D_h = 33.9) and (e) 745 mm (x/D_h = 41.4) downstream from the entrance of this duct. Therefore, we could estimate the wall temperature of the heater plate (T_w) at each position. It could be assumed that the bulk temperature (T_b) of the flow liquid increased linearly with the length of channel, and we were able to calculate the bulk temperature at the each position.

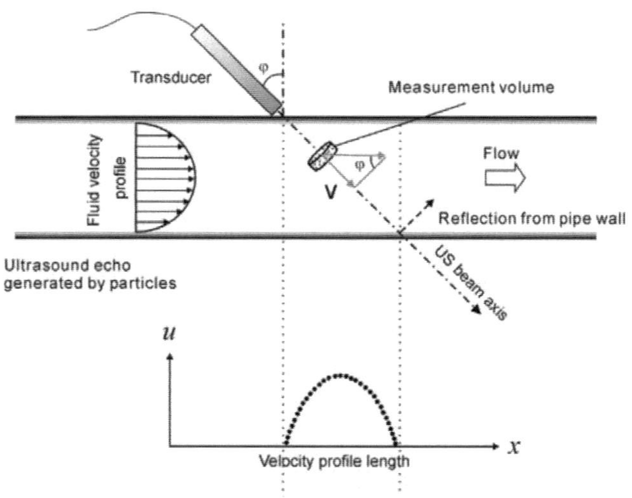

Figure 4. Principle of UVP method.

2.2. UVP method

2.2.1. Principle [18]. Figure 4 shows the principle of the UVP method and the relation between the ultrasonic beam and the flow. The UVP method is based on pulsed ultrasound echography. An ultrasonic pulse is emitted from the transducer along the measuring line and the same transducer receives the echo signal reflected from seeding particles in the flow. All velocity profile information, such as the instantaneous velocity and the positions from where the ultrasound is reflected, can be detected. The position x from where the ultrasound is reflected can be extracted from the time delay τ between the start of pulse burst and reception time as follows;

$$x = c\tau / 2 \qquad (1)$$

where, c is the ultrasonic propagation velocity in the flow liquid. At the same time, the velocity information is derived from the Doppler shift frequency by the following equation.

$$V = cf_D / 2f_0 \qquad (2)$$

where, f_D and f_0 are the Doppler shift frequency and original ultrasonic frequency, respectively. Thus, a velocity profile can be obtained along the ultrasonic beam by analyzing the echo signal to derive instantaneous frequencies at each instant.

2.2.2. Velocity profile measurement by UVP. A UVP probe was set on the center of the rectangular duct and at the position where the magnetic field was applied to the magnetic fluid, as shown in Figs. 2 and 3. Therefore, we could measure the velocity distribution in the x-y plane at the center of the duct. The UVP probe was fixed on the outer wall of the duct at an angle of 14°. In order to obtain the reflected echo signal, we added the poly(methylmethacrylate) particles (MBX-100 produced by Sekisui Plastics Co., Ltd.) as tracer particles. These particles were 115 μm in mean diameter.

2.3. Magnetic fluid. The test magnetic fluid was a water-based magnetic fluid named W-40 produced by Taiho Industries Co., Ltd. W-40 was 40 % weight concentration of fine magnetite particles (Fe_3O_4) in water. However, because W-40 is too dense for this experiment, we diluted this magnetic fluid to 70 volume % with water as a test fluid. The properties of this test fluid are listed in Table 1.

Table 1. Properties of the magnetic fluid (at 25°C).

Test Fluid	Water	Magnetic fluid
Particle material	-	Fe_3O_4
Particle diameter	-	about 10 nm
Carrier liquid	-	water
Density kg/m^3	997	1305
Viscosity mPa·s	0.89	8.0
Specific heat J/kg·K	4110	3080
Thermal conductivity W/m·K	0.58	0.55
Prandtl number	6.2	39.5

2.3.1. Thermo physical properties of test magnetic fluid. Because it is very difficult to obtain the exact values of the thermo physical properties such as specific heat and thermal conductivity, these values are estimated by the following methods:

- *Specific heat:* The specific heat of magnetic fluid c_{pmf} at constant pressure was estimated by the following equation.

$$c_{pmf} = c_{pw} n_w + c_{pm} n_m \tag{3}$$

  where, c_{pw} and c_{pm} are the specific heat of water and magnetite particles, respectively. n_w and n_m are the volume fraction of water and magnetite particles, respectively. The volume fraction can be calculated by the density of magnetic fluid, water and magnetite.
- *Thermal conductivity k:* Takegoshi *et al.*[19] estimated the thermal conductivity of magnetic fluid k_{mf} by Kerner's coating model. The estimated thermal conductivity of magnetic fluid had good agreement with the measured thermal conductivity. Therefore, we also estimated the thermal conductivity of magnetic fluid by Kerner's coating model.
- *Prandtl number:* The Prandtle number can be calculated by the specific heat c_p, thermal conductivity k and viscosity μ by the following equation,

$$\mathrm{Pr} = \frac{c_p \mu}{k} \tag{4}$$

The estimated values of specific heat, thermal conductivity and the Prandtl number are also listed in Table1.

Moreover, when a magnetic field is applied to a magnetic fluid, it is possible that these thermo physical properties will change. Li *et al.* [20] measured the thermal conductivity of magnetic fluid under a magnetic field and they reported that the thermal conductivity of magnetic fluid changed by applying a magnetic field. However, also according to them, when the magnetic field was applied to a magnetic fluid in a direction perpendicular to the temperature gradient, which is the same situation as in our experiment, the thermal conductivity hardly changed. Therefore, in our experiment we treated these values of thermo physical properties of magnetic fluid as being constant under a magnetic field.

2.3.2. Ultrasonic velocity in test magnetic fluid. When the UVP method is applied to measure the velocity profile of the test fluid, the ultrasonic propagation velocity in the test fluid is needed as represented in equations (1) and (2). However, it is very difficult to measure the ultrasonic propagation velocity in a fluid precisely. Moreover, in the case of magnetic fluid, because the magnetic field influences the ultrasonic propagation velocity in a magnetic fluid, it is necessary to investigate the influence of the magnetic field.

In our previous study [21], we measured the ultrasonic propagation velocity in a magnetic fluid and investigated the influence of the magnetic field on the ultrasonic propagation velocity. Details of the measurement technique can be found in Motozawa *et al.* [21]. When a magnetic field is applied to a magnetic fluid, the ultrasonic propagation velocity increases in proportion to the length of time the magnetic field is applied. The inner particles coagulate and form a chain-like cluster under a magnetic field. This change seems to be caused by chain-like cluster formations. However, because the change in ultrasonic propagation velocity is rather small, this change does not influence UVP measurements. We also measured the ultrasonic propagation velocity in the test magnetic fluid by the same technique and ultrasonic propagation velocity of this was c = 1429 m/s.

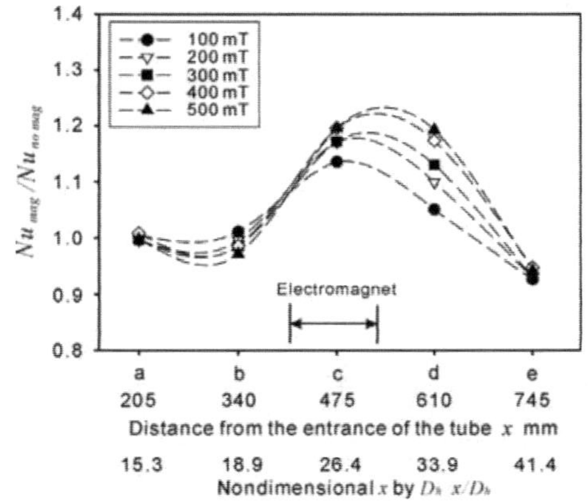

Figure 5. Effect of magnetic field on heat transfer in a laminar flow at Re=960.

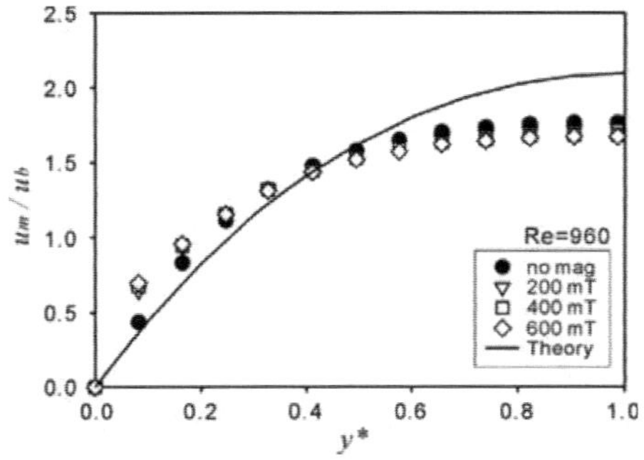

Figure 6. Effect of magnetic field on velocity distribution in a laminar flow at Re = 960.

3. Result and Discussion

3.1. Effect of magnetic fluid on heat transfer and velocity distribution

3.1.1. Laminar flow case. Figure 5 shows the effect of a magnetic field on the heat transfer in the rectangular duct flow of a magnetic fluid. In this experiment, the Reynolds number was set to 960, and the magnetic fluid flow was a laminar flow. The horizontal axis is the distance from the entrance of the duct, and the position of the thermocouples in the duct is also indicated. The flow is from the left to the right side.

Heat transfer coefficient h can be calculated at each position where the thermocouple is attached (i.e. position (a) - (e) in figure 2) by the following equation,

$$h = \frac{\dot{q}}{T_w - T_b} \tag{5}$$

where, $\dot{q}$ is heat flux, T_b is the bulk temperature of magnetic fluid flow which is estimated by T_{in} and T_{out}. Therefore, the Nusselt number is defined by,

$$\mathrm{Nu} = \frac{hD_h}{k} \tag{6}$$

To compare between the Nusselt number with and without a magnetic field, we evaluated the local Nusselt number ratio defined by $\mathrm{Nu}_{mag}/\mathrm{Nu}_{no\ mag}$ at each position. Here, Nu_{mag} and $\mathrm{Nu}_{no\ mag}$ are the local Nusselt number with and without a magnetic field at the same position, respectively.

This figure indicates that when the magnetic field is applied to a magnetic fluid flow, the heat transfer at the position (c) where the magnet field exists greatly increases compared with under no magnetic field. This increment of the Nusselt number becomes larger with increasing magnetic field. In addition, heat transfer is also enhanced just downstream of the position (c) (i.e. position (d)), but slightly smaller than that at the position (c). This seems to show that good heat transfer at position (c) remains but weakens downstream of the area where the magnetic field exists. On the other hand, the heat transfer hardly changes at the entrance and the end of the duct (i.e. position (a) and (e)). This indicates that a magnetic field locally influences the heat transfer of a magnetic fluid flow in the area where the magnetic field is applied.

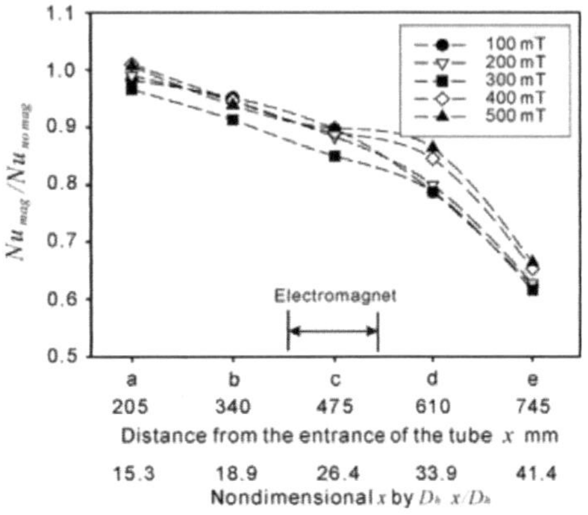

Figure 7. Effect of magnetic field on heat transfer in a turbulent flow at Re=2830.

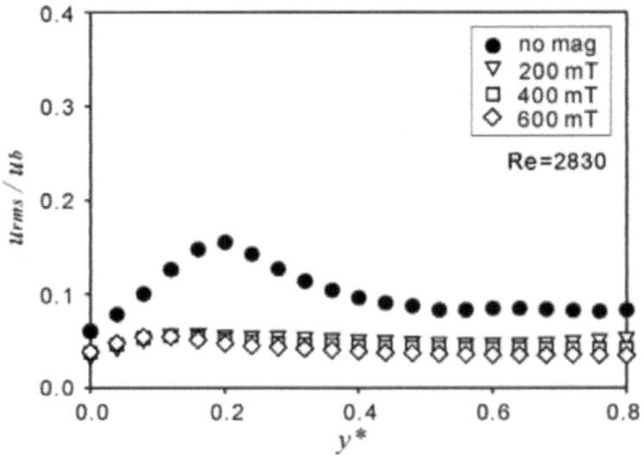

Figure 8. Effect of magnetic field on streamwise velocity fluctuation in a turbulent flow at Re = 2830.

In the case of magnetic fluid flow, there seems to be two prominent reasons why the forced convective heat transfer of magnetic fluid flow changes by magnetic field. One reason is the change in the thermo-physical property of magnetic fluid (e.g. specific heat, thermal conductivity etc.), and the other is the change in the velocity distribution on the basis of the change in the physical properties of magnetic fluid (e.g. apparent viscosity, local density in the region where the magnetic field exists). Regarding the change in the thermo-physical properties of magnetic fluid by magnetic field, as mentioned above, Li *et al.* [20] reported that when a magnetic field is applied to a magnetic fluid, thermal conductivity of a magnetic fluid changes related with the direction of magnetic field. However, if the direction of the magnetic field is perpendicular to the temperature gradient, which is the same direction as our experiment, the thermal conductivity hardly changed. Therefore, the influence of the change in the thermo-physical properties of magnetic fluid is excluded from the reason why the forced convective heat transfer changes.

On the other hand, figure 6 shows the effect of a magnetic field on the velocity distribution of a magnetic fluid flow at Re = 960 (Laminar flow). In this figure, the black plots and solid line indicate the velocity distribution of the magnetic fluid flow under no magnetic field and Newtonian fluid obtained by the theory, respectively. The horizontal axis is the distance from the bottom wall normalized by the half of the duct and the vertical axis is the mean velocity u_m normalized by the bulk mean velocity u_b.

When a magnetic field is not applied to the magnetic fluid, magnetic fluid can be treated as a Newtonian fluid. However, the configuration of the measured velocity distribution is different from that of the Newtonian fluid. The mean velocity of magnetic fluid under no magnetic field near the center of the duct is smaller than that of Newtonian fluid. This is due to a problem in the experimental system. In this experiment, we measured the velocity distribution at the developing region. Therefore, the magnetic fluid flow was not developed enough and the velocity distribution of magnetic fluid under no magnetic field was different from that of the Newtonian fluid.

When the magnetic field is applied to magnetic fluid, this figure indicates that the mean velocity at the center of the duct decreases but the velocity gradient increases near the wall. The increment of velocity gradient gradually becomes larger with increasing magnetic field. In this study, because a uniform magnetic field is applied to magnetic fluid flow, magnetic body force $\boldsymbol{M} \cdot \nabla \boldsymbol{H}$ equals to zero. This means that the velocity distribution does not change by magnetic field. However, because apparent viscosity of magnetic fluid increases by magnetic field and the length of magnetic field region is very short, the velocity distribution changes in the entrance of magnetic field region and magnetic fluid flow is not developed enough. Therefore, the velocity gradient increases in the region where the magnetic field exists.

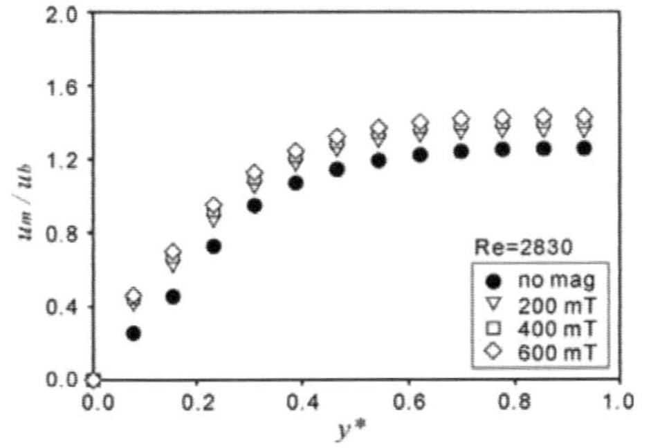

Figure 9. Effect of magnetic field on velocity distribution in a turbulent flow at Re=2830.

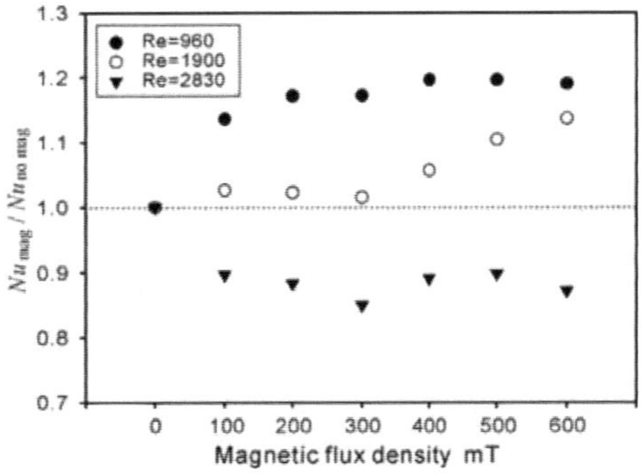

Figure 10. Reynolds number dependence of heat transfer at position (c).

In addition, when a magnetic field is applied to a magnetic fluid, magnetic particles form a micro-scale chain-like structure in the direction of magnetic field [22]. It is a possibility that this clustering structure could inhibit the axial velocity of the flow [23]. This inhibition also leads to the increment of the velocity gradient in the near-wall region because the flow rate is constant when magnetic field is applied to magnetic fluid flow in this experiment. For these reasons, velocity gradient near the heater plate increases in the region where the magnetic field exists and this increment of velocity gradient cause local heat transfer enhancement. However, the increment of velocity gradient indicates the increment of the shear stress. Therefore, it is important to investigate the relationship between heat transfer and flow resistance.

3.1.2. Turbulent flow case. In contrast to the above, figure 7 shows the effect of a magnetic field on the heat transfer for a turbulent flow. In this case, the Reynolds number was set to 2830. This figure indicates that the heat transfer is not enhanced but suppressed. In addition, heat transfer decreases gradually with the distance from the entrance of the duct. Moreover, figure 8 shows the distribution of the streamwise velocity fluctuation normalized by the bulk mean velocity. This figure indicates that when the magnetic field is applied to turbulent magnetic fluid flow, the streamwise velocity fluctuation was largely suppressed. This is because apparent viscosity of magnetic fluid increases by applying magnetic field. This suppression of velocity fluctuation means the suppression of turbulent diffusion. Therefore, heat transfer was also suppressed in the region where the magnetic field exists.

On the other hand, figure 9 shows the results of the measurement of the velocity distribution at the same Reynolds number. This figure indicates that the velocity gradient in the near-wall region greatly increases with increasing magnetic field. As mentioned above, the increment of the velocity gradient leads to the heat transfer enhancement. However, the influence of the suppression of the turbulent diffusion is stronger than the effect of the increment of the velocity gradient. As a result, heat transfer was suppressed in the region where the magnetic field exists.

In addition, regarding the velocity distribution as shown in figure 9, characteristic results were obtained. When the magnetic field is applied to a magnetic fluid, not only the velocity gradient near the wall but also the mean velocity near the center of the duct increases with increasing magnetic field intensity. In this study, experiments were carried out under a constant flow rate when the magnetic field was applied to the magnetic fluid. Therefore, the increment of both velocity gradient and mean velocity near the center of the duct indicates that the velocity distribution in the x-y plane is different from that in the x-z plane, and characteristic anisotropy of the velocity distribution exists in the velocity profile related to the magnetic field direction. In the future, in order to investigate this characteristic velocity distribution, we will measure the velocity distribution in the x-z plane.

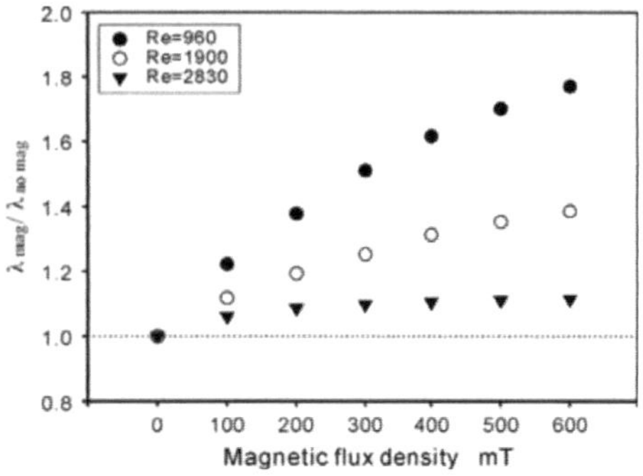

Figure 11. Reynolds number dependence of flow resistance.

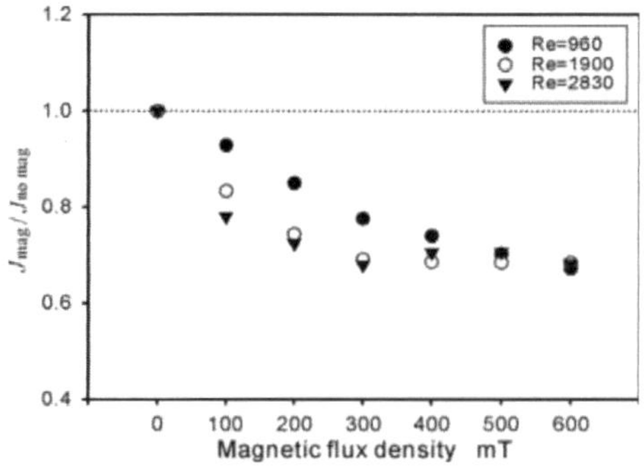

Figure 12. Evaluation of the relation between heat transfer and flow resistance by Colburn's *J*-factor.

3.2. Reynolds number dependence of heat transfer and pipe frictional coefficient

Figure 10 shows the Reynolds number dependence of the heat transfer at position (c) where theagnetic field exists. The horizontal axis represents the magnetic field intensity. As mentioned above, in the cases of laminar flow for Re = 960 and 1900, heat transfer is enhanced with a magnetic field. However, in a flow with a larger Reynolds number, it is necessary for heat transfer enhancement to apply a strong magnetic field. In contrast, heat transfer is not enhanced in a turbulent flow, even if the strong magnetic field is applied.

On the other hand, figure 11 shows the Reynolds number dependence on the increment of the pipe frictional coefficient. To compare between the pipe frictional coefficient λ with and without a magnetic field, we evaluated the increment ratio of the pipe frictional coefficient defined by $\lambda_{mag}/\lambda_{no\ mag}$. This figure indicates that the flow resistance of a magnetic fluid greatly increases with increasing magnetic field in the laminar flow. It is well known that the apparent viscosity increases when applying a magnetic field to a magnetic fluid. As a result of the increase of apparent viscosity, the flow resistance increases with increasing magnetic field intensity. However, the effect of a magnetic field is small in a turbulent flow. Kamiyama *et al.*[24] investigated flow resistance in a circular pipe by experiment, and almost the same result was reported.

3.3. Relationship between heat transfer and flow resistance

As mentioned above, when a magnetic field is applied to a magnetic fluid, heat transfer is greatly enhanced in the laminar flow of the magnetic fluid. However, if heat transfer is enhanced, the flow resistance would generally increases in the same region. Therefore, it is necessary to investigate the relationship between heat transfer and flow resistance. We evaluated the relationship between heat transfer and flow resistance by the following method.

The Stanton number is defined by,

$$\mathrm{St} = \frac{\mathrm{Nu}}{\mathrm{Re}\cdot\mathrm{Pr}} \tag{7}$$

In this study, the Stanton number was evaluated at position (c) where a magnetic field exists. Using the Chilton-Colburn analogy, which can be applied to a wide range of Prandtl numbera, Colburn's *J*-factor is defined by the following equation,

$$J = \frac{\mathrm{St}\cdot\mathrm{Pr}^{2/3}}{(C_f/2)} \tag{8}$$

To compare between the value of J with and without a magnetic field, we evaluated the increment ratio of J defined by $J_{mag}/J_{no\ mag}$. The value of J would be 1 in a water flow theoretically. In this experiment, we obtained about 0.8 of the value of J.

Figure 12 shows magnetic field dependence on the increment ratio of J. This figure indicates that when the magnetic field is applied to a magnetic fluid, $J_{mag}/J_{no\ mag}$ decreases with increasing magnetic field intensity in both the laminar flow case and the turbulent flow case. In the case of a turbulent flow, heat transfer decreases and flow resistance slightly increases when applying a magnetic field. Therefore, it goes without saying that J decreases under a magnetic field. On the other hand, in the case of the laminar flow, this result means that the increment of flow resistance is relatively larger than the heat transfer enhancement. Therefore, although the heat transfer was greatly enhanced in a laminar flow as shown in figure 5, this is unfavorable case in the heat transfer with considering the flow resistance. We will continue to experiment changing in the situation of applying magnetic field in the future.

4. Conclusions

Variation of heat transfer in a rectangular duct flow of a magnetic fluid under a magnetic field was investigated experimentally. Moreover, in order to better understand the heat transfer characteristics of the magnetic fluid flow, we measured the velocity distribution of magnetic fluid flow by the UVP method. We performed experiments on both the laminar flow and the turbulent flow of a magnetic fluid. A magnetic field can be varied from 0 mT to 600 mT. As a result of this study, the following results were obtained.

1. In the case of laminar flow, when the magnetic field is applied to a magnetic fluid, heat transfer is locally enhanced at the region where the magnet field exists. This heat transfer becomes larger with increasing magnetic field but saturated under a strong magnetic field. In addition, the mean velocity at the center of the duct decreases with increasing magnetic field intensity but the velocity gradient increases near the wall.
2. In the case of a turbulent flow, heat transfer is suppressed by the magnetic field. The mean velocity near the center of the duct and velocity gradient near the wall increase with increasing magnetic field intensity. This fact indicates that characteristic anisotropy exists in the velocity profile related to the magnetic field direction.
3. When the magnetic field is applied to a magnetic fluid flow, the flow resistance increases in both laminar flow and turbulent flow. To evaluate the relation between heat transfer enhancement and increment of flow resistance, we calculate Colburn's J-factor. In the case of a laminar flow, increment ratio of J decreases with increasing magnetic field intensity. This means the flow resistance increases more compared with the heat transfer.

Acknowledgements

This study was partly supported by a Grant-in-Aid for Young Scientists (B) of the Japan Society for Promotion of Science.

References

[1] Rosensweig R E 1985 *Ferrohydrodynamics* (Cambridge University Press) chapter 2.
[2] Kamiyama S, Koike K and Oyama T 1983 *J. Magn. Magn. Mater.* **39** 23.
[3] Kikura H, Takeda Y and Durst F 1999 *Exp. Fluids* **26** 208.
[4] Rinaldi C, Chaves A, Elborai E S, He X and Zahn M, *Current Opinion in Colloid & Interface Sci.* **10** 141.
[5] Sudo S, Ohaba M, Katagiri K, Hashimoto H 1993 *J. Magn. Magn. Mater.* **122** 248
[6] Ohno K, Suzuki H and Sawada T 2011 *J. Magn. Magn. Mater.* **323** 1389.
[7] Pankhurst Q A, Connolly J, Jones S K and Dobson J 2003 *J Phys D: Appl Phys* **36** R167.
[8] *Ferrotec Corporation home page*; http://www.ferrotec.com/
[9] Iwamoto Y, Yamaguchi H and Niu X D 2011 *J. Magn. Magn. Mater.* **323** 1378.
[10] Xuan Y and Li Q 2000 *Int. J. Heat and fluid flow* **21** 58.
[11] Sawada T, Kikura H, Saito A and Tanahashi T 1993 *Exp. Therm. Fluid Sci.* **7** 212.
[12] Wu F, Wu C, Guo F and Li D 2001 *Int. J. Heat and Mass Transfer* **44** 4427.

[13] Yamaguchi H, Zhang Z, Shuchi S and Shimada K 2002 *J. Magn. Magn. Mater.* **252** 203
[14] Motozawa M, Chang J, Sawada T and Kawaguchi Y. 2010 *Phys. Procedia* **9** 190
[15] Takeda Y 1986 *Int. J. Heat and Fluid Flow* **7** 313
[16] Sawada T, Kikura H and Tanahashi T 1999 *Exp. Fluids*, **26** 215
[17] Kikura H, Takeda Y and Durst F 1999 *Exp. Fluids* **26** 208
[18] Takeda Y 1995 *JSME Int. J. Ser. B* **38** 8
[19] Takegoshi E, Shibata T, Hirasawa Y and Kosaka A 2000 *Netsu Bussei (Japanese)* **14** 97
[20] Li Q, Xuan Y and Wang J 2005 *Exp. Therm. Fluid Sci.* **30** 109
[21] Motozawa M, Iizuka Y and Sawada T 2008 *J. Phys.: Cond. Matter* **20** 204117
[22] Jeyadevan B and Nakatani I 1999 *J. Magn. Magn. Mater.* **201** 62
[23] Kikura H, Matsushita J, Kakuta N, Aritomi M and Kobayashi Y 2007 *J. Mater. Processing Tech.* 181 93
[24] Kamiyama S, Koike K and Iizuka N 1979 *Bulletin of the JSME* **22** 1205

Visualization on the behavior of nanoparticles in magnetic fluids under the electric field

W-H Lee[1] **and J-C Lee**[2]
[1]Graduate School of Automotive Engineering, Gangneung-Wonju National University, Wonju 220711, Rep. of Korea
[2]School of Mechanical and Automotive Engineering, Gangneung-Wonju National University, Wonju 220711, Rep. of Korea

E-mail: jclee01@gwnu.ac.kr

Abstract. The dielectric breakdown characteristics of magnetic fluids can be influenced by the magnetic nanoparticles included because their properties should be affected by the applied electric field. Based on measuring the dielectric breakdown voltage of magnetic fluids, we found that it is higher than that of the pure transformer oil in the case of the specific volume concentrations of magnetic nanoparticles. It is known from a numerical simulation that the conductive nanoparticles might behavior as electron scavengers in the electrically stressed magnetic fluids and change fast electrons into slowly negative charged nanoparticles for the electrical breakdown. In this study, we focus on the motion of magnetic nanoparticles in the fluids under the electric field applied by the visualization using a microchannel and an optical microscope.

1. Introduction

Magnetic fluid is a colloidal suspension consisting of carrier liquid (water or oil) and magnetic nanoparticles with the size of 10 nm. It is used for the magnetic drug delivery, cooling of electrical system and the rotary shaft seal of a computer disk drive etc. in various fields [1]. Recently, it was found that transformer oil based magnetic fluids are possible to improve the heat transfer property and dielectric performance in electrical power systems such as a high-voltage transformer [2, 3].

Based on measuring the dielectric breakdown voltage of magnetic fluids by Lee *et al.* [4], it is higher than that of the pure transformer oil in the case of the specific volume concentrations of magnetic nanoparticles as shown in Figure 1. It was shown that dielectric breakdown voltage of the pure transformer oil is about 12 kV and the magnetic fluid dielectric breakdown voltage is around 37 kV in the case of the magnetic nanoparticles volume concentrations between 0.08% and 0.6%. This result is accord with true that conductive nanoparticles act as electron scavengers in the electrically stressed transformer oil-based magnetic fluids converting fast electrons to slow charged particles [5]. Also, the magnetic nanoparticles experience an electrical force directed towards the region of maximum electric field strength [5]. This phenomenon is consistent with the dielectrophoresis. It is a phenomenon in which a force is exerted on a dielectric particle when it is affected by a non-uniform electric field. All particles exhibit dielectrophoretic activity in the presence of electric field. If particles are in a non-uniform field, the electric field polarizes the particle and polarizable particles experience a

[2] Corresponding author.

force along the electric field gradient [6]. And the aggregated magnetic nanoparticles like chain by external field affect the dielectric breakdown characteristics [7, 8]. In order to analyze the effect by electric field between the movement of magnetic nanoparticles and dielectric breakdown voltage, the movements of magnetic nanoparticles were observed by microscope and camera.

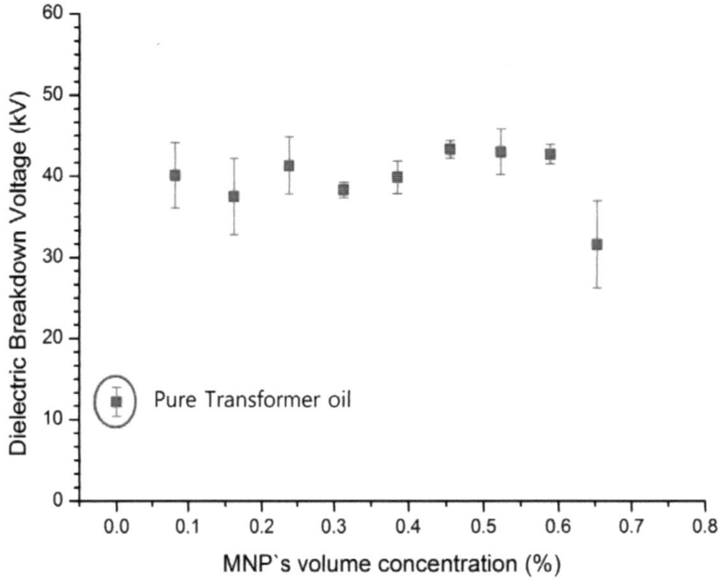

Figure 1. Dielectric breakdown voltage with different magnetic nanoparticles volume concentrations.

2. Experimental method

In this study, the movements of magnetic nanoparticles to applied electric field in a microchannel by an optical microscope were visualized. In order to observe the magnetic nanoparticles, a microchannel with 4 holes for fluid injection and electrode insertion as show in Figure 2(a) was made. In Figure 2(a), No.1 and 4 are holes for fluid injection and No.2 and 3 are holes for electrode insertion. In order to make a microchannel, 50 mL of silicon elastomer base liquid with 5 mL of curing agent were mixed and heated to harden for 3 h at 60°C. After, 4 holes were punched with 6 mm diameter on microchannel for fluid injection and electrode insertion. As shown in Figure 2(b), it has a passage between No. 1 and 4 holes with a length of a 500 μm, a width of 50 μm and a height of a 5 μm.

10 μL of a magnetic fluid (EFH-1, FerroTec, USA) can be diluted, which includes the magnetic nanoparticles, Fe_3O_4, with the size of about 10 nm reported, and 990 μL of transformer oil (OT-4, Michang, Korea) for experiments. The volume concentration of magnetic nanoparticles is 0.065%. A microscope was used (Eclipse Ti-E model) with a camera (HQ2 Monochrome camera) for the observation of magnetic nanoparticles movement. The position of observed photograph is about a height of 2.5 μm near the electrode in a microchannel passage. The applied DC voltage between the electrodes is 2.5 kV. The magnification of a microscope can be controlled into 100 times. The captured photographs have 696x520 pixels where the length of 1 pixel is 900 nm.

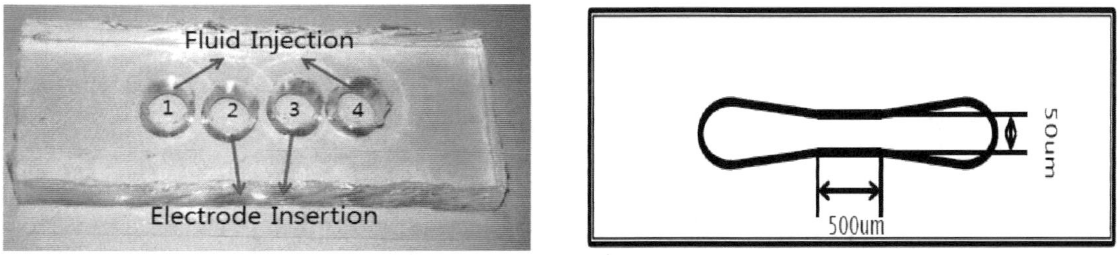

(a) Photograph of the microchannel (b) Schematic diagram of the microchannel

Figure 2. Microchannel with 4 holes for fluid injection and electrode insertion.

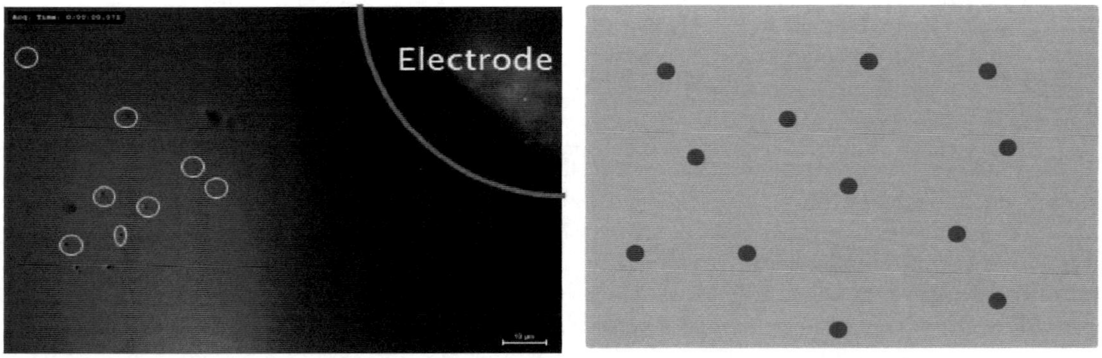

(a) Photograph of magnetic nanoparticles in the microchannel (b) Schematic illustration of magnetic nanoparticles in the microchannel

Figure 3. Dispersion of magnetic nanoparticles in the microchannel when the electric field is not applied.

3. Results

In this study, the magnetic nanoparticle movement according to applied electric field was observed. And the effect by the field between the movement of magnetic nanoparticles in the fluid and the dielectric breakdown voltage of the fluid was analyzed.

In the case of non-applied electric field, the magnetic nanoparticles are uniformly dispersed in the microchannel as shown in Figure 3. In order to measure the size of magnetic nanoparticles in the fluid, NIS Element software to photographs was used. Although the size of particles just above 200 nm by this software can be measured, which is much larger than the original particle size 10 nm, however, the particles can be found to the photographs because it is the shadow of magnetic nanoparticles in the fluid as well as the aggregation of the particles in the initial magnetic fluid. Therefore, the number of magnetic particles visualized in Figure 3(a) should be much less than the real number of magnetic particles in the fluid.

In order to analyze the movement of magnetic particles by electric field, the DC voltage is applied between electrodes up to 2.5 kV. More and bigger particles in Figure 4(a) than the previous in Figure 3(a) can be seen. It is because the magnetic nanoparticles are attracted by the dielectrophoretic force and migrated toward the region of the maximum electric field near electrodes. The particles can be agglomerated by the attraction, the size of particles is 2~3 times larger as shown in Figure 4(a). If consider the particles not visualized in the photograph are considered, the magnetic nanoparticles in the fluid should be attracted near electrode as shown in Figure 4(b).

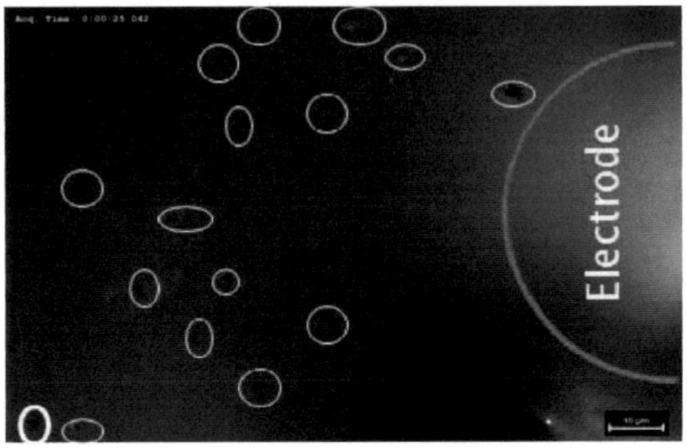

(a) Photograph of magnetic nanoparticle in the microchannel

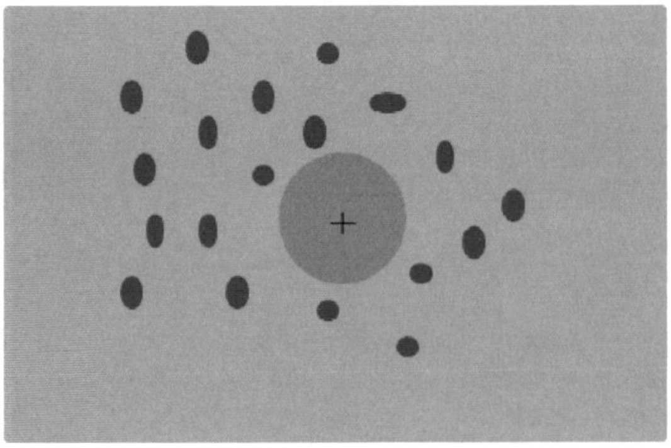

(b) Schematic illustration of magnetic nanoparticles in the microchannel

Figure 4. Dispersion of magnetic nanoparticles in the microchannel when the electric field is applied.

Unfortunately, we have captured pictures in the different positions due to the reflection of light from the electrode, however, we can deduce that what the change of dielectric breakdown voltage of transformer oil based magnetic fluids is primary due to the movement of magnetic nanoparticles by the applied electric field. It is known that the conductive magnetic nanoparticles in the fluids can act as electron scavengers, the dielectric breakdown voltage of the fluid can be changed positively or negatively according to the distribution of magnetic particles near electrodes. The further research on this point will be followed to control the movement of magnetic nanoparticles by dielectrophoretic and magnetophoretic forces.

4. Conclusion

In this study, the movement of magnetic nanoparticles influenced by the applied electric field was observed. From the photographs of magnetic nanoparticles in the fluid with and without the applied electric field, it was found that the magnetic nanoparticles are attracted by the dielectrophoretic force and migrated toward the region of the maximum electric field near electrode. And the particles can be agglomerated by the attraction; the size of particles visualized is 2~3 times larger when the electric field is applied. It can be deduced that what the change of dielectric breakdown voltage of transformer oil based magnetic fluids is primary due to the movement of magnetic nanoparticles by the applied electric field. The further research on this point will be followed to control the movement of magnetic nanoparticles by dielectrophoretic and magnetophoretic forces.

Acknowledgment

This research was supported by Basic Science Research Program through the National Research Foundation of Korea (NRF) funded by the Ministry of Education, Science and Technology (2012-0007808) and by the Agency for Defense Development through Chemical and Biological Defense Research Center.

References

[1] Rosensweig R E 1987 *Ferrohydrodynamics* (New York: Dover)
[2] Das S K, Choi S, Yu W and Pradeep T 2008 *Nanofluids: Science and Technology* (New Jersey: John Wiley& Sons)
[3] Tomco L, Marton K, Herchl F, Kopcansky P, Potocova I, Koneracka M and Timko M 2006 *Phys. Stat. Sol.(c)* **3** 195
[4] Lee J C, Lee W H, Lee S H and Lee S 2012 *Materials Research Bulletin* **47** 2984

[5] Hwang J G, Zahn M, O'Sullivan F M, Pettersson L A A, Hjortstam O and Liu R 2010 *J. Appl. Phys.* **83** 014310
[6] Pohl H A 1978 *Dielectrophoresis* (Cambridge: Cambridge University Press)
[7] Timko M, Kopcansky P, Molcan M, Tomco L, Marton K, Molokac S, Rybar P, Stoian F, Holtescu S and Taculescu A 2012 *ACTA Physica Polonica* A **121** 1253
[8] Kudelcik J, Bury P, Kopcansky P and Timko M 2010 *Physics Procedia* **9** 78

A comparison between micro- and macro-structure of magnetoactive composites

T Gundermann[1], S Günther, D Borin and S Odenbach

Institute of Fluidmechanics, Chair of Magnetofluiddynamics, George-Bähr-Str. 3, 01062 Dresden, Germany

E-mail: thomas.gundermann@tu-dresden.de

Abstract. A possibility to determine the microstructure of magneto active composites using micro-computed tomography (μ-CT) [1] was investigated in this work. The main interest of the current study has been an observation of the magnetic field dependent shift of individual particles from their initial positions inside an elastomeric matrix. For this purpose a μ-CT system has been combined with a sample holder coupled with two permanent magnets, enabling the investigation of the micro-structure under influence of an external homogeneous field. In the experimental investigations samples based on carbonyl iron particles have been used. The particles have been dispersed in a polymeric matrix and the polymer has been created in the presence of a magnetic field driving structure formation of the particles. After the characterization of the sample in its initial state, i.e. without external stimuli, it has been subjected to the magnetic field and its internal structure has been once more studied by μ-CT. As a result a comparison of the macroscopic change of the sample structure and the particle displacement could be undertaken.

1. Introduction

Magneto active composites, also known as magneto-rheological elastomers (MRE) are a kind of smart materials which consists of soft magnetic particles dispersed in an elastomeric matrix. The mechanical properties of this composite material can be changed actively and reversibly by applying a magnetic field. An essential reason for the property changes is an internal magnetodipolarstriction, i.e. a change of the local positions of the particles, which has been widely studied theoretically, e.g. in [2, 3]. The proposed theoretical models show a connection between the local positions of the particles and the behaviour of the matrix. These effects have to be tested using experimental methods.

To investigate different effects of MREs, it is important to distinguish between two scales: A microscopic scale that describes the behaviour of the soft magnetic particles and the interaction between these particles and the elastomeric matrix and a macroscopic scale that describes the behaviour of the composite in a range that is visible to the eye. A typical microscopic effect is the shift of the particles inside the matrix due to the influence of an applied magnetic field. Macroscopic effects are e.g. dynamic-mechanical deformations of the composite due to an applied mechanical stress and/or a magnetic field, e.g. the shape memory effect [4]. Several papers regarding these macroscopic effects were published in the last years [4, 5].

[1] Corresponding author.

Most of these studies were performed on composites, consisting of a silicon rubber as matrix material and carbonyl-iron-powder as particles. Advantage of this composite-system is the liquid state of the silicon components before starting the polymerisation synthesis. This makes it possible to suspend the particles in the silicone and produce samples with an isotropic particle concentration and samples with a chain like structure of the particles parallel to the applied magnetic field, also called anisotropic samples [6].

2. Experimental

2.1. Sample preparation

The particle material that have been used for the preparation were carbonyl-iron powder produced by Höganäs AB with a mean diameter of 35-45μm. Figure 1 shows the volume weighted sum function (circles) and the volume weighted density function (triangles) of the particles, which are used for preparation.

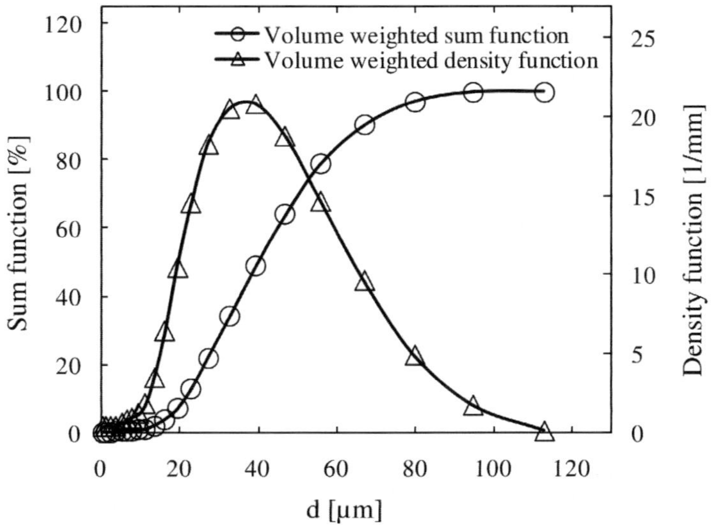

Figure 1. The volume weighted sum function (circles) and the volume weighted density function (triangles) of the carbonyl-iron particles [1].

As silicon matrix a low-molecular vinyl-containing rubber and a hydride-containing cross-linking agent (compound SIEL™) was used [7]. This elastomeric matrix allows Young's modulus smaller than 50 kPa.

The work of D. Günther [1] shows, that the best proportion of soft magnetic particles inside the silicon matrix is around 10 wt.%. A higher proportion of carbonyl iron particles causes that the particles cannot be fully resolved with the μ-CT due to their absorption and a lower content of particles causes a lower macrostructural deformation.

The production of the samples was as follows: The powder was mixed to the silicone and added to the crosslinker. After mixing and degassing under vacuum the sample polymerized in a homogeneous magnetic field. For the polymerisation process, we used a water heating system and placed this in an electromagnetic setup from Bruker Corp. The polymerisation time was 2 hours by 95 °C heating temperature and within a magnetic flux density of 250 mT.The results of the preparation were anisotropic samples with a height of 10 mm and a diameter of 8 mm.

2.2. Experimental setup

The analysis of the samples was performed by the same μ-CT system, which was used in [1]. It is based on a nano-focus tube, a movable sample stage and a detector. Especially for studies under the influence of a magnetic field, a new sample holder was produced. A principle scheme of this new sample holder is shown in Fig. 2. The sample holder consists of a plastic tube with a low absorption coefficient, the sample with a height of 10 mm and a diameter of 8mm on a table and two adjustable permanent magnets. The magnets are neodymium magnets with a remanent flux density of 1.46 T. This configuration enables to vary the magnetic flux density between the magnets from 0 to 250 mT.

To verify this magnetic field, simulations for the magnetic flux density were undertaken. For this purpose, the simulation software COMSOL FEMLAB Multiphysics was used. The simulation parameters were given by the geometrical properties of the permanent magnets and the remanent flux density of $B_r = 1.46$ T. Figure 3 shows the dependence of the magnetic flux density of the radius in the vertical centre of the sample. The minimum distance of 15 mm between the magnets is defined by the combined height of the sample and the sample stage.The magnetic flux density at minimum magnet distance in the centre of the sample is 250 mT and shows a homogeneity across the radius of more than 90%. At the maximum magnet distance of 130mm, the magnetic flux density inside the sample equals zero.

The tomographic measurements where undertaken at two conditions,the first tomogram without field and the second tomogram within a magnetic flux density of 250 mT. After tomography the images have been evaluated with digital image processing software.

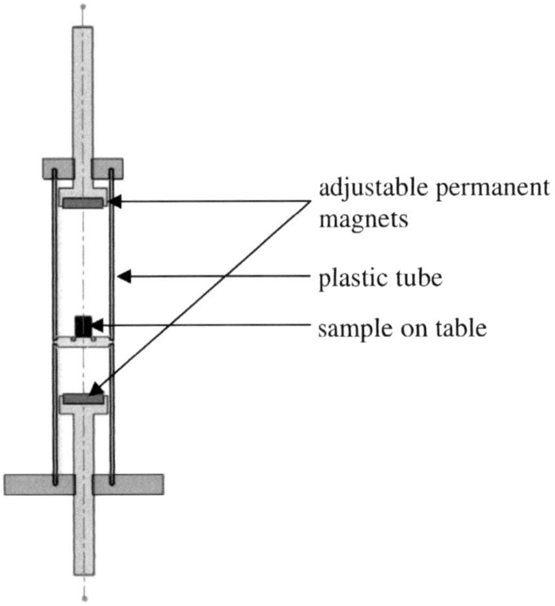

Figure 2. New setup for µ-CT investigations of MRE´s under influence of a magnetic field. It consists of two adjustable permanent magnets, a samplestage and a plastic tube with a low absorption coefficient.

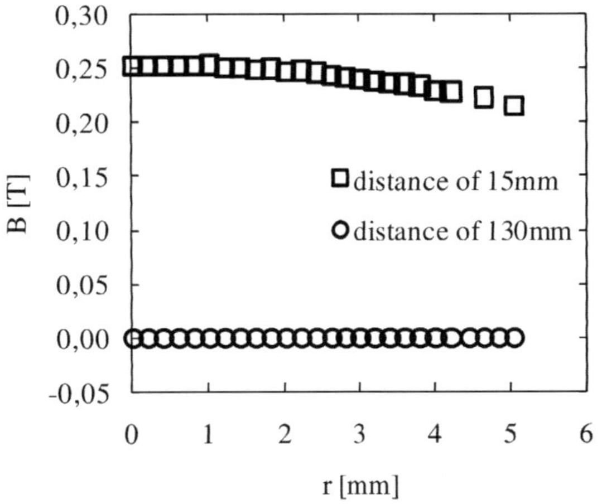

Figure 3. Simulation of the magnetic flux density in the vertical centre of the sample. Shown are the minimal distance (15 mm) and the maximal distance (130 mm) between the magnets.

3. Results and discussion

The undertaken measurements were taken at an X-ray voltage of 90 kV.The magnification of the measurement was nine. This corresponds to a resolution limit of 8 μm which is equal to the measurement error in all measurements

3.1. Macrostructural behavior

Figure 4 shows the tomogram of the sample without the influence of a magnetic field (left) and within a magnetic flux density of 250 mT. Using a digital image software, the dimensions of the sample were determined. The results are also shown in Figure 4. A deformation of the sample in the vertical direction (z-direction), respectively parallel to the applied magnetic field could be detected. In horizontal direction (x-y-plane), it was not possible to detect a deformation due to the spatial resolution of the tomogram. The result shows an extension of the sample of 60 μm. The sample is elongated in the presence of a magnetic field. This effect, also called magnetodeformational effect in magnetoelastics is described in many publications, e.g. in [8] and can be explained by the dipole-dipole interaction of the magnetic particles in the chains [8].

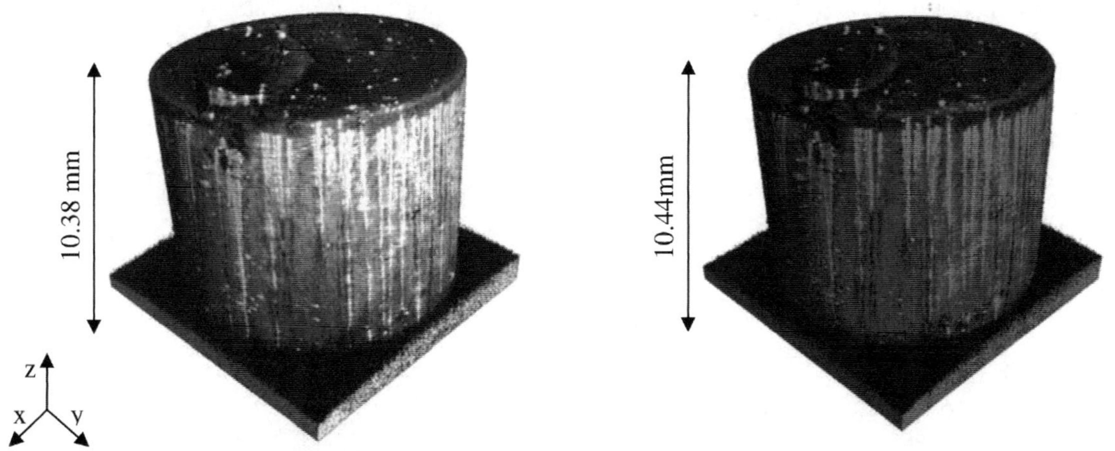

Figure 4. Macrostructural view of the sample without the influence of a magnetic field (left) and within the influence of a magnetic flux density of 250 mT (right).

3.2. Microstructural behavior

Using μ-CT, it is now possible to zoom into the structure and to determine the microscopic behaviour. Figure 5 shows the microstructure of the anisotropic sample for the situation without magnetic field influence (left) and the arrangement with magnetic field influence (right). It can be seen that the particles are arranged in chains, because of the preparation method.

Using the same image processing program as for the evaluation of the macroscopic behaviour it was possible to separate the individual particles from each other. For this step, various filters and the watershed algorithm have been used. The particles were identified by their characteristics such as size, major and minor axes respectively. Using MATLAB with DIPimage it was possible to assign the particles in the case without magnetic influence to the particles within magnetic field influence.

For the associated particles the displacement of the centre points was determined. This shift can be represented by vectors. Figure 6 shows the displacement of the particle centre as vectors. For a better view, the shift in the z-x-planeis shown. Due to the resolution limit, it was found, that there is no difference between the x and the y-direction. This can also explained with the fact that the sample and the magnetic field have been arranged symmetrical with respect to the long axis of the sample cylinder.

This type of analysis requires a large amount of memory. Due to the long computation time, only some particles could be evaluated in the whole sample.Thus only a qualitative description of the particle displacement can be given there. But this result shows a particle shift in vertical direction (z), that is one order of magnitude larger than the particle shift in horizontal direction (x). This could be explained with the parallelism of the vertical direction and the magnetic field. It is also clear that the shift of the particles in vertical direction is higher than the macroscopic deformation of the sample in the same direction.

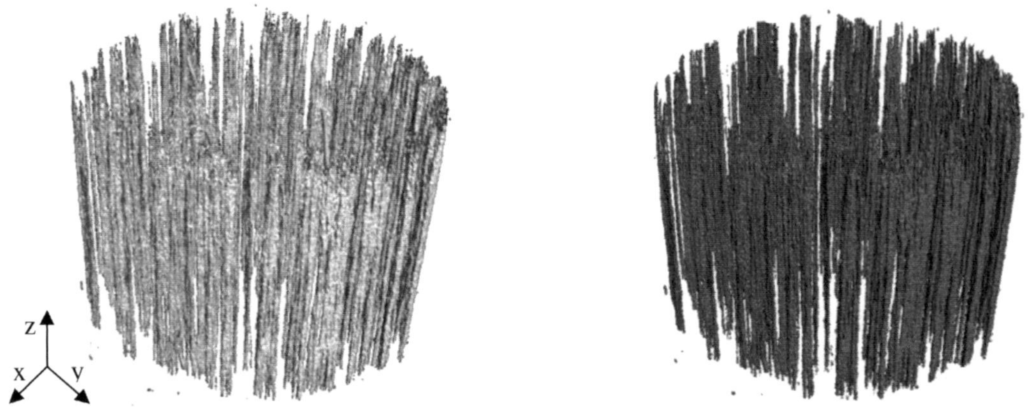

Figure 5. Microstructural view of the sample, specially the chain like structure of the particles without the influence of a magnetic field (left) and within the influence of a magnetic flux density of 250 mT (right).

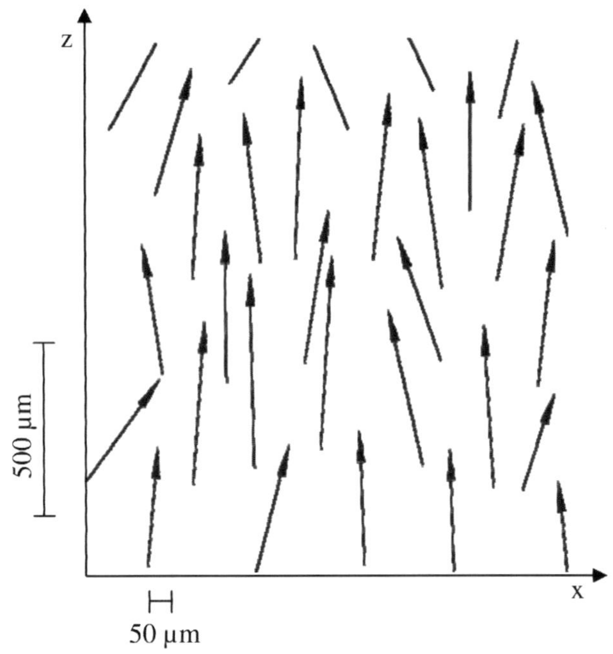

Figure 6. Microscopic change of the structure of a MRE, shown is the displacement of the centre of some particles at different positions in the Matrix. The axes are not scaled.

4. Conclusions

In this work, a new experimental method has been developed to examine the movement of particles in an elastomeric matrix under the influence of a magnetic field and to evaluate the interaction between the matrix and the suspended particles. It could be shown that the particle movement parallel to the applied magnetic field is higher than the particle movement perpendicular to the applied magnetic field. In addition, it has been found that the particle shift is larger than the macroscopical deformation of the sample. This finding leads to the conclusion that the particles in this kind of MR-elastomer can move relative to the matrix with small influence on the matrix deformation only.

Future work will show the overall behaviour of the particles in the matrix. Due to the gradient in the magnetic field, it could be possible that the particle displacement is different between the centre and the edges of the sample. For this purpose, all of the particles have to be segmented and compared. Also the macroscopic deformation at higher content of iron particles will be investigated.

Acknowledgments

This project is funded by the European Union and the Free State of Saxony.

References

[1] Günther D, Borin D Yu, Günther S, and Odenbach S 2012 X-ray micro-tomographic characterization of field-structured magnetorheological elastomers *Smart Mater. Struct.* **21** 015005

[2] Stolbov O V, Raikher Yu L and Balasoiu M 2011 Modelling of magnetodipolar striction in soft magnetic elastomers *Soft Matter* **7** 8484

[3] Davis L C 1999*J. Appl. Phys.* **85**3348

[4] Stepanov G V, Borin D Yu, Raikher Yu L, Melenev P V and Perov N S 2008*J. Phys.: Condens. Matter***20**204121

[5] Böse H and Röder R 2009 *J. Phys.: Conf. Ser.***149** 012090

[6] Borbarth T, Günther S, Borin D Yu, Gundermann Th and Odenbach S 2012 XμCT analysis of magnetic field-induced phase transitions in magnetorheological elastomers, *Smart Mater. Struct.* **21** 105018

[7] Stepanov G V, Borin D Yu, Odenbach S 2009 *J. Phys.: Conf. Ser.* **149** 012098

[8] Nikitin L V, Stepanov G V, Mironova L S and Gorbunov A I 2004 *JMMM* **272-276** 2072-2073

Derivation of stiffness matrix in constitutive modeling of magnetorheological elastomer

D Leng, L Sun[1], J Sun and Y Lin
School of Transportation Science & Engineering, Beihang University, Beijing 100191, China

E-mail: lysunluck@gmail.com

Abstract. Magnetorheological elastomers (MREs) are a class of smart materials whose mechanical properties change instantly by the application of a magnetic field. Based on the specially orthotropic, transversely isotropic stress-strain relationships and effective permeability model, the stiffness matrix of constitutive equations for deformable chain-like MRE is considered. To valid the components of shear modulus in this stiffness matrix, the magnetic-structural simulations with finite element method (FEM) are presented. An acceptable agreement is illustrated between analytical equations and numerical simulations. For the specified magnetic field, sphere particle radius, distance between adjacent particles in chains and volume fractions of ferrous particles, this constitutive equation is effective to engineering application to estimate the elastic behaviour of chain-like MRE in an external magnetic field.

1. Introduction

Magnetorheological (MR) or magneto-sensitive (MS) elastomers are smart materials whose mechanical properties response rapidly with the external magnetic stimulus. These intelligent materials typically consist of micro-sized ferrous particles dispersed in an elastomer matrix. Subjected to uniaxial compression or shear, MREs have been utilized as a single non-linear variable stiffness spring in some applications such as adaptive tuned vibration absorbers, suspensions and automotive bushing [1, 2].

Curing without or with magnetic field, magnetorheological elastomers present isotropic or anisotropic property, respectively. For these two types of MREs, the compression and shear stiffness under uniform magnetic field have been measured by experiment [3-5]. In addition, several

[1] Corresponding author.

Published under licence by IOP Publishing Ltd

mathematical models for isotropic MRE are provided to investigate its constitutive relationship [6-9]. Most of the recent stress energy functions and constitutive equations of MRE were based on the classic work [10] in which the equations of motion for an isotropic non-polar continuum in an electromagnetic field are described by Maxwell's equations and the mechanical and electro-magnetic-structural field for homogenous particle dispersed elastomers. However, it is demonstrated that the chain-like MREs have more obvious magnetic-field response effect. Hence, exploring the constitutive relationship of anisotropic MREs is more significant.

In this paper, based on the anisotropic stress-strain relationships and effective permeability model, the stiffness matrix for elastic deformable chain-like MRE is considered. To valid the constitutive equations of chain-like MRE, the magnetic-structural simulations by finite element method are presented.

2. Constitutive equations of chain-like MRE

For the anisotropic elastic material, the most linear general stress-strain relationships are given by equations of the form [11],

$$\sigma_{\mathrm{ij}} = C_{ijkl} \cdot \varepsilon_{ij} \tag{1}$$

where σ_{ij} is the stress tensor having nine components, C_{ijkl} is the stiffness tensor having eighty-one components, ε_{ij} is the strain tensor having nine components.

As shown in any mechanics of material book, both stresses and strains are symmetric, so that there are only six independent stress components and six independent strain components. This means that the elastic constants must be symmetric, and that the number of nonzero elastic constants is reduced to twenty-one. Hence, the generalized Hooke's law can now be simplified as,

$$\sigma_{\mathrm{i}} = C_{ij} \cdot \varepsilon_j, \quad i, j = 1,2\cdots,6 \tag{2}$$

MRE is a typical material that employs before yield condition, and it exhibits elastic properties. Hence, the constitutive equations of MR elastomers can be expressed as equation (2). Further simplification is possible, as MRE has three mutually orthogonal planes of material properties symmetry (or orthotropic material). Considering a general 3D state of stress consisting of all possible normal and shear stresses for orthotropic materials, the resulting set of equations is given below,

$$\begin{Bmatrix} \varepsilon_1 \\ \varepsilon_2 \\ \varepsilon_3 \\ \gamma_{23} \\ \gamma_{31} \\ \gamma_{12} \end{Bmatrix} = \begin{bmatrix} 1/E_1 & -\nu_{21}/E_2 & -\nu_{31}/E_3 & 0 & 0 & 0 \\ -\nu_{12}/E_1 & 1/E_2 & -\nu_{32}/E_3 & 0 & 0 & 0 \\ -\nu_{13}/E_1 & -\nu_{23}/E_2 & 1/E_3 & 0 & 0 & 0 \\ 0 & 0 & 0 & 1/G_{23} & 0 & 0 \\ 0 & 0 & 0 & 0 & 1/G_{31} & 0 \\ 0 & 0 & 0 & 0 & 0 & 1/G_{12} \end{bmatrix} \cdot \begin{Bmatrix} \sigma_1 \\ \sigma_2 \\ \sigma_3 \\ \tau_{23} \\ \tau_{31} \\ \tau_{12} \end{Bmatrix} \tag{3}$$

For the MR elastomers cured in an external magnet, the chain-like structure of particles is formed. It exhibits specially orthotropic, transversely isotropic material property, as shown in Figure 1.

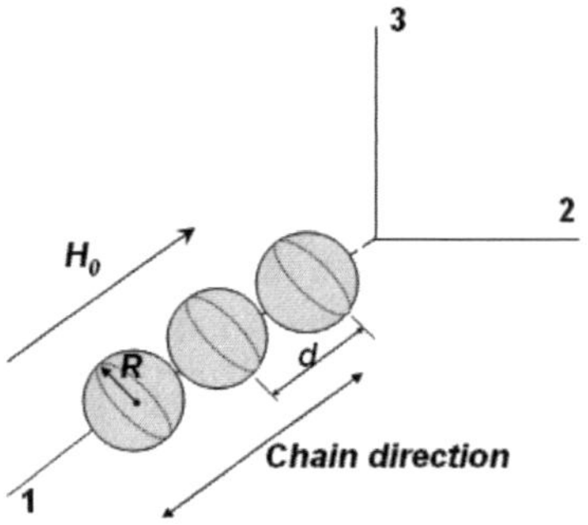

Figure 1. The material plane of chain-like MREs.

For specially orthotropic, transversely isotropic material, the stiffness matrix have twelve nonzero coefficients with five independent constants. The subscripts 2 and 3 in equation (3) can be interchangeable,

$$G_{13} = G_{12} \tag{4}$$

$$E_2 = E_3 \tag{5}$$

$$\nu_{21} = \nu_{31} \tag{6}$$

$$\nu_{23} = \nu_{32} \tag{7}$$

In addition, the familiar relationship among the isotropic engineering constants is now valid for engineering constants associated with the 23 plane, so that [11]

$$G_{23} = \frac{E_2}{2(1+\nu_{32})} \tag{8}$$

Hence, the equation (3) can be simplified as,

$$\begin{Bmatrix} \varepsilon_1 \\ \varepsilon_2 \\ \varepsilon_3 \\ \gamma_{23} \\ \gamma_{31} \\ \gamma_{12} \end{Bmatrix} = \begin{bmatrix} 1/E_1 & -\nu_{21}/E_2 & -\nu_{21}/E_2 & 0 & 0 & 0 \\ -\nu_{12}/E_1 & 1/E_2 & -\nu_{23}/E_2 & 0 & 0 & 0 \\ -\nu_{13}/E_1 & -\nu_{23}/E_2 & 1/E_2 & 0 & 0 & 0 \\ 0 & 0 & 0 & 1/G_{23} & 0 & 0 \\ 0 & 0 & 0 & 0 & 1/G_{12} & 0 \\ 0 & 0 & 0 & 0 & 0 & 1/G_{12} \end{bmatrix} \cdot \begin{Bmatrix} \sigma_1 \\ \sigma_2 \\ \sigma_3 \\ \tau_{23} \\ \tau_{31} \\ \tau_{12} \end{Bmatrix} \tag{9}$$

An effective permeability model of MR elastomers is introduced to estimate the shear modulus in chain direction, G_{23} [12]. This model is based on the Maxwell Garnett mixing rule [13], the field-induced shear modulus of chain direction can be expressed as [12],

$$G` = 6\phi_p \cdot \mu_0 \cdot \mu_m \cdot H_o^2 \tag{10}$$

where ϕ_p is the volume fraction of particles, μ_0 is the permeability of vacuum, μ_m is the permeability of matrix, H_0 is the magnetic field intensity. Before ferrous particle achieves magnetic saturation,

$$H_0 = \frac{B}{\mu_{eff}} \tag{11}$$

where B is the magnetic induction density, and μ_{eff} is the effective relative permeability of MREs,

$$\mu_{eff} = \mu_{ceff} \cdot \phi_c + \mu_m \cdot (1 - \phi_c) \tag{12}$$

where ϕ_c is the volume fraction of chain structure in MREs,

$$\phi_c = \frac{3\phi_p \cdot d}{4R} \tag{13}$$

where d is the distance between adjacent particles in chain structure, R is the radius of ferrous particle, and μ_{ceff} is the effective permeability of chains, it can be predicted as [13],

$$\mu_{ceff} = \mu_m + 2 \cdot \phi \cdot \mu_m \frac{(\mu_p - \mu_m)}{\mu_p + \mu_m - \phi(\mu_p - \mu_m)} \tag{14}$$

where ϕ is the volume fraction of particles in chain-structure,

$$\phi = \frac{4R}{3d} \tag{15}$$

Based on equation (12) to equation (15),

$$\mu_{eff} = \mu_m + 2 \cdot \phi_p \cdot \mu_m \frac{(\mu_p - \mu_m)}{\mu_p + \mu_m - \frac{4R}{3d}(\mu_p - \mu_m)} \tag{16}$$

The shear modulus in chain structure, G_{23}, can be obtained,

$$G_{23} = G_0 + G` \tag{17}$$

where G_0 is the shear modulus of the MRE under zero magnet,

$$G_0 = G_m \cdot \left(1 + 2.5\phi_p + 14.1\phi_p^2\right) \tag{18}$$

where G_m is the shear modulus of the elastomer matrix.

Through equation (9) to equation (13), the shear modulus in chain structure can be obtained,

$$G_{23} = G_m \cdot \left(1 + 2.5\phi_p + 14.1\phi_p^2\right) + 6\phi_p \cdot \mu_m \cdot \left(\frac{B^2}{\left(\mu_m + 2 \cdot \phi_p \cdot \mu_m \dfrac{(\mu_p - \mu_m)}{\mu_p + \mu_m - \dfrac{4R}{3d}(\mu_p - \mu_m)} \right)^2 \cdot \mu_0} \right) \tag{19}$$

According to equation (8), the Young's modulus of chain structure, is

$$E_1 = 2 \cdot (1 + \nu_{12}) \left(G_m \cdot \left(1 + 2.5\phi_p + 14.1\phi_p^2\right) + 6\phi_p \cdot \mu_m \cdot \left(\frac{B^2}{\left(\mu_m + 2 \cdot \phi_p \cdot \mu_m \dfrac{(\mu_p - \mu_m)}{\mu_p + \mu_m - \dfrac{4R}{3d}(\mu_p - \mu_m)} \right)^2 \cdot \mu_0} \right) \right) \tag{20}$$

In the material plane 2 and 3 which are perpendicular to the chain direction, as shown in Fig. 1, the mutual magnetic force between ferrous particles are far less than that of adjacent particles in chain structure [14]. Hence, the shear modulus, G_{12}, is predicted as particle reinforced composites [15].

$$G_{12} = G_m \cdot \left(1 + 2.5\phi_p + 14.1\phi_p^2\right) \tag{21}$$

Based on equation (8), the Young's modulus of chain structure, E_2 , is

$$E_2 = 2G_m \cdot \left(1 + 2.5\phi_p + 14.1\phi_p^2\right) \cdot \left(1 + \nu_{21}\right) \tag{22}$$

Through equation (16) to equation (19), each modulus and the stress-strain relationship of chain-like MREs, under different magnetic field, can be obtained. Based on the Passion's ration, the Young's modulus can be obtained through shear modulus. For the specified magnetic field, sphere particle radius, distance between adjacent particles in chains and volume fractions of ferrous particles, above-mentioned equations can predict the elastic stiffness matrix of MREs with aligned chains of particles, and then the stress-strain relationships can be obtained.

However, the accuracy of the shear modulus in both chain direction and vertical to chain direction is significant to this constitutive equation. Hence, it will be discussed below.

3. Numerical simulations in finite element method

The magnetic-structural simulation of chain-like MREs by finite element method is presented, in order to valid G_{23} and G_{12} in analytical method. The simulation flowchart of magnetic-structural method is illustrated in Figure. 2. In static simulation, ANSYS/EMAG is used to obtain the magnetic forces in an external magnetic field. Then the forces data are used as a boundary condition to simulate the deformation of MRE specimen in ANSYS/STRUC. In this numerical simulation, the unit cell mode is used, and the periodic boundary conditions are applied. The cycle of simulation ends when the deformation of specimen keeps steadily. The parameters of unit cell model are shown in Table 1.

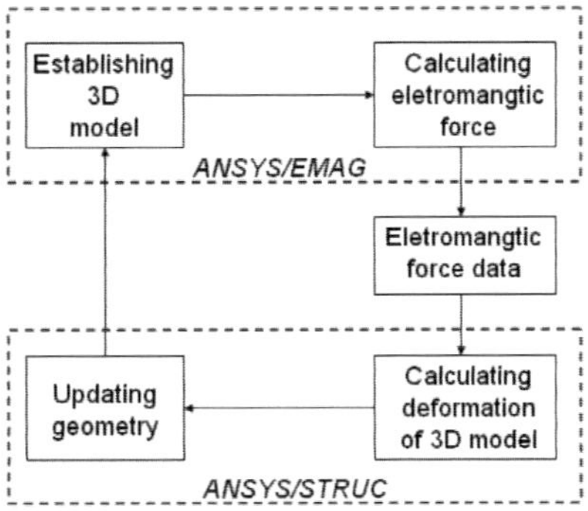

Figure 2. The simulation flowchart of coupling method.

Table 1. The parameters of model.

Part	Young's modulus (G Pa)	Shear modulus (M Pa)	Passion's ration	Density (kg m^{-3})	Size (mm)	Relative Permeability
Particle	210		0.3	7800	4(Radius)	1000
Elastomer		0.0098	0.499	2870		1

In the 3D model, it is assumed that the volume fraction of micro-particles in chain-like MREs, V_p, is 27%. This volume fraction is predicted as the optimum one [15]. As [15] referred, the distance between the adjacent particles in chains, d, can be calculated,

$$d = 3.1 \times R \tag{23}$$

Hence, d is 8.4 mm. It is also assumed that the arrangement of particles in chain-like MREs is 3*3*3 rows, in total twenty-seven particles. The 3D unit cell model is shown in Figure. 3. The Z direction is the chain direction.

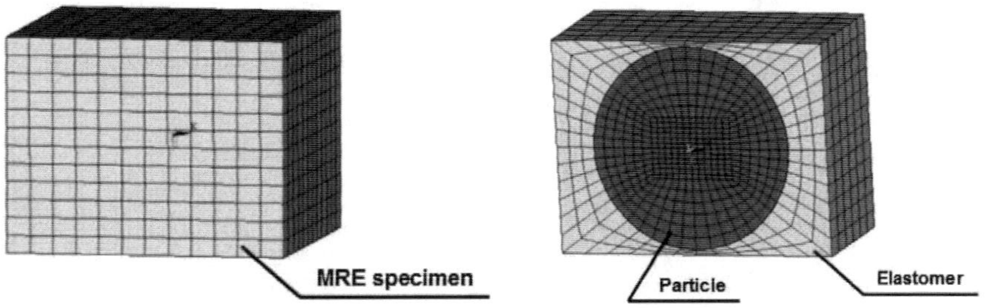

(a) (b)

Figure 3. (a) The 3D MRE specimen and (b) one half of 3D model.

4. Results and discussion

The deformation of MREs in shear condition of both chain direction and vertical to chain direction is shown in Figure 4.

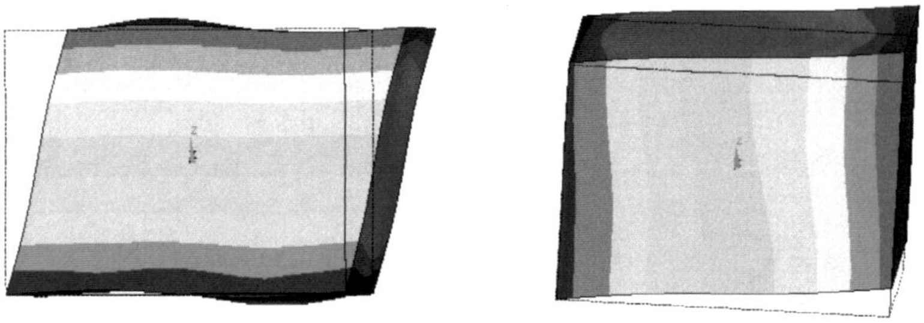

(a) (b)

Figure 4. (a)The shear deformation of MREs in chain direction, (b) the shear deformation in vertical to chain direction.

Through equation (14), the relative effective permeability of MRE is 2.73. According to equation (19), the shear modulus of chain direction can be calculated, as Table 2 shows. The field dependence of shear modulus in chain direction in MREs is shown in Figure 5.

Table 2. Shear modulus of chain direction of MREs.

Magnetic induction density (mT)	Shear Modulus of chain direction (M Pa)	
	Analytical method	FEM
100	0.262	0.166
200	0.273	0.172
300	0.283	0.177
400	0.297	0.194
500	0.315	0.215

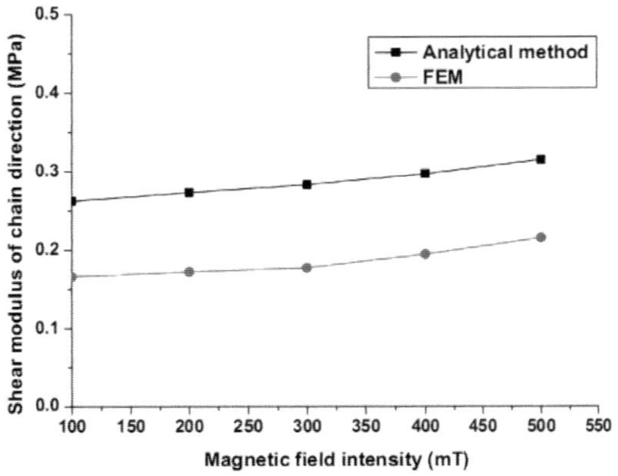

Figure 5. Chain-like shear modulus at different magnetic induction density.

From Table 2 and Figure 5, it can be seen that there is a same tendency of the shear modulus in chain direction of MREs between analytical equations and the finite element method. As the magnetic field intensity increases, the modulus increases correspondingly.

According to equation (21), the shear modulus in the material plane 2 and 3 which are perpendicular to the chain direction can be calculated, as Table 4 shows. The field dependence of shear modulus in material plane 2 and 3 is shown in Figure. 6.

Table 3. Shear modulus in material plane 2 and 3 of MREs.

Magnetic induction density (mT)	Shear Modulus in material plane 2 and 3 (M Pa)	
	Analytical method	FEM
100	0.26	0.162
200	0.26	0.163
300	0.26	0.171
400	0.26	0.174

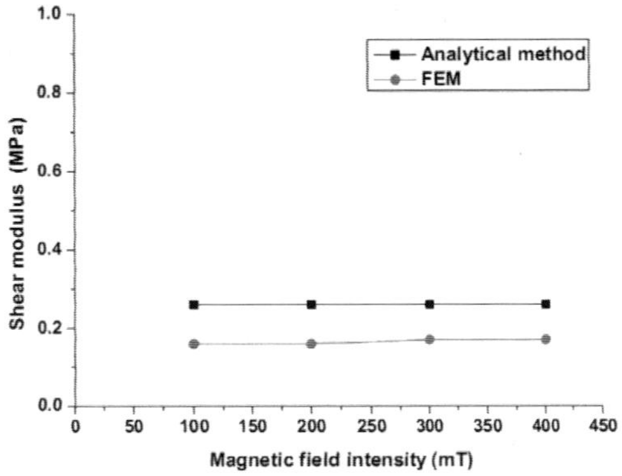

Figure 6. Shear modulus in material 2 and 3 at different magnetic induction density.

From Table 3 and Figure 6, it can be seen that the shear modulus in material plane 2 and 3, in both analytical method and finite element method in low magnetic field intensity, is independent on the external magnetic field, nearly a constant. From Figure 5 and Figure 6, it can be seen that the results of shear modulus in two directions between the analytical method and numerical method have the same tendency.

5. Conclusions

In this paper, it applied the specially orthotropic, transversely isotropic stress-strain relationship and an effective permeability model to predict the stiffness matrix of chain-like MREs. To valid this model, the finite element method of magnetic-structural simulations is presented. An acceptable agreement is obtained between analytical equations and numerical simulations. For the specified magnetic field, sphere particle radius, distance between adjacent particles in chains and volume fractions of ferrous particles, the full system of constitutive equations for elastic deformable chain-like MRE is considered.

In the future, the analyses of constitutive equations at high magnetic field intensity should be studied. Also, the authors will intend to predict the Passion's ratio of MREs to obtain more complete stress-strain equation related.

References

[1] Jolly M R, Carlson J D, Muñoz B C 1996 *Smart Materials and Structures* **5** 607
[2] Kordonsky W 1993 *Journal of Magnetism and Magnetic Materials* **122** 395
[3] Zhou G Y 2003 *Smart Materials & Structures* **12** 139
[4] Shen Y, Golnaraghi M F, Heppler GR 2004 *Journal of Intelligent Material Systems and Structure* **15** 27
[5] Chen L, Gong X, Jiang W 2007 *Journal of Materials Science* **42** 5483
[6] Dorfmann A, Ogden R W 2003 *European Journal of Mechanics-A/Solids* **22** 497
[7] Brigadnov I A, Dorfmann A 2003 *International Journal of Solids and Structures* **40** 4659
[8] Dorfmann A, Ogden RW 2003 *Acta Mechanica* **167** 13
[9] Bustamante R 2009 *Journal of Engineering Mathematics* **64** 285
[10] Pao Y H 1978 *Electromagnetic forces in deformable continua* (*Pergamon Press* vol 4) p 209
[11] Gibson R F 2011 *Principles of Composite Material Mechanics* (*CRC Press* vol 2) p 56
[12] Zhang X., Li W, Gong X 2008 *Communication in Nonlinear Science and Numerical Simulation* **13** 1910
[13] Karkkainen K K, Sihvola A H, Nikoskinen K I 2000 *IEEE Trans Geosci. Remote Sensing* **38** 1303
[14] Kallio M 2005 *The elastic and damping properties of magnetorheological elastomers* VIT publications **3** 87
[15] Davis L C 1999 Journal of Applied Physics **86** 3348

Energy dissipation characteristics of magnetosensitive elastomer under impact loading

D Leng, L Sun[1], J Sun, W Chen, F Ma, W Li and Y Lin
School of Transportation Science & Engineering, Beihang University, Beijing 100191, China

E-mail: lysunluck@gmail.com

Abstract: Magnetosensitive (MS) elastomers are a class of material that ferro-magnetic particles dispersed in rubber or elastomer whose mechanical properties change with the external magnetic fields. To investigate energy dissipation properties of MS elastomers, experimental method is adopted. Firstly, this paper presents a new fabrication method of a magnetosensitive elastomers with particles in millimeter scale distributed in ideal isotropic or in chain. Then, a drop hammer testing setup is developed to measure the energy dissipation and study the impact behaviour of magnetosensitive elastomers (MSEs). For the same volume fraction and size of particle, the dissipated energy per unit length of MSEs increases with the magnetic field increasing, and chain-like structured MSEs dissipate more energy than homogenous MSEs under the same external magnetic field.

1. Introduction

Magnetosensitive elastomer, also referred to as magnetorheological elastomer in the literature, is a class of smart material that consists of non-colloidal, magnetically-polarizable particles dispersed in a non-magnetic polymer medium, typically containing carbonyl iron particles of diameter in several microns [1]. The stiffness of MS elastomers can change rapidly and reversely in the external magnetic stimulus [2-4], which may be useful in impact protection for passenger safety. For instance, some automobile parts such as bumper, headrest, instrument panel and other interior decorations, are required to change stiffness and absorb impact energy adaptively in terms of the variations in mass and velocity of passengers or pedestrians in crash accidents [5]. In order to utilize magnetosensitive elastomers in adaptive energy absorber for car collision protection, it is necessary to understand their energy dissipation properties under impact loading.

[1] Corresponding author

This study investigates the energy dissipation capacity of MS elastomers under two possible influential factors: particle arrangement and magnetic field strength. A new method to fabricate MSE with ideal isotropic and chain-like of macro-particles is introduced. The macro particles (diameter in millimeter) are arranged in elastomers manually, which can greatly avoid the agglomeration of particles. A special drop hammer experimental system is developed that can measure the energy dissipation of MSE under various magnetic fields and impact loadings.

2. Experimental setup

2.1 MSE sample preparation

The sample's ingredients are iron particles, silicone rubber SE901A, and silicone rubber SE901B. The particles are spheres with an average diameter of 8 mm, obtained from Yongda Steel ball Factory (Guangzhou Province, China). The silicone rubber SE901A and B are Kenseer SE900 series from Huitian rubber Corporation, China. Their properties are listed in Table 1.

Table 1. Material properties of SE901A/B.

Type	Density ($g\ cm^{-3}$)	Hardness (Shore A)	Tensile strength (M Pa)	Elongation (%)
SE901A/B	0.98	20	0.5	90

Firstly, silicone rubber SE901A and SE901B are simultaneously poured into a container with 1:1, and then they are mixed by using a stirred bar for five minutes at room temperature. After two ingredients are evenly mixed, the silicon rubber mixture is put into a vacuum to eliminate air bubbles for one hour before it is cured in the mold.

The typical processing method in preparation of MSEs with micro-particles (the particles and elastomers are directly mixed) is not effective for preparing MSEs with macro-particle because the particle in millimeter size will sink in the silicon rubber immediately, therefore the ideal homogenous MSEs or chain-like MSEs would be unable to obtain. Hence, it is necessary to design a new method for processing MS elastomer with macro-particles.

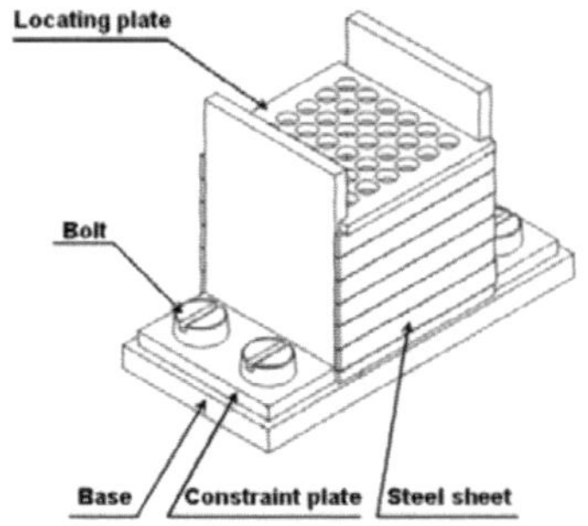

Figure 1. Schematic image of the steel mould.

In this study, a steel mold is designed and manufactured as shown in Figure 1, which is used to prepare the macro-particle MSE composite materials. It includes locating plate, bolt, base, constraint plate and steel sheet. The size and position of the circular holes in locating plate determines the arrangement of macro-particle in MSEs.

Firstly, the base is laid on the desk, and the bolts are used to fasten the constraint plate in L-shape cross section. The first layer of steel sheet is placed to the side of two vertical constraint plates with dual adhesive tape. Secondly, the silicon rubber mixture is poured into the mould until its height reaches the top surface of the steel sheet, being cured for 20 h at room temperature. Thirdly, the locating plate is laid on the top surface of steel sheet. The iron particles are placed on the position of the circular hole in the locating plate, and then the silicone rubber precursor is cured for another 20 h. At this time, the viscosity of silicon rubber mixture can keep the particle not sink. Fourthly, the locating plate is removed, and the second layer of steel sheet is fasten to the constrain plate. The above-mentioned procedure is repeated until the MSE specimen is produced.

The chain of ferrous particles along the external magnetic field is shown in Figure 2. Denoting the radius of particles by R (4 mm), and the gap between the adjacent particles is $2w$. Their relationship can be written as [6],

$$\frac{w}{R} = 0.05 \tag{1}$$

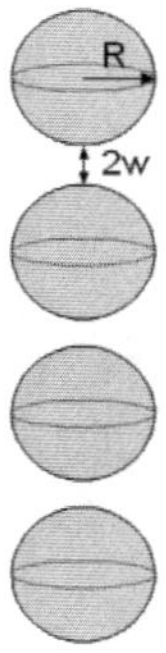

Figure 2. The chain of ferrous particle in MSE.

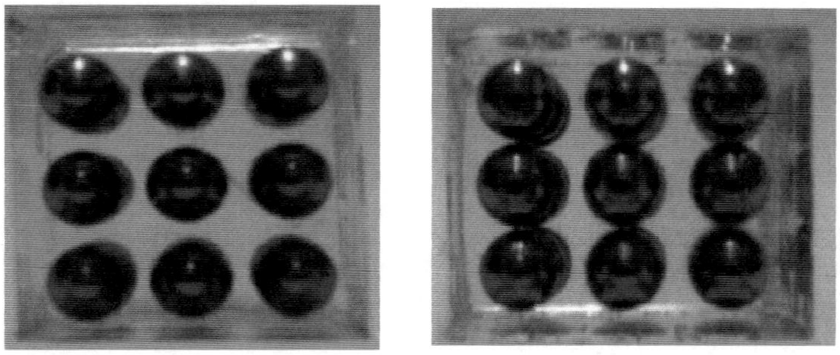

(a) (b)

Figure 3. (a) The homogenous MSEs and (b) the chain-like MSEs.

Hence, w is 0.2 mm. The volume fraction of particles is 27%, which is considered as the optimum volume fraction of MSEs [6]. It is also assumed that the arrangement of particles is 3*3*3, in total twenty-seven particles. The specimen of MSE with homogenous (30 mm in length, 30 mm in width, 30 mm in thickness) or chain-like (35 mm in length, 35 mm in width, 25 mm in thickness) particles are manufactured, as shown in Figure 3(a) and 3(b), respectively.

2.2 Special drop hammer system

The special drop-hammer impact system is developed, as illustrated in Figure 4. The rectangular line in Figure 4 (d) shows the magnetic circuit, which is generated by the electromagnet system. The magnetic intensity is controlled by the electrical current intensity in the coil and measured by a Tesla gauge (model: WT10A, Wei Te magnetoelectric engineering Corp, China). The induced magnetic field is imposed in the direction of particle chains in MSE.

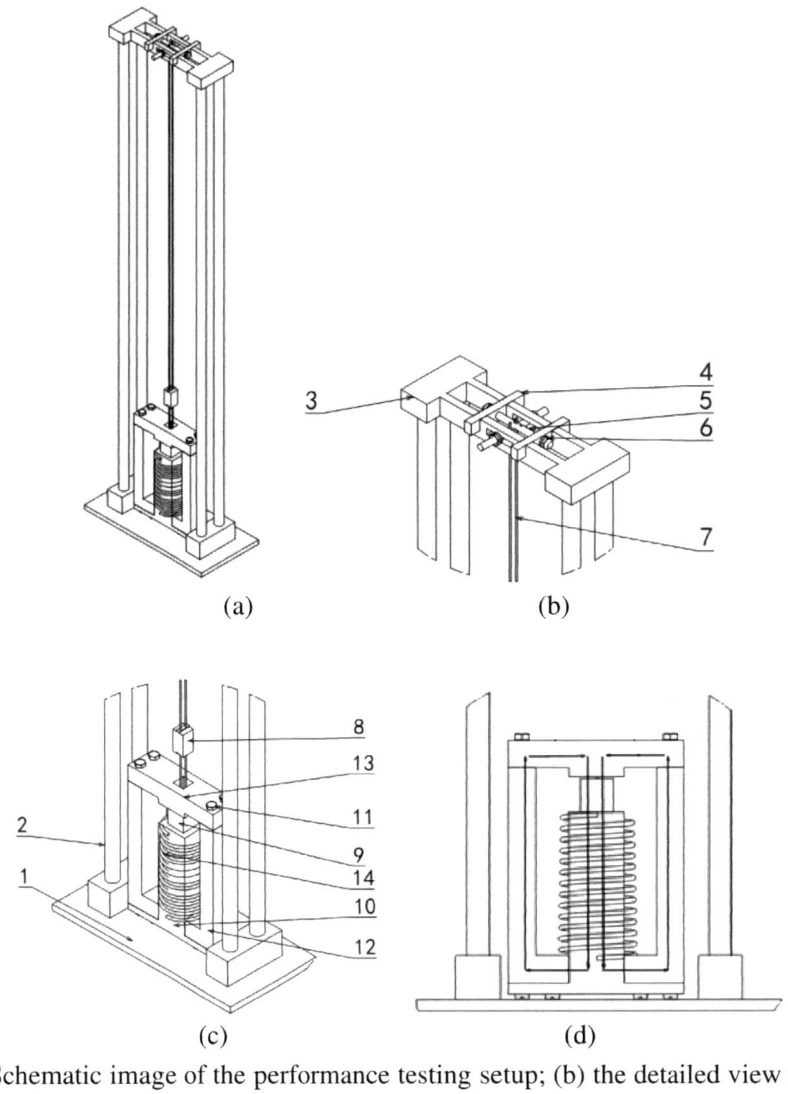

Figure 4. (a) Schematic image of the performance testing setup; (b) the detailed view of upper testing setup; (c) the detailed view of lower testing setup; (d) the detailed view of magnetic route of upper testing setup: (1) base; (2) support rod; (3) cross beam; (4) adjusting plate; (5) cross-shaped shaft; (6) nut; (7) guide bars; (8) hammer; (9) MSEs specimen; (10) T shape iron coin; (11) bolt; (12) L shape iron coin; (13) upper slab; and (14) coil.

The cross-shaped shaft and adjusting plate is used to modify the position of the guide bars, in order to make sure the hammer can pass through the holes in the upper slab smoothly to impact the specimen. The acceleration sensor (model: YJ9A, made by Beijing Oriental Vibration and Noise Institute in China) is laid on the top surface of the hammer with thread connection. The hammer is raised along the guide bars, and its height determines the impact velocity.

3. Experimental results and discussion

For homogenous and chain-like MSEs, the impact behavior is studied at different magnetic intensity from B=0 mT to B=240 mT, with the height of initial position of hammer being 10 mm. At each experimental condition, three identical specimens are conducted, and the final result is an average of these three sets of data. The relationship between acceleration of hammer and time is shown in Figure 5.

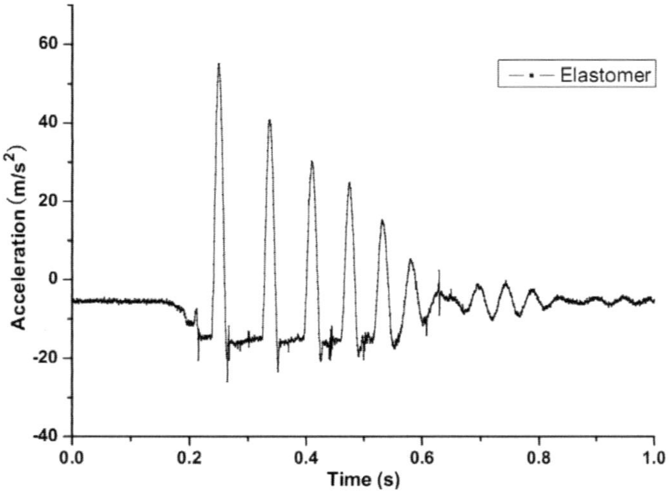

Figure 5. The acceleration-time history of impacting elastomers.

From Figure 5, it can be seen that the hammer rebounds after it impacts the specimen. This rebounding process repeats until the hammer keeps steadily. The amplitude of the acceleration, a_{max}, reduces. Here, only the first impact process is studied.

To analyze the dynamics of the impact system, the impact structure which includes a hammer and MSE can be considered as the single-degree-of-freedom system, as Figure 6 shows.

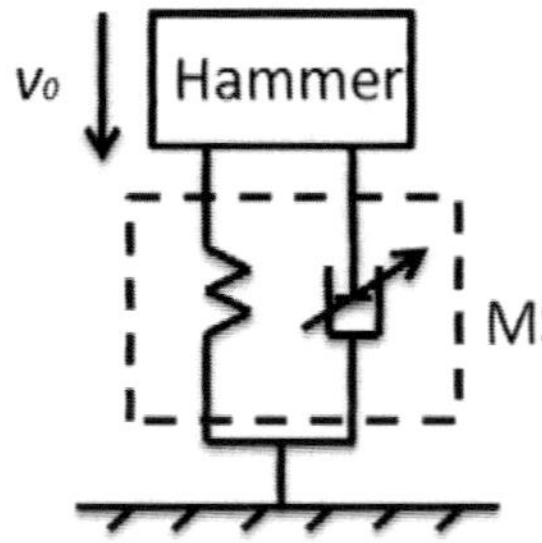

Figure 6. Model of an impact system.

The motion of the hammer is described by the following equations,

$$m\ddot{x} = F_{R1}(\dot{x}) + F_{R2}(x) - mg \tag{2}$$

with the initial conditions,

$$x(0) = 0, \qquad \dot{x}(0) = v_0 \tag{3}$$

where m is the mass of the hammer, F_{R1} and F_{R2} are the damping force and elastic force produced by MSE.

$$F_{R1}(\dot{x}) = c\dot{x} \tag{4}$$

$$F_{R2}(x) = kx \tag{5}$$

where c is the damping coefficient of MSE, and k is the stiffness of MSE. v_0 is the initial impact velocity of hammer,

$$v_0 = \sqrt{2gh} \tag{6}$$

where h is the height of initial position of the hammer, 10 mm. g is the gravity acceleration. Therefore, the initial velocity is 0.447 m/s.

The hammer mass moves along the x direction with decreasing velocity until it reaches zero. Combining the equation (4) and (5) into equation (2),

$$m\ddot{x} = c\dot{x} + kx - mg \tag{7}$$

When the hammer drops down to the maximum displacement, x_{max}, its acceleration also reaches maximum, a_{max}, while the velocity is zero. At this time, the equation (7) can be written as,

$$m\ddot{x}_{\max} = kx_{\max} - mg \tag{8}$$

As the hammer mass is a constant, the maximum acceleration in the first impact process, can directly reflect the stiffness of MSE specimen.

The relationship of acceleration and time for homogenous and chain-like MSEs are shown in Figure 7 and Figure 8, respectively.

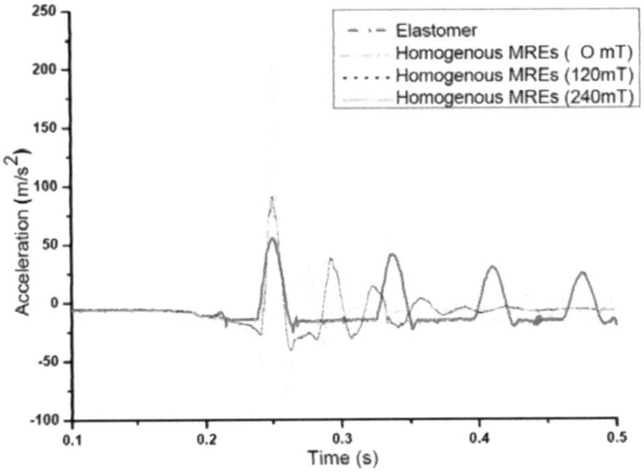

Figure 7. The acceleration-time history of pure elastomer and homogenous MSEs

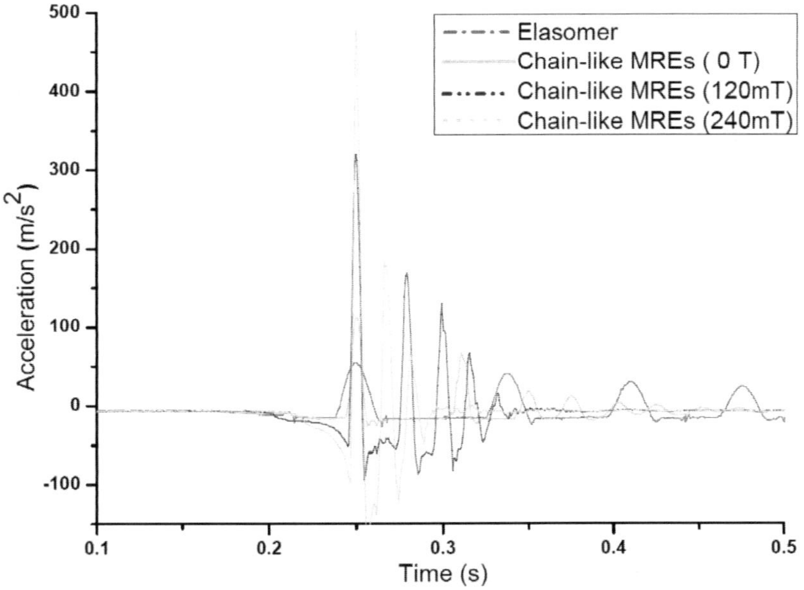

Figure 8. The relationship of acceleration and time of pure elastomer and chain-like MSEs.

From Figure 7 and 8, it can be illustrated that the acceleration amplitude of MS elastomers is greater than that of pure elastomers, and the a_{max} of the MSEs increases with the external magnetic field intensity increasing. It is because that the stiffness of MSE is greater than that of matrix, and the stiffness of MSEs increases with the magnetic field intensity increasing, which is consistent with the results from the published paper [7]. The maximum acceleration of the specimen is listed in Table 2.

Table 2. The maximum acceleration of the specimen in different magnetic field strength.

Type	Magnetic field intensity (mT)	Maximum acceleration (m s^{-2})
Elastomer	0	55.08
Homogenous	0	82.55
	120	91.09
	240	207.06
Chain-like	0	112.46
	120	319.82
	240	478.21

From Table 2, it can be seen that the maximum acceleration in the first impact process of chain-like MSEs is greater than homogenous MSEs. It is because that the stiffness of chain structure is more dependent on magnetic field.

At the time of the maximum acceleration, in Figure 7 and 8, the velocity of hammer drops to zero. Hence, the kinetic energy of the hammer, E, is all dissipated by the specimen.

$$E = \frac{1}{2} m v_0^{\ 2} \tag{9}$$

MSEs dissipate energy in two ways [8]: (1) eliminate or absorb the energy due to its viscosity, E_c; (2) store energy due to its elastic properties, E_k.

$$E_c = \int_0^{x_{\max}} F_{R1}(\dot{x})dx \tag{10}$$

$$E_k = \int_0^{x_{\max}} F_{R2}(x)dx \tag{11}$$

Hence, the energy equilibrium can be written as,

$$E = E_c + E_k \tag{12}$$

Combining equation (4) and (5) into equation (9) to (12),

$$\frac{1}{2} m v_0^{\ 2} = \int_0^{x_{\max}} c\dot{x}dx + \frac{1}{2} k x_{\max}^2 \tag{13}$$

As the hammer mass and the initial dropping height are ensured, the input energy of the impact system is a constant, Q. Therefore, the equation (13) can be illustrated, as Figure 9 shows.

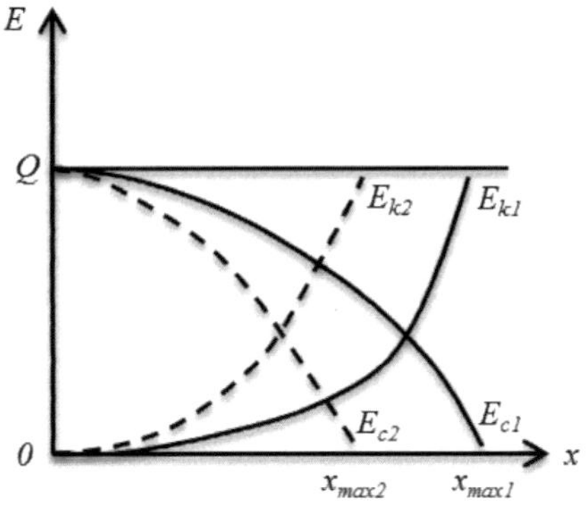

Figure 9. The energy equilibrium curves.

With the stiffness of the specimen increasing, its maximum deformation reduces. The energy dissipation per unit length of specimen is measured as,

$$\overline{E} = \frac{E}{x_{\max}} \tag{14}$$

Hence, the greater the stiffness of MSE specimen is, the greater $\overline{E}$ is. The maximum deformation and energy dissipation per unit length of specimen are shown in Table 3.

Table 3. The energy dissipation characteristic of specimen.

Type	Magnetic field intensity (mT)	δ_{max} (mm)	$\overline{E}$ (J/m)
Elastomer	0	10.7	9.89
Homogenous	0	6.67	18.8
	120	5.39	22.37
	240	4.71	22.48
Chain-like	0	4.76	22.24
	120	4.15	25.50
	240	3.89	27.22

From Table 3, it can be seen that the dissipated energy per unit length of chain-like MSEs is greater than that of homogenous MSEs, and both of them increase with the external applied magnetic field.

4. Conclusion

In this research, a new method to fabricate MSE with macro-particle (size in millimeter) is proposed. By using this method, ideal isotropic and chain-like structured MSEs can be prepared manually. A drop hammer testing setup is developed to investigate the energy dissipation characteristics of MS elastomers under impact loadings and various magnetic fields. It is concluded that, for the same particle volume fraction, particle size and applied magnetic field, the impact acceleration amplitude of chain-like MSEs are greater than homogenous MSEs, and the dissipated energy per unit length of chain-like MSEs is greater than that of homogenous MSEs. The energy dissipation per unit length of MSEs with different particle arrangement both increase with the external applied magnetic field intensity.

Acknowledgments

The authors would like to express the gratitude from the support of National Natural Science Foundation of China (51175016), and appreciate the Vibration and Noise Laboratory of Prof. Xiandong Liu in School of Transportation Science and Engineering, in Beihang Univerisy who provide measurement device and data process software for the experiments. Also, we thank Dr. Xiaojie Wang in University of Nevada, Reno for his valuable comments and proofreading on this manuscript.

References

[1] Carlson J D and Jolly M R 2000 *Mechantronics* **10** 555
[2] Varga Z, Filipcsei G and M Zrinyi 2006 *Polymer* **47** 227
[3] Zhou G Y 2003 *Smart Mater. Struct.* **12** 139

[4] Bellan C and Bossis G 2002 *Int. J. Mod. Phys. B* **16** 2447
[5] Deshmukh S S and Mchinley G H 2007 *Smart Mater. Struct.* **1** 106
[6] Davis L C 1999 *J. Appl. Phys.***86** 3348
[7] Kallio M 2005 *The elastic and damping properties of magnetorheological elastomers* VIT publications **4** 92
[8] Wereley N M 2011 *J. Intell. Mater. Syst. Struct.* **22** 515

Experimental and numerical investigation on damping properties and energy dissipation mechanisms of magnetosensitive rubber

W Li, L Sun[1], J Sun, W Chen, F Ma and D Leng

School of Transportation Science and Engineering, Beihang University, Beijing 100191, China

E-mail: lysunluck@gmail.com

Abstract: This study presents both numerical and experimental investigation on damping properties and energy dissipation mechanisms of magnetosensitive rubber (MSR). Representative volume element (RVE) including particles and matrix was built and periodic boundary condition (PBC) was applied. Various sinusoidal loads, with different frequencies, were applied to RVE under external magnetic field. Considering interaction and complex mechanisms in multi-physics field, finite element method (FEM) based on magneto-mechanical coupling algorithm was adopted. MSR samples were fabricated by aligning iron particles with millimeter level diameter in silicone rubber matrix. The correctness of the numerical method was verified by comparing the results of simulation and quasi-static load test. Dynamic experimental measurement was conducted in material test system. The results demonstrate that the damping properties of MSR are influenced by magnetic induction density and frequency of sinusoidal load. Energy dissipation mechanisms of MSR were explored.

1. Introduction

Magnetorheological elastomers (MREs) are a class of smart materials whose mechanical properties can be controlled by external magnetic field [1]. Besides adjustable elastic properties, it is experimentally demonstrated that MREs also have controllable damping properties [2-4] which possibly can be applied in vibration environment as damper materials. Similar to elastic properties, MREs' damping properties are also probably influenced by both internal factors (shape, size and volume fraction of particle, etc.) and external factors (direction, amplitude, frequency of structural load and magnetic field, etc).

However, the previous experimental and numerical researches mainly focused on studying the relationship between mechanical characteristics of MREs and influencing factors [5-10]. For MREs, complete explanation in terms of the distribution of stress field and magnetic field has not been introduced in detail yet. In order to utilize MREs as damper materials effectively, it is necessary to explore the damping properties and understand the energy dissipation mechanisms by studying their stress field and magnetic field

This research introduced MSR whose structure is very similar to MRE and studied the mechanical characteristics of MSR numerically and experimentally.

[1] Corresponding author.

2. Experiment

2.1. Fabrication of MSR

Traditionally, MREs are fabricated of carbonyl iron powder with micron level particle size, silicone rubber and additives. Because the dimensions of particles and the distance between adjacent particles are rather small, the microstructure of MREs needs to be observed by scanning electron microscopy (SEM). It is difficult to install test probes in the small gap between the magnetized particles to measure magnetic field without destroying the integrity of matrix structure and the distribution of magnetic field. It is also difficult to get stress/strain field of MREs. Hence, iron particles (Puning Iron Particle Factory, China) with millimeter level diameter and transparent silicone rubber (Kenseer Polymer Technology Co., Ltd. China, Model: SE901A and SE901B) are used. This new type of MRE is named as MSR. In the future work, small Hall probes can be installed in MSR. Moreover, photoelastic technology may be used to measure stress field of it. Although there are some differences between MSR and MRE, it will be very helpful to understand mechanical characteristics of MRE by analogical and comparative method.

Particles are aligned in two ways, uniform or chain-like, shown as Figure 1(a) and (b), respectively. Because the elongation is below 10%, the particle arrangement can keep initial configurations in experimental set up well. The diameter of iron particles is 8 mm. The volume fraction of iron particles is 27% which is optimum predicted by Davis [11]. The sizes of samples are about 32mm×32mm×32mm (uniform particles arrangement) and 28mm×35mm×35mm (chain-like particles arrangement), respectively (Because manufacturing error and measurement error, the volume fraction of iron particles is not absolutely 27%). Each sample contains 3×3×3, a total of 27 iron particles. The ratio of distance (W) between adjacent particles in the chain and particle diameter (D) is 0.05, shown as figure 2. The mixing ratio of SE901A and SE901B is 1:1. The mixture is placed in vacuum tank for twenty minutes to remove all air bubbles. During pre-configuration stage, the samples are heated to about 100°C for total of 1 h in a specific mold covered by plastic sheeting, shown as figure 3.

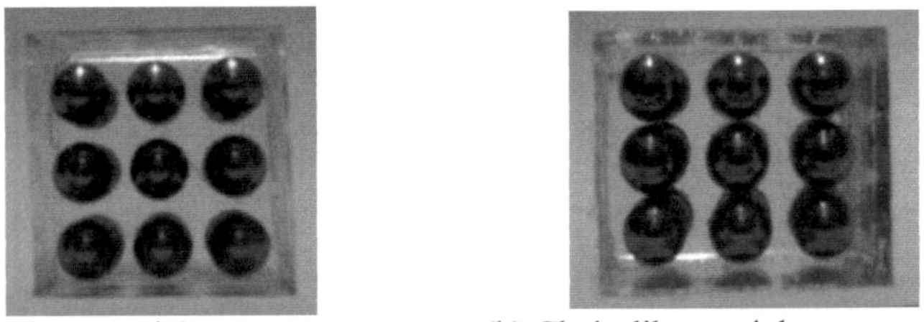

(a) Uniform particles arrangement (b) Chain-like particles arrangement

Figure 1. MSR samples.

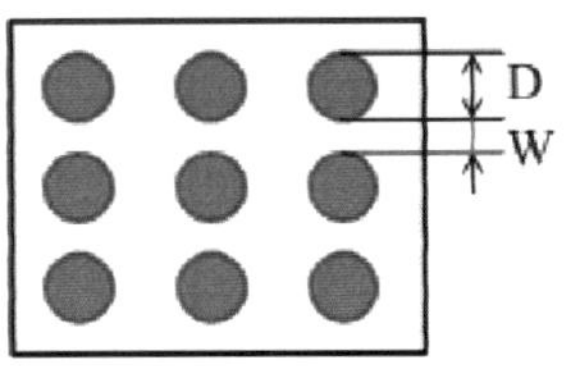

Figure 2. Schematic of MSR.

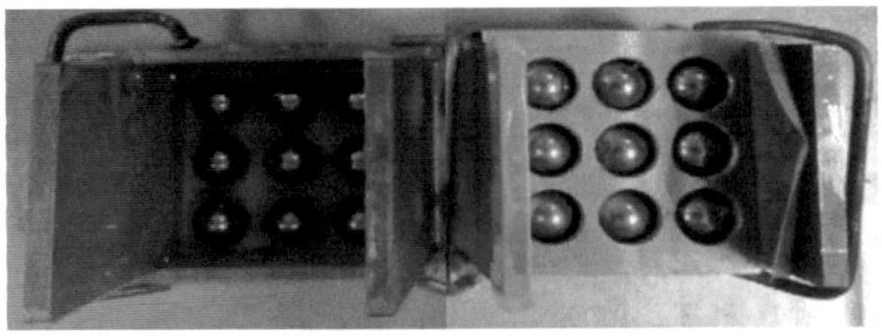

Figure 3. Iron mold.

2.2. Experimental setup

A set of iron fixture including connector base and core is used to install sample and connect to loading device, shown as Figure 4. The magnetic field ranging from 0 mT to 400 mT is generated by the built-in electromagnet coil. Teslameter (Weite Magnetoelectric Engineering Corp, China, Model: WT-10A,) is used to measure the magnetic induction density. MTS-880 material test system is used for both static and dynamic experiments to measure tensile modulus and damping capacity of MRE.

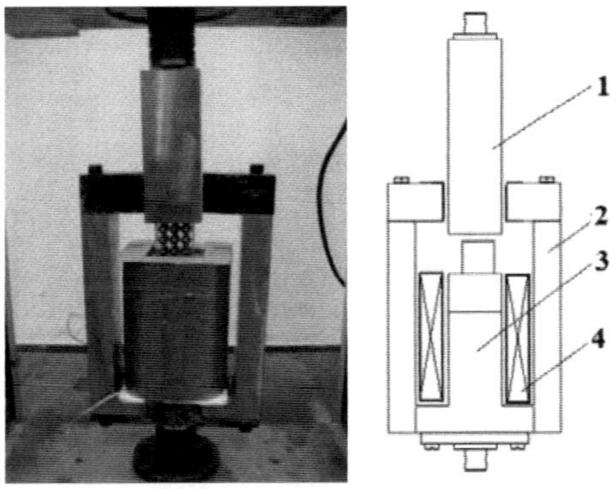

1-connector 2-base 3-core 4-electromagnet coil

Figure 4. Experimental setup.

3. Numerical investigation

3.1. Representative volume element

Besides using iron particles with millimeter level diameter and transparent silicone rubber, finite element techniques can also help to analyze the distribution of magnetic field and stress field.

Each MSR sample contains twenty-seven iron particles. If all particles are included in the 3-D finite element model, it will need so much computational time. Therefore, the characteristics of MSR can be obtained by RVE. For composite materials, the mechanical characteristics of RVE in micro-scale can represent the material properties of structure in macro-scale well. Three typical RVE modes are presented, shown as Figure 5. Type 1 shows that a spherical particle is surrounded by matrix. Type 2 shows that two hemispherical particles embed in the upper side and lower side of matrix. Type 3 shows that eight 1/8 spherical particles distribute in the corners of the cube. Each type of RVE is 2Px width, 2Py length and 2Pz height, shown as Figure 6. The chain is along z-axis direction.

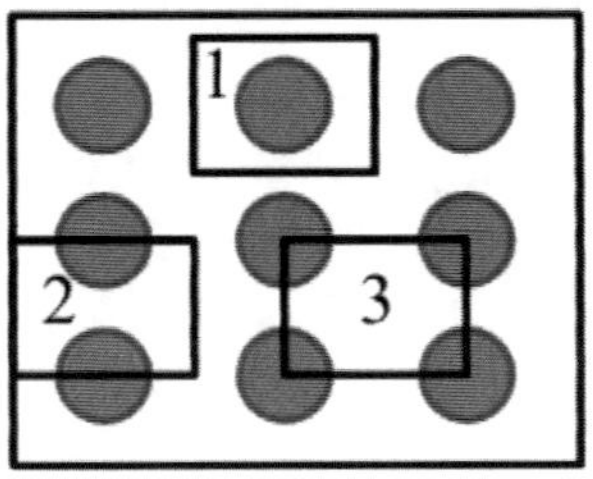

Figure 5. Three typical RVE modes.

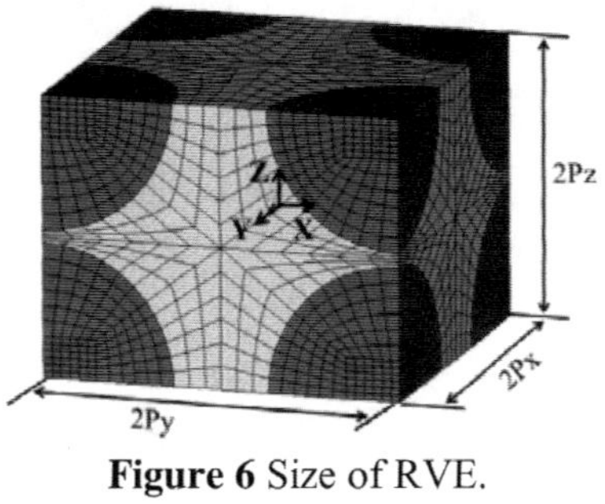

Figure 6 Size of RVE.

When applying magnetic field along the chain, the magnetic force between the adjacent particles within chain and that in different chain both will influence characteristics of RVE, shown as Figure 7. So, type 3 is chosen to analyze.

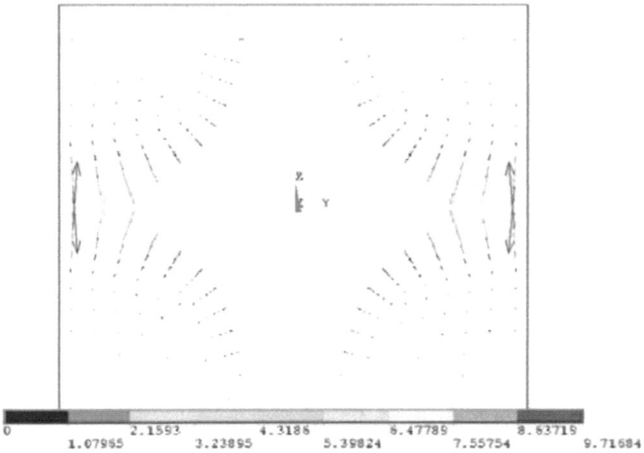

Figure 7. Magnetic force between adjacent particles under 1 T magnetic field.

3.2. Periodic boundary condition

When applying constant or sinusoidal loads along chain, the PBC of RVE is described by equation (1)-(3):

Compatibility equations between left surface and right surface of RVE are:

$$\begin{cases} u(P_x, y, z) - u(-P_x, y, z) = 2\varepsilon_x P_x \\ v(P_x, y, z) - v(-P_x, y, z) = 0 \\ w(P_x, y, z) - w(-P_x, y, z) = 0 \end{cases} \tag{1}$$

Compatibility equations between front surface and rear surface of RVE are:

$$\begin{cases} u(x, P_y, z) - u(x, -P_y, z) = 0 \\ v(x, P_y, z) - v(x, -P_y, z) = 2\varepsilon_y P_y \\ w(x, P_y, z) - w(x, -P_y, z) = 0 \end{cases} \tag{2}$$

Compatibility equations between top surface and bottom surface of RVE are:

$$\begin{cases} u(x, y, P_z) - u(x, y, -P_z) = 0 \\ v(x, y, P_z) - v(x, y, -P_z) = 0 \\ w(x, y, P_z) - w(x, y, -P_z) = \delta_t = A\sin(2\pi f t) \end{cases} \tag{3}$$

Where, u, v and w denotes the displacement along x-axis, y-axis and z-axis, respectively.

ε_x and ε_y represents strain in x-direction and y-direction, respectively.

δ_t is tension-compression loads of z-direction and f is frequency, respectively.

3.3. Finite element method

Sequential magneto-mechanical coupling algorithm which can solve specific static and dynamic problems is written in ANSYS APDL language, the flowchart is shown in Figure 8.

Either magnetic field or mechanical field can be calculated first, and then the numerical information of magnetic field or mechanical field will be transferred to another field. By several calculation cycles, the stress/strain field and the magnetic field will be constant which indicates that RVE is in equilibrium state.

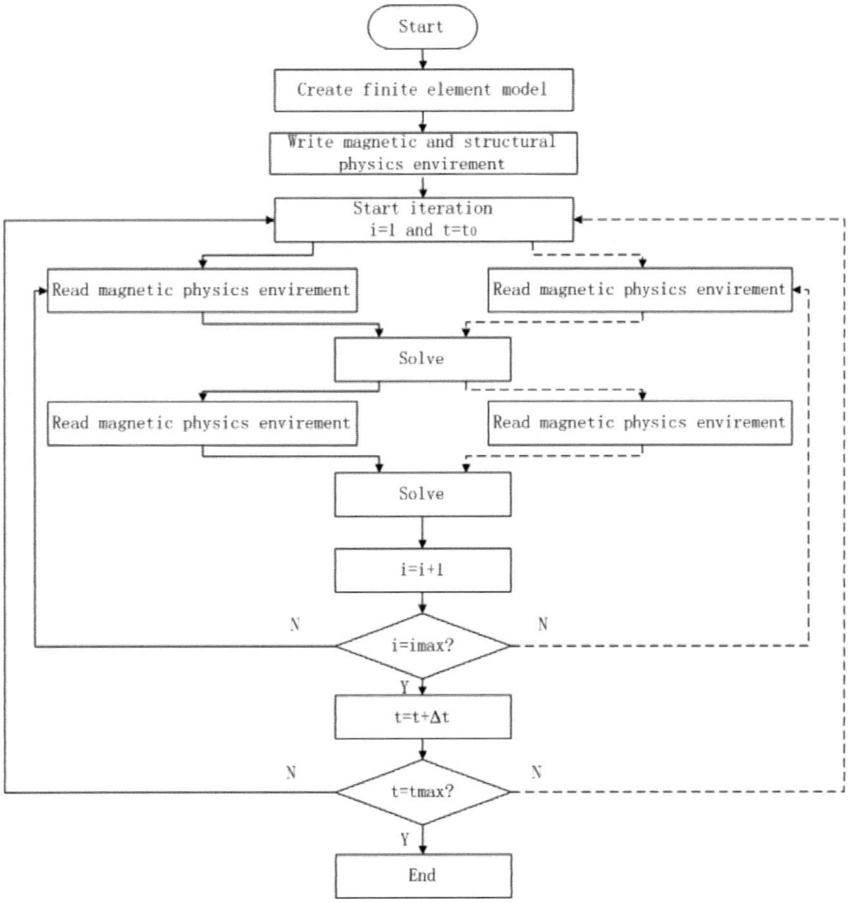

Figure 8. Scheme of magneto-mechanical coupling algorithm.

3.4. Mathematical calculation

The equivalent modulus of RVE can be calculated by equation 4-6 using the numerical results.

$$\bar{\sigma}_{ij} = \frac{1}{V}\int \sigma_{ij} d\Omega \tag{4}$$

$$\bar{\varepsilon}_{ij} = \frac{1}{V}\int \varepsilon_{ij} d\Omega \tag{5}$$

$$Modulus = \frac{\bar{\sigma}_{ij}}{\bar{\varepsilon}_{ij}} \tag{6}$$

Where, $\overline{\sigma}_{ij}$ and $\overline{\varepsilon}_{ij}$ represents the average of stress and strain in volume. V and Ω denotes the volume of the RVE and domain of integration. The superscript bar denotes the volume average of the quantity or effective quantity.

Under cyclic strain loads, energy dissipation is presented in the form of hysteresis loop and stress-strain curves were analyzed to obtain the damping properties of MRE, shown as Figure 9. Specially, damping capacity (φ) is expressed as the ratio of the dissipation energy (ΔW) during a complete cycle to the maximum storage energy (W):

$$\varphi = \frac{\Delta W}{W} \tag{7}$$

$$\Delta W = \int \overline{\sigma}_{ij} d\overline{\varepsilon}_{ij} \tag{8}$$

$$W = \int_0^{\varepsilon_0} \overline{\sigma}_{ij} d\overline{\varepsilon}_{ij} \tag{9}$$

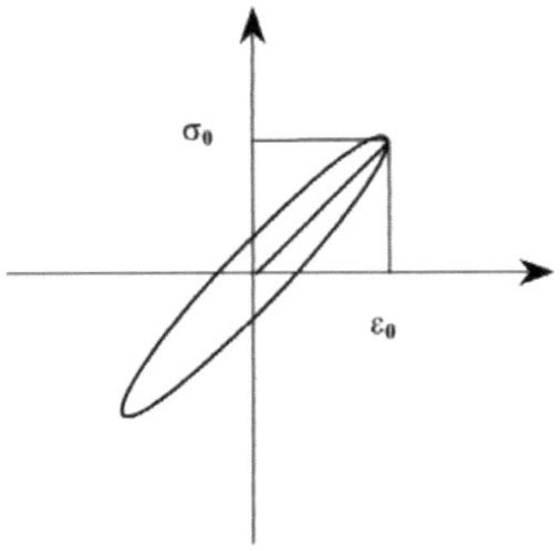

Figure 9. Stress-strain curve (hysteresis loop).

4. Results and discussion

Figure 10 and 11 shows the tensile modulus of MSR with uniform particles arrangement and chain-like particles arrangement under about 6.25% elongation, respectively. Tensile modulus shows an increasing trend with magnetic field. The results suggest that type 3 is accurate enough.

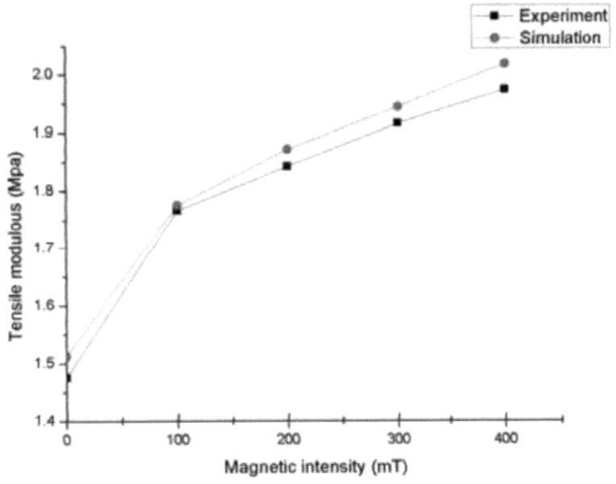

Figure 10. Tensile modulus of MSR with uniform. particles arrangement.

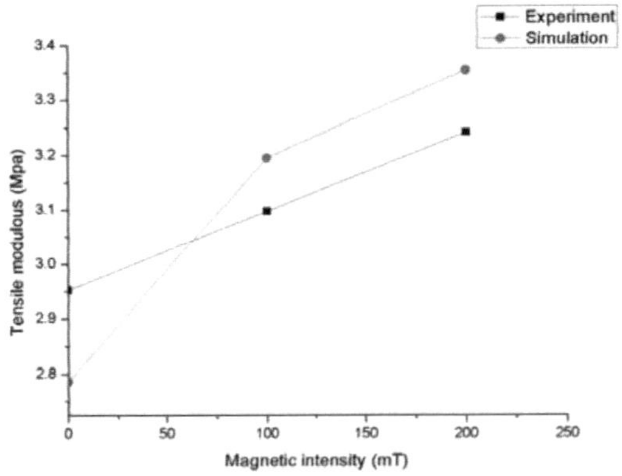

Figure 11. Tensile modulus of MSR with chain-like particles arrangement.

For dynamic experiments, sinusoidal loads with constant strain amplitudes and frequencies were applied to the MSR samples along z-axis. The real-time strain and stress signals were recorded. One strain amplitudes (2 mm) and three frequencies (1 Hz, 5 Hz, 10 Hz) as well as three magnetic fields (100 mT, 200 mT, 300 mT) were selected to study.

Figure 12 and 13 shows the field-dependence of damping capacity of the MSR under sinusoidal loads with different frequencies in experiments. In figure 13, damping capacity of MSR gets a maximum value about 1.12. Because internal defects begin to appear in MSR samples, shown as figure 14, material damping increase sharply.

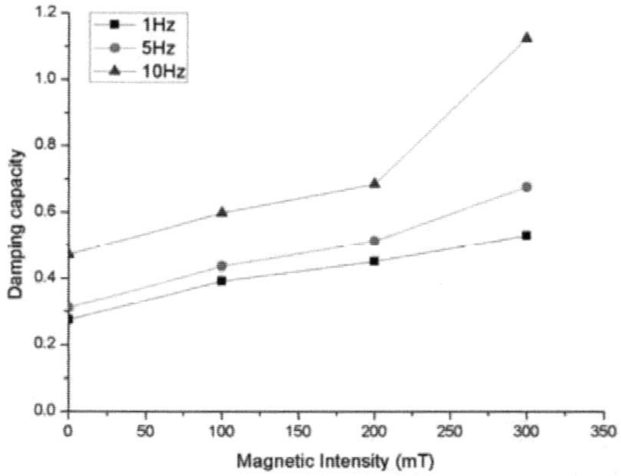

Figure 12. Damping capacity of MSR with uniform particles arrangement.

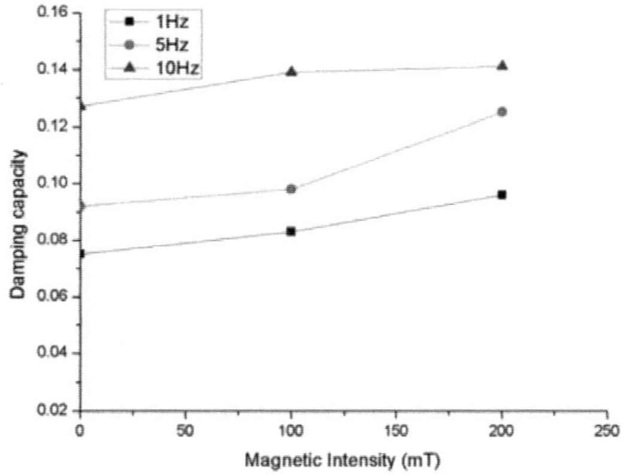

Figure 13. Damping capacity of MSR with chain-like particles arrangement.

(a) Uniform particles arrangement

(b) Chain-like particles arrangement

Figure 14. MSR with internal defects.

The shape of defects domain is similar to cylinder, which mainly locates between adjacent particles within the chain, shown as figure 15(a) and (b). These defects indicate that stress between particles is higher than other place, shown as figure 16. They will influence damping capacity.

The defects will help to explore the energy dissipation mechanisms. In figure 12 and 13, the damping capacity of MRE with uniform particles arrangement is larger than that with chain-like particles. The reason is that cylindrical matrix account for more volume fraction of the former MRE. With the increasing of magnetic intensity, the stress in cylindrical matrix will increase. Moreover, high frequency will enhance the impact of inertia force of particles. Inertia force of particles will compress the matrix. These factors will exacerbate energy dissipation that occurs within MRE through atomic level interactions such as elastic stretching of atomic bonds, vacancy diffusion and dislocation motion.

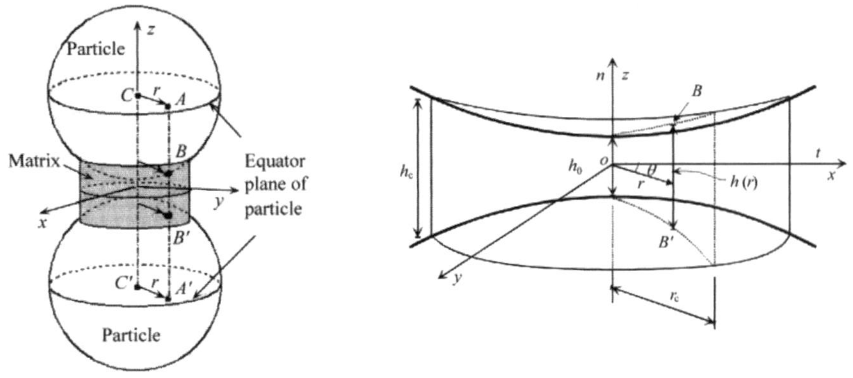

(a) Uniform particles arrangement (b) Chain-like particles arrangement

Figure 15. Adjacent particles [12].

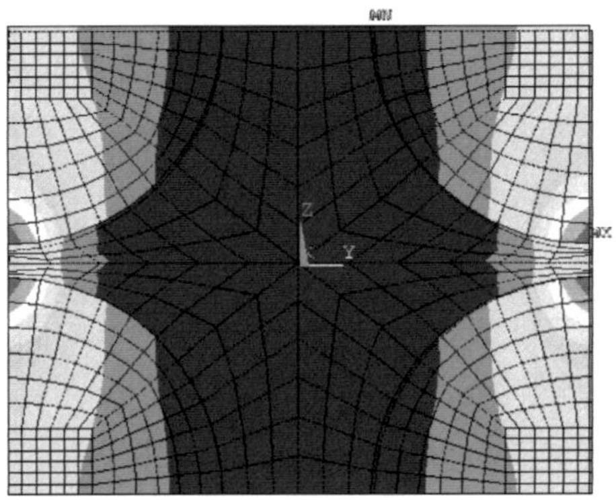

Figure 16. Distribution of stress field of MSR with chain-like particles arrangement.

5. Conclusions

A new type of MRE fabricated of iron particles with millimeter level diameter and silicone rubber is introduced. RVE including particles and matrix is built and PBC is applied. FEM based on magneto-mechanical algorithm is adopted. Both static and dynamic experiments are conducted to measure tensile modulus and damping capacity of MSR. It is concluded that the damping capacity increases with magnetic field intensity and frequency of sinusoidal loads. The cylindrical matrix between adjacent particles within the chain influence energy dissipation and are easily damaged. In the further work, small Hall probes can be installed in MSR to measure magnetic field. Moreover, photoelastic technology can be used to measure stress field of it.

Acknowledgments

This work is financially supported by the National Natural Science Foundation of China (51175016).

References

[1] Chen L, Gong X L and Li WH 2007 *Smart Mater. Struct.* **16** 2645
[2] Shiga T, Okada A and Kurauchi T 1995 *J. Appl. Polym. Sci.* **58** 787
[3] Jolly M R, Carlson J D, Munoz B C and Bullions T A 1996 *J. Intel. Mater. Syst. Str.* **7** 613
[4] Lokander M and Stenberg B 2003 *Polym. Test.* **22** 677
[5] Bellan C and Bossis G 2002 *Int. J. Mod. Phys. B* **16** 2447
[6] Zhou G Y 2003 *Smart Mater. Struct.* **12** 139

[7] Ginder J M, Schlotter W F and Nichols M E 2001 *Proc. of SPIE Smart Structures and Materials 2001: Damping and Isolation* 4331 103
[8] Demchuk S A and Kuzmin V A 2002 *J. Eng. Phys. Thermophys.* **75** 396
[9] Shen Y and Dai H 2004 *Smart Mater. Struct.* **13** 1
[10] Ginder J M, Nichols M E, Elie E D and Clark S M 2000 *Proc. of SPIE Smart Structures and Materials 2000: Smart Structures and Integrated Systems* **3985** 418
[11] L.C. Davis L C 1999 *J. Appl. Phys.* **85** 3348

Magnetodeformational effect of the magnetoactive elastomer and its possible applications

G V Stepanov[1], **E Yu Kramarenko**[2] **and D A Semerenko**[3]
[1] State Scientific Research Institute of Chemistry and Technology of Organoelement Compounds, 105118, Moscow, Russia
[2] Physics Department, Moscow State University, 119992 Moscow, Russia
[3] Moscow State University of Instrument Engineering and Computer Science, 107996 Moscow, Russia

E-mail: gstepanov@mail.ru

Abstract. Magnetodeformational effect (deformation in non-uniform magnetic field) of magnetoactive elastomer has been studied experimentally. It has been shown that in non-uniform magnetic field this material is able to be reversibly deformed by 250%. In dynamic regimes the deformations under elongation are up to 80% while the ones under compression are 2%. Effective frequency range of the material is up to 100 Hz. As a possible application of the material a magneto controlled valve of flow rate (gas, liquid) was developed.

1. Introduction

This work is aimed to a development of new polymer composite materials, namely, magnetoactive elastomers (MAE) based on highly elastic polymeric matrices filled with magnetic microparticles of iron and magnetite. The novel feature of these materials is the ability to change their properties in magnetic fields and thus, perspective for design of magnetic field controlled various devices. New elastomers have low Young's modulus (of the order of tens of kPa) and occupy an intermediate position between rigid magnetic composites and magnetorheological fluids. In the course of magnetic field controlled structuring of the magnetic filler within polymer matrix new materials demonstrate the following properties:

(a) a unique ability to undergo quick and controllable large-scale deformations in external gradient magnetic fields (magnetodeformational effect) [1-3];

(b) a unique ability to undergo quick and controllable large-scale deformations and essential changes in elastic and viscous properties under the action of external homogeneous magnetic fields (magnetostriction and magnetorheological effect) [5-6]. In particular, 100-fold increase of the modulus magnetic field of up to 0.3 T has been observed;

(c) new effect of shape memory or plasticity induced by magnetic field. It is found that in some cases the deformation of the material is virtually fixed by the magnetic field. One may say that the material "remembers" its shape in the presence of magnetic field [3, 5, 6].

In this study, results of experimental studies of magnetodeformational effect of new magnetic elastomers, their dependence on the material composition and strength of magnetic fields were shown, as well as possibilities of their practical applications were demonstrated.

[1] Corresponding author.

2. Experimental

Magnetic field responsive elastomers used in this work consist of a highly elastic polymeric matrix filled with magnetic particles. Polymer matrices were synthesized on the basis of compounds ''SIEL'' produced by GNIIChTEOS. As magnetic filler magnetite particles of 0.2-0.5 μm in diameter and carbonyl iron were used. Homogeneous mixture of all the components was poured in moulds and heated at 120°C for 1 h. Anisotropic samples were obtained by polymerization of composition in homogeneous magnetic field of 30 mT under heating for 1 h. Magnetic particles tend to form chain-like structures under the influence of uniform magnetic field and columnar particle aggregates parallel to the field direction are fixed in the course of polymerization. As a result, materials thus obtained have highly anisotropic properties, namely, their elastic and deformational properties depend on the direction of deformation. In particular, Young's moduli of the material differ by 4 times depending on orientation.

3. Results and discussion

Under the action of non-homogeneous magnetic field samples of magnetic elastomer can be deformed (elongated) by several times. In Figure 1 dependence of the elongation of anisotropic sample synthesized on the basis of magnetite with filler concentration of 50 mass% were plotted. The direction of deformation either coincides with the direction of the magnetic filler structure orientation (parallel orientation) or is perpendicular to it (perpendicular orientation). The sample on the basis of magnetite has very low Young's modulus of around 10 kPa, thus, the deformation in non-homogeneous magnetic field reaches to 250%.

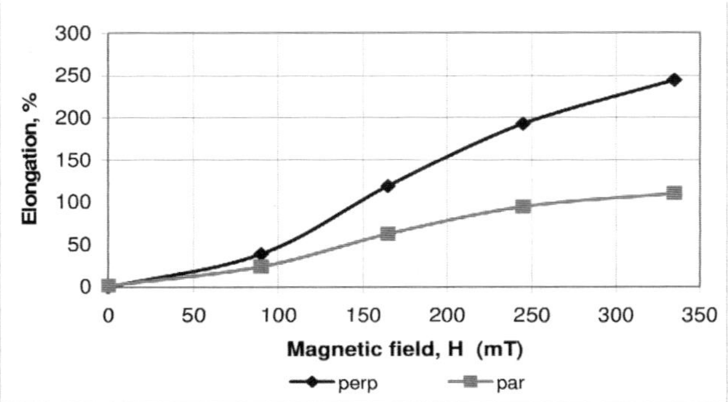

Figure 1. Dependence of the sample elongation in non-homogeneous magnetic field on the field strength at the free end of the sample.

Deformation depends on the direction of elongating magnetic field and direction of magnetic filler columnar structure within samples. The reason for this can be realized from the material structure which is shown on the photo in Figure 2. Concentration of the filler on this photo is low for the structure of the material could be observed in a thin layer by tunneling optical microscope. As one can see from the photo magnetic particles form chains within the material. Elastic properties of polymer composite are defined by the presence of polymer matrix in the space between the filler particles. One can observe on the photo that there is practically no free space between the particles along the filler chains. When the sample is elongated along this direction the composite demonstrates high elasticity because between the particles the amount of polymer capable to elongation is low.

Studies of dynamical properties of magnetic elastomers upon elongation and compression were performed by placing samples above magnetic coil. As one can see in Figure 3 the highest amplitude of deformation is observed at low frequencies of magnetic field. Maximal deformation under elongation was 80%. Sometimes in the vicinity of 50 Hz a resonance was observed along with a decrease of amplitude. In Figure 3, a typical dependence of deformation amplitude under sample elongation on the frequency of alternating magnetic field was plotted. An analogous experiment was performed for a sample of magnetic elastomer which was compressed in non-homogeneous magnetic field. The value of relative compression at low frequency was equal to 2%. Magnetodeformational effect was used for creation of active damper or actuator [7] as well as a magneto controlled valve.

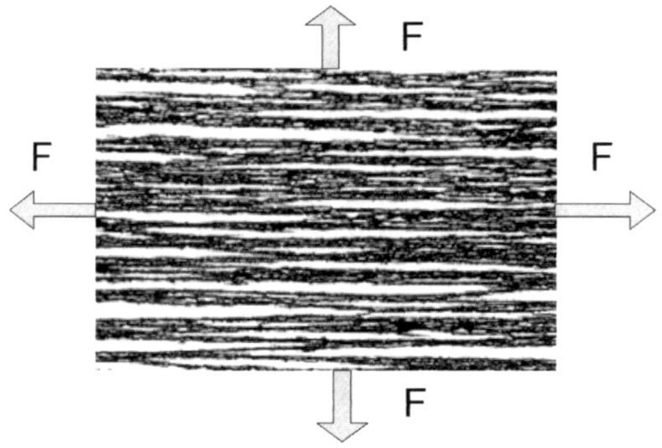

Figure 2. Photo of the structure of the anisotropic sample of magnetic elastomer.

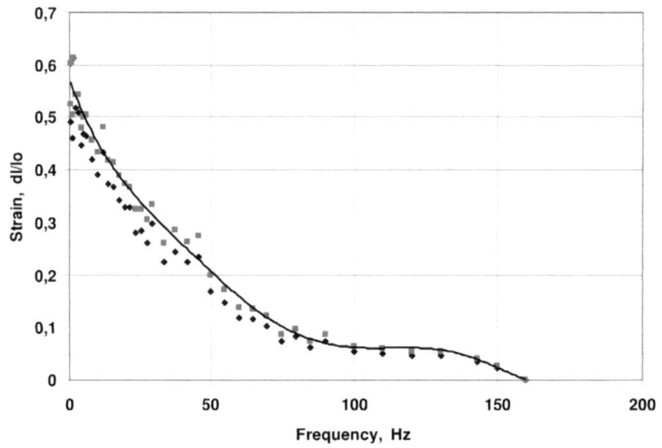

Figure 3. Dependence of elongation of vibrating sample on the frequency of alternating magnetic field.

Magnetic field controlled valve is simple in design; it does not contain plenty of mechanically moving parts as other valves used up to date. It is controlled by an electric current, or a magnetic field, with the minimal overall dimensions and capable to control small gas flow.

In general the valve has the body containing lock-element made of ME composite with the through channel. Outside of the body an electromagnet and/or a constant magnet is placed. ME composite deforms under the action of a non-homogeneous magnetic field. The degree of this deformation can be controlled by the value of the field intensity, or electric current. Thus, the deformed ME composite can either completely lock the channel or only partly decrease its diameter and hence controls gas flow. It is possible to create a system of several electromagnets, besides; the lock-element of the valve can be supplied with the backpressure channel (Figure 4).

In Figure 5 the basic characteristics of the magnetic control valve are shown. Dependence of the valve capacity on magnetic field is considerably non-linear but by special construction it can be corrected to become more linear. It is possible to keep the gas flow in a given interval with the use of an electronic feedback.

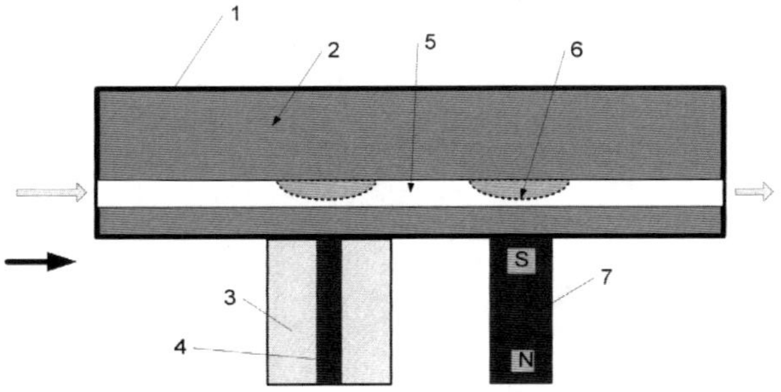

Figure 4. 1-the Case, 2-ME-composite, 3-an electromagnet, 4-the core of an electromagnet, 5-the channel of the valve, 6-A bend of the ME-composite under action of a magnetic field, 7-constant magnet.

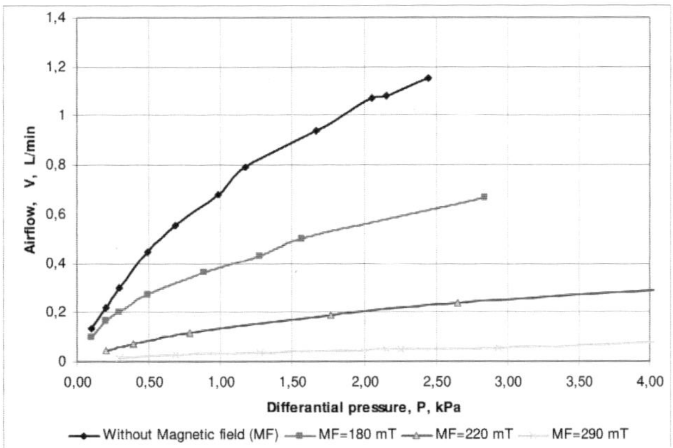

Figure 5. Dependence of air flow from differential pressure for various magnetic fields.

Acknowledgments
This work was supported by the Ministry of Science and Education of Russia

References

[1] Nikitin L V, Mironova L S, Stepanov G V and Samus A N 2001 *Polymer Science Ser. A* **43** 443
[2] Nikitin L V, Mironova L S, Kornev K G and Stepanov G V 2004 *Polymer Science Ser. A* **46** 301
[3] Nikitin L V, Stepanov G V, Mironova L S and Gorbunov A I 2004 *JMMM* **272-276** 2072
[4] Stepanov G V, Abramchuk S S, Grishin D A, Nikitin L V, Kramarenko E Yu and Khokhlov A R 2007 *Polymer* **48** 488
[5] Abramchuk S S, Kramarenko E Yu, Stepanov G V, Nikitin L V, Filipcsei G, Khokhlov A R and Zrinyi M 2007 *Polymers for Advanced Technologies* **18** 883
[6] Abramchuk S S, Kramarenko E Yu, Stepanov G V, Nikitin L V, Filipcsei G, Khokhlov A R and Zrinyi M 2007 *Polymers for Advanced Technologies* **18** 513-518.
[7] Borin D, Stepanov G, Mikhailov V and Gorbunov A 2007 *Magnetogidrodynemic* **43** 437

Magnetorheological elastomer and its application on impact buffer

J Fu[1], M Yu[1,3], X M Dong[2] and L X Zhu[1]
[1] The Key Lab for Optoelectronic Technology and Systems, Ministry of Education, College of Optoelectronic Engineering, Chongqing University, 400044 Chongqing, China
[2] College of Mechanical Engineering, Chongqing University, 400044 Chongqing, China

E-mail: yumiao@cqu.edu.cn

Abstract. In this study, a new magnetorheological elastomer (MRE) based buffer is proposed and its vibration isolation performance is investigated. The MRE buffer with a compact structure is first designed in order to accomplish the maximization of the variable stiffness range. The working characteristics of the MRE buffer are then measured and the model of MRE is established. On the basis of the experimental data, the control model of the MRE buffer is also formulated. A two-degree-of-freedom dynamic model with an MRE buffer is then developed. An intelligent control strategy, human simulated intelligent control (HSIC), is proposed to reduce the impact during the drop crash. Finally, the proposed MRE buffer and controller are validated numerically and experimentally. The results show that the proposed MRE buffer and the control strategy can reduce the impact acceleration effectively.

1. Introduction

Magnetorheological (MR) materials include magnetorheological fluid (MRF), MR foams and MRE, whose rheological properties can be controlled by the application of an external magnetic field[1]. The most common MR material is MRF. There are magnetically polarizable particles suspended in the viscous fluid. In recent years, the MRF technology has made significant advancements [2] and many applications such as the automotive suspension vibration control [3], the earthquake resistance [4], clutch [5] etc. However, MRFs are prone to particle settling with time due to the density mismatch of magnetic particles and the carrier fluid, which may degrade the MR effect. Therefore, another smart material, MRE, has received much attention in recent years.

As solid analogy of MRF, MRE, which includes a wide variety of composite materials, typically consists of magnetically polarizable particles dispersed in a polymer medium with many advantages. Compared to MRF, MRE can avoid some disadvantages such as settling of particles normally associated with MRF. MRE does not require channels or seals to hold or prevent leakage and thus can avoid the particle sedimentation associated with MRF. In addition, the valuable characteristic of MRE is that its mechanical properties such as the storage and loss modulus can be tuned reversibly by the application of an external magnetic field. The field dependence of the mechanical properties enables

[3] Corresponding author: Yu Miao

the construction of controllable elastomeric components. As a result, MRE possesses many potential engineering applications for vibration control in damping and vibration isolation systems such as engine mounts and suspension bushings to reduce noise and vibration.

Due to the little adjustment range of variable stiffness, there are only few application reports about MRE when comparing to the MRF. The salient feature of adjustable stiffness has been studied in vibration isolation. Several researches have shown the shift of resonant frequency can be achieved by the MRE buffer. For example, the tunable automotive mounts and bushings based on MRE were developed by Ginder *et al.* [6]. They found that the suspension resonances excited by torque variation could be suppressed by shifting the resonance away from the excitation frequency. Another similar adaptive tuned vibration buffer based on MRE was studied by Deng *et al.* [7]. Their results show that the natural frequency of the buffer based on MRE can be tuned from 55 to 82 Hz. In their later another work [8], the shift-frequency capability of the adaptive buffer was theoretically and experimentally evaluated. The results demonstrate that the natural frequency of the proposed buffer can be tuned from 27.5 Hz to 40 Hz. Collette *et al.* [9] investigated numerically two systems based on MRE according to the measured characteristics from published studies. The numerical results show that the command ability of the elastomer improves the isolation performance of the MRE isolator under a frequency varying harmonic excitation, and decreases the stress level of the MRE dynamic vibration buffer under a random excitation. Liao *et al.* [10] attached a voice coil motor to a conventional MRE buffer to improve its performance. The results indicate that the resonant frequency of the proposed buffer could be tuned from 11 to 18 Hz whereas the damping ratio would reduce from 0.19 to 0.005. Besides the shift resonant frequency, Opie *et al.* [11] studied a magnetically biased MRE-based vibration isolator in the primary isolation system. The experimental results show that the MRE isolator and the semi-active controller system can reduce resonances and payload velocities by 16-30% when compared to passive systems. To improve the isolation performance of the traditional floating raft system, Sun *et al.* [12] proposed a buffer based on MRE. The results show that the adaptive buffer can improve the isolation performance of floating raft system in a wide frequency range. These researches have shown that tunable vibration buffers based on MRE can achieve better isolation performance than traditional vibration buffers. However, the research focus on steady vibration system and is still in early stage for MRE, most of the study focused on preparation, mechanical tests and simple physical modelling of MRE, and little work has been done in some transient impact isolation systems and control.

Consequently, the main contribution of this study is to propose a new adaptive variable stiffness buffer based on MRE and evaluate its vibration control performance in transient impact isolation systems. To accomplish it, a new vibration buffer based on the fabricated MRE is developed and its control model is formulated on the basis of measured working characteristics. Then a two-degree-of-freedom dynamic model with the proposed MRE buffer is constructed and a human simulated intelligent controller is yet formulated. Numerical and actual drop test are performed to validate the proposed MRE buffer and the proposed control strategy.

2. Development of MRE buffer

2.1. Structure design of MRE

In this study, an adaptive variable stiffness buffer based on MRE is designed and manufactured to realize variable stiffness characteristics of the buffer, which is schematically shown in Figure 1. The buffer consists of a coil, two smart MREs, the outer cylinder and the piston. Two MREs are clung to both the outer cylinder and the piston, respectively. The magnetic field is generated by the coils and can be tuned by tuning the coil current. By the motion of piston, the shear deformation of MRE will appear. The magnetic field exits in the MRE which is perpendicular to the motion of the piston after the current is applied to the coil. The shear modulus of the MRE is altered when subjected to an external magnetic field. Thus, the elastic force of the buffer is controlled by the intensity of the magnetic field.

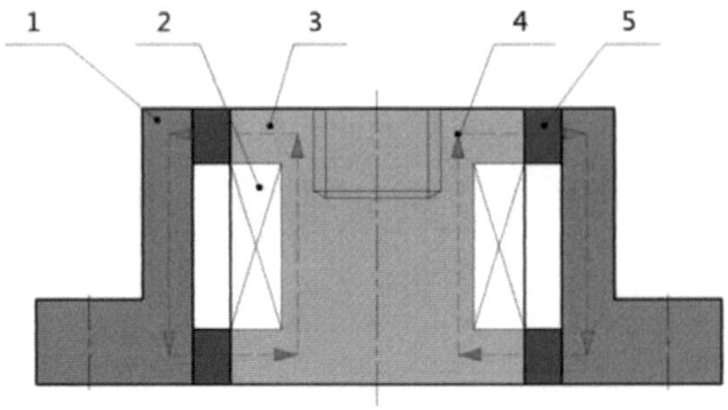

Figure 1. Schematic configuration of MRE buffer (1-Outer cylinder 2-Coil 3-Piston 4-Primary flux path 5-MRE)

In absence of magnetic field, the buffer will produce an elastic force only caused by the MRE's passive shear resistance. If a certain level of magnetic field is supplied to the buffer, the MRE buffer produces an additional and variable elastic force owing to the increased shear modulus of the MRE. The elastic force of the MRE buffer can be continuously tuned by controlling the intensity of the magnetic field.

2.2. Model of MRE

To simplify the analysis of the MRE buffer, a Kelvin-Voigt model is adopted in this study. The model consists of a constant stiffness spring, a variable stiffness spring and a damper with constant damping coefficient in parallel. In the shear direction, k is given by [7]

$$k = \frac{GA}{h} = \frac{(G_0 + \Delta G)A}{h} = \frac{G_0 A}{h} + \frac{A}{h} 36\varphi\mu_r\mu_0\beta^2 H_0^2 (\frac{R}{d})^3 \varsigma = k_e + k_{MRE} \tag{1}$$

where G denotes the shear modulus of MRE and consists of two terms, an initial shear modulus G_0 without a magnetic field and a shear modulus increment ΔG with applying a magnetic field; A is the shear area, and h is the thickness of MRE; μ_r and μ_0 are the relative permeability for the particles and silicon rubber matrix, respectively; φ is volume fraction; R is the average particle radius; d is the particle distance before deflection, H_0 is the applied magnetic field intensity; $\beta \approx 1; \varsigma \approx 1.202$; k_e is the equivalent stiffness coefficient and k_{MRE} is the controllable stiffness part of the MRE.

For simplification, the F_{MRE} can be approximately described as a quadratic function of the input current

$$F_{MRE} = dI^2 + eI + f \tag{2}$$

where d, e, f are constants and obtained by fitting the measured experimental data, I is the controlled current supplied to the coil of MRE buffer.

2.3. Working features test of MRE

To measure working features of the MRE buffer, an experimental platform has been set up, which is shown in Figure 2 and 3. The test system include the MRE buffer, the force sensors, electromagnetic vibration table and data acquisition instrument. A vibrator produced by B&K Co. is used to drive the MRE buffer, and the displacement and force are measured. The velocities are calculated using a central differences approximation. Sinusoidal command signals are used.

In this study, the excitation frequencies include 25, 50 and 75 Hz and the displacement amplitudes have the same value of 0.5mm. Five kinds of constant currents are applied to the MRE buffer to observe the characteristics of the buffer. The applied input current is from 0 to 2 A with the increment of 0.5 A. For brevity, only the force-displacement hysteresis under the excitation frequency of 25 Hz is given in Figure 4. The equivalent stiffness of the MRE buffer is calculated and shown in Figure 5.

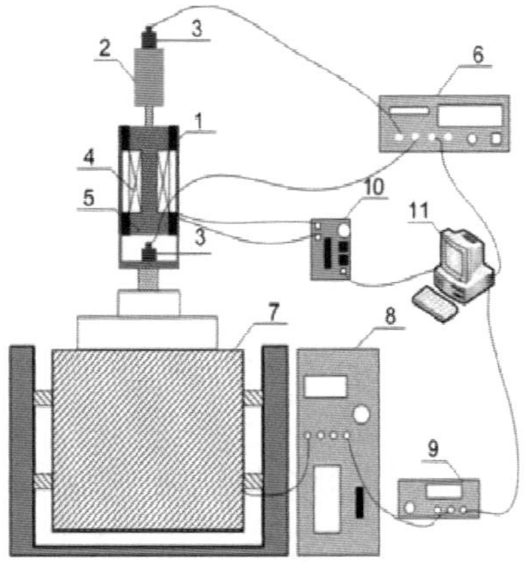

1-MRE , 2-Mass, 3-Sensor, 4-Excitation, 5-Iron Core, 6- Data Acquisition Instrument, 7-Electromagnetic Vibration Table, 8- Power Amplifier, 9-Controller, 10-Current driver, 11-Computer

Figure 2. The test system of MRE buffer.

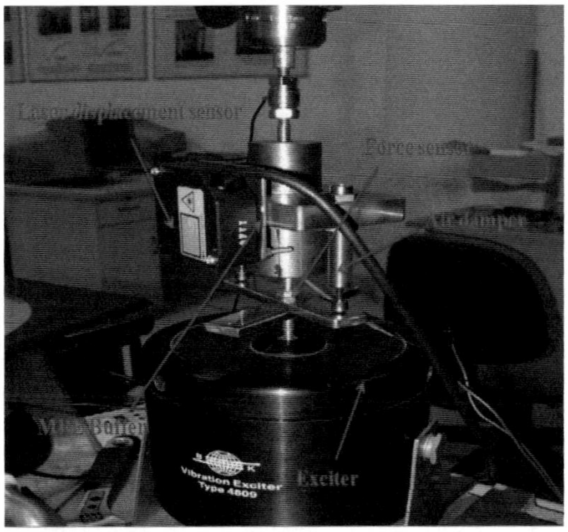

Figure 3. Test set up for MRE buffer.

(1) From the Figures 4-5, it can be seen that the elastic force will increase with the increase of the input current.
(2) It is well known the damping properties can be controlled by the application of magnetic field. However, because the enclosed area of the force-displacement is almost invariable in this study, it is considered the damping characteristic is invariable.
(3) It is noticed that the stiffness is not symmetric with respect to the displacement zero. The tests are repeated for several times. The same phenomenon still existed. The reason is the modulus is different for the case of compression and tension.
(4) In addition, on the basis of the test device in Figure 3, the time response of the MRE buffer under the triangular wave signal excitation is measured, and the excitation is the force. The measured results are shown in Figure 6. When the control current is applied during the back half period of the triangular wave, the force response is different from the shape of the excited signal. There is a time delay, which is about 7 ms. Therefore, it can be found that the response time of MRE buffer is about 7 ms, which is fast enough to formulate the later control strategy.

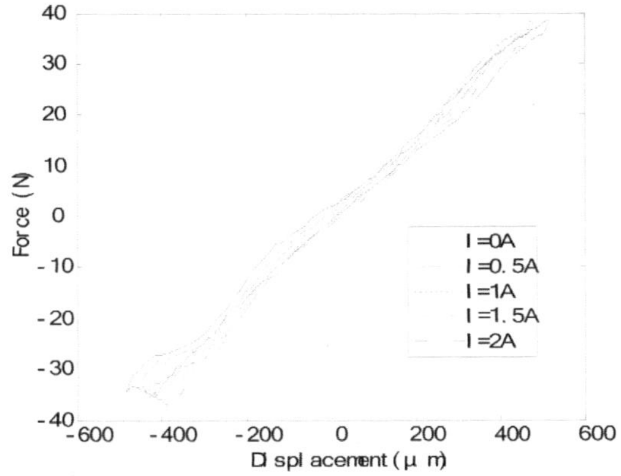

Figure 4. The relationship between force vs. displacement

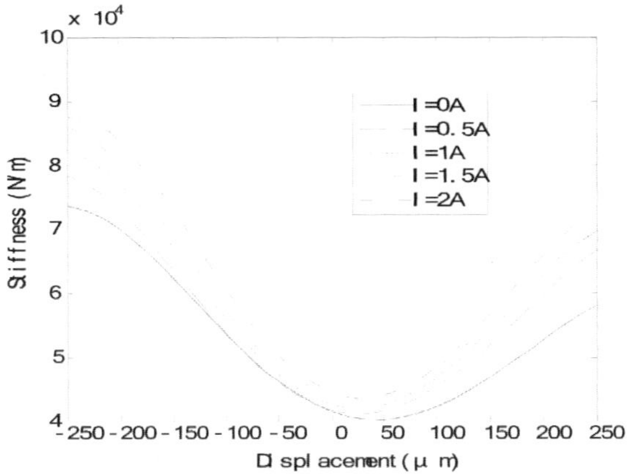

Figure 5. The relationship between stiffness vs. displacement

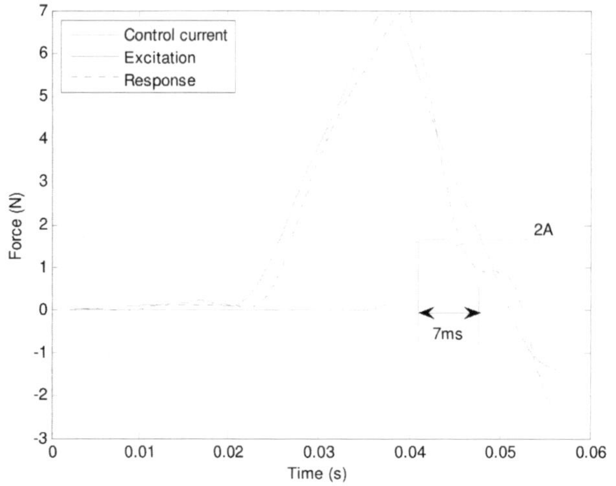

Figure 6. The time response of MRE buffer under triangular wave signal.

3. Dynamic model of MRE buffer system

MRE buffer system is a multi-degree-of-freedom complex vibration system. In order to research vibration performance and design of control law conveniently, it is necessary to simplify the MRE buffer system. Thus the following assumptions are given:

(1) MRE has the advantage of compact structure and light weight. In the work process, the shock of sample is larger than the one in normal operations. Therefore, in the force analysis of spring mass and unspring mass, the influence of gravity force should be ignored.

(2) By the properties of MRE, the shear modulus caused by magnetic field will increase with the exciting current. However, the loss factors have no significant change, and thus the equivalent damping coefficient of MRE may be regarded as constant.

(3) In work process, the shock suffered is mostly vertical without consideration of pitch vibration and tilt angle vibration of MRE.

Based on the above analysis and the designed MRE buffer structure in Figure 1, a two-degree-of-freedom impact dynamic model, which consists of a constant stiffness spring, a variable stiffness spring and a damper with constant damping coefficient in parallel, is constructed to check the performance of the MRE buffer in Figure 7.

According to the Newton's second law, the dynamic model can be derived as

$$M_s\ddot{y}_1 + (C_s + C_e)(\dot{y}_1 - \dot{y}_2) + K_e(y_1 - y_2) + F_{MRE} = 0 \quad (3)$$

$$M_u\ddot{y}_2 - (C_s + C_e)(\dot{y}_1 - \dot{y}_2) - K_e(y_1 - y_2) + K_u(y_2 - w) - F_{MRE} = 0 \quad (4)$$

Where M_s and M_u are the sprung mass the unsprung mass, respectively. C_s is the damping constant of an air damper. C_e is the equivalent damping constant of the MRE buffer. K_e and F_{MRE} are the fixed stiffness coefficient and the controllable stiffness force, respectively. y_1, y_2, w are displace of substrate, displace of sprung mass and displace of unsprung mass, respectively. $\dot{y}_1$, $\dot{y}_2$ are the derivatives of y_1, y_2, respectively, and $\ddot{y}_1$, $\ddot{y}_2$ are the second derivatives of y_1, y_2, respectively.

4. Human simulated intelligent controller development

In order to adjust the variable stiffness of the MRE buffer, an intelligent control algorithm, human simulated intelligent control (HSIC), is proposed, which has been applied successfully in the vibration control of the vehicle MR suspension in our prior works [13-15]. The original human simulated intelligent control (HSIC) algorithm, which was firstly proposed by Zhou *et al.* in 1983 [16], has become a fundamental and systematic method used in resolving some general industrial control problems for the last twenties years. In recent years, the HSIC theory has been further advanced through combing with the schema theory of modern cognitive science in recent years. After having studied for many years [17], HSIC theory based on SMIS has come into being. The theory provides a more effective and systematic method in resolving some complex control problems.

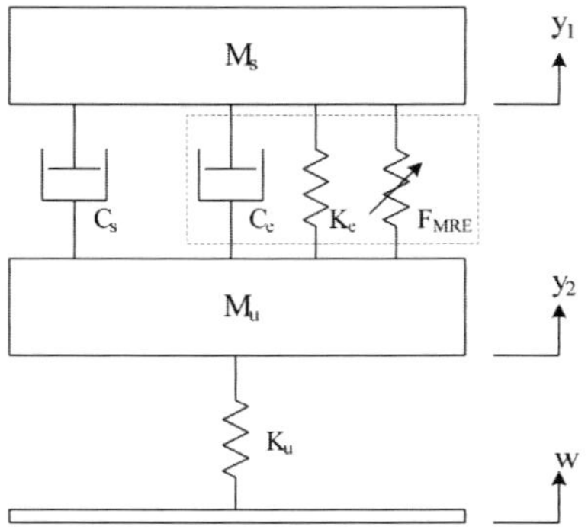

Figure 7. Two-degree-of-freedom impact dynamic model based on MRE buffer.

According to the general design methodology of HSIC theory, the sensed schema, the motion schema and the association schema are formulated in the following.

4.1. Sensed schema

The sensed schema can be described by

$$S_P = < R, Q, K, \otimes, \Phi > \tag{5}$$

in which R, Q, K, $\otimes$ and Φ are the input information set, the characteristic primitive set, the relation matrix, the operational symbol and the characteristic model set, respectively. The characteristic primitive set is selected according to the states of MRE buffer can be selected as

$$Q = [q_1, q_2, q_3, q_4, q_5]^T \tag{6}$$

where $q_1 \Rightarrow a\dot{a} \geq 0$, $q_2 \Rightarrow a\dot{a} < 0$, $q_3 \Rightarrow a > a_{set}$, $q_4 \Rightarrow a \leq a_{set}$, $q_5 \Rightarrow a > 0$, a is the impact acceleration value, a_{set} is the threshold value of acceleration. The relation matrix is set as

$$K = \begin{bmatrix} 1 & 0 & 0 & 1 & 1 \\ 0 & 0 & 1 & 0 & 0 \\ 0 & 1 & 0 & 1 & 1 \end{bmatrix} \tag{7}$$

As a result, the sensed modal can be written as,

$$\Phi = K \otimes Q = \begin{Bmatrix} \phi_1 \mid a\dot{a} \geq 0, a \leq a_{set}, a > 0, \\ \phi_2 \mid a > a_{set}, \\ \phi_3 \mid a\dot{a} < 0, a \leq a_{set}, a > 0 \end{Bmatrix} \tag{8}$$

4.2. Motion schema

The motion schema can be described as

$$S_M = (R, P, L, \Psi, U) \tag{9}$$

in which, $R \in \Sigma^d$, $P = [p_1, p_2, \cdots p_i, \ldots, p_n]^T \in \Sigma^n$, $p_i = f(R)$, $L \in \Sigma^{p \times n}$, $\Psi = \{\psi_1, \psi_2, \cdots, \psi_i, \ldots, \psi_p\} \in \Sigma^p$, $\psi_i = L_i \cdot P = \left\{ u_i = \sum_{j=1}^{n} l_{ij} p_j \right\}$, $U = L \cdot P$ are the input information set, the control mode primitive set, the coordination relation matrix, the control mode set and the control output, respectively.

In this study, the control modal includes three situations:

(1) The first control modal is selected as proportional control during the first phase when the acceleration of sprung mass is below the threshold value.

$$p_1 = f_e k_p a \tag{10}$$

where k_p is the proportional gain and f_e is the adjusting factor.

(2) Once the acceleration is bigger than the threshold value and the second control modal of bang-bang control is activated,

$$p_2 = -f_e^{'} k_m \tag{11}$$

where $f_e^{'}$ is also an adjusting factor, k_m is the maximum value of k_p.

(3) The third control modal is the proportional control during the acceleration of sprung mass is smaller than the threshold value and can be written as

$$p_3 = -f_e k_p^{'} a \tag{12}$$

In which, $k_p^{'}$ is the proportional gain.

The modal selection matrix is $L = \begin{Bmatrix} 1 & 0 & 0 \\ 0 & 1 & 0 \\ 0 & 0 & 1 \end{Bmatrix}$ and the control modal set is written as correspondingly .

$$\Psi_3 = L_3 \cdot P_3^{\ T} = \{p_1, p_2, p_3\} = \{\psi_1, \psi_2, \psi_3\} \tag{13}$$

4.3. Association schema

$$S_A = \{\Omega : \Phi \rightarrow \Psi\}, \Omega_3 = \{w_1, w_2, w_3\} \tag{14}$$

where w_j represents the reasoning procedure of *if* ϕ_j *then* ψ_j ($j = 1,2,3$).

In order to illustrate the HISC idea in the paper, Figure 8 is used to approximate acceleration response of MRE buffer under impact. In section I, states of MRE buffer are satisfied φ_3 in sensed schema, so the control strategy is chosen as p_3 in motion schema. Similarly, in section II (III), states of MRE buffer is satisfied φ_2 (φ_1) in sensed schema and control strategy is chosen as p_2 (p_1) in motion schema.

5. Numerical simulation

Before performing the drop test, the numerical simulation is undertaken. The dynamic model of equations (3)-(4) and HSIC controller are developed on the MATLAB/SIMULINK 7.1. In this study, the model parameters are shown in Table 1. These parameters are coming from the direct measurement of actual drop test system.

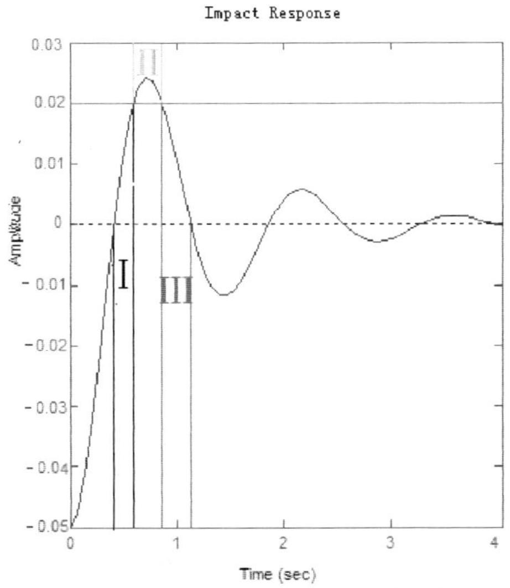

Figure 8. Acceleration response of MRE buffer under impact.

Table 1. The model parameters.

Symbols	Unit	Values
M_s	kg	2.85
M_u	kg	0.25
C_s	N/ms^{-1}	2000
C_e	N/ms^{-1}	200
K_e	N/m	41320
K_u	N/m	1.0×10^6

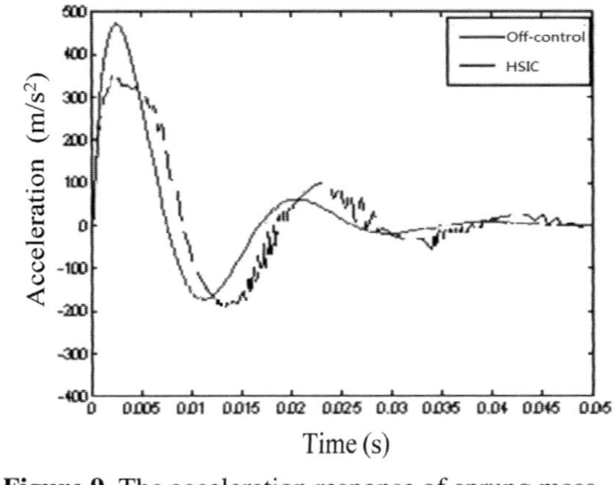

Figure 9. The acceleration response of sprung mass.

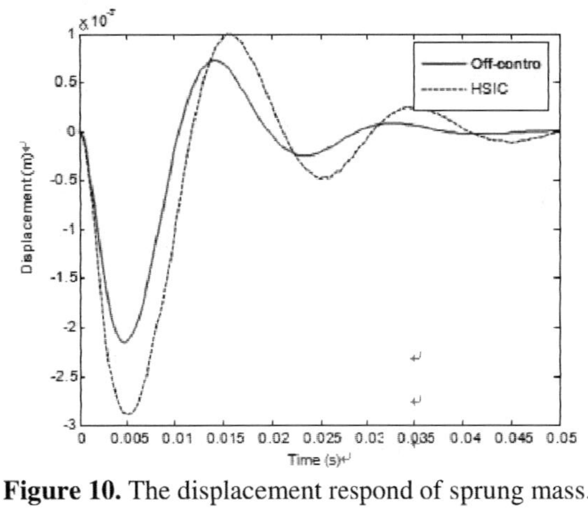

Figure 10. The displacement respond of sprung mass.

To check the control performance, the dynamic system is dropped numerically from a height of 10 cm. The results are shown in the Figures 9-10. From these two figures, it can be found that the maximum impact acceleration of sprung mass of system with control can effectively reduced compared to that without control. However, a slightly longer response time and a bigger displacement are required for the control case. The control output is shown in Figure 11.

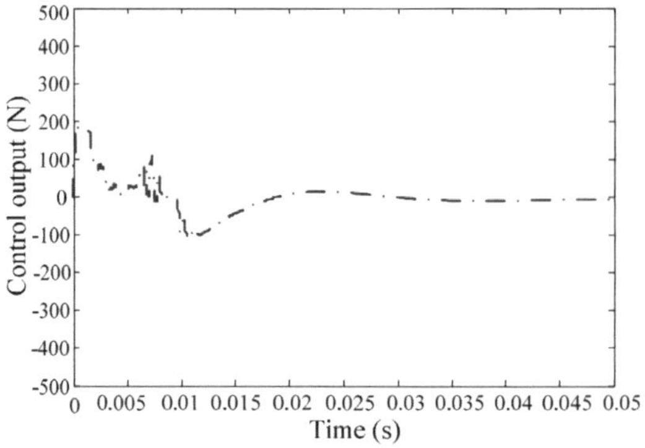

Figure 11. The control output.

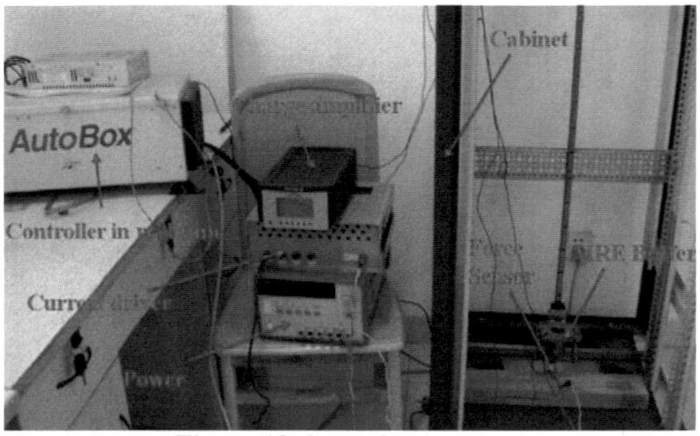

Figure 12. Setup for drop test.

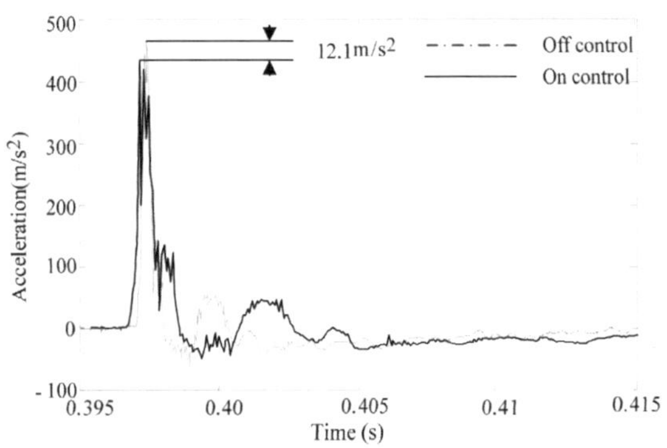

Figure 13. Acceleration time response of sprung mass.

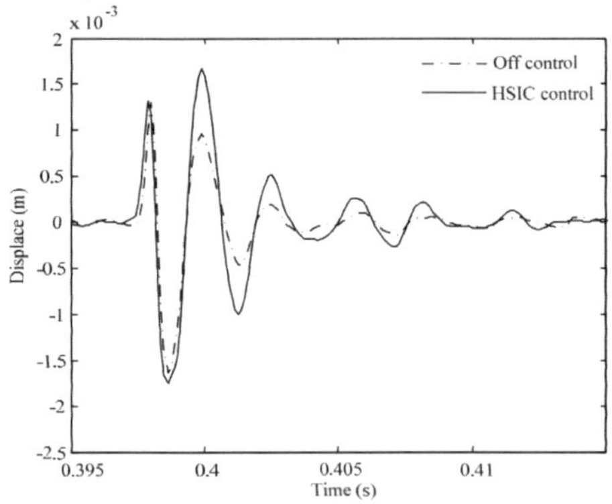

Figure 14. Displacement time response of sprung mass.

6. Drop test and result

To further validate the proposed MRE buffer, a drop test is carried out, which is shown in Figure 12. The spring mass and unsprung mass are connected by the MRE buffer. During the test, the absorbing system is sliding along a guide way vertically. The impact force between the unsprung mass and ground is measured by a force sensor. The acceleration of sprung mass is acquired by the acceleration sensor. In addition, the displacement of sprung mass is also measured by a laser displacement sensor. As comparison, the response of the absorbing system without control is also measured. The height of drop test is 10cm for each test, which is similar to the simulation.

(1) Figures 13 and 14 give the comparison of acceleration and displacement for sprung mass under on-control and off-control, respectively. According to the Figure 13, it can be seen that the maximum acceleration peak of sprung mass under control can be reduced by 12.1 m/s^2 compared to that of the off-control.
(2) Compared to the simulation, the control performance is limited. There are some reasons for this. For example, the time delay or response time of the MRE buffer is not considered and the control range is still very small. On the other hand, some similar phenomena can also be observed in the actual test, a longer adjusting time is required for both the numerical and actual tests.
(3) From Figure14, we can find that the MRE buffer under HSIC can not reduce the displacement response of sprung mass and requires much longer adjusting time, which are similar to that of the simulation. In addition, the amplitude of actual test data is larger than that of simulation. The reason may be that the static friction in absorbing system may be neglected in the simulation.

7. Conclusions

In this study, a variable stiffness buffer based on MRE was developed and its vibration buffer capability, which is described by the relations of force and displacement, stiffness and displacement, was evaluated. The measured results shows time responds is fast enough to formulate the control strategy. Meanwhile, the relation between force and current was approximately described as a quadratic function. A two-degree-of-freedom impact dynamic model is developed to describe the accurate MRE buffer dynamic characteristic. The segment controllers under HSIC strategy realized the impact suppression effectively. Through numerical and actual drop test, it showed that the proposed MRE buffer and controller can reduce the maximum impact acceleration comparing to the off-control case.

Acknowledgment

This research is supported financially by the National Natural Science Foundation of People's Republic of China (No.51073179, 61203098), the China postdoctoral science foundation (20100480043), and the Fundamental Research Funds for the Central Universities (CDJZR11 12 00 03). These supports are gratefully acknowledged.

References

[1] Carlson D J 2000 *Mechatronics* **10** 555-69.
[2] Goncalves F D, Koo J H and Ahmadian M 2006 *Shock and Vibration Digest* **38** 203-19
[3] Ioan B 2002 *J. Magn. Magn. Mater.* **241** (2-3) 196-200
[4] Xu Z D and Guo Y Q 2006 *J. Intel. Mat. Syst. Str.* **17** 871-81
[5] Gordaninejad F K, Barkan M and Wang X J 2007 *Proc. of the SPIE: Active and Passive Smart Structures and Integrated Systems 2007* **6525** 65250C
[6] Ginder J M and Nichols M E 2000 *Proc. of the SPIE* **3985** 418
[7] Deng H X, Gong X L and Wang L H 2006 *Smart Mater. Struct.* **15** N111-N116
[8] Deng H X and Gong X L 2008 *Commun. Nonlinear Sci.* **13** 1938-47
[9] Collette C, Kroll G and Saive G 2009 *Journal of Physics: Conference Series* **149** 1-4
[10] Liao G J, Gong X L, Kang C J and Xuan S H 2011 *Smart Mater. Struct.* **21** 1-11
[11] Opie S and Yim W 2011 *J. Intel. Mat. Syst. Str.* **22** 113-25
[12] Sun H L, Zhang K, Zhang P Q and Chen HB 2010 *Appl. Acoust.* **21** 250-57
[13] Dong X M, Yu M, Li Z S, Liao C R and Chen W M 2009 *J. Intel. Mat. Syst. Str.* **20** (7) 771-86
[14] Dong X M, Li Z S, Yu M, Liao C R and Chen W M 2010 *Control Theory & Applications* **27** (2) 249-56
[15] Yu M, Liao C R and Chen W M 2006 *J. Intel. Mat. Syst. Str.***17** 801-6
[16] Zhou Q J and Bai J K 1983 *Proc. of Multi-National Instrumentation Conf.* Part 1
[17] Li Z S, Zhang H, Wen Y L and Wang G P 2004 *Proc. of the 5th World Congress on Intelligent Control and Automation* 2423-27

On the magnetic field and temperature monitoring of a solenoid coil for a novel magnetorheological elastomer base isolator

Y Li[1], J Li and B Samali

Centre for Built Infrastructure Research, Faculty of Engineering and Information Technology, University of Technology Sydney, NSW 2007, Australia

Email: yancheng.li@uts.edu.au

Abstract. Following a successful experimental validation of a magnetorheological elastomer (MRE) base isolator, this study presents one of the major concerns, the heating of the magnetic coil, in the design and development of the adaptive MRE based isolator. In this research, the MRE materials, with a total thickness of nearly 150 mm, are placed as the magnetic core of the device to best utilize the magnetic energy provided by the coil. A series of tests are undertaken to investigate the magnetic fields inside the coil with or without the MRE materials. Thermocouples are used to monitoring the surface temperature of the coil when it is applied with various currents for 10 min. It is shown that the measurement of field inside the solenoid when no MRE is placed inside agrees with the theoretical analysis. It is also shown that the temperature of the coil increase dramatically when a current is applied. Cooling of the coil may takes even longer, about 4 h, till down to the room temperature. Dropping of the magnetic field is observed when the temperature goes high.

1. Introduction

Finding engineering applications for a new class of smart material, magnetorheological elastomer (MRE), has been a major task for researchers in this field. Novel MRE devices, such as vibration absorbers [1-2] and vibration isolators [3-4], have been proposed and fabricated to pioneer its engineering applications. In civil engineering, Li [5] has proposed a novel MRE based isolator to be used in the base isolation system for mitigating the devastating effects of earthquakes on civil structures.

For any MRE-based device, electromagnetic coil is evitable involved to provide magnetic field for the MRE materials. Comparing with magnetic circuit design in MRF device, i.e. MR damper, MRE devices normally need a larger coil to energise the MRE materials, particularly for a large-scale MRE device. Therefore, investigation of the solenoid on the magnetic field distribution is of great importance. In particular, provision of sufficient and uniform magnetic field is essential towards the success in designing MRF/MRE devices. To understand the mechanism of magnetic field generation in a solenoid is the key for device design and optimisation.

Large electromagnetic coil has high electric resistance. Thus, heating becomes a great concern for the engineering design. Breese and Gordaninejad [6] presented theoretical and experimental studies on heat generation and dissipation of controllable MR fluid shock absorbers. Dogruoz *et al.* [7] performed

[1] Corresponding author.

theoretical and experimental analysis for enhancing heat transfer from MR fluid dampers using fins. Kavlicoglu *et al.* [8] studied heating of a high-torque, multi-plate magnetorheological (MR) fluid limited slip differential (LSD) clutch. These researches focused on the influence of the temperature rising on the performance of the ER/MR devices. Little attention has been paid to the electromagnetic coil itself.

The objective of this research is to investigate the magnetic field distribution and heating performance of a solenoid. Finite element analysis is explored to find the magnetic field distribution in the solenoid. To validate the analytical analysis, an experimental research is conducted to compare with the numerical findings. Finally, temperature monitoring on the solenoid under various applied current inputs is presented.

2. Electromagnetic coil for the magnetorheological elastomer based isolator

The solenoid is designed to provide magnetic field for the MRE based isolator. The sketch map of the solenoid is shown in Figure 1. The solenoid is made of electromagnetic coil and thin non-magnetic support as illustrated in the Figure. The electromagnetic coil is a cylinder shape configuration and is winded by 1.2 mm copper wire. The cylinder shape non-magnetic support is made of epoxy material and was folded into I-shape to protect the coil. Detailed structural parameters of the solenoid are shown in Table 1. The coil is firmly attached with the epoxy support. The total winding number of the coil is 3100 turns. The diameter of the coil wire is 1.2 mm and the total length of the wire is 2.1 km. The wire is made of copper and the resistance of the coil is 32.3 Ω.

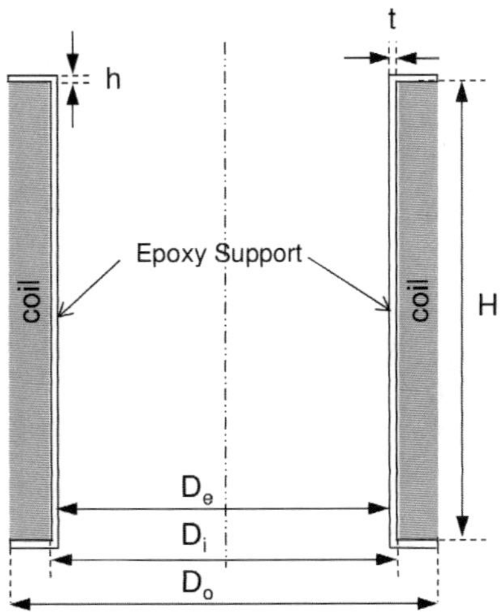

Figure 1. Cross-section of the solenoid.

Table 1. Structural parameters of the solenoid.

Item	**Size** (mm)
Inner diameter of the epoxy support D_e	192
Inner diameter of the coil D_i	196
Outer diameter D_o	232
Height of the epoxy wall H	276
Thickness of the cylinder epoxy wall t	2
Thickness of the top epoxy plate h	2

3. Finite element analysis of the solenoid

To analyse the magnetic field inside the solenoid, a finite element model is developed in Ansoft Maxwell. The FE model is shown in Figure 2. The material properties of the coil and the coil and the epoxy support are properly selected according to the design. Figure 3 shows the magnetic field distribution inside the solenoid when the applied current is 5 A. Figure 4 and 5 are to quantify the magnitude of the magnetic field inside the solenoid, both axially and longitudinally. It can be seen that the magnetic field intensity in the longitudinal middle of the solenoid is higher than in the top and bottom of the solenoid. The magnetic field in the longitudinal-middle is also more uniform than other locations.

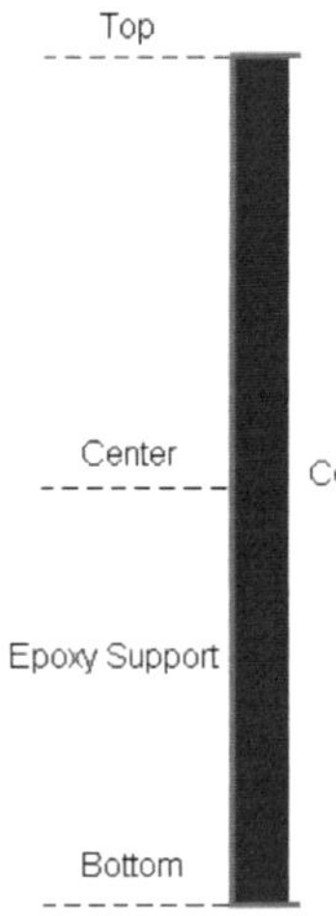

Figure 2. Finite element model in Ansoft Maxwell.

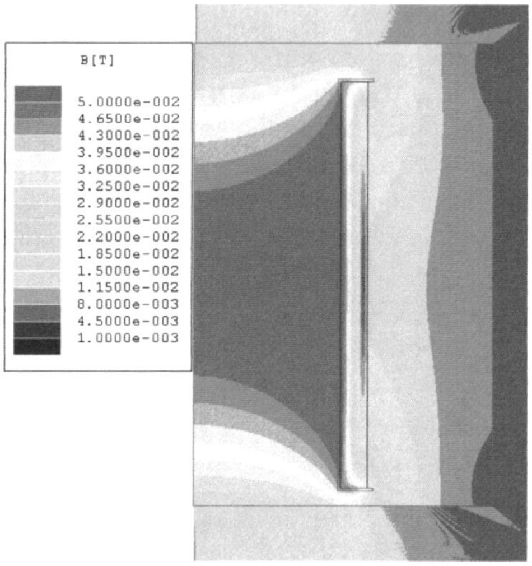

Figure 3. Magnetic field inside the solenoid (I=5A).

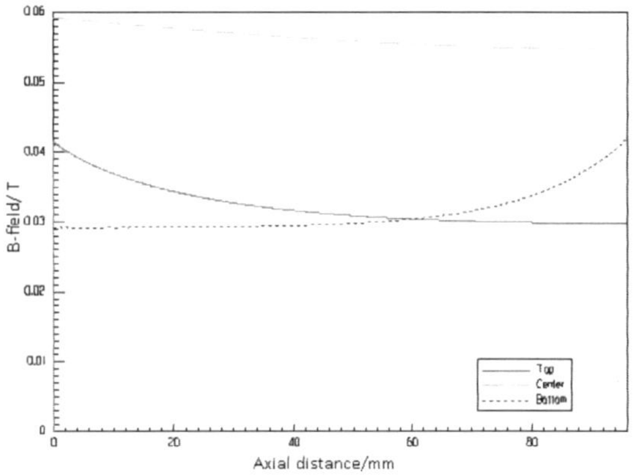

Figure 4. Magnetic field along axial direction (I=5A).

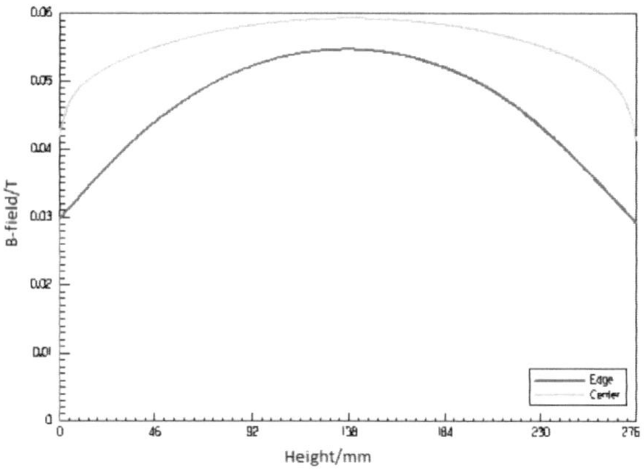

Figure 5. Magnetic field along longitudinal direction (I=5A).

4. Experimental testing

4.1. Experimental setup

To experimentally verify the findings from finite element analysis, an experimental testing is set up as shown in Figure 6. It consists of a solenoid, a DC power supply, a host computer, a magnetic field sensor and a temperature monitor. Details of the solenoid can be found at section 2. The DC power supply, with capacity of 250 V and 10 A, provides the solenoid with required currents from 0 A to 5 A. The magnetic field sensor, IDR-325 Gaussmeter, from Integrity Design and Research Corp, USA, is used to measure the magnetic field inside the MRE layer. The temperature monitor, model YC747UD-k type, is a four-channel thermometer with resolution of 0.1°C and capacity of -100°C to 1300°C. All the data from the magnetic field sensor and the thermometer is recorded by the host computer.

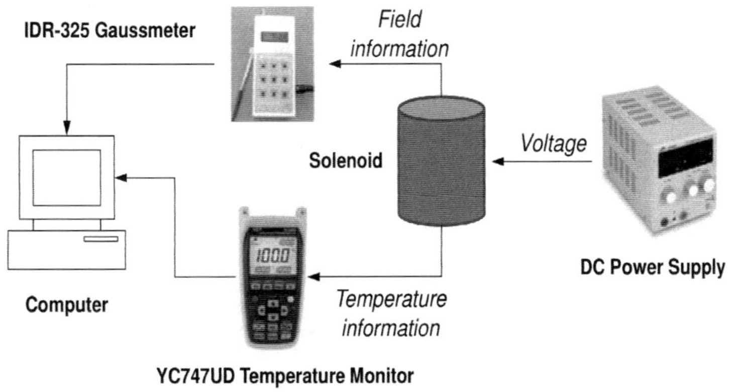

Figure 6. Experimental setup for the magnetic field and temperature measurement.

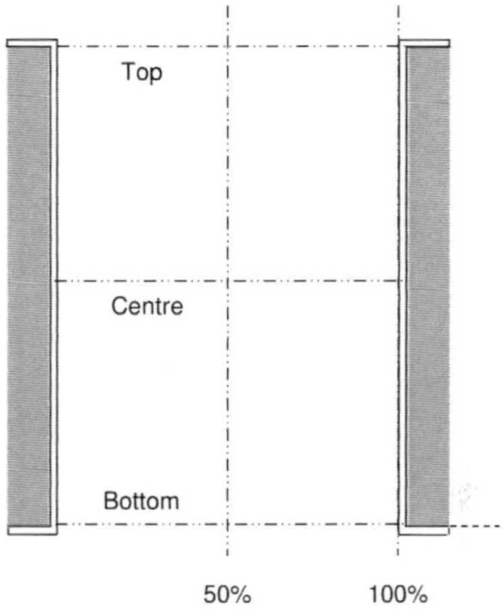

Figure 7. Baselines for the analysis.

4.2. Results and discussion
Figure 8 shows the magnetic field distribution inside the solenoid when the currents applied to the coil are 1 A, 2 A, 3 A and 4 A. The meanings of the terms used in the figures, i.e. "50%", "100%", "Top", "Centre" and "Bottom" are shown in Figure 7. For example, 50% top means the location in the top of vertical centre line inside the solenoid. Comparisons between the finite element analysis and measurements are also listed in Figure 8. It is quite clear to see that the measurements of the magnetic fields follow the simulation results very well. With the increasing of the applied currents, the magnetic field appears a linear increasing trend.

Figure 9 shows the magnetic field distribution along vertical directions inside the solenoid. It can be seen that the maximum magnetic field is in the centre of the solenoid. As indicated from the finite element analysis and the test results, the field intensity here is high and uniform.

Figure 10 is the temperature monitoring results when a certain current pass through for 10 min. When the applied current I = 4 A, the temperature of the coil rises from 25°C to 50°C within 10 min. For the cooling phase, after removal of the current, it will last a long period till back to the room temperature, as shown in Figure11. For the case when a 3 A current is applied to the coil for a certain period of time and the coil temperature is above 40°C, it will last nearly 5 h till fully cooling down. As we know, the MR elastomer will have a weaken MR effect at high temperature, therefore, optimal design of the coil should be considered to find the compromise between the achieved magnetic field and the heating caused by the input current.

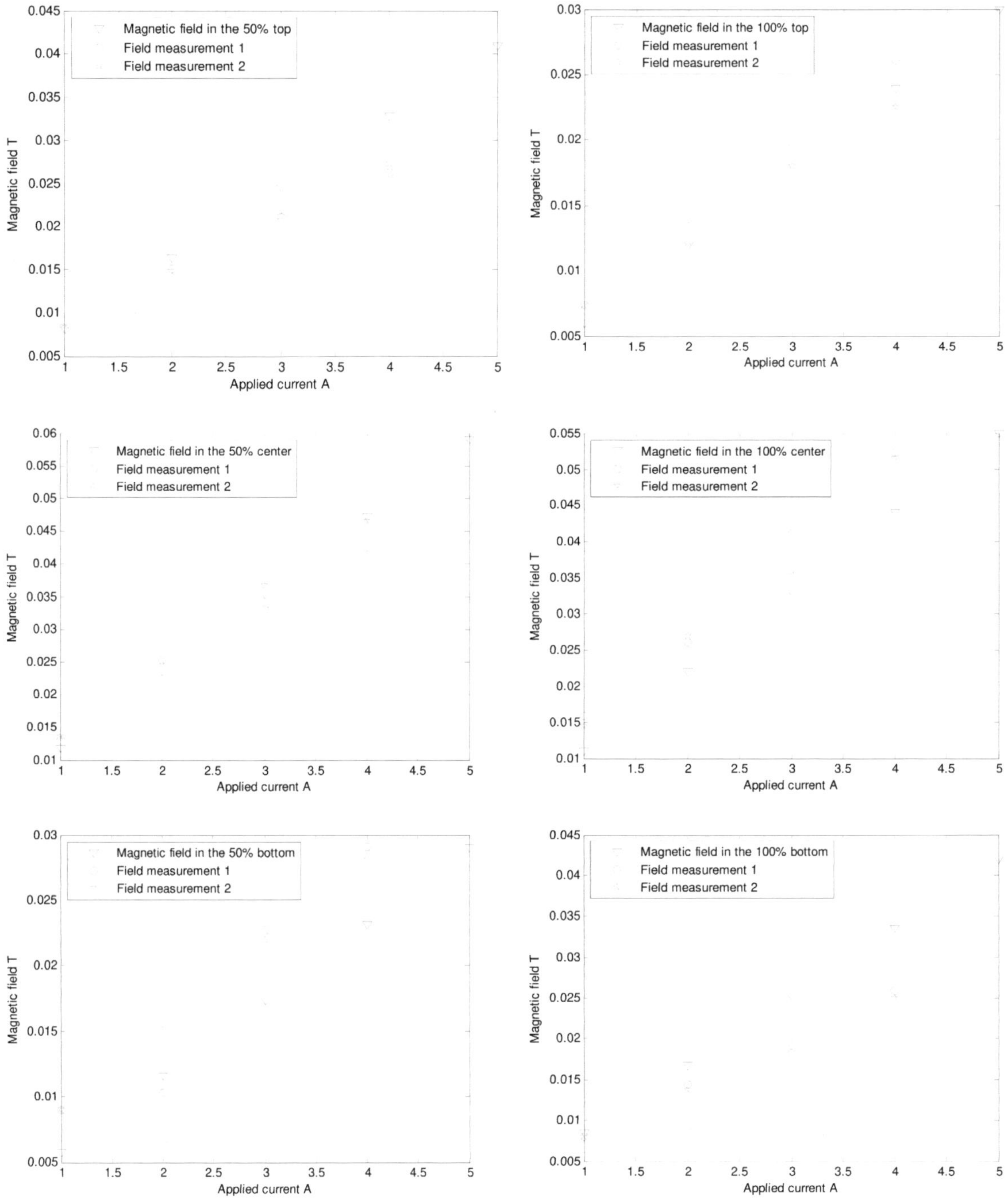

Figure 8. Magnetic field measurements vs. finite element analysis.

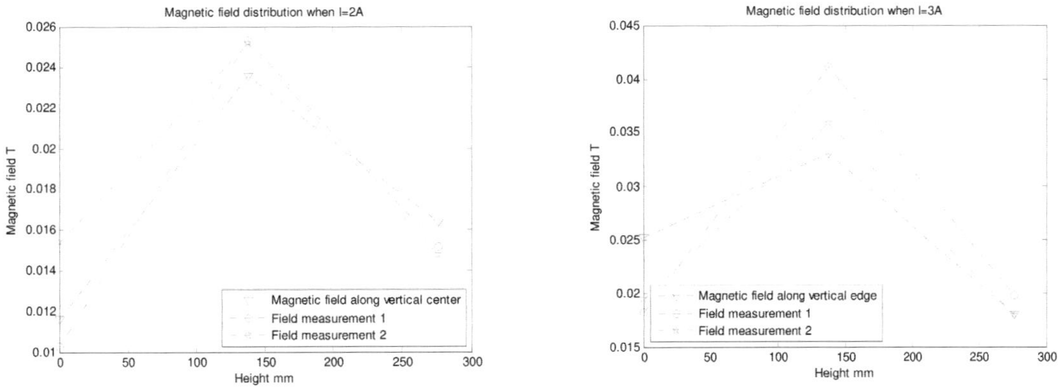

Figure 9. Magnetic field distribution along vertical directions.

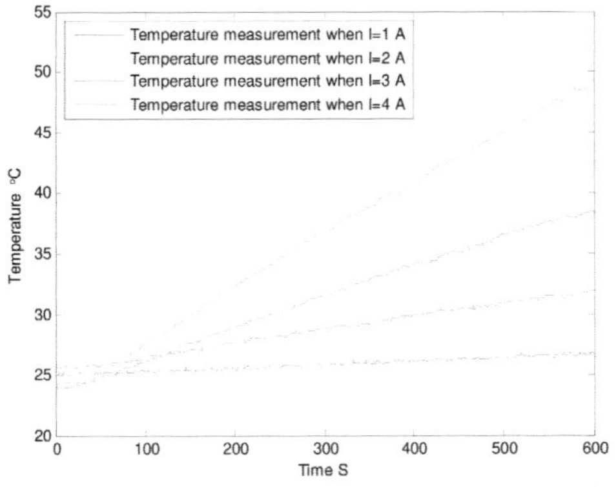

Figure 10. Temperature rising for various current inputs.

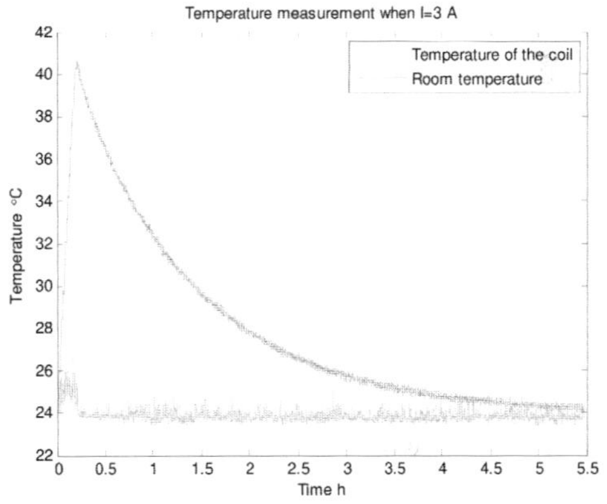

Figure 11. Temperature measurement for I=3A.

5. Conclusions

This study examined the magnetic field distribution and heating of a solenoid for a MR elastomer based isolator. Both finite element analysis and experimental testing were used for the magnetic field analysis in a solenoid. Experimental research on temperature monitoring was also conducted for the solenoid.

Acknowledgement

The authors gratefully acknowledge the financial support by the Early Career Researcher Grant (Grant No. 2032042) and the Chancellor's Postdoctoral Research Fellowship (Grant No. 2032032) from the University of Technology Sydney for this research.

References

[1] Ni Z C, Gong X L, Li J F and Chen L 2009 *J. Intell. Mat. Sys. Struc.* 20 1195
[2] Liao G J, Gong X L, Kang C J and Xuan S H 2011 *Smar. Mat. Struc.* 20 075015
[3] Lerner A A and Cunefare K A 2008 *J. Intell. Mat. Sys. Struc.* 01. 551
[4] Deng H and Gong X 2008 *Comm. Nonlin. Sci. Numer. Simu.* 13 1938
[5] Li Y C, Li J C and Samali B 2012 *Australian Provisional Patent* No. 2012904337
[6] Gordaninejad F and Breese D G 1999 *J. Intell. Mat. Sys. Struc.* **10** 634

[7] Dogruoz MB, Wang E L, Gordaninejad F and Stipanovic A J 2003 *J. Intell. Mat. Sys. Struc.* **14** 79

[8] Kavlicoglu B, Gordaninejad F et al 2003. *Proc. SPIE Smart Struct. Mater. Con.: Indust. and Comm. Appl.* **5054** 341

On the mobility of iron particles embedded in elastomeric silicone matrix

R Rabindranath[1] and H Böse
Fraunhofer ISC, Neunerplatz 2, 97082 Würzburg, Germany

E-mail: raman.rabindranath@iscfraunhofer.de

Abstract. In this contribution the rheological and magnetorheological properties of different polydimethylsiloxane (PDMS) based magnetorheological elastomers (MRE) are presented and discussed. In order to investigate the mobility of the iron particles with respect to the rheological characteristics, the iron particles were silanized with vinyltrimethoxysilane to enable a reaction between the modified particle and the cross-linking agent of the silicone elastomer. In addition, the vinyl-functionalized particles were further modified by the coupling of the superficial vinyl groups with a long-chain hydride terminated PDMS, which enables a reaction pathway with the vinyl terminated PDMS. On the other hand, the iron particles were treated with surfactants such as fatty acids, calcium and aluminum soaps, respectively, prior to vulcanization in order to increase the mobility of the iron particles in the elastomeric matrix. It was found, that both, the modification with the long-chain hydride terminated PDMS as well as the treatment with surfactants lead to an increase of the storage modulus G', the loss modulus G'' and the loss factor tan δ in the magnetic field. It is concluded that both modifications, the coupling with long-chain hydride terminated PDMS as well as the treatment with surfactants, provide a greater mobility of the iron particles and hence a greater friction represented by the increase of the loss factor tan δ. Consequently it is assumed that untreated iron particles are less mobile in the rubber matrix due to covalent bonding with the silicone components, most likely due to the reaction of the hydroxyl groups on the metal surface with the silane groups of the cross-linking agent.

1. Introduction

Magnetorheological elastomers (MRE) represent a new class of smart materials whose rheological properties can be tuned upon exposure to an external magnetic field. MRE displays a composite material consisting of homogeneously distributed or columnar structured ferromagnetic particles embedded in a solid or gel-like, non-magnetic elastomeric matrix. The elastomeric matrix can be polyurethane, natural rubber, synthetic rubber and most of all polydimethylsiloxane [1-3]. The preparation of MRE with chain-like columnar structures requires the alignment of the ferromagnetic particles which is achieved upon exposure to a magnetic field, prior to or during the curing process.

MRE exhibit several interesting properties such as a tunable Young's, storage and loss modulus, which render this material a suitable means for a variety of new applications. Possible applications of MRE comprise adaptive damping, vibration extinction and magnetic field induced actuation [4,5]. In order to gain a deeper understanding on the MRE concepts, which is essential for the tailoring of materials towards a specific application, the mobility of the iron particles within the elastomeric

[1] Corresponding author.

matrix was investigated. For this purpose, the mobility of the magnetic particles within the elastomeric matrix was modified in two different ways. It was assumed that the modification of the particles with a surfactant increases their mobility whereas the attachments of PDMS-linker with functional groups, which enable reaction with the silicone matrix, decrease their mobility. In this study, two different industrial surfactants as well as two different PDMS-linkages were applied and the rheological properties of the MRE made thereof were compared.

2. Experimental

Vinyl-terminated PDMS, hydride-terminated PDMS, platinum-divinyltetramethyldisiloxane complex (Karstedt catalyst) and silicone were purchased from different commercial supplieres. Spherical carbonyl iron powder (CIP), type EM, was supplied by BASF. All the materials were used without further purification. The characterization was carried out on the Rheometer MCR 300 and MCR 501 from Anton Paar, Austria.

The MRE prepared in this work consist of 30 Vol.% on CIP homogenously dispersed in a home-made silicone rubber formulation. The rubber formulation consists of vinyl-terminated PDMS, hydride- functionalized cross-linker and a special plasticizer mixture in order to ensure full solubility of the surfactants. The curing, i.e. hardening process of the rubber formulation and the composite, respectively, is accomplished upon the Pt-catalyzed polyaddition at elevated temperatures using Karstedt's catalyst. The reaction scheme of the cross-linking reaction is shown in Scheme 1.

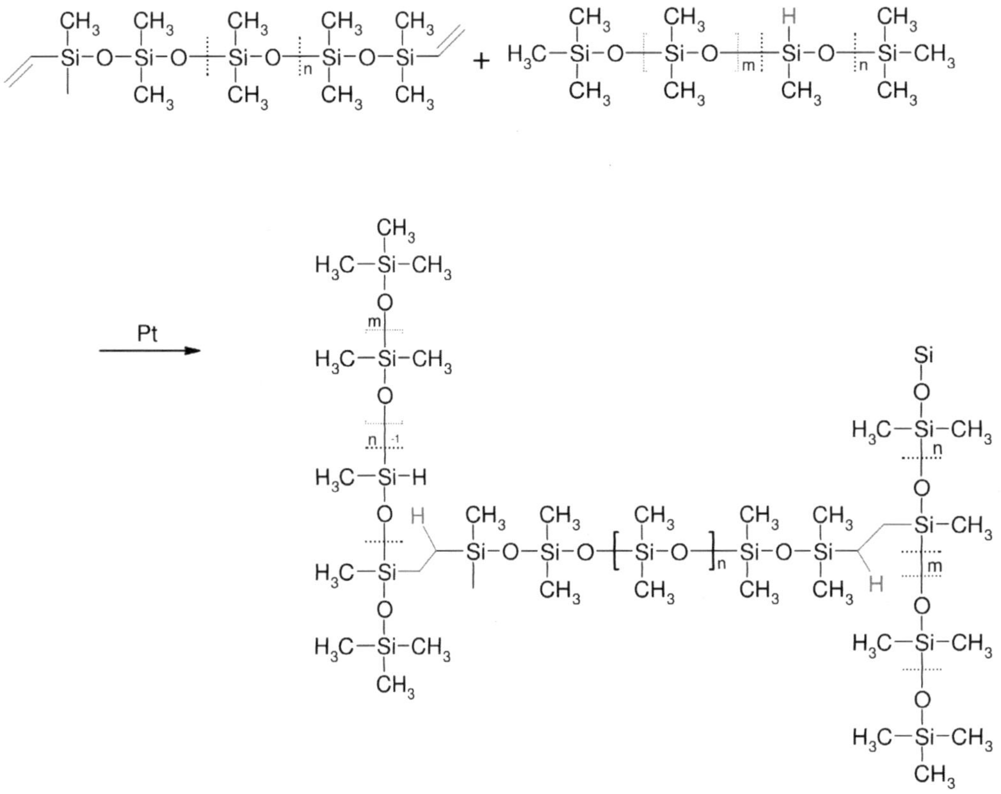

Scheme 1. Cross-linking reaction through Pt-catalyzed hydrosilylation.

The modification via the silanization route of the CIP is depicted in Scheme 2. The silanization is typically carried out in pure ethanol with 1% water added under vigorous stirring of the suspended CIP over a period of six hours at room temperature. At first, the methoxy groups of vinyltrimethoxysilane are hydrolyzed *in-situ* under conversion into silanol groups and release of methanol. In a second step, the silanol undergoes a condensation reaction with the hydroxyl groups on the iron surface leading to the formation of a covalent bond between the iron particle surface and the silicone compound.

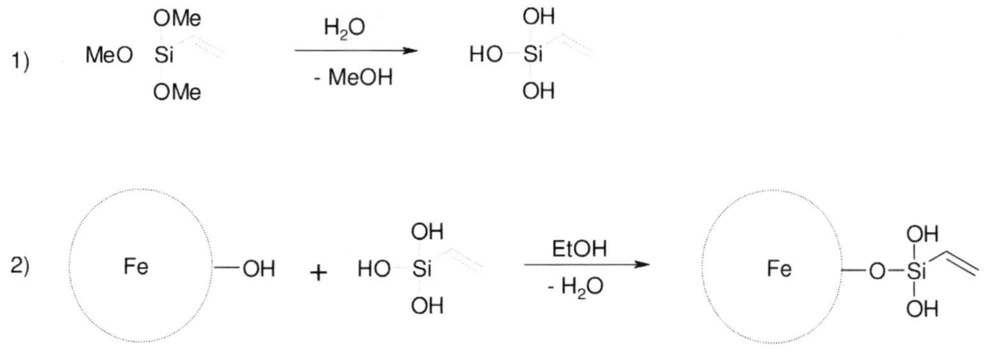

Scheme 2. Silanization of CIP with vinyltrimethoxysilane.

After completion of the silanization, the CIP was simply filtered off, thoroughly washed with ethanol and dried in vacuum. One part of the vinyl-functionalized CIP was then directly processed into MRE containing 30 vol.% of modified CIP.

The other part of the vinyl-functionalized CIP was subjected to a further modification step according to Scheme 3. For this purpose, the vinyl-functionalized CIP was suspended in hexamethyldisiloxane (volatile, low molecular weight silicone oil) and was coupled with a long-chain hydride-terminated PDMS using the Pt-catalyzed hydrosilylation. The reaction was carried out under vigorous stirring with a large excess of hydride-terminated PDMS at 80°C for a period of four hours. After this period, the product was filtered off and thoroughly washed with hexamethyldisiloxane until the filtrate was free of any residues. The final product was then dried under vacuum and processed into MRE.

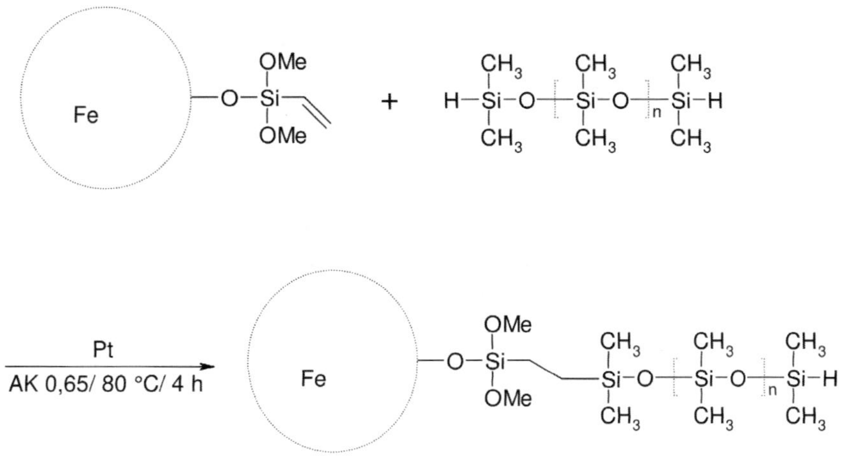

Scheme 3. Modification of vinyl-functionalized CIP with long chain hydride-terminated PDMS.

In contrast to the silanization procedure, the modification of CIP with surfactants is accomplished simply by adding smallest amounts of the surfactant to the rubber formulation. When adding surfactants to the rubber formulation care must be taken in the amount added. It is important to add an effective amount of surfactant for the steric shielding of the particles but the amount must not exceed a certain level at which the degree of cross-linking is altered significantly. The effect of the surfactant addition is displayed by a slightly reduced viscosity of the liquid rubber formulation, shown in Figure 1. In order to ensure full solubility of the surfactants, a solubilizer was incorporated into the rubber formulation. All MRE prepared in this study are based on the same rubber formulation.

3. Results and discussion

Figure 1 displays the storage modulus (G′) and the loss modulus (G″) of modified and unmodified MRE in dependence on the magnetic flux density. Obviously, the dispersant-modified samples and the long-chain modified sample exhibit the same rheological behavior even though the modifications are completely different. Nevertheless, with rising magnetic flux density these modifications lead to increased values of the storage modulus and to a greater maximum of the loss modulus resulting in an increased loss factor (Figure 2).

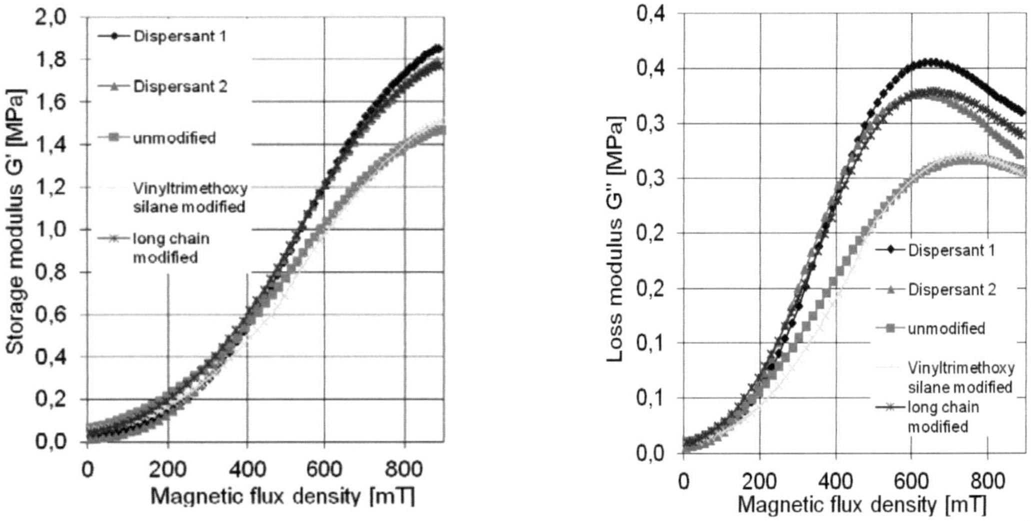

Figure 1. Storage modulus and loss modulus of modified and unmodified MRE samples in dependence of magnetic flux density.

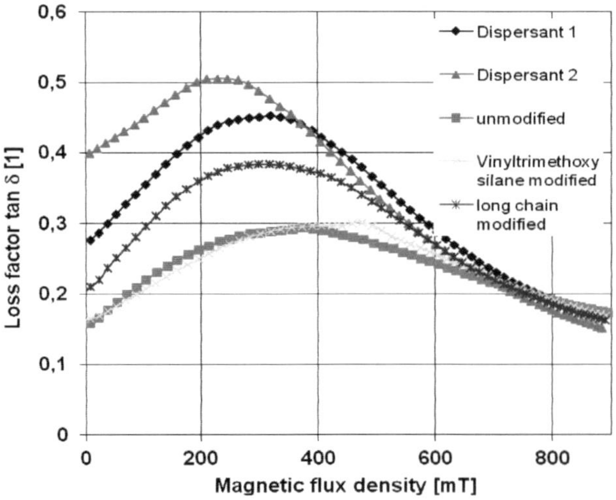

Figure 2. Loss factor of modified and unmodified MRE samples in dependence of magnetic flux density.

On the other hand, the unmodified reference sample and the vinyltrimethoxysilane-modified sample, with a short linkage to the elastomeric matrix, exhibit the same rheological behavior. It is concluded that the mobility of the iron particles in the surfactant-modified sample compares to the mobility in the long-chain modified sample, whereas the mobility in the unmodified reference sample compares to the mobility in the vinyltrimethoxysilane-modified sample. As a consequence, it has to be assumed that even the unmodified particles bond to the elastomeric matrix and hence exhibit a limited mobility comparable to the vinyltrimethoxysilane-modified particles, in which the short silicone linkage does not provide any greater freedom of movement. In reverse, the surfactant- and the long-chain modified sample do provide mobility

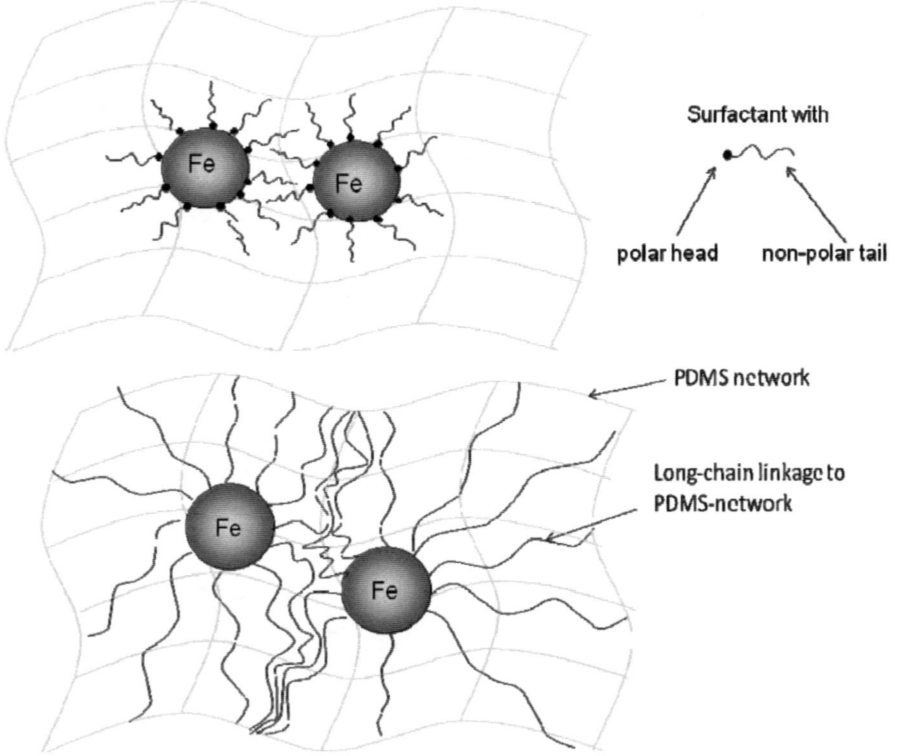

Figure 3. Schematic conditions of modified particles in MRE.

While the surfactants arrange around the particle upon association of the polar surfactant head with the polar hydroxyl groups and thus preventing reaction with the elastomeric matrix, the long-chain modified particles are bond to silicone network through long, coiled and flexible PDMS chains, as schematically shown in Figure3. As a result, both kinds of modifications lead to greater mobility of the iron particles embedded. The arrangement of initially evenly distributed iron particles of different mobility in the magnetic field is illustrated in Figure 4. While the increased storage modulus at a magnetic flux density above 400 mT is attributed to the more pronounced structuring and more dense arrangement of the particles within the magnetic field, the increased loss module represents the dissipative heat loss due the enhanced mobility resulting in greater overall friction. Furthermore, the greater mobility of these modifications results in softer MR elastomers which are revealed by the lower storage modulus without magnetic field. Due to the lower storage modulus in the pristine state and the higher storage modulus in the magnetized state the dispersant and the long-chain modified MR elastomers exhibit an enhanced increase factor in comparison with the unmodified reference sample.

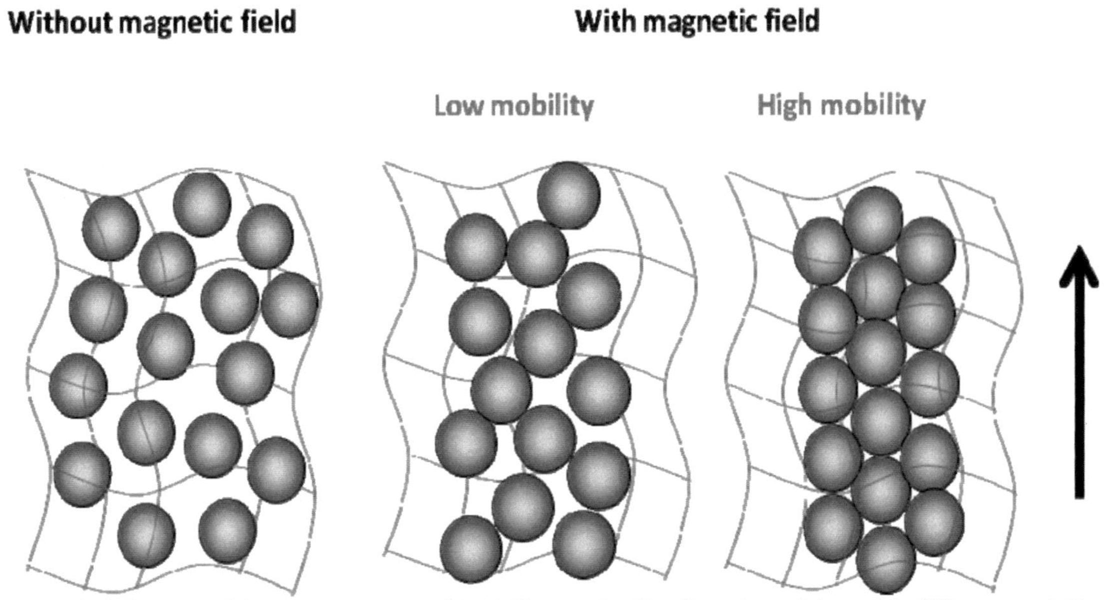

Figure 4. Schemes of the arrangements of initially evenly distributed particles with different mobility in MRE in the magnetic field (arrow).

In order to support the assumption that unmodified particles are bound to the elastomeric network, a common stress-strain experiment using the pure silicone rubber formulation without CIP and a MRE sample made thereof was carried out. With this experiment it is possible to assess the reinforcing properties of a filler material which is displayed by the increase of mechanical work required to break the sample. The reinforcing properties of a filler material are ascribed to the increase of the cross-linking degree due to the bond formation between filler and elastomeric network. Figure 5 shows the stress-strain behavior of the pure elastomer and the corresponding MRE made thereof. The work required to break the sample is represented by the integral of the individual curves. It is clearly shown that the incorporation of CIP leads to a significant increase of work required to break the sample.

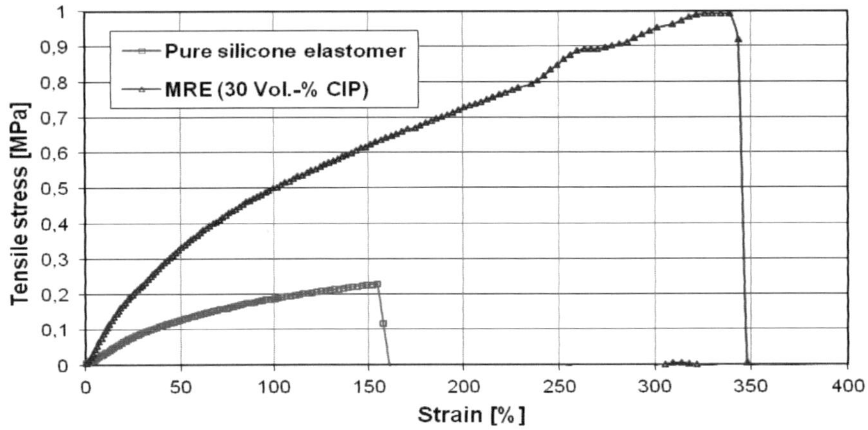

Figure 5. Tensile stress vs. strain of pure and iron particle filled silicone elastomer.

4. Conclusion

The incorporation of surfactants into a MRE admixture as well as the modification of CIP with long-chain hydride-terminated PDMS are suitable means to increase the mobility of the iron particles within the elastomeric matrix. The increased mobility leads to a softer material in the pristine state and to an enhanced stiffening in the magnetized state and hence, to a greater switching factor. Furthermore, it was demonstrated that unmodified CIP particles exhibit reinforcing properties towards addition cure silicones probably due to bond formation between hydroxyl groups on the metal surface and the reactive moieties of the cross-linker. This fact might be considered when designing MRE with a low loss module.

References

[1] Wu J, Gong X, Fan Y and Xia H 2010 *Smart Mater. Struct.* **19** 105007
[2] Chen L, Gong X L, Jiang W Q, Yao J J, Deng H X and Li W H *J. Mater. Sci.* **42** 5483-89
[3] Böse H 2007 *Int. J. Mod. Phys. B* **21** 4790-97
[4] Ginder J M, Schlotter W F and Nichols M E 2001 *Proc. SPIE* p 433
[5] Böse H, Rabindranath R and Ehrlich J 2011 *Proc. 12th Int. Conf. on Electro-Rheological Fluids and Magneto-Rheological Suspensions* (World Scientific) pp 116-122

Predicating magnetorheological effect of magnetorheological elastomers under normal pressure

X Dong[1, 3], N Ma[2], J Ou[2] and M Qi[1]

[1] School of Materials Science and Engineering, Dalian University of Technology, Dalian 116024, People's Republic of China
[2] School of Civil Engineering, Dalian University of Technology, Dalian 116024, People's Republic of China

E-mail: dongxf@dlut.edu.cn

Abstract. Magnetorheological elastomers (MREs) present reversible change in shear modulus in an applied magnetic field. For applications and tests of MREs, a normal pressure must be applied on the materials. However, little research paid attention on the effect of the normal pressure on properties of MREs. In this study, a theoretical model is established based on the effective permeability rule and the consideration of the normal pressure. The results indicate that the normal pressure have great influence on magnetic field-induced shear modulus. The shear modulus of MREs increases with increasing normal pressure, such dependence is more significant at high magnetic field levels.

1. Introduction

Magnetorheological (MR) effect is the magnetic field-induced change of rheological properties that exhibited by MR materials. The effect was first discovered by Rabinow in 1948 [1]. Magnetorheological fluids (MRFs) prepared by dispersing large amounts of micro-sized magnetic particles in a non-magnetic liquid are the most common MR materials [2]. They show a reversible and fast transition from a liquid to a nearly solid state when a magnetic field is applied. Therefore, their yield stress and apparent viscosity can be changed by changing magnetic field. This unique property has made them apply in mechanical systems and civil engineering structures that require the transmission of torque or the control of vibration [3-6]. However, the particles tend to settle out in the liquid due to the density mismatch between the particles and the liquid, and the leakage of the fluids must be avoided by seal during their service [7]. Those shortcomings prevent them from certain applications.

Magnetorheological elastomers (MREs) are solid analogs of MR fluids. The most significant difference between MREs and MRFs is that the matrix of MREs is made of natural or synthetic rubber rather than a liquid. As a new branch of MR materials, the shear modulus of MREs can be controlled rapidly and reversibly by the application of an external magnetic field [8-13]. Compared with MRFs, MREs present several merits. For example, the magnetic particles in MREs do not undergo sedimentation, and the devices based on MREs do not need to be sealed. Those advantages make MREs to be widely used in the field that need controllable stiffness, such as automotive bushings,

[3] Corresponding author.

tuneable stiffness engine mounts, variable impedance surfaces, and adaptive tuned vibration absorbers [14-16]. MREs can be divided into isotropic and anisotropic based on different curing conditions. The anisotropic MREs are prepared by applying a magnetic field during the cure process of the composites, which gives chain-like columnar structures for the particles in the matrix. Generally, the MR effect of the anisotropic MREs is larger than the isotropic ones, the particles in which are homogeneously distributed [17].

Since 1995, many researchers predicted the properties (especially the field-induced modulus) of MREs by theoretical models. Most of the models are established based on the magnetic dipole interactions between two adjacent particles of the chain. Jolly established a model based on the dipolar interaction of particles within an assumed structure, and the magnetic non-linearities as applied field increase was taken into consideration [18]. Davis used finite element analysis method to predict the shear modulus under a varied magnetic field, the results of which indicated the increase in shear modulus due to inter-particle magnetic forces at saturation is about 50% of the zero-field modulus, and the optimum particle volume fraction for the largest relative MR effect at saturation is 27% [19]. Shen *et al.* took all the dipole interaction in a chain and the nonlinear properties of the host composite into account, and established a mathematical model to predict the stress-strain relationship of MREs, which agree well with experimental data [20]. Deng and Gong modified the traditional magnetic dipole model by taking the influences of particles in the same chain and the particles in all chains into consideration to compute the magnetic-induced shear modulus of MREs [21]. Recently, Zhang *et al.* proposed an effective permeability model which taken into account the particle's saturation to predict the properties of MREs with complex structure and components [22]. However, all of those models ignored the effect of the normal pressure on properties of MREs. In fact, a normal pressure must be applied on MREs when they are used or tested, which can not be overlooked; it influences the distance between adjacent particles, and eventually affects the magnetic field-induced shear modulus.

In this study, a model is established based on the effective permeability rule and the consideration of the normal pressure, which is used to predict the magnetorheological properties of MREs under different normal pressure.

2. Theoretical analysis

The chain-like structural construction of an anisotropic MRE is shown in figure1, which contains two domains. One domain is the chains composed by magnetic particles and matrix among them, another one is the pure matrix composed by elastomers.

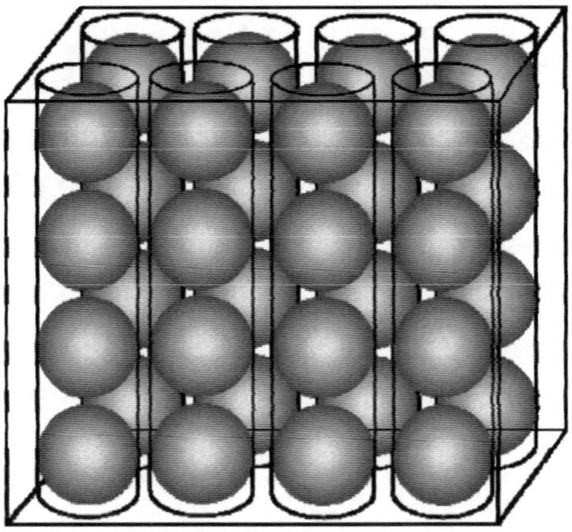

Figure 1. Chain-like structural construction of anisotropic MRE.

For the chains, the effective permeability μ_{ceff} can be calculated by using Maxwell Garnett mixing rule as,

$$\mu_{ceff} = \mu_m + 2\phi\mu_m \frac{\mu_p - \mu_m}{\mu_p + \mu_m - \phi(\mu_p - \mu_m)} \tag{1}$$

where μ_p and μ_m are effective permeability of the particles and the matrix, respectively. ϕ is the particles volume fraction in chains. Based on the hypothesis that all the particles are sphere with the same radius *R*, and the distance between two adjacent particles in a chain is *d*, ϕ can be expressed as,

$$\phi = \frac{4R}{3d} \tag{2}$$

Based on parallel-connection rule, the effective permeability μ_{eff} along the direction of chains is calculated as,

$$\mu_{eff} = \mu_{ceff}\phi_c + \mu_m(1-\phi_c) \tag{3}$$

where ϕ_c is the volume fraction of the chains in the MRE, which can be calculated as,

$$\phi_c = \frac{\phi_p}{\phi} = \frac{3d\phi_p}{4R} \tag{4}$$

where ϕ_p is the particles volume fraction of the MRE. Then the effective permeability along the direction of chains can be obtained by substituting equation (1) and equation (4) into equation (3) as,

$$\mu_{eff} = \mu_{ceff}\phi_c + \mu_m(1-\phi_c) = \mu_m + 2\phi_p\mu_m \frac{\mu_p - \mu_m}{\mu_p + \mu_m - \frac{4R}{3d}(\mu_p - \mu_m)} \tag{5}$$

For shear mode, the magnetic field-induced shear stress can be expressed as,

$$\tau = -\frac{1}{2}\mu_0 \frac{\partial \mu_{eff}}{\partial \gamma} H_0^2 \tag{6}$$

where μ_0 is the vacuum permeability, H_0 is the external magnetic field, and γ is the shear strain. Figure 2 is the shear model of a representative particle chain with two particles; it can be obtained that,

$$\gamma = \frac{x}{d} \tag{7}$$

Then the magnetic field-induced shear stress can be obtained by substituting equation (5) and equation (7) into equation (6) as,

$$\tau = 12\frac{\mu_0\mu_m\phi_p H_0^2}{\gamma}\frac{R}{d}\frac{(\mu_p - \mu_m)^2}{\left[3(\mu_p + \mu_m) - 4\frac{R}{d}(\mu_p - \mu_m)\right]^2} \tag{8}$$

The magnetic field-induced shear modulus ΔG can be expressed as,

$$\Delta G = 12\frac{\mu_0\mu_m\phi_p H_0^2}{\gamma^2}\frac{R}{d}\frac{(\mu_p - \mu_m)^2}{\left[3(\mu_p + \mu_m) - 4\frac{R}{d}(\mu_p - \mu_m)\right]^2} \tag{9}$$

The equation 9 can not be simplified based on the assumption as $\mu_p >> \mu_m$. The reason is the permeability of the particles, which are carbonyl iron particles in most case, is less than 10. Furthermore, the relative permeability of carbonyl iron particles μ_p is a function of the magnetic field intensity H_0, which can be obtained by fitting the experimental data. The empirical equation used as the fitting function is given as,

$$\mu_p(H_0) = \frac{H_0(\mu_{p\max} - 1) + \mu_{p\max} M_s}{H_0(\mu_{p\max} - 1) + M_s} \tag{10}$$

where $\mu_{p\max}$ is the largest relative permeability of particles, and M_s is the saturation magnetization. The dependence of the relative permeability of carbonyl iron particles on magnetic field is shown in figure 3. As it is shown, the empirical equation fits well with the experimental data, and the relative permeability of carbonyl iron particles decreases with increasing magnetic field intensity; especially, at high magnetic field levels, it approaches 2. Therefore, the assumption $\mu_p >> \mu_m$ is invalid. Besides, there is polymer matrix between adjacent particles in chains, which indicates the assumption as $R/d \approx 1/2$ is invalid.

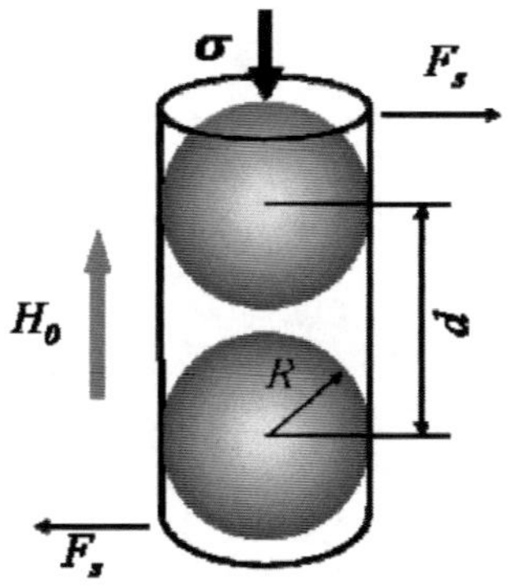

Figure 2. shear model of a representative particle chain with two particles.

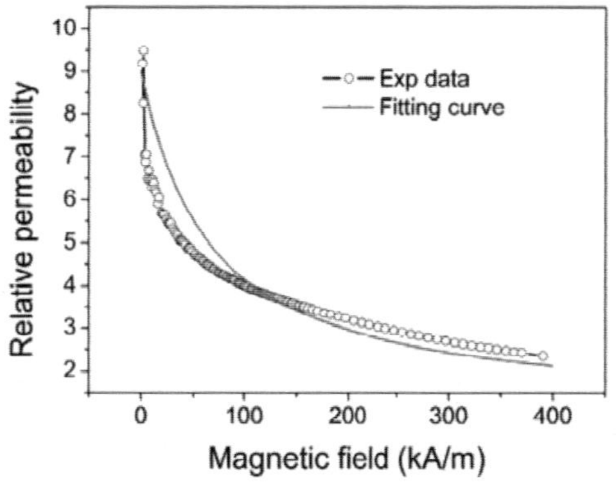

Figure 3. The dependence of the relative permeability of carbonyl iron particles on magnetic field.

To simplify the equation (9), the relative distance between adjacent particles in chains *k* is defined as the ratio of *d* to *R*, *i.e.* $k = d/R$, then the magnetic field-induced shear modulus can be expressed as

$$\Delta G = \frac{12k\mu_0\mu_m\phi_p H_0^2(\mu_p - \mu_m)^2}{\gamma^2[3(\mu_p + \mu_m)k - 4(\mu_p - \mu_m)]^2} \tag{11}$$

Since the Young's modulus of the particles is much lager than that of the polymer matrix, the deformation of the particles under a normal pressure is negligible. Therefore, only the deformation of the polymer matrix is considered in calculating the adjacent particles distance under the normal pressure. The adjacent particle distance can be calculated by,

$$d = d_0(1-\varepsilon) = d_0(1 - \frac{\sigma}{E_m}) \tag{12}$$

where d_0 is the adjacent particles distance under zero pressure, *i.e.* the initial adjacent particles distance, σ is the applied normal pressure, and E_m is the Young's modulus of the polymer matrix. Based on the assumption that the particle size is not changed under the normal pressure, the relative distance between the adjacent particles in chains is expressed as,

$$k = k_0(1 - \frac{\sigma}{E_m}) \tag{13}$$

where k_0 is the initial relative distance between adjacent particles in chains, which is larger than 2.

For an anisotropic MRE with 30% carbonyl iron particles by volume fraction, its magnetic field - induced shear modulus under various normal pressures is calculated and shown in figure 4. As the magnetic field increases, the shear modulus increases. At the same magnetic field, the shear modulus calculated under high normal pressure is larger than those calculated under low normal pressure. The dependence of the magnetic field-induced shear modulus on normal pressure is further shown in figure 5. It is shown that the shear modulus increases with increasing normal pressure, and the influence of normal pressure on shear modulus at high magnetic field levels is more significant than that at low field levels.

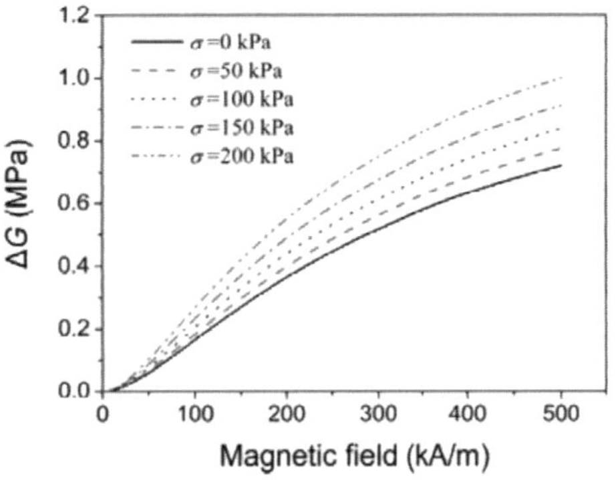

Figure 4. Dependence of shear modulus on magnetic field under various normal pressures for MRE with 30% carbonyl iron particles by volume fraction.

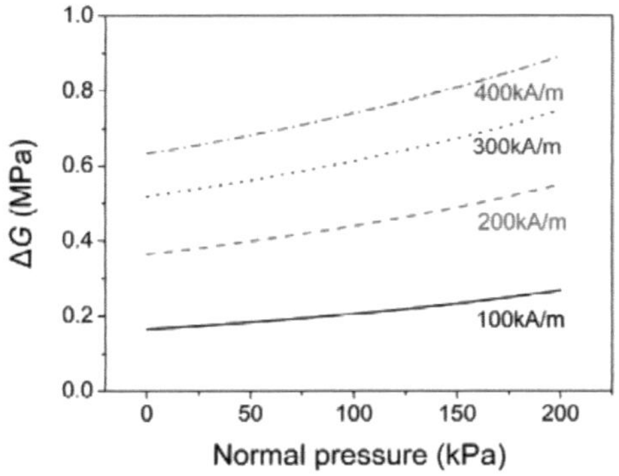

Figure 5. Dependence of the magnetic field-induced shear modulus on normal pressure for MRE with 30% carbonyl iron particles by volume fraction.

3. Conclusion

To predict the influence of normal pressure on magnetorheological properties of anisotropic MREs, a theoretical model is proposed based on the effective permeability rule and the consideration of the normal pressure. The results indicate that the normal pressure have great influence on magnetic field-induced shear modulus; the shear modulus of MREs increases with increasing normal pressure, such dependence is more significant at high magnetic field levels.

Acknowledgements

This research is financial funded by the National Natural Science Foundation of China under the grant number of 51108062, and the National Key Technology R&D Program under the Grant number of 2011BAK02B01.

References

[1] Rabinow J 1948 The Magnetic fluid clutch *AIEE Trans.* **67** 1308–15
[2] Vicente J, Klingenberg D J and Hidalgo-Alvarez R 2011 Magnetorheological fluids: a review *Soft Matter* **7** 3701–8
[3] Ginder J M and Davis L C 1994 Shear stress in magnetorheological fluids: role of magnetic saturation *Appl. Phys. Lett.* **65** 3410–8
[4] Carlson J D and Jolly M R 2000 MR fluid, foam and elastomer devices *Mechatronics* **10** 555–69

[5] Dyke S J, Spencer B F Jr and Carlson J D 1996 Modeling and control of magnetorheological dampers for seismic response reduction *Smart Mater. Struct.* **5** 565–75
[6] Chiriac H and Stoian G 2010 Influence of particle size distributions on magnetorheological fluid performances *Journal of Physics: Conference Series* **200** 072095
[7] Davis L C 1999 Model of magnetorheological elastomers *J. Appl. Phys.* **85** 3348–51
[8] Shiga T, Okada A and Kurauchi T 1995 Magnetoviscoelastic behavior of composite gels *J. Appl. Polym. Sci.* **58** 787–92
[9] Jolly M R, Carlson J D and Munoz B C 1996 A model of the behavior of magnetorheological materials *Smart Mater. Struct.* **5** 607–14
[10] M Lokander and Stenberg B 2003 Performance of isotropic magnetorheological rubber material *Polymer Testing* **22** 245–51
[11] Fan Y, Gong X, Xuan S, Zhang W, Zheng J and Jiang W 2011 Interfacial friction damping properties in magnetorheological elastomers *Smart Mater. Struct.* **20** 035007
[12] Chen L, Gong X, Jiang W, Yao J, Deng H and Li W 2007 Investigation on magnetorheological elastomers based on natural rubber *J. Mater. Sci.* **42** 5483–9
[13] Guan X , Dong X and Ou J 2008 Magnetostrictive effect of magnetorheological elastomer *J. Magn. Magn. Mater.* **320** 158–63
[14] Xu Z, Gong X, Liao G and Chen X 2010 An active-damping-compensated magnetorheological elastomer adaptive tuned vibration absorber *J. Intel. Mat. Syst. Str.* **21** 1039–47
[15] Dwivedy S K, Mahendra N and Sahu K C 2009 Parametric instability regions of a soft and magnetorheological elastomer cored sandwich beam *J. Sound Vib.* **325** 686 –704
[16] Ginder J M, Schlotter W F and Nichols M E 2001 Magnetorheological elastomers in tunable vibration absorbers *Proc. of SPIE* **4331** 103–10
[17] Wu J, Gong X, Fan Y and Xia H 2010 Anisotropic polyurethane magnetorheological elastomer prepared through in situ polycondensation under a magnetic field *Smart Mater. Struct.* **19** 105007
[18] Jolly M R, Carlson J D and Munoz B C 1996 A model of the behaviour of magnetorheological materials *Smart Mater. Struct.* **5** 607–14
[19] Davis L C 1999 Model of magnetorheological elastomers *J. Appl. Phys.* **85** 3348–51
[20] Shen Y, Golnaraghi M F and Heppler G R 2004 Experimental research and modelling magnetorheological elastomers *J. Intel. Mat. Syst. Str.* **15** 27–35
[21] Deng H and Gong X 2007 Adaptive tuned vibration absorber based on magnetorheological elastomer *J. Intel. Mat. Syst. Str.* **18** 1205–10
[22] Zhang X, Li W and Gong X 2008 An effective permeability model to predict field-dependent modulus of magnetorheological elastomers *Commun. Nonlinear Sci.* **13** 1910–6

Prototype of haptic device for sole of foot using magnetic field sensitive elastomer

T Kikuchi[1], Y Masuda, M Sugiyama, T Mitsumata and S Ohori

Graduate School of Science and Engineering, YamagataUniversity, Japan

E-mail: t_kikuchi@yz.yamagata-u.ac.jp

Abstract. Walking is one of the most popular activities and a healthy aerobic exercise for the elderly. However, if they have physical and / or cognitive disabilities, sometimes it is challenging to go somewhere they don't know well. The final goal of this study is to develop a virtual reality walking system that allows users to walk in virtual worlds fabricated with computer graphics. We focus on a haptic device that can perform various plantar pressures on users' soles of feet as an additional sense in the virtual reality walking. In this study, we discuss a use of a magnetic field sensitive elastomer (MSE) as a working material for the haptic interface on the sole. The first prototype with MSE was developed and evaluated in this work. According to the measurement of planter pressures, it was found that this device can perform different pressures on the sole of a light-weight user by applying magnetic field on the MSE. The result also implied necessities of the improvement of the magnetic circuit and the basic structure of the mechanism of the device.

1. Introduction

In Japan, people aged 65 or over account for more than 23% in 2012. This situation is called the super-aging society. Their qualities of life (QOL) are big issues nowadays. Maintenance of health for their mentalities and bodies is an important factor when we consider their QOL. According to the survey on favorite activities of the elderly in Japan [1], the most favorite activity of the elderly is traveling (see Table 1). Actually traveling around the world stimulates their mentality and body. However, a lot number of traveling need a lot of money. Then they love walking around their neighborhood as the second favorite activity. Because the traveling also includes much amount of walking activity, this survey indicates that the walking activity is a favorite physical activity of the elderly. From a viewpoint of their health, walking is a healthy aerobic exercise. Aerobic exercises have a statistically significant positive effect for the elderly with dementia and cognitive impairments [2]. During mild exercises we have increases of blood volume in our brains and these effects are considered to work well to prevent or slow dementia [3]. However, for people suffering from physical and / or cognitive disabilities and who really need the mild aerobic exercises, sometimes it is challenging to go somewhere they don't know well. This is a reason why we are focusing on a virtual walking system which allows them to travel anywhere and anytime they want.

As a system that allows users to do a mild aerobic exercise, we propose a virtual walking system using a haptic device on soles of feet (Figure 1). This system consists of a visual system including a 3D projector and a shutter glass, a sound system, a force plate (Wii Balance Board, Nintendo) as an

[1] Corresponding author

input device, and a desktop computer as a controller. When a user steps on the plate, the visual information and sounds work with the user's motion. A computer graphics of a virtual city was made with computer graphics software and linked with these devices. The users can watch the 3D vision with the shutter glass that switches individual visions on left and right eye. Instead of the 3D glasses, we can use a head mount display for more compact system. The users can travel any places on the earth with staying in their rooms.

Table 1. Favorite activities of the elderly in Japan*.

	Male (%)		Female (%)	
1st	Traveling	56.2	Traveling	53.1
2nd	Walking	53.7	Walking	37.6
3rd	PC	37.1	TV	27.5
4th	TV	29.7	Art	26.0
5th	Golf	28.3	Music	23.3

*Modified from [1]

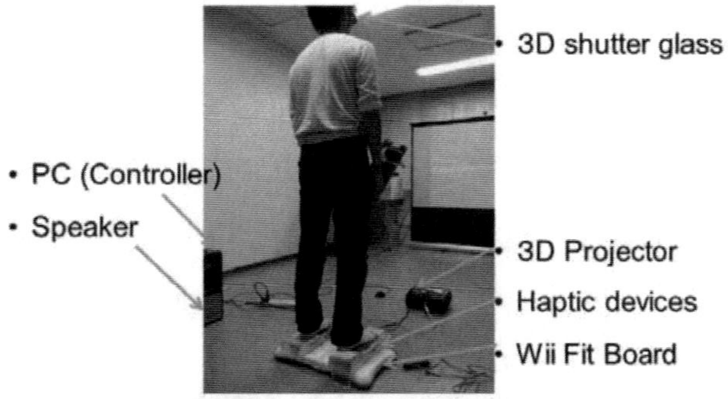

Figure 1. VR Walking System.

In this study, we propose to use a magnetic field sensitive elastomer (MSE) [4], as an interface material of the haptic device on the soles of feet. The MSE is one of functional materials, which have been newly developed in recent years. This material is a composite of a polymer elastomer and magnetic fluids or magnetic particles. Its elastic modulus can be controlled with application of an external magnetic field. There are some mechanical platforms that can display haptic information on users' feet during walking in virtual worlds. For example, Perreault *et al.* [5] developed a cable-driven parallel mechanism to perform reaction force on users' feet and legs. As another example on the virtual walking system, Schmidt developed a large-scale linkage mechanism [6], and Iwata developed a 2 dimensional treadmill system [7]. In any case, they used very large scale system to perform the reaction force on users' feet and legs. However, a compact haptic device on foot with a functional material is a new challenge.

This paper describes the first trial to develop a haptic interface using the MSE. We designed an electromagnet as a magnetic field generator for the MSE. The evaluation of this prototype was also discussed.

2. The first prototype of haptic interface with MSE

We have used a MSE that is a composite material of the magnetic particles (carbonyl iron, 2.8 μm in diameter, 70 wt%) and polyurethane elastomer as a haptic interface of the virtual reality walking system, because of its good endurance for environment and pressure. When we use a haptic interface on sole feet, a high pressure is applied on the material. In such cases, the material should have a high endurance for pressure. As the other special feature of this material, its viscoelasticity is sensitive to the magnetic field. We can easily control the sense of touch only with the control of the external magnetic field without any active element like actuators. This feature helps us to develop a safe device for human.

In this material the magnetic particles disperses evenly in the off-state of the magnet. In this condition, the elastic modulus is less than 10 kPa. On the other hand, when we apply 0.5 Tin the magnetic flux density in this material, the magnetic particles make some clusters that go through the direction of the magnetic flux and become stiff material whose elastic modulus is more than 200 times of that in the off-state [4].

As a resource of the magnetic field, we developed an electromagnet. The sectional view of the magnet is shown in Figure 2. The C-shaped magnetic circuit is made of the magnetic material (structural steel). Two coils at the left and right of the circuit generate magnetic fluxes in it. We put a sheet of the MSE (10mm in thickness) on this device, and control its elasticity with the magnetic field inside it. The result of magnetostatic analysis is shown in Figure 2. The magnetic circuit (geometry and coil) was designed so that they can generate 0.5 T in the magnetic flux density at the MSE. The specification is shown in Table 2. Figure 4 shows the results of compression tests with the magnet and MSE sheet. Handy Digital Force Gauge (HF-50, Japan Instrumentation System Co. LTD) with a probe whose diameter is 15mm was attached on a compression test stand, and used. The stress-strain curves in off-state (current input for the coil: I = 0.0 A) and on-state (I = 1.0 A) were measured. The device performed significantly different stresses between the off-state and on-state of the coil current.

We built 2 magnets on users' heels and toes (Figure 3) in order to control different senses on them. When we use this device, we put the MSE sheets on each magnet and evenly fill normal (non-magnetic) elastomers whose thickness is the same of the MSE at the other place.

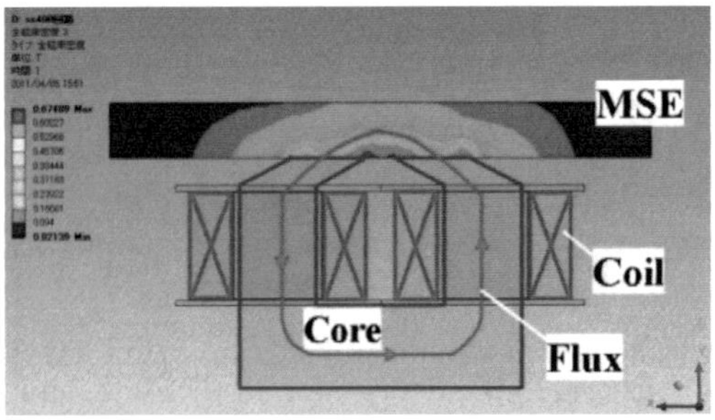

Figure 2. Magnetostatic analysis of the electromagnet for the MSE.

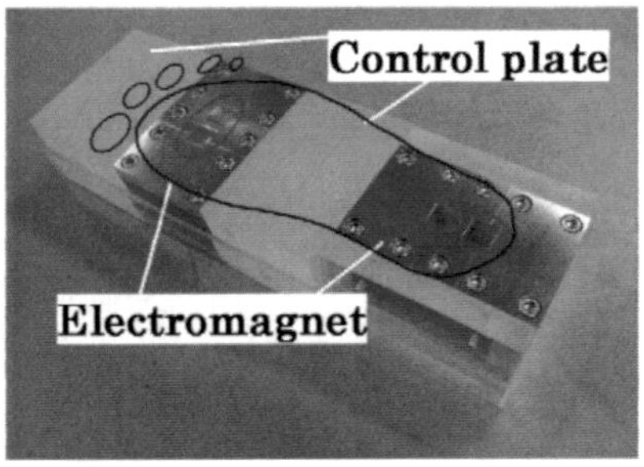

Figure 3. First prototype of the Foot Device (a photo without MSE)

Table 2. Specification of the magnet.

Width	110 mm
Length (Adjustable)	200-300 mm
Height	55mm
Turning number of coil (for each)	840
Diameter of magnetic wire (with coating)	0.45mm
Resistance of coil (for each)	11.4Ω
Maximum electric current (for each)	1A
Maximum and average of Magnetic flux density (@max current)	0.5 T / 0.25T

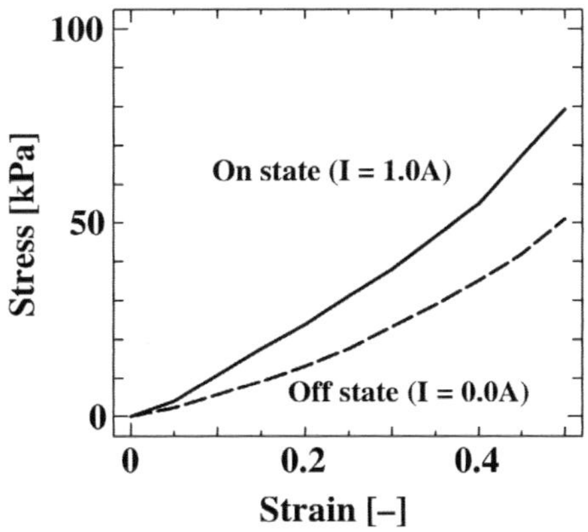

Figure 4. Result of compression test

3. Method

We used an in-sole sheet pressure sensor (Figure 5, F-Scan II, Nitta) to measure the plantar pressure during stepping on the device. The range of measureable pressure is from 50 to 500 kPa. This is a sheet type force sensor and we can put it even in shoes. This sheet has 955 point of the sensor. Maximum sampling rate is 850Hz but we sampled the pressure data at 80 Hz in this study in order to reduce the data number.

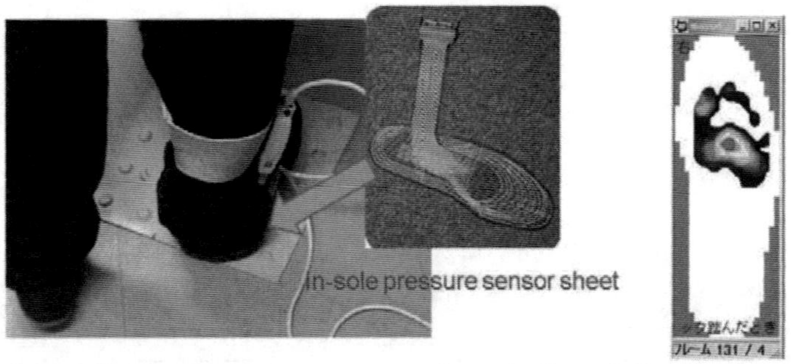

Figure 5. In-sole pressure sensor sheet (left) and distribution of pressure (right)

Two young healthy students were recruited as subjects. The dominant foot of both subjects is right foot. Therefore, the targets are their right feet in this research. Table 3 shows information of them. Subject A is a light weight subject and subject B is a heavy weight subject.

Table 3. Information of target feet (subjects' right feet).

Subject	Mass (kg)	Foot length (m)	Foot width (m)
A	49	0.245	0.083
B	90	0.255	0.105

The prototype can control the stiffness of the MSEs on toes and heels, but we controlled only the MSE on the toe in these experiments. We measured the difference between the plantar pressure when subjects stepped on the off-state (I = 0.0 A) and that on the on-state (I = 1.0 A). As shown in Fig.6, we put the sensor sheet on the MSE device, and the subject was instructed to put his right foot without shoes on the sheet. Then the left foot was stridden from behind the right foot to the forward of it. The pressure profiles during this one stride were sampled at 80 Hz with the pressure sensor.

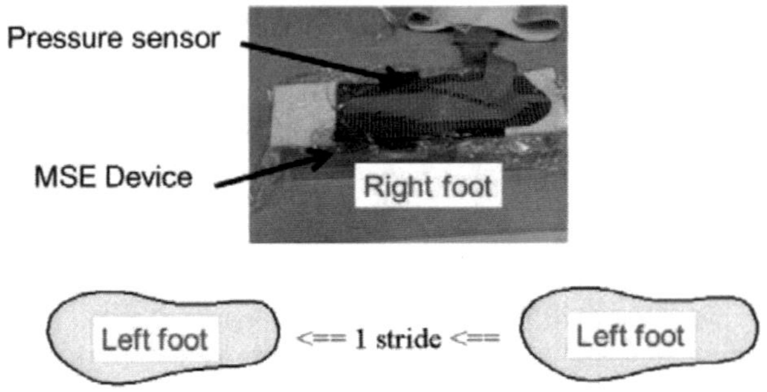

Figure 6. Experimental method to measure plantar pressures during 1 stride.

We measured the pressure profiles of all range of soles every trial. But in this work we discuss the average pressures just under the metatarsal head (MTH, Figure 7), because we put the MSE device on this point and this place is the most stressful point on the sole. The average pressures in 2 cm by 2 cm under the MTH were evaluated below. The target sensors were manually adjusted for every subject.

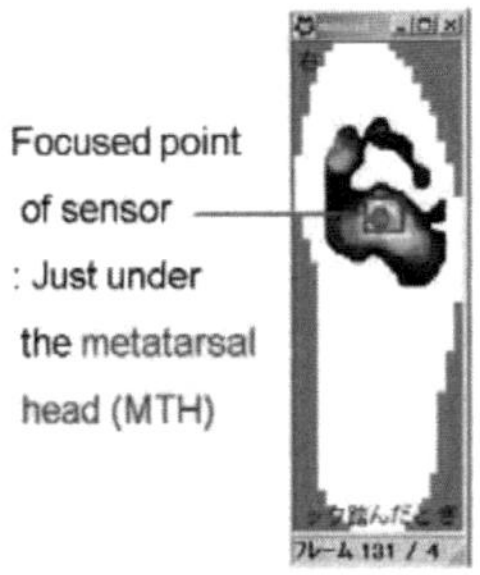

Figure 7. Focused sensing point in this study.

4. Results and discussion

Figure 8 shows the experimental results for subject A and B. For subject A, we can see clear difference of pressure between the on-state and the off-state of the device. However, the difference is not clear for subject B. This means this device has an ability to make stiffness change that can appear clear differences of perceptions under the sole only for light-weight users (~50 kg).

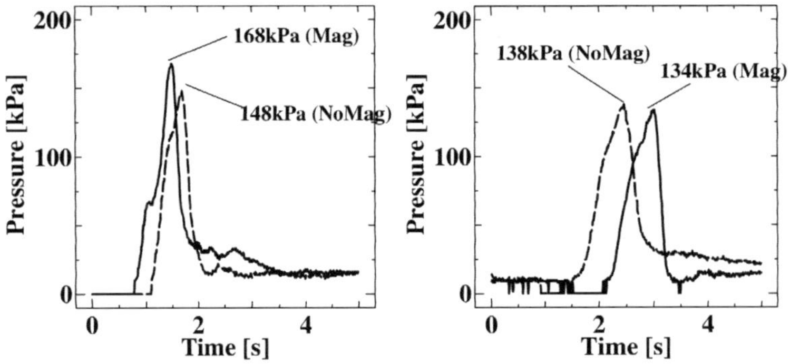

Figure 8. Experimental results for subject A (left) and subject B (right).

As a comparison, we measured a pressure difference between when the subject A (the light subject) stepped on a dot-type braille block (the height of the dots were 5mm, see the left of Fig.9) and a flat floor by the same way of the previous section. As shown in the right of Figure 9, the difference is significantly bigger than the result of previous experiment on the MSE device. This experiment also shows the insufficient performance of the 1st prototype.

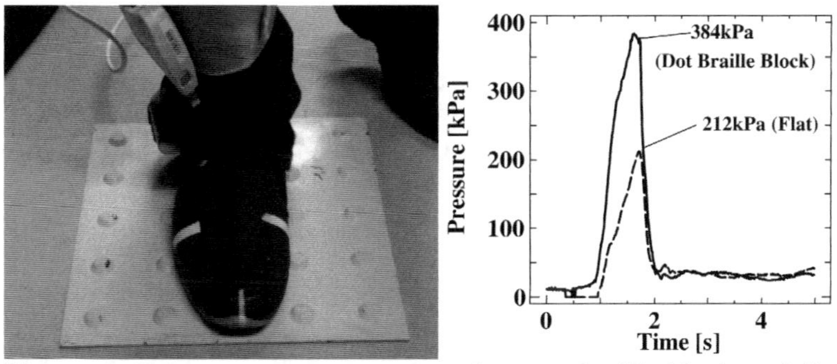

Figure 9. Pressure measurement on dot-type braille block and flat floor for subject A: an experimental scene (left) and results (right).

As shown in Figure 9, the device should support at least 400kPa in 2 cm by 2 cm with a small deformation in order to perform feasible perceptions on foot. We believe that the most critical reason of this insufficient performance of the device is the insufficient magnetic field inside the MSE. We use open air structure in the 1st prototype (Figure 2) and this result in the insufficient intensity of magnetic field. Now we are designing a new version of the device with the MSE. In the new structure, we will sandwich the MSE with magnetic materials and make a close loop of the magnetic flux that completely passes through the MSE and make higher intensity of the magnetic flux density.

5. Conclusion

In this research, we have developed a haptic interface by using a magnetic field sensitive elastomer (MSE). The performance of the 1st prototype is not sufficient but we can find a possibility of the device. We are now designing a new version of the device.

Acknowledgment

This research was supported by Casio Science Promotion Foundation and Strategic Communications R&D Promotion Programme (SCOPE) of Japan.

References

[1] Harada T, Kato K, Oda Y, Uchida H and Ohno T 2011 A survey on living habits of the elderly (2) *Japan. Bull. Nagoya Bunri Univ.* **11** 27-33

[2] Heyn P, Abreu B C and Ottenbacher K J 2004 The effects of exercise training on elderly persons with cognitive impairment and dementia: A meta-analysis *Arch. Phys. Med. Rehab.* **85** 1694-704

[3] Timinkul A, Kato M, Omori T, Deocaris C Custer, Ito A, Kizuka T, Sakairi Y, Nishijima T, Asada T and Soya H 2008 Enhancing effect of cerebral blood volume by mild exercise in healthy young men: A near-infrared spectroscopy study*Neurosci. Res.* **61** 242-8

[4] Mitsumata T and Ohori S 2011 Magnetic polyurethane elastomers with wide range modulation of elasticity *Poly. Chem.* **2** 1063-7

[5] Perreault S and Gosselin C M 2008 Cable-driven parallel mechanisms: Application to a locomotion interface *J. Mech. Des.* **130** 102301-8

[6] Schmidt H 2004 HapticWalker - A novel haptic device for walking simulation *Proc. Eurohaptics* 60-7

[7] Iwata H 1999 Walking about virtual environments on an infinite floor *Proc. IEEE VR Conf.* 286-93

Study of magnetorheology and sensing capabilities of MR elastomers

T F Tian[1], W H Li and G Alici
School of Mechanical, Materials and Mechatronic Engineering, University of Wollongong, Wollongong, NSW 2522, Australia

E-mail: tt436@uowmail.edu.au

Abstract. This study focuses on the magnetorheology and sensing capability of graphite based Magnetorheological Elastomers (Gr MREs). By introducing graphite (Gr) to conventional MREs, the Gr MREs are derived. The anisotropic sample with 20% graphite weight fraction was selected to be compared with anisotropic conventional MREs. The microstructures of anisotropic Gr MREs and conventional MREs were observed. Both steady state tests and dynamic tests were conducted to study rheological properties of the samples. For dynamic tests, the effects of strain amplitude, and frequency on both storage modulus and loss modulus were measured. For sensing capability, the resistance of selected Gr MREs under different magnetic fields and external loadings is measured with a multi-meter. Either higher magnetic field or more external loading results in the resistance increment. Based on an ideal assumption of perfect chain structure, a mathematical model was proposed to investigate the relationship between the MRE resistance with the external loadings. In this model, the current flowing through the chain structure consists of both tunnel current and conductivity current, both of which depends on external loadings. The modelling parameters were identified and reconstructed from comparison with experimental results. The comparison indicates that both experimental results and modelling prediction agree favourably well.

1. Introduction

Magnetorheological elastomers (MREs) are smart materials where polarized particles are suspended in a non-magnetic solid or gel-like matrix [1]. These materials exhibit characteristics that their moduli can be reversely controlled by an external magnetic field. MREs have recently found a variety of applications, such as adaptive tuned vibration absorbers, dampers, sensors, and so on [2-4]. There are two kinds of MREs which are anisotropic and isotropic [5,6]. In anisotropic samples, polarised particles can be arranged as chains in polymer media such as silicon rubbers and natural rubbers.

MREs generally consist of three major components: magnetizable particles, matrix, and additives. Iron particles are generally used as the filler material to fabricate MREs, this is because iron has one of the highest saturation magnetisation values of metallic elements as well as high permeability, low remnant magnetisation and high saturation magnetisation. High permeability and saturation magnetisation are thought to provide high inter-particle attraction and thereby a high MR effect. The effect of particle shape, size and volume fraction on the overall MRE performances have been intensively investigated. Lokander and Stenberg [7] measured the MR effect for isotropic nitrile rubber MRE with varying sizes and content of iron particles. The MR effect was larger for materials

[1] To whom any correspondence should be addressed.

with ASC300 iron (particle size < 60 μm) than for materials with carbonyl iron (particle size 3.9-5.0 μm). Shiga *et al.* [8] measured the increase in shear modulus as a function of the particle volume fraction. For aligned MREs the change in shear modulus increases with an increasing particle volume fraction. When the concentration of filler is higher than 30 vol.%, the mechanical properties of the composite deteriorate rapidly and the stiffening of the material is larger than the increase of the MR effect. Both natural rubber and silicone rubber are used as typical matrixes [9]. Natural rubber is an elastomer. The purified form of natural rubber is the chemical polyisoprene which can also be produced synthetically. Heat is normally required to vulcanise silicone rubber. The silicone rubber and a vulcanising silicon sealant (at room temperature), is mixed with silicon oil to changing its ductility. The silicon oil is selected on the basis of preliminary studies with different elastomers. Additives are used to adjust the mechanism properties or electrical performance of MREs. Silicone oil is an additive to increase the gaps between the matrix molecules and to decrease the gaps between the conglutination of molecules. Apart from increasing the plasticity and fluidity of the matrix, the additives can average the distribution of internal stress in the materials, which makes them ideal for fabricating MRE materials [10]. Graphite powder is also a kind of additive which can affect the magnetorheology and electrical conductivity of MREs [11,12].

By introducing graphite microparticles into the elastic matrix, the Gr MREs is derived that behave a lower electrical conductive characteristic, which has a potential to work as a sensing material for development of force and magnetic field sensors [11]. The Gr MREs show resistance changing when an external load is applied to MRE samples. By this feature of MREs, the external stress signal can be converted to a resistance signal, which can be used potentially in a force sensor.

Kchit and Bossis [13] found that the initial resistivity of metal powder at zero pressure is about 108 Ωcm for pure nickel powder and 106 Ωcm for silver coated nickel particles. The change in resistance with pressure was found to be an order of magnitude larger for a MRE composite than for the same volume fraction of fillers dispersed randomly in the polymer. Wang *et al.* [14] proposed a phenomenological model to understand the impedance response of MREs under mechanical loads and magnetic fields. Their results showed that MRE samples exhibit significant changes in measured values of impedance and resistance in response to compressive deformation, as well as applied magnetic field. Bica [12] found that MRE with graphite micro particles (~14%) is electroconductive. The magnetoresistance has an electric resistance whose value diminishes with both the increase of the intensity of the magnetic field and with the compression force. The variation of resistance with magnetic field intensity is due to the compression of MRE with graphite microparticles. In the approximation of the perfect elastic body, the sums of the main deformations and the compressibility module of MRE with graphite microparticles, depend on the magnetic field intensity. Li *et al.* [11] introduced graphite into conventional MREs and found a MRE sample with 55% carbonyl iron, 20% silicon rubber and 25% graphite powder has the best performance. The test result showed that at a normal force of 5 N, the resistance decreases from 4.62 kΩ without a magnetic field to 1.78 kΩ at a magnetic field of 600 mT. The decreasing rate is more than 60%. This result also demonstrated the possibility of using MREs to develop a sensor for measuring magnetic fields. This result indicates that the detection is very sensitive to the normal force. When the normal force is 15N, the field-induced resistance only has less than 28% change from 0.65 kΩ at 0 mT to 0.47 kΩ at 600 mT. Tian *et al.* [15] proposed a representative volume unit (RVU) to analyze the resistance of Gr MREs theoretically. Many Gr MREs samples with various graphite weight fractions were fabricated and measured for the resistance. They found that the samples' resistance decreases with increasing either the external force or the intensity of applied magnetic field. Meanwhile, the sample with higher graphite concentration shows a lower resistance. Both Bica's [12] and our previous work [11,15] experimentally demonstrated the graphite based MREs exhibit conductive properties; however, the investigation of the mechanism lacks in-depth discussion. For example, very few reports in literature have been developed to investigate the effect of external loading on the electric properties. This is the major motivation of this research.

2. Experimental

2.1. MRE fabrication and microstructure observation

The materials used for the Graphite MR Elastomers are: silicone rubber (Selleys Pty. LTD); silicone oil, type 378364 (Sigma-Aldrich Pty. LTD); carbonyl iron particles, type C3518 (Sigma-Aldrich Pty. LTD); and graphite powder, type 282863 (Sigma-Aldrich Pty. LTD). The particle sizes of graphite powder are about 20 μm, while the iron particles' diameter is between 3 μm and 5 μm.

In this study, a graphite based anisotropic MRE sample (graphite weight fractions Gr = 20% wt) was fabricated to be compared with the conventional anisotropic MRE sample. Table 1 shows the compositions of the two anisotropic MRE samples. All the samples contain the same compositions of 10 g carbonyl iron particles, 3 g silicone rubber and 3 g silicone oil. The only difference is that the graphite based MRE has 20% graphite weight fraction.

Table 1. Components of Gr MRE samples.

Sample	**Carbonyl iron**	**Silicone oil**	**Silicone rubber**	**Graphite**	**Graphite weight fraction (Gr %)**
Conventional MRE	10 g	3 g	3 g	0 g	0%
Graphite based MREs	10 g	3 g	3 g	4 g	20%

After the samples were fabricated, LV-SEM (JSM 6490LV SEM) was used to observe microstructures of MRE samples. Figures 1&2 show the surface imaging for MRE microstructures.

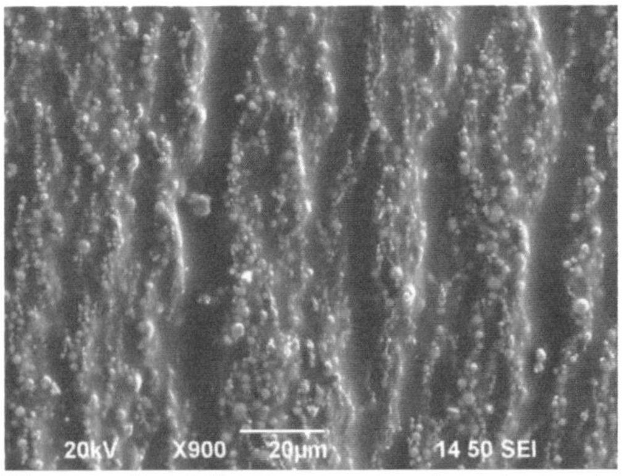

Figure 1. Microstructure of anisotropic conventional MRE (Gr 0%).

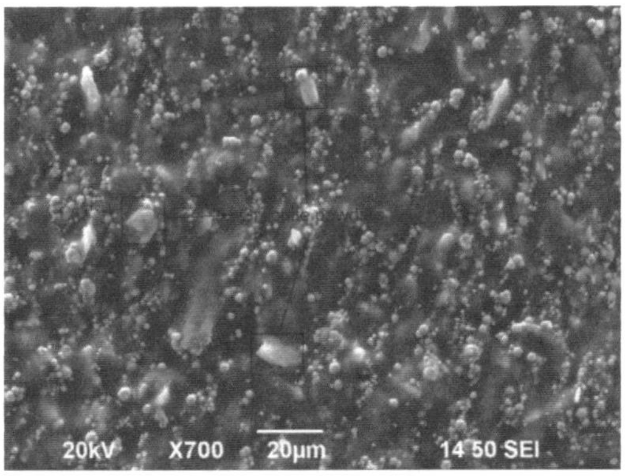

Figure 2. Microstructure of anisotropic Gr MRE (Gr 20%).

In figure 1 that shows the microstructure of anisotropic conventional MRE, the carbonyl iron particles array in chains in the matrix. However, figure 2 demonstrates that in anisotropic Gr MREs, when the carbonyl iron particles array in chains, the graphite powders disperse in the matrix randomly. The reason of this phenomenon is that the magnetic field can not affect on the graphite powders but only on the carbonyl iron particles. So by the magnetism, the carbonyl iron particles move to chains along the same direction as the magnetic field in the matrix.

By comparing figure 1 & 2, we can see that the carbonyl iron chains in the sample without graphite have a better lines performance. The reason is that when the mixed paste of carbonyl iron, silicone rubber, silicone oil and graphite is curing under the magnetic field, the graphite powders in graphite MREs affect the carbonyl iron particles' movement. The more graphite in the mixture, the more effects are applied on to the carbonyl iron chains, which influence the magnetorheology of MREs.

2.2. Rheological experimental device

A rotational Rheometer (MCR 301, Anton Paar Companies, Germany) and a Magneto Rheolgical Device (MRD 180, Anton Paar Companies, Germany) were used to measure the Gr MREs' mechanical properties. The Magneto Rheolgical Device is equipped with an electromagnetic kit which can generate a magnetic field perpendicular to the direction of the shear flow. A 20 mm diameter parallel-plate measuring system with 1 mm gap was used. The samples were sandwiched between a rotary disk and a base parallelly. In this study a steady-state rotary shear and oscillatory shear were both used for the experiments.

In this experiment the magnetic flux density of the sample of MRE (B_{MRE}) in the measuring gap depends not only on the current (I) applied to the samples, but also on the magnetic properties of MRE materials. The relationship between B_{MRE} versus I is found to be: $B_{MRE} = 220\ I$, where the units of B_{MRE} and I are in mT and amp (A), respectively.

In the following test, the test current varies from 0 A to 2 A with the increment 0.5 A, whose intensity of magnetic field is 0 mT to 440 mT with the increment 110 mT.

2.3. Sensing capability experimental measurement

Figure 3 shows a schematic of the experimental device used. In this setup a long plastic plate is used to hold the weight to apply the load to the Gr MRE samples. A Gaussmeter (HT201, Hengtong magnetoelectricity CO., LTD) is used to test the intensity of the magnetic field. A multimeter (Finest 183, Fine Instruments Corporation) measures the resistance in the Gr MRE samples.

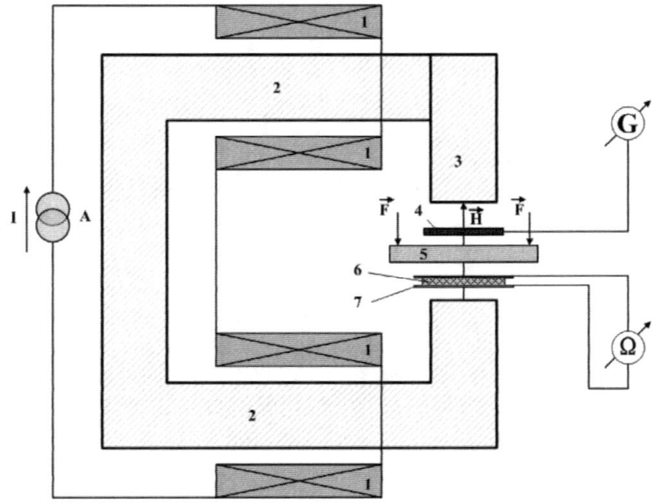

Figure 3. sketch of the experimental device 1- coils; 2- electrical magnetic; 3- moving magnetic pole; 4- Gaussmeter probe; 5- plastic plate; 6- Gr MRE sample; 7- measuring copper plates; G- Gaussmeter; Ω- Multimeter; F- external force applied on Gr MRE samples.

3. Results and discussion

3.1. Steady state rheological test results

Under rotary shear the shear stress and shear strain of MREs under fields varying from 0~440 mT were measured. The MR effect was evaluated by measuring the shear strain-stress curve of the sample with and without a magnetic field applied.

Figure 4 (a) & (b) show the strain-stress curve of the two samples at 5 different magnetic field intensities ranging from 0 to 440 mT. The slope of the strain-stress curve is the shear modulus of the material. As can be seen in the figures, both the samples' shear modulus show an increasing trend with increasing magnetic field intensity before they reach magnetic saturation at high field strength, which proves that both the MRE samples exhibit obvious MR effects. Also from Figure 4, the shear stress shows a linear relationship with the shear strain when the strain is within a range. This means the MRE acts with linear viscoelastic properties when the strain is below a limitation. For conventional MREs, the limitation is around 50% shear strain, which was previously reported [22]. When the graphite weight fraction increases from 0 to 20%, the range of linearity decreases from 50% to around 7%. For the strain over the limitation, the shear modulus reaches a saturation (maximum value) and decrease steadily. This could be due to sliding effect. The higher magnetic field intensity leads to higher steady shear stress.

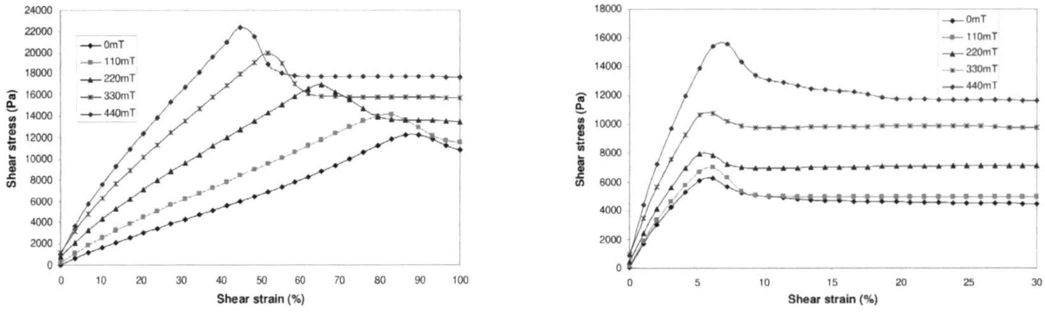

(a) (b)

Figure 4. Strain-stress curve versus magnetic field (anisotropic MRE) (a) Gr 0% and (b) 20%.

For each curve, the slope equals to the ratio of peak shear stress to the relevant shear strain, which is the shear modulus. By analyzing the slopes of the curves, it is easy to see that the more graphite in the material, the less growth of slopes when the magnetic field increase from 0 to 440mT. This is because of the contributions of graphite powders to the stiffness of the samples and to pull down the volume fraction of iron particles in the samples. The graphite increases the initial stiffness of graphite MREs, thus the stiffness changing from the MR effect can not be as the same as the conventional MREs. Meanwhile, the graphite added to the sample downgrades the volume fraction of iron particles in the sample, which plays crucial role in determining the MRE effect of the MR materials. Therefore, the MR effect of Gr MREs is lower than the conventional MREs.

3.2. Dynamic rheological tests results

In order to obtain the dynamic mechanical behavior of MRE, both angular frequency sweep tests and strain amplitude sweep tests were used. Five sets of data were collected for different amplitudes of oscillation, according to the various magnetic fields input to the samples of MR elastomers. Same as the steady state tests, five different magnetic field intensities, 0, 110, 220, 330 and 440 mT, were used in this experiment. The amplitude of shear strain in angular frequency sweep tests is set at 1% and the input frequency was 5 Hz in the strain amplitude sweep tests.

3.2.1. Strain amplitude sweep. In the strain sweep test, the storage and loss moduli were tested by varying strain from 0.01% to 100% at different magnetic fields. Figure 5 & 6 show the changing of storage modulus at the strain amplitude sweep.

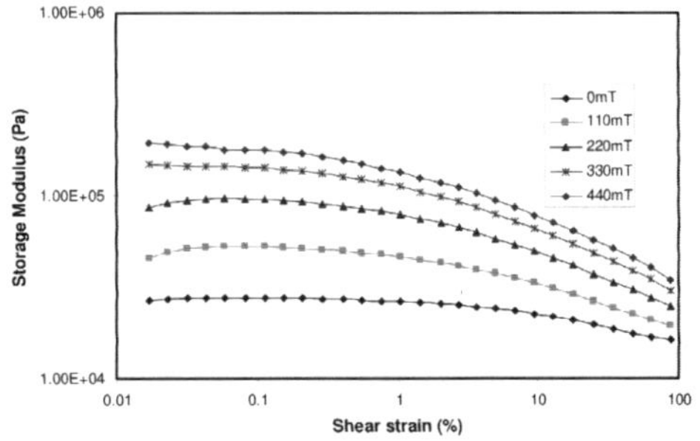

Figure 5. Storage Modulus versus strain amplitude sweep (anisotropic MRE Gr 0%).

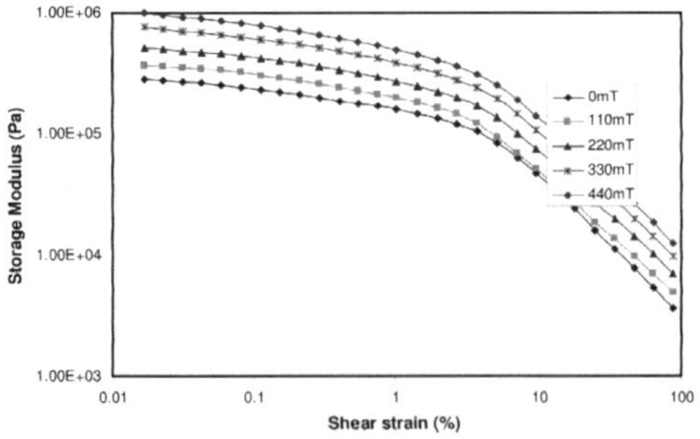

Figure 6. Storage Modulus versus strain amplitude sweep (anisotropic MRE Gr 20%).

In figures 5 & 6, the overall trend of storage modulus is to decrease with the strain amplitude. It goes down smoothly within 10% shear strain and begins to drop significantly over 10% shear strain, which means at the high shear strain, the storage modulus are much smaller than that at low shear strain.

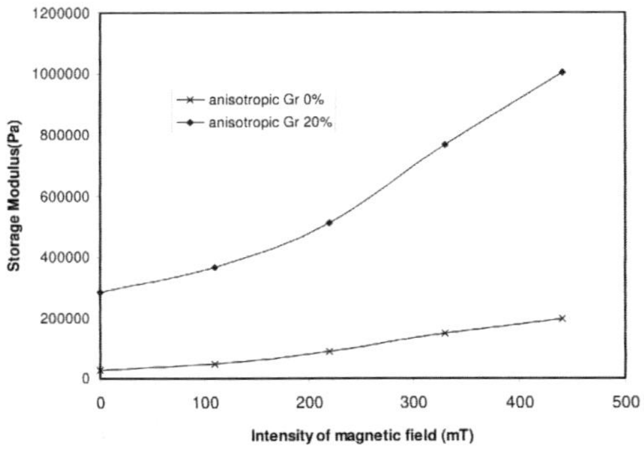

Figure 7. Storage Modulus versus magnetic field at 0.1% shear strain.

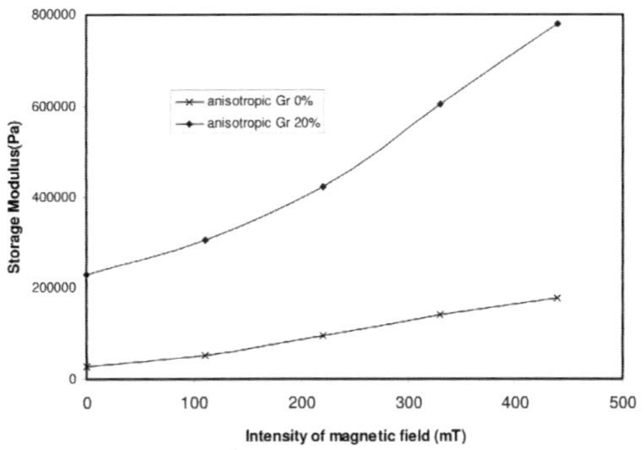

Figure 8. Storage Modulus versus magnetic field at 10% shear strain.

Figures 7 & 8 show the storage modulus versus magnetic field at 0.1% and 10% shear strain, respectively. The two shear strains are the beginning and end of linear range. In these two figures we can see that the storage modulus shows an increasing trend with the intensity of magnetic field. The ratio of storage modulus at 440 mT to that at 0 mT is the MR effect. The MR effect of anisotropic conventional MRE is around 6.3 and 5.4 at 0.1% shear and 10% shear, respectively; when the graphite weight fraction increases to 20%, the MR effect decreases to around 2.5 at 0.1% shear and to 2.38 at 10% shear. This proves that with the growth of graphite, the MR effect decreases.

3.2.2. Angular frequency sweep. In this test, the strain is set at 1%. According to the experimental equipments, the angular frequency was varied from 6 to 100 1/s at different magnetic fields as 0, 110, 220, 330 and 440 mT. The figures 9 & 10 show the storage modulus curves of the MRE samples at frequency sweep.

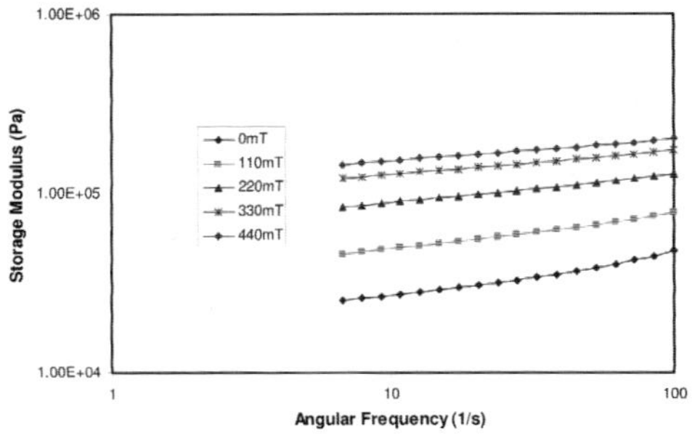

Figure 9. Storage modulus versus angular frequency sweep (anisotropic MRE Gr 0%).

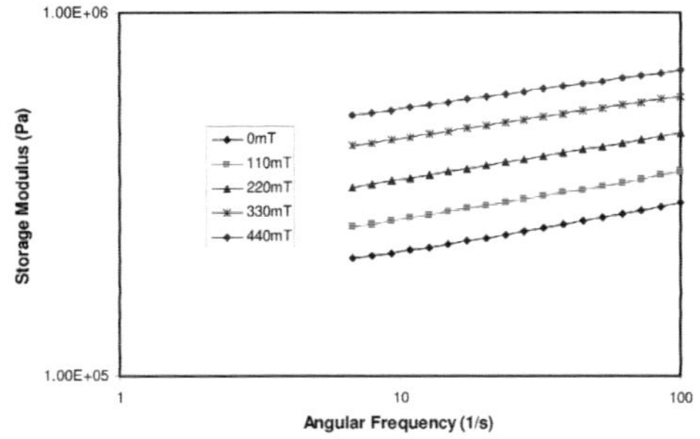

Figure 10. Storage modulus versus angular frequency sweep (anisotropic MRE Gr 20%).

From the figures above, we can see that in the log-log scale, the storage moduli of the two samples are both increasing linearly with the growth of angular frequency. This means with at a higher angular frequency, the samples have bigger storage modulus. This logarithmically linear relationship of the storage modulus to the angular frequency can be used to predict the storage modulus at a certain frequency. The effect of the graphite weight fraction on the MR effect was shown in figures 11 & 12, from which it can be seen that with higher graphite weight faction, the samples have a bigger storage modulus. This also proves that the graphite powders contribute to the initial stiffness of MRE samples.

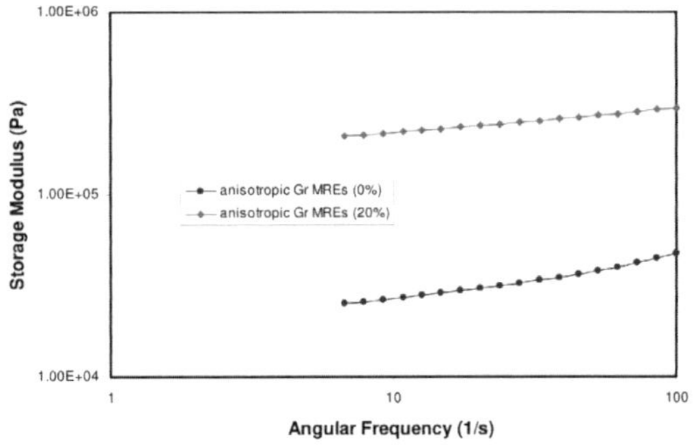

Figure 11. Storage modulus versus angular frequency sweep (without magnetic field).

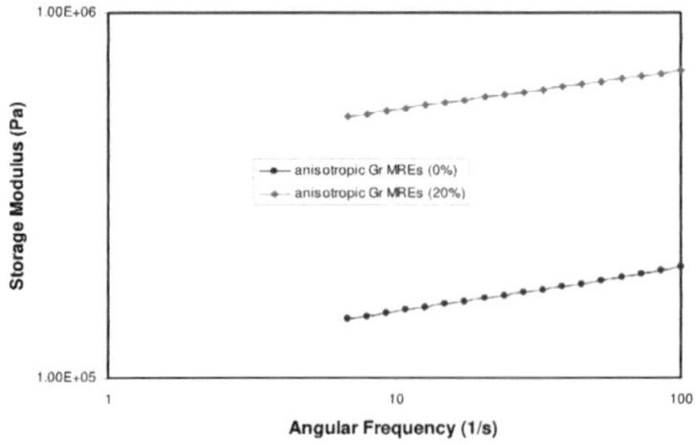

Figure 12. Storage modulus versus angular frequency sweep (440mT magnetic field).

The ratio of storage modulus at 440 mT to the storage modulus at 0mT is the MR effect. The ratio of anisotropic conventional MREs is around 5.2, when the graphite weight fraction increases to 20%, the MR effect is only around 2.4. This again means with the growth of graphite, the MR effect decreases. This phenomenon is because graphite powders contribute to increase the samples' stiffness and to lower iron particles volume fraction in the sample, therefore the MR effect can only have less values on the Gr MRE samples than conventional MREs.

3.3. Sensing capability measurement results and discussion

After the test, the resistance data of anisotropic Gr MRE with 20% graphite weight fraction is shown in figure 13.

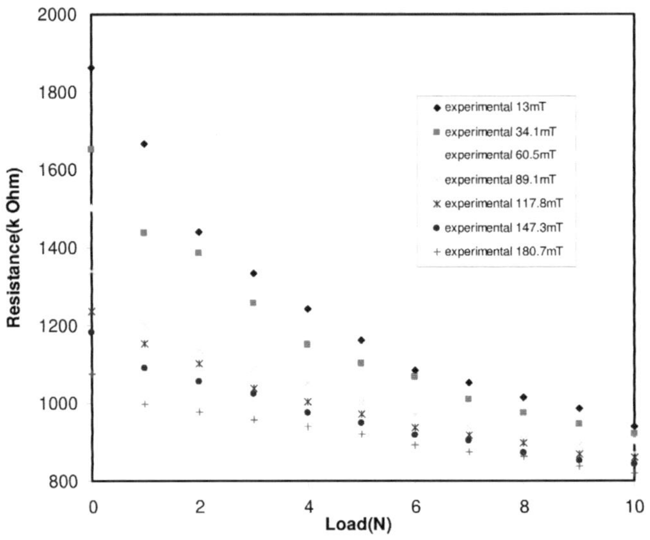

Figure 13. Resistance versus load (anisotropic MRE Gr 20%).

As can be seen from figure 13, in a fixed magnetic field when the external load increases from 0 to 10 N the resistance of Gr MREs reduces at all different magnetic field intensities. With smaller load the resistance changes significantly but it decreases slowly when the load is more than 5 N. Meanwhile it is clear that under a fixed external load, the higher magnetic field intensity leads to a lower resistance value. For instance, at 5 N external force, the resistance decreases from 1086.3 kΩ (without the magnetic field) to 918 kΩ at a 440 mT magnetic field. This trend can be seen in Figure

13. Therefore, The total trend is the decline in resistance with raising the magnetic field or increasing the external load.

To explain how the microstructure affects the resistance of Gr MRE, from the Dipole Model we announced a representative volume unit (RVU) that consists of two neighbouring hemispheres and the surrounding polymer matrix, which can be regarded as the minimum volume element in the conventional MRE. Figure 14 shows the longitudinal section of the unit. [15]

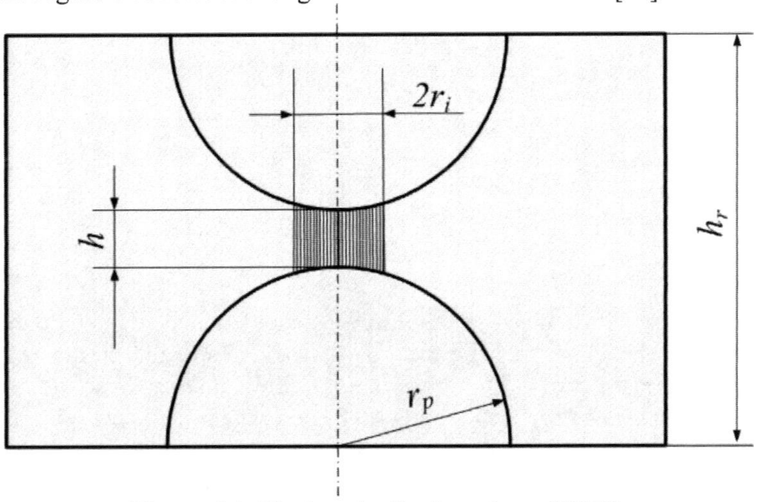

Figure 14. The longitudinal section of RVU.

In the theoretical approach, with the help of Fowler-Nordheim equation, Hertz theory and theory of magnetism [16-21], we got a final equation showing the resistance Rg of Gr MREs.

$$R_g = \frac{\lambda_g \lambda_i l}{3A\phi \cdot \left[\frac{2\alpha}{h^2} E \exp\left(-\frac{h\beta}{2r_p E}\right) + \frac{\sigma_f}{r_p h}\right] \cdot \left[r_{i0} + r_p\left((\sigma_0 + \sigma_1 + \sigma_2)^{1/3} - \sigma_0^{1/3}\right) \cdot \left(\frac{3\pi(1-\nu^2)}{2E_p}\right)\right]} \quad (1)$$

By substituting the known and constant coefficients in to Equation 1, a simplified equation is derived.

$$R_g = \frac{1.78 * 10^{-2}}{10^{-8} + 0.000002\left((2000 + \sigma_1 + \sigma_2)^{1/3} - 2000^{1/3}\right) \times 0.001166} \quad (2)$$

where σ_1 is the compressive stress from the external force and σ_2 is the compressive stress from the external magnetic field.

To show the comparison between the experimental result and the theoretical prediction (anisotropic MRE with graphite weight fraction 20%) clearly, four series of both the experimental result and the theoretical result are selected, as shown in Figure 15. Figure 15 shows that the experimental result and theoretical result couldn't match each other perfectly, but that the trends of both the experimental result and the theoretical result are the same.

Similarly some other comparison can be made from the data. At the fixed external force (which is expressed as load) when the magnetic field intensity increases the resistance of the sample decreases. The data under three external loads such as 1 N, 5 N and 10 N were chosen to compare the experimental and theoretical results in figure 16. With an increase in the magnetic field intensity, the sample's resistance decreases. A higher external applied load leads to lower resistance of Gr MREs.

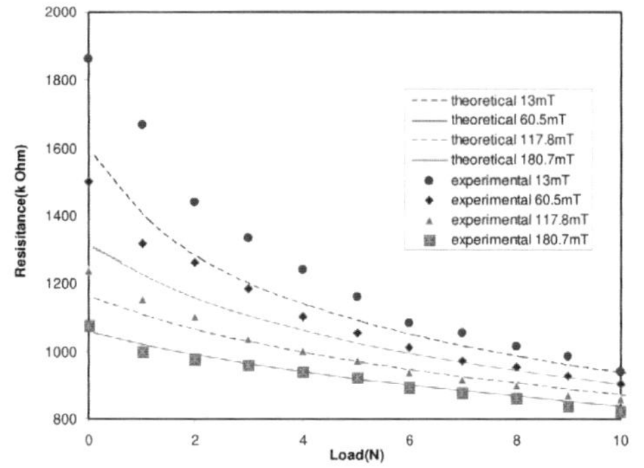

Figure 15. Comparison of experimental result and theoretical result.

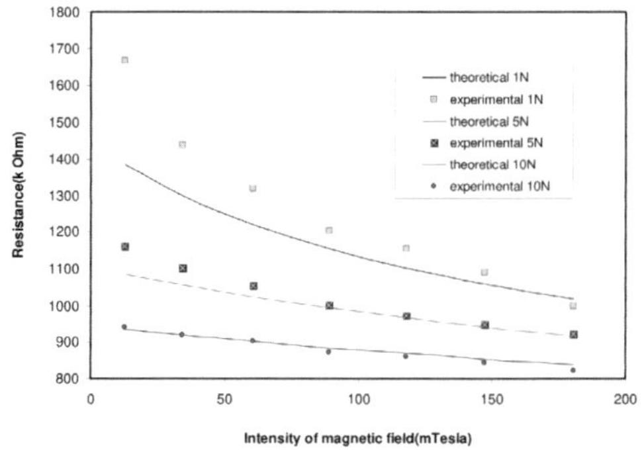

Figure 16. Resistance changing at a fixed load.

4. Conclusion

An anisotropic Gr MREs with 20% graphite weight fractions was selected to compare with anisotropic conventional MREs about rheology and to be studied its sensing capability in this study. LA SEM was used to observe their microstructures. This observation shows that the graphite powders effect the forming of carbonyl iron chains by connecting two iron chains in parallel and connecting the disconnected iron chains the graphite contributes to the conductivity of MREs.

The steady state and dynamic tests such as strain amplitude sweep and angular frequency sweep were used to test the magnetorheology of Gr MREs. With the help of graphite in MREs, the Storage and Loss Moduli are both changed. The steady state tests showed that the graphite decreases or even diminish the viscoelastic linear range of MREs. The dynamic test proved that the samples with higher graphite weight fraction show higher initial storage and loss moduli and lower MR effects.

Based on a Dipole model, a representative volume unit was presented to show the resistance of ideal anisotropic MREs. The current flowing through the ideal chain structure were derived by taking into account both the tunnel current and conductivity current. The exponential parameters were identified and then used to reconstruct modelling predictions. The comparison between experimental results with modelling predictions indicates that the proposed mathematical model performs well in investigating the sensing capabilities of the graphite based MRE elastomers.

Totally compared by the microstructures and experimental data from this study, we got the major effects of graphite in Gr MREs: (a) Graphite powders affect the forming of iron lines when the mixture of iron particles, graphite powders, silicone rubber and silicone oil is curing. The graphite particles lead to an increment of initial mechanical properties and a decrement of the MR effect because of the volume fraction of iron particles was lowered by adding the graphite powders to the sample, (b) For a Gr MREs sample, the graphite powder can connect the off-state lines and link the adjacent lines parallelly, which significantly increases the conductivity of the Gr MREs sample.

References

[1] Carlson J D and Jolly M R 2000 *Mechatronics* **10** 555-69
[2] Ginder J M, Clark S M, Schlotter W F and Nichols M E 2002 *Int. J. Mod. Phys.* B **16** 2412–18
[3] Xu Z B, Gong X L, Liao G J and Chen X M 2010 *J. Intell. Mater. Syst. Struct.* **21** 1039– 47
[4] Zhang X Z and Li W H 2009 *Smart Struct. Syst.* **5** 517–29
[5] Deng H X and Gong X L 2008 *Commun. Nonlinear Sci.* **13** 1938-47
[6] Ni Z C, Gong X L, Li J F and Chen L 2009 *J. Intell. Mater. Syst. Struct.* **20** 1195–1202
[7] Lokander M and Stenberg B 2003 *Polym. Test.* **22** 245-51
[8] Shiga T, Okada A and Kurauchi T 1995 *Appl. Polym. Sci.* **58** 787-92
[9] Gong X L, Zhang X Z and Zhang P Q 2005 *Polym. Test.* **24** 669–76
[10] Leblanc J L 2002 *Prog. Polym. Sci.* **27** 627–87

[11] Li WH, Kostidis K, Zhang XZ and Zhou Y 2009 *Ieee/Asme Int. Conf. Advanced Intelligent Mechatronics* **1-3** 233–38
[12] Bica I 2009 *Mater. Lett.* **63** 2230-32
[13] Kchit N and Bossis G 2009 *Phys. D: Appl. Phys.* **42** 5505
[14] Wang X J, Gordaninejad F, Calgar M, Liu Y M, Sutrisno J and Fuchs A 2009 *J. Mech. Des.* **131** 6
[15] Tian, T F, Li W H and Deng Y M 2011 *Smart Mater. Struc.* **20** 7
[16] Chang W R, Etsion I and Bogy D B 1988 *J. Tribol.* **110** 57-63
[17] Dahlke W E and Sze S M 1967 *Solid State Electron.* **10** 865-73
[18] Etsion I, Levinson O, Halperin G and Varenberg M 2005 *J. Tribol.* **127** 47-50
[19] Jakubovics J P 1994 *Magnetism and Magnetic Materials* 2nd rev. ed. (London: Institute of Materials)
[20] Maxwell J C 1892 *A Treatise on Electricity and Magnetism.* 3rd ed. (Oxford: Clarendon Press)
[21] Weinberg Z A 1982 *J. Appl. Phys.* **53** 5052-56

Journal of Physics: Conference Series **412** (2013) 012038 doi:10.1088/1742-6596/412/1/012038

Study of PDMS based magnetorheological elastomers

T F Tian[1,3], X Z Zhang[1], W H Li[1], G Alici[1] and J Ding[2]
[1] School of Mechanical, Materials and Mechatronic Engineering, University of Wollongong, Wollongong, NSW 2522, Australia
[2] Defence Science and Technology Organization, Human Protection and Performance Division, 506 Lorimer Street Fishermans Bend, VIC 3207, Australia

E-mail: tt436@uowmail.edu.au

Abstract. The fabrication of conventional magnetorheological elastomers (MRE) is usually taken more than 1 day because the conventional matrixes such as natural rubber and silicone rubber need long curing time to become solid state. This study presents a rapid method for fabricating MRE within 90 minutes by using poly(dimethylsiloxane) (PDMS) as the matrix thanks to the rapid curing of PDMS in high temperature. A total of four PDMS based MRE samples were fabricated. Their mechanical and rheological properties under both steady-state and dynamic loading conditions were tested with a parallel-plate rheometer. Additionally, the microstructures of the PDMS based MREs were also observed by SEM and compared with the silicone rubber based MRE.

1. Introduction

Magnetorheological elastomers (MREs) are a group of smart materials where polarized particles are suspended in a non-magnetic solid or gel-like matrix [1]. These materials exhibit characteristics that their moduli can be reversely controlled by an external magnetic field. MREs have recently found a variety of applications, such as adaptive tuned vibration absorbers, dampers, sensors and so on [2-10].

MREs generally consist of three major components: magnetisable particles, matrix and additives. Iron particles are generally used as the filler material to fabricate MREs. Both natural rubber and silicone rubber are used as typical matrixes [11-12]. Additives are commonly used to adjust the mechanical and chemical properties or electrical performance of MR fluids [13-14] as well as MR elastomers [15]. Silicone oil is an additive to increase the gaps between the matrix molecules and to decrease the gaps between the conglutination of molecules. Apart from increasing the plasticity and fluidity of the matrix, the additives can average the distribution of internal stress in the materials, which makes them ideal for fabricating MRE materials [16].

Natural rubber is an elastomer. The purified form of natural rubber is the chemical polyisoprene which can also be produced synthetically. Heat is normally required to vulcanise silicone rubber. The silicone rubber and a vulcanising silicon sealant (at room temperature) are mixed with silicon oil to change its ductility. The silicon oil is selected on the basis of preliminary studies with different elastomers.

PDMS is one example of silicon rubber. The PDMS has a low surface tension and is capable of wetting most surfaces. The stability and chemical neutrality of the system also enables the adhesive to

[3] To whom any correspondence should be addressed.

bond to the metals [17]. PDMS is the most widely used silicon-based organic polymer, and is particularly known for its unique properties such as curing at low temperatures, rapid curing at high temperature, deformation reversibility and, surface chemistry controllable by reasonably well-developed techniques. PDMS is optically clear, and, in general, is considered to be inert, non-toxic and non-flammable [18-20]. Previous studies have focused on the structural properties of PDMS [21] and the kinetic properties under crystallization from solution [22].

In this investigation, PDMS is used as the matrix of MREs and studied the mechanical properties of the new MREs. Carbonyl iron particles can easily blend into PDMS because of its liquid initial state, meanwhile, the rapid curing of PMDS under certain temperature provides a quick method to fabricate MREs.

2. Experimental

2.1. Materials and fabrication

The base and curing agent of PDMS, type Sylgard 184 (Dow Corning Pty. LTD) at weight ratio 10:1 were firstly mixed evenly to be used as the carrying matrix, and then added to the carbonyl iron particles, type C3518 (Sigma-Aldrich Pty. LTD) and stirred sufficiently. The iron particles' diameter is between 3 μm and 5 μm at normal distribution. The density of iron particles and PMDS are 7.86 g/cm^3 and 1.1 g/cm^3, respectively. The final mixtures were placed in a vacuum chamber to eliminate bubbles in 30 min and were placed between two pieces of plastic paper and pressed to be membranes. Some small cylinders with 1 mm thickness were used to ensure the thickness of MREs is 1 mm. Then the mixture was left for curing in an oven at 100°C for 35 minutes and then cut by a punch to disks with 20 mm diameter. The total time to fabricate the PDMS was no more than 100 min, which saves lot of time than the conventional MREs made of natural rubber or silicone rubber that need more than 24 h to curing.

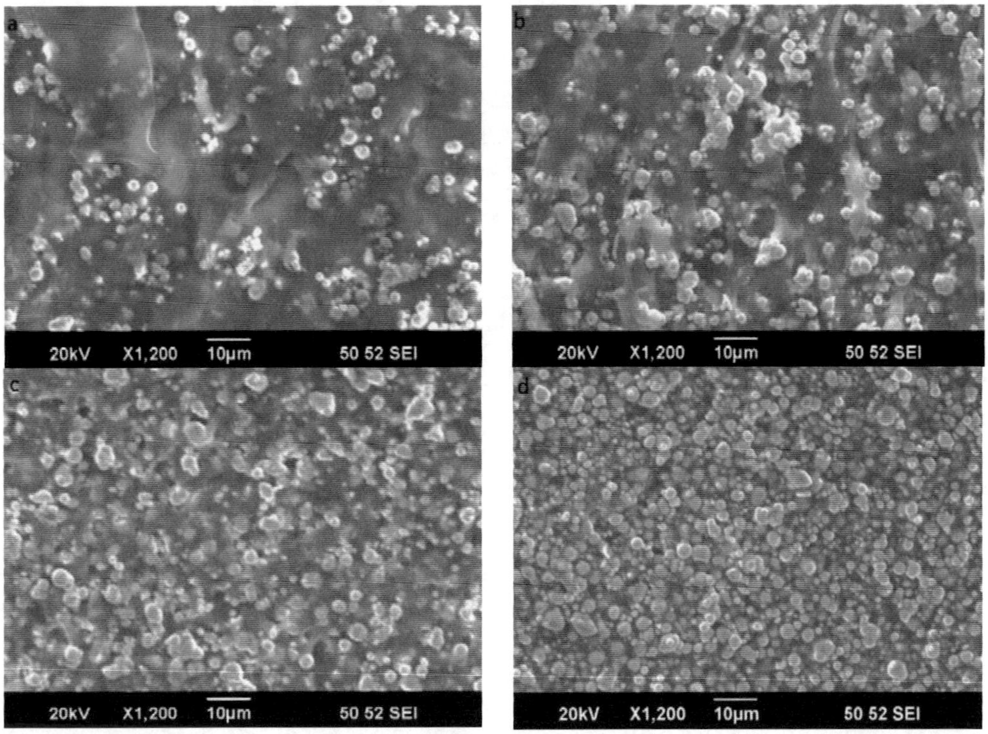

Figure 1. Microstructure of PDMS MREs (a) iron 60%, (b) iron 70%, (c) iron 80%, (d) iron 90%.

In this study, 4 PDMS MRE samples were fabricated whose iron particles' weight fraction (iron wt. %) were 60%, 70%, 80% and 90%. All MRE samples were isotropic.

2.2. Microstructure observation

LV-SEM (JSM 6490LV SEM) was used to observe the microstructure of PDMS MREs. Figure 1 shows the surface imaging for all MRE samples' microstructures and indicates that the carbonyl iron particles disperse randomly in the PDMS matrix due to the isotropy of all samples. The sample having higher iron weight fraction shows more iron particles in the same area.

2.3. Rheological measurement device

The shear-strain dependent rheology of the MREs was measured by a parallel-plate rheometer (MCR 301, Anton Paar, Germany). A Magneto Rheological Device (MRD 180, Anton Paar, Germany) was used to generate magnetic filed in the test area and a temperature control device (Viscotherm VT2, Anton Paar, Germany) was used to control the measuring sample to be at 25 °C. Steady-state shear strain sweeps, dynamic strain sweeps, dynamic frequency sweeps and magnetic field intensity sweeps were carried out using the PP-20 measuring geometry (20 mm diameter) with a gap of 1 mm. The testing procedure for each measurement is illustrated below. Initially, the sample is sheared at a constant shear rate of 100 s^{-1} at zero field for half a minute to distribute the particles uniformly. Next, the desired magnetic field is applied and maintained for more than 30 seconds so that the testing sample forms a static structure. Then, steady shear, dynamic oscillatory shear modes and, magnetic field intensity sweeps were employed to measure rheological properties of the samples under steady and dynamic loading conditions.

In this experiment, the magnetic flux density of the sample of MRE (BMRE) in the measuring gap depends not only on the current (I) applied to the samples and the magnetic properties of MRE materials. As the permeability of MRE samples varies little, an empirical equation, BMRE = 220 I, was employed to predict the flux density for different MRE samples, where the units of BMRE and I are in mT and amp (A), respectively. In the following test, the test current varies from 0 to 2 A with the increment 0.5 A, whose intensity of magnetic field is 0 to 440 mT with the increment 110 mT.

3. Results and discussion

3.1. Steady state

Under rotary shear the shear stress and strain of MREs under fields varying from 0~440 mT were measured at 25°C and 5 rad/s angular frequency. The shear rate range is from 0.0005 to 5 s^{-1}. The MR effect was evaluated by measuring the shear strain–stress curve of the sample with and without a magnetic field applied. Figure 2a–d shows the strain–stress curve of different samples at five different magnetic field intensities ranging from 0 to 440 mT. The slope of the strain–stress curve is the shear modulus of the material.

As can be seen in the figure 2a-d, all the samples' shear modulus show an increasing trend with magnetic field before they reach magnetic saturation at high field strength, which proves that all the MRE samples exhibit obvious MR effects. In figure 2d, it can be seen that the curve at 0 mT, which is the zero-field modulus of the 90 wt.% sample, shows a bigger slop of stress-strain curve compared with the lower graphite concentration samples, which means 90 wt.% sample has a higher zero-field modulus. When the strain is above the limitation, the shear stress reaches a saturation (maximum value) and decrease steadily. This might be due to the sliding effect. Additionally, other factors, such as the sample surface roughness and the normal force, could contribute to the resultant stress. In particular, they influence the static friction between the MRE sample and the upper plate, which consequently result in overshoots, as shown in figures 2a-d.

Also from figure 2a-d, the shear stress shows a linear relationship with the shear strain when the strain is within a range. This means that the MRE acts with linear viscoelastic properties when the strain is below a limitation. Table 1 summarizes the linear ranges of all samples at various magnetic field intensities.

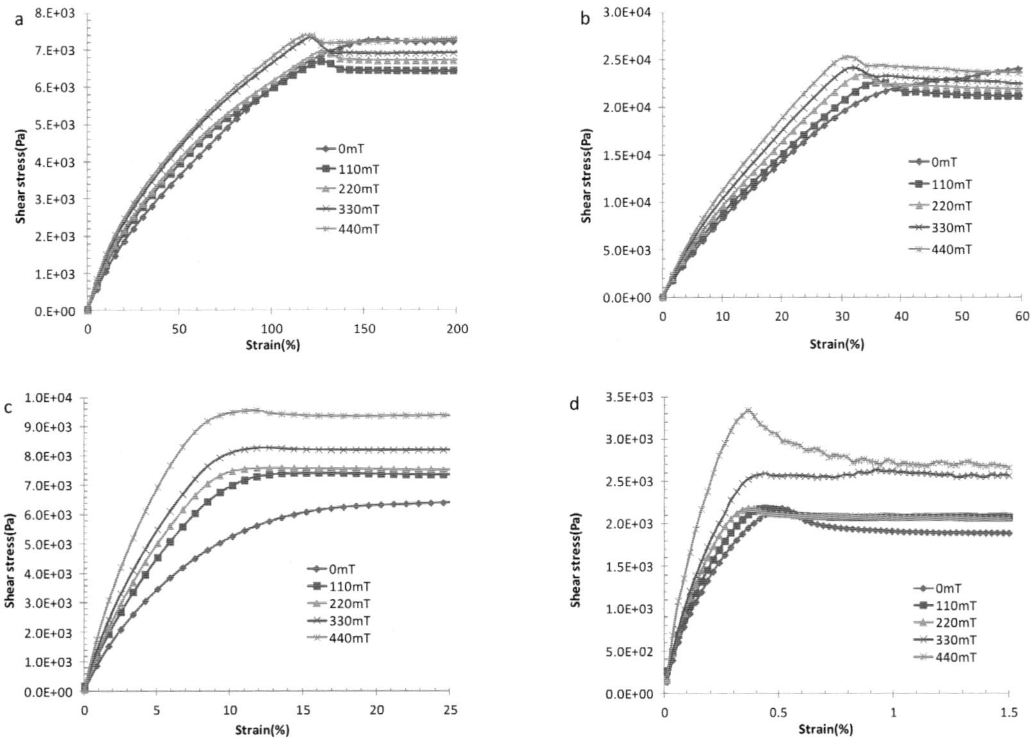

Figure 2. Strain-stress curve versus magnetic field (a) iron 60%, (b) iron 70%, (c) iron 80%, and (d) iron 90%.

Table 1. Linear range of all the samples at different magnetic fields.

Samples	Linear range				
	0mT	110mT	220mT	330mT	440mT
60% iron	150	127	127	123	120
70% iron	80	37	33	31	30
80% iron	40	10	9	9	8
90% iron	0.5	0.4	0.35	0.35	0.35

When the carbonyl iron weight fraction increases from 60% to 90%, the range of linearity decreases from 120%-150% to 0.35%-0.5%. The higher iron particles sample has the lower linear range. Meanwhile, the increasing magnetic field also contributes to the decreasing of the linear range.

3.2. Dynamic tests result

In order to obtain the dynamic mechanical behaviour of MRE, both strain amplitude sweep tests and angular frequency sweep tests were used. Five sets of data were collected for different amplitudes of oscillation, according to the various magnetic field inputs to the samples of MREs. Same as the steady state tests, five different magnetic field intensities, 0, 110, 220, 330 and 440 mT, were used in this experiment. The amplitude of shear strain in angular frequency sweep tests is set at 1% and the input frequency was 5 Hz in the strain amplitude sweep tests.

3.2.1. Strain amplitude sweep. In the strain sweep test, the storage and loss moduli were tested by varying strain from 0.1% to 100% at different magnetic fields and room temperature. Figure 3a-d shows the changing of storage modulus at the strain amplitude sweep.

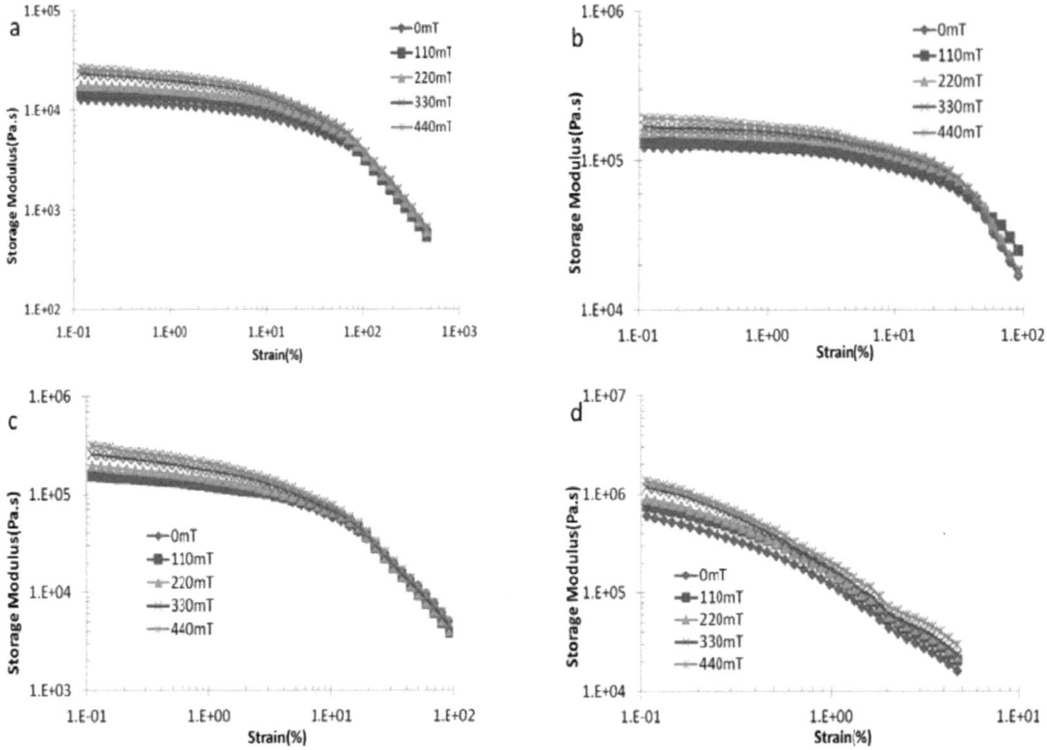

Figure 3. Storage modulus versus strain amplitude sweep (a) iron 60%, (b) iron 70%, (c) iron 80%, and (d) iron 90%.

In figure 3a-d, the overall trend of storage modulus is decreasing with the strain amplitude. It goes down smoothly within a critical shear strain and begins to drop significantly over the critical value and to go crossing with loss modulus, with which we can say that within the critical shear strain, the storage modulus shows approximately a linear relationship with the shear strain. Take sample containing 80% iron as an example, within 10% strain, its storage modulus versus strain as linear. These linear ranges are quite similar to the data in Table 1.

3.2.2. Angular frequency sweep. In this test, the strain is set at 1%. According to the experimental equipments, the angular frequency was varied from 1 to 100 s^{-1} at different magnetic fields of 0, 110, 220, 330 and 440 mT. The figure 4a-d show the storage and loss moduli curves of the MRE samples under frequency sweep and at 25°C.

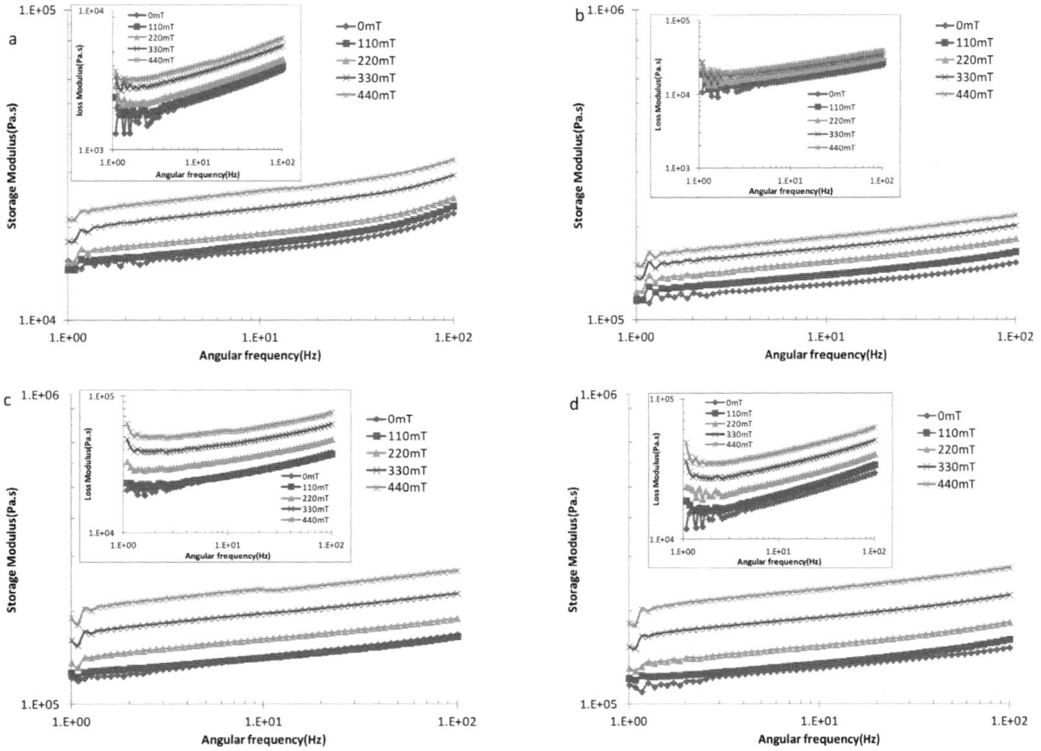

Figure 4. Storage and loss moduli versus angular frequency sweep (a) iron 60%, (b) iron 70%, (c) iron 80%, and (d) iron 90%.

From the figures above, we can see that in the log–log scale, the storage and loss moduli of all the samples are both increasing linearly with the growth of angular frequency. This means that at a higher angular frequency, the samples have bigger storage and loss moduli. This logarithmically linear relationship of the storage and loss moduli to the angular frequency can be used to predict the storage and loss moduli at a certain frequency.

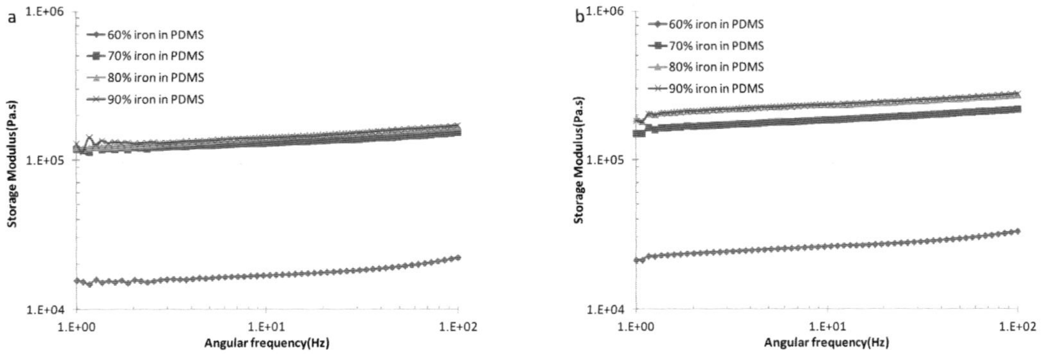

Figure 5. Storage modulus vs. angular frequency sweep (a) 0 mT and (b) 440 mT.

The effect of the iron weight fraction on the MR effect was shown in figure 5a-d. It can be seen from these figures that 60% iron sample has the lowest storage modulus, but the other three samples show very similar storage modulus at both low and high magnetic fields. This means that in PDMS matrix, iron particles contribute to the MRE samples' initial stiffness before a critical weight fraction value between 60% and 70%, after which the stiffness of PMDS dominates in MREs' initial stiffness.

3.3. Magnetic field intensity sweep

The strain and angular frequency are set at 1% and 1 rad/s respectively, in this test. Magnetic field intensity is increasing from 0 mT to 440 mT linearly. This test is mainly used to figure out the relative MR effect (G'_{max}/G'_0) of each sample. G'_0 denotes the MRE samples' zero-field storage modulus, G'_{max} denotes the saturated field-induced modulus, and G'_{max}/G'_0 denotes the relative MR effect. Figure 6a-b shows the storage modulus curves of the MRE samples versus magnetic field intensity at 25°C.

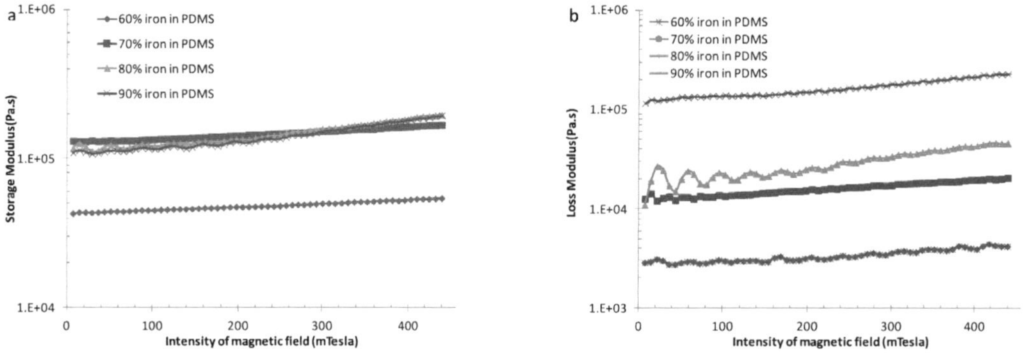

Figure 6. Storage and loss moduli vs. magnetic field intensity of different samples (a) Storage modulus and (b) Loss modulus.

From figure 6a we can see that the relative MR effect is 1.27, 1.30, 1.74 and 1.81 for 60%, 70%, 80% and 90% samples, respectively. The sample with higher iron particles concentration has higher relative MR effect. Also figure 6a shows the similar value of 70%, 80% and 90% sample, which proves that the mechanical properties of PMDS dominates in MREs' stiffness.

4. Conclusion

Four isotropic PDMS based MRE sample with 60%-90% iron particles weight fractions were fabricated in this study. The fabrication time for PDMS MREs is less than 100 min which is much shorter than the time needed for conventional MREs based on natural rubber or silicone rubber. LV SEM was used to observe their microstructures. This observation shows that iron particles disperse randomly in the isotropic samples.

The magnetic field intensity sweep, steady state and dynamic tests such as strain amplitude sweep and angular frequency sweep were used to test the magnetorheology of PDMS MREs. Because of the strong mechanical properties of PDMS, the storage and loss moduli are both changed less than the conventional MREs. The steady state tests showed that the increase of iron particles in the sample would diminish the viscoelastic linear range of MREs. The dynamic and magnetic field intensity sweep test proved that the samples with higher iron weight fraction show higher initial storage and loss moduli and also higher MR effects.

References

[1] Ginder J M, Clark S M, Schlotter W F and Nichols M E 2002 *Int. J. Mod. Phys.* B **16** 2412–18
[2] Deng H X, Gong X L and Wang L H 2006 *Smart Mater. Struc.* **15** 111-6
[3] Ni Z C, Gong X L, Li J F and Chen L 2009 *J. Intell. Mater. Syst. Struct.* **20** 1195–202
[4] Zhang X Z and Li W H 2009 *Smart Struct. Syst.* **5** 517–29
[5] Kim Y K, Koo J H, Kim K S and Kim S H 2011 *Rev. Sci. Instrum.* **82** 035103
[6] Xu Z B, Gong X L, Liao G J and Chen X M 2010 *J. Intell. Mater. Syst. Struct.* **21** 1039–47

[7] Zhang W, Gong X L, Jiang W Q and Fan Y C (2010) *Smart Mater. Struct.* **19** 085008
[8] Chen L, Gong X L and Li W H (2007) *Smart Mater. Struc.* **16** 2645–50
[9] Li WH, Kostidis K, Zhang XZ, Zhou Y 2009 *Ieee/Asme Int. Conf.Advanced Intelligent Mechatronics* **1-3** 233–38
[10] Bica I 2010 *J. Ind. Eng. Chem.* **16** 359–63
[11] Gong X L, Zhang X Z and Zhang P Q 2005 *Polym. Test* **24** 669–76
[12] Li W H, Zhou Y and Tian T F 2010 *Rheol. Acta.* **49** 733–40
[13] Park B J, Song K H and Choi H J 2009 *Mater. Lett.* **63** 1350–52
[14] Fang F F, Choi H J and Jhon M S 2009 *Colloids Surf.* A **35**1 46–51
[15] Zhang X Z, Peng S L, Wen W J and Li W H 2008 *Smart. Mater. Struct.* **17** 045001
[16] Leblanc J L 2002 Prog. *Polym. Sci.* **27** 627–87
[17] De Buyl F 2001 *Int. J. Adhes. Adhes.* **21** 411–22
[18] Sundararajan PR 2002 *Polymer* **43** 1691-93.
[19] McDonald J C, Duffy D C, Anderson J R, Chiu D T, Wu H K, Schueller O J A and Whitesides G 2000 *Electrophoresis* **21** 27-40.
[20] Noonan K J T and Gates D P 2008 *Annu. Rep. Prog. Chem., Sect. A: Inorg. Chem.* **104** 394-413.
[21] Rajan G S, Sur G S, Mark J E, Schaefer D W and Beaucage G 2003 *J. Polym. Sci., Part B: Polym. Phys.* **41** 1897-901.
[22] Feio G, Buntinx G and Cohen-Addad J P 1989 *J. Polym. Sci., Part B: Polym. Phys.* **27** 1-24.

Study on the properties of porous magnetorheological elastomers under shock effect

B X Ju[1,2], M Yu[1,2,1], J Fu[1,2], X Zheng[1,2] and Q Yang[3]
[1] Research Center of Sensing and Instrumentation Technologies, Chongqing University, Chongqing 400044, China
[2]College of Optoelectronic Engineering, Chongqing University, Chongqing 400044, China
[3]State Key Laboratory of Polymer Materials Engineering, Sichuan University, Chengdu 610065, China

E-mail: yumiao@cqu.edu.cn

Abstract. As a safe protector, buffer has been widely applied to engineering applications. The properties of cushion materials play a key role in the performance of the buffer under shock loading. Magnetorheological elastomers (MRE) are a kind of novel smart materials and show to have a controllable, field-dependent modulus, which have attracted increasing attentions and broad application prospects. This paper aims to fabricate a new kind of MRE, named as porous MRE, and study on the properties of porous MRE under shock effect in the presence of an external magnetic field. Three kinds of MRE samples based on polyurethane matrix were prepared without external magnetic field, and ammonium bicarbonate was used as foaming agent with content of 0 wt.%, 0.26 wt.%, 0.67 wt.%, respectively. The microstructures of the sample were observed by using a digital microscope, and image processing and analysis was applied to calculate the parameters of porous MRE. A sleeve structure and mass block were used to test the shock performance of porous MRE under shear mode, and an electromagnetic vibration and shock table was used to provide shock signal with half-sine shock signal. The results show that the content of foaming agent has an obvious influence on the microstructures of porous MRE. The porosity of the porous MRE samples increases with increasing of foaming agent content. Moreover, experimental results show that shock energy dissipation capacity is better than that of traditional MRE. This study is expected to provide guidance in the application of MRE in practical devices, such as in buffer devices.

1. Introduction

Magnetorheological (MR) materials are known as smart materials that have rheological properties that can be changed under a magnetic field, which can be classified into several groups, including magnetorheological fluids (MRF) [1-2], magnetorheological elastomers (MRE) [3-5], magnetorheological gel (MRG) [6], magnetorheological foam [7], magnetorheological plastomer (MRP) [8], and so on. The most common is MRF and MRE. MRF is magnetically polarizable particles suspended in viscous fluids. MRE is a kind of material, which have polarizable particles arranged in chains in polymer media such as silicon rubbers, natural rubbers and polyurethane matrix [5, 9]. There

[1] Corresponding author.

are some problems in MRF, such as particle sediment, stability and leakage in application devices, these problems hinder its further development. To solve these problems, MRE with polymer matrices are developed. Obvious advantages of MRE are that the particles are not able to settle with time and that there is no need for containers to keep the MR material in fixed place.

MR materials have many analogical mechanical behaviors. But, MRE have a unique mechanical performance from other material, MRE have a controllable, field-dependent modulus while MRF have a field-dependent yield stress [10]. The applications of MRE have focused on several areas, such as MRE adaptive tuned vibration absorbers [11], MRE isolators [12], and MRF-MRE dampers [13]. In order to broaden the scope of application of MRE, several aspects of MRE performance have been studied. Chen *et al.* [9] investigated the effects of fabrication conditions (matrix type, external magnetic flux density, and temperature), and materials (plasticizers and iron particles) on MRE performances. When the iron particle weight fraction is 80% and the external magnetic flux density is 1 T, the field-induced increment of shear modulus reaches 3.6 MPa, and the relative MR effect is 133%. Li *et al.* [5] studied the effect of the pre-structure process on MRE performance, the results demonstrated that the pre-structure process strongly affected the magnetic field dependence of shear storage modulus of MRE. The interfacial friction damping properties of MRE is investigated experimentally by Fan *et al.* [14]. The experimental results imply that the interfacial friction damping mainly comes from the frictional sliding at the interfaces between the free rubber and the particles. Moreover, the influence of the matrix type, dynamic strain amplitude, driving frequency, and content of iron particles on damping property was investigated by Li *et al.* [15]. Zhang *et al.* [16] investigated fatigue properties of MRE samples, the results revealed that MR effect, storage modulus, and loss modulus of MRE depended strongly on the strain amplitudes and the number of cycles. The mechanical properties of MRE were tested in compression passively and with increasing magnetic flux density by Kallio *et al.* [17]. It was noted that the dynamic stiffness of aligned MRE increased with increasing testing frequency range, the loss factor of aligned MRE was also tunable with the magnetic flux density but the absolute values also depend on the testing frequency. Therefore, more work should be done to enlarge practical applications of MRE.

This study investigate the shock performance of porous MRE, several kinds of MRE samples based on polyurethane matrix were prepared, the microstructures, dynamic mechanical properties and shock performance were analyzed. We hope this study can provide a good guidance for improving performance of isotropic MRE.

2. Experimental

2.1. MRE samples' fabrication

The material of fabricated MRE samples consist of carbonyl iron particles (type CN, with the size distribution: d_{10} = 3.5 μm, d_{50} = 6.5 μm and d_{90} = 14 μm, provided by BASF corporation) and rubber matrix, stannous octoate (Sinopharm Chemical Reagent Co. Ltd., China) was used as catalyst. The matrix based on polyurethane material was mainly synthesized from castor oil (CO) and diphenylmethane diisocyanate (MDI: 4,4-≈50%, 2,4-≈50%, Yantai wanhua polyurethanes Co. Ltd., China). There are three steps in the synthesis of the MRE samples. Firstly, the mixture was mixed by CO and MDI, and weight ratio of CO and MDI is 100:26. Secondly, ammonium bicarbonate was added to mixture with stirring. At last, the mixture was mixed with carbonyl iron particles, and the mixture was cured at a temperature of 60ºC for 2 hours without external magnetic field. When the samples were cured, ammonium bicarbonate can be decomposed into NH_3, CO_2 and H_2O, the porous MRE samples were prepared as a function of ammonium bicarbonate. In addition, the samples without any ammonium bicarbonate for comparison purposes were also prepared. In this research, three kinds of MRE samples were prepared, the composition of MRE samples were listed in table 1.

Table 1. Composition of the MRE samples (wt.%).

sample name	carbonyl iron particles	matrix	ammonium bicarbonate
Sample 1	70	30	0
Sample 2	70	29.74	0.26
Sample 3	70	29.33	0.67

2.2. Microstructure of MRE

The microstructures of the samples were observed by using scanning electron microscope (SEM, FEI Corporation) and VHX-600 digital microscope (Keyence Corporation). In this paper, three kinds of MRE samples with different foaming agent content were prepared without magnetic field at a certain temperature. The microstructures were investigated and shown in figure 1 and figure 2.

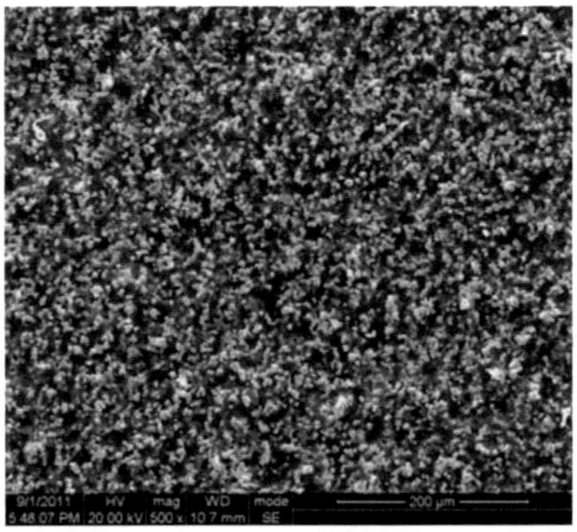

Figure 1. Microstructure of MRE sample (500×).

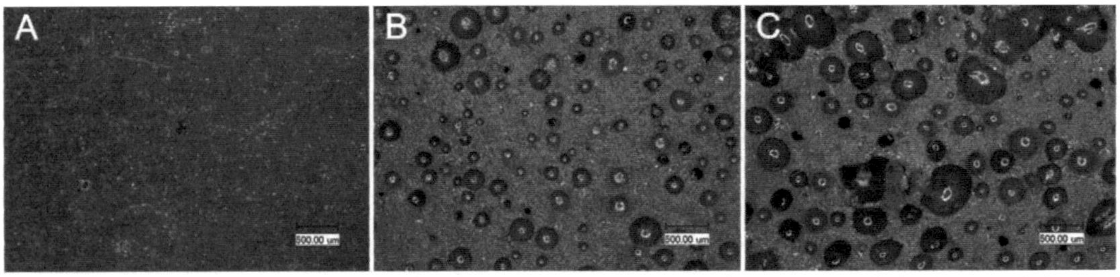

Figure 2. Microstructures of MRE samples (50×, A: sample1, B: sample2, C: sample3).

Figure 1 shows that the carbonyl iron particles are randomly distributed in the rubber matrix; the particles are embedded in and well bound with the rubber matrix. In addition, it can be seen from figure 2 that ammonium bicarbonate has an important influence on the structure of the MRE samples. When ammonium bicarbonate is added to the samples, lots of pores freely distribute in the matrix after curing under a certain temperature, the porosity is very different with different ammonium bicarbonate content.

2.3. Testing of dynamic mechanical properties

Dynamic mechanical properties of MRE samples were tested by using an advanced rheometer (MCR301, Anton Paar), where a magnetorheological device was added on rheometer. The electromagnetic field can generate magnetic fields ranging from 0 to 1.2 T. The picture of testing system is shown in figure 3. This system can applies a fixed oscillatory strain and frequency to samples, shear storage modulus, loss modulus and the damping factor can be calculated. The size of each sample is 20 mm in diameter and 1.2 mm in thickness.

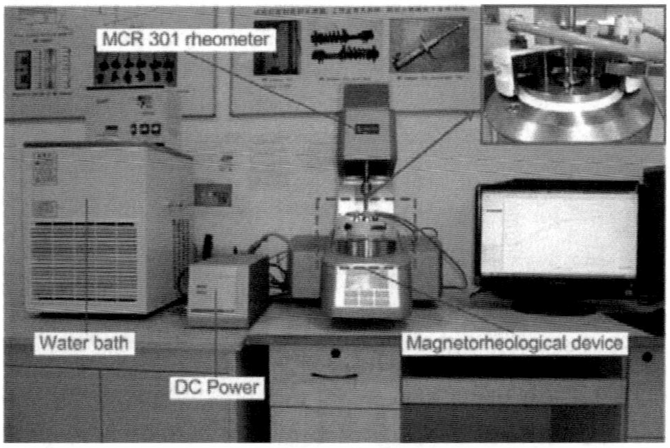

Figure 3. The picture of testing system.

To study the influence of magnetic field on the dynamic mechanical properties of MRE samples, MRE samples were tested under shear mode with constant strain amplitudes of 0.1%, 1% and 2%, respectively. Testing frequency was fixed at 10 Hz, and testing magnetic flux densities of from 0 to 1.2 T were selected. Figure 4 shows the relationship between the shear modulus of MRE and the magnetic flux density under different testing strain amplitude. From this figure, it can be seen that the initial and maximum shear storage modulus of MRE have decreasing trend with increasing of strain amplitude. In additional, magneto-induced shear storage modulus and MR effect are key parameter for evaluating MRE performance. In this study, magneto-induced modulus is defined as the absolute difference between the maximum value of shear storage modulus and the initial modulus. MR effect is the ratio of magneto-induced modulus and initial modulus. The magneto-induced modulus can achieve to 1.16MPa when strain amplitude is 0.1% and MR effect is as high as roughly 209.8%.

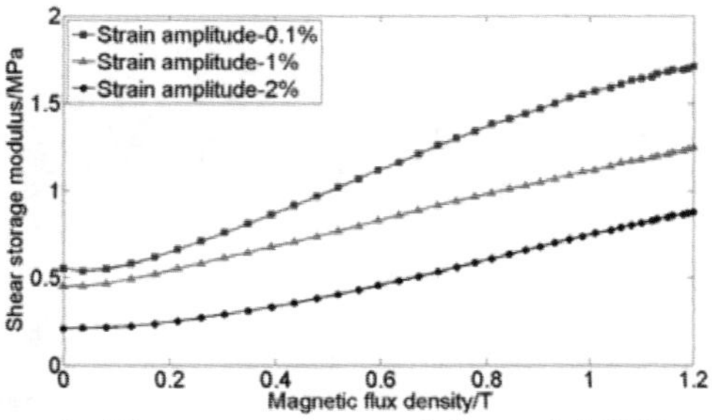

Figure 4. Shear storage modulus of MRE sample (ammonium bicarbonate content is 0 wt.%).

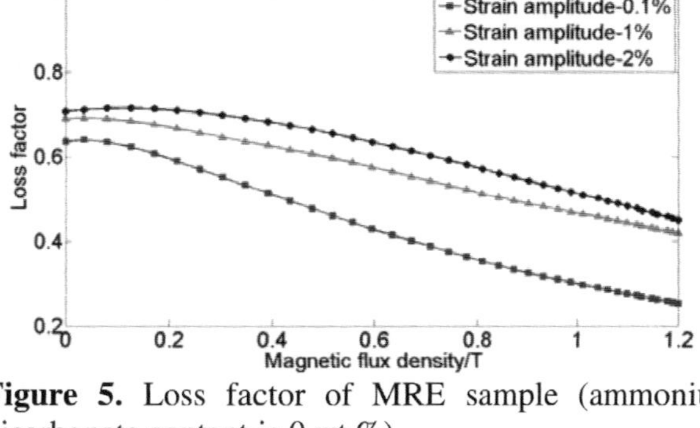

Figure 5. Loss factor of MRE sample (ammonium bicarbonate content is 0 wt.%).

MRE is a new kind of magnetic functional materials, which has been successfully applied to the variable stiffness vibration isolation and vibration absorption device [18-19]. In order to improve vibration absorber and isolation control effect, appropriate damping values of MRE is needed; the relationship of loss factor and magnetic flux density is shown in figure 5. As shown in figure 5, loss factor of MRE sample decrease steadily with increasing of external magnetic field, which represents the ability of dissipating energy of MRE.

2.4 Porosity of MRE

All the samples were cured at a certain temperature, ammonium bicarbonate can be decomposed into several kinds of gas (NH_3, CO_2 and H_2O) when temperature is more than 36ºC, and the porous structure was created due to the gas formation. In order to calculate the porosity of MRE, quantitative image analysis techniques were used to analyze microstructures of the MRE samples. Image segmentation technology was utilized to distinguish the object from the background; it is to distinguish between the rubber matrixes and pores. Threshold methods are widely used in image segmentation because of their simplicity and efficiency. Therefore, this study used threshold segmentation to convert the microstructure image as a 'binary' image. The threshold methods can be expressed as follow:

$$f_t(x,y)=\begin{cases} b_0, & f(x,y)<t \\ b_1, & f(x,y)\geq t \end{cases} \tag{1}$$

Where $f(x, y)$ represent gray scale of pixel point (x, y), t is threshold value, b_0 and b_1 is binary gray level. 0 and 1 was used to instead of and b_0 and b_1. 0 and 1 is represented by black and white, respectively. Once the object has been successfully segmented, threshold segmentation can convert the microstructure image to be a 'binary' image, i.e. white or black, as shown in figure 6.

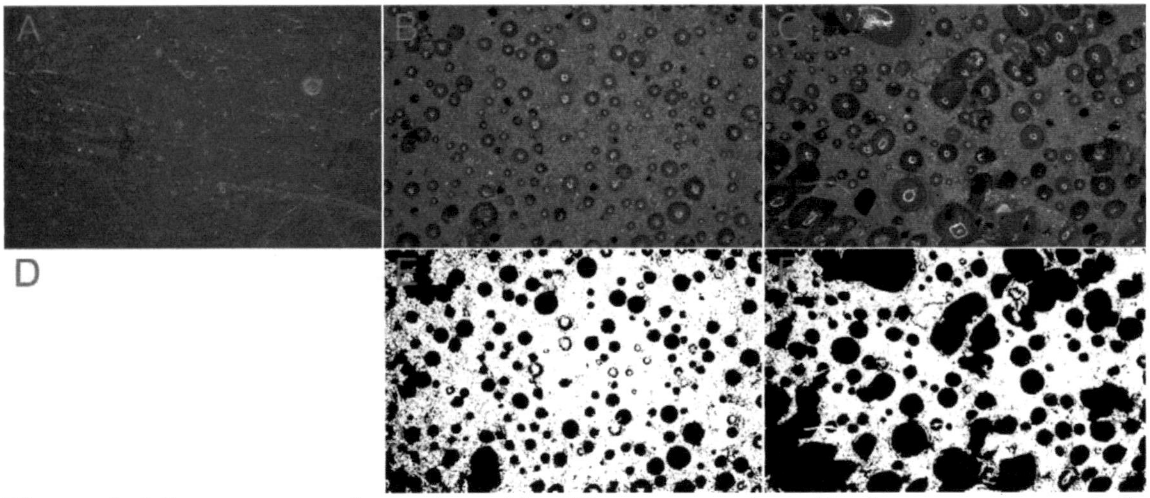

Figure 6. Microstructures of samples with different ammonium bicarbonate contents: A–C respectively corresponds to 0 wt.%, 0.26 wt.% and 0.67 wt.%; D–F corresponds to the results of image threshold segmentation of A–C.

The black represents the pore regions, and the white is the area without pores. So the porosity can be calculated from the numbers of pixels of white and black. The porosity of the MRE samples as a function of ammonium bicarbonate content is shown in table 2.

Table 2. Porosity of MRE samples.

Sample name	Porosity (v/v %)
Sample 1	0
Sample 2	32.5
Sample 3	52.7

It can be seen from table 2 that the porosity of the MRE samples increases with increasing ammonium bicarbonate content. Previous studies have shown that the porosity has a great influence on the performance of the MRE [20].

2.5 Structure of shock experimental device

A MRE-based shock experimental device was constructed as a closed magnetic circuit consisting of iron core, MRE, outer sleeve and excitation coil, as shown in figure 7.

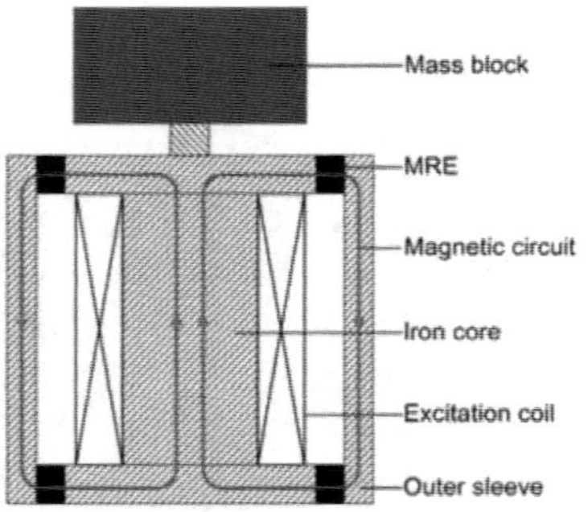

Figure 7. Schematic diagram of shock experimental device.

The metallic portions of the magnetic circuit were constructed of electrical pure iron in order to carry magnetic flux. The magnetic field is adjusted by controlling the current values of the excitation coil, which was parallel with the thickness direction of the MREs. Excitation coil is a kind of copper enamel wire with roughly 500 turns around the iron core. Two MRE samples, worked in shear mode, the mechanical properties of MREs can be adjusted by input electric current to the coil, resulting in variable stiffness and damping characteristics. In this study, the shock experimental device was based on a single degree of freedom system.

3 Results and discussion

In order to test the performance of porous MRE under shock effect, its performance is studied under various shock conditions using electromagnetic vibration table, the photo of equipment for shock experiment is shown in figure 8.

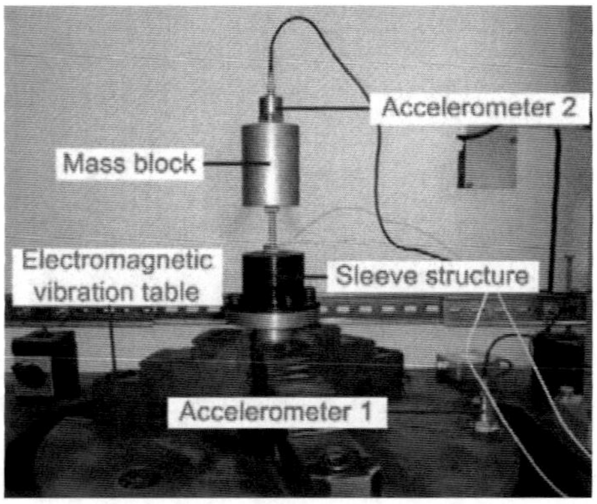

Figure 8. Photo of equipment for shock experiment.

The experiments were carried out using constant acceleration excitation input of 2 g, an electric current ranging from 0 to 2A was applied to the equipment, which can change stiffness and damping of MRE samples during each set of testing, and each testing has been executed at room temperature.

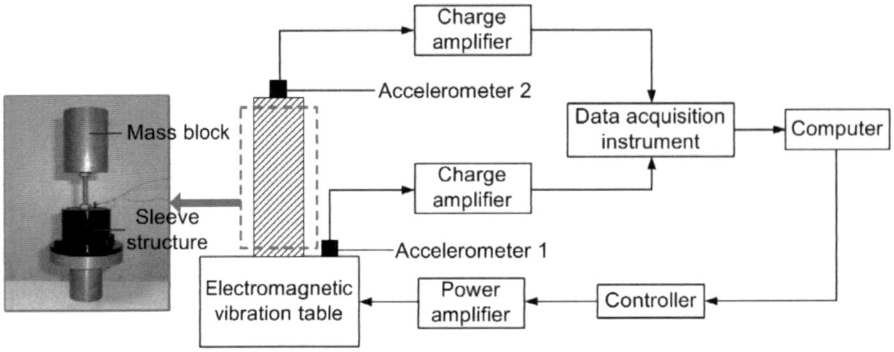

Figure 9. Experimental systems for testing the shock performance of porous MRE.

Figure 9 shows the experimental system for testing the shock performance of porous MRE. In order to investigate the performance of porous MRE under shock effect, the experimental device was mounted on an electromagnetic vibration system (model: MPA406/M232A, ETS Solutions (Beijing) Ltd), this system include power amplifier, controller and electromagnetic vibration table. Half-sine shock signal with different acceleration amplitude and pulse width was generated by this electromagnetic vibration system. A mass block, used as payload, was fixed on the top of the experimental device. Two accelerometers (model: 24108, China Aerospace Science and Technology Corporation) were used to measure the acceleration responses of the base and the payload. The acceleration signals were sent to a data acquisition instrument (model: MDR-05, Beijing Aerostandard New Technology Company), and then sent to a computer for processing. The performance of porous MRE under shock effect can be analyzed through the signals of acceleration excitation and response.

The modulus and damping property of MRE have shown in figure 4 and figure 5, these performances are function of applied electric currents and input strain amplitudes. In order to evaluate the performance of porous MRE under shock effect, the mechanical model of shock can be written as

$$m\ddot{x}(t)+c\dot{x}(t)+kx(t)=P(t) \tag{2}$$

Where $P(t)$ is shock excitation function, m is payload mass, c is viscous damping coefficient, and k is stiffness of MRE.

The duration of shock effect is very short. Therefore, shock excitation can be expressed as

$$P_s=\int_0^{\tau} P(t)dt \tag{3}$$

The maximum response acceleration of the system can be solved through expression 2 and 3

$$\ddot{x}_{\max}=\frac{P_s\omega_n e^{-\xi\omega_n t_{\max}}}{\sqrt{1-\xi^2}}\sin(\omega_n\sqrt{1-\xi^2}\,t_{\max}+\varphi') \tag{4}$$

$$\xi=\frac{c}{2m\omega_n} \tag{5}$$

$$\omega_d=\sqrt{1-\xi^2}\,\omega_n \tag{6}$$

$$\varphi'=arctg\frac{2\xi\sqrt{1-\xi^2}}{1-2\xi^2} \tag{7}$$

Performance of porous MRE under shock effect can be expressed by following equation

$$T = 20\lg \frac{\ddot{x}_{max}}{a} \tag{8}$$

Where $\ddot{x}_{max}$ is maximum response acceleration, a represent excitation acceleration.

The relationship between excitation acceleration signals and excitation acceleration signals is shown in figure 10 and figure 11.

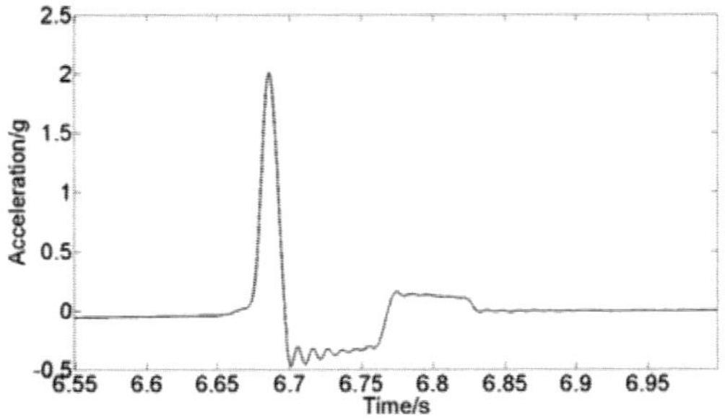

Figure10. Excitation acceleration signal of shock testing.

From figure 10, the excitation acceleration signal response to a half-sine shock signal, where the acceleration amplitude is 2 g. Due to this excitation shock signal, the response acceleration signals of shock testing are shown in figure 11.

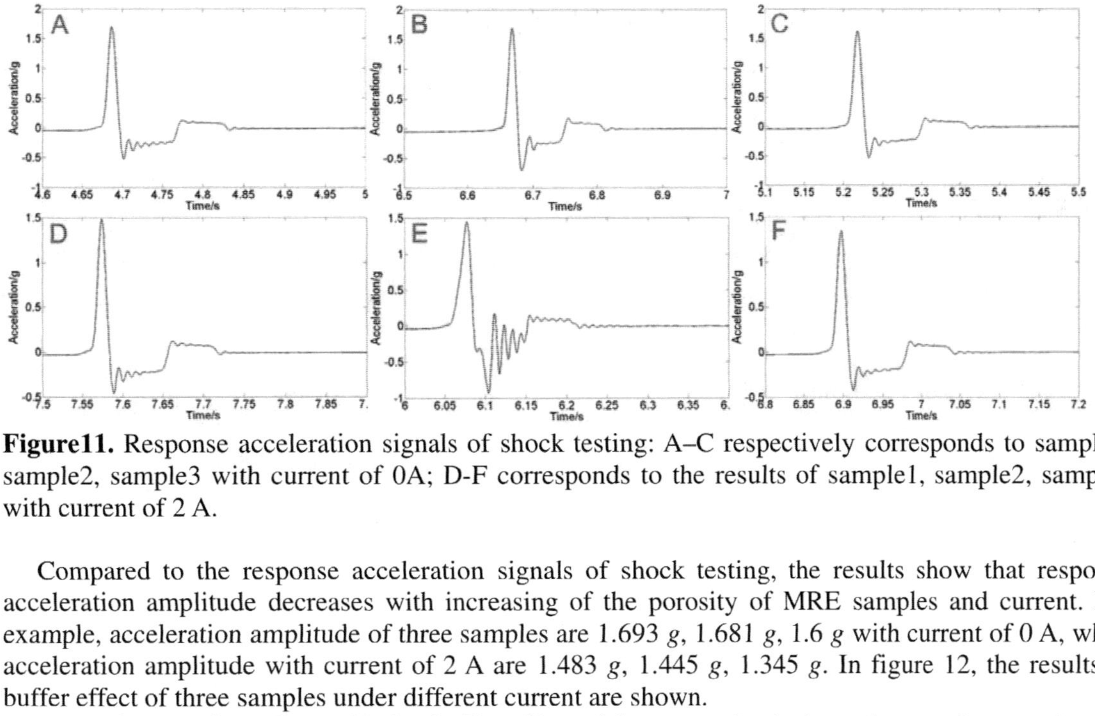

Figure11. Response acceleration signals of shock testing: A–C respectively corresponds to sample1, sample2, sample3 with current of 0A; D-F corresponds to the results of sample1, sample2, sample3 with current of 2 A.

Compared to the response acceleration signals of shock testing, the results show that response acceleration amplitude decreases with increasing of the porosity of MRE samples and current. For example, acceleration amplitude of three samples are 1.693 g, 1.681 g, 1.6 g with current of 0 A, while acceleration amplitude with current of 2 A are 1.483 g, 1.445 g, 1.345 g. In figure 12, the results of buffer effect of three samples under different current are shown.

It can be seen from figure 12 that buffer effect of three samples is dependent on the porosity and current. The porosity of MRE samples has an obviously influence on buffer effect at current of from 0 to 2 A. Sample3 has higher buffer effect than that of sample1 and sample2, Therefore, porous MRE has better buffer effect performance.

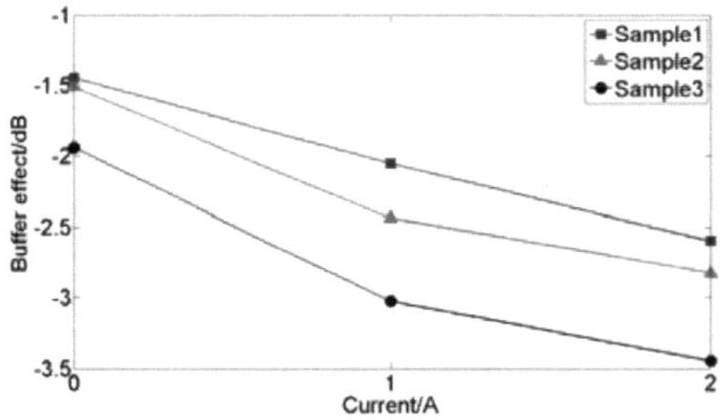

Figure12. Buffer effect of three samples.

4 Conclusions

Three kinds of porous MRE samples based on polyurethane material were fabricated without external magnetic field. Their microstructures and dynamic performance were studied, and image analysis has been applied to analyze the microstructure of porous MRE. It was found that the ammonium bicarbonate content influence the porosity of porous MRE samples greatly and different porous MRE have different buffer effect. The response acceleration amplitude decreases with increasing of the porosity of MRE samples and current, sample3 with higher porosity has higher buffer effect than that of sample1 and sample2. Due to its unique properties, this porous MRE will have broad applications in isolation and buffer device.

Acknowledgement

This work is supported by the National Natural Science Foundation of China (Grant nos. 51073179 and 50830202); The Fundamental Research Funds for the Central Universities (Grant no. CDJZR10120005); The Opening Project of State Key Laboratory of Polymer Materials Engineering (Sichuan University) (Grant no. KF201105); Visiting Scholar Foundation of Key Lab in University (Chongqing University).

References

[1] Rabinow J 1948 *AIEE Transaction.* **67** 1308
[2] Iglesias G R, Lopez-Lopez M T and Duran J D G 2012 *J. Colloid. Interf. Sci.* **377** 153
[3] Jolly M R, Carlson J D and Munoz B C 1996 *Smart. Mater. Struct.* **5** 607
[4] Yu M, Ju B, Fu J, Liu X and Yang Q 2012 *J. Magn. Magn. Mater.* **324** 2147
[5] Li J F, Gong X L, Xu Z B and Jiang W Q 2008 *Int. J. Mater. Res.* **99** 1358
[6] Wei B, Gong X L, Jiang W Q, Qin L and Fan Y 2010 *J. Appl. Polym. Sci.* **118** 2765
[7] Carlson D J and Jolly M R 2000 *Mechatronics* **10** 555
[8] Xu Y, Gong X, Xuan S, Zhang W and Fan Y 2011 *Soft Matter* **7** (5) 246
[9] Chen L, Gong X L, Jiang W Q, Yao J J, Deng H X and Li W H 2007 *J. Mater. Sci.* **42** 5483
[10] Zhou G Y 2004 *Smart. Mater. Struct.* **13** 1203
[11] Ginder J M, Reitberger T and Stenberg B 2001 *Proc.: SPIE* **4331** 103
[12] Behrooz M, Sutrisno J, Wang X J, Fyda R, Fuchs A and Gordaninejad F 2011 *Proc.: SPIE.* **7977** 79770Z
[13] Hu W and Wereley N M 2008 *Smart. Mater. Struct.* **17** 045021
[14] Fan Y C, Gong X L, Xuan S H, Zhang W, Zheng J and Jiang W Q 2011 *Smart. Mater. Struct.* **20** 035007
[15] Chen L, Gong X L and Li W H 2008 *Chinese. J. Chem. Phys.* **21** 581
[16] Zhang W, Gong X L, Sun T L, Fan Y C and Jiang W Q 2010 *Chinese. J. Chem. Phys.* **23** 226
[17] Kallio M, Lindroos T, Aalto S, Jarvinen E and Karna T 2007 *Smart. Mater. Struct.* **16** 506
[18] Liao G J, Gong X L, Xuan S H, Kang C J and Zong L H 2012 *J. Intel. Mat. Syst. Str.* **23** 25
[19] Deng H X, Gong X L and Wang L H 2006 *Smart. Mater. Struct.* **15** N111
[20] Ju B X, Yu M, Fu J, Yang Q, Liu X Q and Zheng X 2012 *Smart. Mater. Struct.* **21** 035001

Tuning the tensile modulus of magnetorheological elastomers with magnetically hard powder

D Yu Borin[1,3], G V Stepanov[2] and S Odenbach[1]

[1]Technische Universität Dresden, Institute of Fluid Mechanics, 01062, Dresden, Germany
[2] State Scientific Research Institute of Chemistry and Technology of Organoelement Compounds, 111123, Moscow, Russia

E-mail: dmitry.borin@tu-dresden.de

Abstract. It has been experimentally determined the tensile modulus of magnetorheological elastomers based on magnetically hard particles. Samples of the elastomer consisting of a soft elastic matrix and micron-sized particles of FeNdB powder have been magnetized in uniform magnetic fields of varying strength in order to provide different remanence magnetizations. The tensile modulus of these samples was measured small and large strain regimes (up to 6.6%) through mechanical elongation with a table top machine. The relative change in the tensile modulus after the sample was magnetized can reach 360%, depending on the remanence magnetization and the strain.

1. Introduction

The most well-known and controllable property of elastic composites with embedded magnetic particles is the so-called MR effect [1], which is in contrast to magnetorheological fluids an increase in elasticity in an external magnetic field. These composites are usually known as MR or magnetoactive elastomers [2, 3, 4]. As previously shown, if the matrix of the composite is sufficiently soft, i.e. if the elasticity modulus does not exceed 200 kPa, an enhanced MR effect exceeding 1000% will be observed, as will magnetically driven shape memory and giant deformational effects [5, 6, 7]. Common ways to develop an MR elastomer with desirable properties are varying the chemical composition of the elastic matrix [8], controlling the parameters of the curing process, or modifying the powder morphology, size, etc. [9, 10]. The magnetic powder is usually made of carbonyl iron, magnetite, or iron oxide. Once cured, MR elastomers have certain viscoelastic properties which are controllable with an applied magnetic field. However, these properties can only be tuned by means of this external stimulus. We propose using a magnetically hard powder which will enable adjustment and control of the elasticity of soft MR elastomers after they are cured. Some studies on the actuation abilities as well as on the electrical properties of composites filled with SmCo particles have been performed [11]. Experiments on the actuation performances of the composite with magnetically hard filler were conducted in [12]. Recently [13], preliminary investigations on elastomers with embedded FeNdB microparticles have been performed. In the present work, we consider the influence of the remanence magnetization on the tensile modulus of an MR elastomer measured by mechanical elongation. This modulus is important parameter in terms of application point of view.

[3] To whom any correspondence should be addressed.

2. Sample preparation

The soft matrix of the composite is composed of a low-molecular, vinyl-containing silicon rubber and a hydride-containing cross-linking agent [4]. The FeNdB-powder is a mixture of small (average size $\sim 4\mu$m) and large (average size of $\sim 50\mu$m) particles which have been mechanically stirred into the liquid polymeric matrix. The overall weight concentration of the powder was $\sim$80%, and the ratio between small and large fractions was 1:4. After the suspension was degassed, it was cast into a mold for further curing at temperatures of about 100°C. In this way, cylindrical samples with a diameter of 10 mm and a height of 15 mm were prepared. After the cross-linking procedure, the samples were magnetized using the uniform magnetic field provided by an electromagnet (Bruker, Germany), in order to obtain a remanence magnetization. The relationship between the intensity of the external field used and the corresponding remanence was determined with a vibrating sample magnetometer (Lake Shore 7407, USA) and is given in Table 1.

Table 1. The relationship between the external field B and the remanence magnetization B_r of the investigated MR elastomer sample.

B (mT)	0	750	1020	1500	1850
B_r (mT)	0	60	83	103	110

3. Experimental

For mechanical testing of the prepared samples, a table-top machine from Dyna-Mess (Germany) was used. The machine is equipped with a force unit with a nominal load capacity up to 5 kN and an efficient servo-pneumatic drive. The class of accuracy for both the load cell and displacement transducer is 0.1, and the best possible resolution for the measurements is 0.01 N and 0.01 mm, respectively.

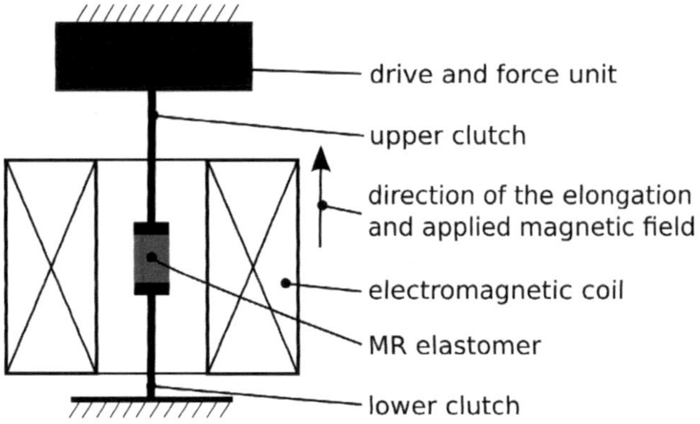

Figure 1. Schematic representation of the experimental setup.

In the experiment (Figure 1), the cylindrical sample is fixed in a movable upper and static lower clutch made of a non-magnetic alloy. Quasi-static elongation tests of the MR elastomer are performed at a speed of 0.05 mm/s using the displacement-controlled mode. The maximum displacement used does not exceed 1 mm, $\sim$6.6% of the initial sample length. This limitation is

necessary to avoid significant influence of the sample deformation in the radial direction on the results of the measurements, since this parameter is not controlled during the experiment.
The force-displacement diagrams for samples magnetized in fields of different strengths are shown in Figure 2. With increasing magnetization, larger forces are needed to elongate the elastomer.

The qualitative explanation of this behavior is similar to the one known for the MR effect, which is driven by an external magnetic field in elastomers containing magnetically soft particles. The particles are oriented and aligned in the direction of the acting field. The higher the field intensity is, the higher the mechanical force needed to overcome interparticle interactions and deform the sample. If a sample is magnetized in a field with a flux density higher than ~1500 mT, no further pronounced changes in the force-displacement dependency can be observed, due to magnetic saturation of the embedded powder.

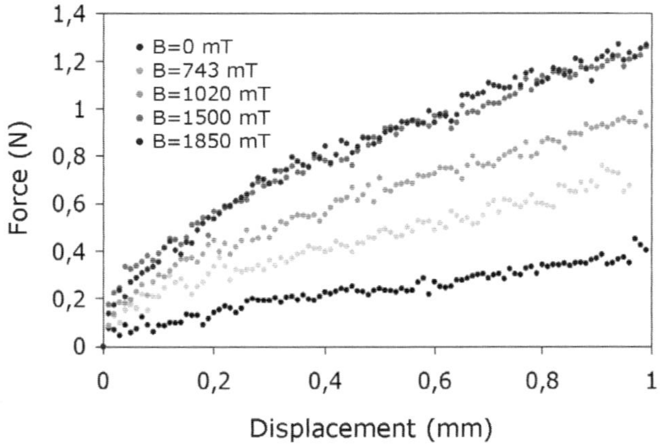

Figure 2. Force-displacement diagrams obtained from the elongation of the MR elastomer samples magnetized in fields of varying induction B.

To determine the tensile modulus, the measured forces and displacements are transformed into stress and strain values, respectively, taking into account the geometry of the sample. The tensile modulus is defined as the slope of the stress-strain curve, and, as shown in Figure 3, one has to distinguish between the moduli for different strains. Generally, the tensile modulus E is calculated as $E = \mathrm{d}\sigma/\mathrm{d}\epsilon$, where σ is the stress and ϵ is the strain, leading to a complex dependence of the modulus E on mechanical deformation. However, it is sufficient to consider the modulus E in two major ranges of strain, as illustrated in Figure 3.
The moduli E obtained this way for small and large strains for samples magnetized in magnetic fields of different flux density B are shown in Figure 4. The modulus in the small strain regime is significantly higher than that obtained for large strain, as expected. For the higher magnetization fields, saturation is observed. Moreover, the dependence of E on the field intensity is similar to other properties of magnetoactive materials [1, 2].

To evaluate the quantitative influence of magnetization on the elasticity of the MR elastomer, the relative change in the tensile modulus is calculated as $R = E(B_r)/E_0 - 1$, where E_0 is the tensile modulus of a non-magnetized sample. The parameter R, given in Table 2 for various remanence magnetizations B_r, can be conventionally designated as a passive MR effect.

Journal of Physics: Conference Series **412** (2013) 012040 doi:10.1088/1742-6596/412/1/012040

Table 2. The relative change of the tensile modulus (R) of the MR elastomer for different remanence magnetizations B_r.

B_r (mT)	60	83	103	110
R, small strain (%)	178	304	354	359
R, large strain (%)	81	133	182	178

4. Summary and outlook

It has been shown that the use of a magnetically hard powder allows tuning of the tensile modulus of an MR elastomer by means of remanence magnetization. The relative change of

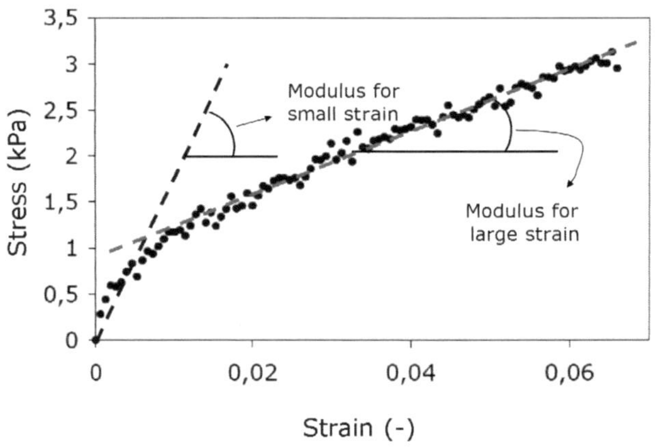

Figure 3. Stress-strain dependence recalculated from the data presented in Figure 2 for the sample magnetized in a field with an induction of 1020 mT. The linear approximations are given to illustrate the difference between the tensile moduli at small and large strains.

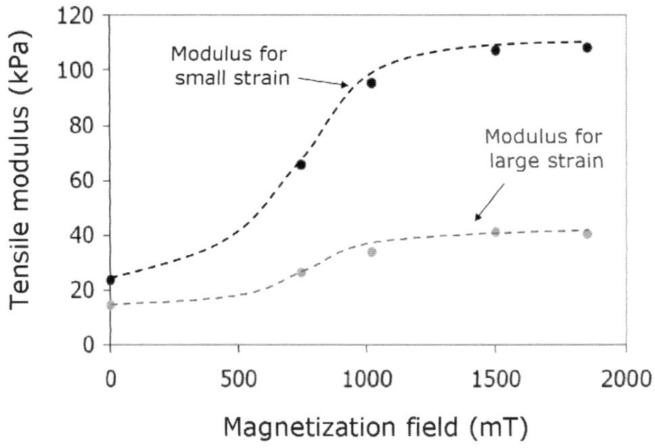

Figure 4. Tensile moduli of the MR elastomer determined for small and large strain versus induction of the magnetic field used for sample magnetization.

the tensile moduli for magnetized samples can reach up to 360%, depending on the remanence magnetization and on the deformation regime considered. Moreover, the same setup allows to perform a compression testing and corresponding results will be published elsewhere.
Further investigations will be devoted to actively controlling this materials properties, as well as to dynamic characterization. Moreover, the basic question of the correlation between observed material performance and microstructural changes will have to be considered for a detailed understanding of the MR effects observed.

Acknowledgments
This project is funded by the European Union and the Free State of Saxony.
G.S. is grateful for the financial support of the Federal Agency for Science and Education of the Russian Federation.

References
[1] Bossis G, Volkova O, Lacis S and Meuner A 2002 *Magnetically Controllable Fluids and Their Applications*, ed. S. Odenbach (Berlin, Heidelberg: Springer-Verlag)
[2] Jolly M R, Carlson J D and Munoz B C 1996 *J.Intell. Mater. Sys. Struct.* **7** 613
[3] Kallio M 2005 *The elastic and damping properties of magnetorheological elastomers* (Espoo: VTT Publications 565)
[4] Nikitin L V, Stepanov G V, Mironova L S and Gorbunov A I 2004 *J. Magn. Magn. Mater.* **272-276** (III) 2072
[5] Stepanov G V, Abramchuk S S, Grishin D A, Nikitin L V, Kramarenko E Yu and Khokhlov A R 2007 *Polymer* **48** 488
[6] Stepanov G V, Borin D Yu, Raikher Yu L, Melenev P and Perov N 2008 *J. Phys.: Condens. Matter.* **20** 204121
[7] Böse H 2007 *Int. J. Mod. Phys. B* **21** 4790
[8] Böse H and Rder R 2009 *J. Phys.: Conf. Ser.* **149** 012090
[9] Padalka O, Song H J, Wereley N M, Filer J A and Bell R C 2010 *IEEE Trans. Magn.* **46** 2275
[10] Li J F, Gong X L, Zhu H and Jiang W Q 2009 *Polym. Test.* **28** 331
[11] Lanotte L, Ausanio G, Hison C, Iannotti V, Luponio C and Luponio C Jn 2004 *J. Optoelectronics Advanced Mater.* **6** 523
[12] Koo J H, Dawson A and Jung H-J 2012 *J Intell. Mater. Syst. Struct.* (in print)
[13] Stepanov G, Chertovich and Kramarenko E Yu 2012 *J. Magn. Magn. Mater.* **324** 3448

Control of haptic master – slave robot system for minimally invasive surgery (MIS)

J S Oh[1], W K Shin[1], C H Uhm[1], S R Lee[1], Y M Han[2] and S B Choi[1,3]
[1]Smart Structures and Systems Laboratory, Department of Mechanical Engineering, Inha University, Incheon 402-751, Korea
[2]Division of Automotive Engineering, Ajou Motor College, Chungnam 355-769, Korea

E-mail: seungbok@inha.ac.kr

Abstract. In this work, 4 degrees-of-freedom (DOF) haptic master featuring electrorheological (ER) fluid is proposed and integrated with a slave. Using an ER fluid, the haptic master is devised which can generate a repulsive force/torque with the 4-DOF motion and provide stimulus information to operator. For realization of master-slave MIS system, an encoder is installed with the ER haptic master to establish the MIS master system. The motion command of the haptic master is realized by slave surgery robot. In order to follow the motion of the haptic master, the mechanism of slave surgery robot is devised. Accordingly, the haptic master-slave system is established by incorporating the slave robot with the haptic master device in which the repulsive force/torque and desired position are transferred to each other. In order to obtain the desired torque trajectories, a sliding mode controller (SMC) is designed and implemented. It has been demonstrated that the desired effective torque tracking control performance is well achieved using the proposed master-slave system.

1. Introduction

Recently, it is very popular to apply robotic technology to modern medical industry, such as minimally invasive surgery (MIS) [1]. Typically, the MIS needs the robot to perform surgery through the usage of long surgical instruments that are inserted through incision points. From the patient's point of view, the advantages of the MIS are less injury to tissue, higher accuracy rate, shorter hospital stay and less pain. But, from the viewpoint of surgeon, the MIS systems only send back visual information and cannot provide information about viscosity and stiffness of the touched tissue or organ. This causes the surgeon not to feel what is happening inside abdomen of patient. Therefore, it is very important to develop the haptic master robot system.

Several medical haptic devices featuring kinesthetic force feedback have been researched and developed for the medical training environment using a simulator and virtual reality, in which a haptic system provides kinesthetic force information of virtual patients to a doctor. Among them, two types of actuation mechanism are common: passive and active types. Although the passive type is known to be safe, the active type using electric motor is widely used as a repulsive force or torque generating actuator. However, it has a complicated mechanism, safety problem and difficulty in continuous force control [2, 3]. In order to adopt a medical haptic master in a real operation room, these problems

[3] To whom any correspondence should be addressed.

should be solved. Subsequently, diverse researches about haptic master using smart materials such as electrorheological (ER) fluid have been researched. ER fluid is able to change their rheological properties such as yield stress by applying electric field. The advantages of ER fluid are also fast response, simple mechanism, continuous force and high stability. These features are ideal properties for haptic master devices in which various different strengths of forces are simulated.

Many master-slave systems have been proposed such as the Polish cardio-robot RobIn Heart (RIH) system [4, 5] and the force-reflecting MC2E system [6, 7]. Among these master-slave systems, the most commercially successful robot master-slave systems are da VinciTM from Intuitive Surgical Inc. [8] and Zeus from Computer Motion Inc. [9]during last decades. These robot systems have large and complex mechanism. So, small and simple slave robot has recently been researched [10, 11].

The main objective of this research is to propose a haptic master device using ER fluid and slave surgery robot. The purpose of haptic master device is not only to generate motion commands for the slave robot but also to generate the physical constraints of the slave robot to the surgeon. Also, in order to follow the motion of the haptic master, the new mechanism of slave surgery robot is devised. Kinematic model of slave surgery robot is analysed. Using the master and robot kinematics, the interface communication between the proposed haptic master and slave robot is discussed. Accordingly, the haptic master-slave system which can transfer the repulsive force/torque and desired position to each other is established by incorporating the slave robot with the master device. In order to obtain the desired force/torque and position trajectories, a sliding mode controller (SMC) is designed and implemented. It has been demonstrated that the desired effective torque tracking control performance is achieved in the time domain.

2. 4DOF ER haptic master

During MIS, the instruments and laparoscope always inserted through fixed small incisions. So, the required motion of MIS is four (three rotation motions and one translation motion). Accordingly, in order to operate the slave surgery robot and generate the repulsive force/torque, the 4 DOF haptic master device is proposed and shown in figure 1. It consists of two actuators: ER spherical joint for 3 DOF rotation motion and ER damper for 1 DOF translation motion. ER spherical joint has been designed and manufactured in the previous study [12]. Detailed geometric shape and modeling of ER damper is discussed.

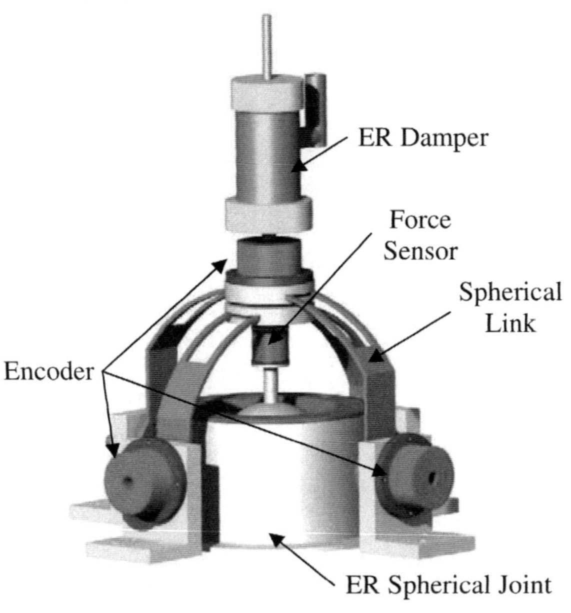

Figure 1. Configuration of 4 DOF haptic master.

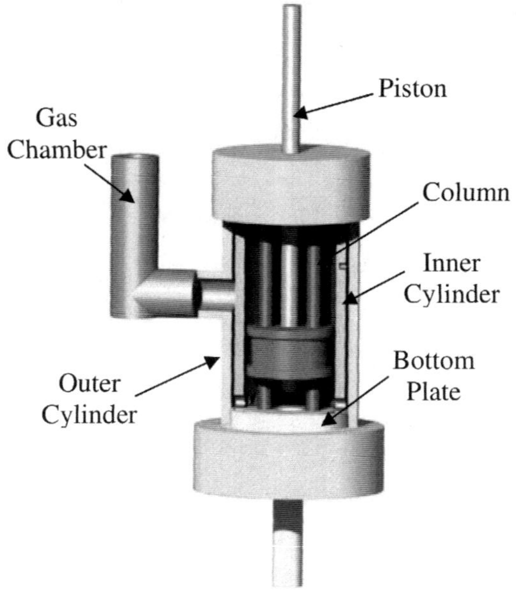

Figure 2. Schematic configuration of ER damper.

In order to realize translational motion, cylindrical ER damper is devised. Figure 2 shows the proposed ER damper consists of columns, piston, inner and outer cylinder. Function of column is to transfer yawing motion of piston to spherical joint. So, if operator rotates the piston, then this motion is transmitted to ER spherical joint via bottom plate. The ER damper is divided into the upper and lower chambers by the piston. These chambers are filled with the ER fluid. When the piston moves, the ER fluid flows from one chamber to the other chamber via the annular duct. The inner cylinder is connected with the positive voltage produced by a high voltage supply unit, playing as the positive (+) electrode. The outer cylinder is connected with the ground playing as the negative (-) electrode. On the other hand, the gas chamber located outside of the annular duct as an accumulator of the ER fluid induced by the motion of the piston. By neglecting the compressibility of the ER fluid, the frictional force and assuming quasi-static behavior of the damper, the damping force can be derived as follows [13]:

$$\begin{gathered} F_z = P_2(A_p - A_f) - P_1(A_p - A_s - A_f) = P_a A_s + c_{vis}\dot{x}_p + F_{ER}\,\mathrm{sgn}(\dot{x}_p) \\ where\ c_{vis} = \frac{6\mu L}{\pi R t_g^{\,3}}(A_p - A_s - A_f)^2,\ F_{ER} = (A_p - A_s - A_f)\frac{cL}{t_g}\tau_y, \end{gathered} \quad (1)$$

where A_p , A_s and A_f are the piston area, the piston-shaft cross-sectional areas and two column area, respectively. P_1, P_2 and P_a are pressures in the upper, lower and gas chamber of the damper, respectively. P_0 and V_0 are initial pressure and volume of the accumulator. γ is the coefficient of thermal expansion which is ranging from 1.4 to 1.7 for adiabatic expansion. x_p is the piston displacement. In this configuration, gas chamber is open. So, P_a is always constant value, atmospheric pressure. In order to simplify the design procedure, stroke of the translation device is fixed by 40mm which is proper value to surgically operate patients and gap size is also fixed by 1 mm. In order to design compact device, the minimum value of r_p is 25 *mm*. With these design values, calculated repulsive force is 98 *N* which people can sense a change of the force enough.

3. Slave surgery robot

In the previous section, an ER haptic master is proposed. The ER haptic master commands slave surgery robot to operate the surgery. From the slave robot, the repulsive force/torque is measured. This force/torque signal is transferred and generated by ER haptic master. In order to follow the motion of the proposed master, a new mechanism for MIS robot is proposed and analysed.

During MIS, the instrument is inserted through fixed small incision of patient. So, it is important to realize the pivot rotation motion, a mechanism of slave robot featuring fixed bar and gimbal structure is proposed as shown in figure 3. Ball joint of the fixed bar is assumed as small incision of patient and gimbal structure can smoothly rotate the instrument around the ball joint. Slave robot is composed of five active joints. Four revolution joints determine a position of end position of slave robot arm. In order to maintain horizontality of gimbal structure, other revolution joint rotates gimbal structure. It is helpful to stable the system. The other revolution joint rotates instrument following the yawing motion of master. The fixed bar can hold the instrument to realize pivot rotation motion. l_1 is the length from bottom to first arm, l_2 is the length of first arm, l_3 is the length of second arm. l_4 is the length of instrument. l_5 is the length from robot to fixed pivot point on X-axis. θ_1 is angle of robot on Z-axis. θ_2, θ_3 are angle of robot arm, respectively. Using forward and inverse kinematics of robot, the rotation angle of each slave surgery robot arm is derived as follows:

$$\begin{gathered} \theta_1 = \arctan\left(\frac{y}{x}\right), \theta_2 = \arctan\left(\frac{c}{\sqrt{a^2+b^2-c^2}}\right) - \arctan\left(\frac{a}{b}\right), \theta_3 = \arctan\left(\frac{x\cdot\cos(\theta_1) + y\cdot\sin(\theta_1) - L_2\cos(\theta_2)}{L_1 + L_2\sin(\theta_2) - z}\right) - \theta_2 \\ \theta_4 = \frac{\pi}{2} - \theta_2 - \theta_3, \quad where\ a = -2L_2(x\cdot\cos(\theta_1) + y\cdot\sin(\theta_1)), b = 2L_2(L_1 - z), c = L_3^{\,2} - (x\cdot\cos(\theta_1) + y\cdot\sin(\theta_1))^2 \end{gathered} \quad (2)$$

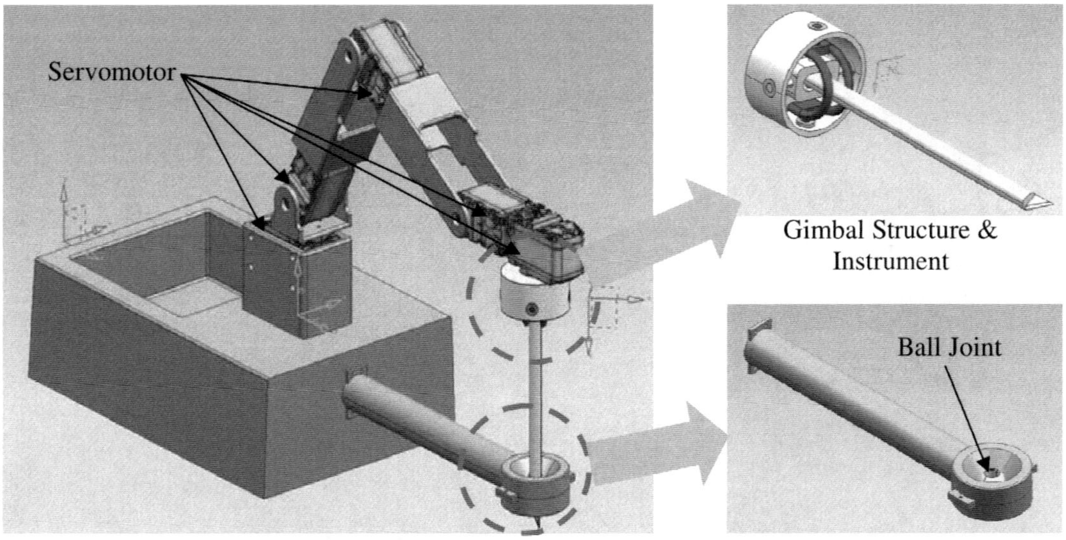

Figure 3. Configuration of slave robot system.

From the MIS robot, the repulsive force/torque such as viscosity and stiffness of the touched tissue or organ is measured and transferred to the haptic master. Then, the haptic master generates a repulsive force/torque using ER fluid.

4. Tele-operation control

ER fluids are known to exhibit hysteretic behaviors. Mechanical friction also exists at the spherical joint. This requires robust control schemes such as a sliding mode controller (SMC). So, SMC for haptic master has been designed in the previous study [14].

$$u = \frac{1}{\omega_n^2 I_i}\left(c\dot{e}_i + \ddot{T}_{di} + 2\varsigma\omega_n\dot{T}_i + \omega_n^2 T_i + ksat(S)\right),\ k > |d| \tag{3}$$

$$u = \frac{1}{\omega_n^2}\left(c\dot{e}_i + \ddot{F}_{di} + 2\xi\omega_n\dot{F}_i + \omega_n^2 F + ksat(S),\quad |k| > |d|\right. \tag{4}$$

Figure 4 shows the photograph of the manufactured haptic master combined with the ER spherical joint and the ER damper. In order to measure positional information to be transferred to the slave system, two spherical link with encoders (Omron) are adopted to the joint. When operator moves the gripper of the master system, position information such as rotational angle is obtained from encoders (Omron), whose resolution is 3600 pulse/rotation, and transferred to the slave system. Figure 5 shows the photograph of the manufactured slave robot. Slave robot is operated and deforms the object upon the position information. Each servomotor of slave robot is controlled by built-in microprocessor. PID controller is programmed in microprocessor. At the same time, the force/torque, which is to be reflected to the user, is measured by force sensor as shown in figure 6. The generated force/torque of haptic master is measured by a 6 axes force sensor (ATI, Nano17) at the mid point of the gripper. The SMC given by Eq. (13) and (15) is then activated to reflect the desired force/torque from slave robot. For palpation test, poched egg is used.

Figure 4. Manufactured 4 DOF ER haptic master.

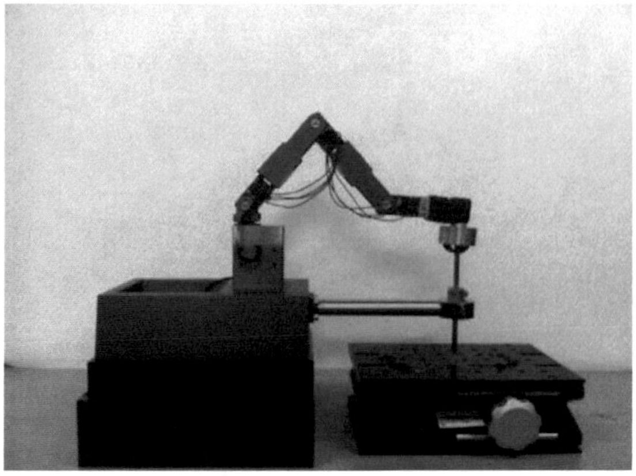

Figure 5. Manufactured slave robot.

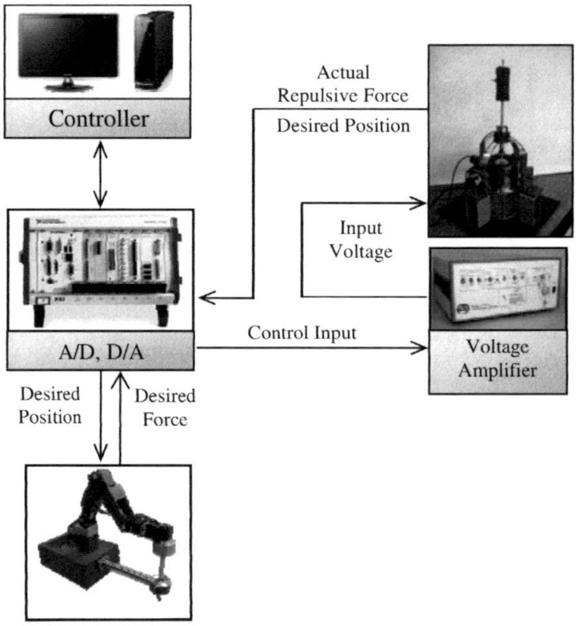

Figure 6. Experimental apparatus of master and slave robot system.

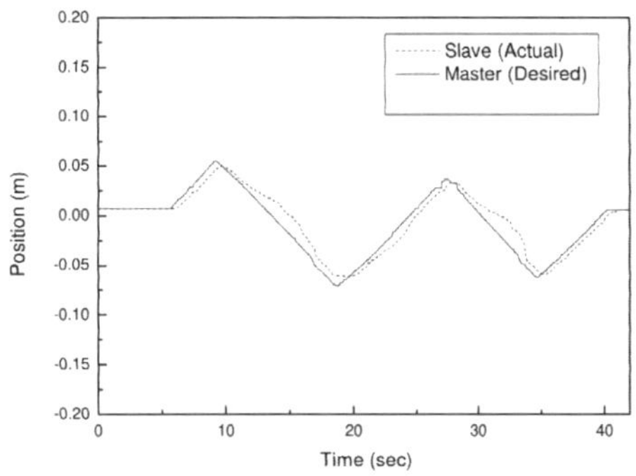

Figure 7. Position tracking control results.

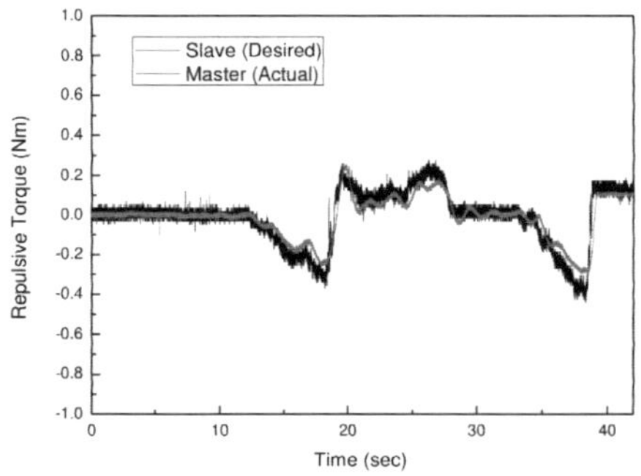

Figure 8. Torque tracking control results.

The operation link was freely moved by a voluntary person in back-and-forth manner and its position information of master was transferred to the slave robot. Figure 7 presents position tracking results of the slave robot about Y axis. At the same time, the measured torque signal is transferred to the master. Figure 8 presents torque tracking control responses of rotation motion about Y axis. From the result, it is clearly seen that the desired torque trajectories from slave robot are well achieved without exhibiting significant tracking errors. From these results, the proposed slave system was successfully operated and communicated with the master without time delay problem.

5. Conclusion

In this study, a haptic master device using ER spherical joint and ER damper for MIS robot was proposed and integrated with the slave robot. The ER damper was designed based on the Bingham model. In order to realize motion of the 4 DOF ER haptic master, a new mechanism of slave surgery robot was proposed. Kinematic analysis of the slave robot was studied. For implementation of MIS robot system, the interface communication between master and slave robot has been analysed. Accordingly, the haptic master-slave system has been established by incorporating the slave robot with the master device in which the repulsive force/torque and desired position are transferred to each other. In order to obtain the desired torque trajectories, a sliding mode controller (SMC) was designed and empirically implemented. It has been demonstrated that the desired effective torque tracking and position control performances are well achieved. It is finally remarked that the haptic master and slave robot developed in this work will be incorporated with the wireless communication for MIS application in the near future.

Acknowledgments

This work was supported by a National Research Foundation of Korea (NRF) grant funded by the Korea government (MEST) (No. 2010-0015090). This financial support is gratefully acknowledged.

References

[1] Intuitive Surgical. Available from: http://www.intuitivesurgical.com.

[2] Pierrot F, Dombre E, Degoulange E, Urbain L, Caron P, Boudet S, Gariépy J and Mégnien J L 1999 Hippocrate: a safe robot arm for medical applications with force feedback *Medical Image Analysis* **3** (3) 285-300

[3] Yamaguchi Y, Furusho J, Kimura S and Koyanagi K 2005 2-D force display system using MR actuators *Journal of the Robotics Society of Japan* **23** (3) 80-7

[4] Nawrat Z and Kostka P 2006 Polish cardio-robot 'Robin Heart'. System description and technical evaluation *Int. J. Med. Robot. Comp.* **2** (1) 36-44

[5] Nawrat Z, Podsedkowski L, Mianowski K, Wroblewski P, Kostka P, Baczynski M, *et al.* 2002 Robin Heart in 2002 - actual state of Polish Cardio-Robot. *Robot Motion and Control, RoMoCo '02 Proc. of the 3rd Int. Workshop on* pp 33-8.

[6] Zemiti N, Morel G, Ortmaier T and Bonnet N 2007 Mechatronic design of a new robot for force control in minimally invasive surgery. *IEEE/ASME Transactions on Mechatronics : A Joint Publication of the IEEE Industrial Electronics Society and the ASME Dynamic Systems and Control Division* **12** (2) 143-53

[7] Shi J, Mederic P, Pasqui V, Morel G and Wang S 2004 Preliminary results on the design of a novel laparoscopic manipulator *Proc. of the 11th World Congress on the Theory of Machines and Mechanisms* 1811-5.

[8] Guthart G S, Salisbury Jr J K 2000 The Intuitive^{TM} telesurgery system: overview and application *Robotics and Automation, 2000 Proc. ICRA '00 IEEE Int. Conf. on.* **2000** (1) 618-21

[9] Chodoussi M, Butner SE, Wang Y 2002 Robotic surgery - the transatlantic case. *Robotics and Automation Proc. ICRA '02 IEEE Int. Conf. on* **2002** (2) 1882-8.

[10] Navarro JS, Garcia N, Perez C, Fernandez E, Saltaren R and Almonacid M 2010 Kinematics of a robotic 3UPS1S spherical wrist designed for laparoscopic applications *The International Journal of Medical Robotics + Computer Assisted Surgery* **6** (3) 291-300

[11] Li J, Wang S, Wang X and He C 2010 Optimization of a novel mechanism for a minimally invasive surgery robot *The International Journal of Medical Robotics + Computer Assisted Surgery* **6** (1):83-90.

[12] Cho HJ, Oh JS and Choi S B 2011 Design and performance evaluation of haptic master device using ER spherical joint *Advanced Materials Research: AMR* **317** (1) 577-80

[13] Nguyen QH and Choi SB 2009 A new approach for dynamic modeling of an electrorheological damper using a lumped parameter method *Smart Mater Struct.* **18** (11) 115020

[14] Oh JS, Cho HJ, Han YM, Choi SB 2012 Design and control of 4 DOF ER haptic master for medical application [8343-15] *Industrial and Commercial Applications of Smart Structures Technologies* **8343** 0G

Control performance evaluation of railway vehicle MR suspension using fuzzy sky-ground hook control algorithm

S H Ha[1], S B Choi[1,4], G S Lee[2] and W H Yoo[3]

[1] Smart Structures and Systems Laboratory, Department of Mechanical Engineering, Inha University, Incheon, 402-751, Korea
[2] RMS Technology, TechpiaBldg, 3F 774, Dujeong-dong, Cheonan-si, Choongchonnam-do, Korea
[3] Korea Railroad Research Institute, Uiwang-si, Kyouggi-do, Korea

E-mail: seungbok@inha.ac.kr

Abstract. This paper presents control performance evaluation of railway vehicle featured by semi-active suspension system using magnetorheological (MR) fluid damper. In order to achieve this goal, a nine degree of freedom of railway vehicle model, which includes car body and bogie, is established. The wheel-set data is loaded from measured value of railway vehicle. The MR damper system is incorporated with the governing equation of motion of the railway vehicle model which includes secondary suspension. To illustrate the effectiveness of the controlled MR dampers on suspension system of railway vehicle, the control law using the sky-ground hook controller is adopted. This controller takes into account for both vibration control of car body and increasing stability of bogie by adopting a weighting parameter between two performance requirements. The parameters appropriately determined by employing a fuzzy algorithm associated with two fuzzy variables: the lateral speed of the car body and the lateral performance of the bogie. Computer simulation results of control performances such as vibration control and stability analysis are presented in time and frequency domains.

1. Introduction

The high speed railway vehicle would cause car body vibration which may induce the various problems such as ride stability, ride quality and track abrasion. Thus the vibration control of high speed railway vehicle is necessary for the improvement of vibration control of car body. In general, the suspension systems of railway vehicle have been proposed depending upon the operation mode: passive, active and semi-active. A passive railway vehicle suspension featuring spring, oil damper and pneumatic damper provides design simplicity, but performance limitations are inevitable in the relatively wide frequency range. The active one generally provides high control performance in wide frequency range. Therefore, many researchers have been proposed active suspension technology for railway vehicle which is utilized by oil valve and pneumatic actuators. However, it requires high power source. And active control suspension would need mechanical power into the system, so the stability of the control system needs to be carefully considered. On the other hand, the semi-active suspension offers desirable performance generally enhanced in the active mode without requiring large

[4] Corresponding author.

power sources and expensive hardware. Furthermore, various semi-active suspension system featuring magnetorheological (MR) fluid have been proposed and successfully applied in the real field, especially in wheel based passenger vehicle suspension system [1, 2]. These previous works have been focused on the effectiveness for vibration control of a MR damper by just implementing exiting control schemes such as sky-hook controller. In general, railway vehicles are exposed to severe working environment at the high speed working. Therefore, high performance suspension systems are required for railway vehicle [3].

Therefore, in this research, the feasibility for improving the vibration control of railway vehicle featuring semi-active suspension system using MR fluid damper was investigated. Firstly, the MR damper system is designed which is incorporated with the governing equation of motion of the railway vehicle which includes secondary suspension. A nine degree of freedom model for a railway vehicle featuring MR damper is established by the fuzzy sky-ground hook controller. This controller takes account for both vibration control of car body and stability of bogie frame. This is achieved by adopting a weighting parameter between two performance requirements, which is appropriately determined via the proposed fuzzy algorithm. Computer simulations are undertaken in order to demonstrate the effectiveness of the proposed control system.

2. Model of railway vehicle with MR damper

In this section, the mathematical model of MR damper was derived in order to evaluating vibration control performance. The governing equation of motion for the railway vehicle, shown in Figure 1, with suspension systems including MR damper can be derived using Newton`s laws. It is expressed by equation 1. A nine degree of freedom passenger railway vehicle models presented to investigate the lateral response on wheel-set motion to measured data. The wheel-set of the railway vehicle is assumed to follow the track perfectly in the vehicle direction; the wheel-set motion is given by the measured data. A bogie frame`s motions have lateral, yaw and roll according to wheel-set motion. The vehicle equations of motion are presented, including the MR damper force term [3].

$$\mathbf{M}\ddot{\mathbf{x}} + \mathbf{C}\dot{\mathbf{x}} + \mathbf{K}\mathbf{x} = \mathbf{B}_a\mathbf{f} + \mathbf{d} \quad (1)$$

where, vector $\mathbf{x}$ is the displacement vectors, respectively. $\mathbf{f}$ is the actuating force of the MR damper. $\mathbf{M}$ is the matrix of the mass. $\mathbf{C}$ is the damping matrix. $\mathbf{K}$ is the stiffness matrix.

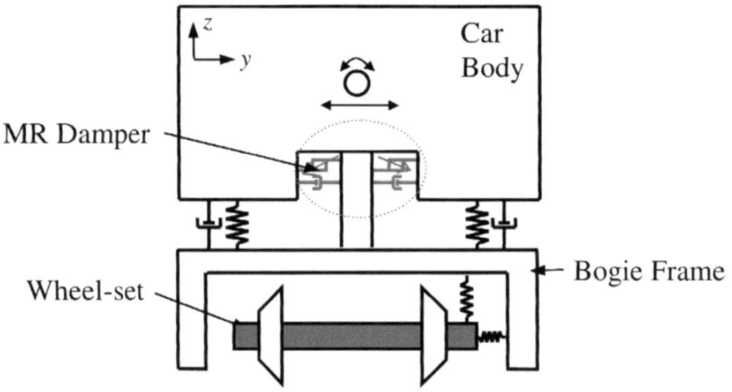

Figure 1. Schematic diagram of railway vehicle.

In this study, the MR damper is employed the cylindrical damper. The outer and inner pistons are combined to form a MR valve structure which divides the MR damper into two chambers: the upper and lower chambers. These chambers are fully filled with the MR fluid. The floating piston incorporated with gas chamber functions as an accumulator to accommodate the piston shaft volume as it enters and leaves the fluid chamber. Using quasi-static behavior of the damper, the damping force can be expressed as follows:

$$F_d = P_2 A_p - P_1 (A_p - A_s) \tag{2}$$

where A_p and A_s are the piston and the piston shaft effective cross-sectional areas, respectively. P_1 is the pressure of upper chamber in the damper. P_2 is the pressureof lower chamber in the damper. The relationsbetween P_1, P_2 and the pressure in the gas chamber, Pa, can be expressed as follows:

$$P_2 = P_a; \quad P_1 = P_a - \Delta P \tag{3}$$

where ΔP is the pressure drops of MR fluid flow through the valve structure. By neglecting minor loss of MR fluid flow, the pressure drops ΔP can be calculated as follows:

$$\Delta P = \frac{6\eta L}{\pi t_d^3 R_d}(A_p - A_s)\dot{x}_p + 2c\frac{L_p}{t_d}\tau_y \tag{4}$$

where τ_y is the yield stress of the MR fluid induced by the applied magnetic field, η is the post-yield viscosity of the MR fluid, L is the length of the inner piston, L_p is the length of the magnetic pole, R_d and t_d are the average radius and gap of the annular duct (the MR valve orifice). The coefficient c depends on the MR flow velocity profile which can be approximately estimated 1.5 to 1.7.where Q is the flow rate of MR fluid flow through the valve orifice. The pressure in the gas chamber can be calculated as follows:

$$P_a = P_0\left(\frac{V_0}{V_0 + A_s x_p}\right)^{\gamma} \tag{5}$$

where P_0 and V_0 are initial pressure and volume of the accumulator. γ is the coefficient of thermal expansion which is ranging from 1.4 to 1.7 for adiabatic expansion. Using Eq. (2), (3), (4) and piston velocity, the MR damping force can be calculated by [4]

$$F_{MR} = P_a A_s + (A_p - A_s)\frac{2cL_p}{t_d}\tau_y \,\mathrm{sgn}(\dot{x}_p) \tag{6}$$

The first term in Eq.(6) represents the elastic force from the gas compliance and the last one is the force due to the yield stress of the MR fluid which can be continuously controlled by the magnetic field across the MR fluid duct. In this work, the commercial MR fluid (RMS technology) is used. The induced yield stress of the MR fluid can be approximately estimated by

$$\tau_y = \alpha H^{\beta} \tag{7}$$

Here, H is the applied magnetic field whose unit is A/mm. The α and β are intrinsic values of the MR fluid to be experimentally determined.

3. Design of control algorithm

The MR damper is semi-active system, it is easily expected that the performance of vibration control of the vehicle motion may be deteriorated if excessive noncontrolled magnetic fields are applied to the MR damper. Thus, it is necessary to use appropriate control scheme to achieve satisfactory control performance for various disturbance. One of the most general control strategies for semi-active suspension system is the sky-hook controller proposed by Karnopp *et al.* [5]. The structure of this control is simple, and hence easy to experimentally realize. However, this controller takes accounts for only ride comfort of vehicle systems. As the railway vehicle, in general, requires the ride comfort as well as the stability to increase vehicle speed. In order to achieve this control target, a new fuzzy sky-ground hook controller is adopted in this work [6]. Unlike the conventional sky-hook controller, the proposed one consists of two ideal dampers; one is fixed to the ceiling while the other fixed to the ground. The fixed one to the ceiling produces a damping force to control the vibration of the bogie frame. Therefore, we may improve both the ride quality and the stability by properly adjusting each component of the damping force. To do this, by assuming independent motion of each bogie frame, the following controller was proposed for each bogie frame.

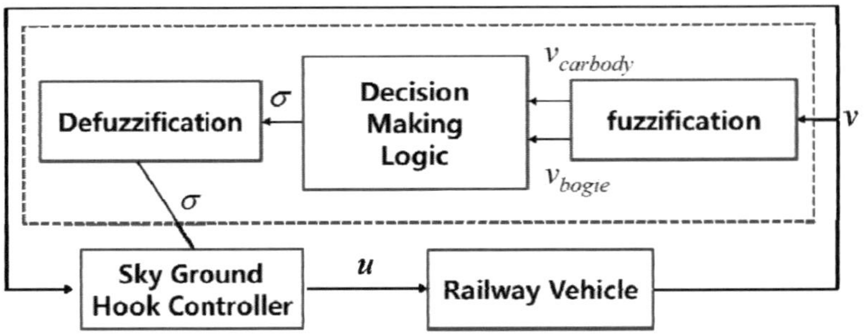

Figure 2. Block diagram for fuzzy sky-ground hook control algorithm.

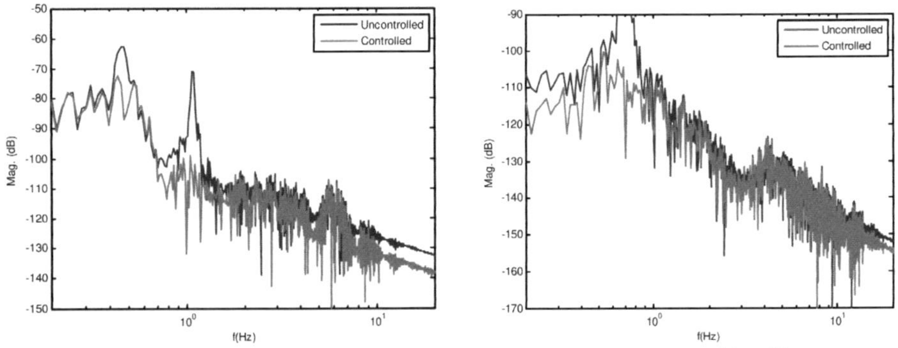

(a) lateral (b) roll

Figure 3. Vibration control results of a car body.

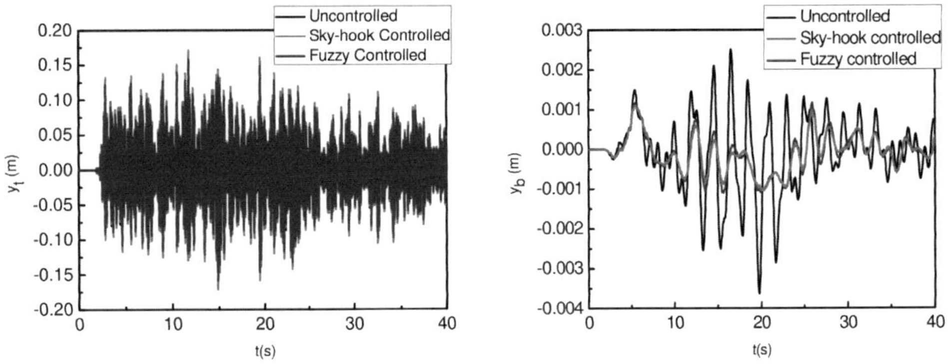

(a) bogie frame (b) car body

Figure 4. Results of fuzzy sky-ground hook controller.

$$u_{Di} = -\sigma_i c_{sky} v_i - (1-\sigma_i) c_{ground} v_i \tag{8}$$

where, c_{sky} is the sky-hook control gain, c_{ground} is the ground-hook control gain, σ_i ($(0 \le \sigma_i \le 1)$ is the weighting parameter between two control input. Figure 2 shows the block-diagram of the proposed fuzzy sky-ground hook control system. The basic configuration of the fuzzy logic composes three components: a fuzzyfication interface, a decision-making logic and a defuzzyfication interface.

4. Performance evaluation of vibration control

For the computer simulation in a subsequent section, the membership functions are adopted by triangular versions. And the fuzzy parameter values are used by 0.1 to 1. Figure 3 presents control response of the car body when the vehicle speed 180 km/h. We are clearly shown that the acceleration power spectral densities (PSD) of the lateral and roll motion are decreased in the primary frequency ranges. As shown in Figure 4 (a) and 4 (b), the time response of sky-hook control algorithm and fuzzy sky-ground hook control algorithm with the same control input. Figure 4 (a) presents the lateral displacement of bogie frame. It is shown that the proposed fuzzy sky-ground hook is decreased vibration of bogie frame.

5. Conclusion

In this work, a semi-active suspension system with MR dampers of railway vehicle has been investigated by considering a lateral response railway vehicle model, which includes three vibration motions such as lateral, yaw and roll of the car body. To illustrate the feasibility and effectiveness of controlled MR dampers on railway vehicle suspension system, the fuzzy sky-ground hook controller was utilized. This controller takes account for both vibration reduction of car body and increasing stability of bogie frame. In order to demonstrate the effectiveness of the proposed control system, computer simulation is undertaken showing vibration control performances. It has been demonstrated that bothvibration control and stability can be improved by employing the proposed control system associated with MR damper.

Acknowledgement

This work was financially supported by the grant from the Technology Development Program (10035438) of the Ministry of Knowledge Economy (MKE) of Korea.

References

[1] Choi S B, Lee H S and Park Y P 2002 H-infinite control performance of a full-vehicle suspension featuring magneto rheological damper *Vehicle Syst. Dyn.* **38** 341-60

[2] Dominguez A, Sedaghati R and Stiharu J 2006 A new dynamic hysteresis model for magnetorheological damper *Smart Mater Struct.* **15** 1179-89

[3] Wang D H and Liao W H 2009 Semi-active suspension for railway vehicles using magnetorheological damper *Vehicle Syst. Dyn.* **47** 1305-25

[4] Lai C Y and Liao W H 2002 Vibration control of a suspension system via a magnetorheological damper *J. Vib. Contol.* **8** 527-47

[5] Karnopp D C 1986 Two contrasting versions of the optimal active vehicle suspension *J. Dyn. Sys. Measurement and Control* **108** 264-68

[6] Choi S B, Suh M S, Park D W and Shin M J 2001 Neuro-fuzzy control of a tracked vehicle featuring semi-active electro-rheological suspension units *Vehicle Syst. Dyn.* **35** 141-62

Control system with magnetorheological fluid device for mitigation of the railway vehicle hunting oscillations

D Baiasu[1, 4], G Ghita[2] and I Sebesan[3]
[1] Atelierele CFR Grivita S.A., Calea Grivitei 359, 010718, Bucharest, Romania
[2] IMS-AR - Institute of Solid Mechanics of the Romanian Academy, C-tin Mille 15, 10141, Bucharest, Romania
[3] Faculty of Transports, Politehnica University Bucharest, Splaiul Independentei 313, 060042, Bucharest, Romania

E-mail: dan.baiasu@grivita.ro

Abstract The paper presents the opportunity of using a magnetorheological device to control the lateral oscillations of a passenger railway vehicle to increase its comfort and speed features. High speed generates large amplitude oscillations in horizontal plan at the axles, bogies and carbody levels which beyond a specific value – the critical speed – leads to unstable movement of the vehicle. The lateral dynamics of the vehicle is simulated using the multibody method. It is build a model with 17 degrees of freedom considering the lateral, yawing and rolling oscillations. An original device with magnetorheological fluid based on the balance logic strategy to control the anti-yaw damper is integrated in the secondary suspension of the railway vehicle. The conception and the design of the magnetorheological device are presented. It is demonstrated that the magnetorheological semi-active suspension improves the safety and the comfort of the railway vehicle.

1. Introduction

High speed trains are more and more present in the landscape of contemporary means of transport, being efficient, economic and environmental friendly. The dynamic response of the vehicle running on rails that have geometric irregularities is determined by the frequency of the random excitation from the rail and by the vehicle's speed [1]. High speeds generate large amplitude oscillations of the axles, bogies and vehicle body which beyond a specific value – the critical speed – leads to unstable movement of the vehicle. To keep under control the oscillations of the vehicle, through the years there were developed different types of suspension systems able to assure stable running at high speeds. The conventional suspension system for a passenger vehicle usually has two levels: the axle's suspension, stiffer, to provide safety and the suspension of the carbody, softer, to offer best ride quality [2]. Those suspension systems usually incorporate passive elements as springs and hydraulic dampers. There are possibilities to increase vehicle's speed by an appropriate design of the passive suspension but this approach proves to have limits. To overcome this issue the present paper considers the possibility of using a semi-active secondary suspension to control the hunting phenomenon of the railway vehicle carbody at high speeds. The semi-active device which will be used is an anti-yaw hybrid magnetorheological damper.

---

[4] Corresponding author.

2. The mathematical model of the railway vehicle

The lateral oscillations of the railway vehicle - the yaw and lateral displacement coupled oscillations - are critical for the vehicle performances because when the speed exceeds a certain value the movement of the vehicle becomes unstable, phenomenon called hunting. A high speed vehicle should be provided with a suspension able to increase the critical speed and in the same time to keep the accelerations of the carbody in certain comfort limits. Having in view the importance of the lateral oscillations phenomenon numerous authors have dedicated studies to the stability issue [3–11]. The mathematical models used in the literature to represent the vehicle differ depending on the number of degrees of freedom taken into account, the vehicle type, the linear or non-linear treatment of the wheel – rail contact phenomenon, of the forces appearing at the wheel – rail contact, as well as the irregularities of the tracks.

In view of studying the vehicle's response in horizontal plan, it was built up a mathematical model which simulates the lateral dynamics of a four axle railway vehicle [12, 13]. It is assumed that all the elastic and damping elements forming the classical suspension systems are weightless and have linear characteristics. Under conditions of geometrical, elastic and inertial symmetry, with identical wheel and rail patterns, the equilibrium position of the coach coincides with its median position in relation to the tracks. The rolling surfaces' contact angles are small and the radii of curvature for the rolling treads remain unchanged. Conicity has been considered as having an equal constant value with the rolling surfaces' effective conicity. The mechanical model's geometrical and elastic symmetry facilitates the decoupling of the lateral movements from the vertical ones [1, 5, 10, 11]. To study the vehicle's lateral oscillations, the mechanical model considers the following degrees of freedom: y_c, ψ_c, φ_c, y_{bj}, ψ_{bj}, φ_{bj}, y_i, ψ_i, where j=1,2 represent the bogies and i=1 – 4 the wheelsets.

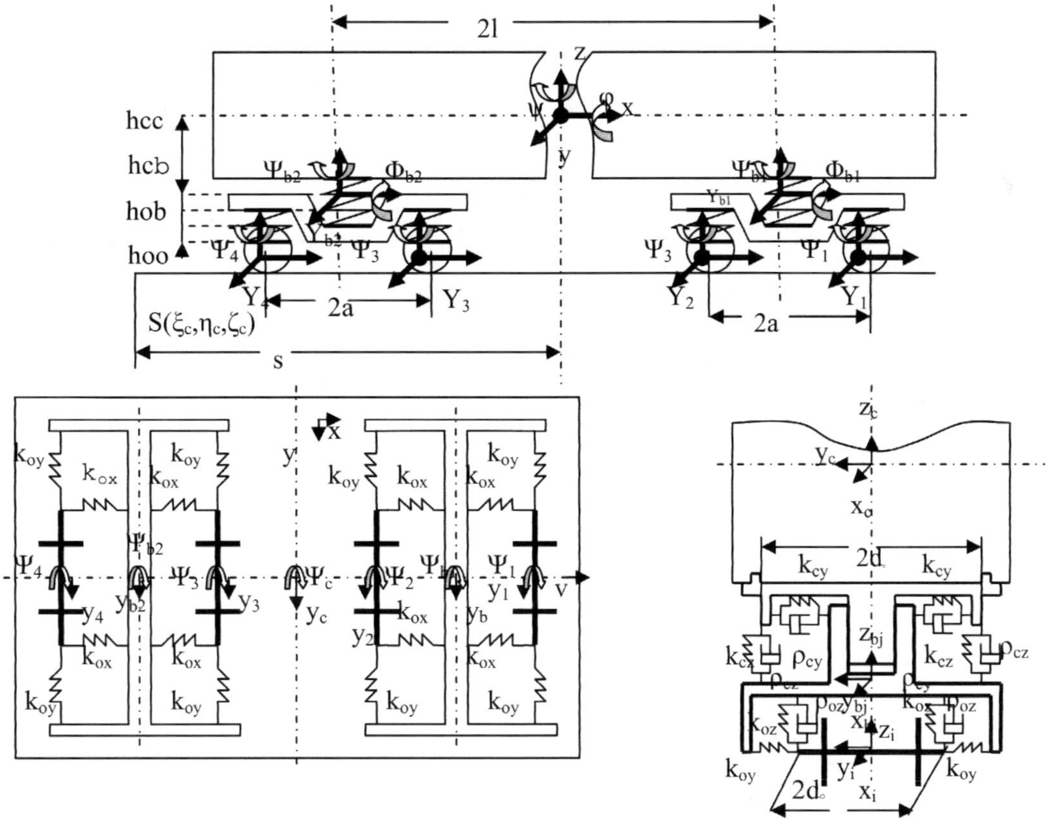

Figure 1. The railway vehicle model for lateral oscillations.

The model highlights the phenomenon of wheel-track contact at large amplitude movement of the axles, characterizing the stability limit cycles, because the flange-rail contact force is considered a linear spring with deadband [3]. The contact forces are expressed according to Kalker's linear theory [14]. The resultant creep force cannot exceed the adhesion force. The nonlinear effect of the adhesion limit is treated according to the Vermeulen - Johnson's nonlinear theory [14] for the case which doesn't consider the spin creepage. In the model, the tangent track's irregularities were considered to be periodical, represented with a sinusoid type expression, according to [5, 10]. The isolate variations of the track geometry type bump, was simulated with a single period sinusoid function which considers the irregularity tangent to the track and provides the variation of its dimensions according to the vehicle's speed.

The vehicle is considered as it is composed of a limited number of rigid bodies, simulating its main parts, connected in between through mechanical weightless linkages: the carbody, the bogies and the wheelsets [15, 16]. Lagrange's equation method was applied in order to establish the movement equations:

$$\frac{d}{dt}\left[\frac{\partial(E-V)}{\partial \dot{q}_k}\right]-\frac{\partial(E-V)}{\partial q_k}+\frac{\partial D}{\partial \dot{q}_k}=Q_k \tag{1}$$

where, q_k is the generalized coordinate, $\dot{q}_k$ is the generalized speed, E is the kinetic energy, V is the potential energy, D is the energy dissipation function; Q_k is the generalized force corresponding to the generalized coordinate q_k.

The generalized forces that are considered refers to y_i and ψ_i degrees of freedom of the wheelsets. The equations of the generalized forces are established using the wheelset efforts diagram presented in [5]. The generalized forces are Q_{y_i} generated by the lateral forces and Q_{ψ_i} due to the yaw moments acting on the wheelsets. Applying Lagrange's equations for the model's degrees of freedom, it can be obtained the motion equations for the body case, bogies and axles [12, 13]:

$$\begin{aligned} m_c\ddot{y}_c &+ 2k_{cy}\left[2(y_c+h_{cc}\varphi_c)-(y_{b1}+y_{b2})+h_{cb}(\varphi_{b1}+\varphi_{b2})\right]+2\rho_{cy}\left[2(\dot{y}_c+h_{cc}\dot{\varphi}_c)-(\dot{y}_{b1}+\dot{y}_{b2})\right. \\ &\left.+h_{cb}(\dot{\varphi}_{b1}+\dot{\varphi}_{b2})\right]=0 \end{aligned} \tag{2}$$

$$\begin{aligned} (I_{cz}/2)\ddot{\psi}_c &+ 2(\rho_{cy}l^2+\rho_{cx}d_c^2)\dot{\psi}_c+2(k_{cy}l^2+k_{cx}d_c^2)\psi_c-\rho_{cy}l(\dot{y}_{b1}-\dot{y}_{b2})+\rho_{cy}l(\dot{\varphi}_{b1}-\dot{\varphi}_{b2})h_{cb} \\ &-\rho_{cx}d_c^2(\dot{\psi}_{b1}+\dot{\psi}_{b2})-k_{cy}l(y_{b1}-y_{b2})+k_{cy}l(\varphi_{b1}-\varphi_{b2})h_{cb}-k_{cx}d_c^2(\psi_{b1}+\psi_{b2})=0 \end{aligned} \tag{3}$$

$$\begin{aligned} (I_{cx}/2)\ddot{\varphi}_c &+ 2(\rho_{cy}h^2{}_{cc}+\rho_{cz}d_c^2)\dot{\varphi}_c+2(k_{cy}h_{cc}^2+k_{cz}d_c^2)\varphi_c+2\rho_{cy}h_{cc}\dot{y}_c-\rho_{cy}h_{cc}(\dot{y}_{b1}+\dot{y}_{b2}) \\ &+(k_{cy}h_{cc}h_{cb}-k_{cz}d_c^2)(\varphi_{b1}+\varphi_{b2})=0+(\rho_{cy}h_{cc}h_{cb}-\rho_{cz}d_c^2) \\ &(\dot{\varphi}_{b1}+\dot{\varphi}_{b2})+2k_{cy}h_{cc}y_c-k_{cy}h_{cc}(y_{b1}+y_{b2}) \end{aligned} \tag{4}$$

$$\begin{aligned} m_b\ddot{y}_{bj} &+ 2\rho_{cy}\dot{y}_{bj}+2(k_{cy}+2k_{oy})y_{bj}-2\rho_{cy}\dot{y}_c-2\rho_{cy}h_{cc}\dot{\varphi}_c-2(-1)^{j+1}\rho_{cy}l\dot{\psi}_c-2(\rho_{cy}h_{cb})\dot{\varphi}_{bj} \\ &-2k_{cy}y_c-2k_{cy}h_{cc}\varphi_c-2(-1)^{j+1}k_{cy}l\psi_c-2(k_{cy}h_{cb}+2k_{oy}h_{ob})\varphi_{bj} \\ &-2k_{oy}(y_{2j-1}+y_{2j})=0 \end{aligned} \tag{5}$$

$$\begin{aligned} (I_{bz}/2)\ddot{\psi}_{bj} &+ \rho_{cx}d_c^2\dot{\psi}_{bj}+(k_{cx}d_c^2+2k_{oy}a^2+2k_{ox}d_o^2)\psi_{bj}-\rho_{cx}d_c^2\dot{\psi}_c-k_{cx}d_c^2\psi_c \\ &-k_{oy}a(y_{2j-1}-y_{2j})-k_{ox}d_o^2(\psi_{2j-1}+\psi_{2j})=0 \end{aligned} \tag{6}$$

$$\begin{aligned}(I_{bx}/2)\ddot{\varphi}_{bj} &+ \left(\rho_{cy}h_{cb}^2 + \rho_{cz}d_c^2 + 2\rho_{oz}d_o^2\right)\dot{\varphi}_{bj} + \left(k_{cy}h_{cb}^2 + k_{cz}d_c^2 + 2k_{oy}h_{ob}^2 + 2k_{oz}d_o^2\right)\varphi_{bj} \\ &+ \rho_{cy}h_{cb}\dot{y}_c + \left(\rho_{cy}h_{cb}h_{cc} - \rho_{cz}d_c^2\right)\dot{\varphi}_c + (-1)^{j+1}l\rho_{cy}h_{cb}\dot{\psi}_c - \rho_{cy}h_{cb}\dot{y}_{bj} + k_{cy}h_{cb}y_c \\ &+ \left(k_{cy}h_{cb}h_{cc} - k_{cz}d_c^2\right)\varphi_c + (-1)^{j+1}lk_{cy}h_{cb}\psi_c - \left(k_{cy}h_{cb} + 2k_{0y}h_{ob}\right)y_{bj} \\ &- \left(k_{cy}h_{cb} + 2k_{0y}h_{ob}\right)y_{bj} + k_{oy}h_{ob}\left(y_{2j-1} + y_{2j}\right) = 0\end{aligned} \tag{7}$$

$$\begin{aligned}I_{oz}\ddot{\psi}_i - 2\rho_{ox}d_o^2\dot{\psi}_{bj} - 2k_{ox}d_o^2\psi_{bj} &+ \left[I_{oy}\frac{v}{r_0}\frac{\lambda}{e} - \frac{2\varepsilon f_{12}}{v}\left(1 + r_0\frac{\lambda}{e}\right)\right]\dot{y}_i + 2\varepsilon f_{33}e\frac{\lambda}{r_0}y_i \\ &+ 2\left[\frac{\varepsilon}{v}\left(f_{22} + e^2 f_{33}\right) + \rho_{ox}d_o^2\right]\dot{\psi}_i + \left[2\varepsilon f_{12} + 2k_{ox}d_o^2 - eQ\lambda\right]\psi_i = 2\varepsilon f_{33}\frac{e\lambda}{r_0}\eta_i\end{aligned} \tag{8}$$

$$\begin{aligned}m_o\ddot{y}_i + 2\rho_{oy}\dot{y}_{bj} + 2k_{oy}y_{bj} &+ (-1)^{j+1}\cdot 2\rho_{oy}a\dot{\psi}_{bj} + (-1)^{j+1}\cdot 2k_{oy}a\psi_{bj} + 2\rho_{oy}h_{ob}\dot{\varphi}_{bj} + 2k_{oy}h_{ob}\varphi_{bj} \\ &+ 2\left[\frac{\varepsilon f_{11}}{v}\left(1 + r_0\frac{\lambda}{e}\right) - 2\rho_{oy}\right]\dot{y}_i + \left(\frac{Q\lambda}{e} - 2k_{oy}\right)y_i + \frac{2\varepsilon f_{12}}{v}\dot{\psi}_i - 2\varepsilon f_{11}\psi_i \\ &= \left(\frac{Q\lambda}{e} - 2k_{oy}\right)\eta_i\end{aligned} \tag{9}$$

The significance of the symbols employed in the equations (2) - (9) is presented in the Table 1. The equation system representing the mathematical model of the lateral motion of the railway vehicle can be written in the following matrix form:

$$[M]\{\ddot{q}\} + [C]\{\dot{q}\} + [K]\{q\} = \{F(t)\} \tag{10}$$

where $[M] \in \Re^{17x17}$, $[C] \in \Re^{17x17}$ and $[K] \in \Re^{17x17}$ are the mass, damping and stiffness matrixes of the vehicle's system, $\{\ddot{q}\}$, $\{\dot{q}\}$ and $\{q\}$ are the accelerations, speeds and displacements vectors and $\{F(t)\} \in \Re^{17}$ is the vector of the periodical excitation given by the tracks.

Through a variable change, the system can be transformed into an ordinary differential equation system [6] which it is treated using numeric methods:

$$\{\dot{y}\} = [E]\{y\} + \{F^*(t)\} \tag{11}$$

where,

$$[E] = \begin{bmatrix} [0] & [I] \\ -[M]^{-1}[K] & -[M]^{-1}[C] \end{bmatrix}$$

$$\{F^*(t)\} = \begin{Bmatrix} \{0\} \\ [M]^{-1}\{F(t)\} \end{Bmatrix} \tag{12}$$

The equations (11) are employed both for the study of stability and the system's response using a numerical integration method of the movement equations, the Runge – Kutta method of 4th order, for a simulation software has been designed. To simulate the vehicle's response, there are used the construction characteristics presented in Table 1. As example, the response of the vehicle with the characteristics in Table 1, launched on a tangent track with periodic irregularities, running with 180 km/h – the maximal testing speed, is presented in the Figures 2 – 4, indicating that the tracks' perturbations effect is slightly felt at the coach case level, as opposed to the bogie and axles where it persists during the coach's circulation. The coach's main suspension acts correspondingly and meets the comfort demands inside the coach

Table 1. Construction characteristics of the vehicle.

Characteristics	Symbols	Values	Units
Body case mass	m_c	30760	kg
Bogie mass	m_b	2300	kg
Wheelset mass	m_o	1410	kg
Carbody moments of inertia	I_{cx}	53596	kg m^2
	I_{cz}	1661732	kg m^2
Bogie moments of inertia	I_{bx}	2240	kg m^2
	I_{bz}	2965	kg m^2
Axles moments of inertia	I_{oy}	980	kg m^2
	I_{ox}	100	kg m^2
Central suspension stiffness	k_{cx}	133	kN m^{-1}
	k_{cy}	133	kN m^{-1}
	k_{cz}	473	kN m^{-1}
Axle suspension stiffness	k_{ox}	256	kN m^{-1}
	k_{oy}	885	kN m^{-1}
	k_{oz}	904	kN m^{-1}
Central suspension damping	ρ_{cx}	0	kN m^{-1} s^{-1}
	ρ_{cy}	25	kN m^{-1} s^{-1}
	ρ_{cz}	18	kN m^{-1} s^{-1}
Damping of the axle suspension	ρ_{oz}	3.67	kN m^{-1} s^{-1}
Wheel tread nominal radius	r_0	0.460	m
The track's gauge	$2e$	1.435	m
The bogie's wheelbase	$2a$	2.560	m
The distance between bogies	$2l$	17.2	m
The distance between the secondary suspension springs	$2d_c$	2	m
The distance between the primary suspension springs	$2d_o$	2	m
The distance case center – central suspension	h_{cc}	1.24	m
The distance axles suspension - bogie center	h_{ob}	0.01	m
The distance central suspension - bogie center	h_{cb}	0.06	m
Load on wheel	Q	51250	N
The longitudinal creep coefficient	f_{11}	9430000	N
The lateral/spin creep coefficient	f_{12}	1200	N m
The spin creep coefficient	f_{22}	1000	N m^2
The lateral creep coefficient	f_{33}	10250000	N
The effective wheel conicity	λ	0.14	-
The maximum testing speed	v_{max}	50	m s^{-1}

The mathematical model of the vehicle it is used to study the vehicle stability. The dynamic system is asymptotically stable if and only if all the eigenvalues of the matrix E have a negative real part [1]. For the vehicle with the characteristics in Table 1, the value of linear critical speed is 245.6 km/h. The non-linear critical speed is determined by simulation [17] with the mathematical model and it has the value of 234.2 km/h. In Figure 5 it is shown the response of the leading axle of the trailing bogie at the stability limit and in Figure 6 the diagram of stability limit.

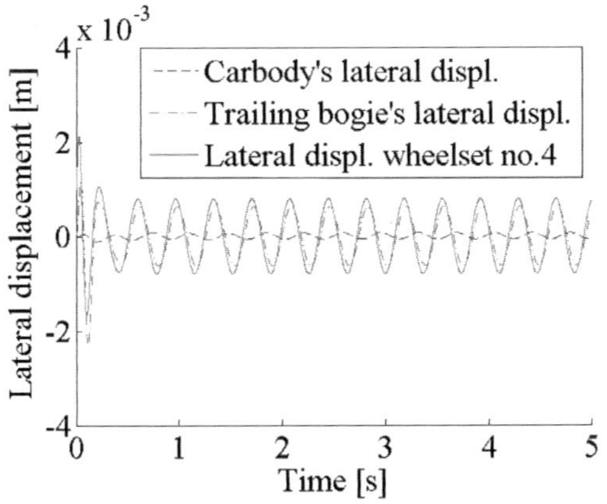

Figure 2. The lateral displacement.

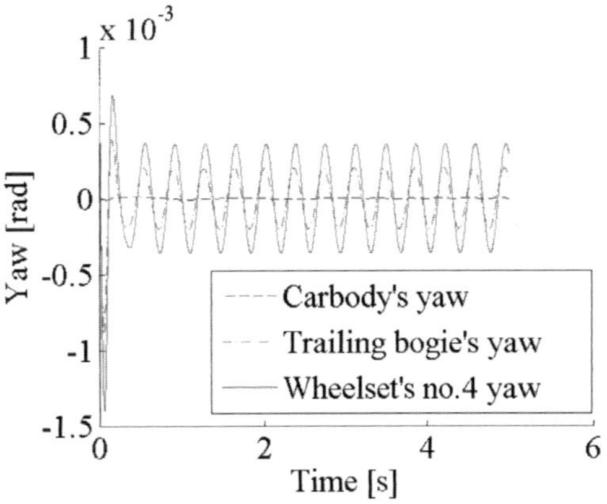

Figure 3. The yaw oscillations.

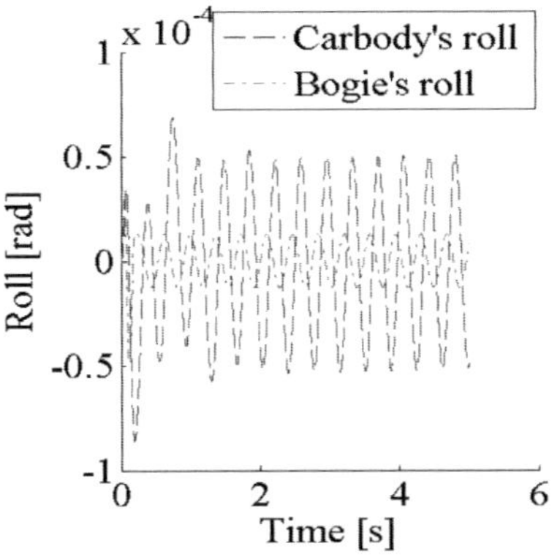

Figure 4. The roll oscillations.

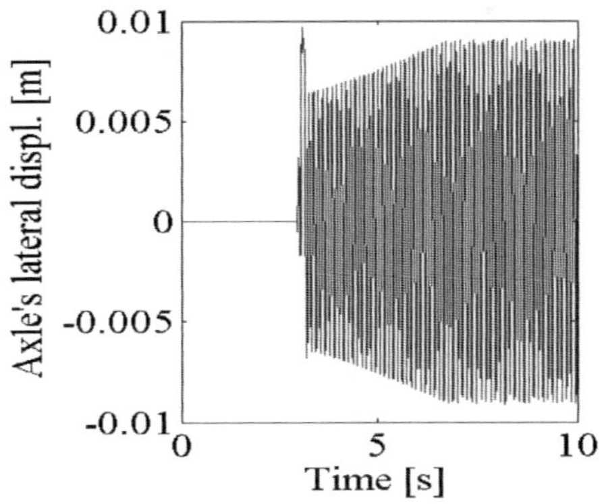

Figure 5. Axle displacement, $v > 234.2$ km/h.

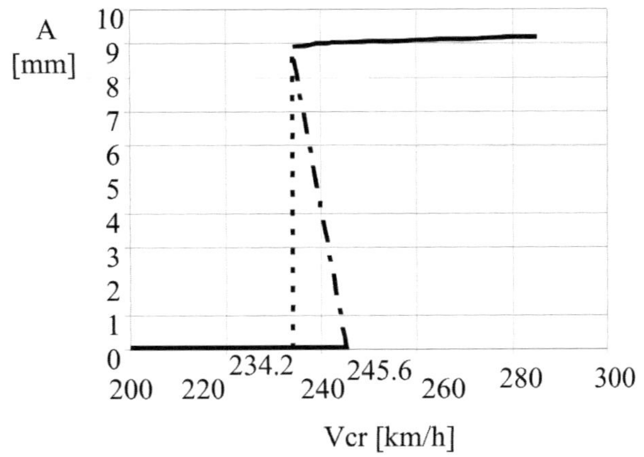

Figure 6. The bifurcation diagram of the railway vehicle.

For the vehicle with the characteristics in the Table 1, the stationary solution of the dynamic system is asymptotically stable for speeds less than 245.6 km/h. For speeds greater than 245.6 km/h the solution of the system is unstable. The point of the diagram with the coordinates (245.6; 0) is consequently a bifurcation point – a point where an unstable periodic solution bifurcates subcritically.

The solution becomes unstable in the saddle-node bifurcation point with the coordinates (234.2; 9) where the non-linear speed is reached and the amplitude of the wheelset's movement equals the rail-wheel clearance.

3. The limits of the classical suspension

The mathematical model of the vehicle previously presented can become a useful tool to improve the suspension design. A parametric study of the vehicle's non-linear stability shows the possibilities of raising the vehicle's speed through the passive suspension construction.

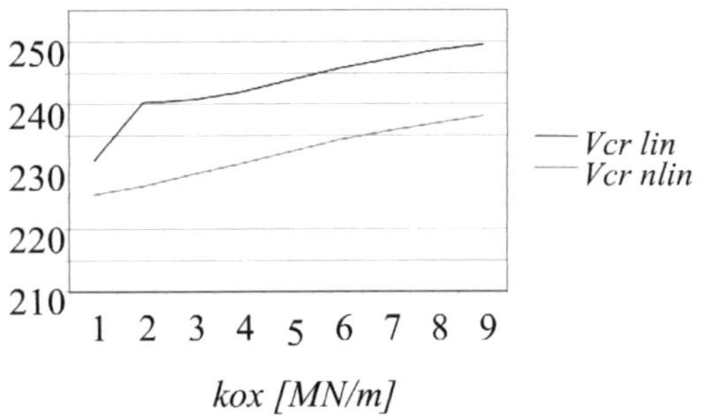

Figure 7. Primary suspension stiffness influence on critical speed.

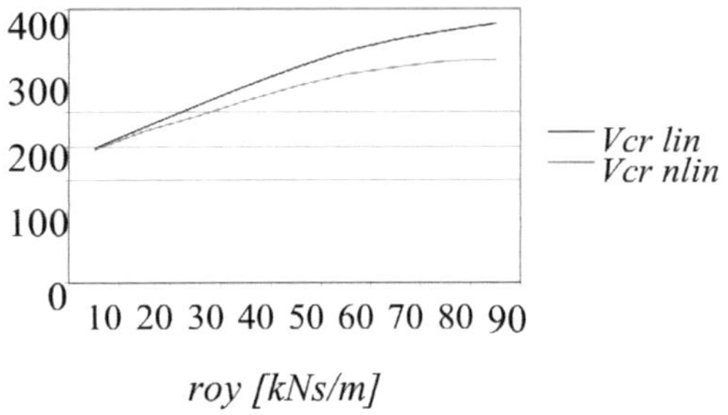

Figure 8. Primary suspension damping influence on critical speed.

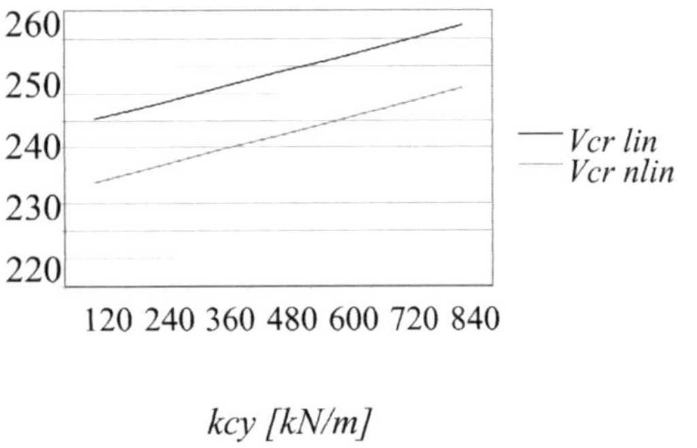

Figure 9. Secondary suspension stiffness influence on critical speed.

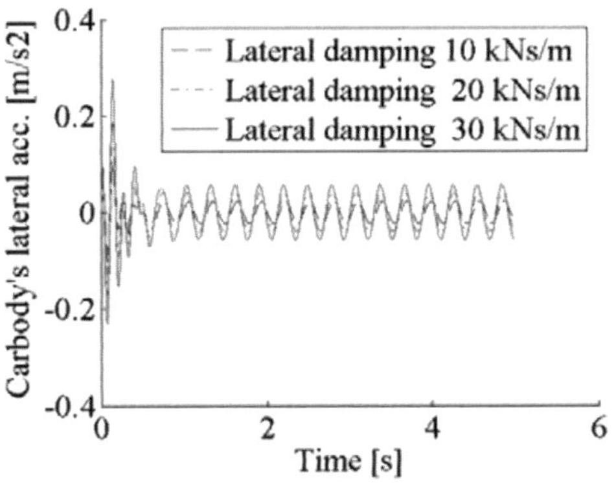

Figure 10. The lateral accelerations of the carbody.

It is investigated the influence of the suspension damping and stiffness lateral and longitudinal components on the critical speed of the vehicle. The critical speed of the vehicle is strongly influenced by the primary suspension parameters k_{ox} , ρ_{oy} and by the lateral damping of the secondary suspension k_{cy} .

In the Figures 7, 8 and 9 are presented the influences of the mentioned parameters on the vehicle's critical speed. The increase of the critical speed is not the unique criteria of the railway vehicle performances. According to UIC 518 leaflet, the assessment of the vehicle ride quality can be made by measuring the accelerations of the carbody on vertical and lateral directions. To establish the lateral acceleration of the carbody it is employed the mathematical model presented previously. The expression of the lateral acceleration is:

$$a_c = -\frac{2}{m_c}\left\{2\left[\rho_{cy}\dot{y}_c + k_{cy}y_c + h_{cc}\left(\rho_{cy}\dot{\varphi}_c + k_{cy}\varphi_c\right)\right] - \left[\rho_{cy}\sum_{j=1}^{2}\dot{y}_{bj} + k_{cy}\sum_{j=1}^{2}y_{bj} - h_{cb}\left(\rho_{cy}\sum_{j=1}^{2}\dot{\varphi}_{bj} + k_{cy}\sum_{j=1}^{2}y_{bj}\right)\right]\right\} \qquad (13)$$

There are determined the accelerations of the carbody for different values of the central suspension lateral damping coefficient. It can be seen that the result of using more rigid dampers in the central suspension is worse in terms of comfort, due to the coupling effect of those between the lateral oscillations of the bogies and of the carbody – Figure 10.

4. The opportunity of a semi-active magnetorheological railway suspension

In the last years the magnetorheological dampers have been employed with predilection in the applications for low frequencies vibrations damping type vibration insulation or earthquake protection of civil constructions. There are also magnetorheological applications in the automotive, aeronautic or military industries.

In the railway field there are not many vehicles using magnetorheological devices. One of the reasons it may be the fact that active technologies are already widely used on high speed trains: the tilting devices which became standard since 1990, the servo-pneumatic and servo-hydraulic active suspensions or the piezoelectric active suspensions. Although extremely effectual, the active suspensions require high powers and sophisticated control technologies. The fact that the active systems bring supplementary energy to the vehicle may lead to system's instability.

Semi-active suspension systems combine the advantages of the construction simplicity and of the low costs specific to the passive devices with those specific to the active systems, resulting very effective devices that assure high performing control with low power consumption, reliable and stable. The magnetorheological dampers valorize the ability of the material to radically modify its rheological properties in milliseconds when in the presence of a magnetic field. The short reaction time becomes another strong point towards the employment of the magnetorheological dampers to railway vehicles, having in view the complexity of the operational conditions of those: variable loads, speeds, track irregularities, weather conditions.

There are few articles treating the magnetorheological suspensions for railway vehicles. In [18] the authors implemented several control strategies used in the automotive industry to a railway vehicle suspension through a computer simulation program which accounts different operational conditions. The results showed that semi-active control improves ride quality of the vehicle. The study, carried on in mid-90th years, evokes the opportunity of using in the railway vehicle suspension electrorheological or ferrofluid dampers. A study of a magnetorheological damper in the secondary suspension of a locomotive is presented in [19]. The paper showed the fact that the semi-active strategy is more efficient than the passive one and the results may be closed to those obtained with an active damper. The design and construction of a magnetorheological double ended damper is presented in [20]. The damper model has been established using a testing bench and the ride quality of a vehicle using such a damper was assessed by means of a simulation software showed to improve. The papers [21, 22] are of interest because it is presented the implementation of a acceleration feed-back control strategy and the positive results of using a magnetorheological damper in vehicle's secondary suspension.

A possibility to continue to increase the vehicle's speed without affecting the comfort is the use of the controllable suspensions. In the present paper, the option for controlling the lateral oscillations of the vehicle's carbody is to use a semi-active control system with a magnetorheological fluid device.

The semi-active control systems have the following advantages:

- the oscillations insulation is accomplished by dissipation of energy, similar to the passive devices;

- the system assures full adaptability as active systems does;
- no additional energy is introduced in the system;
- the energy needed to operate the semi-active device may be supplied by low power sources;
- there are robust and reliable devices;
- the systems are opened to the unconventional technical solutions.

As operational device of the semi-active suspension, a hybrid magnetorheological damper will be introduced in the secondary suspension to control the passive anti-yaw hydraulic damper.

The magnetorheological damper brings forward significant benefits in comparison to the classic versions:

- the damping force is controllable in real time;
- assures great value damping forces at low speeds which is practically impossible with hydraulic dampers;
- the magneto-rheological fluid has its own viscosity which assures a minimal damping force even if the power supply fails;
- the constructive solution is not sophisticated and easy to be implemented to replace classical dampers.

In order to increase the railway vehicle performance, four magnetorheological controlled dampers will be introduced between the bogies and the carbody to replace the passive dampers. The semi-active systems act to reduce the elastic forces resultant across the vehicle secondary suspension. The semi-active suspension system should continuously control the vehicle's response and set the lateral damping force of the secondary suspension to the appropriate values.

If in the equations (10) are replaced the expressions of the lateral damping forces of the secondary suspension with those given by the magnetorheological dampers, it will result the following formulation of the matrix form of the vehicle's model:

$$[M]\{\ddot{q}\}+[C']\{\dot{q}\}+\{F_d(\Delta q,\Delta\dot{q},i)\}+[K]\{q\}=\{F(t)\} \tag{14}$$

where $[C']$ is the passive damping matrix, $\{F_d(\Delta q,\Delta\dot{q},i)\}$ is the controllable damping force of the secondary suspension vector and i is the intensity of the current applied to the driver. The controllable damping force depends on the relative lateral displacement $(\Delta q = \Delta y_c)$and the relative lateral speed $(\Delta\dot{q} = \Delta\dot{y}_c)$in the secondary suspension and on the applied current intensity.

Using the notations according to Figure 1, the relative displacement and speed equations are:

$$\Delta y_c = y_c + h_{cc}\varphi_c + (-1)^{j+1} l\psi_c - y_{bj} + h_{cb}\varphi_{bj} \tag{15}$$

$$\Delta\dot{y}_c = \dot{y}_c + h_{cc}\dot{\varphi}_c + (-1)^{j+1} l\dot{\psi}_c - \dot{y}_{bj} + h_{cb}\dot{\varphi}_{bj} \tag{16}$$

4.1. The hybrid magnetorheological damper

The hybrid magnetorheological dampers are devices that use a hydraulic section to create the damping force and a magnetorheological section to control the hydraulic section. The control element is a small magnetorheological device. In literature this type of damper is known also as magnetorheological piloted hydraulic damper. This type of damper features several advantages as: reduced cost due to the small quantity of magnetorheological fluid required, reduced weight, aspect and overall dimensions similar to those of the passive damper. In fact, the design of the hybrid damper will start from the original passive damper used in the secondary suspension of the vehicle. Figure 11 presents the schematic diagram of a hybrid magnetorheological damper proposed by the authors to be included in the vehicle's semi-active suspension.

The main parts of the hybrid damper, according to Figure 11, are as follows:

- the hydraulic cylinder;
- the control pressure selector;
- the magnetorheological piloting device;
- check valves.

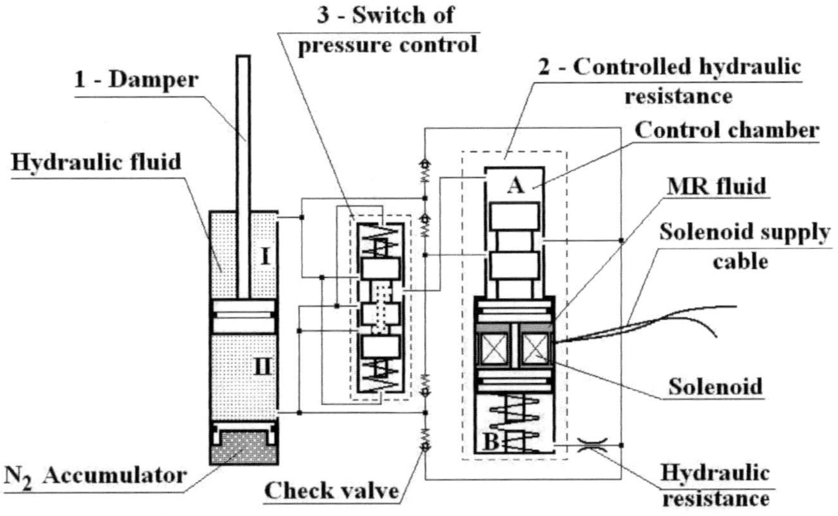

Figure 11. The hybrid magnetorheological damper.

As control strategy for the magnetorheological damper, it is selected a sequential algorithm based on the balance of forces (balance logic) built to control the system's vibrations exclusively on an energy dissipation basis. According to the balance logic, the damping force has to be controlled to balance the elastic force when the carbody is moving away from its equilibrium position – the elastic and damping forces act in opposite directions – and by setting the damping force to a minimal value when the forces have the same direction.

The construction principle of the hybrid magnetorheological damper, presented in Figure 11 fulfils those requirements. The pressures in the chambers I and II of the damper are controlled by the magnetorheological device (the controlled hydraulic resistance) through the pressure switch. The damping force is purely hydraulic. The control system uses speed and movement sensors. When the movement and the speed have opposite signs, the solenoid is supplied with current and the magnetorheological fluid is activated. The current intensity is supplied in such way that the cumulated forces in chamber B of the device and of the magnetorheological fluid resistance allow the pressure in chamber A to move the piston of the device until the flow between the chambers I and II of the damper would determine the pressure gradient which will produce the damping force according to the control algorithm. The solution proposed for the magnetorheological device is a double ended piston with a fixed solenoid mounted in a cylinder. This solution brings benefits because the solenoid supply wires do not have to move in operation.

4.2. The hybrid magnetorheological damper design

The development of the solution for the magnetorheological controlled damper begun from the original passive damper type T50/20x210 with the characteristics presented in the Table 2. Having in view that the active surface of the damper's piston is $A_p = 16.5\,\text{cm}^2$, the pressure drop is $\Delta p_d = 30.31\,\text{daN cm}^{-2}$.

To establish the dimensions of the controlled hydraulic resistance piston it will be calculated the surface of the opening, A_o, necessary to transfer the flow Q with a pressure drop Δp_d.

Table 2. The characteristics of the passive damper

Characteristic	Symbol	Value	Unit
Piston diameter	d_p	0.05	m
Piston rod diameter	d_t	0.02	m
Maximal force	F	5000	N
Velocity	v_p	0.1	m s^{-1}
Fluid flow	Q	1.65x10^{-4}	m^3 s^{-1}

The fluid's flow it is considered turbulent. The following equation will be applied:

$$Q = c_d A_o \left(\frac{2\Delta p_d}{\rho} \right)^{1/2} \tag{17}$$

where $c_d \cong 0.61$ characterize the turbulent flow, $\rho = 880\,\text{kg/m}^3$ is the density of the hydraulic fluid.

The minimal opening surface will be: $A_o^{\min} = 3.26\,\text{mm}^2$.

The maximal surface of the opening will be obtained for free flow between the damper chambers. The minimal pressure gradient is chosen to be 3 bar – the pressure drop estimated as being generated by the hydraulic resistances on the circuit - and the maximal surface of the opening is $A_o^{\max} = 10.36\,\text{mm}^2$.

Considering that the piston of the control device has a diameter of 12 mm, the minimal and maximal forces acting on the piston are:

- $F^{\min} = 34$ N – the force of the spring in chamber B necessary to close the circuit between the chambers I and II by placing the piston of the controllable device in a position that completely closes the hydraulic resistance;
- $F^{\max} = 343\,\text{N}$ – the minimal force that should be produced by the magnetorheological fluid device at the highest value of the applied current.

The opening is considered to be rectangular with the bigger leg, L_o, perpendicular on the hydraulic resistance piston axle equal to 6 mm. The strokes of the two extreme positions of the piston are: $c^{\min} = 0.54\,\text{mm}$ and $c^{\max} = 1.73\,\text{mm}$. The maximal stroke is chosen to be 2 mm.

Table 3. The input data of the design.

MR fluid type	MRF-132LD
Maximal velocity of the piston	$v^{\max} = 0.015\,\text{m s}^{-1}$
Maximal stroke	$c^{\max} = 5 \times 10^{-3}\,\text{m}$
Interior diameter of the device	$d = 22 \times 10^{-2}\,\text{m}$
Diameter of the piston rod	$d_b = 8 \times 10^{-3}\,\text{m}$
Operation temperatures range	$-20^\circ\text{C} \div 150^\circ\text{C}$
A wide range of elastomers compatibility	-

The constructive model is the one of the double ended device. A design method of a double ended magnetorheological damper is presented in [23], [20] and more developed in [24]. The input data are presented in the Table 3.

To continue the dimensional design, following parameters are chosen according to the magnetorheological fluid characteristics supplied by Lord Corp. [25]. The chosen parameters are shown in the Table 4.

Table 4. Magnetorheological fluid parameters

Shear rate	$\dot{\gamma} = 140\,s^{-1}$
Maximal intensity of the magnetic field	$H = 250\,kA\,m^{-1}$
Fluid viscosity	$\eta = 0.25$ Pa s
Yield stress	$\tau_c = 44.1\,kPa$

The aim is to maximize the controllability of the device by maximizing the dynamic range coefficient and to minimize the volume of magnetorheological fluid. Imposing those conditions it results the value of the thickness of the gap between the piston and the inner surface of the device [24].

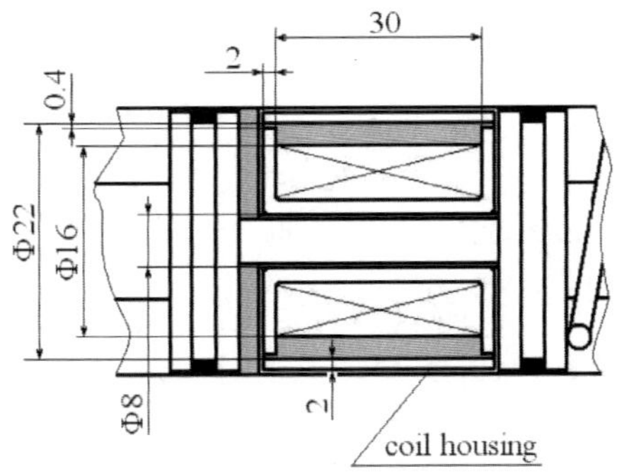

Figure 12. The solenoid of the device

To design the constructive parameters of the solenoid there are calculated the dimensions of the magnetic circuit composed of the poles, the device case between the poles, the support of the solenoid and the magnetorheological fluid [24], considering an applied current of 1.5A.

The resulting parameters of the magnetorheological device shown in Figure 12 are presented in the Table 5.

The functional characteristics of the device with magnetorheological fluid in Figures 13 and 14 were experimentally determined for sinusoidal motion with 1.0 Hz frequency and 0.002 m displacement amplitude.

Table 5. The parameters of the magnetorheologic device.

Thickness of the gap	0.4 mm
Total force of the device	406.7 N
Number of poles	2
Pole width	2 mm
Maximal solenoid diameter	16 mm
Solenoid length	30 mm
Wire diameter	0.6 mm

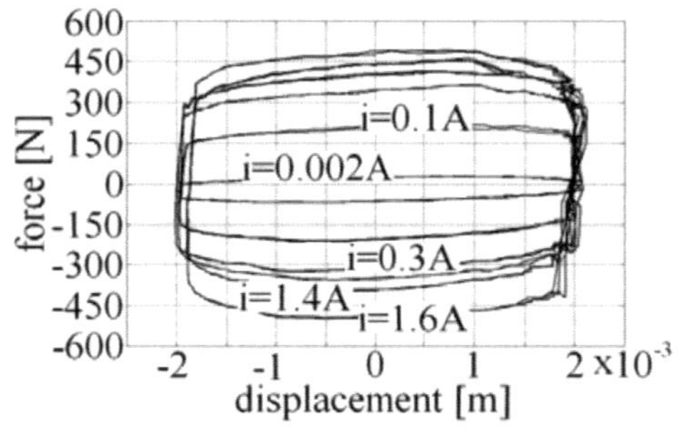

Figure 13. Forces vs. displacement diagram.

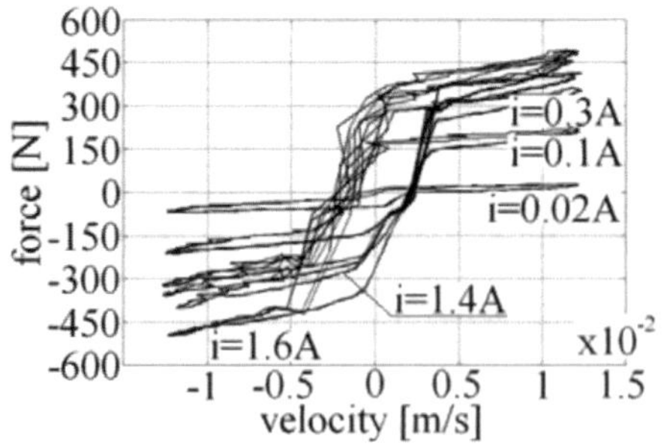

Figure 14. Forces vs. velocity diagram.

As it can be seen in the figures, the maximal value of the force acting on the solenoid piston is bigger than the value presented in Table 5. This fact is generated by the number of windings that for technological reasons was bigger than the calculated one. The maximal value of the current intensity used in the tests, 1.6A is greater than the estimated 1.5A and this has also an increasing effect on the force. The force of the device is greater than the minimal needed value of 343 N and the device is suitable to control the hydraulic damper.

4.3. The semi-active suspension simulation

According to the balance logic control strategy [26], the equation of the semi-active damping force is:

$$F_d(\Delta y_c, \Delta \dot{y}_c, i) = \begin{cases} |F_e(\Delta y_c)| \operatorname{sgn} \Delta \dot{y}_c \,, \Delta y_c \Delta \dot{y}_c < 0 \\ F_{d\min}(\Delta y_c, \Delta \dot{y}_c) \,, \Delta y_c \Delta \dot{y}_c \geq 0 \end{cases} \tag{18}$$

where $F_e(\Delta y_c)$ is the lateral elastic force across the secondary suspension and $F_{d\min}(\Delta y_c, \Delta \dot{y}_c)$ is the minimal damping force, due for example to the inherent energy dissipation when the hydraulic fluid flows through the control pressure selector's orifices, reached when the magnetic field isn't applied ($i = 0$). Having in view the conception of the hybrid damper, in the next pages $F_{d\min}$ will be considered negligible in comparison to the damping forces in semi-active mode. The equation of the damping force will become:

$$F_d(\Delta y_c, \Delta \dot{y}_c, i) = \begin{cases} 2\alpha k_{cy} |\Delta y_c| \operatorname{sgn}(\Delta \dot{y}_c), \Delta y_c \Delta \dot{y}_c < 0 \\ 0 \qquad\qquad\qquad\quad , \Delta y_c \Delta \dot{y}_c \geq 0 \end{cases} \tag{19}$$

where α is a dimensionless gain factor.

To simulate the dynamics of the device with magnetorheological fluid of the hydraulic resistance, it will be used the modified Bouc-Wen model, proposed by Spencer et al. in 1997 [27]. This model improves the simulated answer of the magnetorheological damper in the yield region.

The Bouc-Wen models have been conceived aiming from the start to represent the magnetorheological dampers. The modified Bouc-Wen model, shown in Figure 15, consist in a Bouc-Wen operator which represents the hysteretic behaviour of the damper set in parallel with a viscous damper and a spring simulating the viscous damping and the stiffness of the damper at high velocities, combined with a spring representing the pneumatic accumulator stiffness and a viscous damper to represent the viscous damping effects at high velocities.

The governing equations of the Bouc-Wen model [27] are:

$$F(t) = c_1 y + k_1 (x - x_0) \tag{20}$$

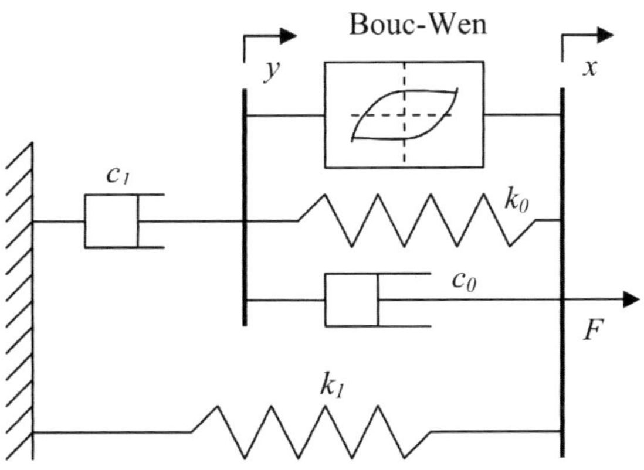

Figure 15. Modified Bouc-Wen model.

$$\dot{y} = \frac{1}{c_0 + c_1}\left[\alpha z + c_0 \dot{x} + k(x - y)\right] \tag{21}$$

$$\dot{z} = -\gamma|\dot{x} - \dot{y}| \cdot |z|^{n-1} \cdot z - \beta(\dot{x} - \dot{y}) \cdot |z|^n + A(\dot{x} - \dot{y}) \tag{22}$$

where y is the internal displacement of the damper, A, γ, β are the parameters that determine the shape of the loop, n is the parameter that determines the smoothness of the force-displacement diagram, x_0 is the initial displacement representing the pneumatic accumulator and $z(x)$ is the hysteretic component that represents a function of the time history of the displacement.

To enlighten the influence of the applied current on the magnetorheological damper behavior, it is assumed that the parameters c_0, c_1, α depend on the voltage applied to the current driver. There are several formulations of this dependence: linear, third grade polynomial or asymmetric sigmoid function. The linear dependence according to [27] can be expressed with the following equations:

$$\alpha(u) = \alpha_a + \alpha_b u \tag{23}$$

$$c_1(u) = c_{1a} + c_{1b} u \tag{24}$$

$$c_0(u) = c_{0a} + c_{0b} u \tag{25}$$

where α_a , c_{1a} and c_{0a} are the values of the parameters at 0 V – free flow of the magnetorheological fluid.

The Bouc-Wen model involves 9 parameters that characterize the damper's behavior. Those parameters should be determined for each value of the applied current in such way that the predicted response of the damper using the model should be as close as possible to the experimental results obtained through tests. This represents an optimization problem that can be solved through genetic algorithms method. In [26] it is recommended a sequential formulation of the voltage control law similar to the balance logic strategy that would simplify the identification process:

$$u(x, \dot{x}) = \begin{cases} \varepsilon K|x|\,\mathrm{sgn}(\dot{x}), & x\dot{x} < 0 \\ 0 & , x\dot{x} \geq 0 \end{cases} \tag{26}$$

where ε is a dimensionless gain factor experimentally determined to obtain a convenient transmissibility or a minimum r.m.s. acceleration output and K is the measuring calibration factor expressed in Vmm^{-1}.

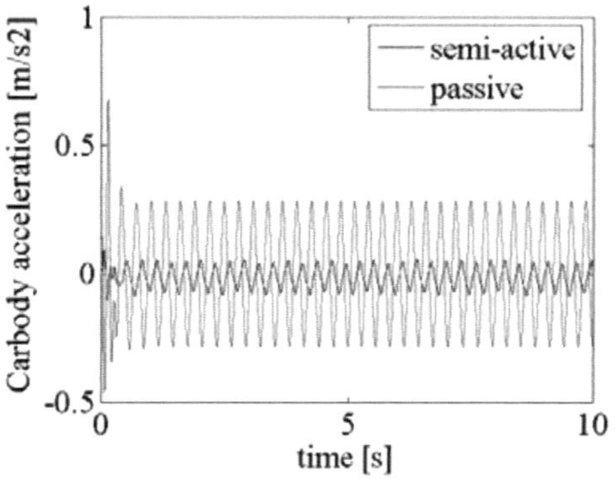

Figure 16. Carbody acceleration (*v=220 km/h*, $\eta_0 = 5mm$).

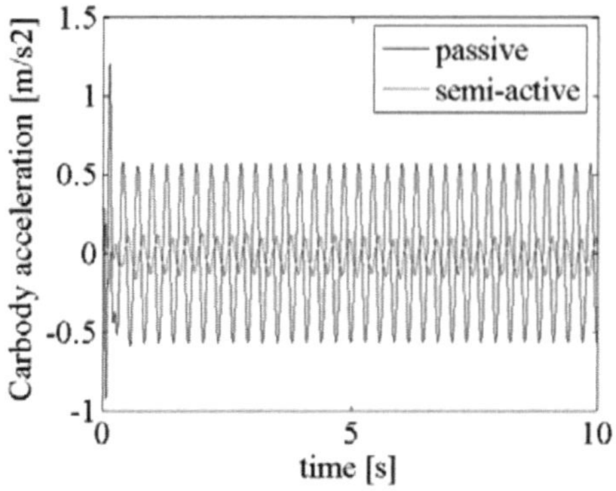

Figure 17. Carbody acceleration (*v=220 km/h*, $\eta_0 = 10mm$).

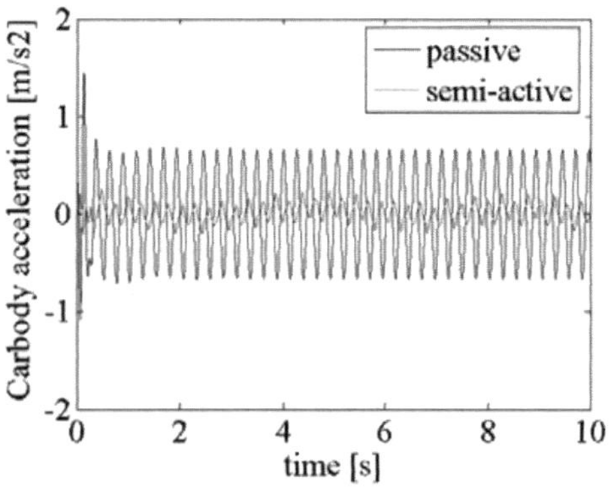

Figure 18. Carbody acceleration (*v=250 km/h*, $\eta_0 = 10mm$).

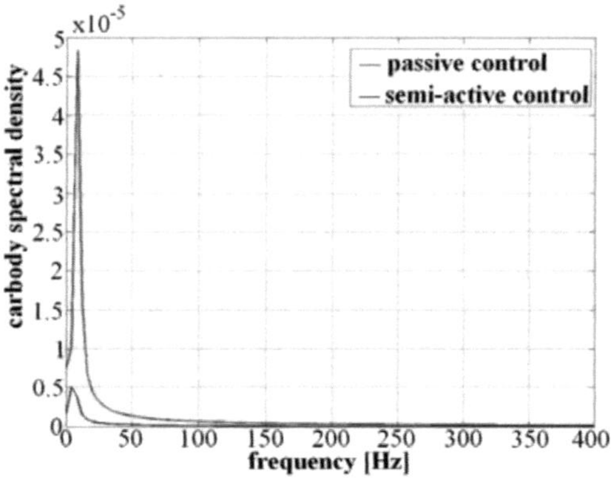

Figure 19. Carbody acceleration PSD.

Several simulations of the employment of the semi-active suspension compared with the use of the classic solution in terms of carbody accelerations are made for different speeds and periodic irregularities amplitudes. To perform those simulations, the controlled damping force generated by the semi-active dampers was included in the mathematical model of the vehicle.

As presented in Figures 16 - 18, the use of a semi-active control strategy reduces significantly the values of the lateral accelerations improving the railway vehicle's comfort.

Figure 19 shows the influence of the semi-active control strategy on the railway vehicle in terms of the p.s.d. value of the carbody acceleration response. There are opened perspectives to study the improvement of passengers comfort using the hybrid magnetorheological damper described previously which offers a simple and reliable control of transmissibility of vehicle's oscillations from the rolling gear to the carbody.

5. Conclusions

Having in view that for the passenger trains the safety and comfort conditions are essential, the paper propose a new type of semi-active suspension controlled by a device with magnetorheological fluid.

The proposed system may transform any hydraulic damper in a damping system with semi-active control. This application shows the capability of the intelligent fluids to develop a control system more reliable and cost effective than the electro-hydraulic devices.

The design of the piloting device with magnetorheological fluid offers possibilities of further developments, being easy adaptable to the specific requirements of the application. For example, if the controlled damper should operate within a wider range of damping forces, the dimensions of the opening and the piston's stroke can be modified accordingly. The control strategy, thought simple, can be very flexible through the dimensionless gain factor which can be adapted according to the specific operational conditions of the vehicle.

The simulation of the employment of a semi-active secondary suspension was performed by means of the mathematical model of the vehicle lateral oscillations developed by the authors. The comparison between the carbody accelerations in passive and semi-active modes shows that the semi-active suspensions represent a reliable option for the improvement of the railway vehicle's comfort and safety.

References

[1] Wickens A H 2003 *Fundamentals of rail vehicles dynamics* (Lisse: Swets& Zeitlinger)
[2] Sebesan I and Hanganu D 1993 *The design of the railway vehicles suspensions* (Bucharest: Technical Publishing House)
[3] Ahmadian M and Yang S 1998 Effect of system nonlinearities on locomotive bogie hunting stability *Vehicle Syst. Dyn.* **29** 365-84
[4] Fan Y T and Wu W F 2006 Stability analysis of railway vehicles and its verification through field test data *ChIEJ* **29** 493-505
[5] Garg V K and Dukkippati R V 1984 *Dynamics of Railway Vehicles Systems* (Toronto: Academic Press)
[6] He Y and McPhee J 2002 Optimization of the lateral stability of rail vehicles *Vehicle Syst. Dyn.* **38** 361-90
[7] Lee S Y and Cheng Y C 2005 Hunting stability analysis of high-speed railway vehicle trucks on tangent tracks *J. Sound Vib.* **282** 881-98
[8] Lee S Y and Cheng Y C 2005 Nonlinear analysis of hunting stability for high-speed railway vehicle trucks on curved tracks *J. Vib. Acoust.* **127** 324-32
[9] Messouci M 2009 Lateral stability of rail vehicles – a comparative study *J. Eng. Appl. Sci.* **4** 13-27
[10] Sebesan I 1995 *Railway Vehicles Dynamics* (Bucharest: Technical Publishing House)
[11] Sebesan I and Copaci I 2008 *Theory of ellastic systems for railway vehicles* (Bucharest : Matrix Rom Editions)
[12] Sebesan I, Baiasu D and Ghita G 2011 Study of the railway vehicle suspension using the multibody method *INCAS Bull.* **3** 89-104
[13] Sebesan I and Baiasu D 2012 Matematical model for the study of the lateral oscillations of the railway vehicle *U.P.B. Sci. Bull.* **74** 51-66
[14] Kalker J J 1991 Wheel-rail rolling contact theory *Wear* **144** 243-61
[15] Shabana A A 2005 *Dynamics of Multibody Systems* (Cambridge: Cambridge University Press)
[16] Shabana A A 2010 *Computational dynamics* (Hoboken: John Wiley & Sons Ltd)
[17] Polach O 2006 On non-linear methods of bogie stability assessment using computer simulations *J. Rail Rapid Transit* **220** 13-27
[18] O'Neal H R and Wale G D 1994 Semi-active suspension improves rail vehicle ride *Comput. Control Eng. J.* **5** 183-8

[19] Shen Y, Yang S, Pan C and Xing H 2006 Semi-active control of hunting motion of locomotive based on magnetorheological damper *Int. J. Innov. Comput. I.* **2** 323-29
[20] Lau Y K and Liao W H 2005 Design and analysis of magnetorheological dampers for train suspension *J. Rail Rapid Transit* **219** 261-75
[21] Wang D H and Liao W H 2009 Semi-active suspension systems for railway vehicles using magnetorheological dampers. Part I: system integration and modeling *Vehicle Syst. Dyn.* **47** 1305-25
[22] Wang D H and Liao W H 2009 Semi-active suspension systems for railway vehicles using magnetorheological dampers. Part II: simulation and analysis *Vehicle Syst. Dyn.* **47** 1439-71
[23] Yang G, Spencer B F Jr, Carlson J D and Sain M K 2002 Large-scale MR fluid dampers : modeling and dynamic performance considerations *Eng. Struct.* **24** 309-23
[24] Sireteanu T, Ghita G and Stancioiu D 2005 *Magnetorheological fluids and dampers* (Bucharest: BREN Editions)
[25] Lord Corporation 1999 Magnetorheological Fluid MRF-132LD *Product Bulletin*
[26] Sireteanu T, Ghita G, Stancioiu D and Stammers C W 2004 Semi-active vibration control using balance-logic strategy: modeling and testing *RJAV* **I** 31-8
[27] Spencer B F Jr, Dyke S J, Sain M K and Carlson J D 1997 Phenomenological model of magnetorheological dampers *J. Eng. Mech. ASCE* **123** 230-8

Damping force control of frictionless MR damper associated with hysteresis modeling

M S Seong[1], **S B Choi**[1,3] **and C H Kim**[2]
[1] Smart Structures and Systems Laboratory, Department of Mechanical Engineering, Inha University, Incheon, 402-751, Korea
[2]Korea Institute of Industrial Technology (KITECH), Cheonan, Korea

E-mail: seungbok@inha.ac.kr

Abstract. This study presents hysteresis modelling and damping force control of the frictionless magnetorheological (MR) damper for semiconductor manufacturing stage. The vibration sources of the semiconductor stage can be classified as two. The one is environmental vibration from the floor, and the other is transient vibration occurred from the stage moving. The transient vibration has serious adverse effect to the process because the vibration scale is quite larger than other vibrations. Therefore, in this research, the semi-active MR damper which can control the transient vibration is devised. In addition, the stage needs to be isolated from tiny vibration to achieve high grade vibration level. At the high frequency range, MR damper acts like a rigid body if the dry friction exists. So the tiny vibration is transferred to the stage directly. Therefore, a dry friction of the MR damper must be removed. In order to achieve this goal, a frictionless MR damper is originally designed. After then, a designed MR damper is manufactured and it's damping force characteristics and hysteresis behaviors are evaluated by experiment. The biviscous hysteresis model of MR damper is formulated and its accuracies are evaluated. Finally, damping force control performances using the hysteresis model is experimentally evaluated.

1. Introduction

The main stream of ultra-precision machining industries like semiconductor is featured by huge scale, integration and multifunction. Therefore, the accuracy limitation of machining, manufacturing and measuring systems is required to be continuously increased [1-3]. The tiny vibration from environment and transient vibration from the moving stage make noticeable problems to the accuracy. Also the transient vibration makes time problem. After a stage moving, transient vibration is occurred and process must be stopped while the vibration is reduced to suitable range. In order to resolve these problems, three types of control systems have been proposed; passive, active and semi-active. The passive system featuring rubber mount or air spring is normally used for vibration absorption. This system provides design simplicity and cost-effectiveness. However, it cannot provides sufficient performance at these days because the accuracy level of the ultra-precision system becomes more important and moving mass on the stage is operated as a huge vibration source. Also it can't reduce the transient vibration fast. Because of the passive system has very low viscosity for vibration isolation at high frequency. On the other hand, the active control system provides high control performance at wide frequency range. However, the performance at resonance frequencies is limited and requires high

[3] Corresponding author.

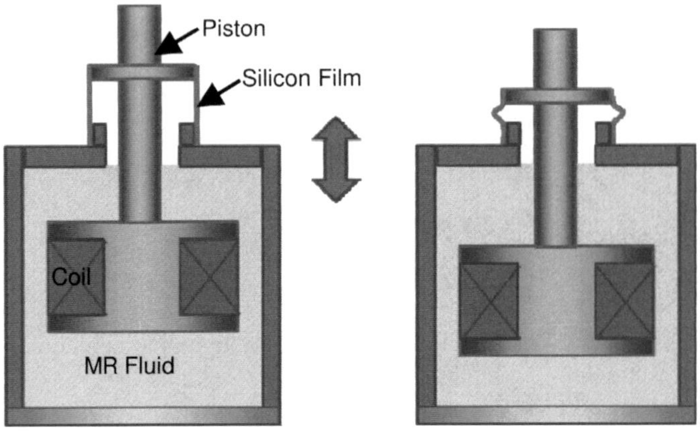

Figure 1. Configuration of the proposed MR Damper.

power source and large space. It is well known that the semi-active control system offers a desirable performance without requiring large power source and space.

Recently, many research works on semi-active actuator and its vibration control strategies have been proposed and successfully implemented. Hong et al. proposed electro-rheological fluid mounts for vibration control of a frame structure [4]. They developed an ER fluid mount and controlled the vibration using optimal controller. Wang et al. developed a new magnetorheological fluid-elastomer mount [5]. They designed a new type of MR-elastomer mount and evaluated its performance. Seong et al. proposed a Preisach hysteretic compensator for the damping force control of MR damper [6]. They evaluated the hysteretic behavior of MR damper and formulated the hysteretic model and compensator using Preisach model. Choi et al. proposed magnetorheological isolators for vibration isolation [7]. They evaluated a new nonlinear hysteresis model with simplicity in form for MR isolator.

In this research, we propose the semi-active MR damper to obtain superior vibration control performance of the semiconductor equipment system. As a first step to achieve this goal, a semi-active MR damper for the stage mount is proposed and designed. The MR damper is then manufactured and it's damping force characteristics are evaluated by experiment. Also the hysteretic behavior of the MR damper is evaluated and modeled using asymmetric biviscous model. Finally, the biviscous hysteretic compensator is formulated and its damping force control performance is evaluated by experiment.

2. MR damper

The vibration sources of a semiconductor manufacturing stage can be classified as two parts : one is environmental vibration from the floor, and the other is transient vibration occurred from the stage moving. The transient vibration occurred from the stage movement affects seriously to the process because the vibration scale is quite larger than other vibrations. Thus, the use of active actuator to attenuate this vibration is not appropriate because it requires a large power source and space. Therefore, in this research, a semi-active MR damper which can control the transient vibration without requiring large power source and space is adopted. On the other hand, micro- or nano-scale tiny floor vibrations need to be isolated from the stage. If the mount has a dry-friction, tiny vibrations are transferred to the stage through friction part. So the dry-friction of the MR damper needs to be eliminated. Therefore, a new type of MR damper which can eliminate the seal friction is proposed in this study.

Figure 1 shows the configuration of the new type MR damper for the semiconductor manufacturing stage. A seal is removed from a piston head. And a gap between a piston head and a cylinder can be used as an orifice. And the silicon film is adopted for sealing between the piston and cylinder. The silicon film is fixed on the piston and the cylinder each side and crumpled according to the piston movement as shown in Figure 1. So the proposed MR damper doesn't make any dry friction. Figure 2

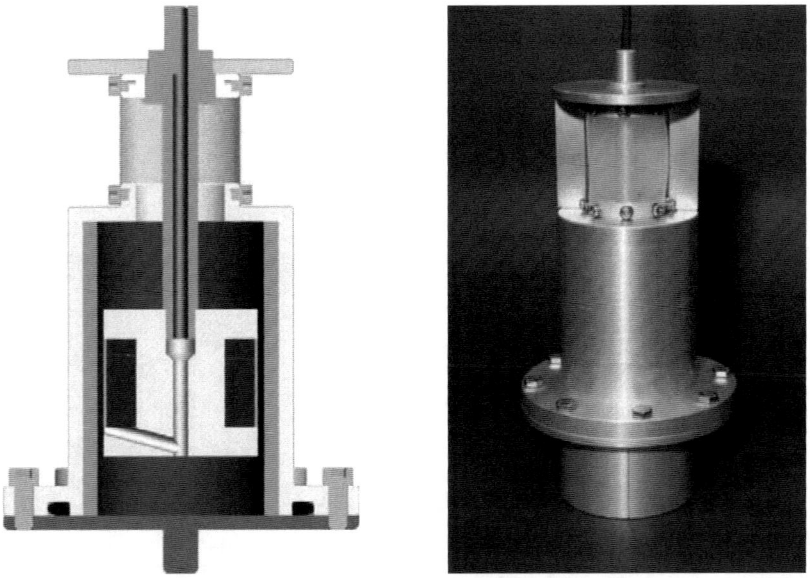

(a) design (b) photograph

Figure 2. Proposed MR damper.

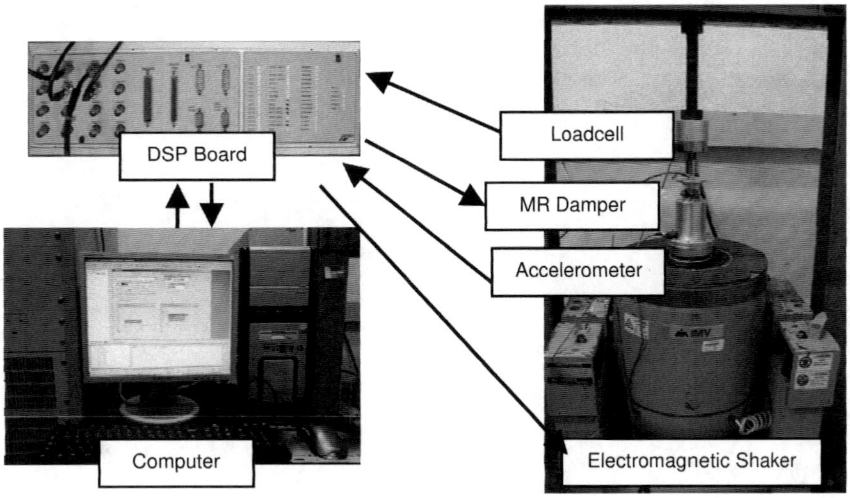

Figure 3. Experimental configuration.

shows the design and manufactured photograph of the proposed MR damper. The gap (orifice) between the piston head and cylinder is 3 mm and the silicon film is adopted. Magnetic field of the piston head is analyzed and maximized by using Ansys^{TM}. Figure 3 shows the experimental configuration for damping force measurement of the manufactured MR damper. The MR damper is excited by amplitude ±4.8 mm and frequency 3.0Hz using a electromagnetic shaker. This excitation condition is similar to the transient vibration from the stage movement. The damping force of the MR damper is measured by a loadcell and MR damper movement is measured by accelerometer. Input signal is generated from a DSP system, and this signal is fed back to the current amplifier and applied to the MR damper.

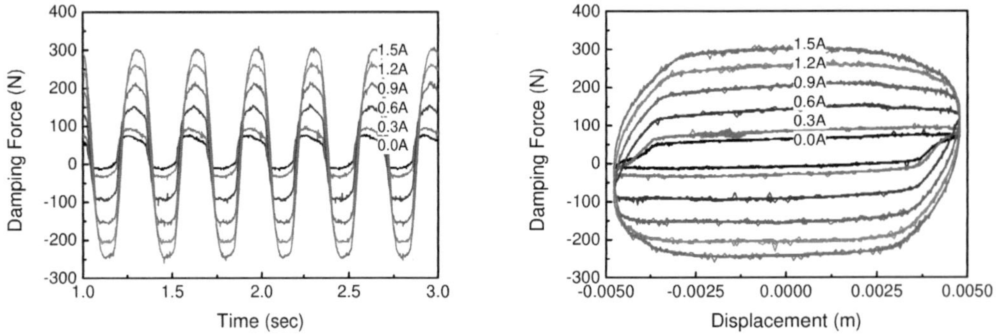

(a) damping force vs. time (b) damping force vs. disp.

Figure 4. Damping force characteristics.

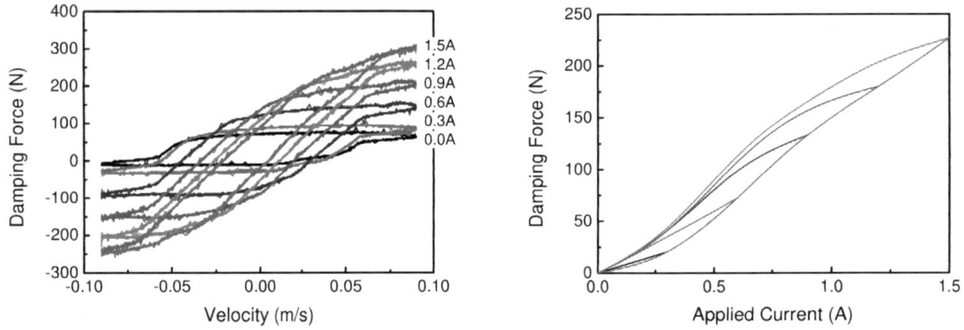

(a) velocity dependent (b) field dependent

Figure 5. Hysteretic behavior.

The measured damping force characteristics of MR damper are shown in Figure 4. The passive damping force range is 107.42 N and the maximum damping force range at 1.5A is 569.05 N from measured data. This damping force range obtained in this test is suitable for vibration control of the semiconductor equipment.

3. Hysteretic model

As shown in Figure 5, MR damper has two types of hysteresis. The one is velocity dependent hysteresis and the other is field dependent hysteresis. If the hysteretic behavior is compensated by model, the control accuracy can be increased. Thus, this behavior should be accommodated in order to achieve more accurate damping force control.

In this research, biviscous model is adopted and modified for represent the hysteretic behavior of the proposed MR damper. Nonlinear hysteretic biviscous model for MR damper proposed by Wereley et al. utilized a set of piecewise linear functions [8]. This nonlinear hysteretic biviscous model considers that the damping force is symmetric at jounce and rebound mode. However, as shown in Figure 5 (a), damping force is not symmetric at jounce mode and rebound mode. Therefore, asymmetric biviscous model is proposed as shown in Figure 6. Asymmetric biviscous model has 10 parameters and every parameter have different values. Also Each parameters is characterized through second- and fourth-order polynomials derived on the basis of parameters identified for each current condition [8]. Therefore, hysteretic behavior of the MR damper can be modelled as follows:

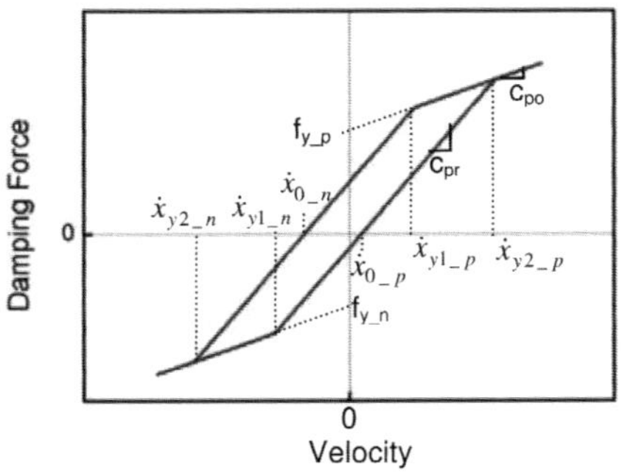

Figure 6. Asymmetric biviscous model.

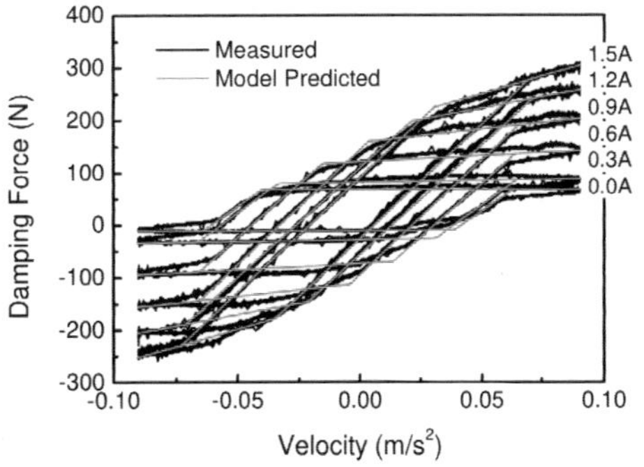

Figure 7. Model prediction.

$$F_d = \begin{cases} c_{po}\dot{x} + f_{y\_n} & \dot{x} \le \dot{x}_{y1\_n} & \ddot{x} > 0 \\ c_{pr}(\dot{x} - \dot{x}_{0\_p}) & \dot{x}_{y1\_n} \le \dot{x} \le \dot{x}_{y2\_p} & \ddot{x} > 0 \\ c_{po}\dot{x} + f_{y\_p} & \dot{x} \ge \dot{x}_{y2\_p} & \ddot{x} > 0 \\ c_{po}\dot{x} + f_{y\_p} & \dot{x} \ge \dot{x}_{y1\_p} & \ddot{x} < 0 \\ c_{pr}(\dot{x} - \dot{x}_{0\_n}) & \dot{x}_{y2\_n} \le \dot{x} \le \dot{x}_{y1\_p} & \ddot{x} < 0 \\ c_{po}\dot{x} + f_{y\_n} & \dot{x} \le \dot{x}_{y2\_n} & \ddot{x} < 0 \end{cases} \tag{1}$$

where $\dot{x}$ is velocity of piston, $c_{po}, c_{pr}, f_{y\_p}, f_{y\_n}, \dot{x}_{0\_p}, \dot{x}_{0\_n}, \dot{x}_{y1\_p}, \dot{x}_{y1\_n}, \dot{x}_{y2\_p}$ and $\dot{x}_{y2\_n}$ are parameters of asymmetric biviscous model. Figure 7 shows the model predicted results. As shown in Figure 8, asymmetric biviscous model is well follow the hysteretic behavior.

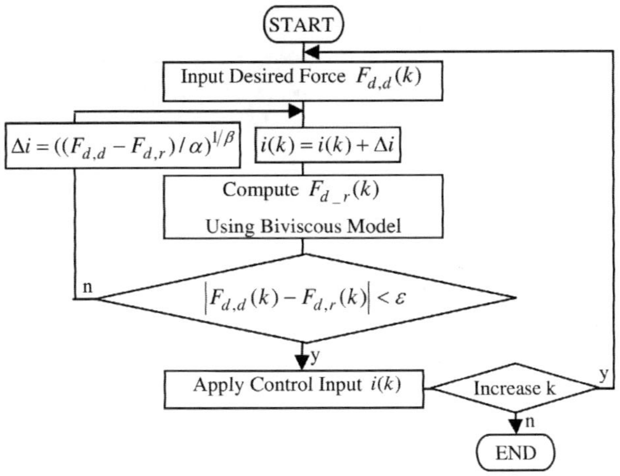

Figure 8. Flow chart of asymmetric biviscous compensator.

4. Damping force control

Because of the asymmetric biviscous model is nonlinear, inverse model cannot be exist. Therefore the compensation algorithm is needed for damping force control. The flow chart for implementation of asymmetric biviscous compensator is shown in Figure 8. The proposed compensation algorithm consists of estimation and linearization of nonlinear hysteretic behavior using asymmetric biviscous model.

Figure 9 presents the damping force control results under sinusoidal decreasing desired damping force. It is seen from the control results that the asymmetric biviscous compensation has overshoot problem. This is occurred because the biviscous model has discontinuous points. This result shows that the asymmetric biviscous model is good for model prediction, but not for damping force control.

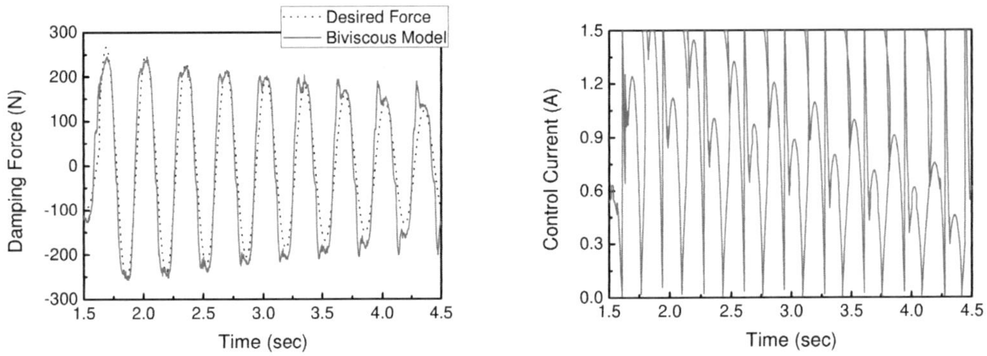

(a) damping force control (b) control current

Figure 9. Damping force control.

5. Conclusion

In this work, a new type of MR damper for the semiconductor manufacturing system was proposed. The dry-frictionless MR damper was originally designed and manufactured. Then, the field-dependent damping force characteristics and hysteretic behaviors were identified. Asymmetric biviscous model was proposed for hysteresis modelling for MR damper and its performance was evaluated. After then asymmetric biviscous hysteretic compensator was formulated and its damping force control performance was evaluated. It is finally remarked that asymmetric biviscous model is good for model prediction of damping force with hysteresis. However, because of its discontinuity, damping force control performance is quite bad compared with inverse Bingham model. From now on, 4 MR dampers with air spring will be manufactured and attached to the real stage. And the vibration control algorithms will be formulated and its performances will be evaluated and compared.

Acknowledgement

This work was financially supported by the grant from the Technology Innovation Program (10040304) of the Ministry of Knowledge Economy (MKE) of Korea.

References

[1] Rahman M, Asad A B M A, Masaki T, Saleh T, Wong Y S and Kumar A S 2010 A multiprocess machine tool for compound micromachining *Int J Mach Tool Manu.* **50** 344
[2] Bringmann B and Maglie P 2009 A method for direct evaluation of the dynamic 3D path accuracy of NC machine tools *Cirp Ann-Manuf Techn.* **58** 343-6
[3] Smith S T and Seugling R M 2006 Sensor and actuator considerations for precision, small machines. *Precis Eng.* **30** 245-64
[4] Hong S R, Choi S B and Han M S. 2002 Vibration control of a frame structure using electro-rheological fluid mounts *Int J Mech Sci.* **44** 2027-45
[5] Wang X J and Gordaninejad F. 2009 A new magnetorheological fluid-elastomer mount: phenomenological modeling and experimental study *Smart Mater Struct.* **18** 095045

[6] Seong M S, Choi S B and Han Y M 2009 Damping force control of a vehicle MR damper using a Preisach hysteretic compensator *Smart Mater Struct.* **18 074008**

[7] Choi Y T, Wereley N M and Jeon Y S. 2005 Semi-active vibration isolation using magnetorheological isolators. *J Aircraft.* **42** 1244-51.

[8] Wereley N M, Pang L and Kamath G M. 1998 Idealized hysteresis modeling of electrorheological and magnetorheological dampers *J Intel Mat Syst Str.* **9** 642-9

Design of a squeeze film magnetorheological brake considering compression enhanced shear yield stress of magnetorheological fluid

C Sarkar[1] and H Hirani[2]
[1] Research Scholar, Mechanical Engineering Department, IIT Delhi, India
[2] Associate Professor, Mechanical Engineering Department, IIT Delhi, India

E-mail:chiranjit.ju@gmail.com

Abstract. A magnetorheological brake, consisting of rotating disks immersed in a MR fluid and enclosed in an electromagnet, is proposed to replace the conventional heavy weight low response hydraulic disk brake. The frictional characteristics of the proposed brake can be controlled by regulating the yield stress of the MR fluid as function of magnetic field and normal compressive force. The controllable yield stress retards the surfaces of rotating disks, thus MR fluid can be used as a brake lining material. The present research work attempts designing a squeeze film MR brake by accounting compression enhanced shear yield stress of magnetorheological fluid. Theoretical calculations indicate that the estimated braking torque of the six plate squeeze film MR brake, under compression, is in the order of 600Nm. To validate the theoretical design and its findings, a prototype of single-plate squeeze film MR disk brake has been developed. Experimental test setup helps to illustrate braking torque under different control currents (0.0 to 1.25 A).

1. Introduction

Conventional hydraulic disc brake, used in high torque application, wears out with time and deteriorates the braking performance. The magnetorheological fluid (MR) brake, having time invariant friction performance, may be considered as a good replacement for hydraulic disk brakes. MR fluid consists of micron sized magnetically permeable particles dispersed throughout the non-magnetic fluid carrier. Iron powder, having high saturation magnetization, is the most popular material to be used as magnetic particles. Under the presence of the magnetic field, magnetic dipole moment within particles induces, causing dipole interactions to form chains in the direction of flux paths. The formed particle-chains restrict fluid movement and increase yield strength of MR fluid. Rotational movement of disk (shear mode) and axial movement of pad (compression mode) affect the particle chains, and therefore the braking torque of MR brake.

Design of MR brake has been studied by many researchers. Huang *et al.* [1] presented theoretical design of a cylindrical MRF brake assuming annular space between circumference of disk and

[1] Corresponding author.

Published under licence by IOP Publishing Ltd

electromagnet filled with MR fluid, and braking torque due to shearing of that MR fluid. Li and Du [2] experimentally investigated torque characteristics of designed disc MR brake operating at low speeds. Bydon [3] described the construction and operation of Lord's MR brake, which offers maximum 5.65 Nm torque at operating speed 1000 rpm. The brake was restricted to be operated within temperature ranging between -30° to 70°C. Park *et al.* [4] presented a theoretical design of MRF brake required for automotive applications. Sukhwani and Hirani [5] designed a MRF brake for high rotational speed up to 2000 rpm. They considered the effect of MR gap on the performance of MR brake. Nguyen and Choi [6] considered minimization of zero-field friction heat in theoretical design of automotive magnetorheological brake to optimize the brake performance. In all aforementioned MRF brake configurations shear phenomenon of MR fluids was accounted. Although shear mode alone has been studied thoroughly by many researchers, shear mode under compression should be investigated to enhance the performance of MRF brake. The advantage of this mode is the attainment of high stresses as compared to shear mode under the same external field strength.

The effect of compression on yield strength of MR fluid was considered by Tang *et al.*[7]. They did experiments on an iron-based MR fluid and found an increase in structure-enhanced static yield stress (~800 kPa) compared to yield stress without compression. Hongyun *et al.* [8] experimentally investigated the compressive and tensile performance of a MR fluid under different magnetic field and compared the test results with the shear yield stress of MR fluids. Mazlan *et al* [9] experimentally investigated the influences of the applied current on two types of MR fluids (MRF 241 ES and MRF 132 DG) under compression mode. Tian *et al.* [10] achieved ultra high yield stress in a general electrorheological fluid under compression. As per their study a lower compressive velocity results in a higher yield stress under the same electric field. The viscous force acting on particles decreases as the compressive velocity increases, while the force acting from the neighboring particles to squeeze particles out increases.

After comprehensive literature review it can be said that mathematical relation expressing compressive resistance of the MR fluid in terms of various variables is missing. There is a strong need to find such expression. Therefore, it has been decided to design compressive magnetorheological fluid brake considering single disc configuration and characterize brake at various speed and control currents to capture its behavior.

2. Theoretical study on squeeze film MR brake

The mechanical response of MR fluids, used in MR brakes, is often modeled using Bingham fluid equation [5]

$$\tau = \tau_{yd} + \eta \frac{\omega r}{h} \tag{1}$$

Where, τ is shear stress of MR fluid, τ_{yd} is yield shear stress of the MR fluid, η is viscosity of the MR fluid, ω is angular velocity of the disk brake, r is the radius of the brake and h is the MR fluid gap. The MRF brake configuration illustrated in Figure 1 indicates that h (MR fluid gap) acts as torque arm. For such configuration, the braking torque produced by MR fluid [2] can be given by

$$T = 2 \times F \times h \tag{2}$$

Where, F is friction force ($F = \text{shear stress} \times \text{area}$). The expression for friction force becomes,

$$F = \int_{r_1}^{r_2} \tau \times 2\pi r dr \tag{3}$$

Substituting equation (3) in equation (2),

$$T = 2\int_{r1}^{r_2} 2\pi rh\left(\tau_{yd} + \eta\frac{\omega r}{h}\right)dr$$
$$\approx 2\pi h\tau_{yd}({r_2}^2 - {r_1}^2) + \frac{4}{3}\eta\pi\omega({r_2}^3 - {r_1}^3) \quad (4)$$

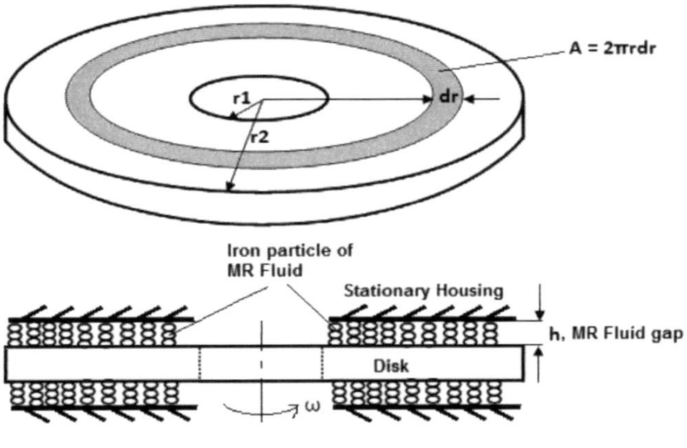

Figure 1. MR fluid gap acts a torque arm in MR brake.

For multidisc, having n disks (friction surface = 2n) MR brake equation (4) becomes

$$T = 2\pi nh\tau_{yd}({r_2}^2 - {r_1}^2) + \frac{4}{3}\eta\pi n\omega({r_2}^3 - {r_1}^3) \quad (5)$$

Where, yield stress (τ_{yd}) is a function of magnetic field intensity (H) and can be expressed [5] as

$$\tau_{yd} = \alpha_1 H^{1.5} \quad (6)$$

For a typical value of yield stress = 69 kPa at magnetic field intensity H = 200 kA/m [11], the value of $\alpha_1$is

$$\alpha_1 = 7.714 \times 10^{-4}\ \text{Pa}\ (\text{m/A})^{1.5}$$

To express yield stress in terms of control parameter (current), one needs to express H in terms of input current I, number of turns in electromagnetic coil N, and MR gap h. With such substitution, the yield stress can be expressed as

$$\tau_{yd} = 7.714 \times 10^{-4}\left(\frac{NI}{2h}\right)^{1.5} \quad (7)$$

Substituting equation (7) in equation (5), expression of torque becomes

$$T = 5.454 \times 10^{-4} n\pi h\left(\frac{NI}{h}\right)^{1.5}({r_2}^2 - {r_1}^2) + \frac{4}{3}\pi\eta n\omega({r_2}^3 - {r_1}^3) \quad (8)$$

Figure 2 shows a configuration of multidisc MR brake considering six plates rotary disk. The model consists of low carbon steel six plates rotary disk brake, separator plates are attached to housing which is made of low carbon steel. Electromagnet has 1000 number of turns of 25 AWG wire. Two radial shaft seals (Nitrile Rubber) are being used in such a manner that they make contact with the separator plates of the housing. Two single row deep groove ball bearing (low carbon steel) support the MR brake structure, w_d is disk width, w_h is width of separator plate of housing, h is MR fluid gap, r_1 is inner radius of the disk and r_2 is outer radius of the disk.

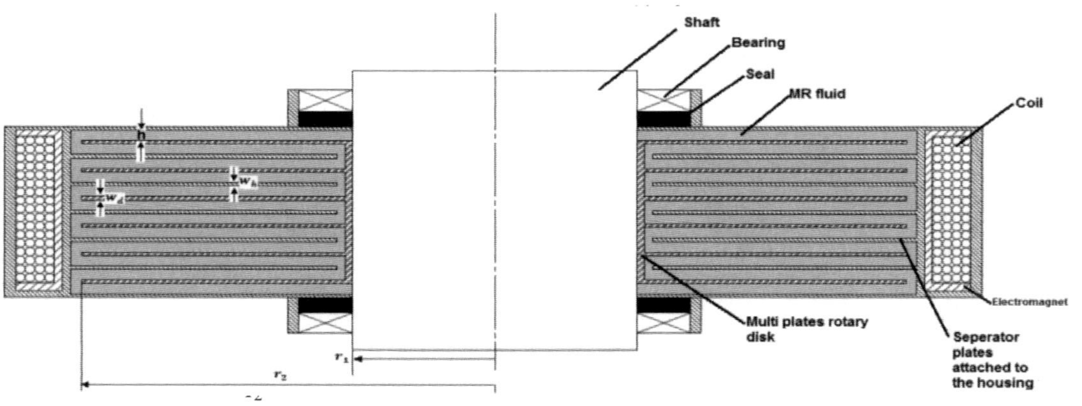

Figure 2. Proposed configuration of six plates rotary disk MR brake.

The magnetic field, induced by electromagnet, generates dipole moment in iron particles of MR fluid. Attractive forces between magnetized particles align them along the field direction as shown in Figure 3.

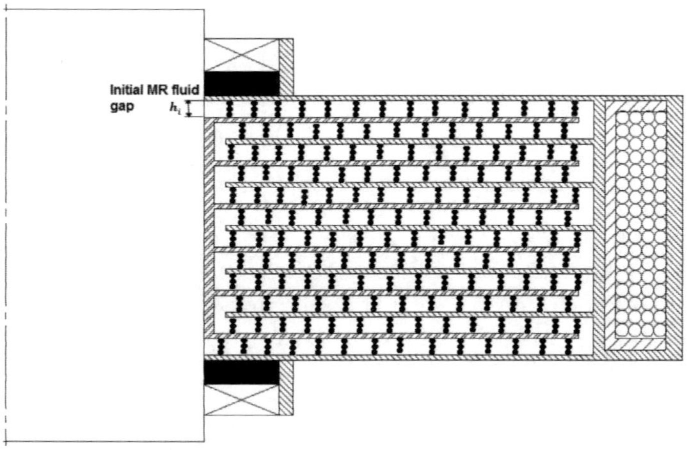

Figure 3. MR fluid particles align along the field direction.

If a rapid compression-assisted-aggregation process is applied to force MR fluids it will form a microstructure that is much stronger than the single-chain structure [12]. MR fluids greatly exceed the theoretical strength predicted from the single-chain structure. Figure 4 shows two zones of six plates MR brake configuration. In compression zone iron particles get compressed and form thick columns with strong and robust end wherever in tensile zone iron particles break their chain structure thus reducing the strength of the chain. The empirical expression of compression-assisted-aggregation model of MR fluid [12] is given by

$$\tau_y(H) = \tau_{yd} + K_H . P_e \tag{9}$$

Where τ_{yd} is the yield stress of the MR fluid without compression-assisted-aggregation. The slope K_H increases with the field H. Yield shear stress $\tau_y(H)$ increases with the normal stress P_e.

The braking torque produced by the MR fluid is given by,

$$T = \left(\int_{r_1}^{r_2}\int_{h_i}^{h_f} 2\pi n r\, \tau_t \, dh\, dr\right) + \left(\int_{r_1}^{r_2}\int_{h_i}^{h_s} 2\pi\, n r \tau_s dh dr\right) \tag{10}$$

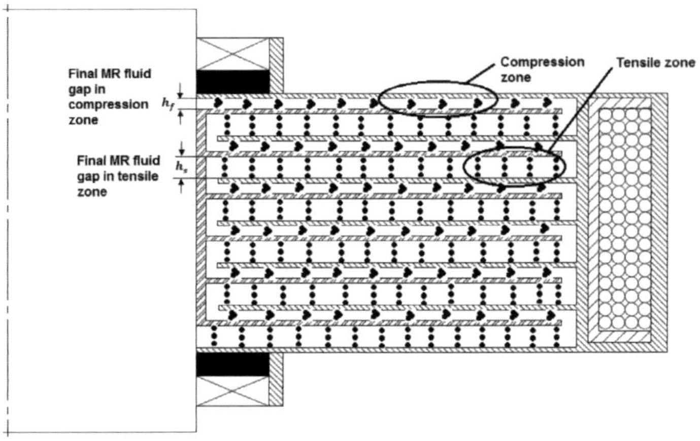

Figure 4. Compression and tensile zone of six plate rotary disk.

Where, $\tau_t = \tau_y(H) + \eta \cdot \frac{\omega^* r}{h}$, τ_t is the stress acting in the compression zone of the disk, h_i is the initial gap of MR fluid as shown in Figure 3 and h_f is the final gap of MR fluid in the compression zone as shown in Figure 4. τ_s is the stress acting in the tensile zone of the disk. Here $\tau_s = \eta \cdot \frac{\omega^* r}{h_s}$ because in the tensile zone the shear stress is only due to viscous force (η, viscosity) of the MR fluid. This is gross approximation, but it will provide a conservation approach (actual torque produced will be more than estimated torque). h_s is the MR fluid gap in the tensile zone of the disk, $h_s = h_i + h_i - h_f = 2h_i - h_f$.

In equation (9), normal stress P_e can be expressed [13] as

$$P_e = \frac{D}{3h}\tau_{yd} \tag{11}$$

Where, D (=2*r) is the diameter of the disk. Therefore equation (9) becomes

$$\tau_y(H) = \tau_{yd}\left(1 + K_H \cdot \frac{2r}{3h}\right) \tag{12}$$

Substituting equation (12) and equation (7) in equation (10) the torque expression becomes

$$T = 2\pi n \left[\begin{array}{c} \left\{2.727 \times 10^{-4} N^{1.5} I^{1.5}\left(\frac{1}{\sqrt{h_f}} - \frac{1}{\sqrt{h_i}}\right)(r_2{}^2 - r_1{}^2)\right\} \\ + \left\{0.101 \times 10^{-4} N^{1.5} I^{1.5}\left(\frac{1}{h_f{}^{1.5}} - \frac{1}{h_i{}^{1.5}}\right) K(H)(r_2{}^3 - r_1{}^3)\right\} + \\ \left\{\frac{\eta\omega(\log h_i - \log h_f)}{3}(r_2{}^3 - r_1{}^3)\right\} \end{array}\right]$$

$$+ \frac{2\pi n \eta \omega}{3}(\log h_s - \log h_i)(r_2{}^3 - r_1{}^3) \tag{13}$$

When normal stress P_e is equal to zero, yield shear stress $\tau_y(H)$ is the yield stress of the MR fluid without compression assisted aggregation (τ_{yd}).

To study the braking torque of MR brake, let $r_1 = 55$ mm, $r_2 = 196$ mm, $h = 1$mm, $w_d = 3$mm, $w_h = 3$mm, $n = 6$, be considered. Viscosity of MR fluid (η) depends on the applied magnetic field (B) in the Bingham model. The approximate polynomial curve of viscosity of MRF-122-2ED [14] is given by

$$\eta = -3.904B^4 + 12.44B^3 - 15.85B^2 + 10.21B \qquad (14)$$

Where,η is in Pas and B is in Tesla. In this theoretical analysis MRF 241 ES [11] is used as MR fluid medium. The magnetic field intensity ($H = \frac{NI}{2h}$) of the MR fluid is in kA/m. The approximate B-H polynomial curve [11] is given by

$$B = -10^{-11}H^2 + 6 \times 10^{-6}H + 0.156 \qquad (15)$$

In equation (12) slope K_H varies with the magnetic field. The polynomial curve of K_H [12] with respect to magnetic field intensity (H) is given by

$$K_H = 0.266 - 4 \times 10^{-7}H + 9 \times 10^{-13}H^2 \qquad (16)$$

In this theoretical analysis speed is kept as 500 R.P.M. To find out the braking torque of compressive MR brake $h_i = 1$mm and final MR gap $h_f = 0.5$mm is assumed. Substituting above values in equation (13), variation of torque T for different magnetic field has been determined and plotted in Figure 5. Figure 5 shows that when the viscosity is function of magnetic field, torque increases with the current because magnetic field (H) varies with current.

To validate the results of braking torque experimentally, small scale (1:5) single plate MR brake configuration has been proposed. Figure 6 shows the schematic drawing of a single plate compressive MR brake and Figure 7 shows the detailed drawing of the compressive MR brakes. This MR brake has inner radius of the disk $r_1 = 10$ mm, and outer radius of the disk $r_2 = 44$ mm, initial MR gap $h_i = 1$mm and final MR fluid gap $h_f = 0.5$mm have been considered. Figure 8 and Figure 9 show the h_i and h_f of the compresive MR brake structure.

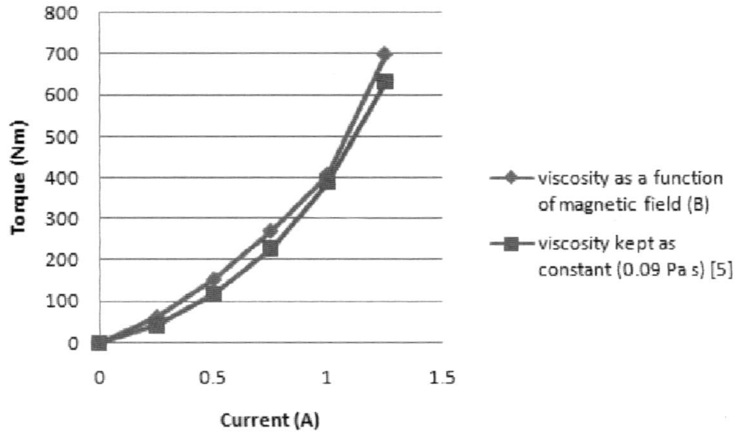

Figure 5. Variation of theoretical torque (T) for different control currents.

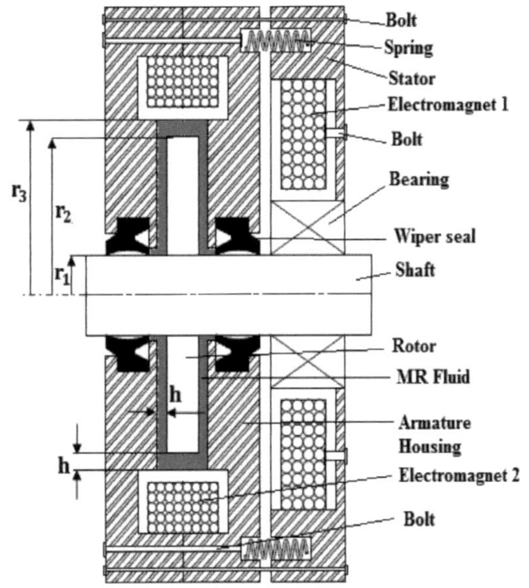

Figure 6. Schematic drawing of MR brake.

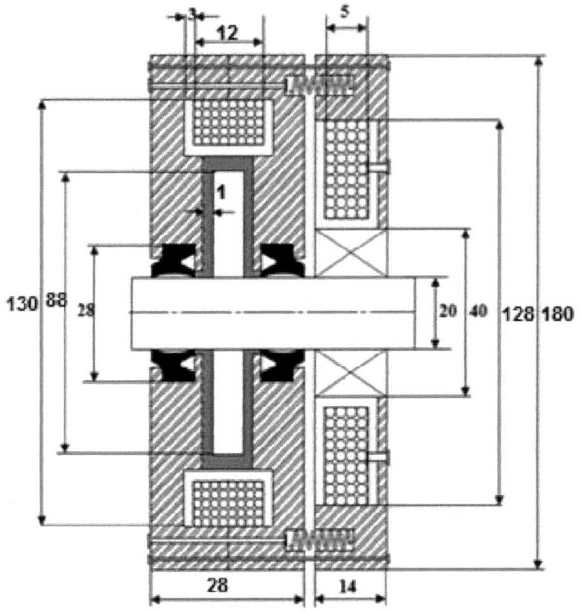

Figure 7. Detailed drawing of MR brake.

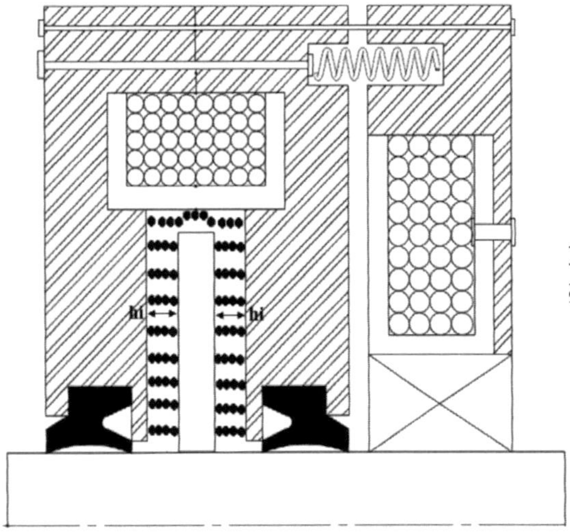

Figure 9. MR fluid gap h_i of the MR brake before after compression.

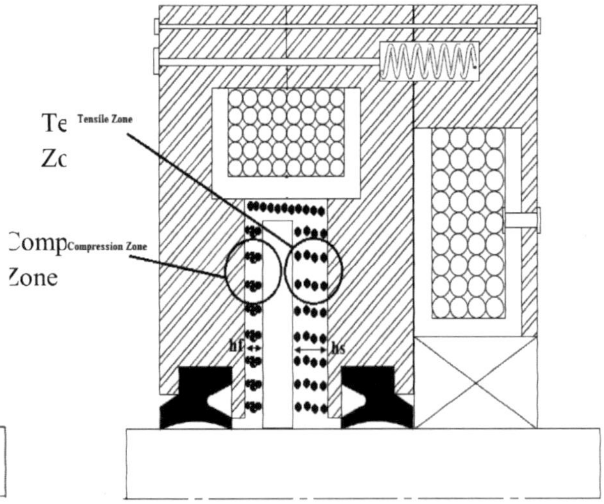

Figure 10. MR fluid gap h_f and h_s in MR brake compression.

3. Experimental study on squeeze film MR brake

In the proposed single disk squeeze film MR Fluid brake accommodating MR fluid in the annular space along the periphery as well as both sides of the rotor have been developed. The required compressive force is applied on the disc through the fluid when the electrical current is supplied to the brake coil. Removing power from the brake coil decompress the armature housing to its initial position. MR brake configuration has been made considering 1000 number of turns of wire on side electromagnet as shown in Figure 6. Figure 6 shows that small gap h between rotor and housing is filled with MR fluid. For ease in manufacturing, shrink fit between brake rotor (radius, r_2) and shaft (radius , r_1) are used as shown in Figure 7. Wiper seals slide along the shaft. It has transition fit to the shaft with 0.8 μm tolerance. Radial ball bearing is fitted to the shaft with -13 μm tolerance interference fit. The bearing has interference fit of 6μm tolerance. Electromagnet1 is attached to the split housing (radius r_3) and electromagnet2 is attached to the stator. Housing is mounted on two wiper seals which will slide along the direction of attractive force by the stator when magnetic field is applied compressing six element springs, which are constrained between housing and stator. On the removal of magnetic field, spring will decompress and housing will go back to its initial position ensuring air gap is equal to 0.5 mm because preload of the springs is zero. Figure 11 shows the fabricated squeeze film MR fluid brake model.

Figure 12 shows the experimental test set up of squeeze film MR brake. It consists of 2 H.P and 1500 R.P.M. D.C. motor with speed controller. A flywheel (20 kg and 0.42 kg m^2inertia) is connected between the D.C. Motor and MR brake through bearing bracket, jaw coupling and flexible coupling. A D.C. power supply is used to apply the D.C. to the electromagnet of squeeze film MR brake. A toggle switch is used to control the motor ON/OFF and MR Brake ON/OFF.

Figure 11. Fabricated model of squeeze film MR brake.

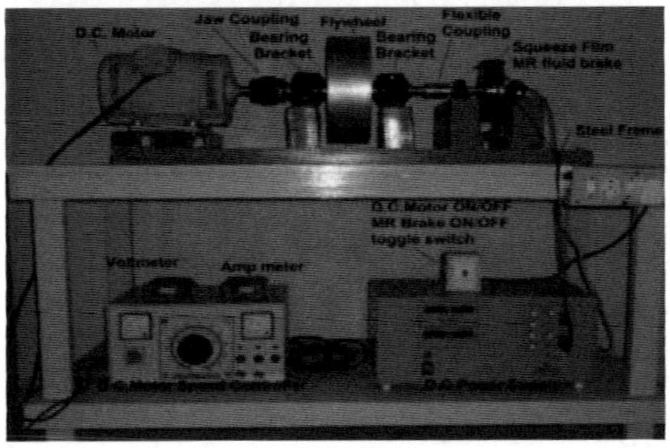

Figure 12. Experimental setup of squeeze film MR brake.

The experimental torque is conducted to explore the torque output of the prototype squeeze film MR brake. In this study, commercially available LORD MRF-241ES [11] is used. The output torque is determined by dividing the power loss of the DC Motor with the angular speed of the shaft measured at different applied coil current.

The transmitted torque is measured for angular speeds of 200 and 400 R.P.M for input currents up to 1.25 A as shown in Figure 13. It implies that the operating speed of squeeze film MR brake does not affect the torque performance, since the viscous torque is relatively small compared to the torque generated by the MR fluid effects.

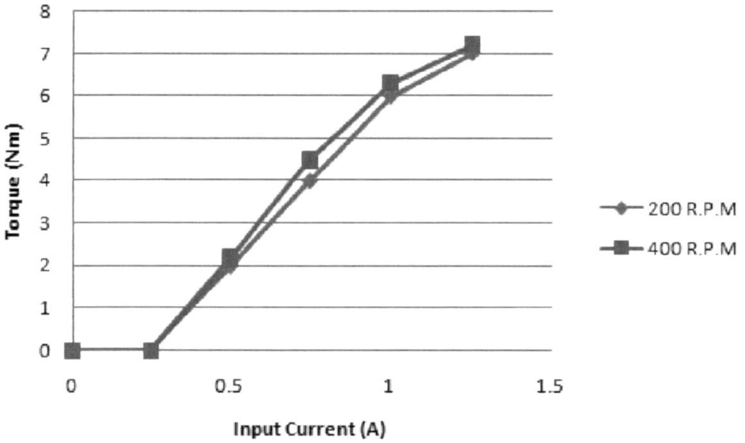

Figure 13. Experimental torque of squeeze film MR brake prototype for different velocities

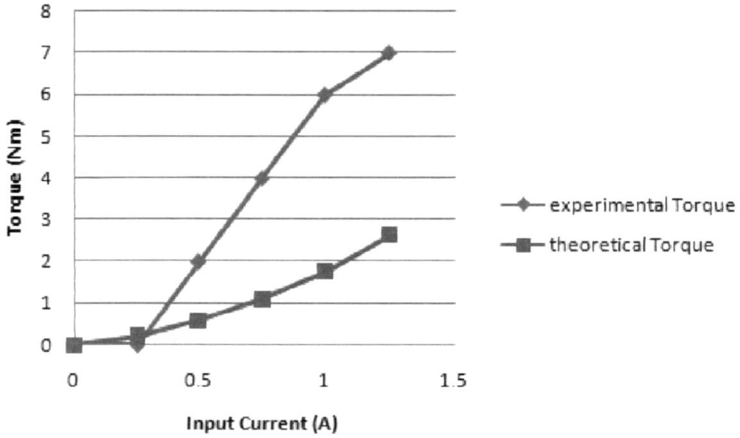

Figure14. Theoretical torque and experimental torque of squeeze film MR brake.

The experimental and theoretical torque output results have been compared in Figure 14. Percentage difference between experimental torque and theoretical torque is shown in Table 1. At 0.25A input current there is no squeezing action due to the high friction force of wiper seal. Friction force of wiper seal plays a dominating role in the squeezing action which is not included in the theoretical calculation.

Table 1. Percentage difference between experimental torque (ET) and theoretical torque (TT).

Current(A)	ET (Nm)	TT (Nm)	% Difference between ET and TT
0.00	0	0	0.00
0.25	0	0.21	-100.00
0.50	2	0.58	239.55
0.75	4	1.09	264.43
1.00	6	1.76	240.63
1.25	7	2.62	167.14

At 0.5A and 0.75A input current percentage differences are 239.55 % and 264.43 % respectively. As current increases, attractive force by the stator increases. Therefore compressive velocity on the disk through the MR fluid increases. A higher compressive velocity results in a shortest test time, corresponding to a more solid like mechanical property of a material [15]. Therefore when the compressive velocity was increased, the compressive strength of a visco-elastic material like MR fluids would increase, which increased the braking torque. The compressive resistance may be described as $P = \eta\dot{\varepsilon} + k\varepsilon$ [16], where η is viscous parameter [11] and k is elastic parameter [9]. These parameters need to be determined experimentally. The ε and $\dot{\varepsilon}$ are compressive strain and compressive strain rate respectively. Table 2 shows the theoretical torque of squeeze film MR brake including strain rate to improve the calculated values. These results have been plotted in Figure 15. On comparing the results of Figures 14 and 15, it can be said that including strain rate improves the results.

Table 2. Theoretical torque (TT) of squeeze film MR brake including strain rate

Current(A)	ET (Nm)	TT (Nm)
0	0	0
0.25	0	3.83
0.50	2	4.03
0.75	4	4.76
1.00	6	6.01
1.25	7	7.77

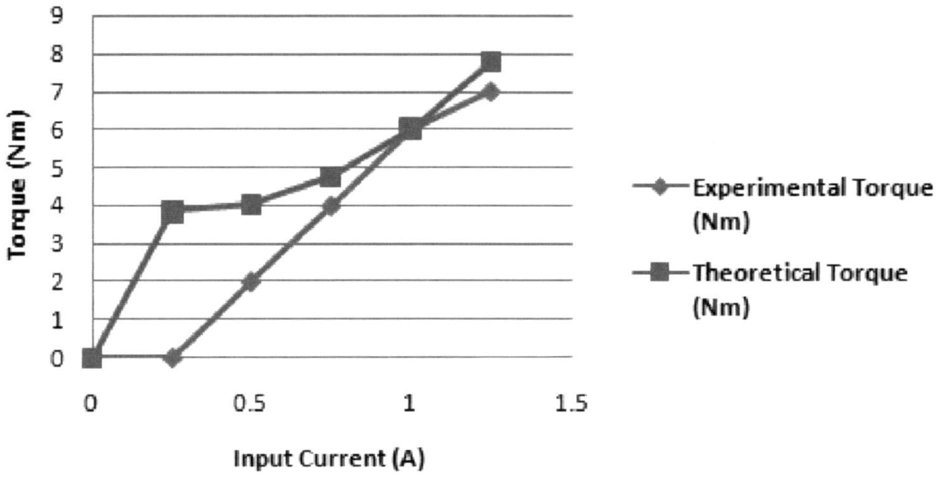

Figure 15. Comparison between experimental torque and theoretical torque including strain rate.

Table 3 shows a comparison of the prototype (described in the present research) and high speed MR brake designed by Sukhwani and Hirani [5]. In both the designs MR gap is kept as 1 mm. The results listed in Table 3, clearly shows far better performance of squeezing action braking compare to high speed MR brake detailed in Sukhwani and Hirani [5]. In other words squeeze film action shall be used for high torque application.

Table 3. Prototype comparison to high speed MR brake.

	High speed MR brake	Squeeze film MR brake
MR fluid	MRF 336AG	MRF 241ES
MR gap (mm)	1	1
Torque (Nm) at I = 0.75A	1	4
Maximum torque (Nm)	4	7

4. Conclusion

A squeeze film MR brake for high torque application is designed and manufactured. Performance of the prototype MR brake is studied using MRF 241ES both theoretically and experimentally. Experimental transient torque analyses show that squeeze film MR brake generates 7 Nm torque for input current of 1.25A which is applicable for high torque application. Dynamic torque is not sensitive to angular velocities up to 400 R.P.M. since the viscous torque is negligible compared to magnetic field torque.

References

[1] Huang J, Zhang J Q, Yang Y and Wei Y Q 2002 *J. Mater. Process. Technol.* **129** 559-62
[2] Li W H and Du H 2003 *Int. J. Adv. Manuf. Technol.* **21** 508-15
[3] Bydon S 2002 *Archiwum Process Control Club* **20**
[4] Park E J, Stoikove D, Falco da Luz L and Suleman A 2006 *Mechatronics* **16** 405-16
[5] Sukhwani V K and Hirani H 2008 *Proc. IMechE Part L: J. Materials: Design and Applications* **222** 73-82
[6] Nguyen Q H and Choi S B 2010 *Smart Material Structure* **19** 115024
[7] Tang X, Zhang X and Tao R 2000 *J. Appl. Phys.* **87** 2634-38
[8] Hongyun W, Chang B, Junwu K, Chunfu G and Wang X 2011 *J. Intel. Mat. Syst. Str.* **22** 811-16
[9] Mazlan S A 2008 The behavior of magnetorheological fluids in squeeze mode *PhD Thesis* 99 -100
[10] Tian Y, Zhang M, Zhu X, Meng Y and Wen S 2010 *Smart Mater. Struct.* **19** 035009
[11] Lord Corporation, http://www.lord.com/, visited on 20.03.2012
[12] Tao R 2001 *J. Phys. Condens. Mat.* **13** R979-R999
[13] Gartling D K and Phan-Thien N J 1984 *J. Non-Newton. Fluids* **14** 347-60
[14] Zubieta M, Eceolaza S, Elejabarrieta M J and Bou-Ali M M 2009 *Smart Mater Struct.* **18** 095019
[15] Larson RG 1999 *The Structure and Rheology of Complex Fluids* (New York: Oxford University Press) pp 3-18
[16] Tian Y, Zhang M, Zhu X, Meng Y and Wen S 2010 *Smart Mater. Struct.* **19** 035009
[17] Nam T H and Ahn K K 2009 *Smart Mater.Struct.* **18** 115029
[18] Kikuchi T and Kobayashi K 2011 *Journal of System Design and Dynamics* **5** 1471-1484

13th Int. Conf. on Electrorheological Fluids and Magnetorheological Suspensions (ERMR2012) IOP Publishing
Journal of Physics: Conference Series **412** (2013) 012046 doi:10.1088/1742-6596/412/1/012046

Design of shear gaps for high-speed and high-load MRF brakes and clutches

D Güth[1], A Wiebe and J Maas

Ostwestfalen-Lippe University of Applied Sciences, Control Engineering and Mechatronic Systems, Liebigstraße 87, 32657 Lemgo, Germany, www.motion-ctrl.de

E-mail: dirk.gueth@hs-owl.de

Abstract. Magnetorheological fluids (MRF) are smart fluids with the particular characteristics of changing their apparent viscosity significantly under the influence of a magnetic field. This property allows the design of mechanical devices for torque transmission, such as brakes and clutches, with a continuously adjustable and smooth torque generation. Especially the use of MRF in devices for applications with the demand on high rotational speeds and high loads can offer advantages, because more dissipation energy and thermal load peaks can be compensated by the torque generating fluid volume. However, high rotational speeds can cause a centrifugation of the particles contained in the MRF especially in idle mode when no magnetic field is applied. This can yield on the one hand to labile and unpredictable behavior of the torque response and on the other hand to a higher wear of the MRF or in the worst case to a destruction of the MRF with respect to a long-term view.For ensuring reliable braking or coupling conditions, in this contribution the development of a Taylor-Couette flow with axisymmetric toroidal vortices in axial shear gaps of MRF brakes and clutches is considered. The developing flow profiles in these shear gaps are modeled and analyzed in detail and finally design rules for axial shear gaps are introduced. Measurements at high rotational speeds up to 6000min^{-1} with a test actuator based on this design prove the positive influence of the vortex flow on the homogeneity of the MRF suspension and on the consistency and predictability of the braking or coupling torque even under long-term conditions.

1. Introduction

The use of magnetorheological fluids (MRF) in rotatory actuators is currently increasing more and more due to their outstanding properties e.g. the smooth adjustable torque, the fast response time, the noiseless operation and the reduced design space of actuators compared to conventional dry friction based systems, [1]. Also the rapid development progress in the electric drive technology with fast rising requirements on the maximum braking power and maximum rotational speeds shows up the performance limits of conventional, friction-based electromagnetic brakes. The required lifetime, peak loads and the dissipation of braking energy cannot be ensured in a sufficient degree by these conventional systems. By using MRF brakes with a scalable fluid volume for braking, new technical capabilities can be opened up. With an appropriate design of the fluidic system of a MRF brake, more dissipation energy and thermal load peaks can be compensated, because of a braking fluid volume and also a better dissipation of energy can be realized, [2, 3]. Further challenges to meet are high rotational

[1] To whom any correspondence should be addressed.

speeds up to 6000min^{-1} and more which are often corresponding to the requirement of high energy brakes. Especially when running a brake at high rotational speeds in idle mode, but no magnetic field is applied, the magnetorheological fluid is exposed to different forces caused by accelerating fields such as gravity, centrifugal forces and inertial force. Due to high differences in the density of the MRF components, carrier fluid and carbonyl iron powder particles, these forces can cause a separation of the two-phase system, whereby the particles will concentrate in the direction of the acting acceleration forces, [4]. For the mentioned applications with high energy dissipation and high rotational speeds especially the shear gap has to be designed in an appropriate manner. Thereby, a distinction will be made mainly between radial (disk-shaped actuators) and axial (drum cylinder-shaped actuators) oriented shear gap designs. Especially for applications with high rotational speeds, axial shear gaps with a narrow fluid gap size in radial direction offer advantages considering a possible particle centrifugation caused by the acting radial accelerations, [5].

In this contribution the developing flow profiles for the non-magnetized case (Off-State mode) in axial shear gaps will be modeled and analyzed in detail. The idea is to take advantage of a developing Taylor-Couette flow in axial shear gaps that can support the homogeneity of MRF when operating with high rotational speeds. Beside a mathematical description of the developing fluid profiles depending on the rotational speed, also analysis are performed with an experimental test actuator for rotational speeds up to 6000min^{-1}.

2. Modelling of fluid flow in cylinder actuators

In the design process of actuators based on MRF, two different states need to be considered: the Off-State mode with no applied magnetic field as well as the On-State Mode with an applied magnetic field for setting up the yield stress. For brakes and clutches, especially at high rotational speeds, it is in particular the Off-State mode when the actuator is driven in an idle state that has to be considered in detail due to the radial accelerations acting on the particles. As mentioned in the introduction, an axial shear gap design (cylinder) offers significant advantages due to the narrow gap size in radial direction.

However, even in these gaps a shift of the particle concentration in radial direction can occur. This can cause on the one hand undesirable effects like an insufficient or undefined behavior of the torque response when switching from idle into braking mode and on the other hand on long-term view an increased wear of the MRF. Therefore, it is necessary to take measures to ensure, even in the case of high rotational speeds, a homogeneous distribution of the particles in the MRF in the idle or Off-State mode. These measures should be based on the developing flow profiles for this type of shear gap that will be studied in detail in the following subsection. For describing the flow profiles occurring in the idle mode, the MR suspension will be treated in the Off-State mode as a continuum with the properties of a Newton fluid, while the solid phase of the suspension will not be considered as discrete particles.

2.1. Flow profiles

The fluid flow of the MRF can be considered in general by common methods of fluid dynamics that are based on the conservation laws for mass, momentum and energy. The continuity equation describes the time variation of mass in a volume element. For the stationary, rotationally symmetric case with an incompressible fluid, the continuity equation in cylindrical coordinates can be written as

$$\frac{\partial v}{\partial r} + \frac{v}{r} + \frac{\partial w}{\partial z} = 0, \tag{1}$$

where r is the radius position and v, w the flow velocity components in radial and axial direction respectively. The Navier-Stokes equation describes the temporal change of momentum in a volume element. With the simplified continuity equation (1) for incompressible fluids, the

Navier-Stokes equation for the azimuthal component φ, the radial component r and the axial component z can be described for the stationary, rotationally symmetric case by

$$\begin{aligned} v\frac{\partial u}{\partial r} + \frac{v \cdot u}{r} + w\frac{\partial u}{\partial z} &= \frac{\eta}{\rho} \cdot \left(\frac{\partial^2 u}{\partial r^2} + \frac{1}{r}\frac{\partial u}{\partial r} - \frac{u}{r^2} + \frac{\partial^2 u}{\partial z^2}\right) + f_\varphi\,, \\ v\frac{\partial v}{\partial r} - \frac{u^2}{r} + w\frac{\partial v}{\partial z} &= -\frac{1}{\rho}\frac{\partial p}{\partial r} + \frac{\eta}{\rho} \cdot \left(\frac{\partial^2 v}{\partial r^2} + \frac{1}{r}\frac{\partial v}{\partial r} - \frac{v}{r^2} + \frac{\partial^2 v}{\partial z^2}\right) + f_r, \\ v\frac{\partial w}{\partial r} + w\frac{\partial w}{\partial z} &= -\frac{1}{\rho}\frac{\partial p}{\partial z} + \frac{\eta}{\rho} \cdot \left(\frac{\partial^2 w}{\partial r^2} + \frac{1}{r}\frac{\partial w}{\partial r} + \frac{\partial^2 w}{\partial z^2}\right) + f_z\,, \end{aligned} \tag{2}$$

where p is the pressure, u the flow velocity component in azimuthal direction, ρ the density and η the dynamic viscosity of the MRF. The considered volume forces f_φ, f_r and f_z take into account gravity, acceleration and other forces.

For low rotational speeds n or consequently low angular velocities ω, the developed flow is a steady and purely azimuthal, circular and especially laminar Couette flow, [6]. The resulting flow velocities can by calculated with the Navier-Stokes equations in equation (2) as a function of the radius r. Therefore it is assumed that the expansion of the fluid gap in axial direction is infinite large. This simplifies the Navier-Stokes equation with $\partial/\partial z = 0$ and $w = 0$. Due to the laminar Couette flow, also the radial velocity component will result in $v = 0$ and the partial derivative $\partial/\partial\varphi = 0$ will vanish. Using these simplifications, the Navier-Stokes equation can be expressed by:

$$\begin{aligned} 0 &= \eta\left(\frac{\partial^2 u}{\partial r^2} + \frac{1}{r}\frac{\partial u}{\partial r} - \frac{u}{r^2}\right), \\ \rho\frac{u^2}{r} &= \frac{\partial p}{\partial r}, \\ \rho f_z &= \frac{\partial p}{\partial z}. \end{aligned} \tag{3}$$

With the help of the first Equation for the azimuthal flow in equation (3), the velocity distribution in the radial direction can be determined. The Equation corresponds to a linear, ordinary, differential equation of second order with variable coefficients of Euler type, [7]. The solution results in the velocity distribution in radial direction for the laminar flow in radial shear gaps

$$u\left(r\right) = C_1 r + \frac{C_2}{r} \tag{4}$$

which is depicted in figure 1a with the coefficients

$$C_1 = \frac{\omega_o - \mathrm{r}_{io}\omega_i}{1 - \mathrm{r}_{io}} \quad \text{and} \quad C_2 = \frac{r_i^2(\omega_i - \omega_o)}{1 - \mathrm{r}_{io}}, \tag{5}$$

wherein ω_i is the angular velocity of the inner rotor, ω_o of the outer rotor and r_{io} the radius ratio defined by

$$r_{io} = \frac{r_i^2}{r_o^2}. \tag{6}$$

For higher rotational speeds n the relative angular velocity ω between the inner ω_i and the outer rotor ω_o has to be considered. If the angular velocity of the inner rotor ω_i is increased compared to the outer rotor ω_o above a certain threshold, the laminar Couette flow, see figure 1b, becomes unstable and a secondary steady state characterized by axisymmetric toroidal vortices emerges as depicted in figure 1c. This behavior is also known as the Taylor-Couette flow.

The change from the Couette flow into the Taylor-Couette flow can be characterized by the Taylor number Ta that is a dimensionless quantity. The beginning of a developing Taylor vortex

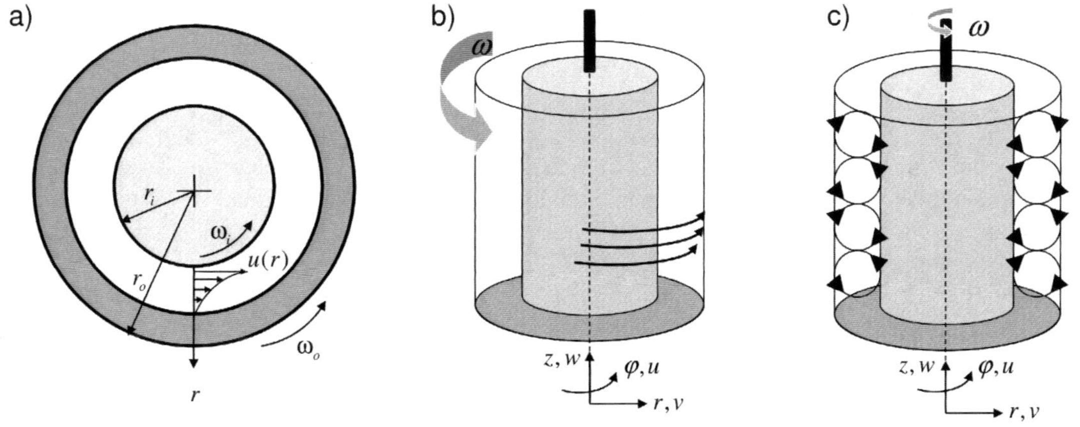

Figure 1. Cross section with velocity profile of an axial shear gap (a), rotating cylinder developing the Couette flow (b) and the Taylor-Couette flow (c).

flow emerges from a threshold of $Ta = 41.3$, [8], and can be calculated for the shear gaps with a narrow height by

$$Ta = \frac{\omega\ r_i\ h_s\ \rho}{\eta} \cdot \sqrt{\frac{h_s}{r_i}}, \tag{7}$$

wherein $\omega = \omega_i - \omega_o$ is the differential angular velocity and $h_s = r_o - r_i$ represents the shear gap height, [9]. This equation can be used for very small shear gap heights $h_s = r_o - r_i \ll r_{i,o}$. The Taylor-Couette flow can be divided into two flow regimes: the Taylor vortex flow and the turbulent Taylor vortices, [10]. Considering the developing flows in axial shear gaps of MRF actuators, it is mainly the Taylor vortex flow that will occur. The turbulent Taylor vortices regime will only be reached at very high rotational speeds or very large heights.

For shear gap sizes $h_s/r_i \leq 1$, as they will be used for MRF brakes and clutches for the announced high energy applications, the transition point can be approximated by defining a critical Reynolds number, [11],

$$Re_{crit} = \omega_i\ r_i \cdot \frac{h_s\ \rho}{\eta} = \left(41.1 + 14.5 \cdot \frac{h_s}{r_i}\right) \cdot \sqrt{\frac{r_i + r_o}{2\ h_s}}. \tag{8}$$

Combining equations (7) and (8), also a critical Taylor number Ta_{crit} can be calculated, which represents a corrected Taylor number describing the transition point for shear gap sizes $h_s/r_i \leq 1$:

$$Ta_{crit} = \left(41.1 + 14.5 \cdot \frac{h_s}{r_i}\right) \cdot \sqrt{\frac{r_i + r_o}{2 \cdot r_i}}. \tag{9}$$

Using equation (8), a critical rotational speed n_{crit} can be calculated that indicates the transition point from the Couette flow to the Taylor-Couette flow for large radial fluid gaps h_s:

$$n_{crit} = \left(41.1 + 14.5 \cdot \frac{h_s}{r_i}\right) \cdot \sqrt{\frac{r_i + r_o}{2\ h_s}} \cdot \frac{\eta}{2\pi\ \rho\ r_i\ h_s}. \tag{10}$$

The axisymmetric toroidal vortices of the Taylor-Couette flow induce a mixing effect in the shear gap, that can cause an inherent effect for the homogeneity of the MR fluid under the influence of centrifugal accelerations due to high rotational speeds.

When designing shear gaps for brakes and clutches for high rotational speeds, this effect should be used. As an essential criterion it has to be considered that the faster rotating boundary is always placed in the inside and the transition from Couette to Taylor-Couette flow is considered by equation (10). A further important property beside geometric aspects that is pointed out by this equation, is the temperature-dependent dynamic viscosity of the MRF. A change in the temperature will result in a change of the developing flow profile and especially in a movement of the transition point from laminar Couette flow to Taylor-Couette flow. By an increase of the temperature of the MRF and the resulting decrease of the dynamic viscosity when running in idle mode, the transition to the Taylor-Couette flow will arise at even lower rotational velocities. This behavior can be an advantage for the safe operation of brakes and clutches based on MRF.

2.2. Methods for verification of the development of Taylor-Couette flow

Common experiments for the investigation of the Taylor-Couette flow are based on optical methods where the investigated fluid is transparent and tracer particles for visual capturing are added. Due to the highly concentrated suspension this method is not suitable for the investigation of the Taylor-Couette flow in MRF actuators. Hence, a different method based on the measured torque in the Off-State mode is used. In case of a laminar Couette flow, the increase of torque T is directly proportional to the increase of the rotational speed respectively angular velocity ω by $T \sim \omega$ following Newtonian fluid behavior, also described as the linear theory in figure 2. In the Taylor-Couette flow (Taylor vortex flow) the behavior changes due to the additional dissipation energy for developing vortices and the torque T increases

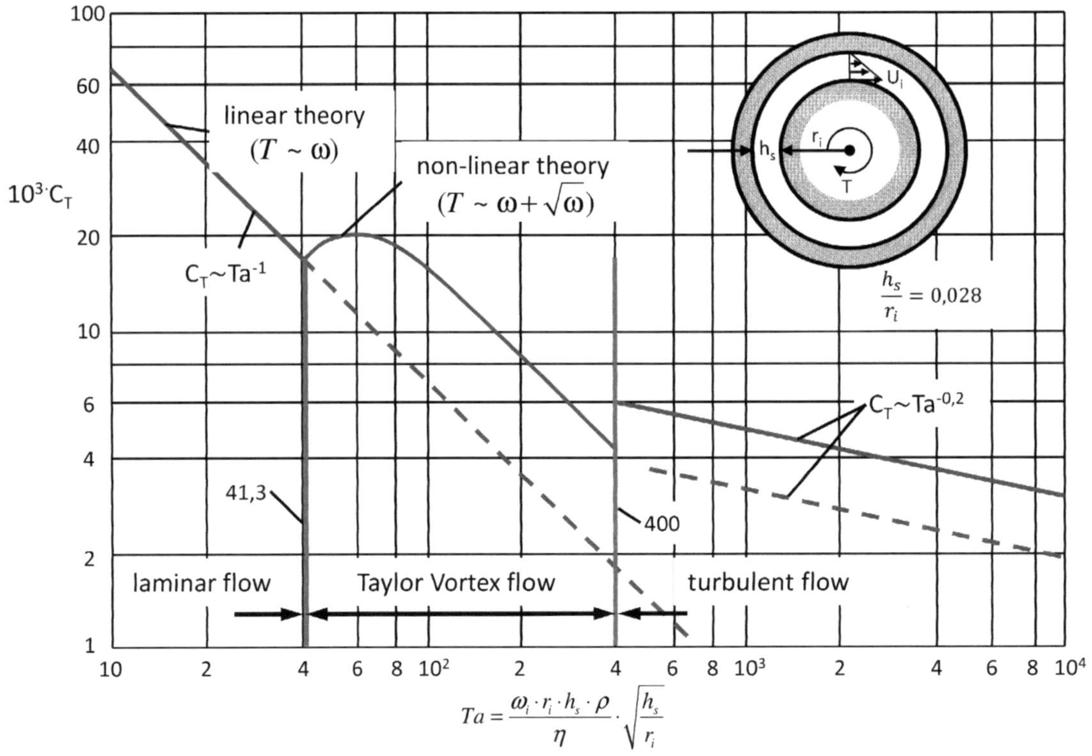

Figure 2. Torque coefficient C_T of the inner cylinder as a function of the Taylor number Ta, [10].

by $T \sim (\omega + \sqrt{\omega})$. This regime is also described as the non-linear theory in figure 2, [12]. Using this relation, an experimental method for the verification of the development of Taylor-Couette flow is pointed out and especially the transition point can be investigated. Therefore, a dimensionless coefficient C_T is introduced that represents the torque. The torque T of the of the inner cylinder of the shear gap can be calculated under consideration of the dimensionless coefficient C_T by

$$T = C_T \cdot \frac{\pi}{2} \cdot \rho \cdot \omega_i^2 \cdot r_i^3 \cdot l \tag{11}$$

under the assumption that the angular velocity of the outer cylinder is $\omega_o = 0$, [10]. Using equation (7), the torque coefficient C_T can be calculated for narrow gaps by

$$C_T = \frac{4\,\eta}{\omega_i\, r_i\, h_s\, \rho} = \frac{4}{Ta} \cdot \sqrt{\frac{h_s}{r_i}}. \tag{12}$$

Considering the correction of the Taylor number Ta expressed by equation (8), the torque coefficient C_T can be calculated as a function of the Taylor number Ta and the transition point and consequently the Taylor-Couette flow can be experimentally verified. Thereby it is important to avoid measuring errors for instant due to changes in the temperature of the MRF.

In the following section an experimental test actuator is introduced for analyzing the flow profiles and especially to validate the Taylor-Couette flow in axial shear gaps of MRF based brakes and clutches.

3. Experimental test actuator

For the experimental investigation, a novel test actuator with the following properties is developed:

- axial shear gap with a variable height
- rotatable inner rotor and stationary outer rotor and vice versa
- capable of high rotational speeds up to 6000min^{-1}
- torque capacity depending on shear gap height of about 30Nm
- rotatable independent excitation system based on one electromagnet
- volume equalization for the compensation of thermal expansion of MRF
- temperature measurement in the shear gap

A design fulfilling these requirements is shown in figure 3. The design is based on a typical axial shear gap with the specific characteristic of two shaft drives: one is for driving the inner rotor and the other for driving the outer rotor. Depending on the operation, a clamping cover has to be installed either on the left side for rotating the inner rotor, see figure 3a, or on the right side, see figure 3b, whereby the other side has to be rotated respectively. By this concept, the basic design of the test system has not to be changed, independent of the kind of operation. This is important for gaining comparable measurement results, e.g. under consideration of the sealing and the resulting friction torque. The dimensions of the shear gap are chosen with respect to the requirements mentioned before. Due to the large amount of dissipation energy generated by the operation at high rotational speeds, also the fluid volume in the shear gap has be chosen in a suitable manner. The basic criterion for dimensioning the shear gap is primarily the resulting torque that has to be achieved.

The torque T of an actuator with an axial shear gap depends on the shear stress characteristics $\tau(B, \dot{\gamma})$ of the MRF, which is a function of the magnetic flux density B in the fluid gap, that is subjected to the magnetic field as well as a function of the fluid viscosity η and the velocity gradient perpendicular to the direction of shear $\dot{\gamma}$ (shear rate). For the considered geometry of

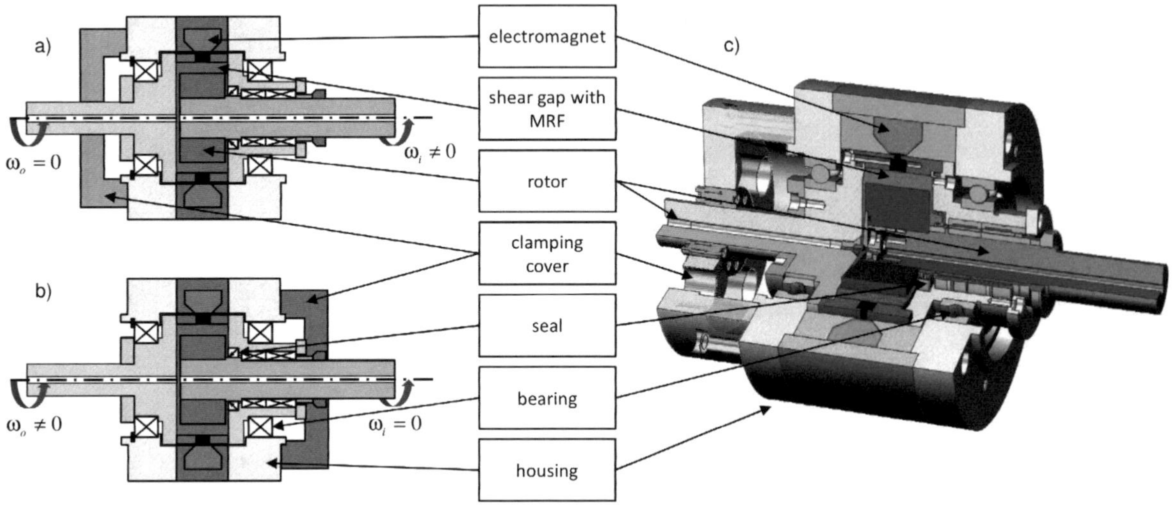

Figure 3. Test actuator for experimental investigations: Half section view with a) inner rotating cylinder as well as b) outer rotating cylinder and c) a 3D model of the actuator.

two coaxially rotating cylinders with ω_i and ω_o, the shear rate $\dot{\gamma}$ for the MRF can be described by the velocity profile $u(r)$ in equation (4) considering $u(r) = \omega \cdot r$ by

$$\dot{\gamma} = -r\frac{d\omega}{dr} = -\frac{du(r)}{dr} + \frac{u(r)}{r} = 2\frac{{r_i}^2\left(\omega_o - \omega_i\right)}{{r_i}^2 - {r_o}^2}. \tag{13}$$

The flow curve describing the rheological behavior of MRF is commonly expressed by the Bingham fluid constitutive model, [13],[14],

$$\tau\left(B, \dot{\gamma}\right) = \tau_0(|B|) \cdot \operatorname{sgn}(\dot{\gamma}) + \eta \cdot \dot{\gamma}, \tag{14}$$

wherein $\tau_0(|B|)$ is the yield point depending on the magnetic flux density B. Due to the focused, high rotational speeds n and resulting high shear rates $\dot{\gamma}$ depending on the shear gap height, see equation (13), the rheological behavior of the MRF is often not well approximated by the Bingham model, [15]. At high shear rates, most MRF show a shear thinning behavior that can be approximated more sufficient by the Herschel-Bulkley fluid constitutive model

$$\tau\left(B, \dot{\gamma}\right) = \left(\tau_0\left(|B|\right) + k|\dot{\gamma}|^n\right) \operatorname{sgn}\left(\dot{\gamma}\right). \tag{15}$$

In this equation n is the power law index and k is the consistency index. The resulting torque T can generally be described by

$$T = \int_r \int_A \tau(B, \dot{\gamma}) \cdot dA \cdot dr, \tag{16}$$

where $\tau(B, \dot{\gamma})$ is the magnetic field and shear rate depending shear stress of the MRF, A is the shear surface at the radius position r in the fluid gap.

The measurements are performed with the BASF fluid *Basonetic 5030* that is an optimized MRF for brake and clutch applications, [16]. The shear stress characteristic $\tau(B, \dot{\gamma})$ is shown in figure 4 and the temperature dependency of the viscosity η is depicted in figure 5. For the investigation of the transition point from Couette to Taylor-Couette flow, two different shear gap

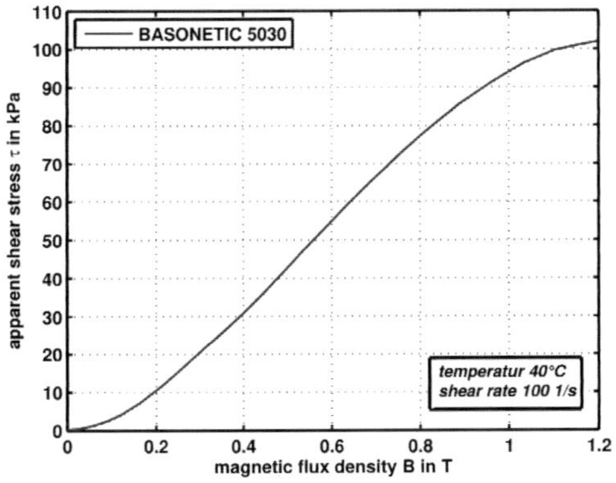

Figure 4. On-State shear stress characteristic τ of Basonetic 5030.

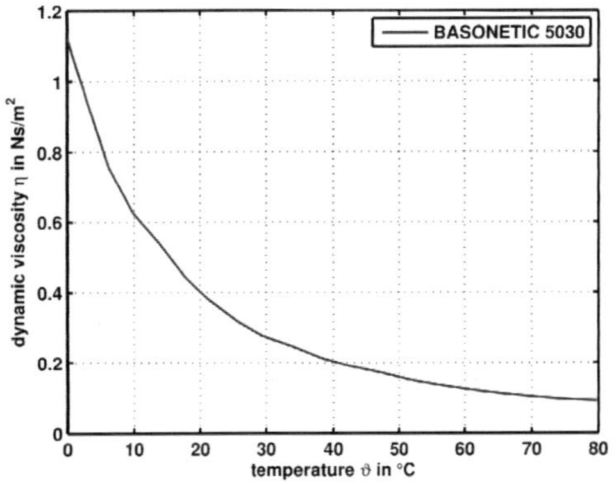

Figure 5. Temperature dependency of the dynamic viscosity η of Basonetic 5030.

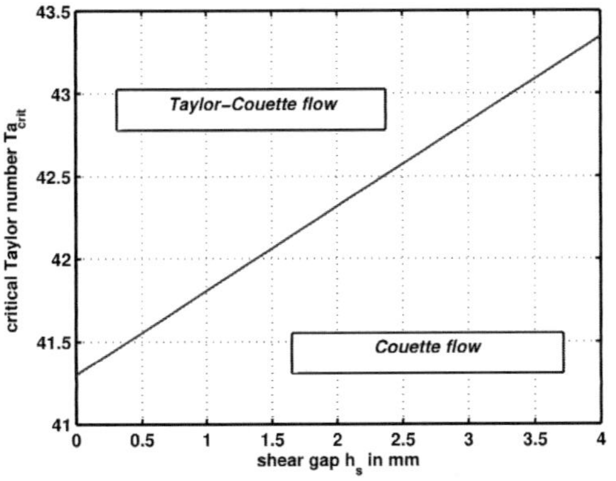

Figure 6. Critical Taylor number Ta_{crit} as a function of the shear gap height h_s.

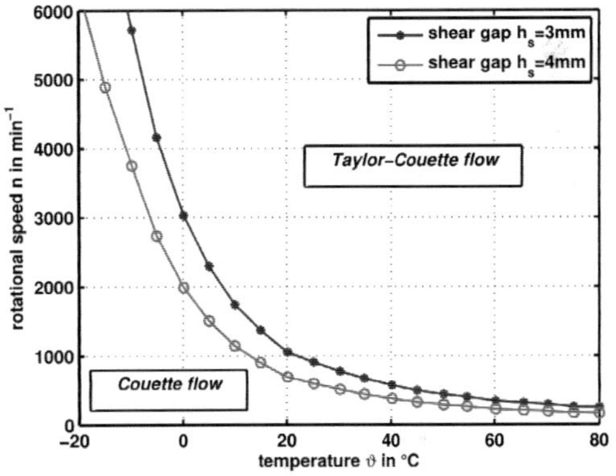

Figure 7. Critical rotational speed n_{crit} as a function of the MRF temperature ϑ.

heights of $h_s = 3$mm and $h_s = 4$mm are considered. The final shear gap design is chosen with a length of $l = 40$mm and a radius of the inner boundary of $r_i = 49$mm. The resulting Taylor number describing the transition point, as defined in equation (9), can be calculated for these shear gap dimensions as depicted in figure 6. Beside the shear gap heights, also the temperature of the MRF has an effect on the transition point. Based on the temperature dependent viscosity, shown in figure 5, the critical rotational speed n_{crit} is influenced as shown in figure 7.

4. Measurements

The test actuator based on the described design has been realized for verifying the occurrence of the Taylor-Couette flow and for investigating the influences of the Taylor-Couette flow on the torque response when applying a magnetic field. Different measurements are preformed that are presented in the following subsections. Furthermore the influences of the Taylor-Couette flow on long-term operation are investigated. The measurements with the test actuator are performed on a test bench that is equipped with an electrical drive and a torque sensor equipped with a real-time processor system for controlling and data capturing.

Journal of Physics: Conference Series **412** (2013) 012046 doi:10.1088/1742-6596/412/1/012046

4.1. Steady state characteristics for torque behavior

Before investigating the occurrence and the influence of the Taylor-Couette flow, the test actuator should be characterized by typical control characteristic lines. These measurements are performed as a function of the rotational speed n for different magnetic excitations due to the applied current i. The characterization is performed for a shear gap height of $h_s = 3$mm, see figure 8, and for $h_s = 4$mm, see figure 9. Due to a non-negligible dependency between the temperature depending dynamic viscosity η and the resulting braking torque, the MRF is thermically preconditioned at a temperature of $\vartheta \approx 40°$C and the measurement is performed step by step to ensure thermal recoveries. By this measurement method thermal influences are almost eliminated. The single points of measurement are marked in the diagrams.

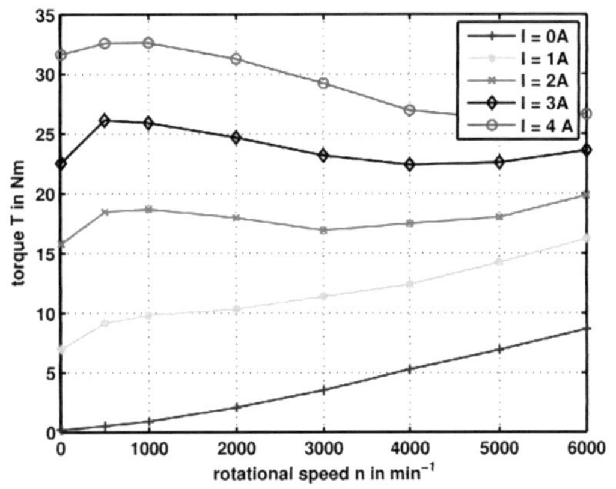

Figure 8. Steady state torque characteristics for a shear gap height of $h_s = 3$mm.

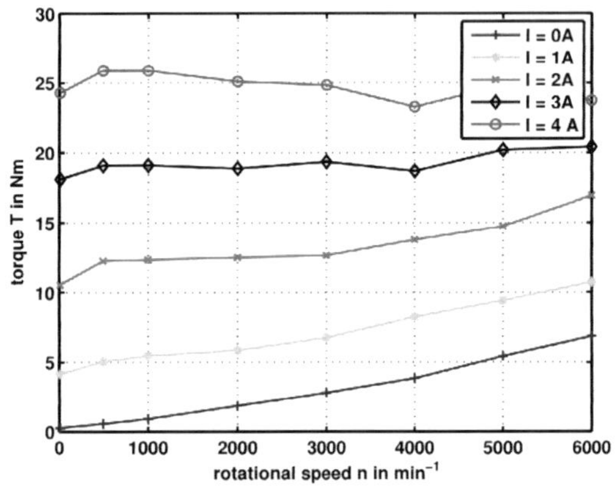

Figure 9. Steady state torque characteristics for a shear gap height of $h_s = 4$mm.

For the Off-State operation with the non-energized excitation system ($I = 0$A), the typical viscous behavior expressed by equation (14) with a proportional increase of the resulting torque T with respect to the rotational speed n is readily identifiable. At higher currents the measured torque T shows a decrease especially at high rotational speeds. This behavior can be explained by the shear thinning behavior of MRF at high shear rates, in this case of up to $\dot{\gamma} \approx 10000\text{s}^{-1}$ at 6000min^{-1}, [15]. The characterization shows that a usage of MRF brakes and clutches up to differential rotational speeds of 6000min^{-1} is possible. But especially the high viscous torque at high rotational speeds has to be considered with respect to the energy efficiency of an application and also to the limited life-time of the MRF, because T_{vis} rises up to about 25% of T_{max}. For applications with high requirements considering efficiency and life-time further measures have to be taken into account like an intelligent fluid control, [17, 18].

4.2. Investigation on the transition from Couette to Taylor-Couette flow

The occurrence of vortices in the developed Taylor-Couette flow should enable a permanent mixing effect of the MRF in the axial shear gap and ensure a homogeneous particle distribution in the suspension. This effect should be verified by measurements that consider the transition from Couette to Taylor-Couette flow at the critical Taylor number Ta_{crit} using the dimensionless coefficient C_T as defined in equation (11). The measurements are performed with the introduced experimental test actuator with an inner rotating rotor and a fixed outer router. The dimension of the shear are varied in the shear gap height with $h_s = 3$mm and $h_s = 4$mm. The related critical Taylor number are $Ta_{crit} = 42.6$ ($h_s = 3$mm) and $Ta_{crit} = 43.3$ ($h_s = 4$mm), see figure 6. Due to the temperature dependency of the viscosity and hence the critical rotational speed n_{crit}, see equation (10), as well as the torque coefficient C_T, see equation (12), the measurement values used for evaluation are captured at a MRF temperature of $\vartheta = 40°\text{C}$. The corresponding critical rotational speeds are $n_{crit} \approx 580\text{min}^{-1}$ ($h_s = 3\text{mm}$) and $n_{crit} \approx 380\text{min}^{-1}$ ($h_s = 4\text{mm}$). The measurement results for the dimensionless coefficient C_T as a function of the Taylor number Ta are shown in figure 10 for the shear gap height of $h_s = 3\text{mm}$ and in figure 11 for the shear gap height of $h_s = 4\text{mm}$. The measured torque results are corrected by the friction torque of the sealing ring.

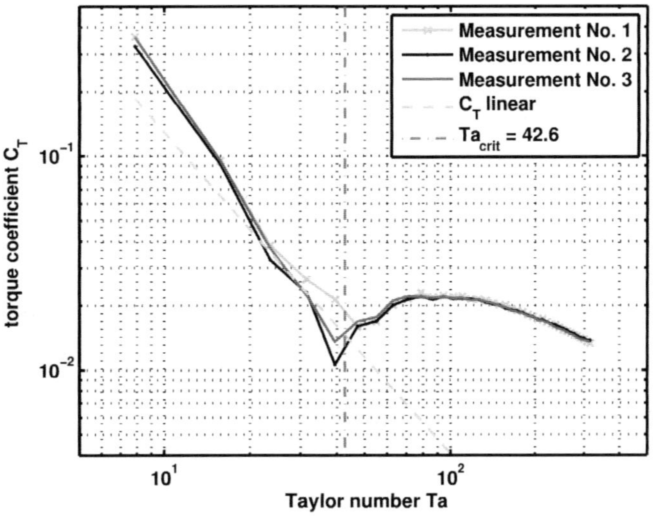

Figure 10. Transition from Couette to Taylor-Couette flow for a shear gap height of $h_s = 3\text{mm}$ at a MRF temperature of $\vartheta \approx 40°\text{C}$.

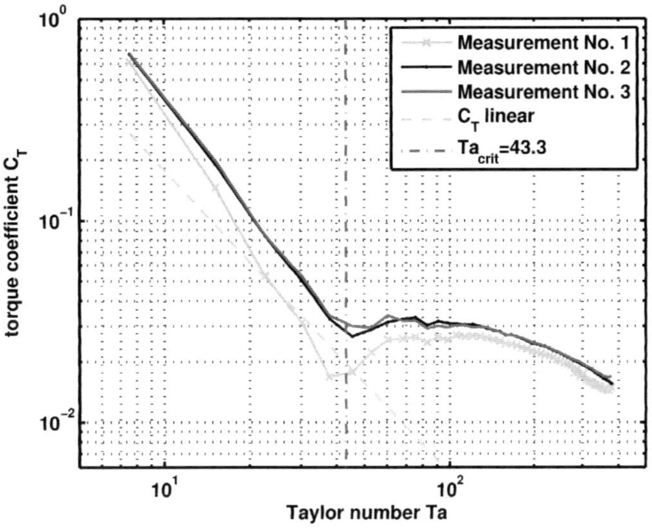

Figure 11. Transition from Couette to Taylor-Couette flow for a shear gap height of $h_s = 4\text{mm}$ at a MRF temperature of $\vartheta \approx 40°\text{C}$.

Journal of Physics: Conference Series **412** (2013) 012046 doi:10.1088/1742-6596/412/1/012046

For both shear heights h_s three measurements for verifying the repeatability are performed. These show a good accordance with the theoretical calculations for the linear regime of the torque coefficient C_T and with the theoretical transition from Couette to Taylor-Couette flow at the corresponding critical Taylor number Ta. These analysis proof the occurrence of a Taylor-Couette flow in axial oriented shear gaps filled with MRF in idle mode. By further measurements the influence and the advantages on the homogeneity of the suspension MRF has to be shown. Especially the advantages on the homogeneity of MRF due to the mixing effect in idle mode needs to be validated.

4.3. Influences of Couette and Taylor-Couette flow on the torque behavior

For analyzing the influences of Couette and Taylor-Couette flow on the torque behavior, both flow regimes should be considered. Therefore, the test actuator should be operated with different constant rotational speeds n in idle mode with no applied magnetic field ($i = 0$A) and after a certain time, a current step I will be applied. For ensuring the both flow regimes, the actuator is operated with a fixed inner rotor ($\omega_i = 0$) for the Couette flow (see figure 3b) and with a fixed outer rotor ($\omega_o = 0$) for the Taylor-Couette flow (see figure 3b). The analysis are performed for rotational speeds of up to $n = 4000\text{min}^{-1}$ and a shear gap height of $h_s = 4$mm.

The measurement results for the Couette flow are shown in figure 12 and for the Taylor-Couette flow in figure 13. When operating the test actuator with the Couette flow, a significant overshoot can be identified for the rotational speeds larger than $n = 2000\text{min}^{-1}$. Also the stationary value is reached for some measurements only after 10 seconds. It is assumed that the slow decrease of torque behavior can be explained by the shearing of the MRF under the influence of a magnetic field which causes magnetic forces acting on the particle and finally the resulting chain formation of particles also results in a homogenization. When utilizing the Taylor-Couette flow with its resulting mixing effect, the torque response shows no overshoot and the stationary torque is immediately reached.

Considering the measurement results the advantages of the Taylor Couette flow can be identified by a more constant and homogenous torque response as predicted by the Bingham model. For ensuring these conditions when applying a sudden current excitation, a homogeneous distribution of particles in the suspension MRF is required.

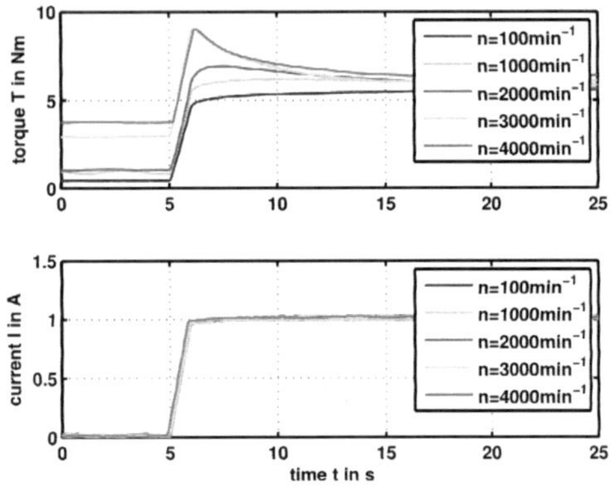

Figure 12. Torque behavior when applying a current step for Couette flow ($\omega_i = 0$).

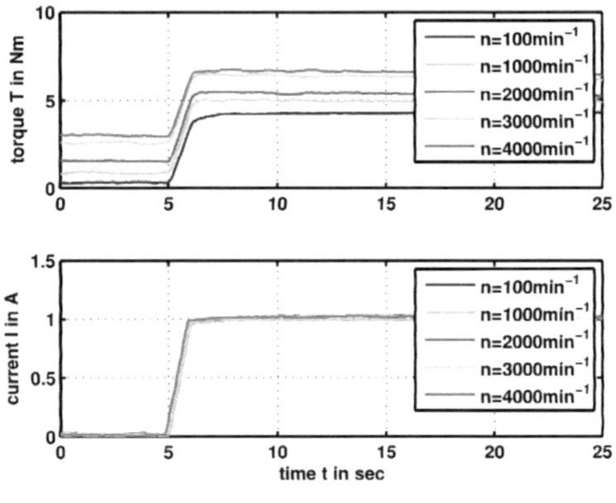

Figure 13. Torque behavior when applying a current step for Taylor Couette flow ($\omega_o = 0$).

4.4. Influences of Couette and Taylor-Couette flow on long-term measurements

The advanced behavior on the torque response using the Taylor-Couette flow should also be analyzed by a long-term run of the experimental test actuator. Therefore an appropriated test cycle is defined as shown in figure 14 with different phases of braking and idle mode. The shear gap height chosen for the measurement is $h_s = 4\text{mm}$. The cycle starts with an On-State measurement with an applied current I at a low rotational speed $n_{stat} = 10\text{min}^{-1}$. The resulting torque is defined as T_{stat}. Subsequently follows an On-State measurement with high rotational speeds n_{dyn} that are varied between rotational speeds from $n_{dyn} = 1500\text{min}^{-1}$ to $n_{dyn} = 4000\text{min}^{-1}$. The resulting torque is defined as T_{dyn}. At the end of the test cycle an idle phase with a constant rotational speed of $n_{idle} = 1000\text{min}^{-1}$ is investigated over a time period of 10s. The measured torque during this phase is defined as T_{idle}. The time durations for measuring T_{stat}, T_{dyn} and T_{idle} are varied for each cycle.

The measurement results are shown in figure 15 for a long-term run with 2000 test cycles for both flow regimes. Therein the three different phases characterized by the torque T_{dyn}, T_{stat} and T_{idle} are presented. For the analysis only stationary torques are evaluated, where acceleration or deceleration phases are not considered. The results for the torque T_{dyn} show a relative constant behavior over all measurement cycles for both flow regimes cases. The significant difference between the results are the variation in torque that is distinctive for the Couette flow. These large variations in the measured torque with peaks up to 5Nm depend on the rotational speed n_{dyn}. With an increase of the rotational speed n_{dyn}, a decrease of the torque T_{dyn} can be identified. When utilizing the Taylor-Couette flow, the decrement in torque is significant lower. Considering the On-State measurements with the low rotational speed $n_{stat} = 10\text{s}$, the resulting torque T_{stat} shows a more stable behavior for the Taylor-Couette flow. The measurement results for the torque T_{idle} at a constant rotational speed $n_{idle} = 1000\text{min}^{-1}$ show a very stable behavior over all measurement cycles for the Taylor-Couette flow. Regarding the torque behavior for the Couette flow, the torque T_{idle} shows a significant degrease in torque after approximately 1000 measurement cycles that can be an indicator for a potential and nonreversible particle centrifugation. A further indicator for this assumption is pointed out by the measurements of the torque T_{dyn} for the Couette flow that is decreasing more compared to the torque T_{dyn} for the Taylor-Couette flow.

The performed mechanical work during these 2000 cycles for each flow regime can be

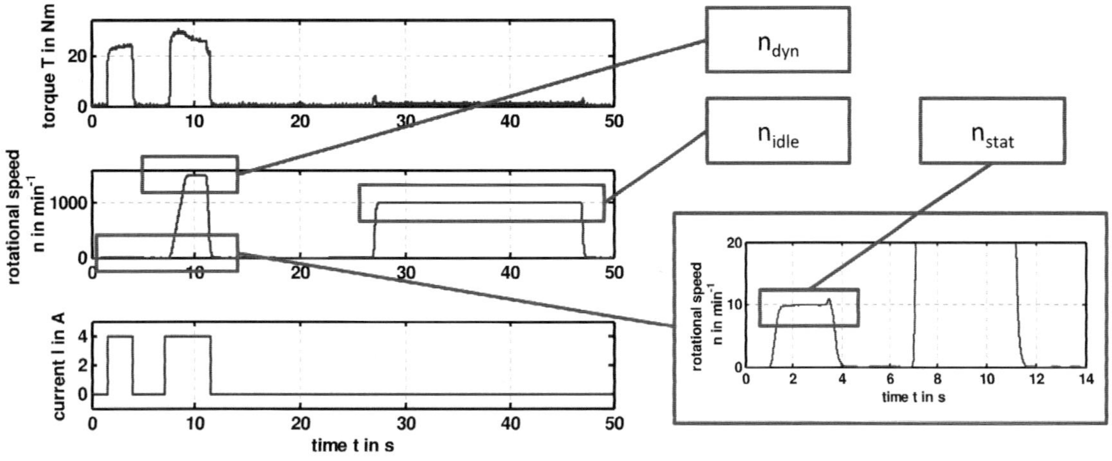

Figure 14. Definition of a test cycle for long-term measurements.

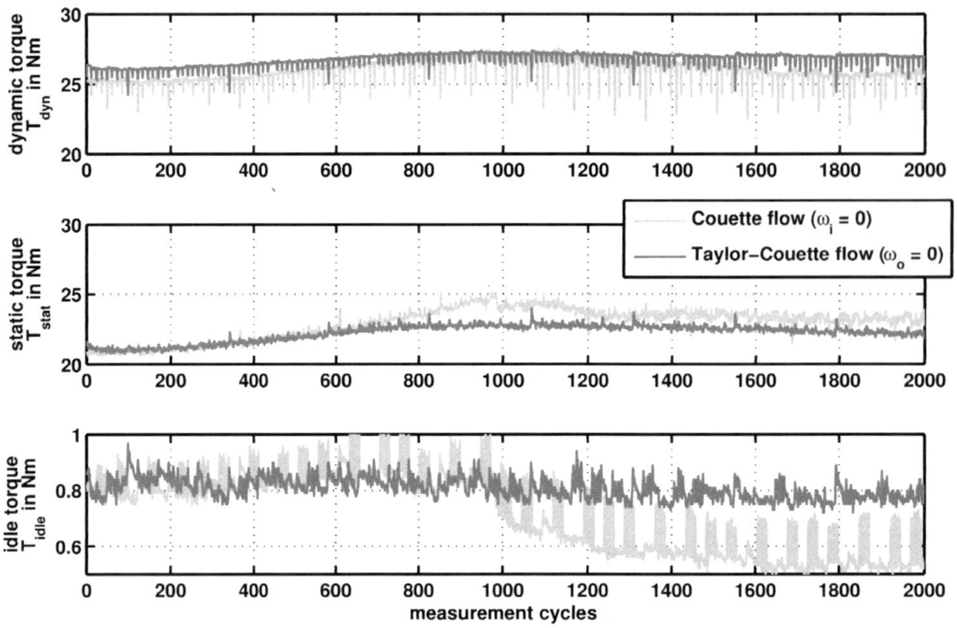

Figure 15. Measurement results comparing the influences of Couette and Taylor-Couette flow on the long-term behavior.

summarized to a total value of 16.2MJ. The total volume of MRF in the shear gap for these measurements was 80ml. Considering a limited **L**ifetime **D**issipated **E**nergy (LDE) of 2.7MJ/ml, [19], the complete test cycle represents a lifetime for the MRF of about 7%.

These measurements compare the influence of the Couette flow and the Taylor-Couette flow and demonstrate advantages considering the consistency and the prediction of the resulting torques when using the mixing effect of the Taylor vortex flow. The measurements also indicates that MRF brakes even for high rotational speeds are realizable with a good performance on long-term view. By considering the limitations like a remaining viscous torque and the torque-speed characteristic shown in figure 8 and figure 9, MRF brakes and clutches can be used for certain applications with requirements like long lifetime, peak loads and a high braking power accompanied by high maximal rotational speeds.

5. Conclusions

In this contribution the different flow profiles in axial shear gaps are modelled in detail. Especially the development of Taylor vortices in axial shear gaps was described and the transition from a simple Couette flow to a Taylor-Couette flow considered. The occurrence of a Taylor-Couette flow should ensure a homogenous particle distribution in the MRF at high rotational speeds by using the mixing effect of the Taylor vortices. A homogenous particle distribution is required for a constant and predictable torque behavior when applying a magnetic field. For proving the transition between these both flow regimes when the rotational speed is increased over a certain threshold, a measurement method based on a dimensionless torque coefficient as a function of the Taylor number was introduced. This transition was investigated by using an experimental test actuator with a variable height of the axial shear gap whose design was based on the theoretical consideration of the development of Taylor vertices. Different measurements are performed with the test actuator that are showing

- the steady state characteristic for torque behavior at high rotational speeds for different shear gap heights,
- the proof of the development of Taylor-Couette flow using the dimensionless torque coefficient for different shear gap heights,
- the advantages of Taylor-Couette flow when operating in idle mode and switching in the On-State mode for setting up a desired and predictable torque and finally
- long-term measurements pointing out the advantage of the Taylor-Couette flow.

Considering the Taylor-Couette flow in the design process of MRF actuators for high energy applications with high rotational speeds, a homogenous suspension with a constant torque behavior also in the long term view can be ensured.

Acknowledgment
This contribution is accomplished within the project HLD-Fluidbrake, funded by the Federal Ministry of Education and Research (BMBF) of Germany under grant number 1713X09.

References

[1] Maas J, Güth D and Wiehe A 2011 Mrf-actuator concepts for hmi and industrial applications *SPIE Smart Structures/NDE* vol 7977 p 797714
[2] Smith A L, Holzheimer J C, Ulicny C and Kennedy L C 2007 *JIMSS* **18** 1131–1136
[3] Wiehe A, Noack V and Maas J 2009 *J. of Phys.: Conf. Series* **149** 012084
[4] Güth D, Wiehe A and Maas J 2010 Modeling approach for the particle behavior in mr fluids between moving surfaces *Proc. of 12th Int. Conf. Electro-rheological Fluids and Magneto-rheological Suspensions* (Singapore: World Scientific) pp 457–463
[5] Neelakantan V A and Washington G N 2005 *JIMSS* **16** 703–711
[6] Constantinescu V 1995 *Laminar Viscous Flow*
[7] Bronstein I N, Semendjajew K A and Musiol G 2007 *Handbook of Mathematics*
[8] Taylor G 1923 *Philosophical Transactions of the Royal Society of London, Series A* **223** 289–343
[9] Kirchgässner K 1961 *ZAMP* **12** 14–30
[10] Schlichting H 1982 *Grenzschicht-Theorie*
[11] Wieghardt K 1974 *Theoretische Strömungslehre* vol 2
[12] Stuart J T 1958 *J. of Fluid Mechanics* **4** 1–21
[13] Bingham E C 1922 *Fluidity And Plasticity*
[14] Phillips R 1969 *Engineering Applications of Fluids with a Variable Yield Stress* Ph.D. thesis The University of California at Berkeley
[15] Lee U, Kim D, Hur N and Jeon D 1999 *JIMSS* **10** 701–707
[16] BASF SE 2009 *Magnetorheological Fluid Basonetic® 5030 - Optimum torque transmission for clutch-type applications* BASF SE www.basonetic.com
[17] Güth D and Maas J 2012 *SPIE Digital Library/Proceedings* **8341** 834121
[18] Güth D, Schamoni M and Maas J 2012 Modeling approach for a fluid movement induced by magnetic forces for viscous loss reduction of mrf brakes and clutches *ASME Conf. Proc.* pp SMASIS2012–8135
[19] Wiehe A, Kieburg C and Maas J 2012 Temperature induced effects on the durability of MR fluids *Proc. of the 13th International Conference Electrorheological Fluids and Magnetorheological Suspensions - in press*

Journal of Physics: Conference Series **412** (2013) 012047 doi:10.1088/1742-6596/412/1/012047

Design of tactile device for medical application using magnetorheological fluid

J S Oh[1], J K Kim[1], S R Lee[1], S B Choi[1,3] and B K Song[2]
[1] Smart Structures and Systems Laboratory, Department of Mechanical Engineering, Inha University, Incheon, 402-751, Korea
[2]School of Mechanical System Engineering, University of Incheon, Incheon, 406-840, Korea

E-mail: seungbok@inha.ac.kr

Abstract. For the tactile recognition of human organ in minimally invasive surgery (MIS), this paper presents a novel tactile device that incorporates with magnetorheological (MR fluid). The MR fluid is contained by diaphragm and several pins. The operator for MIS can feel different force (or stiffness) from the proposed tactile device by applying different magnetic field or current. In order to generate required force from the device, the repulsive force from the human body is measured as reference data and an appropriate size of tactile device is designed and manufactured. It has been demonstrated via experiment that the repulsive force corresponding to the human body can be achieved by applying proper control input current. In addition, it has been shown that we can control the repulsive force by dividing the tactile device by several sections.

1. Introduction

Tactile sensation means to realize contact sensation and stimulate the skin, especially on fingertips. Beginning purpose of such devices is visual information transmission for blind person [1]. However, their potential is exploited, so numerous ranges of new application such as virtual object, training, game, rehabilitation and minimally invasive surgery (MIS) have appeared. Among them, application of MIS is urgently required. MIS is a procedure carried out through the skin or anatomical opening. So, the MIS systems only send back visual information. There is no information about viscosity and stiffness of the touched tissue or organ. This causes the surgeon not to feel what is happening inside patient. Therefore, it is essential to develop the tactile device.

Various mechanisms of tactile device have been researched; pin-arrays utilizing piezoelectric actuators [2], shape memory alloys [3], piezoelectric ceramics [4], ionic conductive polymer gel films [5], polymer fabrics [6], electric motors [7] and MR fluid. The rheological properties of MR fluid are reversibly and instantaneously changed by applying magnetic field to the fluid domain [8]. This behaviour can represent the compliance, damping and sliding feeling of flexible objects of the different materials [9]. Since then, many researches have continued to focus on the MR fluid applied in tactile applications. Scilingo *et al.* proposed a MR application able to mimic the compliance of biological tissues manipulated by surgical tools [10]. In another line of research, Liu *et al.* investigated

[3] To whom any correspondence should be addressed.

the surface force response of a MR fluid-based haptic display with different electro-magnets in single cells, and their results provided a preliminary basis for future developments of this area [11].

The main objective of this research is to propose a new type of tactile device. The purpose of tactile device is to mimic a wide variety of tactile sensations during the minimally invasive surgery. Proposed tactile device consists of diaphragm which filled by MR fluid and movable pins. The magnetic circuit is installed between diaphragm and pin. When diaphragm is compressed by operator, the operator can feel the tactile sensation of MR fluid. The tactile sensation can be controlled by magnetic field. In order to determine design parameters of magnetic circuit, finite elements analysis of magnetic field is conducted and tactile device has been manufactured. In order to evaluate the performance of tactile device, the repulsive force measurement experiment has been conducted. From experimental results, it has successfully been demonstrated that the proposed tactile device can be effectively applied to MIS robot system.

2. Investigation on the human tissue

Recently, several attempts to realize tactile recognition was tried, there has been no successful results. This is because quantification of tactile recognition is very difficult. There are many physical indexes related with tactile recognition, but how these indexes can be affected to tactile recognition has not yet been investigated. So, in order to simplify the procedure, the repulsive force is only utilized for tactile recognition index. Subsequently, the reflection force of human is measured by force sensor. As shown in Figure 1, force sensor is connected with servomotor. So, measurement condition is controlled: 10 mms^{-1}. Measured force is resulted in table 1. From table 1, the objective range of repulsive force is between 0.6 N and 1.3 N.

Figure 1. Measurement equipment for repulsive force.

Table 1. Repulsive force of human body.

Body Part	Repulsive Force [N]
Hand	0.6~0.9
Neck	0.6~1.2
Abdomen	0.4~1.3
Thigh	0.7~1.2
Back	1.0~1.3

3. MR tactile device

In order to realize the tactile recognition, the MR tactile device which consists of MR fluid and several pins is devised. MR fluid which can change its rheological properties by the external magnetic field is utilized to mimic a wide variety of tactile sensations. MR fluid is assumed to behaves as Bingham model given by

$$\begin{aligned} \tau &= \eta\dot{\gamma} + \tau_y(H) \\ \tau_y(H) &= \alpha H^{\beta} \end{aligned} \tag{1}$$

where τ is the shear stress, η is the dynamic viscosity, $\dot{\gamma}$ is the shear rate, and $\tau_y(H)$ is the dynamic yield stress of the MR fluid. Dynamic yield stress of MR fluid is a function of the magnetic field, H, and increases with respect to magnetic field. The parameters α and β are characteristic values of the MR fluid. The measured yield stress of the employed MR fluid (MRF-132DG, LORD Corp.) is $0.0922H^{1.236}$ kPa and shown in Figure 2. The schematic configuration of tactile device is shown in Figure 3. MR fluid is inserted into diaphragm. MR diaphragm is supported by several pins. The pin consists of washer, nut and coil. The washer and nut is utilized for sealing of MR fluid. The bottom of MR diaphragm is connected to coil. By applying input current to coil, rheological properties of MR fluid is changed. Using this smart material, we can control the tactile recognition of surface of MR diaphragm. With this configuration, several sections of tactile device can realize diverse tactile sensations at same times.

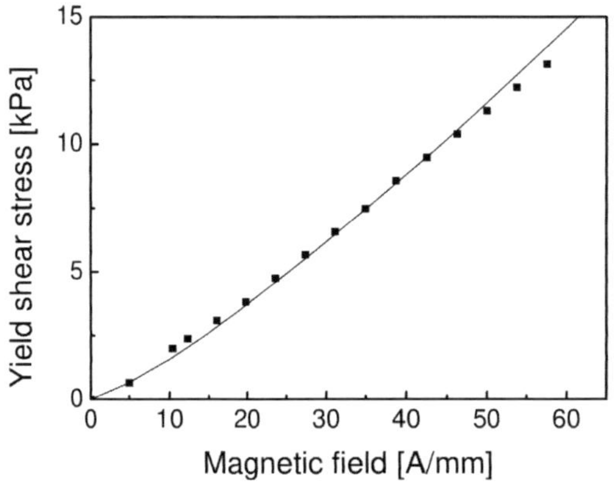

Figure 2. Yield stress of MR fluid.

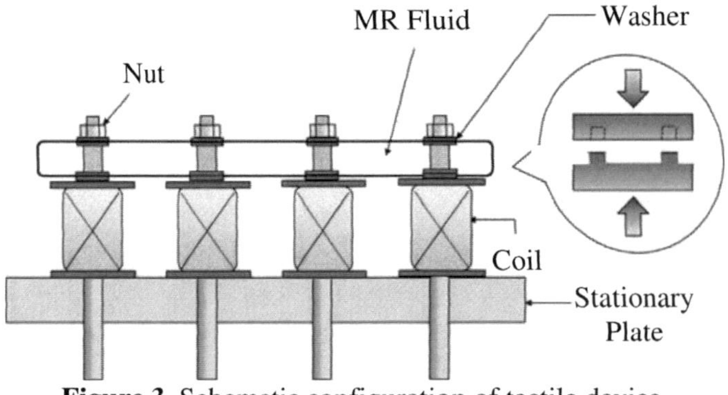

Figure 3. Schematic configuration of tactile device.

However, because magnetic field is developed at the bottom of the MR diaphragm, it is very difficult to dynamic modelling of the tactile device and design process of magnetic circuit. So, in order to get required magnetic flux density, experimental method is implemented as shown in figure 4. When permanent magnet is added at the bottom of the diaphragm, the repulsive force and magnetic flux density is measured. In order to generate 1.5 N, required magnetic flux density is 12 mTesla. Also, the number of coil turns is 400 and maximum input current is assumed as 3 A.

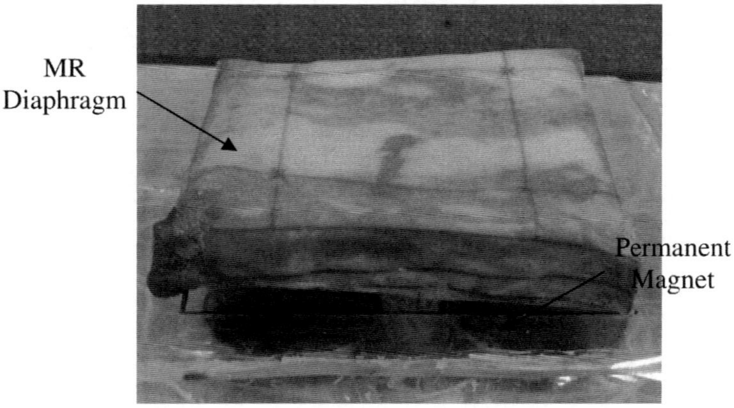

Figure 4. MR diaphragm with permanent magnet.

Next step is to confirm the performance of magnetic circuit. FEM analysis for magnetic circuit is conducted as shown in Figure 5. From Figure 5, it is confirmed that required magnetic flux density is successfully generated. The maximum magnetic flux density is 14 mTesla. As first phase of research, the tactile device has been manufactured with 4 pins and MR diaphragm. As second phase of research, the tactile device will be researched with 16 pins; then, it can represent more diverse tactile recognition.

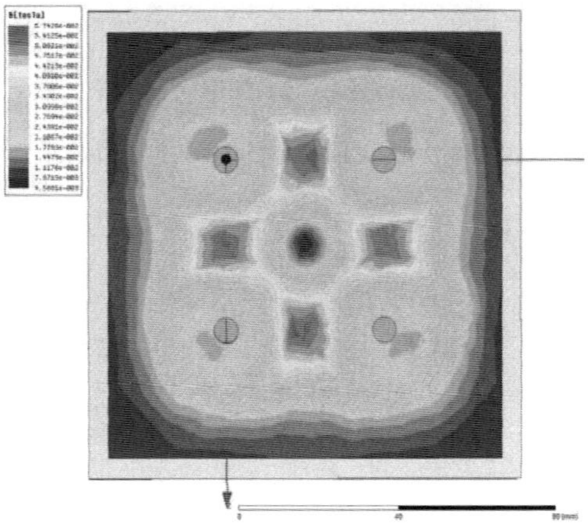

Figure 5. Magnetic analysis results of tactile device.

4. Performance evaluation of MR tactile device

In order to evaluate the performance of tactile device, an experimental setup is established as shown in Figure 6. Because input current is applied to coil, input case is four. So, repulsive force is measured at 2 points (Area1 & Area 5) as shown in Figure 6; Area 1, 2, 3 and 4 are geometrically identical. Figures 7 - 10 present the experiment results of tactile device. From Figure 7, maximum measured force is

1.43 N. So, this result is bigger than repulsive force of human body. But, other results are some smaller than repulsive force of human body.

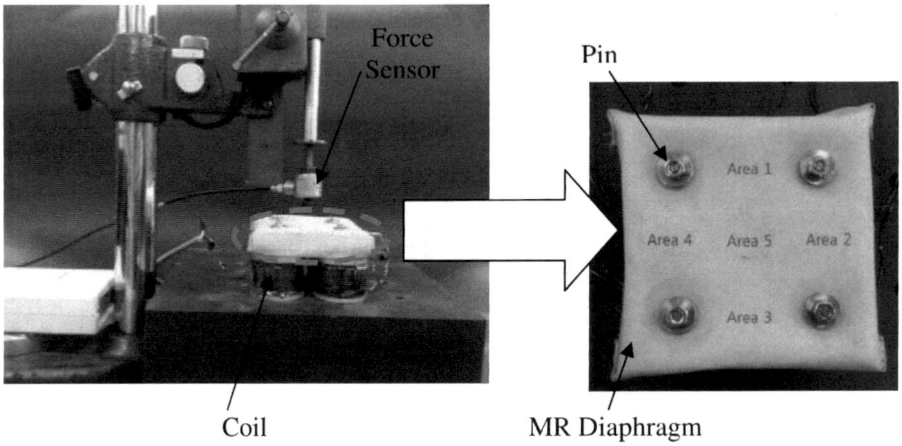

Figure 6. Experimental apparatus for MR tactile device.

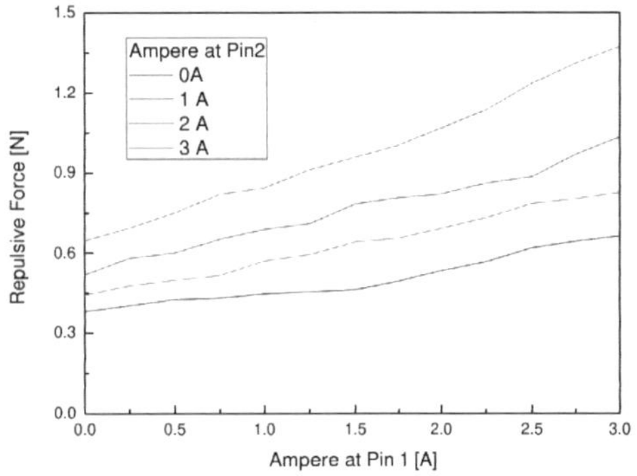

Figure 7. Experiment results at Area 1 (Pin 1&Pin 2).

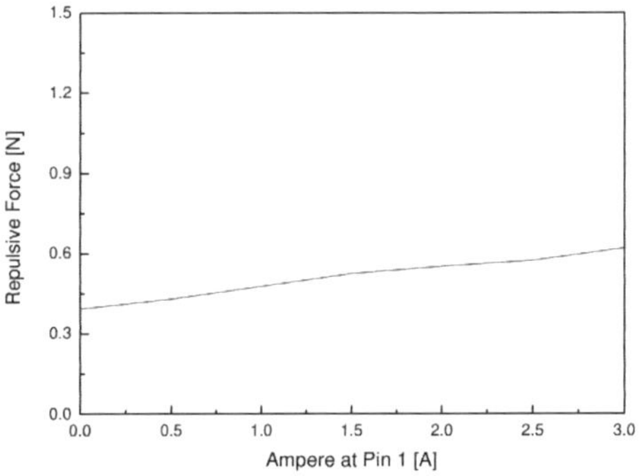

Figure 8. Experiment results at Area 5 (Pin 1).

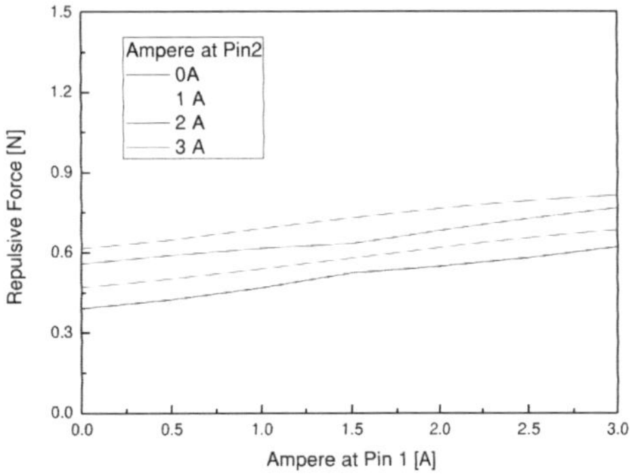

Figure 9. Experiment results at Area 5 (Pin1&Pin 2).

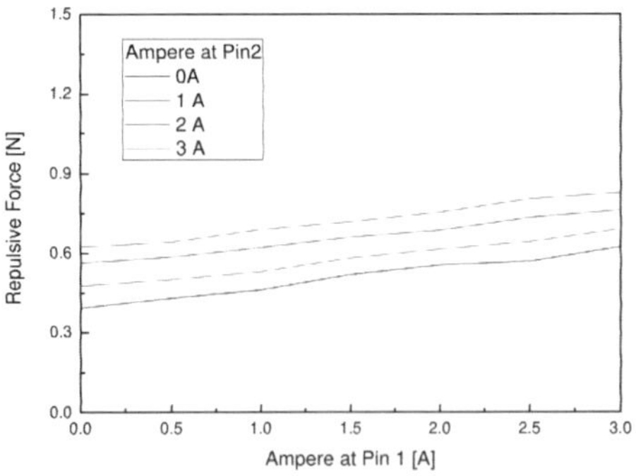

Figure 10. Experiment results at Area 5 (Pin1&Pin 3).

In order to apply experiment results to real applications, the relationship of repulsive force is experimentally obtained as follows:

$$F_1 = 0.4 + 0.04 \cdot I_1 + 0.015 \cdot I_1^2 + 0.024 \cdot I_2 + 0.046 \cdot I_1 \cdot I_2 + 0.022 \cdot I_2^2 \quad (Area\ 1,2,3,4)$$
$$F_2 = 0.4 + 0.09 \cdot I_1 + 0.057 \cdot I_2 + 0.012 \cdot I_1 \cdot I_2 + 0.058 \cdot I_3 - 0.01 \cdot I_1 \cdot I_3 + 0.059 \cdot I_4 - 0.012 \cdot I_1 \cdot I_4 \quad (Area\ 5) \tag{2}$$

where I_1, I_2, I_3 and I_4 are applied current of each pins.

5. Conclusion

In this study, a new type of tactile device was proposed using MR fluid and performance of the proposed tactile device was evaluated. The proposed tactile device consists of diaphragm which filled by MR fluid and several pins. Using MR fluid, tactile device can mimic a wide variety of tactile sensations. Thus, the proposed tactile device can provide repulsive information of the organ to surgeon. In order to determine design parameters of magnetic circuit, finite elements analysis of magnetic field has been conducted and tactile device has been manufactured. Repulsive force experiments have been conducted to evaluate performance on the tactile rendering of the proposed device. From these results, the practical feasibility of the proposed tactile device has been investigated. The proposed tactile device is quite satisfactory for the application of MIS robot system.

Acknowledgments

This work was supported by a National Research Foundation of Korea (NRF) grant funded by the Korea government (MEST) (No. 2010-0015090). This financial support is gratefully acknowledged.

References

[1] Royal National Institute of the Blind. Available from: http://www.rnib.org.uk

[2] Hayward V and Cruz-Hernandez JM 2000 Tactile display device using distributed lateral skin stretch *Proc. Haptic Interfaces for Virtual Environment and Teleoperator Systems Symposium, ASME, IMECE-2000* **DSC-69-2** pp 1309–14

[3] Taylor PM, Moser A and Creed A 1998 A sixty-four element tactile display using shape memory alloy wires *Displays* **18** (3) 163-8

[4] Maucher T, Meier K and Schemmel J 2001 An interactive tactile graphics display *6th Int. Symp. on Signal Proc. and its Applications* **2001** (1) 190-3.

[5] Konyo M, Tadokoro S and Takamori T 2000 Artificial tactile feel display using soft gel actuators. 2000 *Proc. ICRA '00 IEEE Int. Conf. on Robotics and Automation* **2000** (4) 3416-21.

[6] Darden MA and Schwartz CJ 2009 Investigation of skin tribology and its effects on the tactile attributes of polymer fabrics *Wear* **267** (5)1289-94

[7] Gemperle F, Ota N and Siewiorek D 2001 Design of a wearable tactile display *Proc. 5th Int. Symp. on Wearable Computers* **2001** 5-12.

[8] Carlson JD and Jolly MR 2000 MR fluid, foam and elastomer devices *Mechatronics : mechanics, electronics, control* **10** (4) 555-69

[9] Lee CH and Jang MG 2011 Virtual surface characteristics of a tactile display using magneto-rheological fluids *Sensors* **11** (3) 2845-56

[10] Scilingo EP, Sgambelluri N, De Rossi D and Bicchi A 2003 Towards a haptic black box for free-hand softness and shape exploration *Proc. ICRA '03 IEEE Inter. Conf. on Robotics and Automation* **2003** 2412-7.

[11] Liu Y, Davidson RI, Taylor PM, Ngu JD and Zarraga JMC 2005 Single cell magnetorheological fluid based tactile display *Displays* **26** (1) 29-35

Development of 1-DOF manipulator with variable rheological joint for instantaneous force

T Majima[1], S Nagai, H Tomori and T Nakamura
Chuo University, Biomechatronics Laboratory, 1-13-27 Kasuga, Bunkyo-ku, Tokyo, 112-8551, Japan

E-mail: t_majima@bio.mech.chuo-u.ac.jp

Abstract. Highly rigid actuators such as a geared motor or hydraulic actuator are widely used in industrial robots. To obtain high-speed motion, actuators need to increase the actuator output. However, to increase high-rigidity actuators output, it is necessary to make actuators larger. In contrast, humans perform motions with instantaneous force such as jumping or throwing by using muscles. These instantaneous forces are realized by accumulating potential energy to the muscles and the muscles releasing the energy in a short time. Therefore, in this study a 1-DOF manipulator with variable rheological joint for instantaneous force using an artificial muscle and a magnetorheological (MR) brake was developed. In this paper, the method of generating instantaneous force for this manipulator was proposed. Further, the experiment of the proposed method was also conducted. As a result, generating instantaneous force by proposed method was realized.

1. Introduction

High-rigidity actuators such as motors and hydraulic actuators are widely used in industrial robots. To obtain a high-speed motion, actuators need to increase their own output. However, to increase high-rigidity actuators output, it is necessary to make actuators larger. Therefore, it has problem that the robot weight increases as actuators output increases. In contrast, humans perform motions such as jumping or throwing with instantaneous force by using muscles. To obtain instantaneous force, muscles need to increase their own output than usual. These instantaneous forces are realized by accumulating potential energy to the muscles and the muscles releasing the energy in a short time. Since, muscles have high compliance and can change viscosity at high-speed, they can generate such instantaneous force. Therefore, actuators for generating instantaneous force need a high compliance and a high-speed rheological change, which is difficult to achieve with high-rigidity actuators such as motors and hydraulic actuators.

Kovac *et al.* [1] developed a miniature jumping robot using a motor and spring. However, the elastic coefficient of the actuator used by Kovac *et al.* was constant, and the instantaneous force and position could not be controlled independently. Although Niiyama *et al.* [2] developed a jumping robot using the pneumatic artificial muscle, the jumping operation of this robot is designed by trial and error, and the air pressure control capabilities are not adequate. Furthermore, the reported studies do not allow for viscosity control and control of the motion generated by the instantaneous force is not sufficiently developed.

[1] Corresponding author.

Therefore, in this study it was focused on a pneumatic artificial muscle with a high compliance and a MR brake using MR fluid that can change apparent viscosity at high-speed. Applying these features to the joint of the robot, the robot can generate instantaneous force in the same way as human. Then, the 1-DOF manipulator using artificial muscle and MR brake for instantaneous force was developed. Furthermore, since this manipulator can control the viscosity and elasticity independently, it was presumed that this manipulator can control instantaneous force and position independently. Here, a method of generating instantaneous force for the manipulator using the artificial muscle and a MR brake were described. Also an experiment in which instantaneous force generated was described.

2. Straight-fiber-type artificial muscle

Figure 1 shows the diagrammatic illustration of the Straight-fiber-type artificial muscle [3] used in this study. The artificial muscle has a tubular shape and is made of natural latex rubber. The muscle structure includes a carbon fiber layer in the direction of length. Consequently, when air pressure is supplied to the muscle, it will only expand in the radial direction and contract in the direction of length. The contractile force is then used as an actuator. The straight-fiber-type artificial muscle has a high contraction percentage, high contractile force, and a long lifespan compared to the McKibben-type artificial muscle [4]. Straight-fiber-type artificial muscle has non-linear characteristic and variable elasticity.

3. MR brake

In this study, a MRB-2107-3 of the LORD Co. as an MR brake device is used. Figure 2 shows a schematic diagram of the MR brake device. Table 1 shows the specification of the MR brake. This MR brake consisted of a coil, a rotor, and MR fluid. The MR fluid, which is applied to the magnetic field from a coil, produces friction between the rotors. This friction force acts as the brake torque. Applying the MR brake to a manipulator realized the variable rheological joint. Since, MR brake is friction brake, MR brake can hold the rotor not to be rotated. Therefore, it was presumed that the artificial muscle can accumulate potential energy by applying MR brake to a manipulator. Further, the artificial muscle can release the energy quickly because MR brake has a high-speed response.

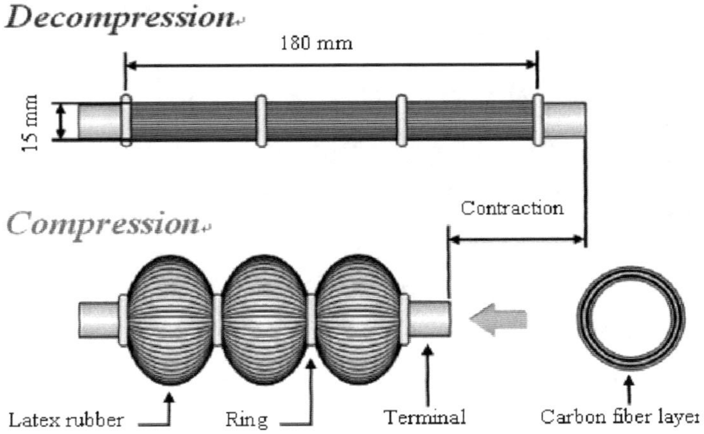

Figure 1. Schematic diagram of straight-fiber-type artificial muscle.

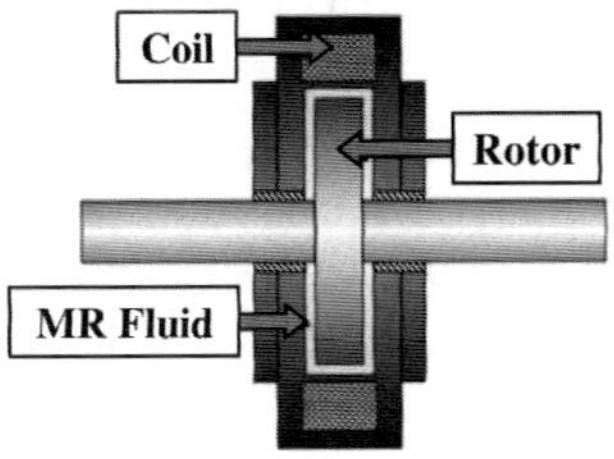

Figure 2. Schematic diagram of the MR brake.

Table 1. Specification of the MR brake

Parameter name	Value
Diameter	920 mm
Length	360 mm
Weight	1.4 kg
Maximum torque	5.65 Nm
Minimum torque	<0.34 Nm
Max current value	1 A
Temperature range	-28.9~71.1

4. 1-DOF manipulator with variable rheological joint

Figure 3 shows the overall view of the developed one-degree-of-freedom (1-DOF) artificial muscle manipulator with variable rheological joint. Table 2 shows the specification of the manipulator. Two artificial muscles are arranged in parallel. The system transmits the contractile force of the artificial muscle to the rotation axis through the belt pulley. Drive torque generated by contractile force of the artificial muscle is very high compared with brake torque of MR brake. Therefore, it cannot be hold the joint rotation only to apply the MR brake directly. For that reason, the MR brake is fixed to the first joint through the gears. As a result, it is possible to apply the brake to the rotation axis and it can be accumulated more potential energy in the artificial muscle. A load cell and a pressure sensor are attached to each artificial muscle. Further, encoder is attached to the joint.

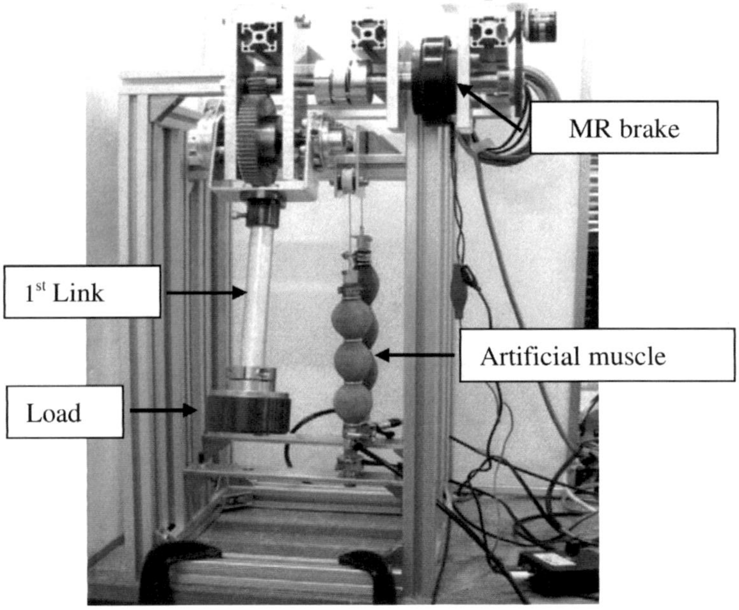

Figure 3. 1-DOF manipulator.

Table 2. Specification of the manipulator.

Parameter name	Value
Mass of 1st link	0.8 kg
Length of 1st link	270 mm
Center of gravity of the 1st link	140 mm
Movable angle	0~120 deg

5. Method for generating instantaneous force

The following steps are proposed as a method for generating instantaneous force in the manipulator. Figure 4 shows overview of the proposed method. (1) Hold the joint rotation with the MR brake, (2) Accumulate potential energy in the artificial muscle by applying initial air pressure and (3) Generate instantaneous force by releasing the arm. This is called the MR method.

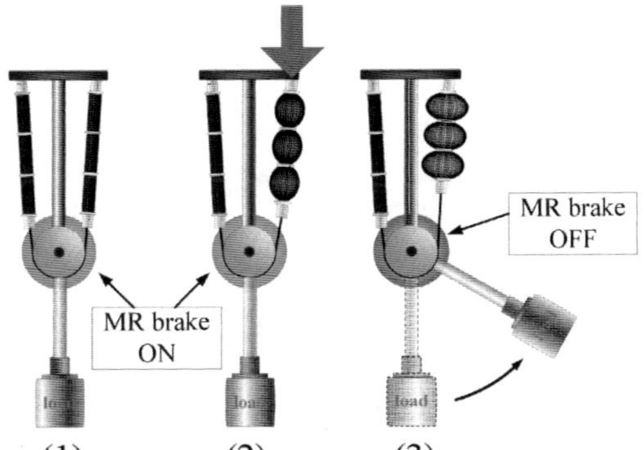

Figure 4. Overview of the MR method.

6. Relationship between volume and pressure of the artificial muscle

In the MR method, No initial pressure is added in the muscle during the manipulator drive. For this reason, the steady-state position of the arm depends on the initial pressure of the artificial muscle. Therefore, it is necessary to calculate the initial pressure to raise the load to the desired arm angle. However, this pressure changes with the volume change and contractile force of the muscle. Thus, experimental relationships between the volume and pressure of the artificial muscle were obtained. In the experiment, the initial pressure is applied to the artificial muscle, and the pressure and contraction are recorded. Figure 5 shows the experimental result. Figure 5 shows that pressure is proportional to volume. This relationship does not depend on the initial pressure. The relationship can be expressed approximately as follows:

$$P_0 = -0.0014\left(V_0 - V_d\right) + P_d \tag{1}$$

Here, P_0 is the initial pressure, P_d is the pressure at the desired angle, V_0 is the volume at the initial angle, and V_d is the volume at the desired angle.

7. Instantaneous force generation experiment

In the instantaneous force generation experiment, the initial pressure was calculated by using equation (1) with the desired arm angle of 60 degrees. Further, we defined instantaneous force as power.
This power is calculated as follows:

$$W = F \cdot r \cdot \theta \cdot s^{-1} \tag{2}$$

Here, W is power of the artificial muscle, F is contractile force of the artificial muscle, r is the radius of belt pulley, θ is the angle of the joint, and s is the sampling time.

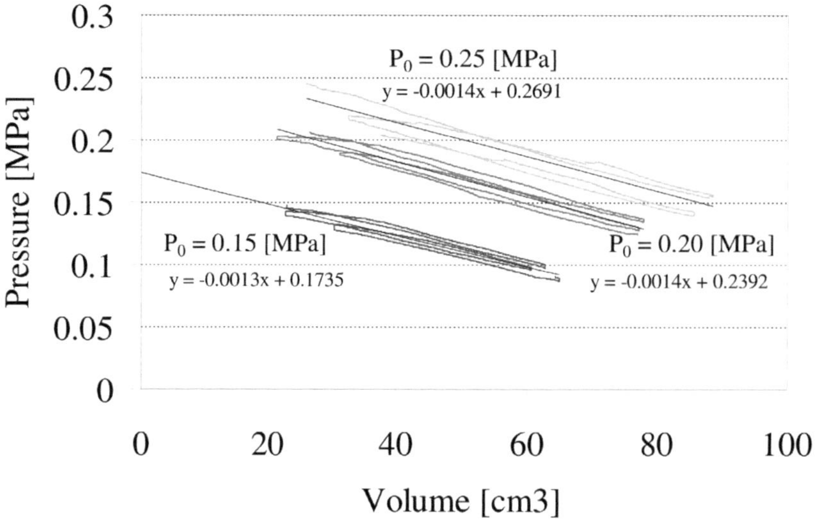

Figure 5. Relationship between volume and pressure of the artificial muscle.

Figure 6 shows the result of response of angle, while Figure 7 shows that of power. Table 3 shows dead and rise times and maximum power of each response. Step response is the response of the arm when air pressure is applied to the artificial muscle by step input. In the experimental result of using the MR method, the dead time decreased by 68%, and the rise time decreased by 11% compared with the step response. Further, maximum power increased by 280% compared with the step response. From that result, instantaneous force generation by MR method was realized. The steady-state position of the arm corresponded to the desired arm angle. Therefore, the initial pressure calculated by equation (1) is useful to control the arm driven by the MR method.

In this experiment, the response of using the MR method oscillated. Because, MR brake was not controlled after starting the manipulator driving. However, applying vibration control using a MR brake [5], vibration of the arm could be controlled easily.

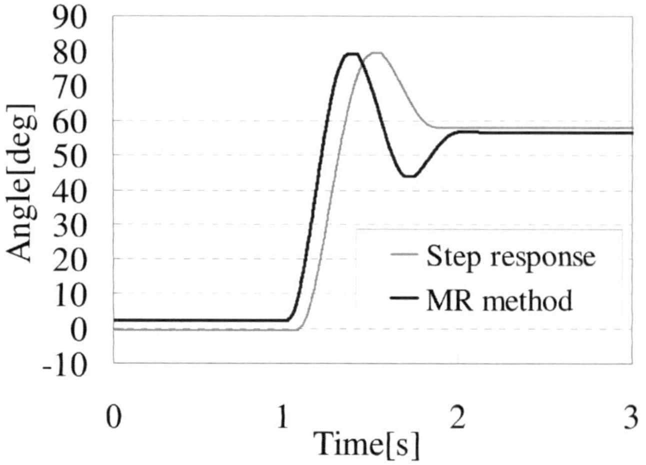

Figure 6. Experimental result of the angle.

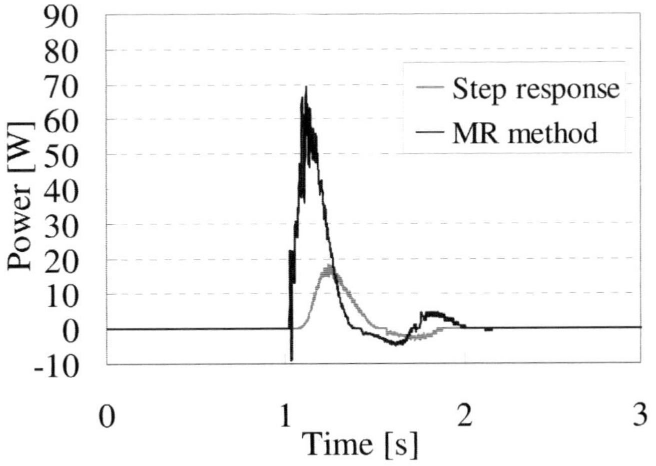

Figure 7. Experimental result of the power.

Table 3. Dead and rise time and maximum power.

	Dead time (s)	Rise time (s)	Maximum power (W)
Step response	0.081	0.176	18.5
MR method	0.026	0.156	69.4

8. Conclusion

In this research, a method of generating instantaneous force for a manipulator with a variable rheological joint was proposed. The proposed method realized the generation of instantaneous force. As a result, the dead and rise times decreased and maximum power significantly increased compared to the step response.

References

[1] Kovac M, Schlegel M, Zufferey J C and Floreano D 2009 A miniature jumping robot with self-recovery capabilities *Proc. IEEE/RSJ Int. Conf. Intelligent Robots and Systems* pp. 583-88

[2] Niiyama R, Nagakubo A and Kuniyoshi Y 2007 Mowgli: A bipedal jumping and landing robot with an artificial musculoskeletal system *Proc. IEEE Int. Conf. Robotics and Automation (ICRA)* pp. 2546-51

[3] Nakamura T and Shinohara H 2007 Position and force control based on mathematical models of pneumatic artificial muscles reinforced by straight glass fibers *Proc. IEEE Int. Conf. Robotics and Automation (ICRA)* pp. 4361-66

[4] Klute G K, Czernieki J M and Hannaford B 1999 McKibben artificial muscles: Pneumatic actuators with biomechanical intelligence *Proc. IEEE/ASME Int. Conf. Advanced Intelligent Mechatronics* pp. 221–26

[5] Nakamura T, Midorikawa Y and Tomori H 2011 Position and vibration control of variable rheological joint using artificial muscles and magneto-rheological brake *Int. J. Hum. Robot.* **8** 205-22

Journal of Physics: Conference Series **412** (2013) 012049 doi:10.1088/1742-6596/412/1/012049

Experimental study on damping characteristics of the tuned liquid column damper with magnetic fluid

H Masuda, T Oyamada and T Sawada[1]

Department of Mechanical Engineering, Keio University, 3-14-1 Hiyoshi, Kohoku-ku, Yokohama 223-8522, Japan

E-mail: `sawada@mech.keio.ac.jp`

Abstract. A magnetic fluid is a smart, colloidal, and strongly magnetizable fluid, and is composed of a base liquid, ferromagnetic particles, and a chemically adsorbed surfactant. A tuned liquid column damper (TLCD) — devised for vibration mitigation of high structures — is a passive damper categorized as a dynamic absorber. Using a magnetic fluid as the working fluid of a TLCD can improve and change the TLCD into a novel semiactive damper. The natural frequency of a TLCD with a magnetic fluid can be changed via a magnetic field. This novel damper can be used as an effective vibration suppression mechanism for building structures. In this study, we investigate the effect of the magnetic field on the natural frequency of a TLCD with a magnetic fluid. In addition, using the experimental results, we demonstrate the relationship between the setting positions of an electromagnet, the strength of the magnetic field, and the changes in the natural frequency.

1. Introduction

The recent development in lightweight high-strength materials and associated construction techniques permits the building of high-rise structures. These structures are very sensitive to wind excitations and long-period seismic waves because of their small structural damping and low natural frequencies. Therefore, low-frequency vibration in high-rise structures is a critical design issue, and the development of an effective vibration suppression mechanism is often necessary. A tuned liquid column damper (TLCD) can mitigate vibrations in high-rise structures and is categorized as a passive damper. Since its introduction by Sakai et al. [1] in 1989, a TLCD has been investigated by many researchers for setting structures [2-4]. It can suppress wind excitations and earthquake-induced vibrations in structures such as tall buildings by absorbing the energy via the motion of the liquid mass in a tube-like container. A TLCD is a reliable and low-cost application. In addition, it yields a simple analyzable motion because it follows a simple mechanism. It consists of a container partially filled with a liquid (water in most cases).

The natural frequency of a TLCD is significant because the oscillation can only be suppressed in a narrow range around its natural frequency. Therefore, it is necessary to control the damping characteristics of a TLCD by using a simple additional mechanism. The semiactive damper is a controllable passive damper. Several researchers have proposed various semiactive TLCDs including a semiactive TLCD with an active orifice, where damping can be controlled by the

[1] Corresponding author.

Published under licence by IOP Publishing Ltd

orifice opening [5,6], and a magneto-rheological tuned liquid column damper (MR-TLCD), which was designed by Wang et al. [7]. An MR-TLCD consists of a magneto-rheological fluid and an orifice, where damping can be controlled by a magnetic field.

In this study, we design a novel semiactive damper using a magnetic fluid as the working fluid. This damper is called the tuned liquid column damper with a magnetic fluid (MF-TLCD). A magnetic fluid is a smart fluid that can be strongly magnetized in the presence of a magnetic field. In addition, it is attracted to magnets strongly depending on the strength of the magnetic field. Because of this attribute, the damping characteristics of an MF-TLCD, especially its natural frequency, can be controlled by the magnetic field. We conduct experiments to analyze the effect of the magnetic field on the natural frequency of the MF-TLCD. We also discuss the relationships among the setting positions of an electromagnet, the strength of the magnetic field, and the changes in the natural frequency.

2. Theoretical approach

2.1. Magnetic fluid column oscillation in a U-pipe in the presence of a magnetic field

Let us consider the magnetic fluid motion in a U-pipe with an electromagnet, as shown in figure 1. Assuming a magnetic fluid is inviscid and incompressible, the equation of motion of a liquid column oscillation in the presence of a magnetic field is

$$\rho A l \ddot{\eta} + 2\rho A g \eta + F_{mag} = 0 \tag{1}$$

where ρ is the density of the magnetic fluid, A is the cross sectional area of the U-pipe, g is the gravitational acceleration, l is the length of the liquid column, η is the displacement of the free surface of the magnetic fluid from the rest position, and F_{mag} is the force generated by the magnetic field. In the absence of the magnetic field, i.e., $F_{mag} = 0$, the natural frequency of this liquid column vibration ω_f is

$$\omega_f = \left(\frac{2g}{l}\right)^{1/2} \tag{2}$$

In the presence of a magnetic field, the magnetic fluid is attracted to the magnet, and the additional pressure p_{mag} appears as follows:

$$p_{mag} = \mu_0 \int_0^{H(s)} M dH' \tag{3}$$

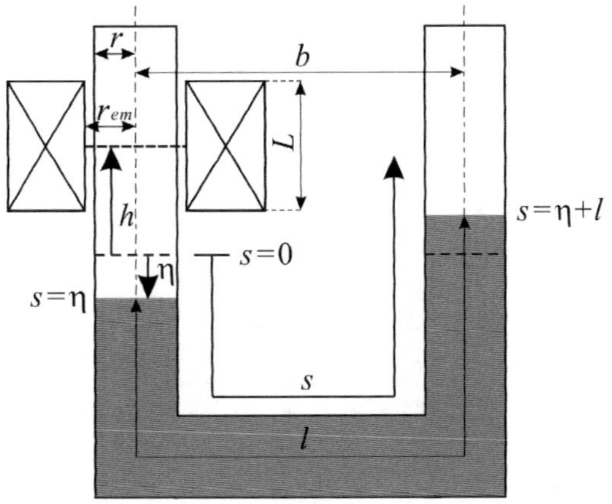

Figure 1. One degree of freedom model of magnetic fluid column oscillation.

where μ_0 is the magnetic permeability in vacuum, M is the magnetization, and H' is the magnetic field, and $H(s)$ is the magnetic field intensity where the position is s. In weak magnetic fields, we can assume that $M = \chi H$, using the susceptibility χ and equation (3) is written as

$$\begin{aligned}\mu_0 \int_0^H M dH' &= \mu_0 \chi \int_0^H H' dH' \\ &= \frac{1}{2}\mu_0 \chi H^2 \end{aligned} \tag{4}$$

Considering a coordinate system s as shown in figure1, the magnetic force dF_{mag} acting on a small line element ds can be written as

$$\begin{aligned} dF_{mag} = A dp_{mag} &= A\mu_0 \int_{H(s)}^{H(s+ds)} M dH' \\ &= \frac{1}{2} A \mu_0 \chi \left[\{H(s+ds)\}^2 - \{H(s)\}^2 \right] \\ &= \frac{1}{2} A \mu_0 \chi \left\{ \frac{d}{ds} \left(H^2 \right) \right\} ds \end{aligned} \tag{5}$$

Since the electromagnet placed on one side of the U-pipe, the magnetic field effects on only one side column part ($h \leq s \leq (l-b)/2$). Integrating equation (5) yields

$$\begin{aligned} F_{mag} &= \frac{1}{2} A \mu_0 \chi \int_{\eta}^{(l-b)/2} \left\{ \frac{d}{ds} \left(H^2 \right) \right\} ds \\ &= \frac{1}{2} A \mu_0 \chi H^2 \Big|_{\eta}^{(l-b)/2} \end{aligned} \tag{6}$$

where b is the distance between two fluid columns. Equation (6) indicates the summation of the magnetic force acting on the magnetic fluid column. According to equations (1) and (6), the

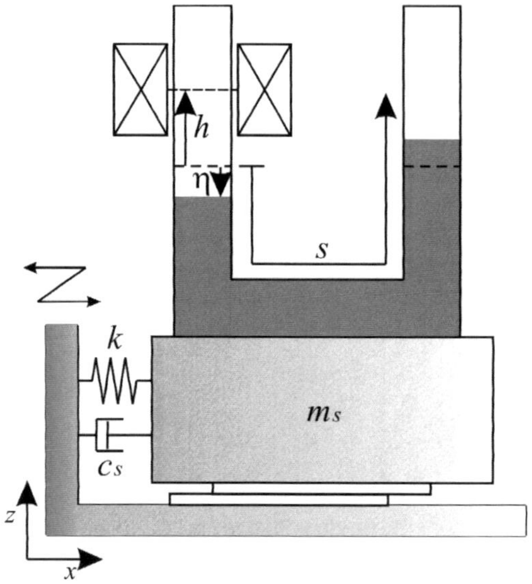

Figure 2. Two degrees of freedom model of the structure with MF-TLCD.

equation of motion of the magnetic fluid column oscillation in the U-pipe in the presence of a magnetic field is

$$\rho Al\ddot{\eta} + 2\rho Ag\eta + \frac{1}{2}A\mu_0\chi\left[\{H(\eta)\}^2 - \left\{H\left(\frac{l-b}{2}\right)\right\}^2\right] = 0 \tag{7}$$

The magnetic field distribution in a solenoidal coil is given by

$$H(\eta) = \frac{NI}{2L}\left[\frac{\eta + h + L/2}{\left\{(\eta + h + L/2)^2 + r_{em}^2\right\}^{1/2}} - \frac{\eta + h - L/2}{\left\{(\eta + h - L/2)^2 + r_{em}^2\right\}^{1/2}}\right] \tag{8}$$

where N is the number of coil turns, I is the supplied electric current, L is the length of the solenoid coil, r_{em} is the inner radius of the solenoid coil, and h is the displacement of the center of the solenoid coil from the free surface of the magnetic fluid.

2.2. Motion of the structure with MF-TLCD

Figure 2 shows two degrees of freedom model of the structure with an MF-TLCD. From equation (7), the equation of motion of the structure with an MF-TLCD is given by

$$\begin{bmatrix} \rho Al & \rho Ab \\ \rho Ab & \rho Al + m_s \end{bmatrix}\begin{bmatrix} \ddot{\eta} \\ \ddot{x} \end{bmatrix} + \begin{bmatrix} c_f & 0 \\ 0 & c_s \end{bmatrix}\begin{bmatrix} \dot{\eta} \\ \dot{x} \end{bmatrix} + \begin{bmatrix} 2\rho Ag & 0 \\ 0 & k \end{bmatrix}\begin{bmatrix} \eta \\ x \end{bmatrix} + \begin{bmatrix} A\mu_0\chi/2 & 0 \\ 0 & 0 \end{bmatrix}\begin{bmatrix} \{H(\eta)\}^2 - \{H((l-b)/2)\}^2 \\ 0 \end{bmatrix} = 0 \tag{9}$$

where x is the displacement of the structure, m_s is the mass of the structure, c_f is the viscous damping coefficient of fluid motion, c_s is the viscous damping coefficient of the structure, and k is the spring constant. In equation (9), we consider that the viscous damping of liquid column vibration works as damping of vibration.

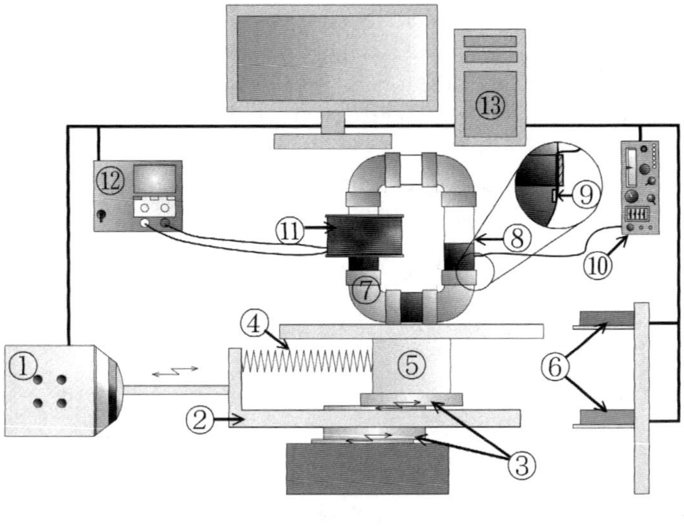

①Actuator ②Vibrator base
③Slider ④Spring ⑤Structure
⑥Laser Sensor ⑦Magnetic fluid
⑧U-pipe ⑨Pressure transducer
⑩Strain amplifier ⑪Electromagnet
⑫DC power supply ⑬PC

Figure 3. Experimental apparatus.

Table 1. Characteristics of magnetic fluid.

Density (25°C)	1350 kg/m^3
Viscosity (25°C)	5.34 mPa·s
Saturation magnetization	48 ± 5 kA/m

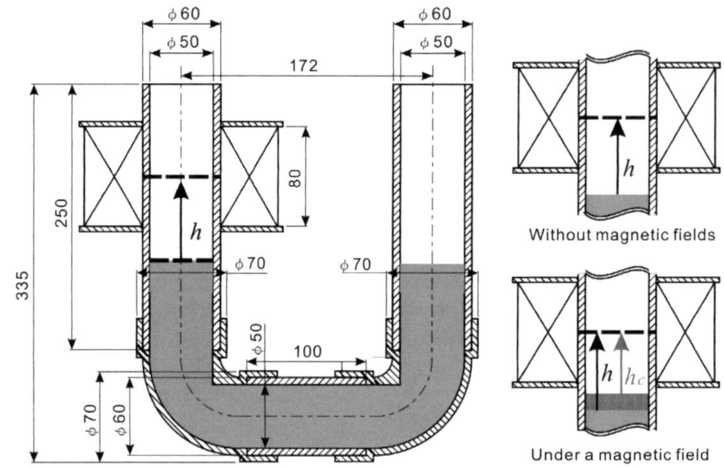

Figure 4. Detailed cross section of the U-pipe and the electromagnet.

3. Experiment

3.1. Experimental setup

Figure 3 illustrates a schematic diagram of the experimental apparatus. The magnetic fluid is placed in the U-pipe container. A detailed cross section of the U-pipe is shown in figure 4. By applying a virtual instrument (VI) LabVIEW, the frequency of the actuator can be controlled. The sine wave vibration is provided by the actuator and is transferred to the structure through the spring and the vibration base between the two smooth slider mechanisms. The vibration amplitude is 0.4 mm. Two laser sensors were used to measure the displacements of the structure and the vibration table, as well as the phases of their vibrations. The pressure of the magnetic fluid in the U-pipe was measured using a pressure transducer. Instead of the displacement of the magnetic fluid's free surface, we monitored the frequency response of the time-dependent pressure. We confirmed that the pressure change corresponds to the free-surface displacement. We used a kerosene-based magnetic fluid, whose characteristics are shown in table 1.

3.2. Arrangement of an electromagnet

An electromagnet was mounted on one side of the U-pipe as shown in figure 4. When the magnetic field was applied, the magnetic fluid was attracted to the electromagnet, and the position of the still free surface changed. As a result, a revised parameter h_c was introduced as shown in figure 4.

3.3. Magnetic field distributions

We measured the spatial distributions of both r and z components of the magnetic field by using a Gauss meter. The intensity of the magnetic field was calculated by using these experimental

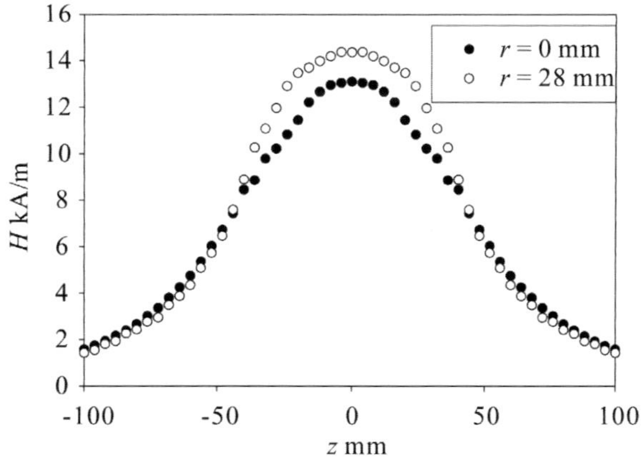

Figure 5. Axial distributions of magnetic field intensity.

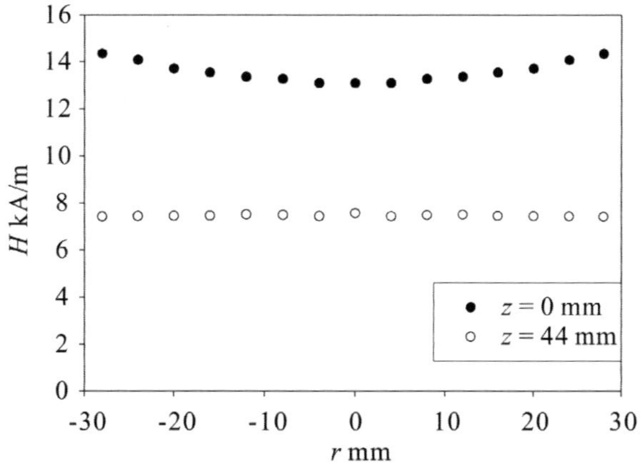

Figure 6. Radial distributions of magnetic field intensity.

magnetic field components. Figures 5 and 6 show the axial and radial distributions of the magnetic field intensity for a 1.0 A electric current, respectively. The magnetic field intensity reaches a maximum value at z = 0 mm, and decreases as $|z|$ increases. Thus, figures 6 and 7 indicate that the magnetic field intensity depends on z rather than r.

4. Results and discussions

Figure 7 shows the frequency responses of the the structure with an MF-TLCD; a_s is the amplitude of the structure; f and a_0 are the excitation frequency and amplitude of the actuator, respectively. The frequency response varies for according to different values of H_0, which is defined as the reference magnetic field intensity obtained at the center point (z = 0 mm, r = 0 mm) of the electromagnet.

Figure 8 shows the relationship between the natural frequency f_n and the magnetic field intensity. These frequency responses correspond to the time-dependent pressure waves obtained from the pressure transducer. Here f_0 is the natural frequency in the absence of a magnetic field. The natural frequency is defined as the frequency where the frequency response is minimal as shown in figure 7. When h is smaller than 0 mm, the natural frequency increases with H_0. In contrast, when h is larger than 0 mm, the natural frequency decreases in the presence of weak

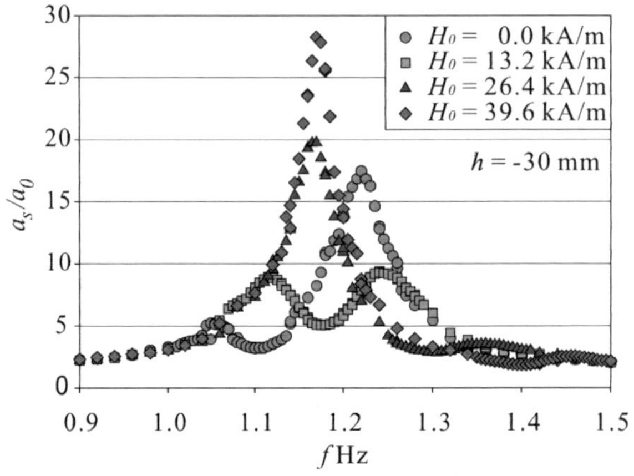

Figure 7. Frequency responses of dimensionless structure amplitude.

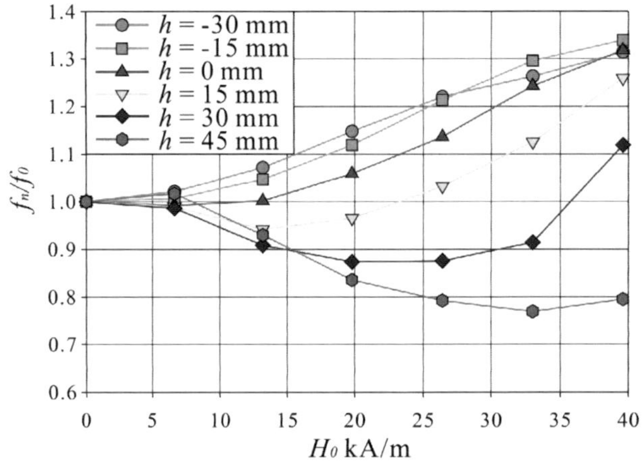

Figure 8. Natural frequency with H_0.

magnetic fields and increases in the presence of strong magnetic fields. Thus, the position of the electromagnet plays an important role in the transition of the natural frequency.

Figure 9 shows the distributions of the natural frequency with h for various values of H_0. The experimental results are compared with those obtained from numerical simulation. The symbol and full line of the same color indicate experimental and numerical results, respectively. We numerically solved equation (7) using the 4th–order Runge–Kutta method. Comparing these results, the experimental and numerical results are found to be qualitatively coincident. Therefore, it is verified that equation (7) expresses the motion of the magnetic fluid column oscillation in a U-pipe in the presence of a magnetic field.

The still magnetic fluid level changes when the magnetic field is applied and h varies according to the magnetic field. Subsequently, we replace h with the revised parameter h_c as shown in figure 4. Figure 10 shows the relation between natural frequency and h_c, and compares the experimental and numerical results. Figure 10 is the revised version of figure 9. When h_c is smaller than approximately -8 mm, all experimental values of f_n are larger than f_0. On the other hand, when h_c is larger than approximately -8 mm, all values of f_n are less than f_0. Thus, it is determined that h_c is a vibration characteristic of the MF-TLCD under in the presence of a magnetic field. However, the zero-cross point in the numerical results is larger than -8 mm.

Figure 11 shows the total magnetic attraction force acting on the magnetic fluid in the entire U-pipe in the presence of various magnetic fields. These results are obtained from the static experiment. We measured the equilibrium position of both magnetic fluid surface levels in the U-pipe with and without a magnetic field. A numerical simulation was carried out ignoring the first term of equation (7). The value of the magnetic force is maximum when h_c is -9 mm, where the natural frequency hardly changes. The electromagnet has a specific position. When the electromagnet is located in this specific position, the natural frequency does not change despite the applied magnetic fields. These experimental results indicate whether the change in the natural frequency is positive or negative, and consequently, estimate their approximate values.

Comparing experimental and numerical results, there exist some differences in h_c for the maximum peak points. The experimental h_c values are smaller than those obtained by numerical simulation. These differences increase with H_0. It is considered that these differences are

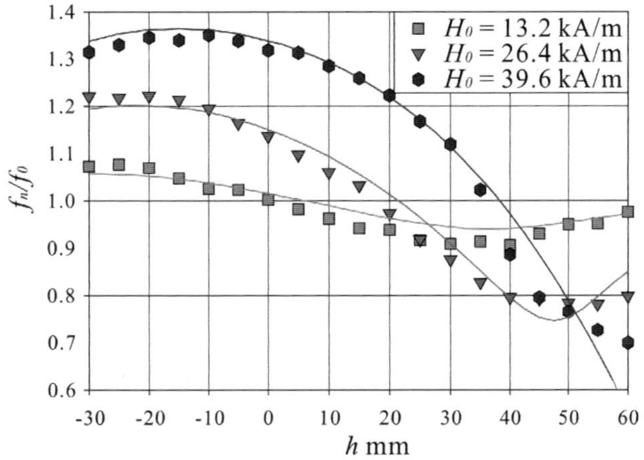

Figure 9. Distributions of the natural frequency with h for various values of H_0.

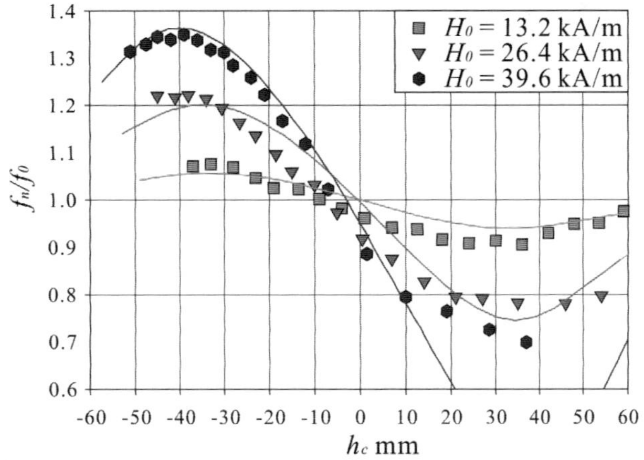

Figure 10. Distributions of the natural frequency with h_c for various values of H_0.

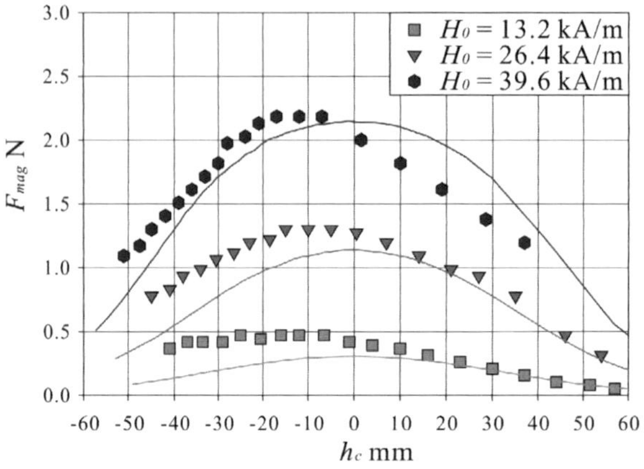

Figure 11. Summations of magnetic attraction forces acting on the magnetic fluid in the U-pipe.

caused by the curvature of the free surface of the magnetic fluid. In numerical simulation, we assumed that the surface of the magnetic fluid is flat in the U-pipe. However, in reality, this is not the case because of the radially distributed gradient of the magnetic field, which results in a surface curvature for the magnetic fluid. This causes h_c to be somewhat different from that in the numerical model, as shown in figure 4. Therefore, it is necessary to consider the radial distribution of the magnetic field to calculate the total magnetic attraction forces more accurately.

5. Concluding remarks

We devised a novel TLCD using a magnetic fluid. It can suppress structural vibrations and change its damping characteristics by changing the magnetic field. Our experimental results showed that the natural frequency was dependent on the position of the electromagnet. These results are in good agreement with those obtained numerically via the 4th order Runge–Kutta method. It is also demonstrated that the natural frequency changes can be estimated from the experimental results.

Acknowledgements

M. Masuda gratefully acknowledges partial financial support from the JSPS Institutional Program for Young Researcher Overseas Visits. This study was partially supported by a Grant-in-Aid for Scientific Research (C) from the Japan Society for the Promotion of Science.

References

[1] Sakai F, Takaeda S and Tamaki T 1989 *Proc. Int. Conf. on High Rise Buildings* Nanjin China 926-931
[2] Balendra T, Wang C N and Cheong N F 1995 *Eng. Structures* **17** 668-675
[3] Hocharainer M J 2005 *Acta Mechanica* **175** 57-76
[4] Wu J C, Shih M H, Lin Y Y and Shen Y C 2005 *Eng. Structure* **27** 1893-1950
[5] Li H M, Jin Q, Song G B and Wang G X 2005 *Adv. Natural Computation* **3612** 1089-91
[6] Yalla S K and Kareem A 2003 *J. Structural Eng.* **129** 960-971
[7] Wang J Y, Ni Y Q, Ko J M and Spencer B F 2005 *Computers and Structures* **83** 2023-34

Magnetorheological torque transmission devices with permanent magnets

H Böse[1], T Gerlach and J Ehrlich

Fraunhofer-Institut für Silicatforschung, Center Smart Materials, Neunerplatz 2, 97082 Würzburg, Germany

E-mail: boese@isc.fraunhofer.de

Abstract. A novel type of magnetorheological (MR) clutch whose magnetic circuit contains a combination of a permanent magnet and an electromagnet is described. Without the support of the electromagnet, the permanent magnet generates a magnetic field in the MR fluid shear gap which enables the MR clutch to transmit a torque without the supply of any electric energy. Hence, the operational states of this clutch are reversed with respect to the common MR clutches equipped with an electromagnet only. Three different MR clutches with hybrid magnetic circuits containing permanent magnet and electromagnet were designed, manufactured and tested. The three clutches differ in their number of mechanical parts which can rotate with respect to each other as well as in their size and weight and in their maximum transmittable torque. The largest MR clutch is capable to transmit torques up to nearly 800 Nm. The designs of the three novel MR clutches and the results of the mechanical tests upon variation of the coil current are presented in this paper.

1. Introduction

Magnetorheological (MR) fluids are suspensions of magnetically polarizable particles in a carrier liquid [1]. When a magnetic field is applied, the particles become magnetic dipoles and attract each other, which results in immediate stiffening of the material. This strong, fast and reversible MR effect known since its discovery by Rabinow more than six decades ago can be exploited for a large number of technical applications.

Most prominent applications of MR fluids have been found in the field of adaptive vibration damping. First automotive shock absorbers entered the market about ten years ago and the use of semi-active suspension systems has been extended to a multitude of vehicle models. Other applications of adaptive vibration damping concern driver seats in trucks and busses, seismic mitigation of buildings and the protection of cable-stayed bridges against wind excitation. A huge number of investigations have been performed on different types of MR dampers considering their design, the damping control and the MR fluids involved.

Less attention has been devoted to the use of MR fluids for controllable torque transmission so far. This lack is surprising, because variable torque transmission offers the opportunity to realize novel clutches and brakes where the torque is controlled by the magnetic field. Such clutches are capable to flexibly distribute the torque from an engine to several consumers. An example for flexible torque distribution is the control of side aggregates in a vehicle which are continuously driven by the

[1] Corresponding author.

combustion engine [2]. A further benefit of MR clutches is the possibility to limit the transmitted torque by the strength of the applied magnetic field. When the input torque reaches a preselected limit, the clutch slips and a higher output torque cannot be transmitted. This is especially worthwhile in applications in which safety issues have to be considered. Moreover, MR clutches bear the advantage of inherent damping in contrast to conventional clutches with hard disks.

Some work on MR clutches has already been published. The first MR clutch was reported by Rabinow who already has been aware of the strong benefit of MR fluids for controlled torque transmission [3]. A large variety of careful investigations was performed by Lampe who studied different clutch designs and basic issues of their performance [4, 5]. Kieburg et al. described a high torque clutch which was successfully tested in the powertrain of a vehicle [6]. A very small and compact MR clutch which can transmit a maximum torque of 6 Nm was developed by Kikuchi et al. [7].

The magnetic field in the MR clutch is usually generated by an electromagnet with a coil. The current in the coil determines the magnetic flux density in the MR fluid and the corresponding torque which is transmitted. This means that without any coil current the clutch works in the disengaged operational state, where only a drag torque due to the base viscosity of the MR fluid and the seal friction is transmitted. In the engaged state, the transmitted torque can only be maintained by the continuous supply of electric energy to the coil. If the clutch is primarily operated in the engaged state, this concept is energy-inefficient.

The reverse operation is possible with the use of permanent magnets in addition to the electromagnet. In such hybrid magnetic circuit with two different sources of the magnetic field the base field is generated by the permanent magnet. Correspondingly, the MR clutch can be operated in the engaged state without electric energy supply. To disengage the clutch, a counter-field is generated by the electromagnet, which weakens the field of the permanent magnet in the MR fluid gap and reduces the transmitted torque. This concept has already been demonstrated for MR dampers, in which a base damping force is generated by permanent magnets [8, 9].

Only few studies have been published regarding MR clutches with permanent magnets until now [10, 11]. Recently the feasibility of the concept of hybrid magnetic circuits for MR clutches was demonstrated [11]. The objective of this paper is to investigate the performance of MR clutches with a combination of permanent magnets and electromagnets. For this purpose, MR clutches with different designs, sizes and transmittable torques were constructed and tested.

2. MR clutch types

MR clutches are usually operated in the shear mode, where the MR fluid in a thin layer mechanically connects or disconnects the input and output shaft of the clutch. Two main designs of MR clutches are known which are designated disk-type and bell-type [4]. In the disk-type clutch, the shear gaps filled with the MR fluid are oriented perpendicular to the rotational axis of the clutch (Figure 1, left). In the bell-type MR clutch, the MR fluid gaps are oriented parallel to the rotational axis (Figure 1, right). Figure 1 also shows the magnetic field lines generated by the coil which penetrate the MR fluid in the shear gaps and attach the output to the input shaft via the stiffening of the MR fluid.

Disk-type and bell-type clutches have different advantages and disadvantages. The disk-type clutch can be made more compact with a large number of parallel disks and corresponding shear gaps, which leads to a large transmittable torque. However, at high rotational speeds the torque is more prone to particle separation due to the centrifugal forces acting on the heavy iron particles in the MR fluid. In contrast, in bell-type MR clutches, the particles can move only over the small distance of the thin MR fluid gap.

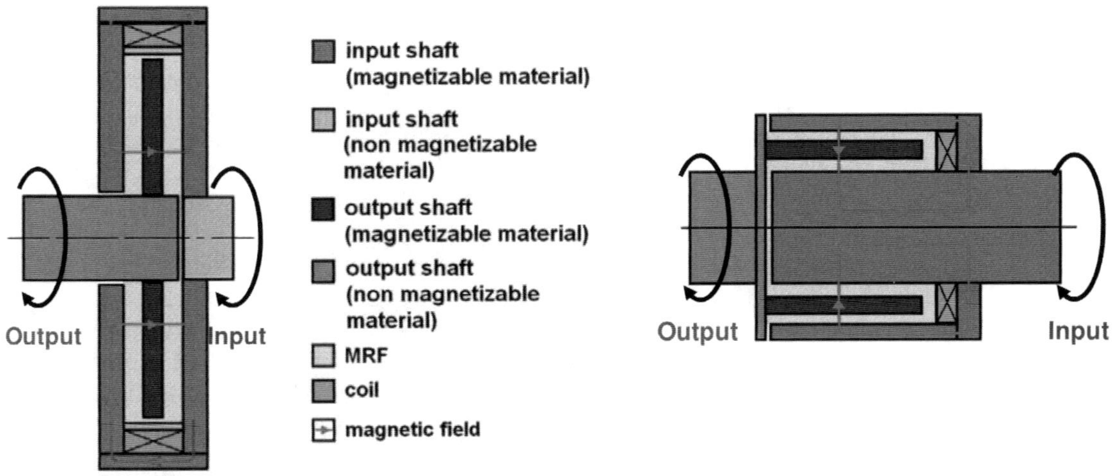

Figure 1. Disk-type MR clutch (left) and bell-type MR clutch (right), each with input shaft and output shaft, the magnetic circuit made from magnetisable material and the magnetic field lines penetrating the MR fluid.

3. MR clutches with hybrid magnetic circuit

In the following, three MR clutches equipped with a hybrid magnetic circuit containing a permanent magnet and an electromagnet are introduced. All three clutches were designed in the bell-type configuration. Their designs especially differ in their number of mechanical parts which can rotate with respect to each other. In the simplest design, the clutch consists only of the input and the output part. In this case, the electromagnet is not stationary in the clutch operation with rotating input and output shaft. Hence, the coil has to be electrically connected by sliding contacts. Other differences between the three presented MR clutches concern their size and weight as well as their maximum transmittable torque.

3.1. Two-part MR clutch

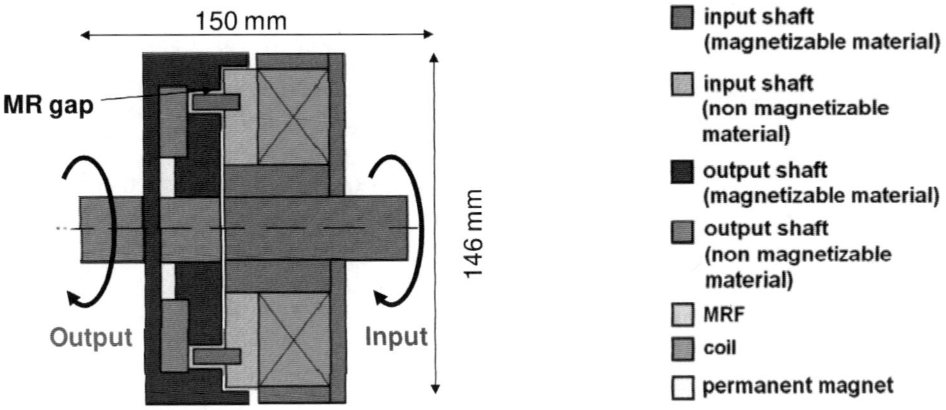

Figure 2. Scheme of the two-part MR clutch with permanent magnet and electromagnet.

The first MR clutch with permanent magnet and electromagnet consists of two parts which are mechanically connected to the input shaft and the output shaft, respectively. Figure 2 exhibits the schematic design as well as the outer dimensions of the clutch. The coil is connected to the input shaft, whereas the other magnetic source, the permanent magnet, is attached to the output shaft. The torque is transmitted by the MR fluid in the two concentric gaps which are oriented parallel to the rotational axis according to the bell-type design. In order to distinguish between the two parts of the clutch in Figure 2, the input part is shown in gray and the output part in blue. The magnetic circuit which guides the magnetic flux through the MR fluid gaps is indicated by darker gray and blue colors.

It becomes apparent that the two magnetic sources generate magnetic fields in different magnetic sub-circuits. However, the magnetic fields superpose in the MR fluid gaps and can weaken and strengthen each other. Without current in the coil, only the permanent magnet generates a magnetic field in the MR fluid. This operational mode is schematically shown in Figure 3, left. Here, the MR fluid in the gaps stiffens without any supply of electric energy and a corresponding torque can be transmitted by the clutch. Figure 3, middle, visualizes the operational state, in which a current applied to the coil causes a magnetic field which counteracts the field of the permanent magnet. As a consequence, the two fields weaken each other and the MR fluid re-liquefies. This means that the engaged and disengaged states of the clutch are reversed with respect to the known MR clutches with an electromagnet only.

Finally, in the third operational mode in Figure 3, right, the magnetic field of the electromagnet strengthens that of the permanent magnet. Such an enhanced magnetic field causes an even higher yield stress in the MR fluid than in the operational mode with the permanent magnet only. This should lead to a higher torque transmission of the clutch.

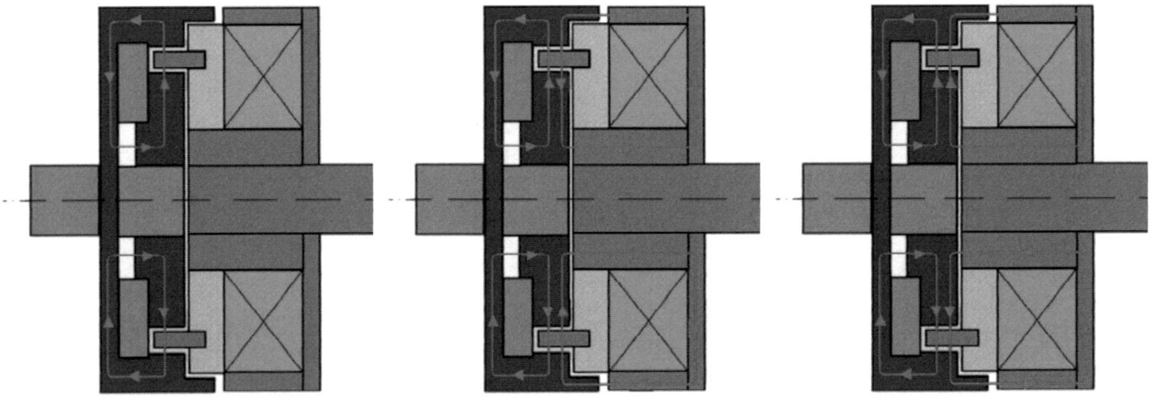

Figure 3. Magnetic field lines in the two-part MR clutch in three different operational modes: Magnetic field generated only by the permanent magnet (left), negative superposition or weakening of the fields (middle) and positive superposition or strengthening of the fields (right).

In order to quantify the magnetic field, FEM simulations of the magnetic flux density in the MR fluid gap were performed. The results are shown in Figure 4. In the operational mode with the permanent magnet only, the magnetic flux density in the MR fluid gap amounts to 450 mT. Decreasing the magnetic field strength by a counter-field caused by a coil current of -7.5 A results in a magnetic flux density of less than 50 mT. When the polarity of the coil current is reversed to +7.5 A, the two magnetic fields strengthen each other and the total magnetic flux density is enhanced to 830 mT. This demonstrates that the magnetic flux density can be varied in a very broad range and a medium flux density is generated by the permanent magnet without electric energy supply.

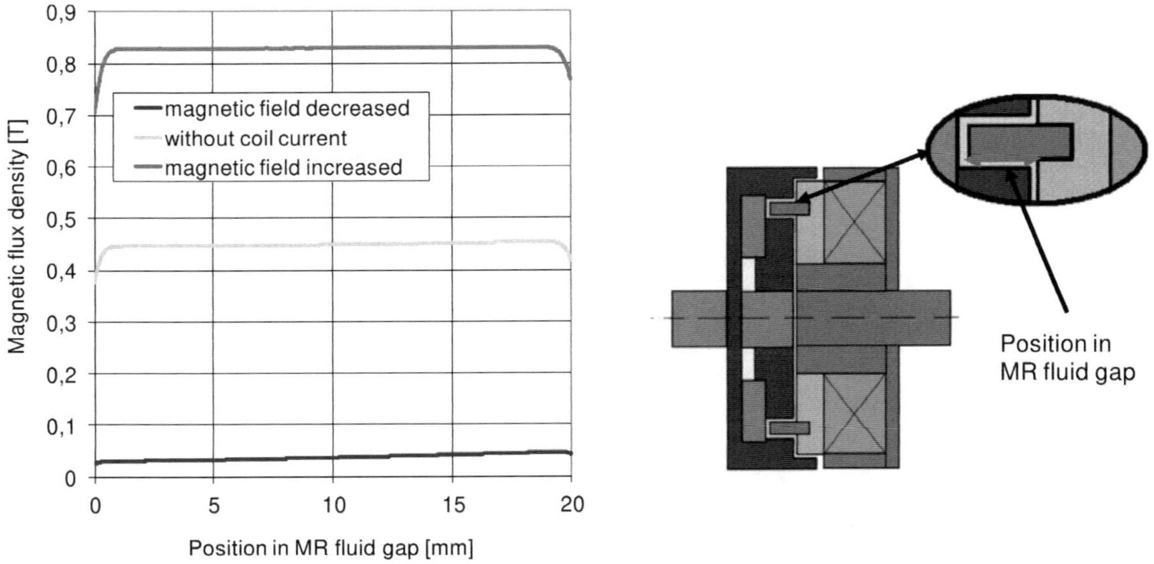

Figure 4. Magnetic flux density in the MR fluid gap for the three operational modes corresponding to coil currents of -7.5 A, 0 A and +7.5 A, received from FEM simulations. The position in the MR fluid gap refers to a line along the gap shown in the right figure.

The described MR clutch with the hybrid magnetic circuit containing permanent magnet and electromagnet was designed in more detail, manufactured and tested. Figure 5 exhibits the design model of the MR clutch and the assembled demonstrator. In both pictures, the electromagnet is integrated in the bottom part and the permanent magnet in the top part of the clutch. The mass of the MR clutch is 11 kg and it can be operated at rotational speeds up to 2000 rpm.

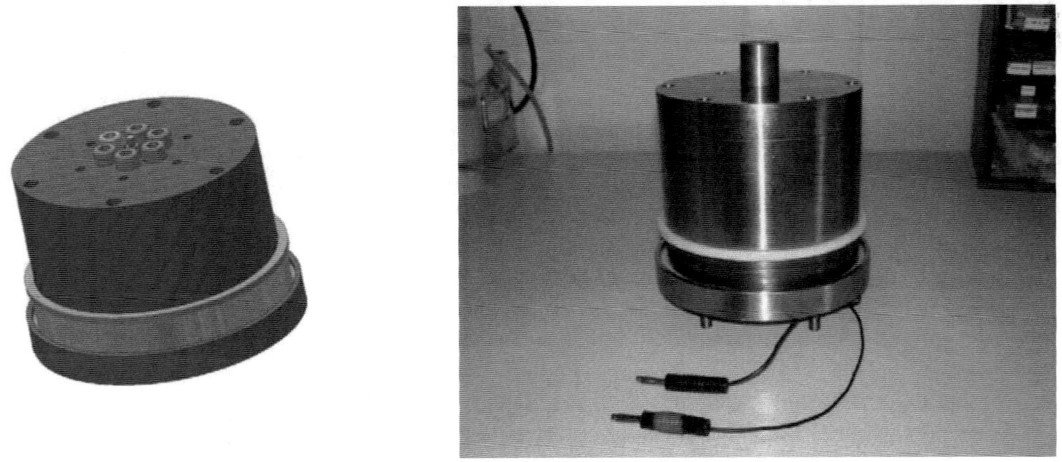

Figure 5. Design model of the two-part MR clutch (left) and demonstrator model (right).

For the evaluation of the performance of the MR clutch, it was filled with a MR fluid which contains 35 vol.% carbonyl iron particles in silicone oil and was developed in the same laboratory. The clutch was mounted in a self-made test-rig which is shown in Figure 6. In these experiments, the torque at different coil currents was measured in the braking mode. This means that the output shaft of the clutch was fixed like in a brake. The input shaft was driven by an electric servomotor at low speed and the braking torque was measured with a torque sensor between the motor and the clutch.

Figure 7 shows the results of the torque measurements in comparison with calculated data. For the calculations, a simple model for the resulting shear force was used, which includes the shear stress of the MR fluid depending on the magnetic flux density in the MR fluid gap. The measurements and the calculations were conducted at various coil currents between -7.5 A and +7.5 A. The comparison in Figure 7 shows that the experimental torqueses are lower than the calculated data. However, the general dependence of the torque on the coil current is in rough agreement with the predicted behavior. Without coil current (0 A), a medium torque of 17 Nm can be transmitted by the clutch without electric energy supply. When a counter-field is generated in the MR fluid gap by a coil current of -7.5 A, the transmitted torque is diminished to 1-2 Nm. With a reversed coil current of +7.5 A, an enhanced transmitted torque of 34 Nm was achieved.

Figure 6. Experimental test setup for torque measurements with the two-part MR clutch integrated.

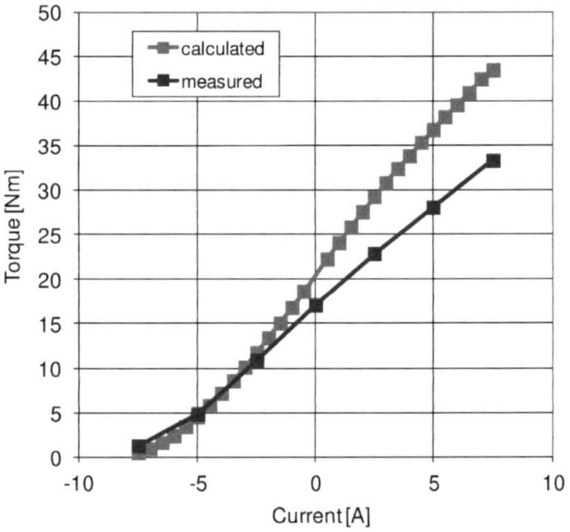

Figure 7. Measured and calculated torque of the two-part MR clutch in dependence on the coil current.

The results demonstrate that the broad variability of the magnetic flux density in the MR fluid gap by the variation of the coil current corresponds to a wide range of torque transmission. Without any coil current and electric energy supply, a medium torque is transmitted by the clutch.

3.2. MR clutch with stationary electromagnet

The second MR clutch with permanent magnet and electromagnet consists totally of four parts which can rotate with respect to each other. Figure 8 exhibits the schematic design of this more sophisticated clutch. The input shaft (blue in Figure 8) is coupled to the part with the permanent magnet. The torque is transmitted to the output shaft (red in Figure 8) via the MR fluid in the axially oriented concentric gaps. In contrast to the two-part MR clutch, the housing of this clutch is stationary and allows a tight attachment to the environment. Finally, the fourth part of the clutch containing the coil is apart from the other three parts. It can be mechanically connected to the housing, which avoids the necessity of sliding electrical contacts. However, the relative motion of the coil with respect to the magnetic circuit in the output part of the clutch could cause disturbing eddy currents. In order to eliminate such eddy currents, alternatively the coil part can be coupled to the output shaft. For this purpose, a commutator ring is attached to the output shaft (Figure 8).

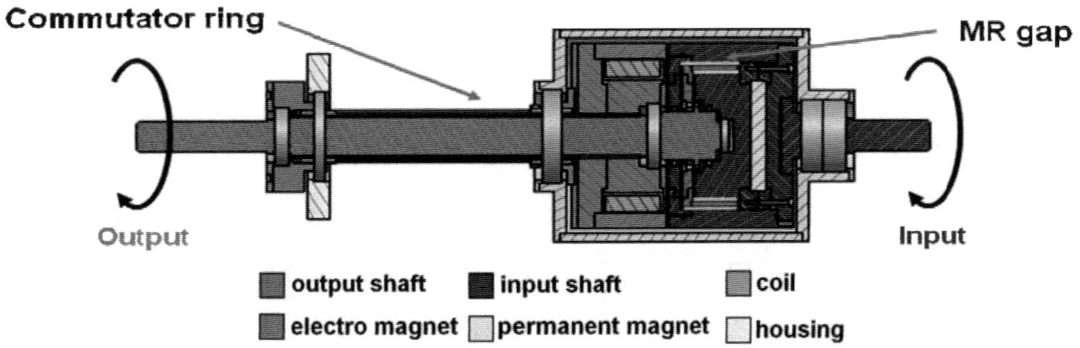

Figure 8. Scheme of the MR clutch with stationary electromagnet consisting of four separate parts: Input shaft, output shaft, coil part and housing.

Figure 9 exhibits the three working modes of the MR clutch in correspondence with the two-part clutch. Again, without any coil current, a magnetic field is generated by the permanent magnet, which causes a torque transmission without energy supply. The magnetic field of the permanent magnet in the MR fluid gaps can be decreased or increased by the electromagnet, depending on the polarity of the coil current.

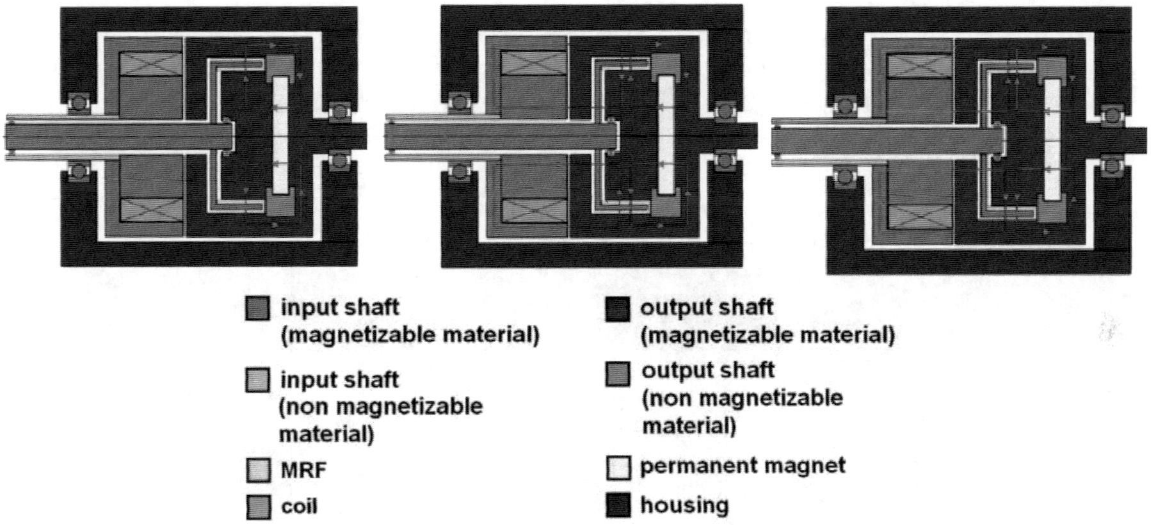

Figure 9. Magnetic field lines in the MR clutch with stationary electromagnet in three different operational modes. Magnetic field generated only by the permanent magnet (left), negative superposition or weakening of the fields (middle) and positive superposition or strengthening of the fields (right).

In Figure 10, maps of the calculated magnetic flux density in the MR fluid clutch received from FEM simulations are revealed. The red ellipses indicate the region of the MR fluid gaps. When only the permanent magnet generates the magnetic field in the gaps, a medium flux density is achieved (Figure 10, left). With the counter-field generated by the electromagnet, the magnetic flux density in the MR fluid gaps nearly vanishes (Figure 10, middle). Finally, a positive superposition of both magnetic sources leads to a very high magnetic flux density in the MR fluid gaps (Figure 10, right).

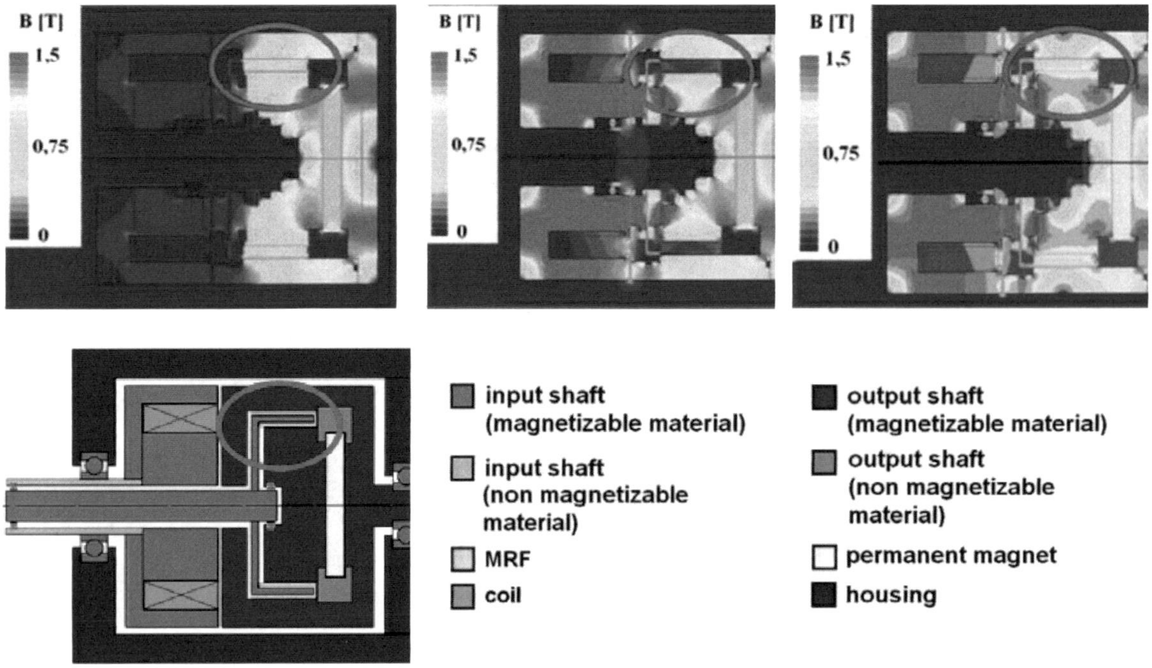

Figure 10. Top: Maps of the magnetic flux density in the MR clutch with stationary electromagnet in three different operational modes (top): Magnetic field generated only by the permanent magnet (left), negative superposition or weakening of the fields (middle) and positive superposition or strengthening of the fields (right). Bottom: Corresponding scheme of the MR clutch. The red ellipses indicate the region of the MR fluid gaps.

Figure 11. Demonstrator model of the MR clutch with stationary electromagnet.

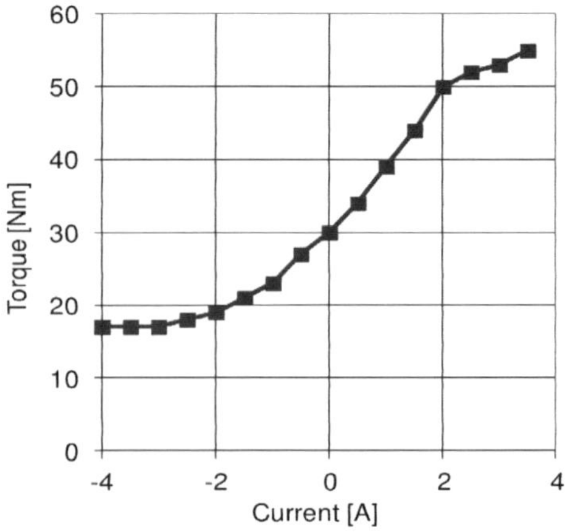

Figure 12. Measured torque of the MR clutch with stationary electromagnet in dependence on the coil current.

Like the two-part clutch, the MR clutch with stationary electromagnet was also designed, constructed and tested. Figure 11 depicts the demonstrator model of the clutch. For the investigations on the same experimental test-rig as before, a MR fluid with 35 vol.% iron particles in silicone oil was prepared and filled into the MR clutch. The measurements were conducted similar to those on the two-part clutch. The results in terms of the transmitted torque in dependence on the coil current are depicted in Figure 12. Without coil current, a medium torque of 30 Nm is transmitted. Depending on the polarity of the coil current, the torque can be decreased or increased up to 55 Nm. However, the decrease of the torque by the counter-field is not as strong as expected. The source of the relatively high minimum torque has been identified in details of the MR fluid gap design. This will be an issue of further improvements of the MR clutch.

3.3. MR clutch with high torque transmission

The third MR clutch with permanent magnet and electromagnet was designed to transmit very high torques. The motivation for this clutch arose from the development of a hybrid vehicle for public transport, whose combustion engine should be variably connected by this MR clutch to the electric generator in the vehicle. Figure 13 exhibits the schematic design as well as the outer dimensions of the new MR clutch. The diameter of the clutch amounts to 350 mm and the length without outer shafts is 304 mm. The permanent magnet with a diameter of about 200 mm was assembled from a large number of small NdFeB magnets in order to avoid the effort of manufacturing a permanent magnet of this size in one piece.

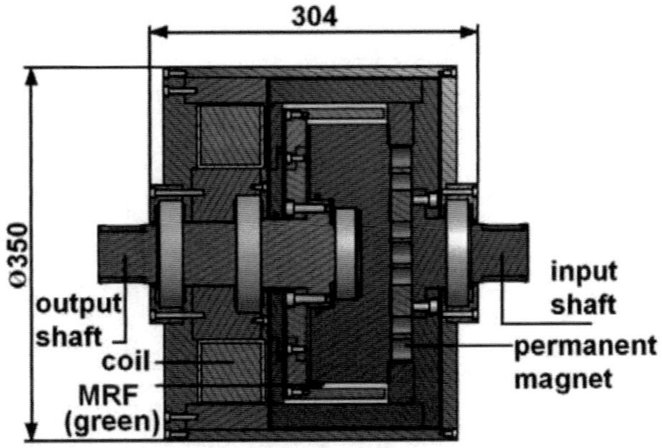

Figure 13. Scheme of the MR clutch with high torque transmission and outer dimensions in mm.

The high-torque clutch was designed in more detail, manufactured and assembled. In order to achieve exceptionally high torques, a MR fluid with 50 vol.% iron particles in silicone oil which reaches a shear stress of 70 kPa at 700 mT magnetic flux density was selected for the clutch and manufactured. The demonstrator model of the clutch is represented in Figure 14. It was integrated in the experimental test setup for torque measurements and its performance was evaluated.

Figure 14. Demonstrator model of the MR clutch with high torque transmission (left) and integrated in the experimental test setup for torque measurements (right).

The results of the torque measurements upon variation of the coil current are shown in Figure 15. Starting from the operational mode without electric energy supply, the torque of 180 Nm can be enhanced by the positive superposition of the magnetic fields of permanent magnet and electromagnet up to nearly 800 Nm. However, the decrease of the transmitted torque by the counter-field of the electromagnet with negative currents down to 120 Nm should be improved by further modifications in the magnetic circuit design in order to extend the range of variability of the transmitted torque.

Figure 16 shows the temporal change of the torque transmitted by the MR clutch upon a stepwise increase of the coil current. The measured torque follows the steps of the current immediately, but requires some seconds to reach the final value. With a coil current of 4.2 A, a maximum torque of 730 Nm is achieved.

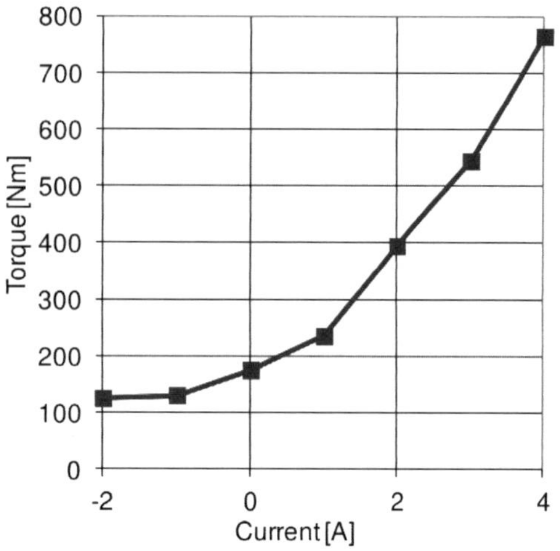

Figure 15. Measured torque of the MR clutch with high torque transmission in dependence on the coil current.

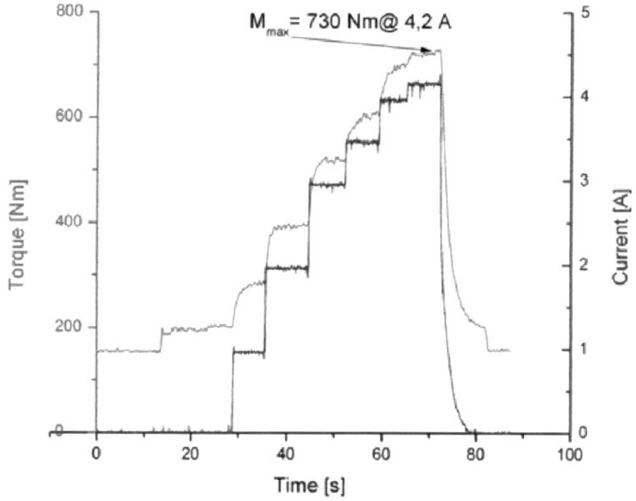

Figure 16. Measured torque of the MR clutch with high torque transmission upon stepwise increase of the coil current.

This high torque would be sufficient for the application in the hybrid vehicle, but it would be even better to accomplish this torque with the use of the permanent magnet only, i. e. without the supply of electric energy. Further work on this high-torque MR clutch shall be focussed on the design of a magnetic circuit, which should generate a stronger magnetic field by the permanent magnet only. Another future goal is the decrease of the minimum transmitted torque which is achieved with the counter-field of the electromagnet by a mutual cancelation of the fields of the two magnetic sources in the MR fluid gaps.

4. Conclusions

Three novel types of MR clutches, each equipped with a hybrid magnetic circuit containing a permanent magnet as well as an electromagnet and designed in the bell-type configuration were described in this paper. The clutch designs especially differ in their number of mechanical parts which can rotate with respect to each other as well as in their size and weight and their maximum transmittable torque.

The most simply designed MR clutch has only two parts which can rotate with respect to each other. Another four-part MR clutch with hybrid magnetic circuit has a stationary housing and allows the mechanical attachment of the electromagnet to this housing in order to avoid any sliding electric contacts. Finally, the third MR clutch with permanent magnet and electromagnet was designed to be operated in a hybrid vehicle for public transport. This clutch is capable to transmit high torques up to nearly 800 Nm.

With all three MR clutch designs, it could be successfully proven, that it is possible to operate the MR clutch in three different modes depending on the coil current in the electromagnet. Without any electric energy supply to the coil, the clutch transmits a medium torque due to the corresponding state of stiffening of the MR fluid. On one hand, when a coil current with negative polarity is applied, the corresponding counter-field in the MR fluid gap weakens that of the permanent magnet and the transmitted torque of the clutch is decreased. On the other hand, with a positive coil current, the fields of both magnetic sources strengthen each other in the MR fluid gaps, which results in an enhancement of the transmittable torque. Hence, a broad range of the transmitted torque which is controlled by the coil current can be achieved with this novel MR clutch principle.

The exploitation of hybrid magnetic circuits offers the possibility to reverse the operational states of MR clutches. If the clutch is engaged in the usual operational state, this engagement can be maintained without any supply of electric energy. Only for a short disengagement of the clutch, electric energy in terms of the coil current has to be supplied. Hence, in such operation the novel MR clutch works very energy-efficient, which offers far-reaching perspectives for a multitude of applications in the automotive sector and in mechanical engineering.

Acknowledgements
Financial support for this work from the German Ministry for education and research as well as from the Bavarian State Ministry for economy, infrastructure, traffic and technology is gratefully acknowledged. The authors thank Ralph Bystricky, Christian Deak and Dominik Fenten for the technical support of this work.

References
[1] Tao R 2011 *Proc. 12th Int. Conf. on Electrorheological Fluids and Magnetorheological Suspensions* (Singapore, World Scientific)
[2] Smith A L, Ulicny J C and Kennedy L C 2007 Magnetorheological fluid fan drive for trucks *Proc. 10th Int. Conf. on Electrorheological Fluids and Magnetorheological Suspensions* (World Scientific, Singapore) pp 479-86
[3] Rabinow J 1948 The magnetic fluid clutch *AIEE Trans.* **67** 1308-15
[4] Lampe D, Thess A and Dotzauer C 1998 MRF-clutch-design considerations and performance *Proc. 6th Actuator* pp 449-53
[5] Lampe D and Grundmann R 2000 Transitional and sold state behaviour of a magnetorheological clutch *Proc. 7th Actuator*
[6] Kieburg C, Oetter G, Lochtman R, Gabriel C, Laun H M, Pfister P, Schober G and Steinwender H 2007 High performance magnetorheological fluids tailored for a 700 Nm automotive 4-wheel-drive clutch *Proc. 10th Int. Conf. on Electrorheological Fluids and Magnetorheo-logical Suspensions* (World Scientific, Singapore) pp 101-7
[7] Kikuchi T, Ikeda K, Otsuki K, Kakehashi T and Furusho J 2009 Compact MR fluid clutch device for human-friendly actuator *J. Phys. Conf. Series* **149** 012059
[8] Böse H and Ehrlich J 2011 Magnetorheological dampers with hybrid magnetic circuits *Proc. 12th Int. Conf. on Electro-Rheological Fluids and Magneto-Rheological Suspensions* (R. Tao, Ed., World Scientific) pp 18-29
[9] Böse H and Ehrlich J 2012 High-Performance Magnetorheological Damper with Hybrid Magnetic Circuit *Proc. 13th International Conf. on New Actuators* pp 109-12
[10] Saito T, Ikeda H 2007 Development of normally closed type of magnetorheological clutch and its application to safe torque control system of human-collaborative robot *Proc. 10th Int. Conf. on Electrorheological Fluids and Magnetorheological Suspensions* (World Scientific, Singapore) pp 557-62
[11] Böse H, Ehrlich J, Gerlach T, Deak C, Janssen C, Trendler A-M 2008 Magnetorheological clutch with reversed torque transmission behavior *Proc. 11th Int. Conf. on New Actuators* pp 847-51

The novel MRF-ball-clutch design – a MRF-safety-clutch for high torque applications

M Jackel[1], J Kloepfer, M Matthias and B Seipel

Division of Adaptronics, Fraunhofer Institute for Structural Durability and System Reliability LBF, Bartningstr. 47, 64289 Darmstadt, Germany

E-mail: marco.jackel@lbf.fraunhofer.de

Abstract. This paper describes the development of a safety-clutch by using magnetorheological fluids (MRF) to switch the transmission torque between a motor and a generator in a bus-like vehicle. The clutch is based on a new design combining an axial MRF-actuator and a ball coupling mechanism. This so called "MRF-ball-clutch" avoids the disadvantages of traditional bell- or disc-MRF-clutch designs where the torque is transmitted by the MRF which leads to a self-heating due to the shearing forces in the fluid and a more or less significant drag torque caused by limitations of the relation between minimal and maximal transmittable torque. The safety clutch based on the new MRF-clutch design requires a minimum of power consumption and allows switching high torsional moments in a very compact, lightweight and robust design.

1. Introduction

A big potential for technical applications based on magnetorheological fluids (MRF), beside adjustable dampers, switchable engine mounts, etc., are so called smart clutches. There are two common designs of these clutches, the disc- and the bell- design (see Figure 1). Both consist of two rotatable parts, the input or drive side and the output or power take off side, with a small gap in between. This gap is filled with MRF that transmits the torsion moment from the input to the output side depending on its viscosity and its interaction with the surfaces. These concepts allow influencing the power transmission and realize an adjustable slip torque.

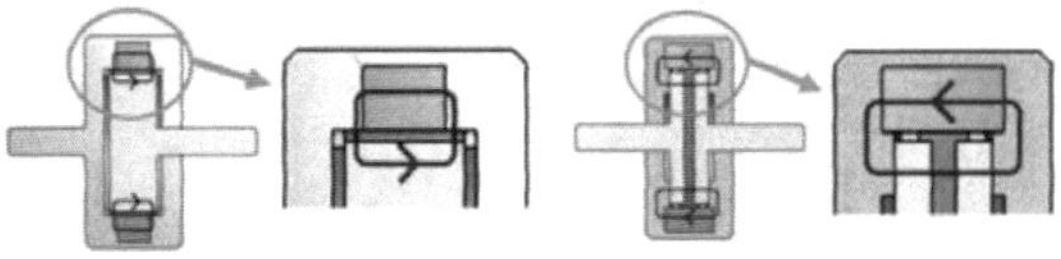

Figure 1. Conventional MRF-clutches, "bell"-design (left) and "disc"-design (right).

To activate the MRF a high magnetic field, generated by an electromagnet is necessary. Therefore a clutch based on these designs has relatively high power consumption and is critical for energy limited applications. Due to the shear stress applied on the fluid both designs have the disadvantage to

[1] Corresponding author.

lead to self-heating effects of the fluid as well as a limitation of the relation between minimal and maximal transmittable torque. Clutches based on these conventional designs are relatively big and heavy in relation to the transmittable torque.

Although the aforementioned designs are very popular, the performance of these clutches has room for improvement. Especially clutches that are designed to transmit higher torques, e.g. some hundred Nm in typical automotive applications, need to have mechanical parts with large surfaces to interact with the MRF. As the corresponding volume of the MRF is also very big, the electromagnetic system has to be really strong to provide a sufficient magnetic field. The minimal torque is, beside the question of the dimension of the clutch, mainly a result of the basic viscosity of the MRF. Caused by this basic viscosity a minimal torsional moment will always be transmitted, which turns out to be the limitation for some applications. Another disadvantage of the traditional MRF clutch designs is caused by the rotation of the mechanical parts that are in contact with the MRF. All passes to the outside have to be sealed very carefully. This is very challenging as the components of the MRF are highly abrasive and therefore bad for the durability of the seals. Caused by these disadvantages a MRF-Clutch based on these classic designs is not in technical or commercial use so far.

For safety critical applications, where a fast reacting clutch to switch the power transmission on or off completely is important and an adjustable slip torque is not needed, a new MRF-clutch-design was developed at the Fraunhofer LBF.

2. Principle of the MRF-ball-clutch

The new design is based on a combination of a ball safety coupling mechanism (see Figure 2) and an axial operating MRF-actuator (see Figure 3) and was named "MRF-ball-clutch". Compared to the described common clutch designs where the torque is transmitted friction locked by the MRF, here mechanical coupling devices, realized with balls which are circularly located in countersinks on the front of two opposite placed discs, act interlocking to transmit the power. By the geometrical interaction of these balls and the two discs an applied torque produces a force in the direction of the rotational axis which increases the distance between input and output disc. If the torque exceeds a certain level the balls are no longer able to rest in their countersinks and the power transmission is stopped immediately. To adjust the maximum transmittable torque, a variable counter force is required that hinders the axial movement of the clutch discs. This force is generated by an axial operating MRF actuator, which is designed similar to a common single-tube damper. A solenoid coil generates a magnetic field in the fluid gap and by this the flow through of the MRF can influenced and the retention forced to open the clutch can be tuned. The principle of the MRF-ball-clutch is patented by the Fraunhofer LBF [1, 2].

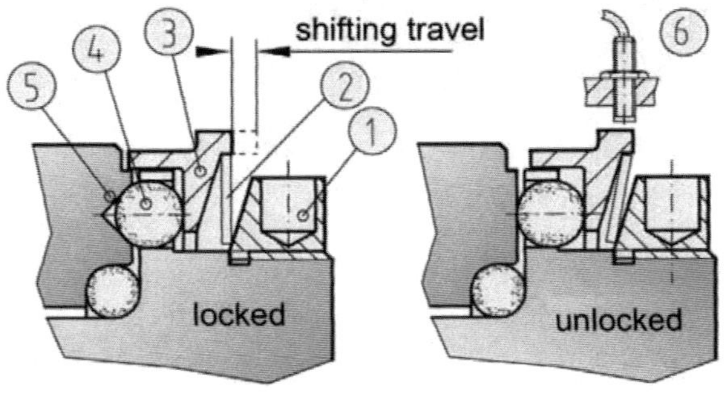

Figure 2. Principle of a ball clutch; locked (left) and unlocked (right) (Source: R+W Antriebselemente GmbH).

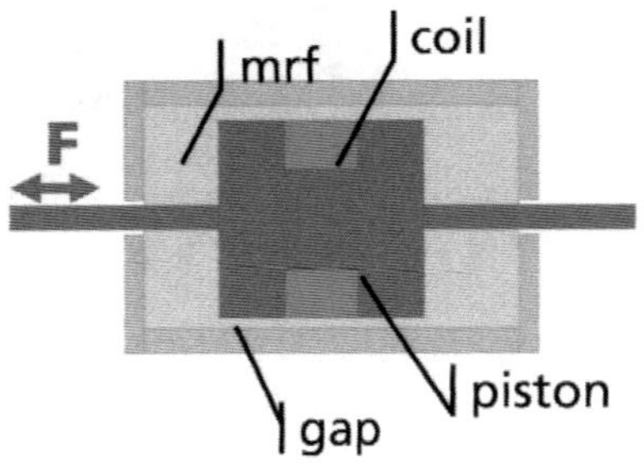

Figure 3. Principle of a MRF single tube damper.

2.1. Example application torque wrench

To show the potential of the developed clutch concept an electrically adjustable torque wrench was chosen as a functional demonstrator to represent a typical application. Figure 4 shows a cross-section of the main parts. On the left side the "ball-clutch" is represented by two hardened steel discs with 90° countersinks and 6 steel balls of 8 mm. The balls are arranged on a circle with a diameter of 36 mm. On the right side the MRF actuator is realized by a low carbon steel piston with a coil of 0.5 mm copper wire and a tube housing also made of low carbon steel. The counterforce of the piston (typically 5.6 kN) is transmitted by a guided rod to the clutch side. To readjust the system, a "weak" retention spring is located at the right end of the rod.

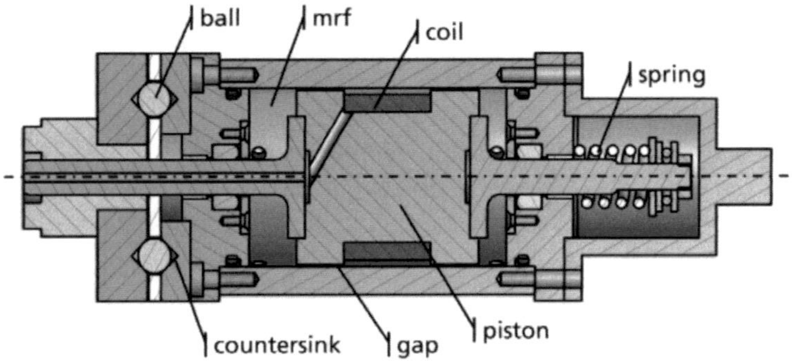

Figure 4. MRF torque wrench, cross section.

The finally realized demonstrator structure (see Figure 5) consists of the MRF actuator, a handle bar with force sensor, a connector for a socket and an electronic unit. With the torque wrench it is possible to tighten a screw exactly with a predefined torque between 5 and 100 Nm.

Figure 5. Prototype of a MRF torque wrench.

With the investigated demonstrator the functional capability of the developed clutch concept was approved [3]. In contrast to mechanical clutch systems, the release torque is online adjustable. The small MRF-ball-clutch is able to transmit up to 100 Nm with very low electrical power consumption.

3. High torque MRF-ball-clutch for E-mobility application

In concepts for future electric vehicles different new innovations are in discussion to solve the main problems as there is for example the relatively small cruising range of electric vehicles. Most of the concepts based on the combination of different engines, electro engines and combusting engines, which increase the complexity of the powertrain and the demand for new coupling elements / clutches.

One of the abandonings of the clutches in future electric mobility is to switch the different components of the powertrain for energy consumption or safety reasons as needed in the actual driving situation.

For a reference application named "AutoTram" [4] (see Figure 6 left) within the scope of "Fraunhofer's System Research for Electromobility – FSEM" [5] a safety clutch has to be developed.

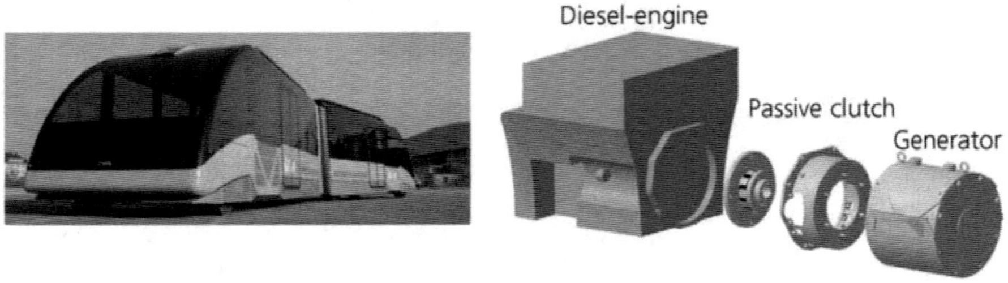

Figure 6. Fraunhofer AutoTram [3] (left) and its power unit (right).

The aim is to replace the passive clutch between the diesel engine and the electric generator within the power unit (see Figure 6 right) of the AutoTram by a MRF clutch. This clutch should work as safety component and should be able to decouple the two main systems even under maximum load conditions. As the power unit consists of a 180 kW V8 Diesel-engine with a maximum torque of typically 560 Nm (operating ratio of 1.2 leads to ~700 Nm) and a maximum speed of 4000 rpm, the clutch design is very challenging.

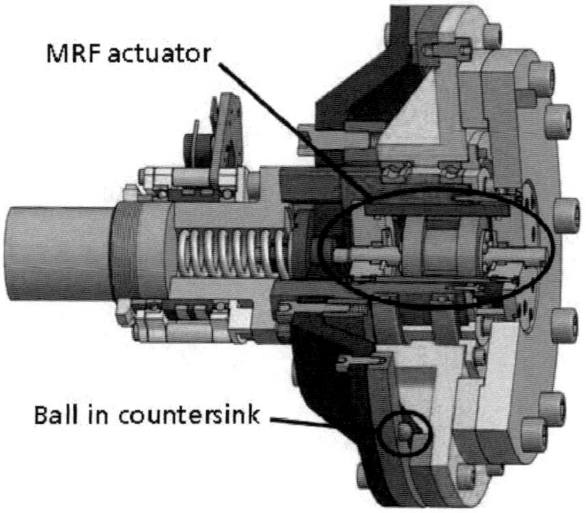

Figure 7. CAD Model of the MRF safety clutch.

Figure 7 shows a CAD model of the clutch. The design of the MRF actuator unit is similar to the design of the torque wrench actuator unit. The size, number and position of the balls are adapted to the different requirements. The clutch has a diameter of 260 mm, an over all length of 300 mm and a mass of about 15 kg. To avoid the axial displacement, generated by decoupling, the torque is transmitted over a splined shaft connection. The power consumption of the electric parts (coil) is only at about 8 W. Compared to traditional active MRF and non MRF clutches these first values show a high potential of the new technology.

3.1. Functional tests of the MRF-ball-clutch

After setting up the MRF safety clutch, initial functional checks were made on a test rig (see Figure 8).

Figure 8. Picture of the MRF safety clutch on a test rig.

First results are promising: it's possible to transmit the torque of 700 Nm and open the clutch on demand. The minimal release torque is about 200 Nm.

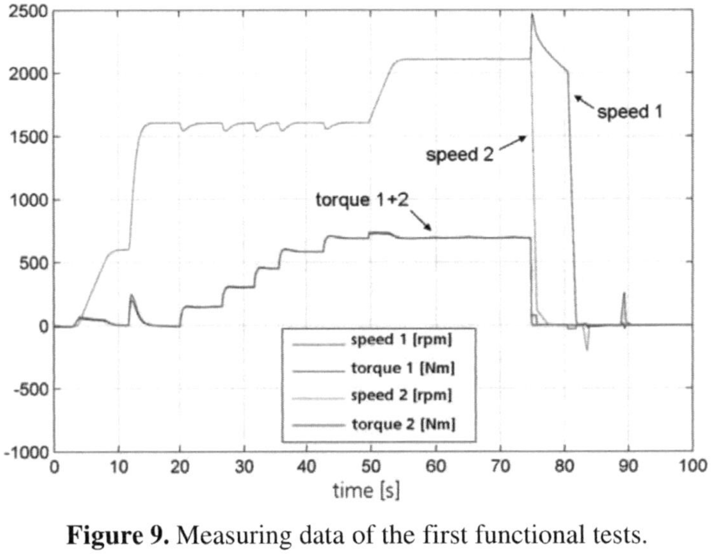

Figure 9. Measuring data of the first functional tests.

The measuring data of a first functional check are shown in figure 9. Torque and rotation speed are increased till 700 Nm at 2100 rpm. After a while of constant torque and rotation speed the current entry of the coil was cut off (t~75s). Consequently the clutch opens and the torque drops down.

3.2. Functional tests of MRF-actuator

After testing the complete MRF-ball-clutch tests with the axial MRF-actuator were performed on a tension-compression testing machine [6, 7]. The goal is that the actuator provides a retention force of 5.6 kN, equivalent to a pressure difference of 3.5 MPa. The test runs were made with different fluids provided by the Fraunhofer ISC [8].

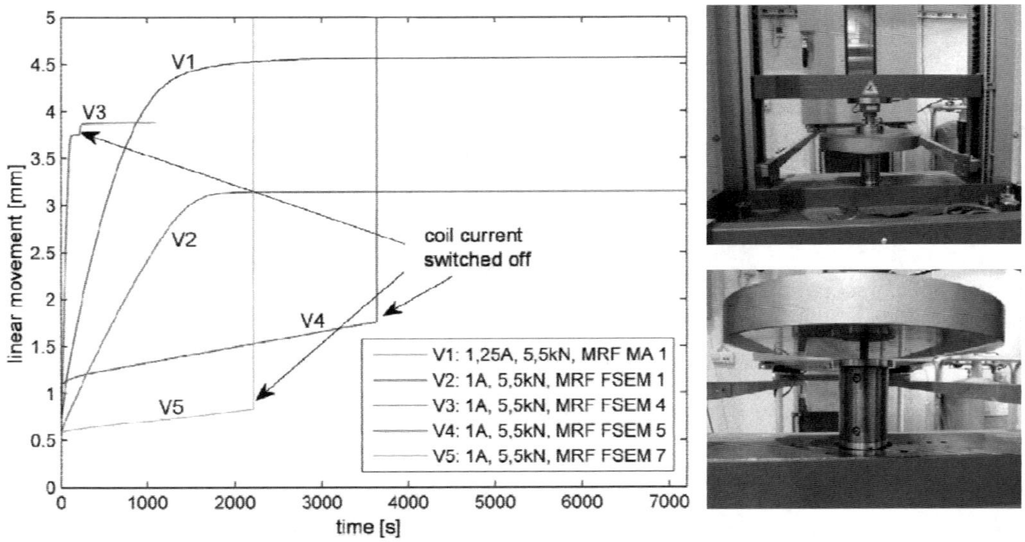

Figure 10. Test results of the linear MRF-actuator.

The results are shown in figure 10. It shows the linear movement over time under a constant load or different fluids. In general the actuator is able to provide the needed force, but in the course of time there is some slow movement under load. The movement at the beginning results from settling effects of the test assembly. After that there is a more or less linear creeping that approaches a final value (see Figure 10 cycle V1 and V2). The reason for this behaviour is the separation of the MRF. The carrier fluid flows from the compression side through the particle chains in the gap to the tension side. The particles on the compression side are resting on their side. Disassembling the MRF-actuator proved that assumption, see Figure 11. But it is still possible that there are other effects like leaky sealings on the compression side.

Figure 11. Disassembled mrf-acuator.

For the MRF-ball-clutch this behavior means, that over time there is an increasing clearance in the drivetrain. This finally may lead to declutching. Even tough the measured data of the MR-fluids FSEM 5 (V4) and FSEM 7 (V5) look much better than the former ones, there may be problems with the MRF-ball-clutch in longterm operation.

The results for some long-term tests are shown in figure 12. The Fluid used in these tests (FSEM 6) is very similar to the fluids FSEM 5 and 7 that used in he former tests. The curves are showing that this fluid is also converging to a final value over time. This is depending on the load after 5 to 25 hours. Curve 4 shows that, after reaching the final value, there is no further linear movement during the duration of the test (140 hours).

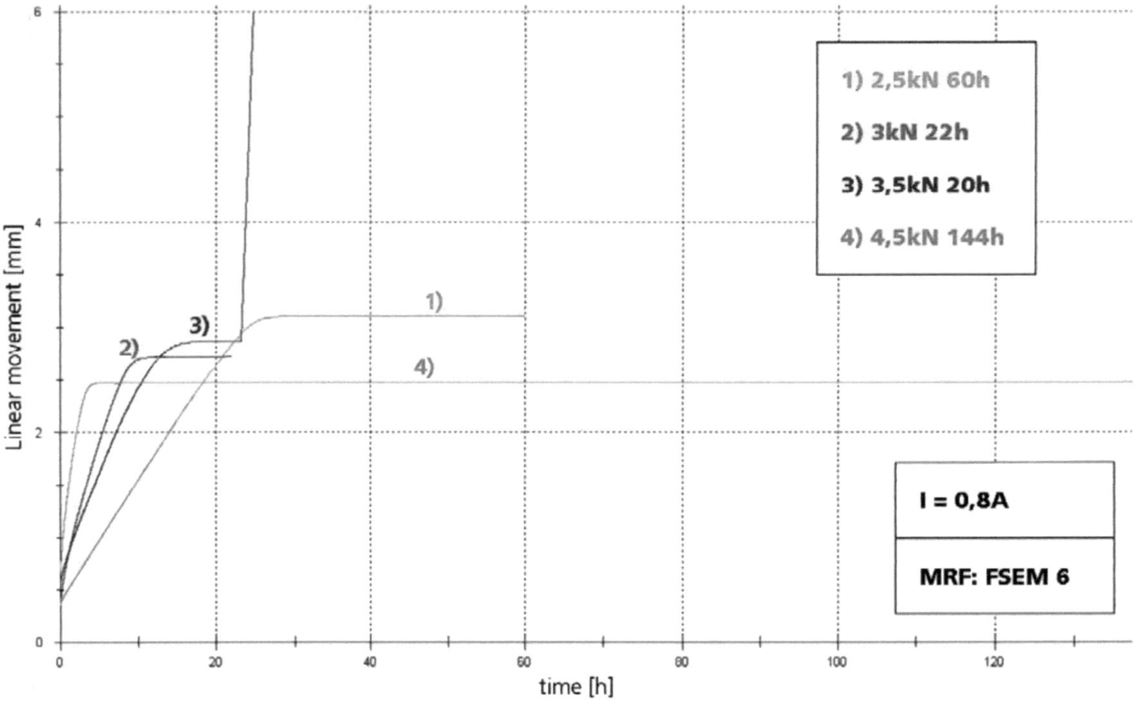

Figure 12. Results of MRF-actuator longterm-tests.

Because of the problems with the creeping of the linear MRF-actuator under high static pressure, some further investigations are necessary. First of all it is intended to perform more tests with the MRF-actuator.

4. Conclusion and outlook

The new concept of MRF based clutches using a combination of a ball safety coupling mechanism and an axial operating MRF-actuator has big advantages compared to clutches based on conventional MRF-clutch designs, which are:

- reduced thermal stress
- lower energy consumption
- reduced complexity of the construction in particular for sealing arrangements
- reduced problems of abrasion
- compact design
- reduced weight
- fast reaction time
- complete couple / decouple without any remaining drag moment

On the other side, the MRF-Ball-clutch does not allow the continuous change of power transmission (slip mode).

With respect to this promising results there seems to be a large field of applications for the described new MRF clutch concept e.g. as a safety clutch for machine tools, hand-operated power tools, in the automotive sector, etc.

Before the technology of MRF-ball-clutches can be implemented in commercial applications further investigations concerning reliability aspects, the creeping phenomena of the fluid etc. have to be done. Therefore it is planned to build up a simple linear mrf-actuator that has no sealings on the compression side of the actuator.

Acknowledgements
The work was done within the LOEWE-AdRIA project founded by the state government of Hessen (Germany) and the Fraunhofer System Research for Electromobility founded by the Federal Ministry of Education and Research BMBF (Germany) [9].

References

[1] Patent DE 102009034055 B4: Drehmomentbegrenztes Kupplungselement sowie Verfahren zum drehmoment-begrenzten Kuppeln, Fraunhofer LBF

[2] Patent WO 2011/009571 A1: Torque-limited coupling element and method for torque limited coupling, Fraunhofer LBF

[3] Matthias M, Jackel M and Seipel B 2010 MRF-Ball-Clutch – A novel magnetorheological clutch design *12th Int. Conf. Electrorheological (ER) Fluids and Magnetorheological (MR) Suspensions* (Philadelphia, PA)

[4] Fraunhofer IVI AutoTram: http://www.autotram.info

[5] Fraunhofer System Research for Electromobility FSEM: http://www.elektromobilitaet.fraunhofer.de/en/

[6] Matthias M, Jackel M, Seipel B and Kloepfer J 2012 Realization of a MRF-safety-clutch for high torsional moments based on a novel ball-clutch design *Active and Passive Smart Structures and Integrated Systems Proc. SPIE* **8341** pp. 83410P-83410P-8

[7] Jackel M, Matthias M, Seipel B, Kloepfer J 2012 The novel MRF-ball-clutch design *Actuator 12 – Int. Conf. Exh. New Actuators and Drive Systems* (Bremen, Germany)

[8] Fraunhofer ISC: http://www.isc.fraunhofer.de/

[9] Federal Ministry of Education and Research BMBF: http://bmbf.de

Unsteady flow damping force prediction of MR dampers subjected to sinusoidal loading

M Yu[1,2,3], S Q Wang[2], J Fu[1,2] and Y X Peng[2]
[1]Key Lab for Optoelectronic Technology and Systems, Ministry of Education, Chongqing University, Chongqing 400044, China
[2]College of Optoelectronic Engineering, Chongqing University, Chongqing 400044, China

E-mail: yumiao@cqu.edu.cn

Abstract. So far quasi-steady models are usually used to design magnetorheological (MR) dampers, but these models are not sufficient to describe the MR damper behavior under unsteady dynamic loading, for fluid inertia is neglected in quasi-steady models, which will bring more error between computer simulation and experimental results. Under unsteady flow model, the fluid inertia terms will bring error calculated upto 10%, so it is necessary to be considered in the governing equation. In this paper, force-stroke behavior of MR damper with flow mode due to sinusoidal loading excitation is mainly investigated, to simplify the analysis, the one-dimensional axisymmetric annular duct geometry of MR dampers is approximated as a rectangular duct. The rectangular duct can be divided into 3 regions for the velocity profile of the incompressible MR fluid flow, in each region, a partial differential equation is composed of by Navier-Stokes equations, boundary conditions and initial conditions to determine the velocity solution. In addition, in this work, not only Bingham plastic model but the Herschel—Bulkley model is adopted to analyze the MR damper performance. The damping force resulting from the pressure drop of unsteady MR dampers can be obtained and used to design or size MR dampers. Compared with the quasi-steady flow damping force, the damping force of unsteady MR dampers is more close to practice, particularly for the high-speed unsteady movement of MR dampers.

1. Introduction

It is known that magnetorheological (MR) and electrorheological (ER) fluids are a type of non-Newtonian fluids, and the fluid flow obey the governing Navier–Stokes (NS) equations. Due to its analysis simplicity, Quasi-static analysis model is usually used to provide the required force-stroke profile as an analytical method for the design of MR /ER fluids device, the quasi-static model neglects the effect of fluid inertia on the MR fluid flow [1-3]. However, fluid inertia term acts on the fluid flow in practical application and should be incorporated in the governing NS equation for oscillatory or unsteady fluid flow [4]. For example, under the sinusoidal excitation, it is not significant for the fluid inertial effect on the unsteady flow of MR fluid in the very low range of the oscillatory frequency, but it is obvious for the higher oscillatory frequency [5]. Chen et al. proposed and solved the velocity profile and pressure gradient of the unsteady state unidirectional flow of Bingham fluid between parallel plates by the Laplace transform method [6]. Based on the unsteady flow analysis of an ER

[3] Corresponding author: M Yu, Phone: +86 23 65111016; Fax: +86 23 65111016.

fluid, Nguyen and Choi developed a dynamic modeling of the ER damper without experimental results for parameter identification considering the unsteady behavior of ER fluid in a high frequency and large stroke damper. In addition, experiments were conducted to validate the proposed model and showed the effect of unsteady behavior is significant and should be taken into account at high frequency [7].

It is necessary to study dynamic damping force of the MR/ER dampers for its geometric parameter configuration or control modelling parameter identification without experimental results. The main objective of this paper is to introduce a theoretically analyze method of unsteady flow of MR fluids, based on Bingham-plastic (BP) [8] and Herschel–Bulkley (HB) model [2], the MR unsteady fluid flow within flow mode MR dampers is described using partial differential equations (PDE). According to the PDE, numerical velocity solution of MR fluid unsteady flow is solved, and then the dynamic damping force of MR dampers can be predicted.

2. Simplified mathematical formulations

2.1. Constitutive equations

In this paper, both BP model and HB model are used to describe the incompressible MR fluid behavior. The constitutive equations of the two models are described as

BP model
$$\begin{cases} \tau = \tau_y \operatorname{sgn}(\dot{\gamma}) + \eta\dot{\gamma}, |\tau| > \tau_y \\ \dot{\gamma} = 0, |\tau| < \tau_y \end{cases} \tag{1}$$

where τ is shear stress, τ_y is the critical yield stress, η is zero field viscosity of MR fluids, $\dot{\gamma}$ is the rate of stress strain.

HB model
$$\begin{cases} \tau = \tau_y \operatorname{sgn}(\dot{\gamma}) + \kappa(\dot{\gamma})^m, |\tau| > \tau_y \\ \dot{\gamma} = 0, |\tau| < \tau_y \end{cases} \tag{2}$$

where m , κ are the power-law exponent and the consistency index, respectively. The HB model accounts for shear-thinning/thickening behavior displayed by MR fluids at higher strain rates. The BP model can be seen as a generalization of the HB model that takes into account shear-thinning/thickening behavior of MR fluid by changing in the effective viscosity with the applied shear rate through a power-law behavior. There is a substitution relationship between BP and HB models, namely:

$$\eta = \kappa(\dot{\gamma})^{m-1} \tag{3}$$

2.2. Governing equations

In this section, a flow mode MR damper with an approximate parallel-plate geometry model instead of the actual annular damping channel is investigated, for the damping channel width (d) is smaller than damping channel inner radius (R_1) and effective length (L). The pressure gradient between piston head sides $\Delta P(t)$ is assumed to evenly distribute along the damping channel length (L). In addition, the Cartesian coordinate system (x, y, z) is used, x-axis is taken as the centre line direction in the two parallel plates, y is the coordinate normal to the plate, z presents the perimeter of the annular damping channel. According to NS equations and constitutive equation of MR fluid, here, a one-dimensional governing equation of the parallel-plate model can be confirmed and written as

$$\rho\frac{\partial u(y,t)}{\partial t} = -\frac{\Delta P(t)}{L} + \frac{\partial \tau(y,t)}{\partial y} \tag{4}$$

where ρ is the fluid density, t is the time variable. $u(y,t)$ is the fluid velocity, $\tau(y,t)$ is the shear stress.

2.3. Boundary conditions

Figure 1 shows a typical velocity profile of MR fluids flowing through annular damping channel of flow mode MR dampers. According to velocity profile, boundary of parallel-plate gap, inner boundary of the plug region (y_{pi}) and outer boundary of the plug region (y_{po}) divide parallel-plate gap into three regions, region1 (post-yield) ($0 \le y \le y_{pi}$), region2 (pre-yield) ($y_{pi} \le y \le y_{po}$) and region3 (post-yield) ($y_{po} \le y \le d$). Each region has different boundary conditions, which are listed in table1.

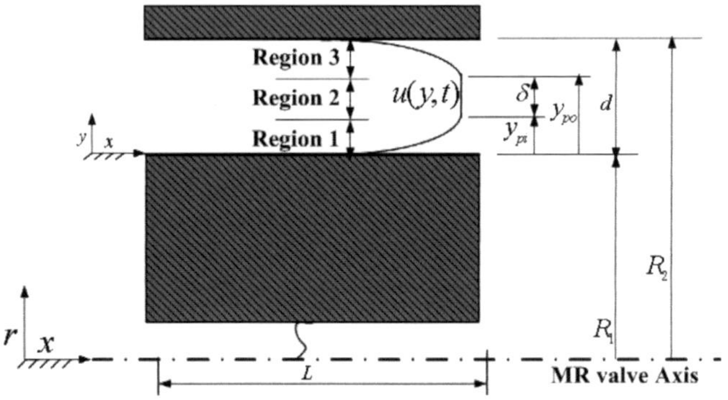

Figure 1. Typical MR fluids velocity profile in the annular damping channel of flow mode MR dampers.

Table 1. Boundary conditions of annular damping channel.

Regions	Region1	Region2	Region3
Boundary conditions	$u(0,t)=0$,	$\tau(y_{pi},t)=\tau_y$,	$u_y(y_{po},t)=0$,
	$u_y(y_{pi},t)=0$	$\tau(y_{po},t)=-\tau_y$	$u(d,t)=0$

2.4. Mathematical equation

The question of MR fluids flow through damping channel can be described by the partial differential equation (PDE) incorporating the governing equation, boundary conditions, initial condition in each region. Velocities of every regions $u_i(y,t)$ satisfies a independent PDE, namely:

In the region 1,
$$\begin{cases} u_{1t}(y,t) = v \cdot u_{1yy}(y,t) + \Delta P(t)/L \\ u_1(0,t) = 0, u_{1y}(y_{pi},t) = 0 \\ u_1(y,0) = \varphi(y) \end{cases} \tag{5}$$

In the region 2,
$$u_2\left(y_{pi} \le y \le y_{po}, t\right) = u_1(y_{pi},t) = u_3(y_{po},t) \tag{6}$$

In the region 3,
$$\begin{cases} u_{3t}(y,t) = v \cdot u_{3yy}(y,t) + \Delta P(t)/L \\ u_3(d,t) = 0, u_{3y}(y_{po},t) = 0 \\ u_3(y,0) = \varphi(y) \end{cases} \tag{7}$$

where subscript i is a representative of numbers 1, 2 and 3, $v = \eta/\rho$, $u_{it}(y,t) = \partial u_i(y,t)/\partial t$, $u_{iy}(y,t) = \partial u_i(y,t)/\partial y$ $u_{iyy}(y,t) = \partial u_i^2(y,t)/\partial y^2$.

3. Methodology of solution

3.1. Velocity solutions

To solve the PDE, the pressure gradient between piston head sides $\Delta P(t)$ should be given by a known functional form. The nonlinear constitutive law for the MR fluid dampers is assumed to be

$$F(t) = c\left|\frac{dX(t)}{dt}\right|^{\alpha} \mathrm{sgn}(\frac{dX(t)}{dt}) \tag{8}$$

where $F(t)$ represents the damping force on the piston, $X(t)$ is the displacement of the piston, c is the damping coefficient and α is a fractional exponent that accounts for the nonlinearity inherent in a MR fluid damper with the range $0 \le \alpha \le 1.5$ [9]. It can be seen from equation (8) that there is a determinate $\Delta P(t)$ for an excitation velocity at the certain time t. In this paper, to simply the analysis, assume the pressure gradient $\Delta P(t)$ is an approximate linear function of load velocity on the MR damper, namely

$$\Delta P(t) = \varepsilon V_p(t) \tag{9}$$

where ε is the undetermined constant, $V_p(t)$ is the velocity of the piston motion. The PDE can be solved with variable separation method, and the velocity of MR fluid through the annular damping channel $u(y,t)$ can be obtained. In addition, according to equation (3), the velocity solutions of BP model $u_B(y,t)$ is also used to calculate the effective viscosity of HB model.

Assume $\Delta P(t)/(\rho L) = f(t)$, the velocity solutions of BP model are:

$$\begin{aligned} u_{B1}(y,t) &= -\sum_{n=1}^{\infty}[\frac{4}{(2n-1)\pi}\int f(t)e^{K_{B1}t}dt + C]e^{-K_{B1}t}\sin\frac{(2n-1)\pi y}{2y_{pi}} \\ u_{B2}(y,t) &= -\sum_{n=1}^{\infty}[\frac{4}{(2n-1)\pi}\int f(t)e^{K_{B1}t}dt + C]e^{-K_{B1}t} \\ u_{B3}(y,t) &= \sum_{n=1}^{\infty}[\frac{4}{(2n-1)\pi}\int f(t)e^{K_{B3}t}dt + C]e^{-K_{B3}t}\sin\frac{(2n-1)\pi(y-d)}{2(d-y_{po})} \end{aligned} \tag{10}$$

where $K_{B1} = [\frac{(2n-1)\pi}{2y_{pi}}]^2 v$, $K_{B3} = [\frac{(2n-1)\pi}{2(d-y_{pi})}]^2 v$.

The velocity solutions of HB model are

$$u_{H1}(y,t)=-\sum_{n=1}^{\infty}[\frac{4}{(2n-1)\pi}\int f(t)e^{K_{H1}t}dt+C]e^{-K_{H1}t}\sin\frac{(2n-1)\pi y}{2y_{pi}}$$
$$u_{H2}(y,t)=-\sum_{n=1}^{\infty}[\frac{4}{(2n-1)\pi}\int f(t)e^{K_{H1}t}dt+C]e^{-K_{H1}t} \tag{11}$$
$$u_{H3}(y,t)=\sum_{n=1}^{\infty}[\frac{4}{(2n-1)\pi}\int f(t)e^{K_{H3}t}dt+C]e^{-K_{H3}t}\sin\frac{(2n-1)\pi(y-d)}{2(d-y_{po})}$$

where $K_{H1}=[\frac{(2n-1)\pi}{2y_{pi}}]^2\frac{\kappa}{\rho}(\frac{du_{B1}(y,t)}{dy})^{m-1}$, $K_{H3}=[\frac{(2n-1)\pi}{2(d-y_{pi})}]^2\frac{\kappa}{\rho}(\frac{du_{B3}(y,t)}{dy})^{m-1}$, the arbitrary constant C can be confirmed based on the initial conditions.

Based on the above velocity solutions, it can be seen that there is a comprehensive characteristic function affected by the piston excitation velocities for the velocity solution, that is

$$\phi(t)=[\frac{4}{(2n-1)\pi}\int f(t)e^{K_{Bi}t}dt+C]e^{-K_{Bi}t} \tag{12}$$

It is a function of the time variable and concerned with the input excitation mode (piston motion).

3.2. Dynamic damping force

To calculate $\Delta P(t)$, there are at least three equations for three unknown quantities $\Delta P(t)$, y_{pi} and y_{po}. One equation is about volume flow. For incompressible continuous MR fluids, it is known that the total volume flux through the damping channel Q_c equals the volume flux displaced by the piston Q_p, namely

$$Q_c=Q_p=S_pV_p(t) \tag{13}$$

where S_p is the active cross-section of piston head. The total volume flux composes of the volume fluxes in each region obtained by velocity-annular gap width integral, namely

$$Q_c=|Q_1+Q_2+Q_3|=\left|b\int_0^{y_{pi}}u_1(y,t)dy+b\int_{y_{pi}}^{y_{po}}u_2(y,t)dy+b\int_{y_{po}}^{d}u_3(y,t)dy\right| \tag{14}$$

where b is the average perimeter of the annular damping channel, $u_1(y,t)$ is general term for $u_{B1}(y,t)$ and $u_{H1}(y,t)$, $u_2(y,t)$ is general term for $u_{B2}(y,t)$ and $u_{H2}(y,t)$, $u_3(y,t)$ is general term for $u_{B3}(y,t)$ and $u_{H3}(y,t)$.

The plug flow region 2 is rigid and the total yield stress ($\tau(y_{pi})-\tau(y_{po})=2\tau_y$) linearly varies along the plug thickness. Based on force equilibrium of plug region 2, the plug thickness can be obtained and can rewritten as

$$y_{po}-y_{pi}=2\tau_y\left|\frac{\Delta P(t)}{L}-\rho\frac{\partial u_2(y,t)}{\partial t}\right|^{-1} \tag{15}$$

Another equation, according to equations (6), (10) or (11), the system of equations can be obtained, that is

$$y_{po}+y_{pi}=d \tag{16}$$

Incorporating equations (14), (15) and (16), the numerical solution of pressure drop $\Delta P(t)$ can be worked out, and the damping force can be confirmed by the equation

$$F(t)=\Delta P(t)S_p \tag{17}$$

4. Illustration of examples

For the unsteady analysis of MR dampers, four piston excitation motion cases are discussed. The flow due to linearly accelerated piston motion and suddenly piston motion proposed by Das and Arakeri are analyzed [10], in addition, the piston velocity $V_p(t)$ moves with an exponential velocity, and the oscillatory piston motion are also considered. For a certain excitation load, if the comprehensive function of equation (12) is confirmed, its damping force can be calculated. Hence, only the comprehensive functions of four piston excitation motion is solved.

4.1. Constant acceleration piston motion

For a piston with constant acceleration (a_p, where $V_p(t) = a_p t$), the corresponding characteristic function is

$$\phi(t) = [\frac{4Ba_p(K_{Bi}t-1)e^{K_{Bi}t}}{(2n-1)\pi K_{Bi}^{\ 2}} + C]e^{-K_{Bi}t} \tag{18}$$

where B is a constant.

4.2. Suddenly started piston motion

For a piston motion with a suddenly started velocity, the motion equation is

$$\begin{aligned} V_p(t) &= 0, (t \le 0) \\ V_p(t) &= V_0, (t > 0) \end{aligned} \tag{19}$$

where V_0 is the constant velocity. The corresponding characteristic function is

$$\phi(t) = [\frac{4BV_0 e^{K_{Bi}t}}{(2n-1)\pi K_{Bi}} + C]e^{-K_{Bi}t} \tag{20}$$

4.3. Exponential velocity piston motion

The piston motion with exponential velocity can be described as

$$V_p(t) = V_1 e^{-\lambda t} \tag{21}$$

where λ is the attenuation coefficient of velocity, V_1 is the initial velocity. The corresponding characteristic function is

$$\phi(t) = [\frac{4BV_1 e^{(K_{Bi}-\lambda)t}}{(2n-1)\pi(K_{Bi}-\lambda)} + C]e^{-K_{Bi}t} \tag{22}$$

4.4. Oscillatory piston motion

The sinusoidal displacement excitation is usually applied on MR dampers to characterize their properties, such as force-stroke and force-velocity profiles. Here, the sinusoidal displacement excitation is

$$X(t) = A\sin(\omega t) \tag{23}$$

where A , ω are the sinusoidal excitation displacement amplitude and angular velocity, respectively. The corresponding characteristic function is

$$\phi(t) = \frac{4B}{(2n-1)\pi({K_{B1}}^2+\omega^2)}\left[(K_{B1}\cos\omega t+\omega\sin\omega t)e^{K_{Bi}t}+C\right]e^{-K_{Bi}t} \tag{24}$$

According to the solution analysis procedure in section 3, damping forces for each piston motions can be numerically solved. In this work, force-stroke and force-velocity profiles of flow mode MR dampers are investigated, in addition, assuming the initial condition is satisfied the following equation

$$u(y,0) = -u(y,T/2) \tag{25}$$

Assume there is an MR damper with the configuration parameters listed in table 2 referring to design methods of ER or MR dampers in the literatures [11, 12].

Table 2. Simulation parameters.

Parameters	Value
Sinusoidal excitation displacement amplitude	0.005m
MR fluid yield stress	20kPa
Sinusoidal excitation frequency	40Hz
Damping channel length	0.02 m
Damping channel inner radius	0.0185 m
Damping channel outer radius	0.02 m
MR dampers piston rod radius	0.01 m
Average perimeter of the damping channel	0.1209 m
damping channel width	0.0015 m
MR fluids mass density	2.650×103 kgm^{-3}
MR fluid zero field viscosity	1.000 Pa s

The relationships of the MR damper response between force versus displacement and force versus velocity with yield stress 20 kPa and frequency 40 Hz are shown in Figure 2(a) and 2(b), respectively. The consistency index κ in equations (2) and (3) is assumed to equal zero field viscosity of BP model η , in addition, the power-law exponent of HB model m of 0.9 and 1.0 are discussed, since most of the MR fluids show the shear thinning behaviours ($m<1$). For the non-stationary motion of MR damper piston, the dynamic damping force of MR damper can be predicted by the numerical analysis solutions of MR fluid unsteady flow. Compared with the damping force of MR damper for quasi-static analysis, the dynamic damping force can well present the effect of fluid inertia on the unsteady fluid flow with high velocity rates, shown in figure 2.

5. Conclusions

In this paper, a numerical solution procedure of dynamic damping force is given for unsteady flow of MR fluid within flow mode MR damper for the controlled non-stationary piston motion. According to the governing NS equations, the boundary and initial conditions, the analytical solutions of velocity profile in the annular damping channel can be derived and used to confirm the dynamic damping force. The numerical solution method can be applied to arbitrary non-stationary piston of MR damper.

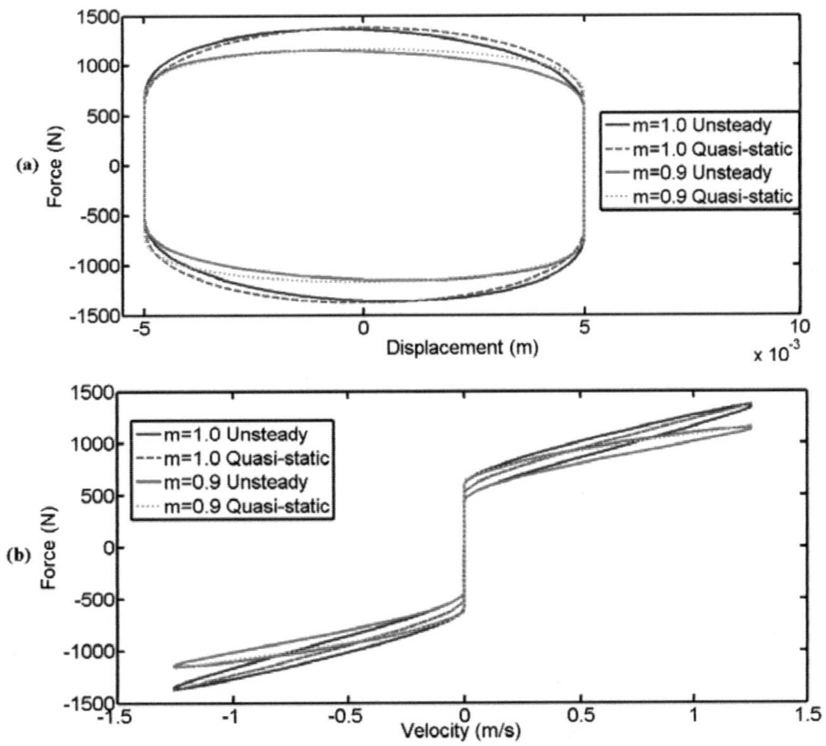

Figure 2. Relationships of (a) force versus displacement (b) force versus velocity for different power-law exponent of HB (0.9, 1.0).

Acknowledgments
The work of this paper was supported by the National Natural Science Foundation of China (No.50830202, No.51073179), the Fundamental Research Funds for the Central Universities (Grant No. CDJZR11128801) and the China postdoctoral science foundation (20100 480043).

References
[1] Yang G, Spencer Jr B F, Carlson J D and Sain M K 2002 *Eng. Struct.* **24** 309
[2] Lee D Y and Wereley N M 1999 *J. Intell. Mater. Syst. Struct.* **10** 761
[3] Hong S R, John S, Wereley N M, Choi Y T and Choi S B 2008 *J. Intell. Mater. Syst. Struct.* **19** 959
[4] Wereley N M and Li P 1998 *Smart Mater. Struct.***7** 732
[5] Li W H, Yao G Z, Chen G, Yeo S H and Yap F F 2000 *Smart Mater. Struct.* **9** 95
[6] Chen C I, Chen C K and Yang Y T 2004 *Appl. Math. Modelling* **28** 697
[7] Nguyen Q H and Choi S B 2009 *Smart Mater. Struct.* **18** 055016 (8pp)
[8] Stanway R, Sproston J L and El-Wahed A K 1996 *Smart Mater. Struct.* **5** 464
[9] Gordaninejad F and Breese D G 1999 *J. Intell. Mater. Syst. Struct.* **10** 634
[10] Das D and Arakeri J H 2000 *ASME J. Appl. Mech.* **67** 274
[11] Rosenfield N C and Wereley N M 2004 *Smart Mater. Struct.* **13** 1303
[12] Wang D H, Ai H X and Liao W H 2009 *Smart Mater. Struct.* **18** 115001

Vibration control of an artificial muscle manipulator with a magnetorheological fluid brake

H Tomori[1,3], Y Midorikawa[1] and T Nakamura[2]
[1]Dept. of Science and Engineering, Chuo University, 1-13-27 Kasuga, Bunkyo-ku, Tokyo, Japan
[2]Chuo University, 1-13-27 Kasuga, Bunkyo-ku, Tokyo, Japan

E-mail: h_tomori@bio.mech.chuo-u.ac.jp

Abstract. Recently, proposed applications of robots require them to contact human safely. Therefore, we focus on pneumatic rubber artificial muscle. This actuator is flexible, light, and has high-power density. However, because the artificial muscle is flexible, it vibrates when there is a high load. Therefore, we paid attention to the magnetorheological (MR) fluid. We propose a control method of the MR brake considering energy of the manipulator system. By this control method, MR brake dissipates energy leading to vibration of the manipulator. In this paper, we calculated the energy and controlled the MR brake. And, we deliberated the proposal method by simulation using the dynamic model of the manipulator, and experiment.

1. Introduction

Recently, there have been advances in the development of partner robots aimed at providing life support and robots aimed at providing medical treatment and care such as power assistance [1]. These robots are expected to be safe to human when these robots contact human. Then, we developed the straight-fiber-type artificial muscle [2] as actuators. It has a high contraction percentage and high contractive force, and this muscle has a long life compared with past McKibben type rubber artificial muscle [3-6]. Moreover, these artificial muscles are lightweight, have high output, and are flexible.

However, because the artificial muscle is flexible, it vibrates when there is a high load. Moreover, the artificial muscle manipulator has a slow response because it is operated by air pressure, and there is a limit to the response of the momentary power. Therefore, we are interested in the use of magnetorheological (MR) fluid [7], because its apparent viscosity can be changed very quickly (in milliseconds) by applying a magnetic field.

Until now, we have been developed a 1-degree-of-freedom artificial muscle manipulator using MR brake and controlled vibration of the artificial muscle by simple control of the MR brake. However, this control does not consider the characteristic of the manipulator. As a result, the MR brake interfered with the artificial muscle excessively.

Thus, in this paper, we propose a control method of the MR brake considering energy of the manipulator system. By this control method, MR brake dissipates energy leading to vibration of the manipulator. And we examine this proposal method by experiment and simulation using dynamic characteristic model of the manipulator.

---

[3]Corresponding author

2. The straight-fiber-type artificial muscle

The schematic of the artificial muscle developed at our laboratory is shown in figure 1. Table I shows the specifications of the artificial muscle used in this study. The form of this artificial muscle is tubular, and the material used is natural-rubber-latex liquid and carbon fiber. Moreover, since its structure includes a carbon fiber layer in the direction of an axis, if air pressure is supplied to this artificial muscle, the rubber membrane will expand, but the fiber stops the growth in the direction of an axis so that the membrane is mostly not extended. As a result, the muscle expands only in a radial direction and contracts in the direction of an axis. This artificial muscle is lightweight, has high output, and is flexible.

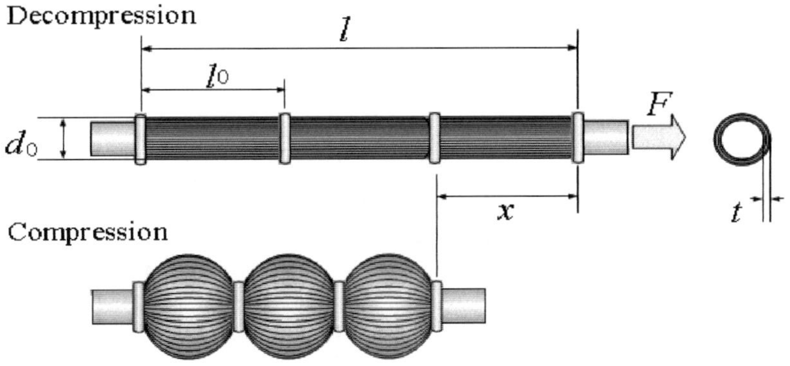

Figure 1. A schematic diagram of the straight-fiber-type artificial muscle.

Table 1. Specifications of artificial muscle.

Length l [mm]	187.0
The length between cap and ring l_0 [mm]	62.3
Diameter d_0 [mm]	12.3
Thickness t [mm]	0.23
Elastic modulus K [Pa]	800000
Number of rings	2

3. MR brakes

We focused on the MR brake as a way to apply MR fluid to the joint. We used the MRB-2107-3 of the LORD Co. as MR braking equipment. The overall view and the diagrammatic illustration of the MRB-2107-3 are shown in figure 2. And, the characteristics of the MR brake are shown in figure 3. As figure 3, the MR brake changes apparent viscosity by magnetic field. And relationship between its brake torque and rotational speed is low that it can ignore.

This MR brake has high speed response in units of several milliseconds. Thus, we guess MR brake can control vibration of the arm of the manipulator. This device generates friction on the surface of the disk as it changes in response to the magnetic field by arranging the MR fluid in the internal disk surroundings. It is possible to apply brakes to a rotation operation according to this mechanism. The MR brake has the following features:

- Since torque density is high, it is small, and it is high output.
- Since the structure is simple, it is strong.
- The speed of response is in units of several milliseconds.
- It can be changed continuously.
- It is stable.
- It is an energy-saving model.

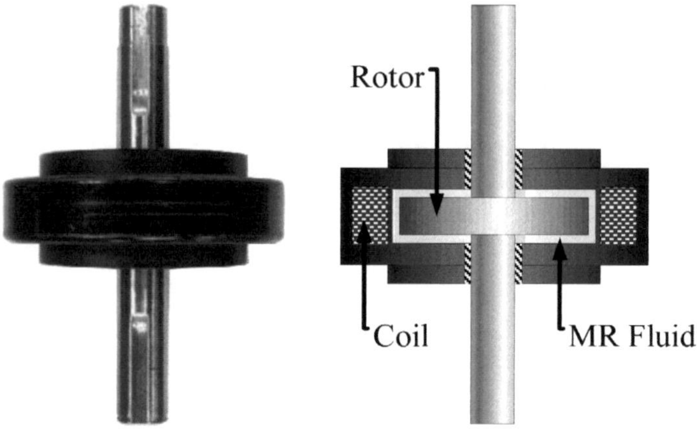

Figure 2. A configuration of MR brake.

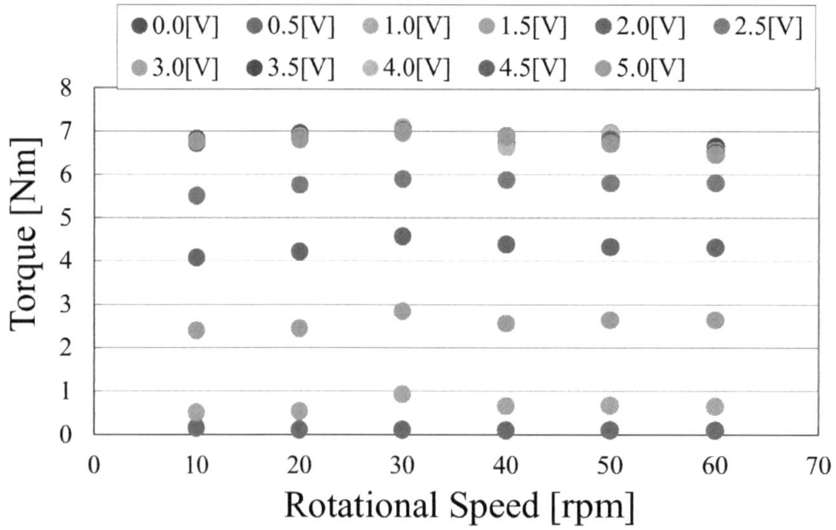

Figure 3. Relationship between the rotational speed and the torque of MR brake.

4. The artificial muscle manipulator with the MR brake

Figure 4 shows the overall view of the developed one degree of freedom artificial muscle manipulator. And Table 1 shows parameters of this manipulator. The artificial muscle manipulator is arranged as two artificial muscles in parallel. Further it transmits the contractile force of the artificial muscle to the rotation axis through the belt pulley. The MR brake equipment is fixed to the first link side, and it can apply braking power to the axis of rotation. The manipulator is installed in the load cell to detect the load torque that hangs to the joint and the encoder to detect the joint angle.

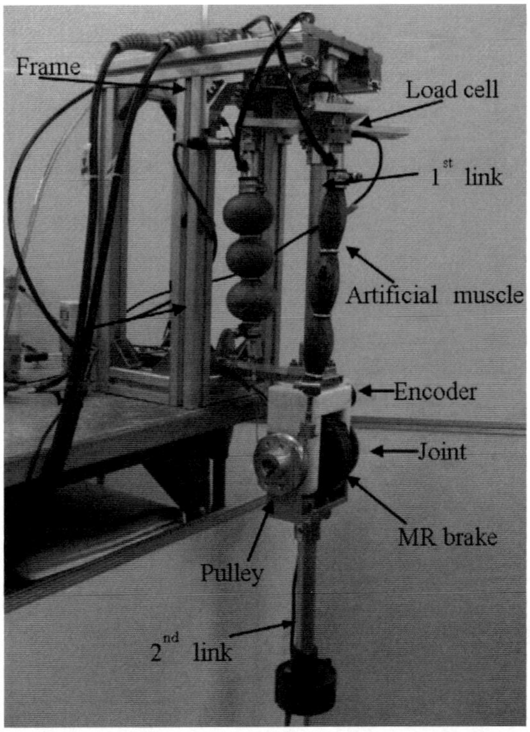

Figure 4. The artificial muscle manipulator with the MR brake.

Table 2. Parameters of the artificial muscle manipulator.

Mass of 2nd link M_2 [kg]	0.168
Length of 2nd link l [mm]	250
Center of gravity of 2nd link l'_G [mm]	240
Movable range θ [rad]	$-\pi/2 \leqq \theta \leqq \pi/6$
Mass moment of inertia of 2nd link I_1 [kgm^2]	4.80×10^{-6}
Parallel number of artificial muscles n_p	1

5. Manipulator model

In this study, we built a manipulator model for approach against control from theory. However, this system is difficult to reproduce as simple mass-spring-damper system because there is lateness of air pressure in this system. Therefore, we built a manipulator model while considering the dynamic characteristics of artificial muscles.

The schematic view of this manipulator model is shown in figure 5. This model consists of three parts. The first part is controller using a mechanical equilibrium model [2] treating the static characteristic of an artificial muscle. Since the artificial muscles are highly nonlinear in contraction, contractile force, and pressure, they are difficult to control. This controller can linearize the characteristic of artificial muscles.

The second part is a dynamic characteristics model containing the elements related to the dynamic characteristics of the air artificial muscle system. As these elements, the speed response of a solenoid valve, the ease of passing of the air in an air tube, the pressure change in an artificial muscle, and the volume change in an artificial muscle are mentioned [8].

The third part is a load system model of the manipulator. This model is based on forward dynamics. These models were combined and considered as the manipulator model.

Journal of Physics: Conference Series **412** (2013) 012053 doi:10.1088/1742-6596/412/1/012053

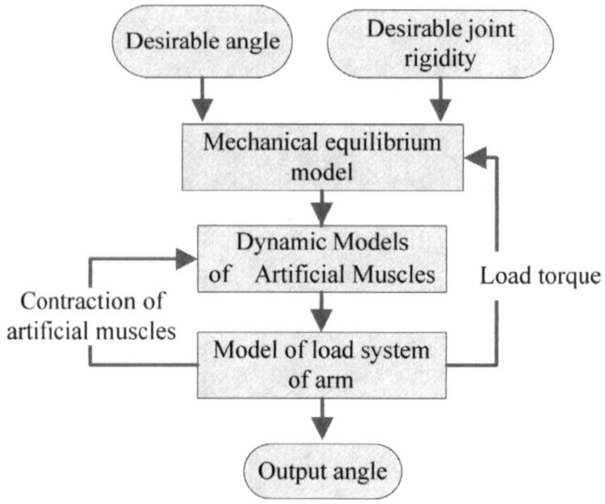

Figure 5. A schematic of the manipulator mode.

6. Calculation of energy for dissipation

In this section, we propose a control method considering mechanical energy in the manipulator system. And a schematic diagram of joint manipulator is shown in figure 6. In this system, load cells detect forces F_1 and F_2. These forces contain force for joint stiffness F_0, force by load F_L and elastic force Fe by inertia force of the arm.

By the proposed method, the MR brake dissipates energy leading to vibration of the arm of the manipulator. Therefore, we calculate kinetic energy of the arm and elastic energy of the artificial muscle. Then, we calculate brake torque of MR brake for dissipation of this mechanical energy.

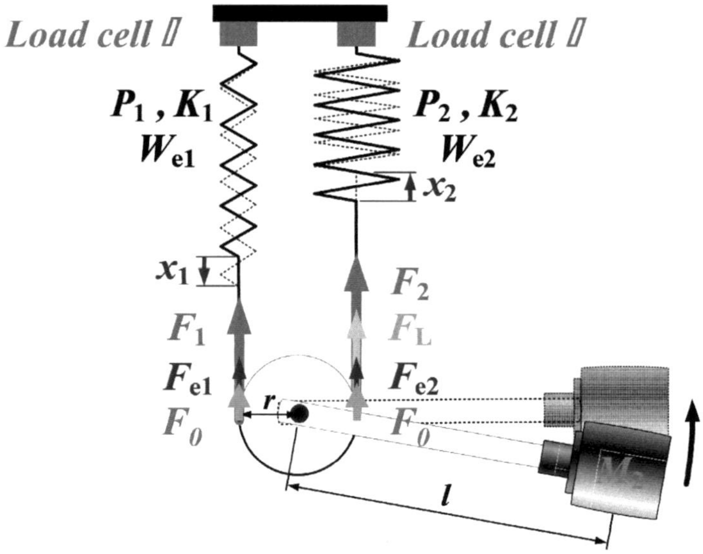

Figure 6. Schematic diagram of joint manipulator.

6.1. Kinetic energy

Kinetic energy of the manipulator W_k is calculated by following equation. Where, I is Mass moment of inertia of the arm, and θ' is angular velocity.

Journal of Physics: Conference Series **412** (2013) 012053 doi:10.1088/1742-6596/412/1/012053

$$W_k = \frac{1}{2} I \dot{\theta}^2 \tag{1}$$

6.2. *Elastic energy*

The artificial muscles generate vibration because of own flexibility. Therefore, we control the vibration by dissipating elastic energy using MR brake. Elastic energy of artificial muscles leading to vibration of the arm W_e calculated by following equation. Where, K_i is a spring constant of artificial muscles.

$$W_e = W_{e1} + W_{e2} \tag{2}$$

$$W_{ei} = \frac{1}{2} K_i x_i^2 \tag{3}$$

Next, we calculated K_i. The spring constant of artificial muscles K_i changes with pressure values linearly [2]. Therefore, each spring constant is calculated by the following formulas. Where, P_i is pressure, and K_{ai} are coefficients between stiffness and pressure of artificial muscles.

$$K_i = K_{ai} P_i \tag{4}$$

Next, we calculated displacement of artificial muscle x_i from the spring constant of artificial muscles and elastic force F_{ei}.

$$x_i = \frac{F_{ei}}{K_i} \tag{5}$$

Where, the artificial muscle manipulator is arranged two artificial muscles to parallel. Further it transmits the contractile force of the artificial muscle to the rotation axis through the belt pulley. Therefore, force F_1 and F_2 detected by load cell contain ①,force for joint stiffness, ② load and ③ elastic force. Thus, for detection of elastic force, we calculated difference of force F_1 and F_2. Thereby, force for joint stiffness is eliminated.

$$F_e = F_2 - F_1 - F_L \tag{6}$$

$$F_{e1} = F_e \quad (F_e > 0) \tag{7}$$

$$F_{e2} = -F_e \quad (F_e < 0) \tag{8}$$

Where, F_L is force of load calculated from mass of load M_2, length of arm l, radius of pulley r and angle of the arm θ. The elastic force F_e is pull force to the artificial muscles by inertia force of the arm. This elastic force acts on one of artificial muscles or another by the movement direction of an arm as F_{e1} or F_{e2}.

$$F_L = \frac{M_2 g l \sin\theta}{r} \tag{9}$$

And, we calculated W_e from (2) to (9).

$$W_e = \frac{1}{2}\left(\frac{F_{e1}^{\ 2}}{K_{a1} P_1} + \frac{F_{e2}^{\ 2}}{K_{a2} P_2} \right) \tag{10}$$

6.3. *Dissipation energy of the MR brake*

In this section, we calculated the dissipation energy by MR brake W_m. Figure 7 shows the schematic view of the energy dissipation by MR brake. Where, θ_d is desirable angle of the arm. In the proposed method, we dissipate energy using MR brake in the area named dissipation section $\theta\alpha$. Changing $\theta\alpha$, we can adjust dissipated energy. From this figure, the energy W_m calculated from brake torque τ_m and angular velocity.

$$W_m = \tau_m \int \dot{\theta} dt \tag{11}$$

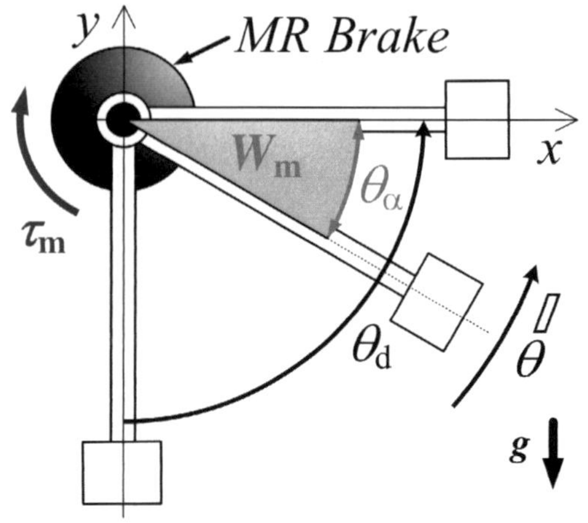

Figure 7. A schematic of the energy dispersion.

And, output torque of the MR brake τ_m is controlled for dissipation mechanical energy leading to vibration. Therefore, τ_m is calculated by following equation.

$$W_m = W_k + W_e \tag{12}$$

$$\tau_m = \frac{\left\{ I\dot{\theta}^2 + \frac{F_{e1}^{\ 2}}{K_1} + \frac{F_{e2}^{\ 2}}{K_2} \right\}}{2\int \dot{\theta} dt} \tag{13}$$

7. Vibration control experiment by the MR brake

7.1. Simulation for decision of energy dissipation section

In this proposed method, energy dissipated by MR brake depends on width of dissipation section θ_α. Thus, we need to decide width of energy dissipation section. Therefore, we deliberated the section using the dynamic model of the manipulator. In this simulation, desirable angle is 60 [deg], joint stiffness is 0.1 [Nm/deg], load is 9.6[N] and dissipation section is 10 [deg] to 40 [deg].

Figure 8 shows a result of simulation, and figure 9 shows the amount of vibration and deviation area in each dissipation section. In figure 9, Integration value of speed is defined as Integration value of the magnitude of the velocity vector of the arm. This value becomes high when vibration of the arm increases. Thus, we used this value as evaluation of vibration of the arm. Therefore, we decided dissipate section so that integration value of speed and Difference area of position to step input become low. In this study, we attached weight to Integration value of speed. Therefore, we chose 30 [deg] as dissipation section.

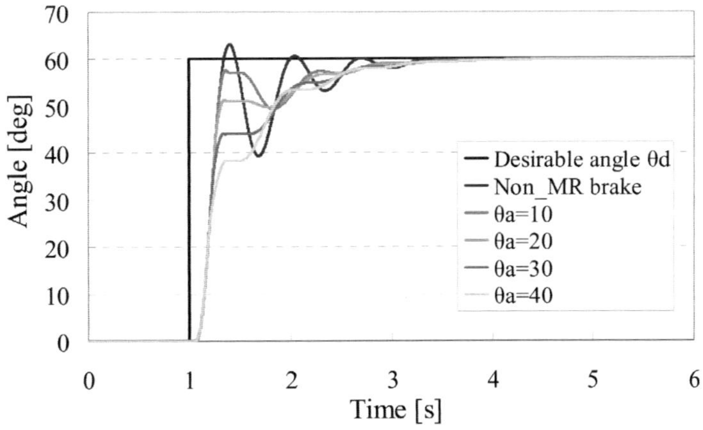

Figure 8. The simulation result (desirable angle is 60 [deg], joint stiffness is 0.1 [Nm/deg], and load is 9.6[N]).

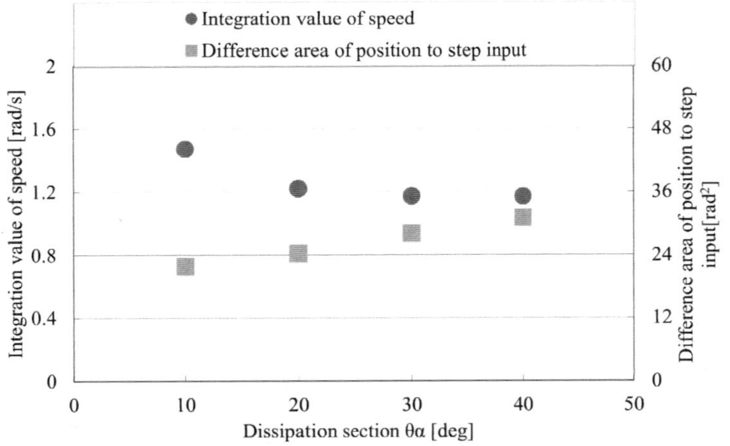

Figure 9. Relationship between dissipation section and integration value of speed and difference area of position to step input.

7.2. Vibration control experiment

In this section, we experiment to lift the arm with vibration control by proposed method. In this experiment, desirable angle θ_d is 60 [deg], joint stiffness is 0.1 [Nm/deg], load is 9.6[N] and dissipation section θ_α is 30 [deg]. First, the arm raises the load to the desirable angle. Then, energy dissipation is started 30 degrees before from the desirable angle. Dissipated energy is calculated from sensor value in real time.

A result of vibration control experiment and simulation is shown in figure 10. From This figure, the dynamic model of the manipulator reproduces system of the manipulator in points of dead time, rise time and motion of the arm. And we conclude that the MR brake controls vibration of the arm successfully by proposed method, from experimental result and simulation result. And from figure 11, we confirmed that the mechanical energy in the vibration domain was dissipated.

However, there is response delay of MR brake because of coil in experimental result. Thus we need to develop a driver for the MR brake.

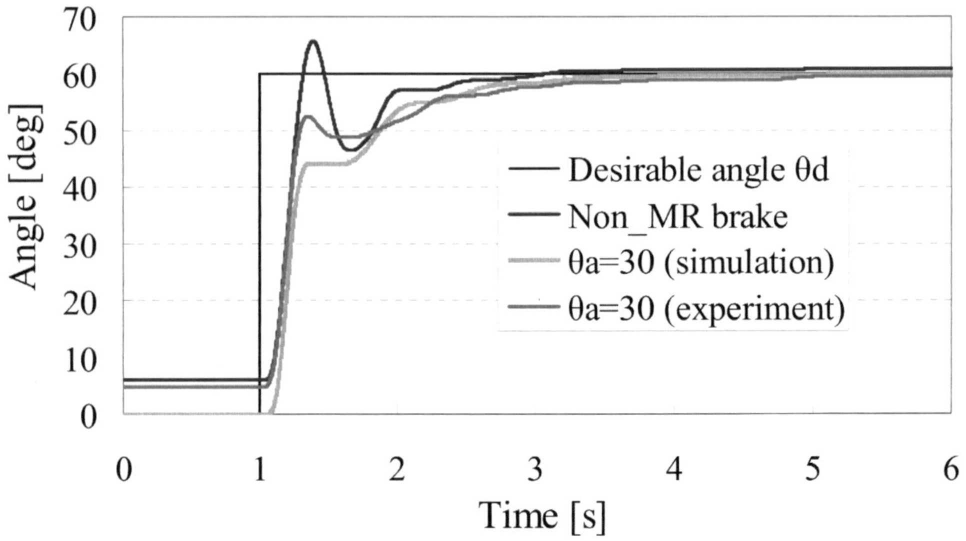

Figure 10. The experimental result (the dissipation section is 30 [deg]).

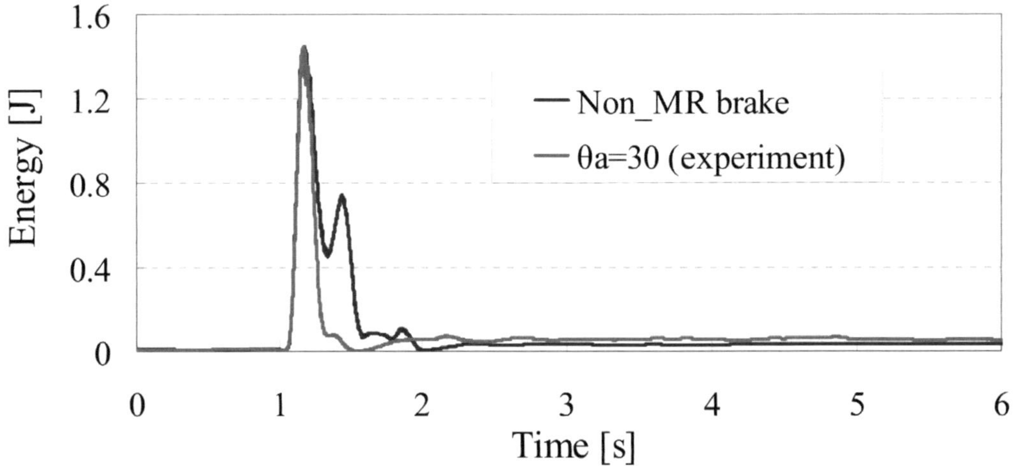

Figure 11. The energy change of the manipulator when the energy dissipation is applied.

8. Conclusion

We proposed the energy dissipation method for control of the MR brake. We calculated the energy leading to vibration and controlled the MR brake. Then, we applied the proposed method numerically using the dynamic model of the manipulator. As a result, we have shown that the MR brake controls vibration of the arm successfully by the proposed method. We confirmed that the mechanical energy in the vibration domain was dissipated faster.

For future works, we plan to develop a MR brake driver to improve response of the MR brake. In addition, we develop a multi-DOF manipulator using MR brake.

References

[1] Satoh H, Kawabata T, Tanaka F and Sankai Y 2010 *Transferring-care assistance with robot suit HAL* (Transactions of the Japan Society of Mechanical Engineers, Series C) **76** 227-35

[2] Nakamura T and Shinohara H 2007 Position and force control based on mathematical models of pneumatic artificial muscles reinforced by straight glass fibers *Proc. of IEEE Int. Conf. on Robotics and Automation ICRA* (Roma, Italy) pp 4361–66

[3] Gavrilovic M M and Maric M R 1969 Positional servo-mechanism activated by artificial muscles *Medical and Biological Engineering* **7** 77–82

[4] Klute G K, Czernieki J M and Hannaford B 1999 McKibben Artificial Muscles: Pneumatic Actuators with Biomechanical Intelligence *Proc. of the IEEE/ASME Int. Conf. on Advanced Intelligent Mechatronics* (Atlanta, GA, USA) pp 221–26

[5] Chou C P and Hannaford B 1994 Static and dynamic characteristics of McKibben pneumatic artificial muscles *Proc. of IEEE Int. Conf. On Robotics and Automation* (San Diego, CA, USA) pp 281–86

[6] Nakamura T 2006 Experimental comparisons between McKibben type artificial muscles and straight fibers type artificial muscles *SPIE Int. Conf. on Smart Structures, Devices and Systems III* (San Diego, CA, USA)

[7] Park B J, Park C W, Yang S W, Kim H M and Choi H J 2008 Core-shell typed polymer coated-carbonyl iron suspension and their magnetorheology *Int. Conf. on Electrorheological Fluids and Magnetorheological Suspensions ERMR08* (Dresden, Germany) p 102

[8] Tomori H, Midorikawa Y and Nakamura T 2010 Nonlinear dynamic characteristic model of artificial rubber muscle manipulator using MR brake *Proc. of the 12th Int. Conf. on Electrorheological Fluids and Magnetorheological Suspensions ERMR2010* (Philadelphia, PA, USA) pp 136-41

force, good durability, low energy consumption and strong reliability, which can be treated as a passive control devices even the control system failures. MR damper as a semi-active control device combines the reliability of passive control systems and the strong adaptability of active control system. It can achieve the similar control performance of the active control system by means of a certain control law. It has a good prospect of engineering application as semi-active damping device. Therefore, it has received more and more attention in recent years [3-6].

The performance of a semi-active control system is dependent on the choice of control strategy [7], such as skyhook control [8-10], groundhook control [11], neuro-fuzzy control [12], LQG control [13], adaptive control and nonlinear control [14] and H_∞ control [15-17]. The parameter uncertainties, caused by the structural vibration and load changes, affect the stability of the control system when modelling the seat suspension system. Therefore, the seat suspension control system requires a certain degree of robustness to deal with the uncertainty of the model and external incentives. However, previous studies often neglected the effect of the uncertainty on the control performance in the modelling process, whereas the H_∞ control can takes account of the effect of uncertainty and guarantee the robust stability of the system [18]. On the other hand, the damping force will increase with increasement of excitation. However, the damping force will not increase longer when the excitation current reaches a certain degree because of the internal structure of the damper, and the actuator saturation appears. Robust H_∞ control can effectively deal with the problem of the actuator saturation when exerting the control force over the seat suspension [18-19]. So, an appropriate robust H_∞ controller for the semi-active seat suspension system is designed to provide a trade-off of the two main performance requirements (ride comfort and suspension deflection).

In this research, the dynamical model of a seat suspension system with parameter uncertainties and actuator saturation is established. Then, combined with the damping force characteristics of MR damper, a state feedback controller is designed which guarantees H_∞ performance index and the constraints of the actuator saturation and the suspension deflection. Finally, some necessary comparisons between the H_∞ control strategy and the passive and skyhook control strategy are given, and the simulations are used to validate the effectiveness of the proposed control strategy.

2. Dynamic model of the seat suspension with MR damper

2.1. The characteristic of MR damper

The MR damper is illustrated schematically in Figure 1 and the Bingham model is adopted in this study for the MR damper (shown in Figure 2). The expression of the damping force is shown as follow:

$$F = c_e \dot{x}_d + f_d \operatorname{sgn}(\dot{x}_d) \tag{1}$$

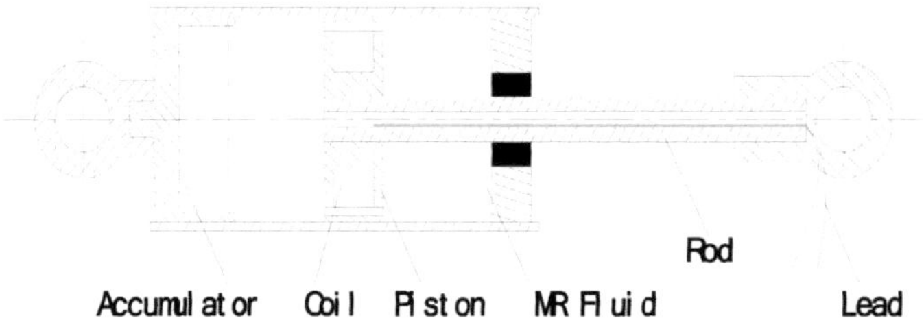

Figure 1. Schematic drawing of MR damper.

Where c_e is the viscous damping coefficient of the MR damper, $\dot{x}_d$ is the relative velocity between the piston and the cylinder. Therefore, the MR damper can be seen as the sum of a passive viscous damper and a semi-active coulomb damping device. The passive viscous damping is not controllable and coulomb damping force is controllable. By designing appropriate control algorithm, the current is changed, and the shear yield strength of the MR fluid is changed accordingly. Then, desired control force is achieved.

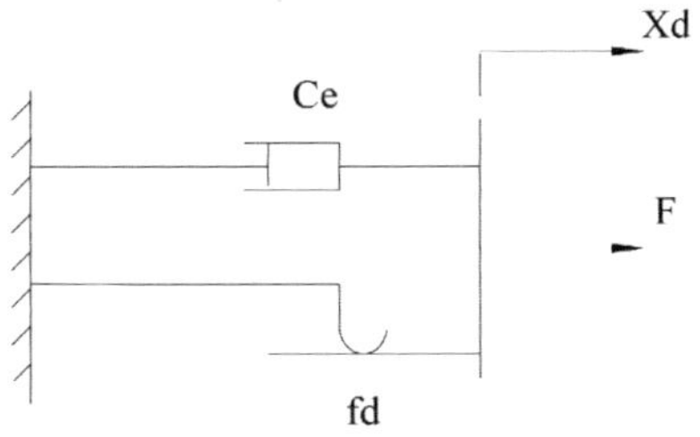

Figure 2. The Bingham model of MR damper.

RD 8040-1 MR damper from Lord Corporation is used in this study. By applying sinusoidal excitation with a fixed frequency of 1Hz and constant amplitude of 15mm, the response of force is changed with different input current from 0 to 2A with increments of 0.5A. Thus, the force versus displacement loops and the force versus velocity loops under various electric currents were obtained as shown in Figure 3 and Figure 4. It can be seen that the damper force will remarkably increase with the increasement of the input current.

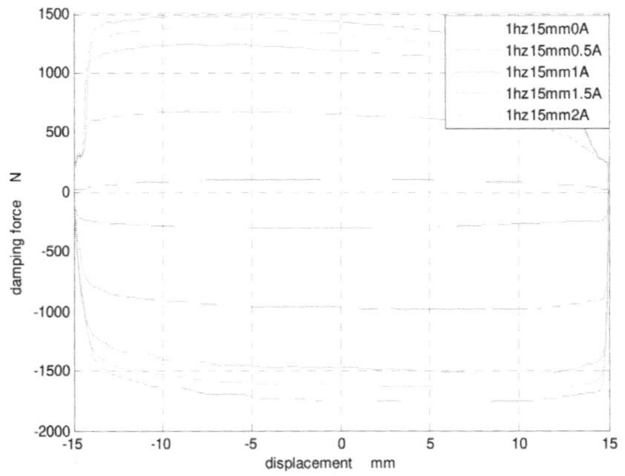

Figure 3. Force versus displacement loops under various currents .

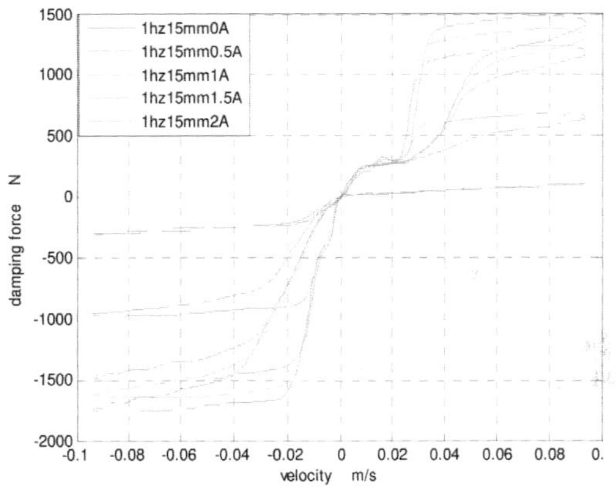

Figure 4. Force versus velocity loops under various current.

By expression 1, we can put viscous damping force into the system damping matrix. So, the MR damper can be seen as a simple controlled friction damper.

2.2. The dynamical model of a seat suspension system with MR damper

A physical model of the semi-active seat suspension system with a passive spring and a controllable MR damper is shown in Figure 5, whose dynamical model is abstracted in Figure 6.

Make $c_e = c_s$, $\dot{x}_d = \dot{z}_s - \dot{z}_r$, $f_d = u$ and define $w(t) = \dot{z}_r$

The dynamic equations of motion of the seat suspension system are given by:

$$m_p \ddot{z}_p + c_p(\dot{z}_p - \dot{z}_s) + k_p(z_p - z_s) = 0$$

$$m_s \ddot{z}_s + c_s(\dot{z}_s - \dot{z}_r) + k_s(z_s - z_r) - c_p(\dot{z}_p - \dot{z}_s) - k_p(z_p - z_s) = -u \quad (2)$$

Where m_p , m_s are the mass of the driver and the seat, respectively; z_r is the road displacement input, z_p , z_s are the displacements of the driver and the seat, respectively; k_p , k_s are stiffnesses of the driver and the passive suspension system, respectively; c_p , c_s represent the damping of the driver and the MR damper when the input current is zero, respectively; u is the active control input of the seat suspension.

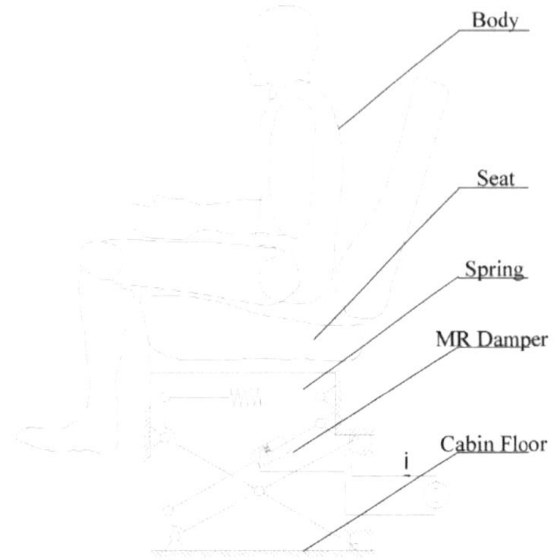

Figure 5. Physical model of the semi-active seat suspension with MR damper.

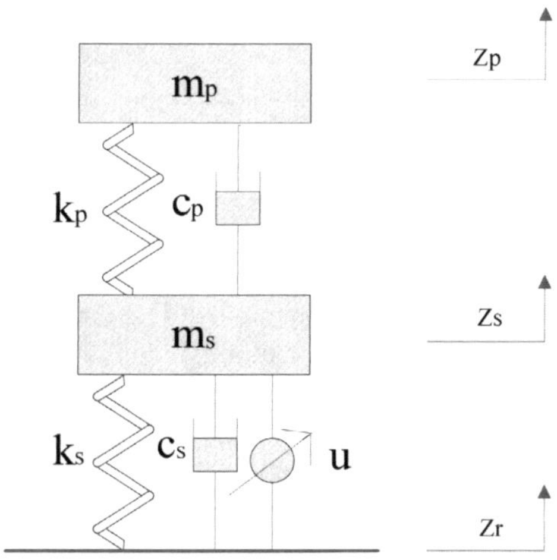

Figure 6. A dynamical model of the seat suspension system with MR damper.

Define the set of states:

$$x = [x_1, x_2, x_3, x_4]^T$$

where the components of state variables are given as follows:

$$x_1 = z_p - z_s \quad , \quad x_2 = \dot{z}_p \quad , \quad x_3 = z_s - z_r \quad , \quad x_4 = \dot{z}_s$$

Therefore, the dynamic equations in (2) can be written into a state space equation form as:

$$\dot{x}(t) = Ax(t) + Bu(t) + B_w w(t) \tag{3}$$

where

$$A = \begin{bmatrix} 0 & 1 & 0 & -1 \\ -\frac{k_p}{m_p} & -\frac{c_p}{m_p} & 0 & \frac{c_p}{m_p} \\ 0 & 0 & 0 & 1 \\ \frac{k_p}{m_s} & \frac{c_p}{m_s} & -\frac{k_s}{m_s} & -\frac{c_p + c_s}{m_s} \end{bmatrix}, B = \begin{bmatrix} 0 & 0 & 0 & -\frac{1}{m_s} \end{bmatrix}^T, B_w = \begin{bmatrix} 0 & 0 & -1 & \frac{c_s}{m_s} \end{bmatrix}^T$$

taking the parameter uncertainties and actor saturation into account, the seat suspension model can be expressed as:

$$\dot{x}(t) = A_1 x(t) + B_1 u(t) + B_w w(t) \tag{4}$$

The parameter uncertainties considered here are norm-bounded of the form:

$$A_1 = A + \Delta A\,,\ B_1 = B + \Delta B$$

$$\begin{bmatrix}\Delta A & \Delta B\end{bmatrix} = L\varepsilon(t)\begin{bmatrix}E_A & E_B\end{bmatrix}$$

where ΔA, ΔB are unknown real matrices, reflecting the parameter uncertainties in the system model. L, E_A, E_B are known constant real matrices of appropriate dimensions. $\varepsilon(t)$ is an unknown unreal matrix function of appropriate dimensions and satisfies $\varepsilon^T(t)\varepsilon(t) \le I$, I represents the unit matrix. And the actuator saturation nonlinearity is described by:

$$|u(t)| \le u_{\max} \tag{5}$$

where $u_{\max}$ is the maximum controlled damping force.

Ride comfort is closely related to the vertical acceleration of the body. Denote body acceleration as control output:

$$z_1(t) = \ddot{z}_p \tag{6}$$

Meanwhile, in order to ensure the ride comfort and prevent the seat mechanical structure from damaging, the controller should be able to inhibit suspension disturbance from exceeding the limits of its vibration. In order to satisfy the performance constraint, denote the

$$z_2(t) = \left|\frac{z_s - z_r}{z_{\max}}\right| < 1 \tag{7}$$

where $z_{\max}$ is the maximum limit of suspension deflection.

Therefore, the seat suspension control system can be described by equation 4 and the following equations:

$$z_1(t) = C_1 x(t) + D_1 u(t) \tag{8}$$

$$z_2(t) = C_2 x(t) \tag{9}$$

where $C_1 = \begin{bmatrix} -\dfrac{k_p}{m_p} & -\dfrac{c_p}{m_p} & 0 & \dfrac{c_p}{m_p}\end{bmatrix}$, $C_2 = \begin{bmatrix} 0 & 0 & \dfrac{1}{z_{\max}} & 0\end{bmatrix}$, $D_1 = 0$, C_1, C_2 are known constant real matrices of appropriate dimensions for describing the system model.

According to the performance requirements, the design goal can be summed up as follows:

(1) Define $T_{zw}(s)$ as the transfer function from the disturbance inputs $w(t)$ to the controlled output $\ddot{z}_p(t)$. The H_∞ controller is designed for improving the ride comfort such that the closed-loop system guarantees:

$$\|T_{zw}(s)\|_\infty < \gamma \tag{10}$$

where γ is a prescribed scalar.

(2) Taking the actuator saturation into account, equation 5 should be satisfied.

(3) Taking the safety of mechanical structure into account, equation 7 should be satisfied.

3. Robust H_∞ controller design for the seat suspension system with MR damper

In order to get the controller parameters, the following lemmas are necessary.

Lemma 1 [20]. Given appropriately dimensioned matrices D, $F(t)$ and E. For all $F(t)$ satisfying $F^T(t)F(t) \le I$, we have

$$DF(t)E + E^T F^T(t)D^{\mathrm{T}} \le \varepsilon^{-1}DD^{\mathrm{T}} + \varepsilon E^T E,$$

where ε is any scalar.

Lemma 2 [20]. For any matrices (or vectors) U and V with appropriate dimensions, we have

$$U^T V + V^T U \le \alpha U^T U + \alpha^{-1} V^T V,$$

where α is any scalar.

Lemma 3 [21]. For given symmetric matrix $S = \begin{bmatrix} S_{11} & S_{12} \\ S_{21} & S_{22} \end{bmatrix}$, the following three conditions are equivalent to each other, where S_{11} has the dimension of $r \times r$.

(i) $S < 0$; (ii) $S_{11} < 0, S_{22} - S^T{}_{12} S^{-1}{}_{11} S_{12} < 0$; (iii) $S_{22} < 0, S_{11} - S_{12} S^{-1}{}_{22} S^T{}_{12} < 0$.

In order to achieve the effect of the semi-active control, assume the MR damper as an active control device firstly, and the ideal force is derived from the active control algorithm, obtained control force is constrained by the semi-active control law, and then the force of MR damper is approximate to the ideal force to perform the semi-active control.

In this study, it is assumed that all the state variables can be measured, and we are interested in designing a state feedback controller:

$$u(t) = Kx(t) \tag{11}$$

where K is the state feedback gain matrix to be designed, such that the H_∞ norm of the closed-loop system is minimized, while satisfying the constraints in (5) and (7) for all nonzero $w(t) \in L_2[0,\infty)$ with the initial condition.

Theorem 1: Let positive scalars γ, η and ρ be given. If there exist matrices $X > 0$, Z, satisfying inequalities (12), (13), (14), such that the closed-loop system (4) is asymptotically stable for the state feedback controller (11), $\|T_{zw}(s)\|_\infty < \gamma$ is satisfied and the constraints in (5) and (7) are guaranteed for all nonzero $w(t) \in L_2[0,\infty)$. Moreover, the state feedback gain matrix is obtained as $K = ZX^{-1}$.

$$\begin{bmatrix} XA^T + Z^T B^T + AX + BZ & L & XE_A{}^T + Z^T E^T{}_B & B_w & XC_1{}^T \\ L^T & -\eta^{-1} I & 0 & 0 & 0 \\ E_A X + E_B Z & 0 & -\eta I & 0 & 0 \\ B_w{}^T & 0 & 0 & -\gamma^2 I & 0 \\ C_1 X & 0 & 0 & 0 & -I \end{bmatrix} < 0 \tag{12}$$

$$\begin{bmatrix} -I & \sqrt{\rho}Z \\ \sqrt{\rho}Z^T & -u^2_{\max}X \end{bmatrix} < 0 \tag{13}$$

$$\begin{bmatrix} -I & \sqrt{\rho}C_2X \\ \sqrt{\rho}XC_2^T & -X \end{bmatrix} < 0 \tag{14}$$

Proof: Define a Lyapunov function for system as

$$V(x(t),t) = x^T(t)Px(t) \tag{15}$$

where P is a positive definite matrix.

By differentiating (15), we obtain

$$\begin{aligned} \dot{V}(x(t),t) &= \dot{x}^T(t)Px(t) + x^T(t)P\dot{x}(t) \\ &= x^T(t)\left[(A+BK)^T P + P(A+BK)\right]x(t) \\ &+ x^T(t)\left[(E_A+E_BK)^T \varepsilon^T(t)(PL)^T + PL\varepsilon(t)(E_A+E_BK)\right]x(t) \\ &+ \left[B_w w(t)\right]^T Px(t) + x^T(t)P\left[B_w w(t)\right] \end{aligned} \tag{16}$$

By using **Lemma 1**, we obtain

$$\dot{V}(x(t),t) \le x^T(t)\Gamma x(t) + \left[B_w w(t)\right]^T Px(t) + x^T(t)P\left[B_w w(t)\right] \tag{17}$$

where $\Gamma = (A+BK)^T P + P(A+BK) + \eta^{-1}PL(PL)^T + \eta(E_A+E_BK)^T(E_A+E_BK)$, and η is any positive scalar.

Define H_∞ performance index $J = \int_0^\infty \left[z_1^T(t)z_1(t) - \gamma^2 w(t)^T w(t)\right]dt$.

By using inequality (17) and J

$$\begin{aligned} &\dot{V}(x(t),t) + z_1^T(t)z_1(t) - \gamma^2 w(t)^T w(t) \\ &\le \left[x^T(t) \quad w^T(t)\right]\Pi\left[x(t) \quad w(t)\right] \end{aligned} \tag{18}$$

where,

$$\Pi = \begin{bmatrix} \Gamma + C_1^T C_1 & PB_w \\ B^T_w P & -\gamma^2 I \end{bmatrix} \tag{19}$$

By using lemma 3, if $\Pi < 0$, we obtain

$$\begin{bmatrix} (A+BK)^T P + P(A+BK) & PL & (E_A+E_BK)^T & PB_w & C_1^T \\ L^T P & -\eta^{-1}I & 0 & 0 & 0 \\ E_A+E_BK & 0 & -\eta I & 0 & 0 \\ B_w^T P & 0 & 0 & -\gamma^2 I & 0 \\ C_1 & 0 & 0 & 0 & -I \end{bmatrix} < 0 \tag{20}$$

It is now deduced from the above that if $\Pi < 0$, then $\dot{V}(x(t),t) < 0$. Therefore, the closed-loop system (4) with the controller (11) is asymptotically stable for the parameter uncertainties and actuator saturation of the seat suspension system. And if $\Pi < 0$, $\dot{V}(x(t),t) + z_1^T(t)z_1(t) - \gamma^2 w(t)^T w(t) < 0$. Therefore, the system satifies the H_∞ performance index for all nonzero $w(t) \in L_2[0,\infty)$.

By using equation (16) and lemmas 2, we have

$$\dot{V}(x(t),t) \le \gamma^2 w(t)^T w(t) \tag{21}$$

and by integrating both sides of inequality(21), we obtain

$$V(x(t),t) - V(x(0),0) \le \int_0^t \gamma^2 w(t)^T w(t)dt \le \gamma^2 \|w(t)\|_2^2 = \gamma^2 w_{\max} \tag{22}$$

where $w_{\max}$ is the disturbance energy.

This shows that,

$$x^T(t)Px(t) \le V(x(0),0) + \gamma^2 w_{\max} = \rho \tag{23}$$

Define $\bar{x}(t) = P^{\frac{1}{2}}x(t)$, from inequality (23), it follows that $\bar{x}^T(t)\bar{x}(t) \le \rho$. Hence

$$\begin{aligned}
\max_{t\ge0} |u(t)|^2 &= \max_{t\ge0} \|Kx(t)\|_2^2 = \max_{t\ge0} \left\|x^T(t)K^T Kx(t)\right\|_2 \\
&= \max_{t\ge0} \left\|\bar{x}^T(t)P^{-\frac{1}{2}}K^T KP^{-\frac{1}{2}}\bar{x}(t)\right\|_2 \le \rho\lambda_{\max}\left(P^{-\frac{1}{2}}K^T KP^{-\frac{1}{2}}\right) \\
\max_{t\ge0} |z_2(t)|^2 &= \max_{t\ge0} \left\|x^T(t)C_2^T C_2 x(t)\right\|_2 \\
&= \max_{t\ge0} \left\|\bar{x}^T(t)P^{-\frac{1}{2}}C_2^T C_2 P^{-\frac{1}{2}}\bar{x}(t)\right\|_2 \le \rho\lambda_{\max}\left(P^{-\frac{1}{2}}C_2^T C_2 P^{-\frac{1}{2}}\right)
\end{aligned} \tag{24}$$

where $\lambda_{\max}(\cdot)$ represents the maximum eigenvalue.

Then, the constraints in (5) and (7) hold if

$$\rho P^{-\frac{1}{2}}K^T KP^{-\frac{1}{2}} < u^2_{\max} I \tag{25}$$

$$\rho P^{-\frac{1}{2}}C_2^T C_2 P^{-\frac{1}{2}} < I \tag{26}$$

By using lemma 3, inequalities (25) and (26) are equivalent to

$$\begin{bmatrix} -I & \sqrt{\rho}K \\ \sqrt{\rho}K^T & -u^2_{\max}P \end{bmatrix} < 0 \tag{27}$$

$$\begin{bmatrix} -I & \sqrt{\rho}C_2 \\ \sqrt{\rho}C_2^T & -P \end{bmatrix} < 0 \tag{28}$$

Since expressions like (20) cannot be handled directly by LMI optimization. In order to solve the nonlinear problem, pre- and post-multiply by $J_1 = diag(P^{-1}, \quad I, \quad I, \quad I, \quad I)$ and its transpose to (20) and pre-multiply and post-multiply by $J_2 = diag(I, P^{-1})$ and its transpose to (27) and (28). Define $X = P^{-1}$ and $Z = KP^{-1}$, inequalities (12), (13) and (14) are derived.

Remark 1. It is noticed that inequality (12), (13) and (14) are LMIs to γ^2. In order to obtain the lower limit of H_∞ performance index, we can minimize the variable γ. Therefore, the controller design problem can be changed to a problem of finding a solution to:

$$\min \quad \gamma^2 \quad s.t \ (12), (13) \text{ and } (14) \tag{29}$$

This minimization problem can be solved by using the Matlab LMI Toolbox. And the solution to (29) will be dependent on the values of η and ρ. The choice of values for η and ρ is a trial and error process. In general, the use of small values of η and ρ may mean that a high gain controller design could be obtained.

The controller gain matrix, K ,obtained by solving the problem of (29), will be used to generate a desired control force. Then desired damping force of MR damper could be obtained by the semi-active control law, and the expression is as follows

$$f_d = \begin{cases} u(t)_{\max} & u(t)(\dot{z}_s - \dot{z}_r) \geq 0, \ |u(t)| \geq u(t)_{\max} \\ u(t) & u(t)(\dot{z}_s - \dot{z}_r) \geq 0, \ |u(t)| < u(t)_{\max} \\ 0 & u(t)(\dot{z}_s - \dot{z}_r) \leq 0 \end{cases} \tag{30}$$

The actual damping force can be obtained by applying the corresponding current to the MR damper.

4. Simulations

In order to validate the effectiveness of the control method proposed in the above section, simulations are conducted in this section. The parameters that describe seat suspension system with MR damper are listed in table 1 for reference. And the maximum suspension deflection is defined as $z_{\max} = 0.1m$ [19], the maximum control force is assumed as $u_{\max} = 3000N$. The norm-bounded parameter uncertainties are expressed as:

$$L = \begin{bmatrix} 0 & 0 & 0 & 1 \end{bmatrix}^T, \ E_A = \begin{bmatrix} 0 & 0 & -\dfrac{0.1k_s}{m_s} & \dfrac{0.1c_s}{m_s} \end{bmatrix}$$

According to the Table 1, constant matrices for describing the seat suspension system model are as follows:

$$A = \begin{bmatrix} 0 & 1 & 0 & -1 \\ -\dfrac{44100}{50} & -\dfrac{1420}{50} & 0 & \dfrac{1420}{50} \\ 0 & 0 & 0 & 1 \\ \dfrac{44100}{26} & \dfrac{1420}{26} & -\dfrac{17085}{26} & -\dfrac{2420}{26} \end{bmatrix}, \ B = \begin{bmatrix} 0 & 0 & 0 & -\dfrac{1}{26} \end{bmatrix}^T,$$

$$B_w = \begin{bmatrix} 0 & 0 & -1 & \frac{1000}{26} \end{bmatrix}^T, C_1 = \begin{bmatrix} -\frac{44100}{50} & -\frac{1420}{50} & 0 & \frac{1420}{50} \end{bmatrix}, C_2 = \begin{bmatrix} 0 & 0 & 10 & 0 \end{bmatrix}$$

Table 1. Parameter values of the seat suspension system model.

Mass (kg)	Stiffness (N/m)	Damping coefficient (Ns/m)
m_p 50	k_p 44100	c_p 1420
m_s 26	k_s 17085	c_s 1000

By choosing $\rho = 0.0009$, $\eta = 1$ and solving the minimization problem of (29), we obtain the controller gain matrix by using the theorem 1 with a guaranteed H_∞ performance index and constraints in time domain, $K = 10^6 \times [-1.0690 \quad -0.0333 \quad -0.0160 \quad 0.0361]$
Substitude K into (30), we can get the semi-active control force.

To show the advance of the proposed method, some necessary comparisons with the passive and skyhook control strategy are given.

(1) Passive control

There is no controllable control force for passive control system, it can only play certain a vibration reduction on the system, relying on elastic elements of the fixed stiffness and damping components of the fixed damping force.That is

$$u = 0 \tag{31}$$

(2) Skyhook control

Skyhook control strategy assumes that the imaginary damper is mounted between the sprung mass m_s and virtual inertial space (Sky), and the skyhook control model shown in Figure 7.

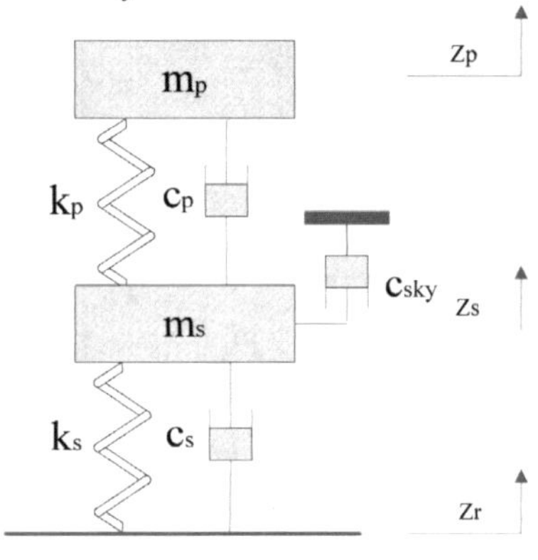

Figure 7. The skyhook control model

The control strategy for the skyhook is given as:

$$u = -C_{sky}\dot{z}_s \tag{32}$$

$$f_d = \begin{cases} u(t)_{\max} & \dot{z}_s(\dot{z}_s - \dot{z}_r) \geq 0,\ |u(t)| \geq u(t)_{\max} \\ u(t) & \dot{z}_s(\dot{z}_s - \dot{z}_r) \geq 0,\ |u(t)| < u(t)_{\max} \\ 0 & \dot{z}_s(\dot{z}_s - \dot{z}_r) \leq 0 \end{cases} \tag{33}$$

where C_{sky} is the skyhook damping coefficient. It can be determined according to the suspension system parameter optimization [22], and here $C_{sky} = 1000Nm/s$.

To validate the seat suspension performance in the time domain, two typical types of road disturbance, i.e., bump road disturbance and random road disturbance, will be considered in the simulation.

4.1. Simulation of bump road disturbance input

Road disturbances can be generally assumed as shocks. Shocks are discrete events of relatively short duration and high intensity, such as a convex bag or pothole on a smooth road surface. It is assumed that the bump road disturbance input has the following form: (the step input in time domain is shown in Figure 8)

$$w(t) = \begin{cases} -0.1 & 0 \leq t < 0.1 \\ 0 & 0.1 \leq t \end{cases} \tag{34}$$

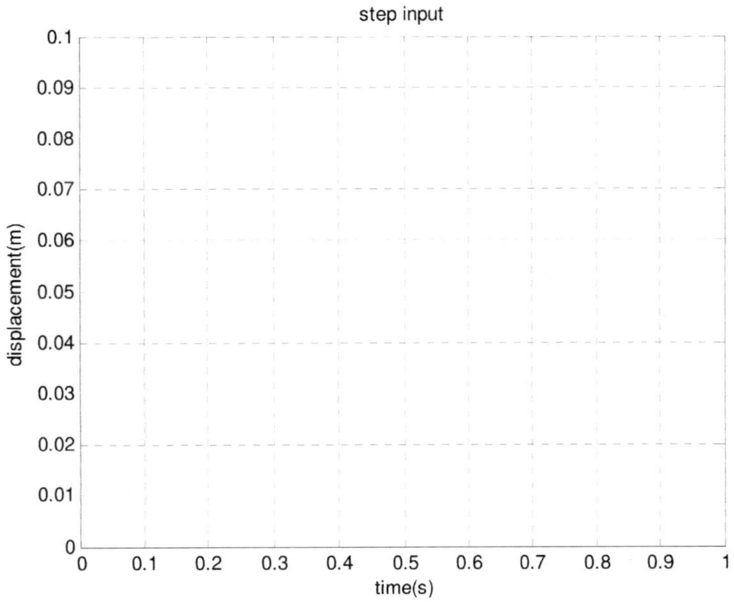

Figure 8. The step input in time domain

Table 2. The comparison of the body acceleration under the step input

Control method	Max peak-to-peak value (ms^{-2})	Attenuation time (s)
Passive	17.6486	0.50
Skyhook	15.9290	0.28
H_∞	14.4778	0.04

Table 3. The comparison of the suspension deflection under the step input

Control method	Max suspension deflection (m)	Attenuation time (s)
Passive	0.0155	0.78
Skyhook	0.0150	0.37
H_∞	0.0146	0.38

Tables 2 and Table 3 are the comparison results of the body acceleration and the suspension deflection under the step input. As can be seen from the tables, the acceleration under the H_∞ control attenuates faster than the other two control methods (passive and skyhook control) and the settling time is reduced more quickly. The max peak-to-peak value of the body vertical acceleration under the skyhook control and the H_∞ control reduces by 9.74% and 17.97%, respectively. The time domain responses of body vertical acceleration under the step input is shown in Figure 9. The time domain responses of seat suspension deflection under the step input is shown in Figure 10.

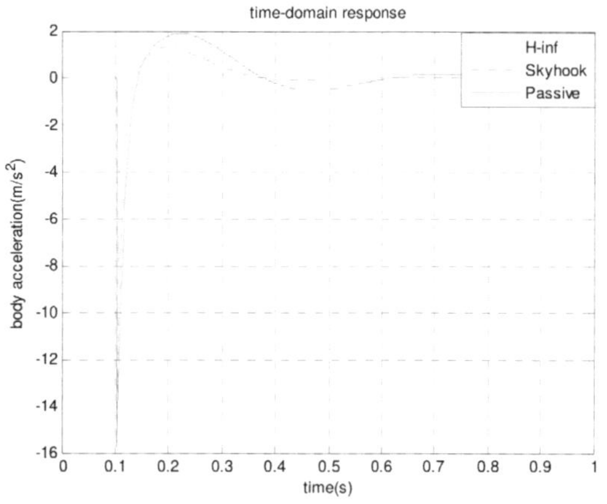

Figure 9. The time domain responses of the body vertical acceleration under the step input

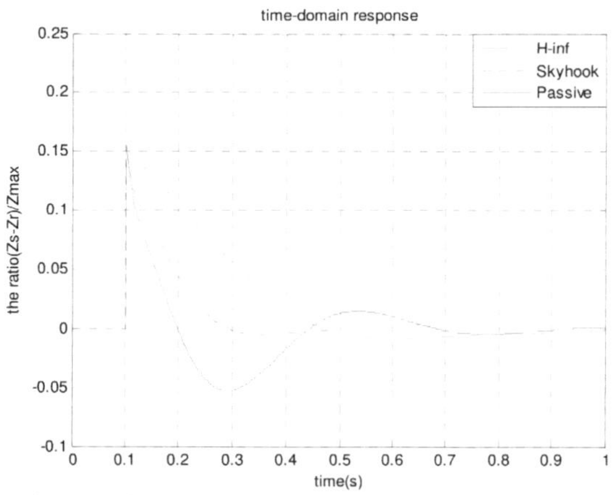

Figure 10. The time domain responses of the suspension deflection under the step input

4.2. Simulation of the random road input

Random disturbance of the road is a continuous vibration, it refers to the continuous excitation along the road length direction, such as asphalt pavement and washboard road. For the continuous and random road surface, it is generally described by spatial frequency power spectral density function and the corresponding time domain representation:

$$G_q(n) = G_q(n_0)(\frac{n}{n_0})^{-w} \tag{35}$$

where n is the spatial frequency and n_o is the reference spatial frequency of $n_o = 0.1(1/m)$; $G_q(n_o)$ stands for the road roughness coefficient; $w = 2$ is the road roughness constant.

In order to facilitate the analysis, the spatial frequency functions need to be translated into time frequency function, set the vehicle forward velocity as v, then we have :

$$f = nv \tag{36}$$

where, f is the time frequency, and its unit is Hz.

Therefore, the ground displacement power spectral density (PSD) is

$$G_q(f) = G_q(n_0)n^2{}_0 \frac{v}{f^2} \tag{37}$$

which is only related with the vehicle forward velocity. When the vehicle forward velocity is fixed, the ground displacement can be viewed as a white-noise signal. Select the road roughness as $G_q(n_0) = 64 \times 10^{-6} m^3$ [23], which is corresponded to B Grade. Set the vehicle forward velocity as $30km/h$ to generate the random road profile as Figure 11.

Table 4 is the comparison results of RMS value of the body acceleration and the suspension deflection under the random road input. As can be seen from the Table 4, the RMS value of body vertical acceleration under the skyhook control and the H_∞ control reduces by 10.13% and 39.25%, respectively. It shows that the H_∞ control outperforms the passive control and the skyhook control and realizes a better ride comfort.

Remark 2. Reducing the body acceleration and reducing the suspension deflection are a pair of contradiction, so that the improvement of ride comfort is at the expense of sacrificing the operational stability. Therefore, the suspension deflection has an increase for the H_∞ control. However, the time domain constraint is still guaranteed for the seat suspension system.

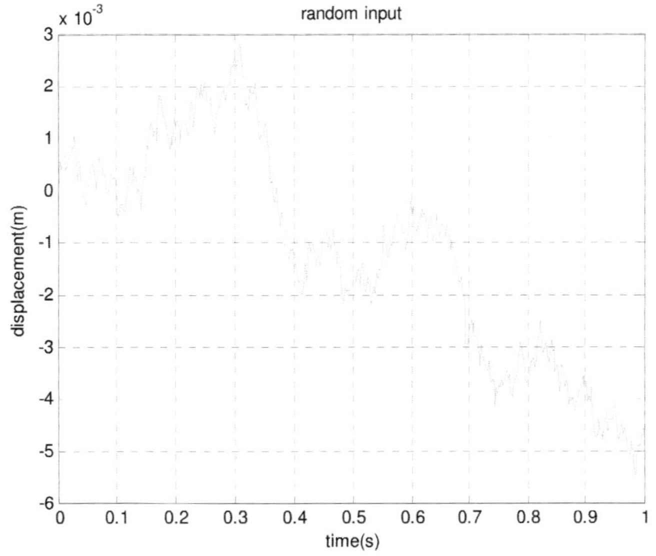

Figure 11. The random road input in time domain.

Table 4. RMS of driver body acceleration and suspension deflection for a B grade road (30Km/h) .

Control method	RMS (A) ($m\ s^{-2}$)	RMS (D) (m)
Passive	0.4283	8.5803e-004
Skyhook	0.3849	7.6642e-004
H_∞	0.2602	9.9959e-004

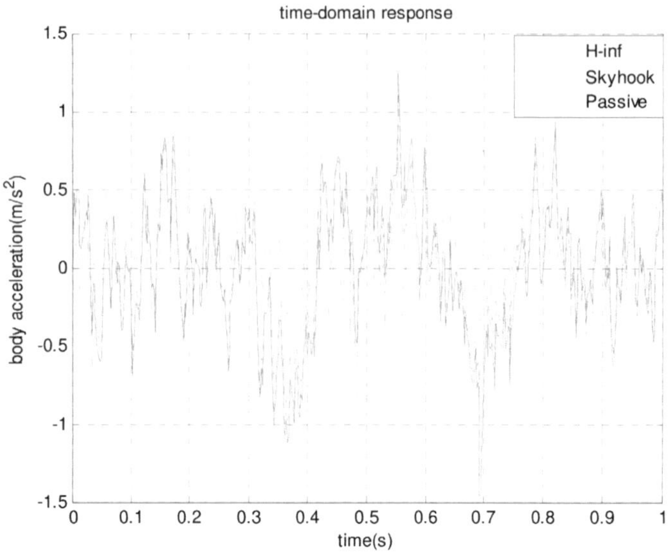

Figure 12. The time domain responses of body vertical acceleration under the random road input.

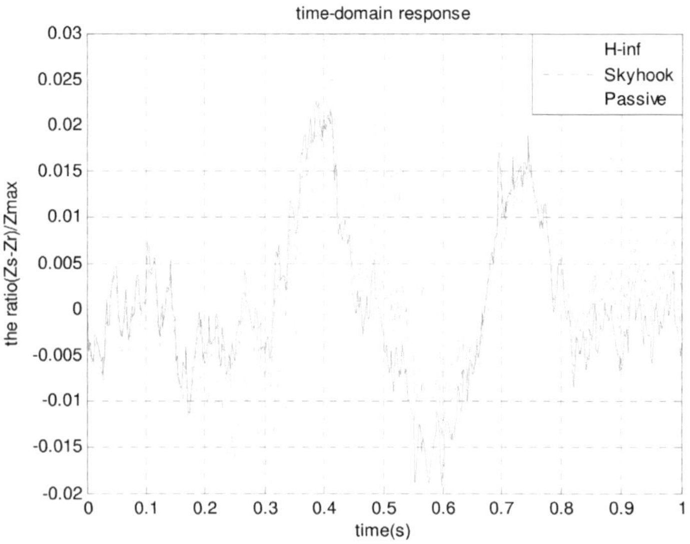

Figure 13. The time domain responses of seat suspension deflection under the random road input.

The time domain responses of body vertical acceleration under the random road input is shown in Figure 12. The time domain responses of the seat suspension deflection under the random road input is shown in Figure 13. The frequency domain responses of body vertical acceleration under the random road input is shown in Figure 14. There are two distinct peaks, the first one is the corresponding resonance frequency for the body (6Hz), and the second one is the corresponding resonance frequency for the seat (12Hz). It can be seen that, compared with the passive control and the skyhook control, the H_∞ control greatly reduces the second-order resonance peak of the seat suspension system. The power spectral density of body acceleration under the the random road input is shown in Figure 15. It is observed from Figure 15 that the system for H_∞ control has lower PSD of body acceleration and smaller PSD of body acceleration value results in better ride comfort.

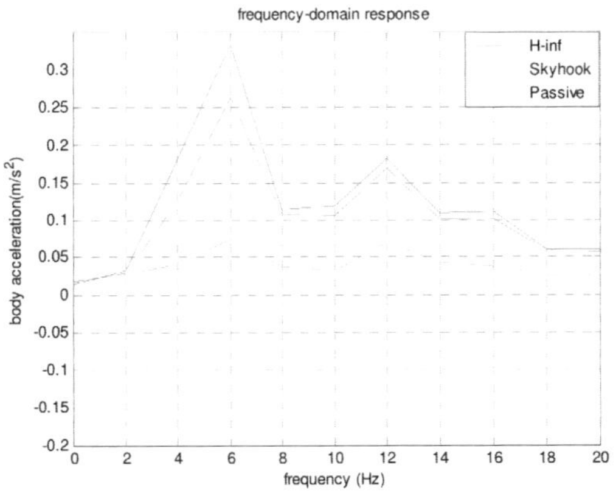

Figure 14. The frequency domain responses of body vertical acceleration with the random input.

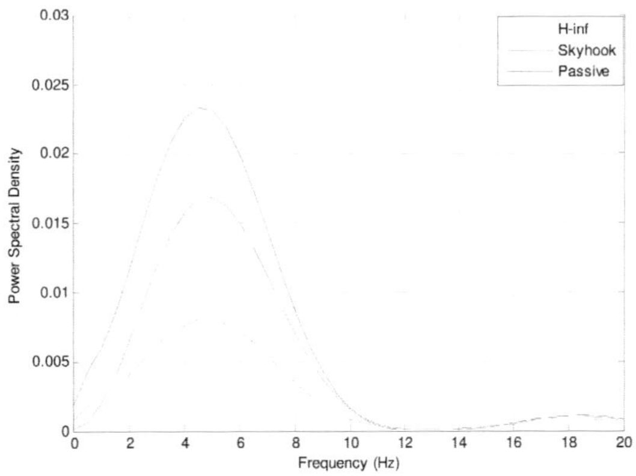

Figure 15. The power spectral density of body acceleration with the random input.

5. Conclusions

This study presents the robust H_∞ control method for the semi-active seat suspension system with MR damper with parameter uncertainties and actuator saturation. The solution of the state feedback controller is given in terms of LMIs, which guarantees H_∞ performance index and the constraints. The simulation results in time and frequency domain demonstrate that it is closer to the actual system by considering the model parameter uncertainties and actuator saturation. And the proposed approach can achieve better vertical vibration attenuation and improve ride comfort for seat suspension.

Acknowledgment

This research is supported financially by the National Natural Science Foundation of People's Republic of China (No.51073179, 61203098), the China postdoctoral science foundation (20100480043), and the Fundamental Research Funds for the Central Universities (CDJZR11 12 00 03). These supports are gratefully acknowledged.

References

[1] Sun W C, Li J F, Zhao Y and Gao H 2010 *J. Mechatronics* **21** 250-60
[2] Du H P, Li W H and Zhang N 2011 *Smart Mater. Struct.* **20** 105003 (10pp)
[3] Du H P, Li W H and Zhang N 2011 *Smart Mater. Struct.* **20** 1-10
[4] Yao G Z, Yap F F, Chen G, Li W H and Yeo S H 2002 *Mechatronics* **12** 963-973
[5] Maciejewski I 2010 *Vibrations in Physical Systems* **24** 898-912
[6] Ballo I 2007 *J.Vehicle Syst. Dyn.* **45** 1065–1073
[7] Ying Z G, Zhu W Q and Soong T T 2003 *Journal of Sound and Vibration* **259** (1) 45-62
[8] Ahmadian M, Song X B and Steve C 2004 *J. Transactions of the ASME* **126** 580-584
[9] Poussot-Vassal C, Sename O, Dugard L, Ramirez-Mendoza R and Flores L 2006 *Mechatronic Systems* **4** (1) 243-249
[10] Sammier D, Sename O and Dugard L 2003 *Vehicle System Dynamics* **39** (4) 279 -308
[11] Guglielmino E, Sireteanu T, Stammers C W, Ghita G and Giuclea 2008 *Semi-active suspension control* (Springer)
[12] Schurter K C and Roschke P N 2001 *Proc. of the American Control Conf.* 1097-1102
[13] Hrovat D 1982 *Automatica* **18** 117-119
[14] Fialho I, Balas G J 2002 *IEEE Trans. Control Syst. Technol.* **10** 43-54
[15] Du H P, Sze K Y and Lam J 2005 *Journal of Sound and Vibration* **283** 981-996
[16] Du H P, Lam J and Sze K Y 2003 *Engineering Applications of Artificial Intelligence* **16** 667-680

[17] Gao H J, Sun W C and Shi P 2010 *IEEE Trans. Control Syst. Technol.* **18** 238-245
[18] Zhao Y B, Sun W C and Gao H J 2010 *Journal of Sound and Vibration* **329** 4335-4353
[19] Sun W C, Gao H J, Kaynak O 2011 *IEEE Trans Control Syst Technol.* **19** 416-422
[20] Xue A K 2008 *The robust optimal control theory and its application* (Science Press)
[21] Yu L 2002 *Robust control-the linear matrix inequality processing method* (Tsinghua University Press)
[22] Caponetto R, Diamante O, Fargione G, Risitano A and Tringali D 2003 *IEEE Trans. Control Syst. Technol.* **11** 786-798
[23] Yu Z S 2011 *Car theory* (Mechanical Industry Press)

Experimental and numerical determination of the static critical pressure in ferrofluid seals

W Horak and M Szczęch[1]

AGH University of Science and Technology, Krakow, Poland

E-mail: szczech@agh.edu.pl

Abstract. Ferrofluids have various engineering applications; one of them are magnetic fluid seals for rotating shafts. There are various constructions of this type of seals, but the main difference is the number of sealing stages. The development of this construction is a complex process which requires knowledge of ferrofluid physical and rheological properties and the magnetic field distribution inside the sealing gap. One of the most important parameters of ferrofluid seals is the critical (burst) pressure. It is the pressure value at which a leak will occur. This study presents results of numerical simulation of magnetic field distribution inside the seal gap and calculations of the critical pressure value. The obtained pressure values were verified by experiments.

1. Introduction

Rotating shaft seals with magnetic fluid belong to the class of contactless technical seals. The principle of operation of such seals is a formation of a tight barrier by means of a magnetic fluid ring, which is kept in the gap by the magnetic field. This type of design is characterized by a small resisting moment, high reliability and durability. Such seals can operate, in a wide range of work parameters under static and dynamic conditions, maintaining an absolute tightness.

Ferrofluids are colloidal suspensions of magnetic particles of dimensions: 5-10 nm. Usually oil or water is applied as a carrying fluid, and in order to improve their stability ferrofluids contain additions of surface active substances. In contrast to magnetorheological fluids ferrofluids are highly stable.

Constructions of magnetic fluid seals consist of some basic components. An example of the design used in numerical calculations and in the presented hereby investigations is shown in Figure 1. A magnetic field source constitutes either the permanent magnet or an electromagnet (pos.3) polarized in the axial direction. The rotating shaft (pos.2) and pole pieces (pos.1) constitute the main part of the magnetic circuit. The whole is placed in a non-magnetic housing. The magnetic fluid (pos.4) is kept by the magnetic field in the sealing stage (pos.7). For research reasons this is an asymmetrical seal with stages situated on one side only.

2. Critical pressure

The crucial parameter characterising each sealing is the pressure value of the sealing factor, at which an abrupt leak will occur (critical pressure). The seal burst is of a point character and occurs in the weakest place in the circuit of the magnetic fluid ring. The fluid is displaced and thrown out from the sealing in the axial direction.

[1] Corresponding author.

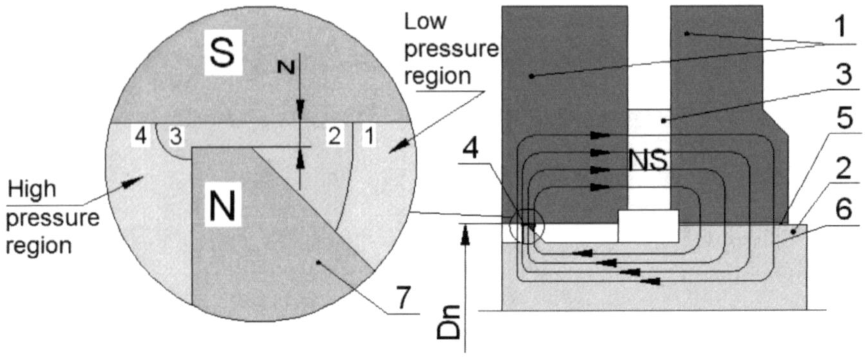

Figure 1. Geometry of the magnetic fluid seal; 1– Pole pieces, 2– Sealing bush, 3– Permanent magnet, 4– Ferrofluid, 5– Air gap, 6– Magnetic field, 7- Sealing stage, z - Gap size, D_n –Nominal diameter.

In case of an excessive pressure increase and a leak, the magnetic fluid sealing can restore itself when the pressure will decrease below its critical value. This effect is known as 'Reseal'. The critical pressure value can be calculated on the bases of the Bernoulli's theorem [1]:

$$P + \frac{1}{2}\rho v^2 + \rho g h - \mu_0 M H = \text{const.} \quad (1)$$

where P is total pressure in a magnetic fluid (Pa s), ρ is density of a magnetic fluid (kg/m^3), g is gravitation constant (m/s^2), M is magnetisation (A/m), μ_0 is vacuum magnetic permeability (V·s/A·m), H is magnetic intensity (A/m), and h is reference height (m).

When the influences of gravitation and centrifugal forces are omitted the following equation is obtained [1]:

$$\Delta P_{kr} = \mu_0 (M_s H_3 - M_s H_2) = \mu_0 M_s \Delta H = M_s \Delta B_{3-2} \quad (2)$$

where M_S is magnetic saturation in A/m, $\Delta B_{3\text{-}2}$ is difference of the magnetic induction value in points 3 and 2 (Figure 1).

Magnetic saturation M_S is a constant value at the assumption that the magnetic fluid is of an intensity higher than 800 kA/m. This condition is fulfilled in the discussed cases. Typical M_s values of magnetic fluids are within the range: 20-60 kA/m. In order to assess the critical pressure value the knowledge of the magnetic field distribution in the sealing is necessary. In the majority of seals due to a small height of the gap z, in which the magnetic fluid is placed, there is no possibility of an accurate, direct measurement of the magnetic field distribution. The smallest thickness of probes in the equipment for measuring the magnetic field are of the order of 1-1.5 mm, while the heights of working gaps are within the range: 0.1-0.5 mm. Due to a non-linear character of the problem and sometimes complicated sealing geometry it is difficult to assess this distribution by means of mathematical formulae.

3. Numerical simulations of the magnetic field distribution

Analyses of magnetic circuits in seals with the magnetic fluid were performed by the finite element method. Due to the circular symmetry of the analysed geometry in numerical simulations two-dimensional, axisymmetric model was assumed. The grid size was compacted in the gap zone. Dirichlet's zeroing boundary conditions were introduced, which means that it was assumed that the magnetic flux was parallel to the boundaries of the analysed area and that the flux value at the boundary would be zero. Simulations were performed for the magnetic fluid sealing of a nominal diameter: D_n= 50 mm. Simulations as well as experiments were carried out for three values of a gap height: 0.1, 0.2, 0.3 mm and three volumes of ferrofluid: 0.05, 0.1, 0.2 mL applied for sealing with one stage. For sealing of two or three stages the gap of 0.2 mm and the fluid amount of 0.1 mL were selected. As the magnetic field source the permanent magnet was used. This was the magnet of a trade name N38 and a coercive force H_c=937.4 kA/m and B_r = 1.241T. For the ferromagnetic elements (shaft, pole pieces) the curve B-H, characteristic for structural steel of a carbon content up to 0.27% [2], was assumed. The magnetic fluid BM-30 Batch 257 [4], for which the saturation magnetisation equals M_S = 37.5±0.5 kA/m was used in investigations. On the grounds of the references the magnetisation curve B-H for this type of fluid was assumed according to [5].

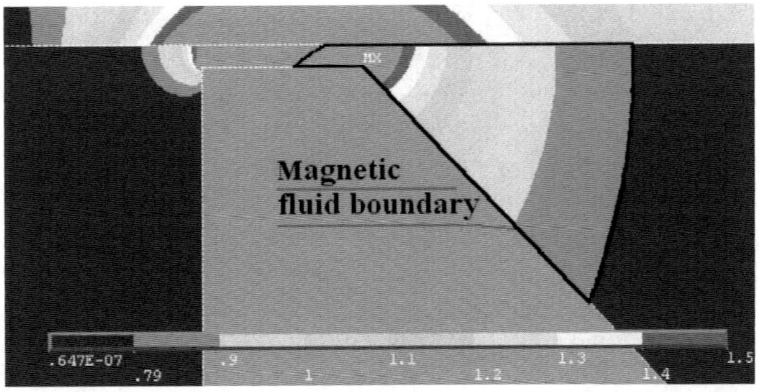

Figure 2. Contour distribution of the magnetic induction T with the modelled fluid-0.1 mL.

The shape of the investigated geometry was modelled according to the real dimensions of the structure. The position and shape of the magnetic fluid in the seal were modelled with maintaining the constant fluid volume rule and the fluid boundary occurrence along the constant magnetic induction rule [6]. The typical contour distribution of the magnetic induction in the analysed seal and the assumed ferrofluid shape is presented in Figure 2. For the needs of the simulation the fluid was placed in the position in which it would have been if the critical pressure had occurred.

On the bases of the numerical analysis the curves of the magnetic induction distribution in the gap - were determined. An example of the diagram for the geometry applied in investigations is presented in Figure 3.

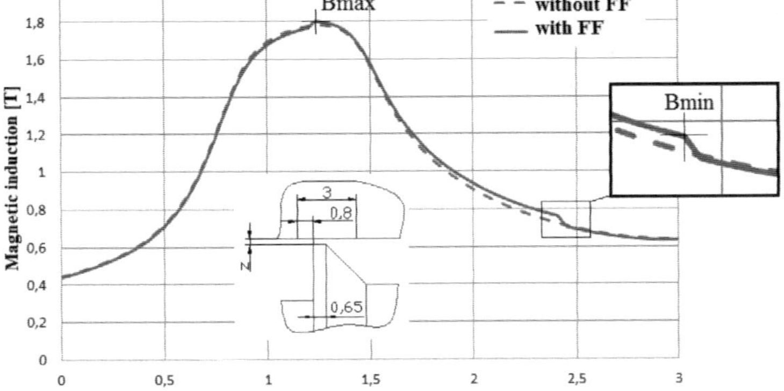

Figure 3. Magnetic induction distribution in the analysed geometry of the cases with and without the magnetic fluid [3].

Journal of Physics: Conference Series **412** (2013) 012055 doi:10.1088/1742-6596/412/1/012055

The characteristic stage, which appears in the magnetic fluid presence and is not seen when there is a lack of this fluid, is emphasised in this figure. The stage appears at the interface boundary. Its value depends on magnetic fluid properties and on the magnetic field intensity in the gap.

The critical pressure value was calculated on the basis of equation (2), as a product of the ferrofluid saturation magnetisation and the difference of the magnetic induction $B_{max\text{-}Bmin}$.

Accuracy of the simulation method was verified with use additional templet with gap width 1.5 mm and constant magnetic induction. This has provided the possibility of direct measurement magnetic induction inside gap by the Hall Effect sensor. Measurement results were compared with numerical simulation results. On this basis it has been shown that the difference between the simulation and the measured value of the magnetic induction does not exceed 2% of the measured value.

4. Experimental determination of the critical pressure value

Experimental investigations verifying the correctness of the calculated critical pressure value were performed on the research stand for testing seals with the magnetic fluid MAST1 [7]. This stand allows for the precise measurements of the pressure, speed, torque and temperature. Investigations were performed for the same geometry of seals as in the case of numerical simulations. Measurements were carried out under static conditions (system rotational speed = 0 rpm). The measurement consists in introducing the linearly increasing pressure and observing the moment of its abrupt drop.

In seals with magnetic fluids the static critical pressure is a time function. With the stabilisation time increasing obtains higher and higher values - up to a maximum. In order to eliminate this effect influence the special measuring procedure was applied. Each measurement was preceded by 5 minutes of the stabilising period after adding the fluid, followed by successive 5 minutes when the tested seal operated at the peripheral speed equal 1 m/s, after which the right measurement was performed. An example of the critical pressure diagram is presented in Figure 4.

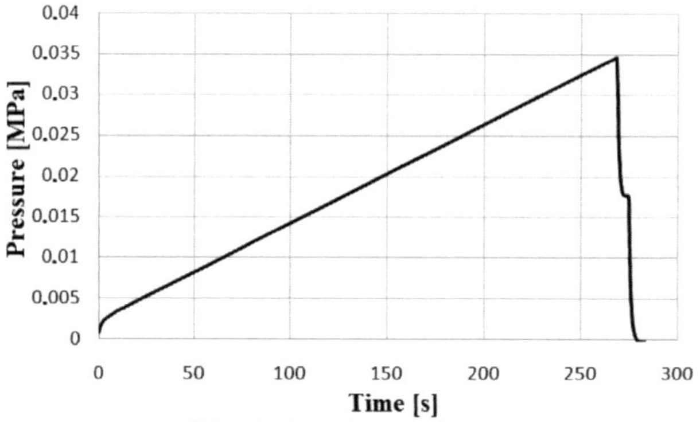

Figure 4. Burst pressure curve for BM-30 Batch 257 fluid; working gap: 0.1 mm, rotational speed: 0 rpm, critical pressure: 0.0346 MPa [3].

The critical pressure results obtained by means of the simulation and their comparison with experimentally obtained values are presented in Fig. 5. They concern seals with one sealing stage. The largest error was obtained for the smaller volume of the applied fluid (0.05 mL) and the largest gap z = 0.3 mm. The pressure obtained on the simulation basis was by 34.3 % higher than the measured one. The smallest error was obtained for the largest gap z = 0.3 mm and the largest volume of the magnetic fluid (0.2 mL) -1.4 %. Positive values of the relative error, it means that the measured critical pressure was higher than the one numerically obtained, appeared only for the smallest working gap size (0.1 mm).

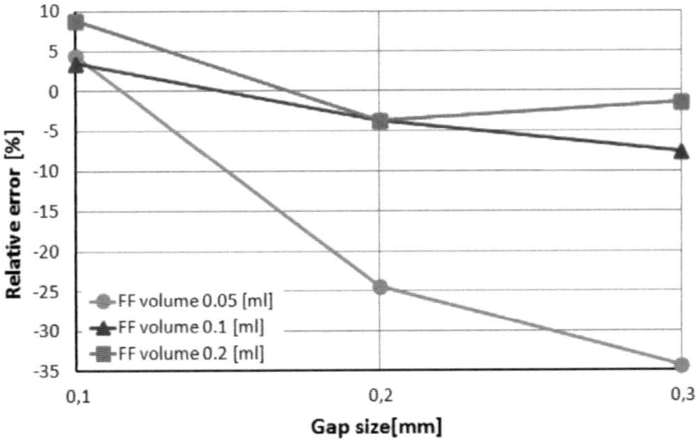

Figure 5. Relative error (%) of the critical pressure between measurement and simulation results, for various working gap sizes and various fluid volumes applied for one sealing stage.

The relative error results and critical pressure values for seals with various sealing stages are presented in Figure 6 (a,b). The highest critical pressure value equal 0.097 MPa was obtained in tests for three sealing stages. The relative error for one and two sealing stages does not exceed -4%. On the other hand for the three sealing stages this error has been already –26.3%. In each case simulations overstated the critical pressure value.

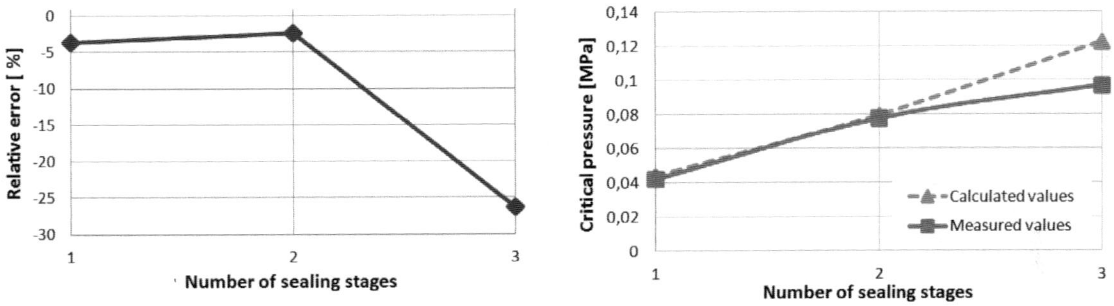

(a) (b)

Figure 6. (a) Relative error (%) of the critical pressure between the measured values and calculated on the basis of simulations for various sealing stages, working gap z =0.2 mm, fluid volume applied for one sealing stage 0.2 mL, (b) Critical pressure values.

5. Conclusions

The determination method of the critical pressure of the sealing with the ferrofluid with the application of the numerical simulation of the magnetic field distribution and on the basis of experiments, was presented in the hereby research.

On the bases of the performed considerations, it can be stated that the proposed method allowed to determine the static critical pressure of the sealing with the magnetic fluid on the bases of analytical equations and simulation calculations.

The relative error of the determined pressure value depended on the ferrofluid volume modelled in numerical simulation, in addition to which the modelling of small fluid volumes leads to under-assessing the critical pressure value, first of all for seals with a larger gap size. In case of multistage seals the simulation results should be corrected, since for more than 3 stages the error was above 25%.

Differences between the critical pressure calculated values and values assumed as the real ones are caused by an inaccuracy in producing and assembling elements of the research stand, surface roughness, misalignment of sealing bushes and pole pieces and others. Divergences of the obtained results were also influenced by the ferrofluid surface shape assumed in simulations as well as the selection of curves describing magnetic properties of the numerically modelled geometry.

References

[1] Rosensweig R E 1985 *Ferrohydrodynamics* (Cambridge University Press)
[2] Gieras J F, Piech Z J and Tomczuk B 1999 *Linear synchronous motors: Transportation and automation systems*
[3] Horak W and Szczęch M 2011 Wyznaczanie statycznego ciśnienia krytycznego w uszczelnieniach z cieczą ferromagnetyczną na podstawie obliczeń numerycznych *Mechanik* **12** 988
[4] Salwiński J, Szydło Z, Horak W and Szczęch M 2011 Badanie zmian lepkości cieczy ferromagnetycznych poddanych działaniu stałego pola magnetycznego *Tribologia* **2** 143
[5] Ivanov A O, Kantorovich S S, Reznikov E N, Holm Ch, Pshenichnikov A F, Lebedev A V, Chremos A and Camp P J 2007 Magnetic properties of polydisperse ferrofluids: A critical comparison between experiment, theory, and computer simulation *Phys. Rev. E* **75** 6
[6] Balac S and Caloz G 2002 Magnetostatic field computations based on the coupling of finite element and integral representation methods *IEEE Transactions on Magnetics* **38** 2
[7] Salwiński J, Szydło Z, and Szczęch M 2010 Experiments on magnetic fluid seals operating in water condition *17th Int. Colloquium Tribology* Technische Akademie Esslingen 93

Journal of Physics: Conference Series **412** (2013) 012056 doi:10.1088/1742-6596/412/1/012056

Brownian dynamic simulations and experiments of MR fluids

JP Segovia-Gutiérrez[1], J de Vicente[1,3], R Hidalgo-Alvarez[1] and AM Puertas[2]
[1]Department of Applied Physics, Faculty of Sciences, University of Granada, C/ Fuentenueva s/n, 18071-Granada, Spain
[2]Department of Applied Physics, University of Almería 04.120-Almería, Spain

E-mail: jvicente@ugr.es

Abstract. The use of computational techniques in magnetorheology is not new. I general, these approaches assume dipolar magnetic interactions, hard sphere repulsions, and no-slip conditions. In this contribution we focus on the dynamics of the equilibrium state in the presence of uniaxial DC fields. To achieve this goal we make use of Brownian Dynamic Simulations. We highlight the importance of the Brownian forces versus magnetic dipolar interaction in the range of low magnetic field strengths. We monitor the formation of columnar structures and their dynamics, in competition with the Brownian motion, until a hexatic crystal phase appears at high field strengths for monodisperse systems. The shear viscosity is computed from the Einstein relation and eventually compared with experimental data at very low-shear rates. A reasonably good agreement between both data sets is observed.

1. Introduction

The use of particle-level dynamic simulations techniques in magnetorheology is not new. Initially, these techniques were employed to interrogate the static/structural properties under no flow conditions [1]. Then, researchers moved to the investigation of MR fluids out of equilibrium under shearing [2-5] and squeezing flows [6]. Generally speaking, these approaches assume dipolar magnetic interactions, hard sphere repulsions, and no slip conditions. Most of the research efforts have assumed that Brownian motion is negligible [2-7] but there are also some reports where thermal motion is considered in spite of the large size of the constituents [8]. In this contribution we focus on the dynamics of the equilibrium state in the presence of uniaxial DC fields. To achieve this goal we make use of Brownian Dynamic Simulations [9]. We highlight the importance of the Brownian forces versus magnetic dipolar interaction in the range of low magnetic strengths for the formation of columnar structures and its dynamics. Upon increasing the field strength, we observe the formation of columnar structures, in competition with the Brownian motion, until a hexatic crystal phase (monodisperse systems) appears at high field strengths. It is also shown that random forces introduce defects in particle aggregates along the field direction.

2. Simulations

In our simulations, we measure structure parameters to identify the formation of columnar structures and crystals. The dynamics of the system were measured with the mean square displacement and bond correlation functions. The rheology of the system was studied by means of the non-diagonal terms of the stress tensor and its correlation function. The method applied in this work presents advantages with

[3] Corresponding author.

respect to the classical procedure as we are capable to extract more information about the system in the equilibrium state.

Our simulation system consists in 1000 monodisperse spherical particles interacting via exponential repulsive hard-core and dipolar magnetostatic forces without bounding surfaces. Dipolar magnetic interaction between two particles is described below:

$$\vec{F}_{ij}^{mag} = 3U_0\left(\frac{1}{|\vec{r}_{ij}|}\right)^4 \left[\left(2cos^2\theta_{ij} - sin^2\theta_{ij}\right)\hat{r} + sin2\theta_{ij}\hat{\theta}\right] \quad (1)$$

Here, U_0 controls the external magnetic field strength, $\vec{r}_{ij}$ denotes the relative position between two particles and θ_{ij} is the angle that $\vec{r}_{ij}$ forms with the field direction. We pay special attention to the calculation of the low-shear viscosity (at *zero*-shear rate) following the Einstein relation [10]:

$$\eta = \lim_{t\to\infty}\eta(t) = \frac{\beta}{6V}\lim_{t\to\infty}\frac{1}{t}\langle\Delta A^2(t)\rangle \quad (2)$$

where $\beta=1/(k_BT)$, T is the temperature and k_B is the Boltzmann constant, V is the volume of the simulation box, t is the time and $\langle\ldots\rangle$ indicates the average over all particles. Finally, $\Delta A(t)$ is defined as follows:

$$\Delta A(t) = A(t+s) - A(s) = \int_s^{t+s}\sum_{\alpha<\beta}\sigma^{\alpha\beta}(t')dt' \quad (3)$$

Here, $\sigma^{\alpha\beta}(t)$ are the non-diagonal components of the stress tensor. In our case, the magnetic field is in the Z-direction, and accordingly we take the 13 and 23 components for the calculation of the viscosity. A typical calculation of the viscosity is shown in Figure 1. The shear viscosity is read from the long time plateau value.

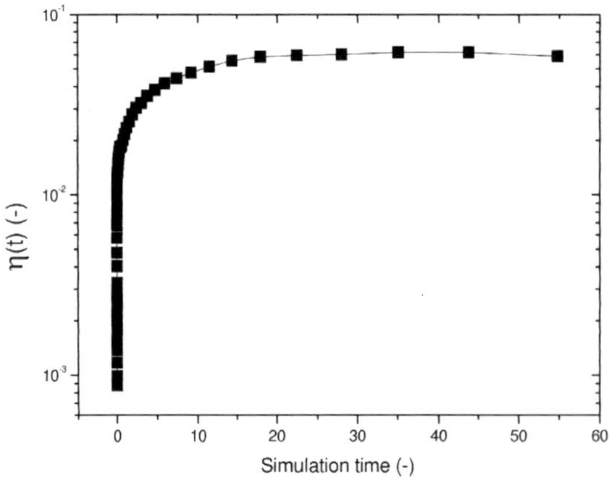

Figure 1. Evolution of *η(t)* in an equilibrium state. The viscosity is given by the long time plateau.

3. Experiments

Conventional MR fluids were formulated by dispersing carbonyl iron spheres (HQ grade, BASF) in Newtonian silicone oils of different viscosities (20 and 500 mPa·s, Sigma-Aldrich). The particle concentration was 5 vol.%.

Low-shear viscosity data were obtained from steady simple shear flow curves using a magnetorheometer. To attain this goal, a custom-built fixture was designed to achieve sufficient torque resolution at low magnetic field strengths (below 10 kA/m). To ensure that experimental measurements were carried out in equilibrium, prior to the test, the system was stabilized under magnetic fields during a long time period (~ $2 \cdot 10^3$ s). Furthermore, we took a sufficient time interval length over each measured data point to achieve a steady shear viscosity value at low-shear rates.

4. Discussion

Figure 2 shows a typical snapshot of the system for moderate field strengths (Lambda ratio is around $\lambda = 30$). The system is still in the fluid state as observed by the mean square displacement (see Figure 2) where we can distinguish both ballistic and diffusive regimes at very short and long times, respectively.

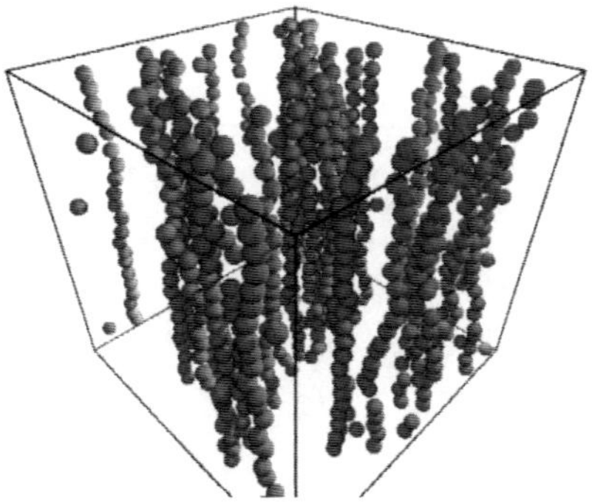

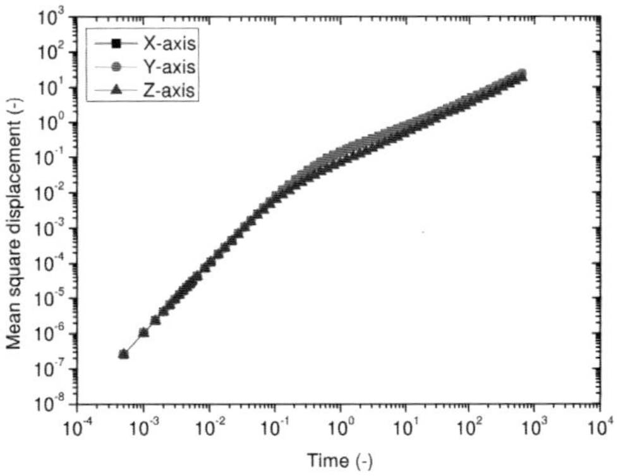

(a) (b)

Figure 2. (a) Microscopic structure for a monodisperse system at $U_0 = 35$. Particle columns are aligned in the field direction; (b) Mean square displacement for the three coordinate axes at the same value of U_0.

Thus, the columns in the snapshot are not stable over long times, but dissolve and reform, as noted in the bond correlation function (see Figure 3). This behavior is observed up to crystallization, in the case of this monodisperse system, where $\lambda \sim 50\text{-}60$.

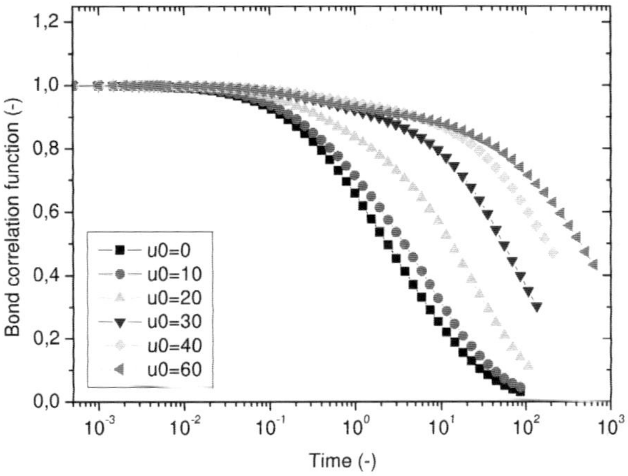

Figure 3. Bond correlation functions for different values of the control parameter U_0.

To carry out a comparative study between experiments and simulations, the viscosity was rescaled as follows,

$$\frac{\eta-\eta_S}{\eta_{H=0}-\eta_S} \tag{4}$$

where $\eta_{H=0}$ corresponds to the shear viscosity at zero magnetic field and η_S is the solvent viscosity. In this way, we correct for i) the different solvent viscosity in the simulations and experiments, and ii) the non-magnetic residual interactions in the experimental system. Importantly, using this scaling function, experimental curves collapsed for different solvent viscosities and shear rates ranging from 0.01 up to 1 s^{-1} (see Figure 4).

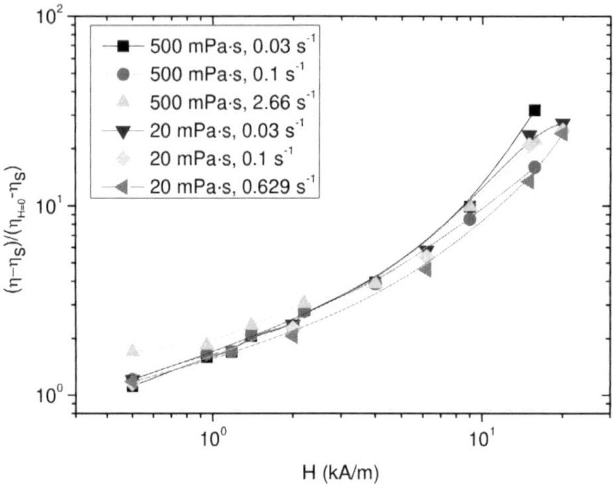

Figure 4. Experimental normalized viscosities as a function of the magnetic field.

The comparison between experiments and simulation results is shown in Figure 5. Here, the X-axis is rescaled to collapse all curves. The reason for this rescaling is probably the remnant magnetization of the particles, or the simplification implied in the dipolar interaction. The agreement between the experimental and simulation data sets is reasonably good and deviations appear only at high magnetic fields, where the crystallization starts in the simulated monodisperse system. At this point it is worth to stress that carbonyl iron particles employed in the formulation of the MR fluids are very polydisperse and as a consequence crystallization is prevented. As expected, by introducing a slight polidispersity in the simulation system the deviation is reduced (see Figure 5).

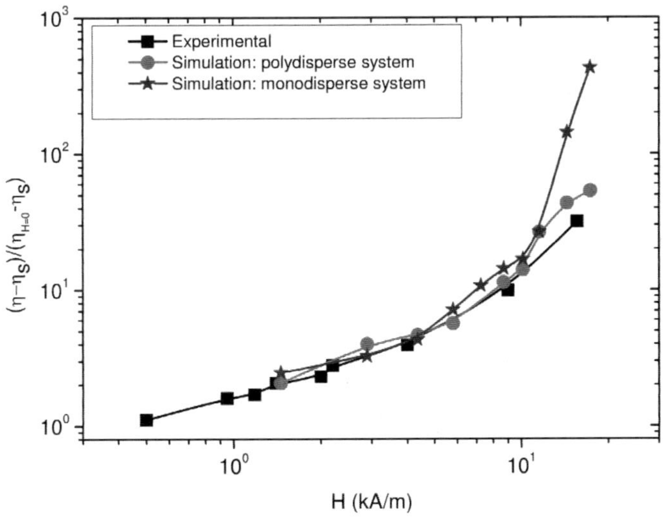

Figure 5. Simulation and experimental viscosities as a function of the magnetic field. Shear rate was fixed at 0.03 s^{-1} in experimental measurements.

5. Conclusions

Despite the many approximations there is a reasonably good agreement between the low shear viscosity obtained from experiments and simulation results. Polydispersity improves the behavior of the simulation system at high magnetic fields and avoids crystallization. Also, valuable information is obtained from the calculation of order parameters such as potential energy and the average number of neighbors per particle, which achieve a plateau when the system is equilibrated. The present results show that the simulation, if properly rescaled, reproduces not only qualitatively the experimental trend but also quantitatively. Also, we found a nondependent behavior of the rescaling with shear rate in at least two decades.

Acknowledgments

This work was supported by MICINN MAT 2010-15101 project (Spain) and by Junta de Andalucía P10-FQM-5977, P10-RNM-6630 and P11-FQM-7074 projects (Spain). J. P. Segovia-Gutiérrez acknowledges financial support by the "Ministerio de Educación: Becas del Programa de Formación del Profesorado Universitario (FPU)" (AP2008-02138). J. de Vicente and R. Hidalgo-Alvarez thank the financial support received from CEI-Biotic 20F12/16.

References

[1] Tao R 1999 *Rheology Series* **8** 659-76
[2] Klingenberg DJ, van Swol F and Zukoski CF 1989 *J. Chem. Phys.* **91** 7888
[3] Klingenberg DJ, van Swol F and Zukoski CF 1991 *J. Chem. Phys* **94** 6160
[4] Parthasarathy M and Klingenberg DJ 1999 *J. Non-Newtonian Fluid Mech.* **81** 83
[5] de Vicente J, Vereda F, Segovia-Gutiérrez JP, Morales MP and Hidalgo-Álvarez R 2010 *J. Rheol.* **54** 1337
[6] de Vicente J, Ruiz-López JA, Andablo-Reyes E, Segovia-Gutiérrez JP and Hidalgo-Álvarez R 2011 *J. Rheol.* **55** 753
[7] Heine M, de Vicente J and Klingenberg DJ 2006 *Physics of Fluids* **18** 023301
[8] Martin JE 2000 *Phys. Rev. E* **63** 011406
[9] Paul W and Yoon DY 1995 *Phys. Rev. E* **52** 2076
[10] Puertas AM, De Michele C, Sciortino F, Tartaglia P and Zacarelli E 2007 *J. Chem. Phys.* **127** 144906

Continuous media theory for MR fluids in non-shearing flows

JA Ruiz-López, R Hidalgo-Alvarez and J de Vicente
Department of Applied Physics, Faculty of Sciences, University of Granada, C/ Fuentenueva s/n, 18071-Granada, Spain

E-mail: jvicente@ugr.es

Abstract. The enhanced mechanical response of magnetorheological fluids under slow compression has been investigated by means of experiments, theory and particle-level simulations. A wide range of magnetic field strengths (0–354 kA/m), dispersing medium viscosities (20–500 mPa·s) and particle concentrations (5–30 vol%) were investigated. Plastic media theory in compressive flow was in good agreement with experimental data. Slight deviations from the theory were associated to the so-called strengthening effect as the yield shear stress could increase during compression. Particle-level simulations were in good agreement with both experiments and simulations.

1. Introduction

In general, available devices using magnetorheological (MR) fluids can be classified according to their flow mode as direct-shear flow mode, pressure-driven flow mode and squeeze-film flow mode. Among the three modes, it is well known that the squeeze flow mode provides the largest yield stress under the same field [1]. The rheological properties of MR fluids under shearing flows have been extensively investigated in the literature. However, the understanding of the behavior of MR fluids under non-shearing elongational flows, and particularly in squeeze flow mode is still far to be complete mainly because of the lack of both a thorough understanding of the basic MR mechanisms and reliable experimental data [2, 3].

First reports on squeeze flow magnetorheology were devoted to investigate the enhancement of MR performance by the so-called compression-assisted aggregation process [4-6]. Later, in a study, a series of low-strain tests on MR fluids where the behaviors under constant velocity squeezing flow and shear flow were compared [7]. A field dependence of $H^{0.91}$ was found for compression in contrast to the $H^{1.4}$ dependence observed under shearing.

In this work, we follow previous papers where we experimentally demonstrated the appearance of normal forces under no-slip compression in the filtration dominated regime [3, 8]. In the present study the long-standing question of whether a single theory for model plastic fluids is suitable to deal with the unidirectional compression problem in MR fluids was addressed. In this work, an extensive experimental and particle-level simulation investigation of the performance of MR fluids in slow-compression, no-slip, constant-volume squeeze mode under different magnetic field strengths (0–354 kA/m), dispersing medium viscosities (20–500 mPa s) and particle concentrations (5–30 vol%) were investigated.

2. Theory

Plastic media theory in squeeze flow mode was first developed by Covey and Stanmore [9]. The solution of the motion equation for a Bingham fluid depends of the plasticity number, defined as:

$$S = \frac{\eta_p v R}{h^2 \tau_y} \quad (1)$$

where η_p is the plastic viscosity, v is the compression velocity, R is the sample radius, h is the plate-plate gap and τ_y is the shear yield stress. For low plasticity numbers, in the so-called filtration dominated regime, both Bingham and biviscous theories are applied that reduce to the following analytical expression in the case of constant volume tests [8]:

$$F = \frac{2\tau_y V^{3/2}}{3\sqrt{\pi} h_0^{5/2} (1-\varepsilon)^{5/2}} \quad (2)$$

where V represents the total volume of the sample ($V = \pi R^2 h$) and the elongational strain ε is defined here as the ratio of the moving distance of the upper plate to the initial distance between the plates $\varepsilon = (h_0 - h)/h_0$.

3. Simulations

A three-dimensional particle-level simulation method to understand the effect of particle concentration was also employed. Magnetic forces were approximated by the point-dipole limit and high-order magnetic multipoles were neglected. A free draining approximation was used for including hydrodynamic interactions. Short-range repulsive interactions between particles, and particles and walls were taken as an exponential interaction [3, 8].

Finally, the normal force F^* acting on the plate is calculated by simply differentiating the total magnetic energy U^* according to:

$$F^* = -\frac{dU^*}{dh^*}, \quad (3)$$

4. Results and discussion

Uniaxial monotonic compression tests were carried out in the presence of a uniaxial magnetic field at low compression rates under constant volume conditions. In all cases, the resulting normal force shows an initial 'plateau' in the low-strain region related to the compressive yield stress. This initial plateau is in agreement with equation 2 [3].

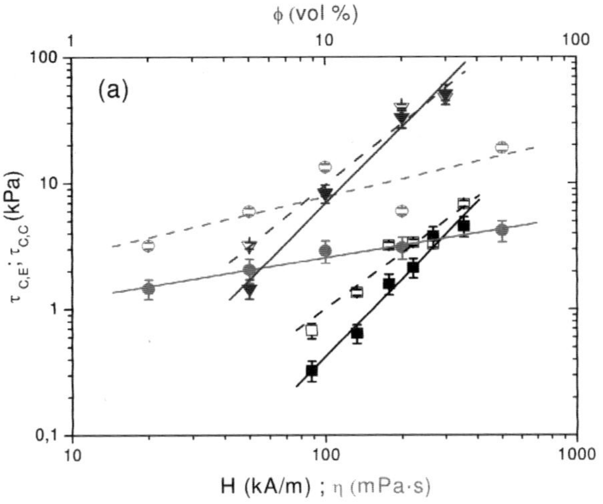

Figure 1. Low-strain compressive stresses as a function of magnetic field strength H (squares), dispersing medium viscosity η (circles) and particle volume fraction ϕ (triangles).

The compressive yield stress was obtained from experimental data by dividing the initial normal force plateau value by the MR fluid area. For a comparative analysis, in Fig. 1, experimentally measured compressive yield stresses $\tau_{C,E} = A/\pi r_0{}^2$ (solid symbols) as well as calculations $\tau_{C,C}$ using $\tau_{C,C} = \frac{2\tau_y V^{3/2}}{3\pi^{3/2} h_0^{5/2} r_0^2} = \frac{2\tau_y r_0}{3h_0}$ (open symbols) were shown. As demonstrated, calculated compressive yield stresses compare reasonably well with experimental measurements especially taking into account that there are not free parameters in these calculations. The yield compressive stress does increase with increasing the magnetic field strength, medium viscosity and particle content. A power law dependence is found in the three cases. Particle-level simulations show a similar dependence of the compressive yield stress with the particle concentration as shown in Fig. 2.

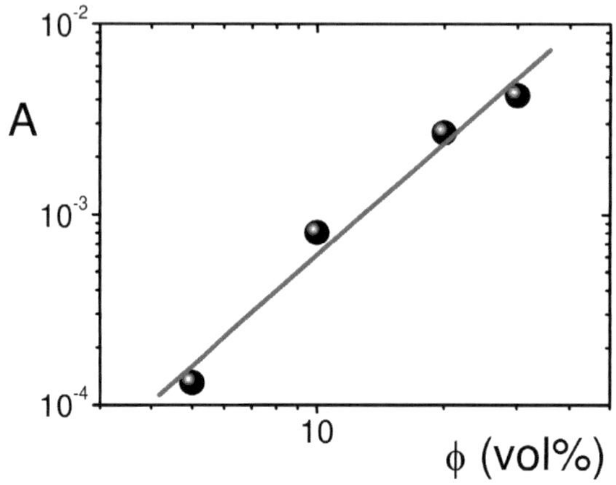

Figure 2. Simulation results for the low strain plateau value A as a function of the particle concentration ϕ. Red line is the best fit and reveals a quadratic dependence (1.94 ± 0.22).

In order to compare theory, experiments and particle-level simulations, a normalization of the normal force with the low-strain plateau value was carried out. According to equation 2, this normalized normal force must only depend on the elongational strain. As shown in Fig. 3, experimental data reasonably collapse in a master curve and are in good agreement with both theoretical plastic models and simulations [3, 8].

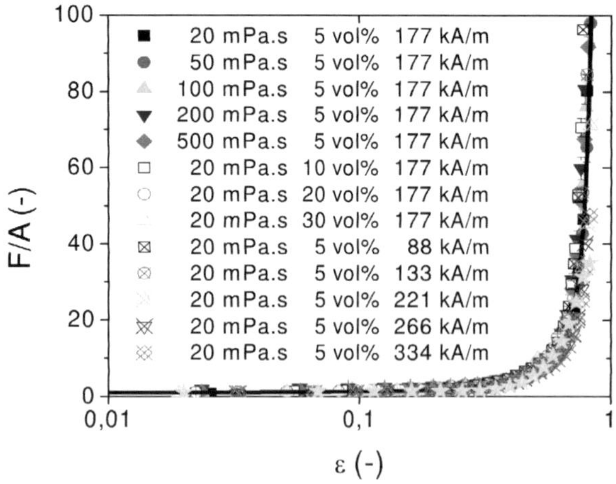

Figure 3. Dimensionless normal force as a function of compressive strain ε for different magnetic field strength, dispersing medium viscosity and particle volume fraction. Solid line corresponds to the prediction by the continuous media theory for plastic materials equation 2. Stars correspond to particle level dynamic simulations.

Deviations from the theory can be associated to the so-called strengthening effect [10]. Under slow compression, particles may restructure in shorter, thicker and eventually stronger structures. As a consequence, the shear yield stress may enhance during compression. To further investigate the structural changes during squeeze, compression tests by superposition of a Small-amplitude oscillatory shear flow were carried out where the upper plate was made to undergo rotary oscillations about a mean position $5x10^{-6}$ rad at a given frequency f = 10 Hz in the viscoelastic linear region. This kind of test would allow us to monitor viscoelasticity changes at the same time the sample is being compressed [3].

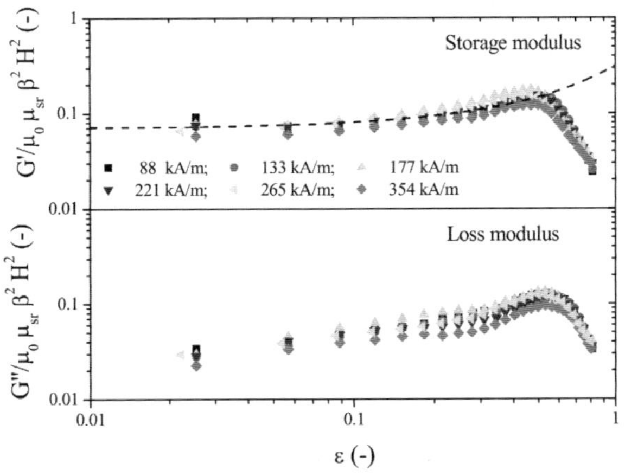

Figure 4. Strain dependence of the normalized shear viscoelastic moduli for different magnetic field strengths. Dashed line corresponds to $G'/(\mu_0\mu_{sr}\beta^2H^2) = 0.07e^{1.5\varepsilon}$

As shown in Fig. 4, viscoelastic moduli increased under compression following an exponential law. Moreover, viscoelastic moduli scale with a Mason-like number in a master curve. Strengthening effect is also suggested by rheomicroscopic structural observations and simulation results where the structure appears to become thicker under compression [3].

5. Conclusions

A unified description of the behavior of MR fluids in terms of a continuous media theory for plastic materials were proposed. This allowed us to obtain collapsed compression curves for a wide range of magnetic field strengths, medium viscosity and particle concentration. Deviations from the theory were explained in terms of the squeeze strengthening effect. On the one hand, a quadratic dependence with the magnetic field strength (2.0±0.1) and particle concentration (2.0±0.2) was found. On the other hand, a negligible dependence with the continuous phase viscosity was found. Experiments reported here validated another procedure to determine static yield shear stresses when slowly compressing the MR fluids. Particle-level dynamic simulations at large $1-\varepsilon$ suggest a quadratic volume fraction dependence (1.94 ± 0.22) in good agreement with experimental data.

Acknowledgments

This work was supported by MICINN MAT 2010-15101 project (Spain), and by Junta de Andalucía P10-FQM-5977, P10-RNM-6630 and P11-FQM-7074 projects (Spain). J.A.R.-L. acknowledges financial support by the "Ministerio de Educación: Becas del Programa de Formación del Profesorado Universitario (FPU)" (AP2010-2144).

References

[1] Havelka K O and Pialet J W 1996 *CHEMTECH* **26** 26
[2] de Vicente J, Klingenberg D J and Hidalgo-Álvarez R 2011 *Soft Matter* 7 3701
[3] de Vicente J, Ruiz-López J A, Andablo-Reyes E, Segovia-Gutiérrez J P and Hidalgo-Álvarez R 2011 *J. Rheol.* **55** 753
[4] Tang X, Zhang X, Tao R and Rong Y 2000 *J. Appl. Phys.* **87** 2634
[5] Zhang X Z, Gong X L, Zhang P Q, Wang Q M 2004 *J. Appl. Phys.* **96** 2359
[6] Zhang M L, Tian Y, Jiang J L, Zhu X L, Meng Y G, Wen S Z 2009 *Chin. Phys. Lett.* **26** 048301
[7] See H 2003 *Rheol. Acta* **42** 86
[8] Ruiz-López J A, Hidalgo-Álvarez R and de Vicente J 2012 *Rheol. Acta* **51** 595
[9] Covey G H and Stanmore B R 1981 *J. Non Newtonian Fluid Mech.* **8** 249
[10] Tang X, Zhang R, Tao R and Rong Y 2000 *J. Appl. Phys.* **87**(5) 2634

Formation of a layer of magnetorheological fluid on the surface of the moving object in the gradient magnetic field

A M Rusetski, A A Mokeev and E V Korobko[1]

A.V. Luikov Heat and Mass Transfer Institute of NAS Belarus, 220072,
15 P. Brovka str., Minsk, Belarus

E-mail: evkorobko@gmail.com

Abstract. The main forces that form the free surface of a layer of magnetorheological fluid (MRF) in contact with the surface of a moving object in the gradient magnetic field are determined. These forces are the force associated with the pressure caused by an external magnetic field, the force, determined by the ability of MRF material to be magnetized and magnetostrictive force. It is found that MRF layer in an inhomogeneous magnetic field retains its shape when the above mentioned forces are balanced by the elastic forces and surface tension.

1. Introduction

Devices for mechanical polishing of surfaces from different materials in the magnetic field are based on the use of fluid abrasive materials or pastes, which are modified by introduction of magnetically sensitive particles. For a more rough treatment using the lapping technology, magnetic fluids (MF) are usually applied [1, 2], and for the finishing of precision optical and semiconductor components using fluidic technology – magnetorheological fluids (MRF) are used [3-6]. Magnetic fluids contain colloidal ferromagnetic particles with the size, $d < 10^{-8}$ m characterized by their noticeable Brownian motion and in experimental studies and physical-mathematical description they are considered [7] as a continuum medium. It is assumed that the external forces are applied to the centers of mass of the representative volumes in which a short-range order exists. The momentum, obtained by the particles of this volume for the time greater than the relaxation time (the time to reach an equilibrium state) is transmitted to the carrier fluid.

Unlike magnetic fluids magnetorheological fluid suspensions include a large multi-domain structure of ferromagnetic particles (size range 0.1–10 μm), has a low coercivity and low residual magnetization that allow one to classify them as superparamagnetic. Brownian motion in them has a much weaker impact on the particles then interaction between each other and with the external magnetic field. Approaching rigorously, a continuum approximation is inapplicable to them. However, up to the present time, due to the lack of data for accurate mathematical description of the kinetics of particle-particle interactions and the absence of any other reasonable physical and mathematical tools for the correct description of such media at shear deformation in the magnetic field, they are, especially in the case of medium- and highly concentrated suspensions are considered quasi-homogeneous and are subjected to a continuum approximation [8].

---

[1]Corresponding author.

Thus, it is assumed that the center of mass of the particle or aggregate is moving at an average speed of the fluid layer. Structures formed in the flowing MRF by the impact of the field of the dispersed particles is continuously broken by shear flow and are carried away by it, and are also continuously recovered in a short time from other fragments downstream. Kinetic processes, depending on the energy ratio of electrical and shear forces determine the overall performance characteristics of MRF fluidity.

As for the considered objectives soft ferromagnetic hysteresis-free particles are mainly used as a dispersed phase, and as a dispersion medium – oil or water; the degree of MRF magnetization is uniquely determined by the intensity of the magnetic field. Assuming that MRF during the time of experiment that is longer than the relaxation time, the rate of which is determined by the Stokes damping, but shorter that the time to establish a full balance, has the properties of the magnetic fluid and to its movement a steady-state approximation can be applied [7].

In the design of polishing devices it is necessary to keep the MRF layer as a ribbon on a moving support with a known cross section of the layer. The solution to this problem is possible by considering the equilibrium conditions of the bulk forces and the support reaction and Maxwell forces and surface tension, as in [9].

2. Forces impacting the layer of magnetorheological suspension in the magnetic field

Let us consider the peculiarities of formation of magnetorheological fluid layer on the surface of the machining tool-support in the finishing devices for various objects. Let MRF flow in a stream at a constant rate from the nozzle and get into the hydraulic system of a moving object-ribbon support of finite size in a region of a high gradient magnetic field. Its further movement takes place at a constant speed of the support as the layer with a free outer surface. Geometric parameters (the thickness, the shape of the free surface), and the adhesion to the surface to the support and tension on the free surface of the layer are defined by the ratio of support movement conditions, leakage rate in the initial pattern, gradient and impact area of the magnetic field. In our case, the non-uniform magnetic field created by a horseshoe magnet, placed so that the magnetic field lines cover the support and the entire MRF layer within it in planes perpendicular to the movement, as shown in figure 1.

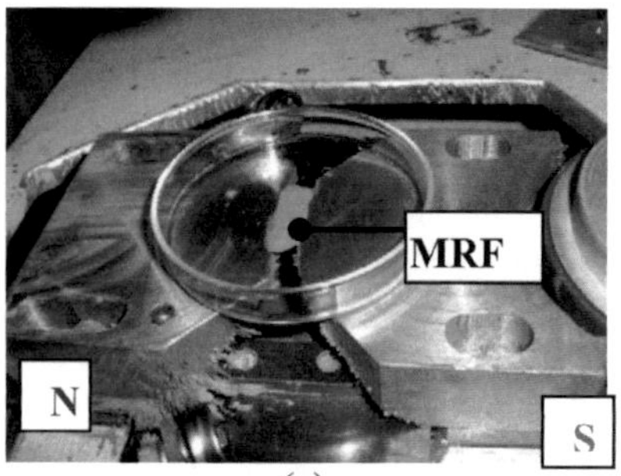

(a)

Figure 1. Experimental cell for studying the peculiarities of MRF layer formation in the polishing device (a) MRF layer is in between magnet poles on the steady support; (b) MRF layer and its fragment and (c) are in between magnet poles on the moving (rotating) support.

(b)

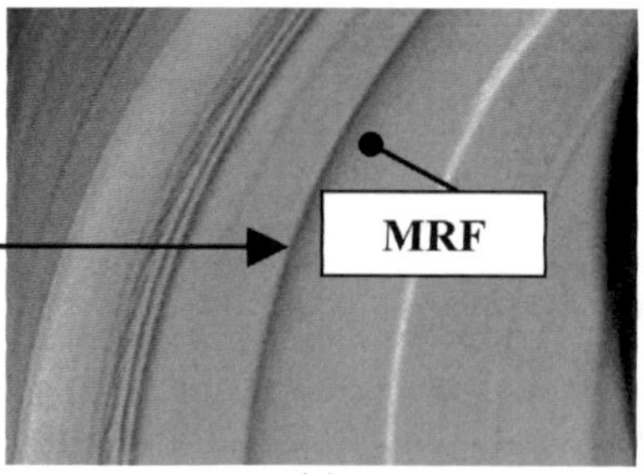

(c)

The shape of the free surface of MRF layer in cross section relative to the speed of movement in the gradient magnetic field and especially its interaction with the bearing support are determined by the bulk and surface Maxwell forces induced by the magnetic field.

The induction $\vec{B} = \mu_0\left(\vec{H} + \vec{J}\right) = \mu_0 \mu \vec{H}$ appearing in MRF is equal to the sum of the induction of the external field $\vec{H} = \dfrac{\vec{B}_0}{\mu_0}$, and the induction of the magnetization field, that is proportional to the magnetization $\vec{J}$ of the magnetic (MRF), and μ_0 – the coefficient of the magnetic permeability of vacuum, $\mu = 1 + \chi$ – magnetic permeability of the magnetic, χ – its magnetic susceptibility. It is known that the intensity of the magnetic field, expressed through the induction and magnetic permeability of vacuum μ_0, does not depend on the properties of the magnetic, in particular on its density ρ [10]. For practical applications as particle's material a superparamagnetic material with a narrow hysteresis loop (or its absence) is used, having a very small coercetive force and residual magnetization.

Holding MRF layer on the support is provided by the bulk force, applied to it from the magnet.

The induction creates in MRF a field of bulk forces [10], density of which is described by the function:

$$\overline{f\left(\vec{r}\right)} = -\nabla\left[\int v\frac{\mathrm{d}B}{\mathrm{d}v}dH + \int \mu_0 J\mathrm{d}H\right] + \mu_0 J\nabla H\,, \tag{1}$$

where v – the volume of MRF in the unit of mass, in which the particle material is superparamagnetic.

It can be expressed through the gradient of the magnetic pressure as:

$$\vec{f}(\vec{r}) = -\nabla(P_S) - \nabla(P_{JH}) + \mu_0 J\nabla(H) \tag{2}$$

The first term in (2) – is the gradient of magnetic striction pressure, which can be represented considering the average value of the induction as:

$$P_S = \int v\frac{\mathrm{d}B}{\mathrm{d}v}dH = v\frac{\mathrm{d}}{\mathrm{d}v}\frac{B(H)}{2}H = \mu_0 v\frac{\mathrm{d}}{\mathrm{d}v}\frac{J(H)H}{2} = -\frac{1}{2}\mu_0 H\rho\frac{\mathrm{d}\mu(H)}{\mathrm{d}\rho}, \tag{3}$$

where according to [9] for a weakly magnetic material we have:

$$\rho\frac{\mathrm{d}\mu(H)}{\mathrm{d}\rho} = \chi$$

The second term in (2) – is the gradient of magnetic pressure:

$$P_{JH} = \int \mu_0 J\mathrm{d}H = -\frac{1}{2}\mu_0 JH\,. \tag{4}$$

Regarding (2) and (3) the impacting force of the external magnetic field on the element of MRF is determined as:

$$\vec{f}(\vec{r}) = -\frac{1}{2}\mu_0 H^2(\rho\frac{\partial}{\partial\rho}\nabla\mu - \nabla\mu) + \frac{1}{2}\rho\frac{\mu_0\partial\mu}{\partial\rho}\nabla H^2\,. \tag{5}$$

In quasi-linear representation this force expressed through magnetic permeability, depending on the intensity of the magnetic field is written as:

$$\vec{f}(\vec{r}) = -\frac{1}{2}\mu_0 \nabla\left(H^2 \rho \frac{\partial}{\partial \rho}\mu \right) - \mu_0 H \nabla(J) \cdot \tag{6}$$

And like gravity force it presses MRF layer to the region with the bigger value of the intensity of the field.

According to [9] bulk forces could be brought to the surface Maxwell tensions Tn relative to the unit normal $\vec{n}$ to the surface element of the layer $d\vec{S} = \vec{n} \cdot dS$:

$$\vec{T}n = \left(\vec{B}\cdot\vec{n}\right)\vec{H} + \vec{n}\left[\int v \frac{dB}{dv} dH + \int B dH \right] \tag{7}$$

In MRF like in other superparamagnetics

$$\vec{T}n = \left(\vec{B}\cdot\vec{n}\right)\vec{H} + \frac{1}{2}\vec{n}\left(1+\hat{b}\right)\vec{B}\cdot\vec{H}, \qquad \hat{b} = -\rho\frac{\partial}{\partial\rho}, \tag{8}$$

where $\hat{b}$ determines the magnetic striction stress in MRF.

From this relation according to [10] it follows that in equilibrium the vector of intensity $\vec{H}$ bisects the angle between the normal $\vec{n}$ to the surface and tension vector $\vec{T}n$.

The condition of equilibrium of forces on the surface of the layer together with the boundary conditions of electrodynamics assumes the equality of the forces acting on each element of the surface, it means the equality of the sum of the Maxwell tensions, or the difference between the external pressures from the air P_0 and from the magnet P_1 [10] to the surface tension force of P_c .

Given that

$$P_1 = P_0 + P_{JH} + P_s + \frac{1}{2}\mu_0 J_n^2 ,$$

where P_{JH} – magnetic pressure (4), P_s - striction pressure (3) , surface tension force $P_c = 2\sigma k$, and σ- coefficient of surface tension, κ - surface curvature, we have

$$P_1 - P_0 = 2\sigma\kappa \quad . \tag{9}$$

In MRF besides the above mentioned forces there is also an impact of magnetization pressure:

$$P_J = -\frac{1}{2}\mu_0 J_n^2 , \tag{10}$$

where J_n – normal to the surface of the layer component of magnetization which in the presence of hysteresis in ferromagnetic particles does not disappear after switching off the magnetic field.

From condition (9) after substituting expressions for the pressure we have the expression for the curvature of the free surface of MRF layer:

$$\kappa = -\frac{\mu_0}{4\sigma}\left(\mu\chi - \rho\frac{\partial\mu}{\partial\rho} \right) H^2 \quad . \tag{11}$$

From it follows that the curvature of the surface is maximum there, where the maximum value of the field intensity is.

Resultant force of Maxwell tensions on the surface of the layer is equal to the resultant bulk forces, i.e.

$$\vec{F}=\oint \vec{T}n dS=\int \vec{f}(\vec{r})dV=\int \frac{\mu_0}{2}\nabla\left(H^2\rho\frac{\partial}{\partial\rho}\mu\right)dV+\int \mu_0 H^2\nabla(\mu). \qquad (12)$$

It is directed to the area of a strong magnetic field and presses the layer to the surface of the object-support.

Consideration of the expressions for bulk and surface forces of Maxwell tensions, as well as the pressures applied to MRF layer in the gradient magnetic field, can further define the cross-sectional shape of the layer.

3. Shape of the cross section of MRF layer in the gradient magnetic field

When designing devices for finishing of precision optical or semiconductor components there is a need not only to keep MRF layer as a ribbon on a moving support, but also to satisfy the desired shape of the cross section of the layer.

The shape of the cross section of MRF layer in the gradient magnetic field of a permanent magnet under the conditions of applicability of the quasi-static approximation of a continuous medium is determined by the terms of the balance of forces on the surface (9). From (6), (12) according to [10] it is concluded that in the equilibrium state the vector of Maxwell tension $\vec{T}n$ near the surface of the layer is directed at the angle equal to twice the angle between the normal $\vec{n}$ to the surface and the direction of the magnetic field intensity vector $\vec{H}$. It can be represented as a vector sum of the normal and tangential components:

$$\overrightarrow{Tn} = \overrightarrow{Tnn} + \overrightarrow{Tnt} . \qquad (13)$$

Let us consider the features of formation of the surface of MRF layer under conditions corresponding to the system of polishing device. According to the equilibrium condition (9) and (13), if the direction normal to the surface of the layer is close to the direction of the field intensity vector and the angle between the normal vector and the Maxwell tension $\vec{T}n$ is less than the right angle, its tangential component $\overrightarrow{Tnt}$ causes a shift of MRF in the direction of the slope of the field intensity vector $\vec{H}$ to the surface layer. In this zone, the volume of fluid increases and the normal to the surface of the layer deviates more than that of the initial extent from the direction of the field.

If the direction of the normal vector $\vec{n}$ to the surface layer is close to the intensity vector $\vec{H}$ of the perpendicular direction, the angle between the normal vector and the tension vector is more than that of the right angle. Then the tangent vector component of Maxwell tension causes the movement of MRF volume in the direction opposite to the slope of the intensity vector to the surface of the layer. This leads to an increase in the volume of fluid in this area and the deviation of the normal vector $\vec{n}$ to the surface of the layer in the direction perpendicular to the field intensity vector $\vec{H}$. This situation is more stable and the free surface of MRF layer takes a position along the magnetic field intensity vector, as shown in figure 2.

If the layer has a large volume, in which the point of contact of the object's surface by the layer is in an area where the intensity vector $\vec{H}$ crosses the support at an angle greater than the right angle relative to the normal to the support, than near this point on the surface of contact the normal to the surface of MRF layer is directed to the nearest pole of the magnet. Tension vector $\overrightarrow{Tn}$ forms an angle with it that is less than the right angle, leading to compression of the layer along the surface of the object.

At the closest point on the free surface of the layer the vector $\overrightarrow{Tn}$ is directed into the layer and compresses it in a direction opposite to the pole of the magnet. As a result the free surface of the layer extends along the lines of magnetic field intensity.

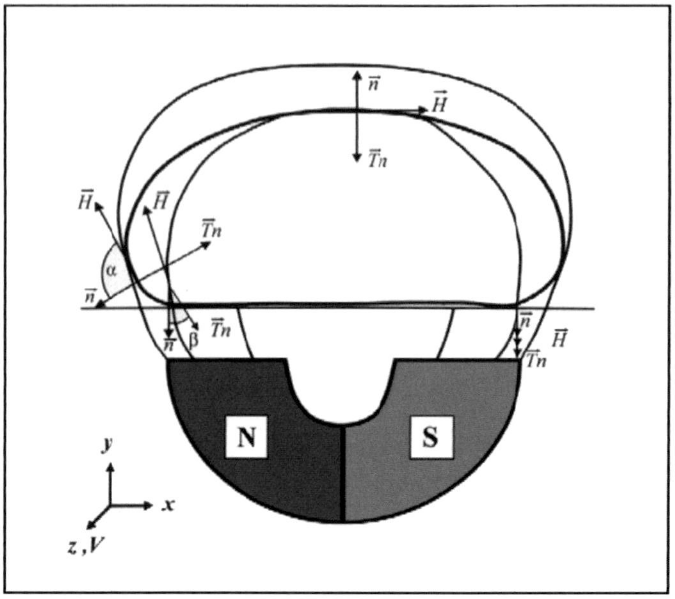

Figure 2. The shape of the cross section of MRF layer in the gradient magnetic field, the field lines of which intersect the surface of the layer and the object-support at an angle greater than the right angle.

The compression of the layer in the side direction prevents constriction in its middle, which could occur under the influence of the Maxwell tension, compressing the layer in the middle section in the direction to the poles of a magnet. When the tension is balanced by the surface tension, an equilibrium shape of the section layer is settled in the form of a truncated oval with tops extended to the poles of the magnet (in the regions of the most powerful magnetic field), like the one shown in figure 2.

Thus, the analysis of the ratio of orientation patterns of the magnetic field lines, the surface tension forces and magnetic forces in the polishing device allows us considering the equilibrium conditions to determine the main factors affecting the features of the formation of the settled cross-section of MRF layer of the certain thickness in a gradient magnetic field, moving together with object-ribbon support. Knowing the shape of the free surface of the layer, the value of the surface tension force and the degree of its pressing to the object-support, depending on the magnitude and direction of the gradient of the magnetic field will optimize the design of the processing unit and operating data for the implementation of high-precision machining of the surfaces of various products.

4. Conclusion

The considered model of formation of magnetorheological fluid layer allows us to determine the shape of its cross section, if the distribution of the magnetic field intensity is known. Received laws define methodology for further calculations of MRF layer characteristics in different shear conditions, as well as in contact of the free surface with the surface of the outer layer (machining) object.

References

[1] Taketomi S, Tikadzumi S 1993 *Magnetic fluids* (Moskow: Mir)
[2] Grinchik N N, Kagoroda O P and Khomich N 2010 *J. Eng. Phys. and Therm.* **83** (2) 638–49
[3] Kordonski W I and Gorodkin S R 2011 *J. Applied Optics* **50** (14) 1964–94
[4] Kordonski W I and Jacobs S 1996 *Int. J. Mod. Phys.* B **10** 2837–48
[5] Rusetski A, Novikova Z, Gorodkin G and Korobko E 2011 *Proc. National Academy of Sciences*

of Belarus **55** (5) 97–104
[6] Gorodkin G and Novikova Z 2012 *J. Intell. Mat. Syst. and Struct.* **23**(9) 959–62
[7] Berkovsky B M, Medvedev V F and Krakov M S 1993 *Magnetic Fluids:Engineering and Applications* (Oxford: Oxford University Press)
[8] Shulman Z P, Kordonski V I , Zalzgendler E A, Prohorov I V, Husid B M and Demchuck S A 1983 *Structure, Physical Properties and Dynamics of Magnetorheoogical Fluids* Preprint № 19 (Minsk: HMTI NASB Press)
[9] Tamm I E 2003 *Electricity Theory Bases* (Moskow: Fismatlit)
[10] Rosensweig R 1985 *Ferrohydrodynamic* (Cambridge: Cambridge University Press)

AUTHOR INDEX

AUTHOR INDEX